Emerging Technologies in Wireless LANs

Wireless LANs have become mainstream over the last few years. What started out as cable replacement for static desktops in indoor networks has been extended to fully mobile broadband applications involving moving vehicles, high-speed trains, and even airplanes. An increasing number of municipal governments around the world and virtually every major city in the United States are financing the deployment of 802.11 mesh networks, with the overall aim of providing ubiquitous Internet access and enhanced public services. This book is designed for a broad audience with different levels of technical background and can be used in a variety of ways: as a first course on wireless LANs, as a graduate-level textbook, or simply as a professional reference guide. It describes the key practical considerations when deploying wireless LANs and equips the reader with a solid understanding of the emerging technologies. The book comprises 38 high-quality contributions from prominent practitioners and scientists, and covers a broad range of important topics related to 802.11 networks, including quality of service, security, high-throughput systems, mesh networking, 802.11/cellular interworking, coexistence, cognitive radio resource management, range and capacity evaluation, hardware and antenna design, hotspots, new applications, ultra-wideband, and public wireless broadband.

"Benny Bing has created a masterful, horizon-to-horizon compendium covering the foundations, functionality, implementation, and potential-for-the-future of IEEE 802.11 wireless LAN communications. Whether your interests are in QoS, security, performance and throughput, meshing and internetworking, management and design, or just the latest in Wi-Fi applications, you will find an in-depth discussion inside these covers. *Emerging Technologies in Wireless LANs: Theory, Design, and Deployment* is an excellent resource for anyone who wants to understand the underpinnings and possibilities of the Wi-Fi offerings we see evolving in the marketplace today."
– Robert J. Zach, Director, Next Generation Broadband, EarthLink, Inc., USA

"Over the past 20 years, wireless LANs have grown from technical curiosity to a mainstream technology widely installed across residential, enterprise, and even municipal networks. The mobility and convenience of wireless has been augmented by the advanced throughput and range performance available in today's products, extending the reach of wireless LANs to a broad array of applications. This book explores all aspects of contemporary wireless LANs, from the basics through wireless security, meshes, QoS, high throughput, and interworking with external networks. The broad range of topics and perspective make this the ideal reference for experienced practitioners, as well as those new to the field."
– Craig J. Mathias, Principal, Farpoint Group, USA

"This book is a wonderful resource for anyone who works with Wi-Fi wireless technologies. It provides an excellent overview for the newcomer and an extensive and up-to-date reference for the expert. This book is a crucial tool for everyone involved in this exciting, fast-paced field. Everyone will learn from it!"
– Professor David F. Kotz, Director, Center for Mobile Computing, Dartmouth College, USA

"The ability of Wi-Fi technology to expand in so many directions while maintaining backwards compatibility has been one key to its success and the technology will certainly continue to evolve. This book has hopefully given you some insights into where we have been and where we may be headed."
– Greg Ennis, Technical Director, Wi-Fi Alliance

Benny Bing is a research faculty member with the School of Electrical and Computer Engineering, Georgia Institute of Technology. He is an IEEE Communications Society Distinguished Lecturer, IEEE Senior Member, and Editor of the *IEEE Wireless Communications* magazine.

Emerging Technologies in Wireless LANs

Theory, Design, and Deployment

Edited by

BENNY BING

Georgia Institute of Technology

FLORIDA GULF COAST
UNIVERSITY LIBRARY

CAMBRIDGE UNIVERSITY PRESS
Cambridge, New York, Melbourne, Madrid, Cape Town, Singapore, São Paulo, Delhi

Cambridge University Press
32 Avenue of the Americas, New York, NY 10013-2473, USA

www.cambridge.org
Information on this title: www.cambridge.org/9780521895842

© Cambridge University Press 2008

First published 2008

Printed in the United States of America

A catalog record for this publication is available from the British Library.

ISBN 978-0-521-89584-2

Contents

Authorship by Chapter

Chapter 1   Luke Qian, Cisco Systems

Chapter 2   Wildpackets, Inc

Chapter 3   Mathilde Benveniste, Avaya Labs

Chapter 4   Giuseppe Bianchi, Universita di Roma Tor Vergata
Sunghyun Choi, Seoul National University
Ilenia Tinnirello, University of Palermo

Chapter 5   Yiannis Andreopoulos, Queen Mary University of London
Nicholas Mastronarde, University of California at Los Angeles
Mihaela van der Schaar, University of California at Los Angeles

Chapter 6   Motorola, Inc

Chapter 7   Dorothy Stanley, Aruba Networks
Joshua Wright, Aruba Networks

Chapter 8   Richard VanNee, Qualcomm, Inc

Chapter 9   Bjorn Andre Bjerke, Qualcomm, Inc
Irina Medvedev, Qualcomm, Inc
John Ketchum, Qualcomm, Inc
Rod Walton, Qualcomm, Inc
Steven Howard, Qualcomm, Inc
Mark Wallace, Qualcomm, Inc
Sanjiv Nanda, Qualcomm, Inc

Chapter 10   Stephen Rayment, BelAir Networks

Chapter 11   Ambatipudi Sastry, PacketHop

Chapter 12   John Macchione, Juniper Networks

Chapter 13   Josef Kriegl, Tranzeo Wireless Technologies USA
William Merrill, Tranzeo Wireless Technologies USA

Chapter 14	Devabhaktuni Srikrishna, Tropos Networks
Chapter 15	Francis Dacosta, Meshdynamics
Chapter 16	Jan Kruys, Cisco Systems Luke Qian, Cisco Systems
Chapter 17	Srinivasan Balasubramanian, Qualcomm, Inc
Chapter 18	Sandro Grech, Nokia Corporation Henry Haverinen, Nokia Corporation Vijay Devarapalli, Nokia Corporation Jouni Mikkonen, Nokia Corporation
Chapter 19	Enrique Stevens-Navarro, The University of British Columbia Chi Sun, The University of British Columbia Vincent Wong, The University of British Columbia
Chapter 20	Steve Shellhammer, Qualcomm, Inc
Chapter 21	Eldad Perahia, Intel Corporation
Chapter 22	Richard Paine, Boeing
Chapter 23	Nestor Feses, Bandspeed, Inc
Chapter 24	John DiGiovanni, Xirrus, Inc Perry Correll, Xirrus, Inc
Chapter 25	Srenik Mehta, Atheros Communications David Weber, Atheros Communications Manolis Terrovitis, Atheros Communications Keith Onodera, Atheros Communications Michael Mack, Atheros Communications Brian Kaczynski, Atheros Communications Hirad Samavati, Atheros Communications Steve Jen, Atheros Communications Weimin Si, Atheros Communications MeeLan Lee, Atheros Communications Kalwant Singh, Atheros Communications Suni Mendis, Atheros Communications Paul Husted, Atheros Communications Ning Zhang, Atheros Communications Bill McFarland, Atheros Communications David Su, Atheros Communications Teresa Meng, Stanford University

Bruce Wooley, Stanford University

Chapter 26	Frank Caimi, Skycross, Inc
	David Wittwer, Intel Corporation
	Anatoliy Ioffe, Intel Corporation
	Marin Stoytchev, Rayspan Corporation

Chapter 27   Jasbir Singh, Pronto Networks

Chapter 28   James Keeler, Wayport, Inc

Chapter 29   Boingo Wireless®, Inc

Chapter 30   William Merrill, Tranzeo Wireless Technologies USA
Dustin McIntire, Tranzeo Wireless Technologies USA
Josef Kriegl, Tranzeo Wireless Technologies USA
Aidan Doyle, Tranzeo Wireless Technologies USA

Chapter 31   Ekahau, Inc

Chapter 32   Appear

Chapter 33   Timothy Brown, University of Colorado at Boulder
Brian Argrow, University of Colorado at Boulder
Eric Frew, University of Colorado at Boulder
Cory Dixon, University of Colorado at Boulder
Daniel Henkel, University of Colorado at Boulder
Jack Elston, University of Colorado at Boulder
Harvey Gates, University of Colorado at Boulder

Chapter 34   Kai Siwiak
Yasaman Bahreini

Chapter 35   Dagnachew Birru, Philips Research North America
Vasanth Gaddam, Philips Research North America

Chapter 36   British Telecom

Chapter 37   Philip Marshall, Yankee Group

Chapter 38   Becca Vargo Daggett, Institute for Local Self-Reliance

Foreword

Every now and then, a technology comes along which changes everything. Wi-Fi is one of those technologies.

Although wireless LAN technology has been around for close to 20 years, what we think of today as Wi-Fi has really existed for less than a decade. The IEEE 802.11b standard was ratified in 1999, enabling the then unheard of speed of 11Mbps. Shortly thereafter, the Wi-Fi Alliance was formed to focus on product interoperability certification and the development of the ecosystem and market. The combination of the right industry standard, unprecedented industry cooperation, and the novel utilization of unlicensed spectrum, created a new paradigm in terms of how people could connect to the Internet without wires.

Today, with the advent of draft 802.11n technology, we are able to deliver data rates in the multi-hundred Mbps range. We can now reliably cover most homes with a single access point using sophisticated MIMO techniques. We can connect large cities using advanced mesh architectures. With these developments, Wi-Fi is no longer confined to just the PC and networking application segments. Rather, Wi-Fi is now becoming a must-have feature in the latest consumer electronics products and handsets, ushering in new applications like voice and video. In a short period of time, Wi-Fi has moved from a cool, niche technology to one that is a mainstream, global phenomena.

I hope this book gives you a better appreciation for the power of Wi-Fi and stimulates your thoughts on where it can go in the future. Enjoy!

Frank D. Hanzlik
Managing Director
Wi-Fi Alliance

Preface

Wireless Fidelity (Wi-Fi) networks have become mainstream over the last few years. What started out as cable replacement for static desktops in indoor networks has been extended to fully mobile broadband applications involving moving vehicles, high-speed trains, and even airplanes. Perhaps lesser known is the proliferation of unique Wi-Fi applications, from Wi-Fi mosquito nets (for controlling malaria outbreaks) to Wi-Fi electric utility and parking meters to Wi-Fi control of garden hose sprinklers. The global revenue for Wi-Fi was nearly $3 billion at the end of 2006 and will continue its upward trend in the coming years.

When Wi-Fi wireless LANs were first deployed, they give laptop and PDA users the same freedom with data that cellphones provide for voice. However, such networks need not transfer purely data traffic. It can also support packetized voice and video transmission. People today are spending huge amounts of money, even from office to office, calling by cellphones. With a Wi-Fi infrastructure, it costs them a fraction of what it will cost them using cellphones or any other equipment. Thus, voice telephony products based on 802.11 have recently emerged. A more compelling use of Wi-Fi is in overcoming the inherent limitations of wireless WANs. An increasing number of municipal governments around the world and virtually every major city in the U.S. are financing the deployment of Wi-Fi mesh networks with the overall aim of providing ubiquitous Internet access and enhanced public services. Cheap phone calls using voice over IP may turn out to be one of the biggest benefits of a citywide Wi-Fi network, benefiting residents, businesses, tourists, and government agencies. This has led some technologists to predict that eventually we are more likely to see meshed Wi-Fi cells that are linked together into one network rather than widespread use of high-powered WAN handsets cramming many bits into expensive and narrow slices of radio spectrum.

I first edited a Wi-Fi book, *Wireless LANs*, in 2002. The book was well received by both academia and industry and was extensively reviewed by the *IEEE Network*, the *ACM Networker*, and the *IEEE Communications Magazine*, the first time a book has been featured by all 3 journals. This edited book comprises 38 new chapters covering a wide range of interesting Wi-Fi developments, including mesh networking, sensors, real-time tracking, cellular interworking, coexistence, hotspots, high-throughput multiple antenna systems, cognitive radio resource management, hardware and antenna design, ultra-wideband, and new 802.11 initiatives focusing on some of the areas mentioned above.

Organization of the Book

This book is designed to be accessible to a broad audience with different levels of technical background. It is not a collection of research papers that only specialists can understand nor is it collection of articles from trade magazines that give general overviews. Rather, it aims to strike a balance between technical depth and accessibility. To achieve this goal, the book is organized into a mix of chapters that cover fundamental tutorials, standards and case studies, mathematical analysis and modeling, and emerging technologies. Many chapters are written by prominent research scientists and industry leaders.

Part I: Introduction to 802.11

The original 802.11 standard is celebrating its 10th birthday this year and has progressed with a number of amendments since 1997. However, understanding the family of 802.11 amendments, including the acronyms, can be a daunting process. To this end, the first two chapters attempt to equip the reader with the necessary background for the rest of the book. The first chapter gives an overview of the emerging 802.11 amendments while the second chapter provides a detailed guide to 802.11 functionality and deployment issues. Chapter 2 also contains a list of basic 802.11 acronyms used throughout the book and it is highly recommended that these terms be familiarized before proceeding to other chapters.

Part II: 802.11 Quality of Service

The ratified IEEE 802.11e amendment will serve as a benchmark for servicing time-sensitive traffic such as voice and video and will become a major component of many home entertainment systems and set-tops, including Slingboxes that now come equipped with Wi-Fi connectivity. In the future, 802.11e may assume a more important role in mobile entertainment with the growing trend of Wi-Fi enabled portable devices such as iPod®s and smartphones. Chapter 3 covers the fundamental aspects of 802.11e namely, channel access, admission control, and power management mechanisms, with an emphasis on voice transmission. This is followed by a chapter on 802.11/802.11e modeling, written by a lead author (G. Bianchi) who developed the first analytical model for the 802.11 MAC protocol. The final chapter in this section presents an analytical framework for video transmission over multi-hop 802.11 networks. I am confident these three chapters will provide a solid foundation for engineers and researchers to evaluate the performance of voice, video, and data transmission over single-hop and multi-hop 802.11 networks.

Part III: 802.11 Security

Mobile client devices are becoming increasingly smarter and can easily act as an authorized Wi-Fi station. They can also move to different locations and shut off at any time. As such, soft access points involving client devices are becoming harder to detect, identify, and locate than hard-wired rogue access points. More recently, "evil twin" hotspots are becoming a rising danger for users who rely on public hotspots for Internet access. A hacker simply creates a hotspot with the same or similar name to a legitimate hotspot nearby. There are powerful features in 802.11i/WPA2 that can effectively counter security breaches related to intentional and accidental association. Thus, there has been a gradual migration from captive portals (often employed by Wi-Fi hotspot service providers) and VPNs to security architectures built around these standards. Unfortunately, Wi-Fi devices conforming to these standards can potentially add latencies in the order of hundreds of milliseconds and this can be very disruptive to voice connections as mobile users roam between networks. New methods such as key caching may be needed to support real-time traffic and the emerging 802.11r amendment is addressing secure mobility (and mobile QoS) with reduced handoff delays between 802.11i (and 802.11e) access points. Since different levels of Wi-Fi security lead to different levels of convenience for the end-user, the Wi-Fi Alliance's Wi-Fi Protected Setup (WPS) standard was designed to ease the set-up process of Wi-Fi networks. The first chapter in this section equips the reader with a clear understanding of Wi-Fi security basics while the second chapter focuses on a more in-depth coverage of Wi-Fi security issues, including handshaking and advanced encryption mechanisms, practical intrusion detection methods, analysis and countermeasures, and secure mesh networking.

Part IV: High Throughput 802.11

Wi-Fi data rates have continued to increase from 2 to 54 Mbit/s with current rates in the 802.11n draft amendment topping 600 Mbit/s. This development, coupled with the emergence of the 802.11s mesh amendment, may eventually render wired Ethernet redundant in the enterprise network. Despite the impressive progress in data rates, 802.11n products are backward-compatible with legacy 802.11b/g devices that operate in the 2.4 GHz unlicensed frequency band, even though the underlying physical layer transmission for 802.11 has changed dramatically over the last 7 years. Spread spectrum transmission that was used in first-generation 802.11 networks has given way to OFDM while multiple antenna MIMO-OFDM promises higher data rates, improved range performance, and better reliability for the future. To achieve higher speeds, channel bonding of two 20 MHz channels is allowed 802.11n. However, since there are only 3 non-overlapping 20 MHz channels in the 2.4 GHz band, this means that only one adjacent network operating in the remaining 20 MHz channel can co-exist. Hence, most 802.11n deployments in the 2.4 GHz band are not likely to include channel bonding. Because there are more non-overlapping channels in the 5 GHz band, the ratification of 802.11n may result in more widespread deployments of 5 GHz 802.11 networks, especially high-speed backhaul/backbone mesh deployments for enterprises and public municipal networks. Besides MIMO-OFDM and channel bonding, frame aggregation is another key feature of 802.11n. This feature allows the throughput efficiency to be improved by reducing the number of backoff delays required for frame transmission, thereby reducing the overheads per frame. Although 802.11n has yet to be ratified, dual-radio (2.4/5 GHz) products based on the draft amendment have started penetrating the WLAN market. The two chapters in this part describe the features and performance of this important amendment.

Part V: 802.11 Mesh Networks

Wi-Fi mesh networking will transform both enterprise and public networks. Because the same MAC and PHY layers can be used throughput the span of the network, such networks may see the distinction between WANs and LANs blurring for the first time in the history of computer networking. In addition to widespread municipal deployment, the multipoint capability of Wi-Fi mesh networks has been widely used in outdoor fairs and carnivals. Mesh networks are highly flexible networks with the ability to self-form and self-heal, thereby reducing the cost for backhaul deployment, system engineering, and network management. Wi-Fi access points in a mesh network not only deliver wireless coverage to end-user devices, they also act as routing nodes for other access points in the network. Obstruction, noise, and interference can be avoided dynamically by a reroute to the next best possible route. Unlike long-range wireless solutions such as 3G, the shorter hops in a Wi-Fi mesh network lead to lower variations in throughput and channel fading. Moreover, proprietary mesh protocols can sometimes provide a form of information security for wireless packet routing. While mesh networks are scalable in deployment, throughput scalability poses a huge challenge, even with multiple radio nodes. In addition, municipal Wi-Fi networks face a variety of challenges: the need to ensure high quality end-user experience, to meet guaranteed connectivity from first responders and emergency services, and to offer committed service level agreements with business and home users in an interference-prone public environment. Unlike traditional telecommunication systems, a multi-layered architecture is typically required: backhaul, capacity injection, mesh, and access. Bandwidth management and traffic policing are crucial in determining smooth operation and acceptable quality of user experience. The six industry contributions in this section cover different aspects of Wi-Fi mesh networking and offer many useful tips on network design and deployment. Additional insights on the development of the 802.11s amendment are provided in Chapter 16.

Part VI: 802.11/Cellular Interworking

Broadband cellular technologies such as 3G were originally targeted to compete with Wi-Fi. However, like unified wired Ethernet/wireless Wi-Fi switches, cellular and Wi-Fi convergence with single number access (regardless of device make) has now become mainstream. During the last two years, the Wi-Fi Alliance has certified about 100 Wi-Fi phones, the majority of which are dual-mode cellular handsets. Such handsets offer users the ability to transfer calls between home, office, and cellular phones seamlessly. Although Wi-Fi operates on unlicensed spectrum, the higher data rates afforded by a Wi-Fi connection can result in better voice quality, in addition to solving the notorious cellular signal fade inside buildings. An interesting alternative to Wi-Fi/cellular convergence is the use of femtocells, which are essentially simplified cellular base stations that act like personal access points for the home or office. With the ability to work with an existing cellular handset, femtocells can be very attractive when compared to VCC and GAN/UMA-based Wi-Fi services that require a new dual-mode handset. This makes some sense since the cellular phone of today is a much more innovative (and expensive) device compared to the cellular phone of yesteryear. With advances in computing power and storage, many cellular smartphones now come equipped with the ability to take and store photos, view TV programs, share real-time video, play games, provide navigation, act as a remote monitoring device, in addition to voice transmission. Nevertheless, I believe that by integrating with Wi-Fi in a dual-mode handset, the reach and affordability of a cellular connection can become more attractive. The first chapter in this section provides a very detailed coverage of the underlying issues associated with Wi-Fi/cellular interworking. The second chapter proposes an architecture for Wi-Fi/cellular integration. The third chapter presents a comprehensive analytical framework for evaluating the performance of Wi-Fi/cellular networks.

Part VII: Coexistence

The success of Wi-Fi operating in unlicensed frequency bands has provided the impetus for regulatory bodies around the world to open up more radio spectrum for unlicensed use. For instance, the white space in the TV bands can create a new market for Wi-Fi in future. It has been demonstrated that a Wi-Fi device using this unlicensed spectrum can co-exist with high-definition TV operation. However, interference must be carefully managed in any unlicensed environment, more so when incompatible devices operate in the same radio band. Dynamic spectrum access will play a critical role in these environments. The 802.11n draft 2.0 amendment allows co-existence with legacy 802.11 devices as well as non-802.11 devices such as Bluetooth. Such co-existence can sometimes result in serious restrictions on an 802.11n network and one may not be able to use the 40 MHz bandwidth for higher speed operation on the 2.4 GHz band. The first chapter in this section gives a comprehensive overview of co-existence issues for a wide range of radio bands, from UHF to microwave to millimeter bands. It also recommends cognitive sensing solutions for secondary devices operating in the 802.22 TV bands. The second chapter focuses on 802.11n and Bluetooth coexistence, which will become increasingly important as 802.11n networks are deployed in the coming years.

Part VIII: 802.11 Network and Radio Resource Management

Radio bandwidth resource management is key to the success of any wireless network deployment. A dynamic resource allocation method is needed to assign bandwidth, channel, and power levels depending on current interference, propagation, and traffic conditions. Currently, Wi-Fi access points need proper setup and maintenance in order to perform optimally. However, with the emerging 802.11k amendment, tedious configuration procedures may be a thing of the past. In addition, one can increase the number of access points to improve reliability and capacity without having to consider frequency planning or conduct detailed site surveys. The two chapters in this section discuss the state of the 802.11k amendment, and the features and benefits of a cognitive WLAN architecture.

Part IX: 802.11 Range

In general, the range of a network determines its utility. Currently, there are long-range Wi-Fi solutions as well as Wi-Fi mesh, both allowing the deployment of Wi-Fi networks that cover a large area. The chapter in this section compares the tradeoffs between the range of Wi-Fi and the data rates, coverage, and capacity. It also discusses how range can be improved using advanced MIMO technologies.

Part X: 802.11 Hardware Design

The two chapters in this section focuses on 802.11 chip and antenna design for portable computers. The first chapter describes an integrated single-chip system-on-a-chip (SoC) that can meet both the cost and form factor requirements by implementing all of the functions of an IEEE 802.11g WLAN system in a single 0.18-μm CMOS die. The integrated SoC combines the RF transceiver, analog baseband filters, data converters, digital baseband, physical layer, and medium access controller. A brilliant chapter on antenna fundamentals and design for portable devices follows. The authors describe various antenna diversity methods for practical implementation and recommend that the best design at lowest possible cost should take into account the radiation characteristics of the antenna elements and their interaction, and the channel characteristics.

Part XI: Wi-Fi Hotspots

Currently, there are nearly 150,000 free and paid Wi-Fi hotspots around the world. A more recent trend is the strong emergence of community hotspot providers offering Wi-Fi routers for users to share their broadband Internet connections with others in exchange for being able to use other users' connections for free when they are away from home. The cost savings associated with this concept of network sharing have some parallels with content sharing (e.g., peer-to-peer file sharing). Unlike past Wi-Fi community networks that are solely operated by residential users, the scale of these new Wi-Fi community networks has become much more extensive and are no longer limited to a single community or even a single country. Similar to the current applications of the Internet, such networks now serve as a powerful platform for social networking and are becoming increasingly integrated with commercial public hotspots. The chapters in this section are written by technical authorities from some of the leading hotspot providers, including Pronto Networks, Wayport, and Boingo Wireless®.

Part XII: Wi-Fi Applications

Wi-Fi sensor networks have become more pervasive. An example is the city of Cambridge in Massachusetts, which is building a Wi-Fi-based sensor network that will monitor the weather and pollution. The CitySense network will eventually support some 100 sensors around the city. A Wi-Fi real-time location system (RTLS) allows an organization to track high-value assets in a fast and efficient manner. Besides increasing asset visibility, it also enhances device security, simplifies IT management, and tightens control on the network environment. Context-aware computing enables applications to discover and exploit contextual information (such as user location, time of day, nearby people and devices, and user activity). Wi-Fi can be invaluable in such a mobile computing platform. The chapters in this section have been carefully selected to cover some of the most unique Wi-Fi applications. These include sonobuoy sensor network deployment, RTLS, context-aware computing, and unmanned aerial vehicles.

Part XIII: Ultra WideBand (UWB)

UWB technologies promise to deliver data rates in the order of a gigabit/sec. Being more focused on the end-user device, UWB can serve as an excellent complement to longer range Wi-Fi network deployments. The first chapter in this part describes the fundamental concepts of UWB, including emerging standards, pulse radiation and reception, channel models, and new applications. The second chapter describes the Multiband OFDM Approach (MBOA) to UWB and offers key insights into emerging UWB technologies.

Part XIV: Public Wireless Broadband

Public wireless broadband promises to revolutionize many facets of our lives, specifically in pervasive content access and mobile entertainment, while having the added benefit of affordable subscription. The Internet and Wi-Fi have both become the defacto media for entertainment and social networking. Many Wi-Fi radios not only stream audio from Internet radio stations, but also stream music files from the computer. Wi-Fi Internet TV and video game systems have also become prevalent. Just like bundling computer sale with Internet access, many popular game systems are now bundled with free Wi-Fi access in public hotspots, allowing young male gamers to go online and challenge each other in multi-player games in public locations. For example, over 5 million unique users have logged on to the Nintendo® Wi-Fi Connection (http://www.nintendowifi.com) and played over 200 million game sessions since the service was first launched in November 2005. Other massively multiplayer online games (e.g., World of Warcraft) have attracted millions of subscribers. However, with the mobility afforded by Wi-Fi, this will add to the complexity of the backend systems already limited by the number of simultaneous users and what players can do in the virtual world. The final chapters of this book comprise three interesting essays that cover the impact of Wi-Fi on public networks, the mobilization of the Internet, and the solutions to the broadband problem in the U.S.

In compiling the chapters for this book, I humbly admit that I have gained invaluable knowledge from a group of highly accomplished contributors. Through my experience in interacting and collaborating with industry, I recognize the importance of practical perspectives. I believe the many chapters from our industry colleagues will enable the reader to appreciate some of the major engineering considerations when designing and deploying Wi-Fi networks. As a researcher, I also value the insights provided by theoretical analysis, simulation, and proof-of-concept prototypes and testbeds. To this end, I hope the chapters from academia will adequately address the key problems associated with current and emerging Wi-Fi technologies and applications.

I take the opportunity to thank all contributors for generously investing their time and efforts, their co-operation in observing the deadlines, and their patience in seeing this book put to print. In addition, I wish to express my sincere thanks to the following individuals:

- Frank Hanzlik and Greg Ennis (both from the Wi-Fi Alliance) for taking time from their busy schedules to write the thought-provoking foreword and the epilogue.
- Jay Botelho (WildPackets, Inc) for contributing a superb chapter on 802.11 analysis. Ronnie Holland (also from WildPackets, Inc) was instrumental with the logistics.
- Ed Tan (Motorola, Inc) for furnishing the excellent Wi-Fi security writeup.
- Lorna Pierno (Xirrus, Inc) for co-ordinating the chapter on the range of Wi-Fi.
- Mia Falgard (Appear) for supplying the interesting chapter on context awareness using Wi-Fi.
- Judson Vaughn (Ekahau, Inc) for contributing the chapter on Wi-Fi tracking systems.
- Nicole Iannello (British Telecom) for contributing the Wireless Cities chapter in this book. Elliott Grady and Suzannah Ritch (both Fishburn Hedges) also deserve special mention for ensuring that this chapter was finalized on time.

- Phil Meyler, Emma Collison, and Anna Littlewood (all from Cambridge University Press) for their excellent support in ensuring the timely publication of the book. Much credit for the excellent layout of this book goes to Anna for her diligence in checking the manuscript and for providing numerous comments.

Sometime ago, I read about this initiative called One Laptop Per Child or OLPC (http://laptop.org). The first phase of this project aims to provide a rugged laptop to 100 million children in underdeveloped countries in the next few years at a cost of at least $10 billion (i.e., roughly $100 per laptop). This is about twice the current worldwide annual laptop sales at a much lower price per laptop. Each laptop is equipped with a screen that is readable in darkness or full daylight and has far more capabilities than commercial units costing ten times more, including the ability to function as a game console, a home theater, or an e-book. Although the power and display innovations are interesting, the most prominent feature in these laptops is the use of specially designed external antennas for Wi-Fi mesh networking (based on the draft 802.11s amendment). The flip-up antennas act as a switch to turn on the Wi-Fi radio without waking the CPU and provide better gain and range than the internal antennas in a typical laptop. Thus, Wi-Fi plays a vital role in this project and is a clear winner in the world of mobile computing.

On a more forward-looking note, research developed by scientists from the Massachusetts Institute of Technology allows electricity to be transferred wirelessly using magnetically coupled resonant objects. The design comprises two copper coils, each a self-resonant system. One of the transmitting coils is attached to a power source. Instead of flooding the environment with electromagnetic waves, the transmitting coil fills the space around it with a non-radiative magnetic field oscillating at MHz frequencies. The non-radiative field facilitates the power exchange with the other coil at the receiving unit, which is specially designed to resonate with the field. The overall effect is a strong interaction between the transmitting and receiving units, while the interaction with the rest of the environment is weak. For laptop-sized coils, power levels more than sufficient to run a laptop can be transferred efficiently over room-sized distances in virtually any direction, even with objects completely obstructing the line-of-sight path between the two coils. If this technology matures, then a truly wireless world beckons where both electric power and information can be exchanged over the airwaves.

I hope the technical depth and breadth of the chapters in this book will serve many readers well. The book can be used in a variety of ways: as a first course on WLANs, as a graduate-level textbook, or simply as a professional reference guide. I would greatly appreciate any feedback from readers regarding the book as well as suggestions of new topics for future editions. Please feel free to contact me via email (bennybing@ieee.org).

Benny Bing
Atlanta, GA, USA

1

Emerging IEEE 802.11 Standards

Luke Qian[a]

As the popularity of IEEE 802.11 wireless LANs (WLANs) grows rapidly, many new 802.11 wireless standards are emerging. New 802.11 standards are being developed in two major categories: specifications that make use of advanced wireless technologies in Radio Frequency (RF) and Physical layer (PHY), such as 802.11n, and specifications that address the needs in wireless network management, performance measurements, inter-networking, fast roaming, and the needs in other various specific applications and use scenarios. These include 802.11k, 802.11p, 802.11r, 802.11s, 802.11T, 802.11u, 802.11v, 802.11w and 802.11y. In this chapter, we discuss briefly the goals and scopes of these emerging standards. Emphasis will be given on 802.11n standard because of the significance in the technology advances it brings in.

1.1 IEEE 802.11n: Enhancements for Higher Throughput

802.11n is a long anticipated upgrade to the IEEE 802.11a/b/g wireless local-area network standards. It is expected to bring significant increase in MAC throughput of over 100 megabits per second (Mbps) and an enhanced communication range in the 2.4 and 5 GHz bands. 802.11n is also required to make efficient use of the unlicensed spectral resources by achieving at least 3 bits per second per Hz at the highest 802.11n rate.

The first draft of 802.11n supports PHY rates as high as 270 Mbps or five times that of a 802.11a/g network, which runs at 54 Mbps. The PHY rates can increase even more, up to 600 Mbps with four spatial streams and 40 MHz bandwidth, in the longer term when more receiver and transmitter antennas are employed. Currently, many chip vendors have already shipped pre-11n devices and have delivered the performance enhancements in both higher throughput and longer range promised by 802.11n.

High throughput (HT) devices are compliant with 802.11a/b/g standards. 802.11n defines a number of modes for backwards compatibility and interoperability with 802.11a and/or 802.11g.

802.11n builds upon existing 802.11 standards with enhancements to the MAC and the use of multiple-input multiple-output (MIMO) technologies. MIMO employs multiple transmitter and receiver antennas to allow for simultaneous data streams. The technology is

---

[a] *Cisco Systems, Inc*

capable of increasing data throughput via spatial multiplexing and increasing range via spatial diversity.

1.1.1 802.11n PHY

802.11n PHY is a high throughput orthogonal frequency division multiplexing (HT OFDM) system operating in the unlicensed 2.4 GHz and 5 GHz bands. The 802.11n PHY accommodates the required high throughput using MIMO, channel binding, beam forming, and Space-Time Block Coding (STBC). A number of other PHY features are also included in 802.11n to help meet the requirements.

The 802.11n PHY operates in one of three modes: Non-HT mode, Mixed mode, and the optional Green Field mode. In the Mixed mode, packets are transmitted with a preamble compatible with the non-HT receivers followed by a HT specific part for estimation of the MIMO channel. The Mixed mode enables non-HT devices to detect the transmission of a Mixed mode HT packet. In the Green Field mode, the non-HT compatible part of the Mixed mode preamble is omitted for higher efficiency. The two HT PLCP packet formats are illustrated in Figure 1.1.

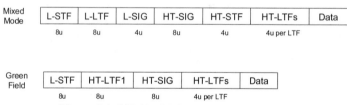

Figure 1.1: 802.11n PLCP Packet Formats.

1.1.1.1 MIMO

802.11n employs a mandatory basic MIMO of space division multiplexing. With MIMO, multiple spatial streams transmitted at the same time significantly increase the data rate (Figure 1.2).

1.1.1.2 40 MHz Channel Binding

Optionally, 802.11n allows for two adjacent 20 MHz channels to be combined into a single 40 MHz channel. This enables twice the amount of data to be carried in a single OFDM symbol. One of the tradeoffs is that it will reduce the number of total available channels. Spectrum use may not be optimal when the interference between devices in the various channels is taken into account. This is especially a concern in the 2.4 GHz band where the available channels are less than in 5 GHz.

1.1.1.3 Beam Forming

Beam forming is an optional technique adopted in 802.11n in which the transmitter utilizes the knowledge of the MIMO channel to generate a spatial mapping matrix that will improve reception in the receiver. There are two flavors of beam forming: implicit beam forming and explicit beam forming.

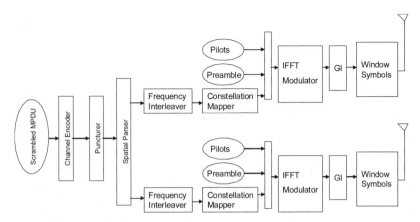

Figure 1.2: A Simplified MIMO Transmission Block Diagram.

Implicit beam forming uses the fact that the channel between STA A and STA B is the transpose of the channel between STA B and STA A. In reality, the actual channel includes the transmit chain in STA A and receive chain in STA B. So a calibration procedure is needed to correct for the difference in the measured channel.

In explicit beam forming, a transmitting STA needs the receiving STA to measure and send the channel matrices or the mapping matrices. The matrices can be either uncompressed (which can be used directly by the transmitter) or compressed (which requires further processing by the transmitter).

1.1.1.4 STBC

Space-Time Block Coding (STBC) is an optional PHY feature in 802.11n. STBC transmits multiple copies of a data stream across multiple antennas. On the receiving side, STBC combines all the received copies of signals optimally. The various received versions of the data provides the redundancy that results in a higher probability for one or more of the received copies of the data to be correctly decoded.

1.1.1.5 Other 802.11n PHY Features

There are a number of other optional PHY features introduced in 802.11n to help achieve the high throughput and increased range. These include:

- Green Field (GF) mode: an optional HT mode which provides high efficiency by omitting the backward-compatible portion in the preambles of packets operating in the HT Mixed mode.
- Short guard interval (GI): use of a short guard interval of 400 ns rather than a regular long GI of 800 ns.
- Low density parity check (LDPC) codes: an advanced error correcting code introduced along with an iterative probability-based decoding algorithm developed by Gallager in the early 1960's. Sparse random parity check matrices are used in constructing the codes.

- Reduced Interframe Space (RIFS): Allows a transmitter to transmit a sequence of PPDUs each separated by 2 μs.

1.1.2 802.11n MAC

The major function of 802.11n MAC is to meet the high throughput requirement under the constraint of the 802.11e QoS specifications. It also provides legacy compatibility protection and interoperability. In addition, new frames (such as sounding frames for channel measurements) are introduced to support HT PHY capabilities.

1.1.2.1 High Throughput Support

802.11n introduces extra PHY overheads to accommodate the transmission of 802.11n frames with the advanced PHY technologies. The overheads, along with the shortened transmission time of the payload due to the higher data rates, greatly reduces the MAC efficiency. The following plot in Figure 1.3 shows the MAC efficiency for an example of a point-to-point setting at a PHY rate of 243Mbps. For a packet size within the range of regular frame sizes, the MAC efficiency is found to be well below 20%.

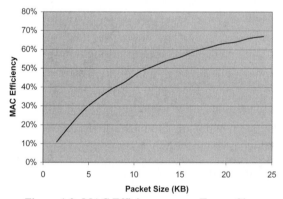

Figure 1.3: MAC Efficiency versus Frame Size.

To better understand the cause of this low MAC efficiency, Figure 1.4 breaks up the transmission time for an example of a frame and its acknowledgement. Apparently, channel access, inter-frame spacing, and PHY headers in particular, take up a considerable amount of the entire packet transmission time. As a result, the transmission time taken by the payload portion becomes very small.

Increasing the MAC efficiency to realize the high data rates provided by HT PHY therefore becomes one of the major objectives in the development of 802.11n MAC enhancements. Since the MAC efficiency increases with the frame size, the major technique to achieve HT throughput in the 802.11n MAC is to perform frame aggregations. These include aggregated MSDU (A-MSDU), aggregated MPDU (A-MPDU), and RIFS Bursting.

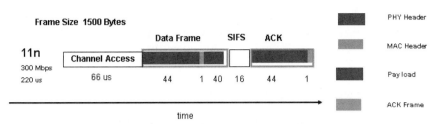

Figure 1.4: Breakup of a Frame Transmission Time.

A-MSDU defines an efficient MAC frame format as illustrated in Figure 1.5. A-MSDU aggregates multiple MSDUs in a single MPDU. Each sub-frame in an A-MSDU consists of a sub-frame header followed by a MSDU and 0-3 bytes of padding. The maximum allowed A-MSDU size is extended to around 4 KBytes and an optional 8 KBytes for MAC efficiency improvement.

All sub-frames in an A-MSDU must be addressed to the same receiver address because they share one single MAC header. In the same light, all sub-frames in an A-MSDU must be of the same access category (AC). Since there is only one single FCS for an A-MSDU, error recovery can be expensive and an error in any of the sub-frames will require the re-transmission of all the sub-frames in the A-MSDU.

A-MSDU Sub-frame 1	A-MSDU Sub-frame 2	...	A-MSDU Sub-frame n

Figure 1.5: Frame Format of A-MSDU.

A-MPDU is another form of aggregation in 802.11n. A-MPDU aggregates multiple MPDUs in a single PPDU as shown in Figures 1.6 and 1.7. The maximum allowed A-MPDU size is extended to around 64 KBytes, much longer than that of A-MSDU.

A-MPDU Sub-frame 1	A-MPDU Sub-frame 2	...	A-MPDU Sub-frame n

Figure 1.6: Frame Format of A-MPDU.

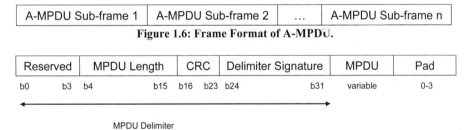

Figure 1.7: Format of A-MPDU Sub-frame.

A-MPDU is purely a MAC function. The PHY has no knowledge of the MPDU boundaries and treats the A-MPDU as if it were a regular MPDU. Each sub-frame in an A-MPDU has its own FCS. When an error occurs to a sub-frame, it is possible to continue parsing the rest of the PPDU and recover them with the A-MPDU delimiters and the unique A-MPDU signature pattern embedded in the delimiters. This error recovery mechanism in A-MPDU makes it less expensive than A-MSDU when an error in a sub-frame occurs. On the other hand, since there are multiple MAC frames in a single A-MPDU, the Block ACK (BA) to acknowledge multiple frames has to be involved to accommodate the A-MPDU and this adds significant complexity and buffer requirement in the implementation of the A-MPDU. To offset the overheads relating to the use of BA, a number of BA enhancements are introduced in the 802.11n MAC. For example, it allows implicit BA without a BA request and a compressed BA bitmap with its size reduced significantly.

1.1.2.2 Legacy Protection, Coexistence, and Interoperability

With the introduction of multiple HT modes, co-existence of devices running at various modes can impose a major challenge in the deployment of 802.11n networks. On one hand, 802.11n devices have to co-exist with the 802.11 a/b/g devices in previously deployed wireless networks. For example, an inappropriately configured 802.11n device that runs in the Green Field mode can seriously affect the performance of a legacy 20 MHz device nearby since the latter is not able to recognize and yield to the GF signals properly. On the other hand, achieving interoperability is not trivial even among the 802.11n devices. There are quite a few 802.11n features that are optional. As a result, interoperability with other draft 802.11n devices is challenging. Compatibility with the future ratified standard can be another big concern.

L-SIG TXOP protection is one of the new protection mechanisms introduced in 802.11n to help the coexistence. As shown in Figure 1.8, in the L-SIG field of HT frames with a Mixed mode PHY header, the rate is always set to 6 Mbps and the length field can be set to some value that causes non-HT devices to defer transmission for a period of time. This effectively provides a protection mechanism for non-HT devices to defer transmission when a HT device is transmitting.

Figure 1.8: Basic Concept of L-SIG Protection.

To ensure the coexistence, it is required in 802.11n that GF mode operation must be protected when there are non-GF mode devices present nearby. It is also required that 40 MHz operation must be protected in the presence of non-HT devices.

1.2 IEEE 802.11k: Radio Resource Measurement

Emerging technologies and wireless applications (such as voice over IP, video over IP, location services, large scale WLAN deployment and management) impose many new requirements over the capabilities of WLANs. These advancements demand standardized facilities to acquire and exchange statistics and measurements to better deploy and manage the WLAN, to better utilize the wireless bandwidth, to automatically optimize network performance, and to improve the reliability of the WLAN. Such facilities are of key importance to the pre-eminence of 802.11 wireless networks.

802.11k Radio Resource Measurement (RRM) specifies the facilities to meet the requirements of information about the radio environment. The specification defines Radio Resource Measurement enhancements by specifying a list of standardized measurements for radio resources and providing mechanisms to higher layers in the network stack for consistent radio and network measurement reports. The mechanisms include measurement requests and reports as well as the MIB with an Object Identifier (OID) interface to upper layers.

The provided radio measurements can be used for various benefits such as enabling simplified and automatic radio configuration, achieving better performance for the WLAN, optimizing the use of the client's radio resources, alerting a network administrator to issues, notifying end user radio status, etc.

The 802.11k RRM measurements are exchanged with measurement pairs of requests and reports. The measurements include: Beacons, Measurement Pilots, summary of received packets, Noise Histograms, STA Statistics, Location Configuration Information, Neighbor Report, Link Measurements, QoS, QBSS Loads, Access Delay, etc.

802.11k adopts a layer management request/response model to collect statistics and perform measurements. In general, 802.11k only contains measurements that nearly all vendors can support via a driver or firmware upgrade without requiring hardware modifications.

1.3 802.11p: Wireless Access for the Vehicular Environment

IEEE 802.11p defines enhancements to 802.11 required to support Intelligent Transportation Systems (ITS) applications, which includes data exchange between high-speed vehicles and between the vehicles and the roadside infrastructure in the licensed ITS band of 5.9 GHz (5.85-5.925 GHz). 802.11p is also referred to as Wireless Access for the Vehicular Environment (WAVE).

802.11p provides the lower layers of the Dedicated Short Range Communications (DSRC) solution and will be used as the groundwork for DSRC. DSRC is a U.S. Department of Transportation project. It targets at vehicle-based communication networks, especially for applications such as vehicle safety services, toll collections, and commerce transactions via cars. Its ultimate vision is a nationwide network that enables communications between vehicles and other vehicles or roadside access points. The higher layers of the DSRC solution are provided by standards outside of IEEE 802.11 family, such as IEEE 1609, IEEE 1556 for beacon authentication, and NEMO for mobility.

802.11p uses 5 and 10 MHz channels of 802.11 OFDM PHY at 5.9 GHz, with a spectral mask that cannot be easily met by 802.11a devices. It also requires a substantially extended MAC and uses only very few 802.11 facilities such as a basic access mechanism of EDCA.

1.4 802.11r: Fast BSS-Transitions

Prior to 802.11r, BSS transitions are supported under 802.11a/b/g, but only good enough for the best-effort data, not for QoS data. With the emergence of QoS applications, such as Voice over IP (VoIP), a satisfactory BSS transition solution for QoS data is required.

VoIP mobile phones are designed to work with wireless Internet networks. These VoIP devices must be able to disassociate from one access point and associate to another rapidly. The handoff delay typically cannot exceed a threshold of about 20 msec. There are several issues with the BSS transitions. For one thing, the handover latency is too long (often average in the hundreds of milliseconds) to support QoS traffic. This excess delay can lead to loss of call connectivity or degradation of voice quality. Another problem is that a VoIP device cannot know if necessary QoS resources are available at a new access point until after a transition. It is therefore impossible to know beforehand whether a transition will lead to satisfactory VoIP performance. In addition, it is also problematic for secure 802.11 connections using WPA2 or WPA.

802.11r defines enhancements to 802.11 required to provide a solution for fast BSS transition. It provides a faster handoff solution to address the needs of security, a minimal latency, and QoS resource reservation, which is essential to widely deployed VoIP applications. 802.11r will permit connectivity aboard vehicles in motion, with fast handoffs from one access point to another seamlessly.

802.11r allows a wireless client to establish a security and QoS state at a new access point before making a transition, which leads to minimal connectivity loss and application disruption. The overall changes of the roaming process do not introduce any new security vulnerabilities. This preserves the behavior of current stations and access points.

802.11r will govern the way roaming clients communicate with candidate APs for instance in establishing security associations and reserving QoS resources. Under 802.11r, clients can use the current AP as a passage to other APs, allowing clients to minimize disruptions caused by the roaming transition.

1.5 802.11s: Wireless Mesh Networks

802.11s defines enhancements to 802.11 required for a new topology of 802.11 wireless LANs, the Mesh Networks, which supports frame delivery in a hop-by-hop fashion. The work of 802.11s started in 2005 and its publication is expected by late 2008 or early 2009. Currently, the specification has been drafted and is in the stage of refinement.

802.11s inherits from existing 802.11 standards many features including security, QoS. and device power-saving mechanisms. For example, secure mesh links are set up based on 802.11i and push/pull key distribution using the key hierarchy and a mesh KDC, with fall back to pre-shared keys for small or home networks. It also adopts a variety of concepts such as Beacons and Probe/Response to advertise Mesh ID, Routing protocol, Security Capability etc.

802.11s defines a 6-address scheme to accommodate mesh tunneling. New mesh related features include route discovery, route maintenance, route recovery or re-establishment, and mesh routing functionality for frame routing and forwarding. In particular, 802.11s supports a layer 2 routing protocol for small and mid-size mesh networks called hybrid wireless mesh protocol (HWMP), which is a hybrid of two wireless routing protocols: the Tree Based Routing and the AODV Routing. Both fixed and mobile mesh applications are supported by HWMP.

The major frame exchanges employed by HWMP include:

- PANN – portal announcement that enables mesh segmentation by allowing nodes to choose a portal as their gateway.
- RANN – root announcement that enables passive and active mesh formation (registration).
- RREQ – routing request that builds forward paths and enables registration of STAs at node
- RREP – routing response that builds reverse paths
- RRER – route error that signals the breaks-up of a path

1.6 802.11T: Wireless Performance Prediction

802.11T is developed to meet the need of the 802.11 industry for an objective means of evaluating functionality and performance of the increasingly sophisticated 802.11 products. Its goal is to define a set of testing methods and conditions and a set of performance metrics as a recommended practice to measure and predict performance in a consistent and uniform manner. These metrics are valuable to all involved in 802.11n products, including developers, end-users and product reviewers. 802.11T defines test metrics in the context of use cases, which are classified into three major groups: data, latency sensitive and streaming media.

The data applications include Web downloads, file transfers, file sharing, e-mails. Such traffic typically belongs to the best-effort access category in 802.11e. They do not put strict requirements on networks. The metrics important for data applications include throughput vs. range, AP capacity, and AP throughput per client.

Latency sensitive applications, such as voice over IP, are time-critical, whose traffic usually belong to the voice access category in 802.11e. Those applications impose a strict requirements of Quality of Service (QoS) over networks, including limits on voice latency, jitter and packet loss vs. range, network load, and admitted calls, etc.

Streaming-media applications include real-time audio/video streaming, stored content streaming, and multicast high-definition television streaming. These applications require even stricter QoS over networks than voice applications, including guarantees of bandwidth and latency. The related performance metrics include video quality (throughput, latency, jitter) vs. range and network load.

In addition to the above, there are other metrics such as throughput vs. path loss, fast BSS transition, receiver sensitivity, and AP capacity and association performance, etc.

The metrics are further classified as primary and secondary. The primary metrics (e.g., voice quality) directly affect the user experience. Secondary metrics, such as latency, jitter, and packet loss, affect the primary metrics and therefore indirectly affect the user experiences.

The recommended metrics are tested in the settings of either conducted (which provide RF isolation often in a shielded chamber and emulate controlled motion) or over-the-air. For motion emulation and measurement repeatability, most of the tests require a conducted environment.

1.7 802.11u: Wireless Inter-working with External Networks

802.11u is an amendment to the 802.11 to add features that improve inter-working with external networks. 802.11u is still in its early stage of development. Its scope covers improved enrollment, network selection, emergency call support, user traffic segmentation, and service advertisement.

Current 802.11 assumes that a user is pre-authorized to use the network. 802.11u covers the cases where user is not pre-authorized. With 802.11u, a network will be able to allow access based on the user's relationship with an external network (e.g., hotspot roaming agreements). A network can also indicate that online enrollment is possible or allow access to a strictly limited set of services such as emergency calls. This capability in 802.11u will also greatly improve the user experience of traveling users who will be able to select access to an external network based on the provided services and conditions for example.

There are a number of issues under considerations within IEEE 802.11 TGu, which include: wireless inter-working with external networks, requirements of address changes within the 802.11 PHY and MAC and enabling inter-working with non IEEE 802 networks.

802.11u has identified a set of mandatory requirements in the following clusters:

- Network Selection:
  - way to determine if a network in hotspot supports a particular SSPN (Subscription Service Provider Network) without authentication.
  - method for a client with multiple credentials to choose a proper one.
  - way for an AP to support multiple SSPNs.
  - method for a client to determine inter-working services before association.
- SSPN Interface: method to define the authorization information to be provided to the MAC and associated functionality.
- QoS Mapping: to define the mapping from external QoS information to 802.11 specific parameters.
- Media Independent Handover (MIH in 802.21).
- Emergency Sequence: to define the functionality for e911 call.

1.8 802.11v: Wireless Network Management

The existing mechanism in 802.11 for wireless network management is mostly via SNMP. However, there are quite a few issues for this approach. For example, not all wireless clients on the market possess SNMP capabilities. In the case that a wireless client cannot get IP connectivity, management of a device may be required, but use of SNMP is impossible then. In addition, the complexities of 802.11 APs nowadays require more management capabilities than what current SNMP MIBs can provide. Therefore we need to create more MIBs and seek a new management approach more advanced than SNMP.

802.11v provides the wireless network management enhancements to 802.11. It extends prior work of 802.11k in radio resource measurements to form a complete and coherent upper layer interface for management of STAs by APs and WLANs by higher layers in wireless networks. While 802.11k provides the messages to retrieve station information, the 802.11v provides the ability to configure stations.

1.9 802.11w: Management Frame Protection

802.11w is an amendment of 802.11 designed to protect selected uni-cast management frames (MF), such as action frames, disassociate and de-authenticate frames, etc, and multicast/broadcast management frames for their data integrity, data authenticity, replay protection, and data confidentiality. 802.11w extends data frame protection in 802.11i to management frames. These extensions will have interactions with 802.11r and 802.11u.

Prior to 802.11w, system management information is transmitted in unprotected management frames. The wireless network can be vulnerable because of network disruptions caused, for example, by malicious devices that forge disassociation requests that appear to be from a valid station. Protection of MF is therefore required in a robust security network association (RSNA) to protect against forgery, eavesdropping and unauthorized disclosure attacks on selected uni-cast management frames, and to protect against forgery attacks on selected broadcast management frames.

Selected uni-cast MF is protected with the negotiated RSNA data frame protection protocol (CCMP or TKIP). Forgery protection for broadcast/multicast management action frames (MAF) is achieved through the Broadcast Integrity Protocol (BIP) with AES-128-CMAC for message integrity. The protocol provides replay protection, as well as forgery protection against insider attacks by authenticated stations.

1.10 802.11y: Contention Based Protocol

The FCC opened up the 3.65 - 3.7 GHz band for public use in July 2005. This band is previously reserved for fixed satellite service networks. The FCC has developed a light licensing scheme for the 3.65 - 3.7 GHz band. Wireless clients use contention-based protocols (CBP) to minimize interference while the APs are lightly licensed. The mobile clients must receive an enabling signal from an AP before transmitting to avoid interfering with the FSS and Radiolocation Services, who are the primary users in the 3.65 GHz band.

IEEE 802.11 TGy is working on the necessary amendments, the IEEE 802.11y, for operation with contention-based protocols to minimize interference in this 3.65 - 3.7 GHz band. 802.11y also streamlines the adoption of new frequencies in the future.

The APs are allowed to operate with a peak power of 25W/25MHz while the mobile clients will be allowed to operate with a peak power of 1W/25MHz. The high powers allowed in this band can practically be used to support long range 802.11 infrastructure and long range mesh networks.

While 802.11y is defining CBP for operating in 3.65 - 3.7 GHz, there are various companies pushing for changing the rules to allow 802.16 and other license friendly protocols.

1.11 Conclusions

Evidently, consumer demands, industry interest, and technology advances will lead to more IEEE WLAN standards in future. In the March 2007 IEEE 802.11 plenary meeting, 802.11 formed a new study group, >1 Gbps Study Group (Very High-Throughput), to address the ITU-T IMT-2000 throughput requirement for the nomadic wireless interface. We expect the work initiated in such study groups to result in new exciting standards in the near future.

Acknowledgement

The author would like to thank Benny Bing, the editor of this book, for his valuable comments when drafting this chapter.

Disclaimer

THE INFORMATION HEREIN IS PROVIDED "AS IS," WITHOUT ANY WARRANTIES OR REPRESENTATIONS, EXPRESS, IMPLIED OR STATUTORY, INCLUDING WITHOUT LIMITATION, WARRANTIES OF NONINFRINGEMENT, MERCHANTABILITY OR FITNESS FOR A PARTICULAR PURPOSE.

2

Guide to Wireless LAN Analysis[a]

The market for 802.11 wireless local area networks (WLANs) continues to grow at a rapid pace. Business organizations value the simplicity and scalability of WLANs as well as the relative ease of integrating wireless access with existing network resources. WLANs support user demand for seamless connectivity, flexibility and mobility. This chapter provides an overview of wireless networks and the 802.11 WLAN standards, followed by a presentation of troubleshooting wireless network problems with the types of analysis required to resolve them.

2.1 Introduction

802.11 is no longer a "nice-to-have." It is a critical element in all enterprise networks, whether by design, by extension or by default. Office workers expect to have a wireless option as part of the overall network design. Mobile users extend their reach by using wireless networks wherever they are available, including in public places, in a prospect's conference room, or at home. Even when the policy states "No Wireless," wireless networking is alive and well as a built-in default on most laptops today. 802.11 enables tremendous mobility, and is becoming the foundation for other technologies, like campus-wide wireless voice.

Maintaining the security, reliability and overall performance of a wireless LAN requires the same kind of ability to look "under the hood" as the maintenance of a wired network - and more. Wireless networking presents some unique challenges for the network administrator and requires some new approaches to familiar problems. In order to see what these are - and why they are - we need to know something about how WLANs work.

2.2 Overview of Wireless LANs

WLANs gain great flexibility by the use of radio waves instead of wires to carry their communications. This freedom comes at a cost in network overhead and complexity however, as the radio medium is inherently less reliable and less secure than a wired network.

---

[a] *WildPackets, Inc*

For example, 802.11 WLANs are able to transmit and receive at a variety of data rates and switch between them dynamically. They step down to a lower rate when transmission conditions are poor, and back up again when signal reception improves. They can also dynamically impose their own fragmentation, reducing packet size to reduce data loss in poor conditions.

In a free-form network, stations must create an explicit association with one another before they can exchange unicast data traffic. In a medium where reception can be problematic, each unicast data packet is separately acknowledged. Because stations cannot detect collisions created by their own transmissions, special rules are needed to control access to the airwaves.

The public nature of radio transmissions and the desired flexibility of network membership create special challenges for security, requiring special authentication and confidentiality measures. This section provides overviews of each of these aspects of 802.11 WLANs.

2.2.1 WLAN Physical-Layer Standards

The first 802.11 standard was published by the Institute of Electrical and Electronics Engineers (IEEE) in 1997. Since that original standard, many amendments and corrections have been published, and versions of the standard (and amendments) have been adopted as standards by the ISO.

It is conventional to refer to various aspects of the standard by the name of the revision document in which they were first introduced - for example: 802.11b, 802.11i, and so forth. We follow this convention when distinguishing between physical medium specifications (802.11a, b, and g). It is important to recognize, however, that all of these documents form a single integrated set of specifications for wireless networks, some parts of which are optional and others mandatory.

Table 2.1 shows the development of the primary physical layer specifications for 802.11 WLANs, including the band in which they operate, the encoding methods used, and the mandatory and optional data rates achieved by those encoding methods. (For more about encoding methods, see "Encoding and data rates")

2.2.1.1 802.11n

802.11n promises to be one of the most exciting changes in 802.11 technology for many years. 802.11n leverages an existing modulation technology, "multiple input, multiple output," or MIMO, creating a standard implementation for employing this technology for 802.11 purposes. It uses multiple antennas on both APs and clients to make the transmission of multiple, simultaneous data streams possible. The resulting effect is both increased throughput and increased range. Theoretical throughput increases are impressive, with the capability of achieving a theoretical throughput rate of up to 600 Mbps. It is this type of throughput that is making 802.11n such an exciting technology.

Though there is already tremendous pressure on vendors to begin delivering 802.11n equipment, and some already have, ratification of the IEEE 802.11n standard has been slow in coming. This has had an adverse effect on the widespread adoption of 802.11n to date. However, the Wi-Fi Alliance recently announced a certification program for 802.11n hardware based on a draft standard of the IEEE specification, and this is likely to bring about stability and increased adoption. The IEEE has also recently approved a draft version of the 802.11n specification, making a significant step towards ratification. Currently, it is estimated that the 802.11n specification will be ratified by the IEEE sometime in the first half of calendar year 2008.

Table 2.1: Standards, band, encoding, and data rates.

Year	Standard or revision	Band	Encoding	Data Rates (Mbps) Mandatory, optional	Comments
				Mandatory, optional	Mandatory data rates shown in **Bold**, other rates are optional.
1997	**802.11**	IR (infrared)	PPM	**1, 2**	Never implemented. (PPM = Pulse Position Modulation)
1997	**802.11**	2.4 GHz	FHSS	**1, 2**	Commercially insignificant. (FHSS = Frequency Hopping Spread Spectrum)
1997	**802.11**	2.4 GHz	DSSS	**1, 2**	Distributed Sequence Spread Spectrum (DSSS) methods also supported by later 802.11b and 802.11g revisions for backward compatibility. (Original standard had an insignificant installed base.)
1999	**802.11b**	2.4 GHz	DSSS/CCK	**1, 2, 5.5, 11**	The first widely deployed WLAN hardware. Added complementary Code Keying (CCK) to original DSSS methods to achieve 5.5 and 11 Mbps rates.
1999	**802.11b**	2.4 GHz	DSSS/PBCC	**1, 2, 5.5, 11**	Added Packet Binary Convolutional Coding (PBCC) as an optional approach to achieving 5.5 and 11 Mbps data rates.
1999	**802.11a**	5.0 GHz	OFDM	**6**, 9, **12**, 18, **24**, 36, 48, 54	Introduced Orthogonal Frequency Division Multiplexing (OFDM) to achieve significantly higher data rates. Ratified in 1999, but hardware was not available until 2002.
2003	**802.11g**	2.4 GHz	DSSS/CCK	**1, 2, 5.5, 11**	Included for backward compatibility with 802.11b nodes operating in the same band.
2003	**802.11g**	2.4 GHz	OFDM	**6**, 9, **12**, 18, **24**, 36, 48, 54	Pure 802.11g mode (no 802.11b nodes present).
2003	**802.11g**	2.4 GHz	DSSS/OFDM	**6**, 9, **12**, 18, **24**, 36, 48, 54	Optional hybrid mode using DSSS preamble/header, OFDM payload.
2003	**802.11g**	2.4 GHz	PBCC	**22**, 33	Optional additional PBCC data rates.

2.2.2 WLAN Regulation

In addition to official standards bodies such as IEEE and ISO, three other classes of entities have an impact on wireless networks: regulatory agencies, industry groups, and major vendors.

Radio frequency (RF) spectrum use is regulated by the Federal Communications Commission (FCC) in the United States, and by other agencies in other jurisdictions. This document reflects usage in the United States, but notes those areas in which usage in other jurisdictions may vary.

The Wi-Fi Alliance is the most significant industry trade group for 802.11 equipment manufacturers in North America. The group provides certification of equipment manufactured by its members, indicating the equipment meets various named sets of specifications published by the group. These specifications are based very closely on the IEEE standards, but are not absolutely identical with them in all respects. In particular, options permitted by the IEEE standard but absent from the Wi-Fi Alliance certification programs may be implemented rarely, if at all. When they are implemented, features from different vendors not covered by Wi-Fi Alliance certification may not be interoperable.

Individual manufacturers, seeking to differentiate their products, may add sets of features neither covered nor prohibited by the published standards or certification programs. In general, this document makes only passing mention of such features.

2.2.3 Wireless LAN Topologies

WLANs are designed for flexibility and mobility. The standards refer to the nodes of a wireless network as stations (STAs). A special type of station called an access point (AP) is connected to both the wired and the wireless network and bridges communications between the two. The AP (sometimes called a base station) also provides synchronization and coordination and forwarding of broadcast packets for all the associated STAs. The area of operation of an AP is sometimes referred to generically as a radio cell.

The standard distinguishes between Infrastructure topologies (those with an AP and a connection to a wired network) and Independent topologies, made up of STAs with no access point and no direct connection to the wired network.

Figure 2.1: Stations in an independent Basic Service Set (IBSS) or Ad Hoc group are able to communicate with one another without connection to a wired network.

The simplest arrangement is an ad hoc group of independent wireless nodes communicating on a peer-to-peer basis, as shown in Figure 2.1. (Ad hoc is a Latin phrase meaning "for this (purpose)," indicating a temporary arrangement.) The standard refers to this topology as an Independent Basic Service Set (IBSS) and provides for some measure of coordination by electing one node from the group to act as the proxy for the missing access point.

The fundamental unit of the Infrastructure topology is the Basic Service Set (BSS), consisting of a single AP (connected to the wired network) and the STAs associated with it (shown in Figure 2.3). The user configures the AP to operate on a single channel.

To cover a larger area, multiple access points are deployed. When multiple BSSs are connected to the same wired network (Figure 2.2), the arrangement is called an Extended Service Set (ESS). Each access point is assigned a different channel wherever possible to minimize interference. If a channel must be reused, it is best to assign the reused channel to the access points that are the least likely to interfere with one another.

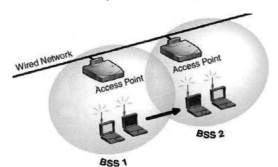

Figure 2.2: Extended Service Set (ESS) supports roaming from one cell to another.

When users roam between BSSs, they will find and attempt to connect with the AP with the clearest signal and the least amount of network traffic. This can ease congestion and help a roaming STA transition from one access point in the system to another without losing network connectivity.

An ESS introduces the possibility of forwarding traffic from one BSS to another over the wired network. This combination of APs and the wired network connecting them is referred to as the distribution system (DS). Messages sent from a wireless device in one BSS to a device in a different BSS by way of the wired network are said to be sent by way of the DS.

The 802.11 WLAN standards attempt to ensure minimum disruption to data delivery, and provide some features for caching and forwarding messages between BSSs. The 802.11i revision provides some support for optional fast transitions for stations moving between BSSs within a single ESS. Particular implementations of some higher layer protocols such as TCP/IP may be intolerant of dropped and restored connections. For example, in a network where DHCP is used to assign IP addresses, a roaming node may lose its connection when it moves across cell boundaries and have to reestablish it when it enters the next BSS or cell.

Additional specifications are also being developed, including 802.11r, to standardize "fast roaming" by reducing latency during handoffs between APs.

2.2.4 Establishing a Wireless Connection

Because the physical boundaries and connections within a radio cell are not fixed, there is no guarantee that a radio source is who or what it claims to be. Security requires some means of authentication. Because the physical arrangement and membership of any group of WLAN stations is purposely fluid, stations must be able to manage their own connections to one another. The WLAN standard refers to this logical connection between two nodes as an association. Because radio signals are inherently public, confidentiality requires the use of encryption.

These three functions - authentication, association, and confidentiality - are all a part of making a connection in a WLAN. We add a fourth function, discovery (not explicitly named in the standards), to construct a general picture of the process of creating a connection in a WLAN.

2.2.4.1 Discovery

A station or access point discovers the presence of other stations by listening. Access points (and their equivalents in ad hoc networks) can periodically send out management packets called beacon packets containing information about their capabilities (data rates, security policies, BSSID, SSID, and so forth). Stations can send a probe request packet to elicit a probe response containing similar information. A probe request can be sent to a particular station, or to the broadcast address (in which case, any response will come from access points, or the equivalents in an IBSS).

2.2.4.2 Authentication and Deauthentication

The first step in creating an association between two stations is authentication. If a station receives an association request from a station that is not authenticated with it, it sends a deauthentication notice to the requester.

Authentication is achieved by an exchange of management packets. The standards support several types of authentication.

The original standard provided only two forms: open and shared key. In open authentication, any standards-compliant node is automatically authenticated. In shared key authentication, the node must prove it knows one of the Wired Equivalent Privacy (WEP) keys in use by the network.

These original methods are still supported, but the 802.11i revision added additional steps for networks using the newer encryption methods. To avoid making complex changes to the original protocol, these newer methods first use the older open authentication method, then create a new security association between the two nodes during the association phase immediately following. The security association encompasses both authentication and encryption, and is described in more detail below. Briefly, authentication in a security association can be handled by a separate 802.1x authentication server, or be based on demonstrating possession of the correct pre-shared key (PSK).

A station can be authenticated with multiple other stations at any one time. The standard also supports an optional measure of pre-authentication in support of roaming within an ESS by stations already authenticated with the network.

2.2.4.3 Association, Disassociation, and Reassociation

In order to exchange unicast data traffic, stations must create an association between them by an exchange of management packets. A station can send an association (or reassociation) request to any station with which it is authenticated. If the association response is positive, the association is created.

In an IBSS, each station must create a separate association with each of the others in the group.

In a BSS, each station has a separate association with the access point. Stations can only be associated with one access point at any given time. To move between access points within an ESS, the roaming station sends a reassociation request to the new access point in order to seamlessly join the new BSS and leave the old one. Any station can terminate an association by sending a disassociation notice.

2.2.4.4 Confidentiality

Confidentiality is achieved by protecting transmitted information with encryption. The standards offer several options for encryption, each of which is a part of a larger security policy. APs (and their equivalent in an ad hoc network) advertise these security policies in beacon and probe response packets. An AP can enforce security policies for all the nodes in its BSS. Each node in an ad hoc network must enforce its own security policy. This topic is covered in detail in the next section, Security.

2.2.5 Security

Secure communication is problematic in all radio networks. A wired network can be secured at its edges - by restricting physical access and installing firewalls, for example. A wireless network with the same measures in place is still vulnerable to eavesdropping. Wireless networks require a more focused effort to maintain security.

This section presents a few basic concepts of communications security, then describes the main generations of security enhancements to 802.11 WLANs.

2.2.5.1 Concepts of Secure Communications

Communications security is often described in terms of three elements:

- Authentication ensures that nodes are who and what they claim to be.
- Confidentiality ensures that eavesdroppers cannot read network traffic.
- Integrity ensures that messages are delivered without alteration.

Authentication is typically based on demonstrating knowledge of a shared secret, such as a username and password pair. In more complex systems, possession of the shared secret may be demonstrated by proving possession of a token that is more difficult to steal or forge, such as a certificate or a smart card.

Confidentiality is typically protected by encrypting the contents of the message. Encryption applies a known, reversible method of transformation (called a cipher or encryption algorithm) to the original message contents (called the plaintext), scrambling or disguising them to create the ciphertext. Only those who know how to reverse the process (decrypt the message) can recover the original text. The most common forms of encryption are mathematical transformations which use a variable called a key as a part of their manipulations. The intended receiver must know both the correct method and the value of the key that was used, in order to be able to decrypt the message. For commercial encryption schemes, the method will be public knowledge. Protecting the secrecy of the key becomes crucial.

Integrity, in the context of communications security, refers to the ability to make certain that the message received has not been altered in any way and is identical to the message that was sent. The frame check sequence (FCS) bytes are one example of an integrity check, but they are not considered secure. The ordinary FCS bytes are not calculated over the plaintext message and protected by encryption. Instead they are calculated over the ciphertext, using a known method and sent in the clear (unencrypted). The FCS bytes help to identify packets that have been accidentally damaged in transit. An attacker, however, could recalculate the ordinary FCS (for example, to hide their deliberate alteration of a packet they captured and retransmitted). The harder it is for an attacker to correctly recalculate the integrity check sequence or security hash function, the more reliable a test of message integrity it is.

The concept of integrity is sometimes extended to include verifying that the source of the message is the same as the stated source. Timestamps and message sequence numbers can protect against "replay attacks," but, again, they are not considered secure unless they are protected by encryption.

Security is always relative, never absolute. For every defense, there is (or will soon be) a successful attack. For every attack, there is (or will soon be) a successful defense. Only time and effort are really at issue. The better the defense, the more time and effort it takes to breach.

The right defense is the one that is balanced and that matches the expected range of attacks. Balance is important in two senses. First, the weakest link must be secure enough. Second, the passive elements of authentication, encryption, and integrity check must be backed up by active elements such as monitoring and pursuing attempted breaches, maintaining security discipline, and so forth. The right defense is one in which a breach requires just slightly more time and effort from attackers than they are willing to invest. Security measures impose costs and constraints on the defender. Like any other business decision, these trade-offs must be made with eyes open.

2.2.5.2 Confidentiality and Encryption

Confidentiality (preventing unauthorized access to message contents) is achieved by protecting the data contents with encryption. Encryption is optional in 802.11 WLANs, but without it, any similar standards-compliant device within range can read all network traffic.

There have been three major generations of security approaches for WLANs. In chronological order of introduction, these are:

- WEP (Wired Equivalent Privacy)
- WPA (Wi-Fi Protected Access)
- 802.11i / WPA2 (Wi-Fi Protected Access, version 2)

To address vulnerabilities in WEP, the IEEE established the 802.11i working group in 2001. Based on early drafts from the working group, the Wi-Fi Alliance trade group established WPA at the beginning of 2003. WPA was intended as an interim solution that could be achieved with existing equipment, using only firmware and software updates. The Wi-Fi Alliance refers to their implementation of the more robust security features defined in the final 802.11i document (July, 2004) as WPA2. The more powerful encryption requires hardware acceleration, and is not supported by older WLAN equipment. The 802.11 standard now defines multiple alternative security arrangements for WLANs. For the sake of simplicity we use the terminology of the Wi-Fi Alliance to group the various alternatives, presented in Table 2.2.

Table 2.2: Encryption Methods in 802.11 WLANs.

Wi-Fi name	Authentication	Key distribution	Encryption	Algorithm
(none)	open	none	none	none
WEP	open or shared key (WEP)	out of band	WEP	RC4
WPA – Personal	open, followed by shared secret = PSK	out of band (PSK=PMK)	TKIP	RC4
WPA – Enterprise	open, followed by 802.1x, in which shared secret = certificate or other token	PMK from Authentication Server	TKIP	RC4
WPA2 – Personal	open, followed by shared secret = PSK	out of band (PSK=PMK)	CCMP	AES
WPA2 – Enterprise	open, followed by 802.1x, in which shared secret = certificate or other token	PMK from Authentication Server	CCMP	AES

TKIP is the Temporal Key Integrity Protocol. It uses a message integrity check called "Michael." Like WEP, TKIP uses the RC4 stream cipher encryption algorithm.

CCMP stands for CTR (Counter mode) with CBC-MAC (Cipher Block Chaining Message Authentication Code) Protocol. CTR is an attribute of the encryption method. CBC-MAC is used for message integrity and authentication. CCMP uses an AES (Advanced Encryption Standard) block cipher encryption algorithm.

WPA represents a significant improvement over the older WEP standards. The final 802.11i standard (implemented by the Wi-Fi Alliance as WPA2) defines even stronger security methods, but the greater computational burdens of CCMP/AES require specific network hardware. For many networks, WPA with TKIP continues to be a viable choice.

The expense and complex administration required for a full implementation of 802.1x can be beyond the reach of smaller networks, making the alternative of pre-shared keys (PSKs) more welcome there.

802.1x is a separate IEEE protocol used in support of the Extensible Authentication Protocol (EAP). In WLANs, 802.1x is used with EAP over LAN (EAPoL). The 802.11 standard specifies the use of 802.1x, but many details of the authentication services and methods used are left to the implementor. In general, 802.1x involves the use of a separate authentication server (such as Remote Access Dial-In User Service (RADIUS)) and valid certificates (or other secure tokens of authenticity) for each network node.

When encryption is in use, only the 802.11 headers of data packets are sent in the clear (that is, unencrypted). Management and control packets are not encrypted.

WEP uses a set of up to four static keys that must be installed manually on every station and access point. Different implementations of WEP support different key lengths. The revised 802.11 standard supports two WEP key lengths: 40-bit (expanded to 64 by the addition of a 24-bit initialization vector (IV)) and 104-bit (expanded to 128 with the IV). Other proprietary systems support longer key lengths. The unencrypted portion of the packet header can show which of the four WEP keys was used to encrypt the payload.

TKIP and CCMP use a separate Pair-wise Master Key (PMK) for each pair of peers - a pair of stations, or a station and an access point. This master key is used to derive other keys which are the ones actually used to encrypt and decrypt different elements of the traffic between the pair of nodes. This approach keeps the master key less exposed and allows for frequent rekeying.

The standard provides for two different methods of distributing PMKs. When an 802.1x authentication server (such as RADIUS) is in use, the PMK is derived when a station authenticates with the server. For networks that do not use an 802.1x server, a pre-shared key (PSK) is distributed out of band to every station and access point. This PSK is the PMK.

The security association between the two nodes is created during an exchange of four EAPoL packets called a four way handshake. During this transaction, the nodes derive a pair-wise transient key (PTK), which is then partitioned to provide the individual keys the pair will use for encryption, data integrity, and so forth. The PTK is derived from the PMK and a random value from both the station (the SNonce) and the access point (the ANonce).

When TKIP or CCMP are in use, broadcast and multicast traffic is also protected by encryption, using a Group key shared by all members of the BSS or IBSS. The Group Temporal Key (GTK) is distributed during the four-way handshake, or can be distributed in a separate group key handshake.

2.2.6 Collision Avoidance and Media Access

One of the most significant differences between Ethernet and 802.11 WLANs is the way in which they control access to the medium, determining who may talk, and when. Ethernet uses CSMA/CD (carrier sense multiple access with collision detection). An Ethernet device can send and listen to the wire at the same time, detecting the pattern that shows a collision is taking place. When a radio attempts to send and listen on the same channel at the same time, its own transmission drowns out all other signals. Collision detection is impossible. Because they cannot be reliably detected, collisions must be avoided. 802.11 WLANs use CSMA/CA (carrier sense multiple access with collision avoidance).

802.11 WLAN standards provide two basic methods for gaining access to the radio medium: the mandatory Distributed Coordination Function (DCF) and the optional Point Coordination Function (PCF).

Under DCF, stations listen to make sure the medium is clear, wait for a specified length of time, wait an additional random backoff interval, then attempt to send. The period during which stations are waiting their respective random backoff intervals is known as the contention period. Data and management packets also contain a Duration/ID field. Stations within range use this to determine how long the current transaction will take, deferring contention until it is complete.

Under PCF, the access point acts as the Point Coordinator (PC) for all associated stations, polling each in turn to ask if it would like to send. The PC acts as the reservation system for air time within the group. Under PCF, the group alternates between contention free periods (during which access is controlled by the PC) and contention periods, during which access is controlled exactly as in DCF. PCF was designed to support voice, video, and other time-sensitive transmissions. It is not implemented by most vendors. The standards leave room for interpretation, and interoperability among equipment from different vendors that do support PCF may be problematic.

DCF has one significant weakness, addressed in the standard. This is known as the "hidden node" problem. In a wireless network, a device can be in range of two others, neither of which can hear the other, but both of which can hear the first device. For example, the access point in Figure 2.3 can hear both node A and node B, but neither A nor B can hear each other. This creates a situation in which the access point could be receiving a transmission from node B without node A sensing that node B is transmitting. Node A, sensing no activity on the channel, might then begin transmitting, jamming the access point's reception of node B's transmission already under way.

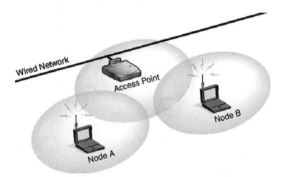

Figure 2.3: Basic Service Set (BSS), showing the hidden node problem.

To solve the hidden node problem, the standard specifies an optional method in which use of the medium is reserved by an exchange of control packets called request to send (RTS) and clear to send (CTS). A station sends an RTS to its intended unicast recipient. If the recipient receives the RTS and can accept the proposed transmission, it replies with a CTS. When it receives the CTS, the first station begins to send. This has two advantages and one drawback. First, the packets are small, and any collision caused by the transmission will be brief. Second, both parties to the proposed communication send a packet whose Duration/ID field covers the whole proposed transaction. That allows all stations within range of either station to defer use of the medium until the transaction is complete. The disadvantage, of course, is that the overhead represented by the RTS/CTS exchange must be added to each transaction.

A special case can occur between 802.11b and 802.11g stations using the same channel. Because 802.11b nodes cannot interpret the higher-speed OFDM-encoded transmissions of 802.11g nodes, additional steps must be taken to minimize contention between them. The standard refers to these steps as protection, invoked whenever 802.11b and 802.11g nodes are both associated with the same access point, or part of the same IBSS.

One protection option is for all stations to use the full RTS/CTS method for every unicast exchange, but this imposes significant costs to 802.11g throughput. As an alternative, 802.11g nodes can send a single CTS packet at 802.11b rates addressed to themselves (CTS to Self) to reserve the medium. This does not solve the hidden node problem, but it does allow 802.11g nodes to provide all 802.11b nodes within range with the information they need to defer using the medium until the 802.11g transaction is completed.

The use of RTS/CTS can be set to be always on, always off, or be invoked automatically when fragmentation reaches a preset level (for example, a data packet length of 500 bits). The precise methods are dependent on the implementation of the equipment vendor. Note that RTS/CTS is never used with broadcast or multicast traffic, nor for other control packets (such as an ACK).

2.2.7 Physical Layer

The 802.11 WLAN standards specify the lowest layer of the OSI network model (physical) and a part of the next higher layer (data link). A stated goal of the initial IEEE effort was to create a set of standards which could use different approaches to the physical layer (different frequencies, encoding methods, and so forth), and yet share the same higher layers. They have succeeded, and the Media Access Control (MAC) layers of the 802.11a, b, and g protocols are substantially identical. At the next higher layer still, all 802.11 WLAN protocols specify the use of the 802.2 protocol for the logical link control (LLC) portion of the data link layer. In the OSI model of network stack functionality (Figure 2.4), such protocols as TCP/IP, IPX, NetBEUI, and AppleTalk exist at still higher layers. Each layer utilizes the services of the layers underneath. This section describes the nature of the RF medium, some problems particular to it, and the solutions to those problems embodied in the 802.11 standards.

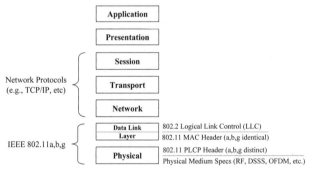

Figure 2.4: 802.11 and the OSI Model.

2.2.7.1 Radio frequencies and channels

Where Ethernet sends electrical signals through wires, WLANs send radio frequency (RF) energy through space. Wireless devices are equipped with a special network interface card (NIC) with one or more antennae, a radio transceiver set, and circuitry to convert between RF signals and the digital pulses used by computers.

Depending on the design of the antenna, radio waves may emanate more or less equally in all directions (the most common design), or be stronger in one direction than in others. Radio waves broadcast on a given frequency can be picked up by any receiver within range tuned to that same frequency.

Effective or usable range depends on a number of factors. In general, higher power and lower frequency increase the range at which a signal can be detected. Distance from the signal source and interference from intervening objects or other signals all tend to degrade reception. Filtering, accurate synchronous timing, and a variety of error correcting approaches can help distinguish the true signal from reflections, interference, and other noise.

Low output power limits 802.11 WLAN transmissions to fairly short effective ranges, measured in hundreds of feet indoors. Signal quality, and hence network throughput, diminishes with distance and interference. The higher data rates rely on more complex encoding methods. These in turn require an ability to distinguish very subtle modulations in the RF signals.

The WLAN standards for physical media (802.11a, b, and g) define the full set of channels for each type of network. Each channel is defined as a range of frequencies within a narrow band around a center frequency. When a WLAN radio uses a channel, it actually transmits or receives on multiple frequencies around that center frequency. The particular pattern of frequency use is determined by the encoding method, which also determines the nominal data rate.

RF spectrum is a limited resource which must be shared by competing users. While the standards define the range of possible channels, the actual channels used and the power outputs permitted on them are set by each regulatory agency for all 802.11 devices operating within its jurisdiction. The 802.11d and 802.11h revisions provide additional generalized methods for complying with the particular requirements of these agencies with respect to RF frequency use and power output in 802.11 devices. The 802.11j revision adds specifications particular to Japan.

The 802.11b and 802.11g WLAN standards both use the same 2.4 GHz band. Taking 2412 MHz as the center frequency of the first channel, the standard describes 14 channels, 5 MHz apart, numbered 1 to 14. In the United States, the FCC has allocated bandwidth to support the first 11 channels. Regulatory bodies in other jurisdictions have made different allocations from within this same band.

The 802.11a WLAN standard uses the 5.0 GHz band. The standard defines channels 1-199, starting at 5.005 GHz, with their center frequencies spaced 5 MHz apart. The FCC in the United States has allocated bandwidth in three parts of the spectrum, as shown in Table 2.3. The ETSI and ERM in Europe, MKK in Japan, and other regulatory agencies in other jurisdictions have made their own allocations within this band.

Notice that the FCC channel numbers for 802.11a WLANs appear in a gapped sequence, with 20 MHz separating the center frequency of one allocated channel from the next. This is a recognition of the fact that the encoding methods used in all 802.11 WLAN standards actually take up far more spectrum than 5 MHz. In fact an active channel using OFDM (whether in 802.11a or 802.11g) fills more than 16 MHz.

Table 2.3: FCC Channels for 802.11a WLANs.

Band	Center frequency	Channel number	Maximum power
U-NII low band (5150 MHz to 5250 MHz)			
	5180 MHz	36	40 mW
	5200 MHz	40	40 mW
	5220 MHz	44	40 mW
	5240 MHz	48	40 mW
U-NII medium band (5250 MHz to 5350 MHz)			
	5260 MHz	52	200 mW
	5280 MHz	56	200 mW
	5300 MHz	60	200 mW
	5320 MHz	64	200 mW
U-NII high band (for outdoor use) (5725 MHz to 5825 MHz)			
	5745 MHz	149	800 mW
	5765 MHz	153	800 mW
	5785 MHz	157	800 mW
	5805 MHz	161	800 mW

2.2.7.2 Signal and Noise Measurement

Electrical energy in radio waves is typically measured in the unit of power, Watts, or (in the case of 802.11 WLANs) milliWatts (mW). A typical 802.11b WLAN card might have a transmit power of 32 mW. The energy detected at the receiving antenna would be several orders of magnitude less than this. The wide range of values encountered in radio engineering could be expressed with exponential notation (for example, 3.2×10^{-5} mW), but radio engineers came up with a simpler solution. They measure signal strength with a unit called the decibel-milliWatt (dBm).

The decibel is a unit of relationship between two power measurements, and is equal to one tenth of the exponent of ten. That is, 10 decibels denotes an increase by a factor of 10, 20 decibels an increase by a factor of 100, and 30 decibels an increase by a factor of 1,000. These correspond to 10 raised to the power of one (10^1), 10 raised to the power of two (10^2), and 10 raised to the power of three (10^3), respectively.

Decibels are dimensionless. By associating decibels with a particular unit, it is possible to write and compare a wide range of power values easily. By the definition of the decibel milliwatt, 0 dBm = 1 mW. Power values larger than 1 mW are positive numbers. Power values smaller than 1 mW are expressed as negative numbers. Remember, this is an exponent. For example, the power output of 32 mW mentioned above could be written as

15 dBm. A typical lower limit of antenna sensitivity for an 802.11b WLAN card might be expressed as -83 dBm. A more practical lower limit might be -50 dBm, or 0.00001 mW.

Not all 802.11 WLAN cards report signal strength in dBm. The 802.11 WLAN standard itself calls for makers to implement their own scale of received signal strength, and report that within the protocol as a value called Received Signal Strength Indicator (RSSI). While one manufacturer might use a scale of 0-31, another might use 0-63.

Noise is also a form of electrical energy, and is reported in the same way, either as a percentage or in dBm. The signal to noise ratio is simply the difference between signal and noise. Noise is present in all 802.11 deployments, and can take many forms. Regardless of its source, determining the overall noise measurement is very important in determining both the quality of the signal and the expected data rate that can be received. To maintain a given data rate, a certain signal to noise ratio (SNR) must be achievable, which is of course based on the specific noise measurement. Table 2.4 provides some rule-of-thumb guidance for the SNR that is required to maintain certain data rates. For example, assuming a noise level of -80 dBm, a signal level of -61 dBm must be achievable at any point within the WLAN to ensure that all users can operate at the maximum data rate of 54 Mbps (S = SNR + N; -61 dBm = 19 dB + -80 dBm). With the knowledge that -61 dBm is the lowest signal that should be measured anywhere in the WLAN to achieve maximum throughput, AP placement can now be quantitatively assessed, and the minimum number of APs to be deployed can be determined.

Table 2.4: SNR for desired data rate.

Data Rate	Required SNR (dB)
6 Mbps	2
9 Mbps	5
12 Mbps	5.5
18 Mbps	7.5
24 Mbps	10.5
36 Mbps	12.5
48 Mbps	17
54 Mbps	19

2.2.7.3 Encoding and data rates

WLAN stations communicate by manipulating radio signals in agreed-upon ways. These manipulations encode the information using various combinations of frequency modulation, frequency hopping or spreading, and pulsing the energy on and off - all in a particular pattern. The most commonly used encoding methods are Direct Sequence Spread Spectrum (DSSS) and Orthogonal Frequency Division Multiplexing (OFDM).

DSSS (in particular configurations appropriate to the desired data rate) is used by 802.11b networks, and by 802.11g devices for backward compatibility with them.

Complementary Code Keying (CCK) is used in conjunction with DSSS to achieve the higher data rates of 5.5 Mbps and 11 Mbps.

OFDM (again, in particular configurations for each data rate) is used by 802.11a networks, and by 802.11g networks when operating in an "802.11g-only" environment.

An additional encoding method, Packet Binary Convolutional Coding (PBCC) is an option in both the 802.11b and the 802.11g standards.

The 802.11g standard also defines an optional hybrid method, combining DSSS for packet preambles and headers and OFDM for the body. This is intended to allow older 802.11b stations to follow the conversation, even though they cannot interpret the OFDM part of the transmission.

Table 2.1 shows the development of the primary physical layer specifications for 802.11 WLANs, including the band in which they operate, the encoding methods used, and the mandatory and optional data rates achieved by those encoding methods.

Stations must use the same encoding methods in order to communicate with one another. The nominal data rate of a WLAN is directly related to the encoding method used. In general, more complex encoding methods are used to create a more dense information flow for higher data rates. More complex encoding and decoding takes longer to perform. The more complex encoding can also be more susceptible to signal degradation.

2.2.8 Packet Structure and Packet Types

Like the rest of the 802 family of LAN protocols, 802.11 WLAN sends all network traffic in packets. There are three basic types: data packets, network management packets and control packets. The first subsection describes the basic structure of 802.11 WLAN data packets and the information they provide for network analysis. The second subsection describes the management and control packets, their functions and the role they play in network analysis.

2.2.8.1 Data Packet Structure

All the functionality of the protocol is reflected in the packet headers. RF technology and station mobility impose some complex requirements on 802.11 WLAN networks. This added complexity is reflected in the long physical layer convergence protocol (PLCP) headers as well as the data-rich MAC header.

802.11 WLAN data packet structure

OSI Physical (PHY) layer		OSI Data Link layer		Higher OSI layers	Packet trailer	
PLCP preamble header		MAC Header	LLC (opt)	Network data	FCS	End Delimiter

Because 802.11 WLANs must be able to form and re-form their membership constantly, and because radio transmission conditions themselves can change, coordination becomes a large issue in WLANs. Management and control packets are dedicated to these coordination functions. In addition, the headers of ordinary data packets contain a great deal more information about network conditions and topology than, for example, the headers of Ethernet data packets would contain.

802.11 MAC header (WLAN)

Frame Control	Duration ID	Address 1	Address 2	Address 3	Sequence Control	Address 4
2 Bytes	2 Bytes	6 Bytes	6 Bytes	6 Bytes	2 Bytes	6 Bytes

802.3 MAC header (Ethernet)

Dest. Address	Source Address	Type or Length
6 Bytes	6 Bytes	2 Bytes

A complete breakout of all the fields in the packet headers and the values they may take is beyond the scope of this chapter. For this overview, Table 2.5 below presents a list of the types of information 802.11 WLAN data packet headers convey. The table also shows the types of information carried in management and control packets.

2.2.8.2 Management and Control Packets

Control packets are short transmissions which directly mediate or control communications. Control packets include RTS, CTS and ACK packets used in the four way handshake, as well as power save polling packets and short packets to show (or show and acknowledge) the end of contention-free periods within a particular BSS or IBSS.

Management packets are used to support authentication, association, and synchronization. Their formats are similar to those of data packets, but with fewer fields in the MAC header. In addition, management packets may have data fields of fixed or variable length, as defined by their particular sub-type. The types of information included in management and control packets are shown in Table 2.5, along with the related information found in data packet headers.

2.3 Wireless Network Analysis

Wireless networks require the same kinds of analytical and diagnostic tools as any other LAN in order to maintain, optimize and secure network functions, with one notable exception. In a LAN environment, all signals are conducted over fixed, well-defined and "electrically stable" network of cables. This is in stark contrast to wireless networks, where signals are transmitted using radio frequency (RF) technology. Radio frequency waves propagate outwardly in all directions from their source, and are very sensitive to disruption or interference. The quality of the transmitted signal varies over time and space, even if the source and destination remain fixed. The path between the source and destination also has a very significant impact on the quality of the resulting communication. Open propagation of data means that anyone can receive the data, even those not "connected" to the network, making security a far bigger issue for WLANs. The use of unlicensed spectrum by 802.11 also increases its vulnerability to interference, as it must share its available bandwidth with non-802.11 devices, including Bluetooth, cordless telephones, and microwave ovens.

Fortunately, the 802.11 WLAN standard offers even more data to packet analysis than any of the other members of the 802 family of protocols. Also, new technologies are being developed to simplify the identification and mitigation of interference sources by analyzing the 802.11 physical layer - the actual RF environment that is the transmission network.

This section describes four broad areas in which wireless network analysis solutions can be of particular use in network planning, management, troubleshooting, and administration.

Table 2.5: Protocol Functions in 802.11 WLANs.

Authentication/Privacy	
The first step for a device in joining a BSS or IBSS is authentication. This can be an open or a shared key system, If WEP encryption of packet data is enabled, shared key authentication should be used. Authentication is handled by a request/response exchange of management packets.	
Authentication ID	This is the name under which the current station authenticates itself on joining the network.
Security enabled	If this field is true, then the payload of the packet (but not the WLAN headers) will be encrypted.
Network membership/Topology	
The second step for a device joining a BSS or IBSS is to associate itself with the group, or with the access point. When roaming, a unit also needs to disassociate and reassociate. These functions are handled by an exchange of management packets. The current status is shown in the packet headers.	
Association	Packets can show the current association of the sender. Association and Reassociation are handled by request/response management packets. Disassociation is a simple declaration from either an access point or a device.
IBSSID or ESSID	The ID of the group or its access point. A device can only be associated with one access point (shown by the ESSID) or IBSS at a time.
Probe	Probes are supported by request/response management packets used by roaming devices in search of a particular BSS or access point. They support a roaming unit's ability to move between cells while remaining connected.
Network conditions/Transmission	
The 802.11 WLAN protocol supports rapid adjustment to changing conditions, always seeking the best throughput.	
Channel	The channel or radio frequency used for the transmission.
Data rate	The data rate used to transmit the packet. 802.11 WLAN nodes can rapidly adjust their transmission data rate to match conditions.
Fragmentation	802.11 WLANs impose their own fragmentation on packets, completely independent of any fragmentation imposed by higher level protocols such as TCP/IP. A series of short transmissions is less vulnerable to interference in noisy environments. This fragmentation is dynamically set by the protocol in an effort to reduce the number, or at least the cost, of retransmissions.
Synchronization	Several kinds of synchronization are important in WLANs. Network management packets called "beacon" packets keep members of a BSS synchronized. In addition, devices report the state of their own internal synchronization. Finally, all transmissions contain a timestamp.
Power save	Laptops and handheld devices need to conserve power. To facilitate this, the protocol uses a number of fields in data packets plus the PS-Poll (power-save poll) control packet to let devices remain connected to the network while in power save mode.
Transmission control	
While the protocol as a whole actually controls the transmission of data, certain header fields and control packets have this as their particular job.	
RTS, CTS, ACK	Request to send (RTS), clear to send (CTS), and acknowledgement (ACK) – these control packets are used in the four way handshake in support of collision avoidance.
Version	The version of the 802.11 protocol used in constructing the packet.
Type and sub-type	The type of packet (data, management, or control), with a sub-type specifying its exact function.
Duration	In support of synchronization and orderly access to the airwaves, packets contain a precise value for the time that should be allotted for the remainder of the transaction of which this packet is a part.
Length	Packet length.
Retransmission	Retransmissions are common. It is important to declare which packets are retransmissions.
Sequence	Sequence information in packets helps reduce retransmissions and other potential errors.
Order	Some data, such as voice communications, must be handled in strict order at the receiving end.
Routing	
Again, many fields are related to routing traffic, but the following are most directly related.	
Addresses	There are four address fields in 802.11 WLAN data packets, instead of the two found in Ethernet or IP headers. This is to accommodate the possibility of forwarding to, from, or through the distribution system (DS). In addition to the normal Destination and Source addresses, these fields may show the Transmitter, the Receiver, or the BSSID. The type of address shown in each address field depends on whether (and how) the packet is routed by way of the DS. Control and management packets need only three address fields because they can never be routed both to and from (that is through) the DS.
to/from DS	In an ESS, traffic can be routed from a device using one access point to a device using a different access point somewhere along the wired network. These fields describe routing through the distribution system (DS) and tell the receiving device how to interpret the address fields.
More data	Access points can cache data for other devices. This serves both roaming across BSS or cell boundaries and the power save features. When a device receives a message from an access point, it may be told the access point has more data waiting for it as well.

2.3.1 Planning and designing a WLAN

One of the advantages of 802.11 WLANs is their ability to dynamically adjust to changing conditions and to configure themselves to make the best use of available bandwidth. These capabilities work best, however, when the problems they address are kept within limits. To do this, you must understand the limits of the RF environment in the areas where wireless is to be deployed. This is best done by assessing the overall area over space and time to get a quantifiable baseline of your environment.

2.3.1.1 Predeployment

When developing your baseline, it is imperative to assess two specific areas - interference sources from non-802.11 devices and signals from existing 802.11 equipment. Interference sources are often ignored when planning a WLAN deployment, yet this information is critical in designing AP placement, spacing, and channel selection. For example, where interference is high, 802.11 WLAN nodes will continue to increase fragmentation, simplify spectrum spreading techniques, and decrease transmission rates in an attempt to best use the available bandwidth. In addition, physical layer interference increases retransmissions, especially when they occur despite high fragmentation. While some network applications may show no ill effects from a given source of interference, others may begin to lag with too many retransmissions of packets already reduced well below their most efficient transmission size. Remember that 802.11 WLAN packet headers are quite large. This means high overhead and a low usable data rate when packet fragmentation and retransmissions are both high. If only one or two network applications seem to be affected, it may not be immediately obvious that there is a more general problem.

2.3.1.2 Initial deployment

Once the environment is understood and an initial layout is determined, it is time to test it out. Analyzers can be used to see if new interference sources have been introduced, and to see the interfering effects that each AP in your design may be having on its neighboring APs - otherwise known as adjacent channel interference (ACI), or co-channel interference when the APs are on the same channel (Figure 2.5). ACI occurs because each 802.11 b/g channel occupies approximately 20 MHz of bandwidth, while the actual channel separation is only 5 MHz (for example, Channel 1 is assigned at 2412 MHz while Channel 2 is assigned at 2417 MHz). ACI is the reason that it is often stated that there are only three non-overlapping b/g channels in the US: 1, 6, and 11, and this is why most multi-AP deployments use only these channels while making every effort to keep neighboring APs off the same channel.

Analyzers can be used to assess overall throughput and signal strength at key locations in your network, and to troubleshoot both the wireless and wired side of your network, simultaneously, should problems be identified. The ability to troubleshoot both the wired and wireless side of the network simultaneously is critical, and this is illustrated in Figure 2.6. Ongoing problems with packet retransmission on the wireless side of the network are clearly demonstrated.

Analyzers can also be used to test the interaction of clients and APs in multi-AP deployments. 802.11 WLAN BSSs and ESSs have the ability to dynamically configure themselves, associating and reassociating roaming nodes, first with one access point and then with another. The physical location and RF channel used by each access point should be optimized, and these choices can lead to smooth network functioning or to unexpected problems. To help evaluate network topologies, a packet analyzer must be able to display signal strength and transmission rate for each packet found on a given channel. Furthermore, the user must have control over what channel - better still, which base station - the packet analyzer will scan. With these tools, a packet analyzer can be used to build a picture of conditions at the boundaries between cells in an ESS.

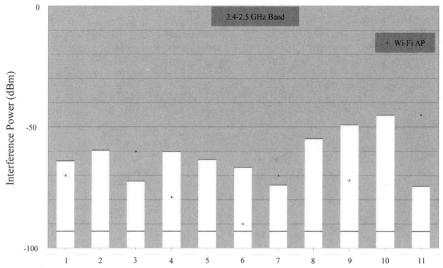

Figure 2.5: Analyzers can illustrate the impact of adjacent channel interference.

Performing a survey with the analyzers may find dead spots in a particular configuration or identify places where interference seems to be unusually high. Solving the problem may require changing the channel of one or more access points, or perhaps moving one or more to a new location.

2.3.2 Managing a WLAN

2.3.2.1 Managing Signals

Management of your WLAN begins with simple "dashboard" views that you can use to quickly assess the overall health of your network. Analyzers can provide an accurate display of signal and noise on your WLAN by showing a continuously updated bar graph of the most recently reported signal strength, noise, or signal to noise comparison on every channel on which traffic is detected.

2.3.2.2 Managing Users

Wireless networks are made up of one or more radio cells, centered on Access Points (APs). Unlike wired networks, the precise topology of the WLAN changes as clients roam from one AP to the next. The topology can be expressed as a hierarchical tree, with the ESSs (all APs connected to the same DS) at the top, then individual BSSs (individual APs and their clients), then the individual client nodes or stations (STAs).

Figure 2.6:   The dual capture and compare feature allows simultaneous wired and wireless analysis.

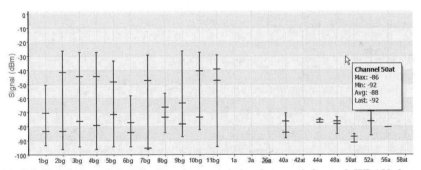

Figure 2.7: Detailed graphical display of reported signal strength for each WLAN channel.

An analyzer can display the wireless devices on your network in a hierarchical tree (Figure 2.8). Individual devices are identified by their ESSID, BSSID, or MAC address (as appropriate). An ESSID identifies a group of access points. This is the identifier sent out as "SSID" from the access point. The BSSID is the specific identifier of the access point, naming its MAC address. The view tracks dozens of 802.11 characteristics for each node, including encryption state, authentication method, channel, data rate, signal and noise statistics (dBm or %), and throughput statistics. Trust values can also be assigned to each node, allowing you to quickly distinguish friend from (potential) foe.

In addition to a hierarchical view of network users, some analyzers can also be used to represent the network as it is physically deployed. This is illustrated in Figure 2.9.

2.3.3 Administering a WLAN

2.3.3.1 Securing the WLAN

Because they use radio transmissions, WLANs are inherently more difficult to secure than wired LANs. Simple encryption and authentication techniques such as WEP prevent outsiders from casually or inadvertently browsing your WLAN traffic, but they cannot stop a deliberate attack. WPA, and particularly WPA2 is quite secure today, and meets the need of the most demanding security officer. Even the best passive defenses, however, must be paired with an active defense in order to really work. First, attempted breaches must be identified and stopped. Second, networks must be monitored to ensure that security policies are followed.

Analyzers can be used to monitor compliance with security policies, and to identify, intercept, log, and analyze unauthorized attempts to access the network. They can automatically respond to security threats in a variety of ways, making them ideal both for monitoring and for more focused analysis. Expert, real-time analysis of all traffic on the network identifies anomalies and sub-optimal performance. Some analyzers can provide a set of expert troubleshooting and diagnostic capabilities and problem detection heuristics based on the network problems found. Some examples of security related expert diagnoses include:

- Denial of Service (DoS) attacks
- Man-in-the-Middle attacks
- Lapses in security policy (such as wrong or default configurations)
- Intrusion detection
- Rogue access point and unknown client detection
- Adherence with common wireless network policies

Figure 2.10 shows an example of the Expert ProblemFinder Settings.

With some analyzers, you can assign levels of trust to any node, making it easy to tell at a glance who is who. Keeping a current list of your own network's members is easy, and allows you, for example, to automatically identify and easily locate rogue access points (see Figure 2.8). Assign a value of Trusted to the devices that belong to your own network. The intermediate value of Known lets you segregate sources that are familiar, but beyond your own control, such as an access point in a neighboring office. Nodes classified as Unknown (the default) can be quickly identified.

Some analyzers also ship with a security audit template, which you can use as is or extend or modify to meet particular requirements. The template makes use of special filters, alarms, and pre-configured capture sessions to create a basic WLAN security monitoring system. The security audit template scans network traffic in the background, looking for indications of a security breach. When it finds one, it captures the packets that meet its criteria and sends a notification, keeping you informed of suspicious activity on your WLAN.

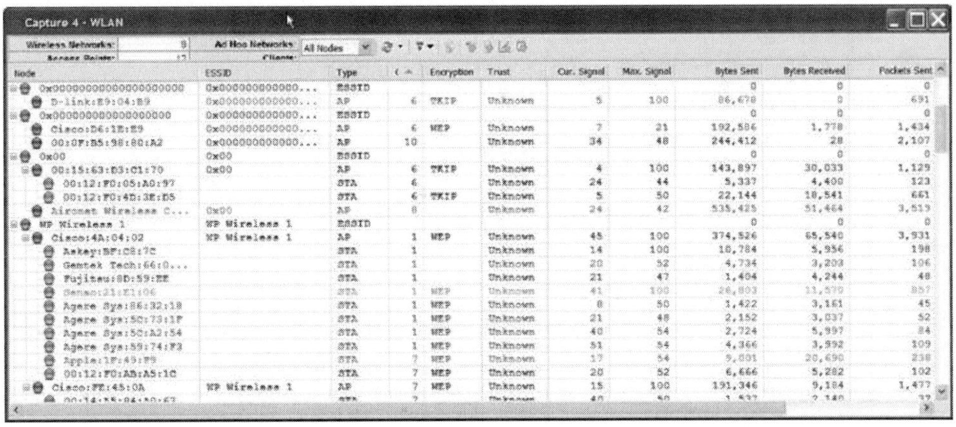

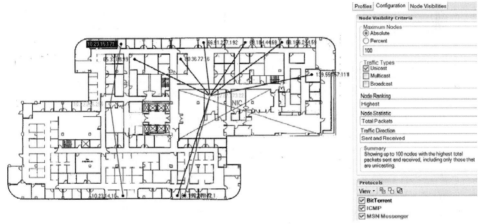

Figure 2.8: Hierarchical tree view of WLAN.

Figure 2.9: A physical representation of network users.

Security issues are not always malicious. Even with well-established security policies in place, well-intentioned users can be inadvertently violating these policies due to misconfigured security settings or even just an overall lack of knowledge of wireless security. With some analyzers, security policies established around common operating procedures like those illustrated in Figure 2.11, can be monitored in real time, providing instantaneous alerts when a single client is in violation of the policy.

2.3.4 Troubleshooting - Analyzing Higher Level Network Protocols

Managing a network is more than just managing Ethernet or the WLAN. It also means making sure all the resources users expect to access over the network remain available. This means troubleshooting the network protocols that support these resources. When WLANs are used to extend and enhance wired networks, there is no reason to expect the behavior of higher level protocols on these mobile clients will be any more or less prone to problems than on their wired equivalents.

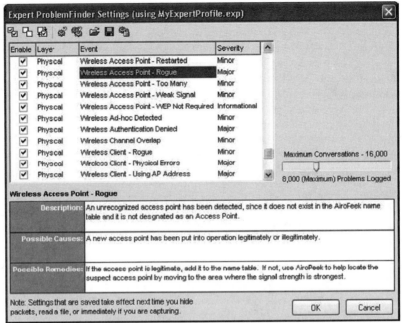

Figure 2.10: A sample of Wireless Expert Events.

Although part of this work can be done by capturing traffic from the wired network alone, some problems will yield more quickly to analysis of wireless-originated traffic captured before it enters the DS. To determine whether access points are making errors in their bridging, or if packets are being malformed at the client source, you must be able to see the packets as they come from the client node, as shown in Figure 2.12.

In an all-wireless environment, the only way to troubleshoot higher level protocols like TCP/IP is to capture the packets off the air. In smaller satellite offices in particular, this all-wireless solution is increasingly common. It offers quick setup and can cover areas that would be awkward to serve with wiring, such as non-contiguous office spaces on the same floor. The only wired part of such networks may be the connection from the DSL modem, through the router to the access point.

Event	Severity	Ena...
⊕ VoIP		☐
⊕ Wireless		▣
⊟ Network Policy		☑
Network Policy Violation - Vendor ID	Minor	☑
Network Policy Violation - Channel	Minor	☑
Network Policy Violation - ESSID	Severe	☑
Network Policy Violation - WLAN Encryption	Minor	☑
Network Policy Violation - WLAN Authentication	Minor	☑
⊕ Client/Server		☐
⊕ Application		☐
⊕ Session		☐
⊕ Transport		☐
⊕ Network		☐
⊕ Data Link		☐
⊕ Physical		▣

Figure 2.11: Wireless Network Policies.

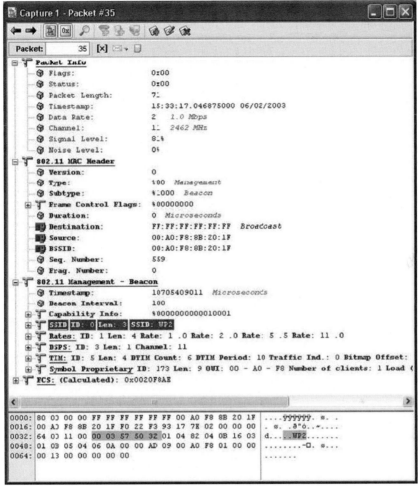

Figure 2.12: Partial Example of a Detailed Packet Decode.

The actual troubleshooting of these higher level protocols is no different on a wired or a wireless LAN, provided the network analysis software can read the packets fully. If security is enabled, the protocol analyzer must be able to act like any other node on the wireless network and decode the packet payloads using the shared keys. The ability to use WEP in the same way as all other nodes on the network must be built into the analyzer.

2.3.4.1 Leveraging Existing Assets with AP Capture Adapters

One of the most significant issues that exists in WLAN troubleshooting today is access to packets at the source of the trouble. Overlay networks, a deployment of wireless sensors that can monitor all wireless traffic from existing APs, is an effective but very costly means of having instantaneous access to wireless packets. A far more attractive solution is to be able to capture packets using the existing wireless network - after all, the hardware is designed to both transmit and receive. An AP capture adapter allows existing APs to be put into a "listen-only" mode, and directs them to forward all of the packets they receive to an analyzer over the wired network. No additional hardware, or expense, is required to implement this solution. Access to information for troubleshooting from any location on the network is only a few clicks away.

2.4 Conclusion

The demand for wireless networks is strong and increasing. The technology continues to evolve rapidly. Improvements in throughput, reliability, security, and system interoperability consistently add to this demand. Both the security of the new WLANs and their performance depend on active, informed network management. Effective network management requires the right tools.

Appendix: Wireless Terms

Access Point	Provides connectivity between wireless and wired networks
Ad Hoc Network	Peer-to-Peer network of roaming units not connected to a wired network
Base Station	Access Point
BSS (Basic Service Set)	Wireless network utilizing only one access point to connect to a wired network
Cell	The area within range of and serviced by a particular base station or access point
CSMA/CA	Carrier Sense Multiple Access with Collision Avoidance
CSMA/CD	Carrier Sense Multiple Access with Collision Detection
CTS	Clear To Send
DHCP	Dynamic Host Configuration Protocol, used to dynamically assign IP addresses to devices as they come online
DS (Distribution System)	Multiple access points and the wired network connecting them
DSSS	Direct Sequence Spread Spectrum
ESS (Extended Service Set)	A wireless network utilizing more than one access point
Frame	A packet of network data, framed by the header and end delimiter
FHSS	Frequency Hopping Spread Spectrum
IBSS	Independent Basic Service Set or Ad Hoc Network
IEEE	The Institute of Electrical and Electronics Engineers
Infrastructure	Wireless network topology utilizing access points to connect to a wired network
LLC	Logical Link Control
MAC	Media Access Control
NIC	Network Interface Card
OFDM	Orthogonal Frequency Division Multiplexing
Roaming	Traveling from the range of one access point to another
RF	Radio Frequency
RTS	Request To Send
WEP	Wired Equivalent Privacy
WFA	Wi-Fi Alliance, an industry organization specializing in interoperability and promotion of 802.11 WLAN equipment
WLAN	Wireless Local Area Network

3

WLAN QoS

Mathilde Benveniste[a]

With both the enterprise and residential sectors embracing voice over IP (VoIP) at an accelerating pace, and the pervasive use of wireless local area networks (WLANs), the natural requirement emerged for a technology to support VoIP over WLANs without degradation of its quality of service (QoS). QoS requirements for WLANs are imposed also by video and multimedia applications tailored for use with WLANs. A QoS-focused MAC Layer standard, IEEE 802.11e, was developed to meet the QoS requirements of a range of applications. In addition to VoIP/ multimedia QoS, the new standard serves mission-critical functions by reducing latency across a WLAN. This chapter discusses the enhancements the new standard adds to WLAN technology with respect to QoS performance, channel use efficiency, and power management of battery-based wireless devices.

3.1 Introduction

Since the initial emergence of the 802.11 network interface card for laptop computers and access points, the appeal of 802.11 technology has been so strong worldwide that it is now appearing in a wide range of devices, including consumer electronics devices and VoIP phones. Enterprises wish to extend VoIP over wireless LANs for the convenience wireless service brings to the mobile user throughout the building, campus, quad and warehouse, as well as anywhere a WLAN is accessible. Residential users purchasing VoIP service for cost savings, look to the WLAN to enable them to make their telephones cordless. The installation of WLANs in public spaces, backed up by a ubiquitous Internet, makes the case of VoIP over WLANs even more compelling. Users can have telephone service portability free of any wires anywhere a WLAN is present. The new trends in the expansion of WLAN use include consumer electronics appliances generating multi-media traffic streams from applications such as video streaming and interactive gaming.

"All this could happen if wireless LANs could support QoS adequately in a congested WLAN" was typically the reaction to the above observations prior to the adoption of the new standard for IEEE 802.11 WLANs, known as IEEE 802.11e. The new standard enables frames from QoS-sensitive applications to be transmitted sooner than other frames, thus minimizing latency. It also introduces new power management features that will prolong the life of mobile devices powered by battery. The channel-use efficiency gains

---

[a] *Avaya Labs*

introduced by the new standard make it worth pursuing even in situations where all traffic is of the same type, thus allowing privileged treatment to none. The lower latency achievable with 802.11e enables also the prioritization of time-critical data. Devices observing the new standard can co-exist with 802.11-compliant devices.

This chapter gives a high level overview of the major 802.11 mechanisms that have been modified and the new mechanisms introduced in 802.11e. They cover specifically the areas of channel access, admission control, and power management. QoS challenges that remain specifically in 802.11 mesh networks are also discussed.

3.1.1 Terminology and Abbreviations

For the reader reviewing the IEEE 802.11e standard [1], we note in this section some relevant naming conventions used in the standard. The QoS-aware contention-based random access is referred to as Enhanced Distributed Channel Access (EDCA). In the early standard drafts, and in much of the published literature, the same access approach had been called 'enhanced distributed coordination function' (EDCF), following the naming convention of the 802.11 standard, where Distributed Coordination Function (DCF) referred to contention-based random access [2]. Polled access was called Point Coordination Function (PCF) in the 802.11 standard. The enhanced polled access mechanism in 802.11e is called HCF Controlled Channel Access (HCCA). An access point (AP) supporting 802.11e features is called a 'QAP', and a station equipped to use 802.11e features is called a 'non-AP QSTA'. A QoS-aware WLAN, i.e., the group of stations supported by a QoS-aware AP, is called 'QBSS' in the 802.11e standard as compared to 'BSS' (basic service set), the group of stations supported by a legacy AP.

Prioritization for the various functions of the channel access protocol is achieved by imposing waiting requirements of variable durations after the channel becomes available. The different durations are known as interframe spaces, with the shortest, SIFS (short inter-frame space), used when a transmission is acknowledged, PIFS (priority inter-frame space) used for PCF, and DIFS (DCF inter-frame space) required of stations using DCF.

Abbreviations or acronyms used in this chapter are defined below.

AP	access point
AIFS	arbitration inter-frame space
APSD	automatic power save delivery
CSMA/CA	carrier sense multiple access/collision avoidance
CW	contention window
CWMax	contention window maximum
CWMin	contention window minimum
DCF	distributed coordination function
DIFS	DCF inter-frame space
EDCA	enhanced distributed channel access
HCCA	HCF controlled channel access
PIFS	priority inter-frame space
PCF	point coordination function
S-APSD	scheduled APSD
SIFS	short inter-frame space

TCMA	tiered-contention multiple access
TSPEC	traffic specification
TXOP	transmit opportunity
U-APSD	unscheduled APSD
WLAN	wireless local area network

3.2 Channel Access

A Wireless LAN operates on either the 2.4 GHz ISM band or the 5 GHz UNII band, each containing multiple radio channels. The IEEE 802.11 standard specifies procedures for WLAN stations by which they share a single radio channel for asynchronous data transfer. Two channel access mechanisms are specified, contention-based and polled access. With contention-based access, stations transmit to peers and to the AP by accessing the channel using the distributed random access method that employs the CSMA/CA (Carrier Sense Multiple Access/Collision Avoidance) MAC protocol [3]. If an AP is present in a WLAN, peer-to-peer communication is not allowed independently of the AP. With polled access, the AP transmits frames to a station and polls for its transmissions.

The IEEE 802.11e standard amendment also provides contention-based and polled access mechanisms, both of which represent enhancements of the 802.11 mechanisms. The latter are referred to in this chapter as the 'legacy' mechanisms. 802.11e aims at reducing access delay and jitter in delivering QoS-sensitive frames from the source to the destination through enhanced functionality at the MAC Layer. The access delay comprises over-the-air time and queuing delay plus time consumed in retransmissions, when they occur. To achieve this goal, channel access in an 802.11e-compliant WLAN distinguishes among priorities of individual frames as introduced by IEEE 802.1D [4]. QoS-sensitive traffic is typically assigned higher priority than best effort data. Stations may transmit/ receive traffic streams of different priorities concurrently. The channel access mechanisms are described in detail later in this section.

The 802.11e amendment also introduces features to improve channel use efficiency. It allows stations in WLAN served by an AP to communicate directly with one another. Signaling must be exchanged between the stations, through the AP, according to the Direct Link Setup protocol.

A block acknowledgement mechanism is introduced in 802.11e, which improves channel efficiency by aggregating several acknowledgements into a single frame. Special signaling is needed between stations to negotiate this type of acknowledgement.

Another efficiency enhancing feature introduced in 802.11e is the transmit opportunity (TXOP). In a contention based TXOP, a station may transmit a sequence of frames without having to contend for the channel, following a successful channel access attempt. Because all but one frame in a TXOP is transmitted without contention, TXOPs help reduce the frequency of collisions and thus increase channel use efficiency. A station granted a 'polled TXOP' when polled by the AP may transmit several frames to the AP, thus obviating the need for additional polls.

The remainder of this section describes the 802.11e channel access mechanism after first presenting the channel access protocols employed in 802.11, namely contention-based and polled access.

3.2.1 Legacy Channel Access Methods

3.2.1.1 Legacy Contention-Based Channel Access

According to the legacy contention-based access mechanism, DCF, each station listens to the channel and, if busy, postpones transmission and enters into the 'backoff procedure' [2]. This involves deferring transmission by a random time, which facilitates collision avoidance between multiple stations that would otherwise attempt to transmit immediately after completion of the current transmission.

The length of time for which a station will postpone its transmission depends on the 'backoff value', a number chosen randomly from a range of integers known as the Contention Window (CW). The backoff value expresses, in time slots, the cumulative time the channel must be idle before access may be attempted, excluding an additional time interval DIFS that the channel must remain idle following each period the channel is busy. In other words, transmission occurs after initially setting a 'backoff timer' to the backoff value, and then counting down once for every slot of time that the channel remains idle following a busy period excluding an initial idle period of length DIFS. A transmission may not be attempted until the backoff timer expires. CW is set initially at the value CWmin, and is doubled after each collision involving its transmitted frame, until reaching the value CWmax; after which it remains constant for further retries. The frame is dropped if it cannot be transmitted successfully after a specified number of retries. CW is reset to CWmin following a successful transmission. When the backoff timer expires, it is reset to a new backoff value, drawn randomly from the contention window CW, regardless of whether there is a frame queued for transmission.

Following anyone's transmission on the channel, a station is allowed to transmit only after the channel remains idle for at least a DIFS time interval. The value of DIFS is selected in order to enable DCF to share the same channel with the centralized protocol PCF of the 802.11 standard. In PCF, the channel is accessed at the end of a transmission after a PIFS idle interval, which is shorter than DIFS.

3.2.1.2 Legacy Polled Channel Access

At pre-specified regular time intervals, an AP engaged in polled access starts a 'contention free period' by transmitting a beacon frame. The AP can access and maintain control of the channel, once the channel becomes idle for the duration of the contention-free period by transmitting after a shorter time interval, PIFS, than a station. In addition, a station hearing the beacon will refrain from transmitting if not polled until it receives notice from the AP that the contention-free period is over.

The AP first transmits all broadcast and multicast frames and frames addressed to power-saving stations that subscribe to polled access. The transfer of frames to and from non-power-saving stations follows. The AP maintains a 'polling list' of stations to be polled, and polls each of them at least once during a contention-free period. At the same time the AP transfers frames to the stations on the polling list. Every poll elicits a single frame from the polled station. The AP stops polling a station in a contention-free period if it receives a Null frame (frame containing a header but no body) in response to a poll. An acknowledgement, data and/or poll can be combined into a single frame in order to save overhead.

3.2.2 802.11e Contention-Based Channel Access

The IEEE 802.11e contention-based access mechanism, EDCA, extends the contention-based access mechanism of the 802.11 standard to provide frame prioritization [1]. That is, given a collection of contending entities, prioritized access enables frames of higher priority to access the channel sooner. IEEE 802.11e uses the TCMA MAC protocol, a variant of CSMA/CA designed for priority differentiation [5, 6].

A key consideration in formulating EDCA was fairness. Because certain stations will transmit frames of different priorities, while others will transmit frames of a single priority, it was important for the channel access mechanism to provide the same performance to all frames of a given priority regardless of their source. Thus, instead of buffering all frames in a single queue (as with 802.11 stations), an 802.11e station employs four queues, one for each 'access category' based on the frame's priority [7]. The mapping of user priorities to access categories specified for a WLAN must be observed by all the stations. Traffic in an access category mapped to higher priorities will access the channel more readily than lower-priority access categories.

The different queues of a station contend for channel access independently of one another, almost as if they resided in different stations. The only difference is that any internal collisions between two queues of the same station are resolved by allowing the higher priority frame to be transmitted, while the lower priority frame is treated as if it had experienced a collision.

A contending queue in a station with multiple access categories behaves just like a station with traffic of a single access category with respect to accessing the channel. For simplicity of presentation, therefore, both are referred to as 'a station' in the description of prioritized access that follows.

The underlying MAC protocol in 802.11e contention-based access is CSMA/CA, which was described above. In 802.11e, the protocol's parameters CWmin and CWmax values are allowed to vary with the access category [8]. Assigning lower CWmin or CWmax values to an access category causes the contention window to be shorter when transmitting, or retransmitting following a collision. This indeed could offer priority differentiation, as shorter backoff values would lead to shorter average access delays, provided that the number of contending stations in the access category in question is small. The user should be cautioned, however, that when the number of contending stations in the access category is large, a short contention window would cause multiple stations to draw the same backoff, leading to collisions and consequently longer rather than shorter delays [9]. When differentiation with respect to CWmin or CWmax is pursued, the AP must be equipped to adjust these parameters dynamically in response to traffic conditions. The 802.11e standard permits such adjustments.

Another priority differentiation mechanism for contention-based access is through differentiation of the arbitration time, referred to as AIFS in the 802.11e standard. This concept was introduced as part of the TCMA MAC protocol, which is described in the next section [5, 6].

Like the contending stations, the AP also differentiates between traffic of different priorities by using different access parameters. The access parameters used by the AP for a given access category may be different from those used by the stations. The AP is thus allowed to use higher-priority access parameters than the stations, a prudent measure since the AP typically transmits more traffic than a station.

An important-efficiency enhancing feature introduced in 802.11e is the transmit opportunity [10]. Following a successful channel access attempt, a station may transmit a sequence of frames without having to contend for the channel. That is, a station is allowed consecutive transmissions of frames from the same access category without the need for backoff by using spacing between consecutive frames of SIFS, which is the shortest inter-frame spacing. The station thus maintains control of the channel for the entire TXOP by waiting a shorter time between transmissions than any other station. A good portion of a TXOP is also protected from collisions with hidden terminals. The acknowledgement to a frame indicates in the Duration field the length of the following frame in the TXOP or the remaining TXOP duration. This length is derived from the Duration field of the transmitted frame that is acknowledged. Because all but one frame in a TXOP is transmitted without contention, TXOPs help reduce the frequency of collisions. This increases channel use efficiency.

Another useful efficiency measure introduced in 802.11e is the expiration of frames based on the time queued, waiting for transmission. The time limit for the expiration of a frame, known as MSDULifetime, varies with access category, as overly delayed frames may not be useful in some applications but useful in others. For instance, applications with low latency tolerance, like voice, drop excessively delayed frames on the receiving end. Excessively delayed frames are dropped in this case without further transmission attempts, thus making room for other transmissions. Naturally, one must be careful when setting MSDULifetime limits for different access categories to separate traffic types with different tolerance for packet loss. For instance, VoIP signaling and VoIP media should be in separate access categories in such a case.

The impact of dropping excessively delayed frames has been studied in [5]. Figure 3.1 shows the effect of dropping voice frames if the time spent in the MAC layer exceeds the MSDULifetime limit. The average over-the-air delay experienced by a single station engaged in a voice call without dropping frames appears in Figure 3.1(a). Figure 3.1(b) shows the same resulting delay as a consequence of dropping any frames delayed by 20 milliseconds.

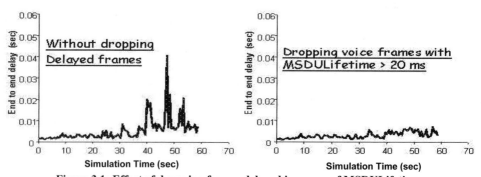

Figure 3.1: Effect of dropping frames delayed in excess of MSDULifetime.

It is important to note that, because the contention-based access mechanism of 802.11e is based on access prioritization, part of the advantage it brings to a WLAN over the legacy CSMA/CA protocol is lost when the traffic load consists primarily of one type of traffic – as for instance in the case of call centers, where much of the traffic comprises voice calls. The channel overhead penalty introduced by the longer MAC headers of 802.11e frames may counteract efficiency gains introduced in EDCA. The benefit of choosing EDCA over legacy DCF in such a case relates to the greater flexibility found in the former, as illustrated in Section 3.4.

3.2.2.1 TCMA MAC Protocol

According to the CSMA/CA protocol, as implemented by 802.11, a station engaged in backoff countdown must wait while the channel is idle for time interval equal to DIFS before decrementing its backoff immediately following a busy period, or before attempting transmission. According to the TCMA (Tiered Contention Multiple Access) protocol, variable lengths of this time interval, which is called Arbitration-Time Inter-Frame Space (AIFS), lead to varying degree of accessibility to the channel [5,6]. A shorter AIFS will give a station an advantage in contending for channel access. Differentiation between different access categories is achieved by assigning a shorter AIFS to a higher priority access category. An example is shown in Figure 3.2.

The effectiveness of priority differentiation of access categories is only partly due to allowing the station with the shortest waiting requirement to access the channel first, given two or more stations with expired backoff. This mechanism was used in 802.11 to give priority to stations engaged in PCF to access the channel before any other station. For instance, an AP would wait an idle time period of length PIFS, which is shorter than the length of DIFS required of a station. When a legacy AP has to engage in backoff, however, it uses the same backoff countdown rules as a station. It must wait for an idle interval of DIFS duration before decrementing its backoff timer.

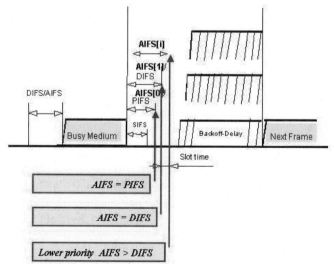

Figure 3.2: AIFS differentiated contention-based access.

The benefit from shortening the waiting time for transmission is small relative to the effect of the different AIFS values when decrementing the backoff timer. Since countdown of the backoff timer following a busy period may not occur unless the channel has been idle for a time period equal to AIFS, backoff countdown of lower priority frames slows down, and even freezes, in the presence of higher-priority frames with expired backoff. This is because a transmission will occur and the channel will be busy again before the lower-priority station, with the longer AIFS, has a chance to decrement its backoff timer. This would occur commonly in congestion.

Hence, in congestion conditions, the priority mix of stations with expired backoff timers favors higher priorities. In general, high priority stations will have lower backoff values than lower-priority stations when one looks at the residual backoff values of a mix of stations at any point in time. This desirable result is achieved without shortening the contention window from which the backoff value is drawn, which if pursued would increase the likelihood of collisions among the high-priority stations. Given any mix of initial random backoff values, the tendency of high-priority frames to reduce their backoff faster than lower-priority frames under TCMA leads to lower delay and jitter than without AIFS differentiation. Finally, the same tendency also reduces the likelihood of collisions between frames of different priorities, thus leading to a lower collision rate and higher throughput. These observations, which lead to the adoption of AIFS differentiation into the 802.11e standard, have been confirmed by subsequent performance evaluation studies [11 – 12].

3.2.3 802.11e Polled Channel Access

The IEEE 802.11e standard improved the PCF polled channel access mechanism of the earlier 802.11 standard to achieve better delay and jitter performance and greater channel use efficiency. The enhanced mechanism, called HCF controlled channel access (HCCA) in the 802.11e standard, resembles PCF, but with the following modifications. Polling is not limited to the contention-free period, but instead it can occur any time. The polling schedule is tailored to the time profile of the individual traffic streams, thus reducing both overhead, delay, and jitter. Overhead, delay, and jitter are also reduced through uplink TXOPs, which cause frames to be transmitted sooner than would have been possible otherwise.

In general, a *service period* is a time interval of continuous communication between the AP and a station, comprised of downlink transmissions and/or a poll and the station's response to the poll. Polled-access service periods occur periodically at a negotiated service interval subject to limited time slippage. The AP transmits downlink frames to stations as single frames or as TXOPs. A downlink frame may be combined (or piggybacked) with a poll. With the poll, the AP grants a polled TXOP to the station. That means a response to a poll may consist of multiple uplink frames. An uplink frame can be combined with the acknowledgement to a downlink frame. The station can request extension of the TXOP by indicating the desired duration in a special QoS control subfield: TXOP Duration Requested. Uplink transmissions are protected from contention from other stations in the WLAN for the value of the Duration subfield in the downlink frame(s) sent to the station during the station's service period.

By allowing multiple frames to be transmitted uplink without contention, in response to a single poll, a lot of the signaling frames that would otherwise be required are eliminated. TXOPs reduce contention when employed by either access method. TXOPs that are secured by the AP and granted to a station employing polled access give the station priority over any station using contention-based access, regardless of their respective priorities.

To match polling frequency to the traffic, a station that starts a new traffic stream exchanges signaling with the AP to establish the schedule by which the station will be polled. A station may have several traffic streams going on at once. An ADDTS frame is submitted for each traffic stream associated with the station, describing various aspects of transmission/delivery in the TSPEC element. These include the following: the nominal size of data frames (Nominal MSDU Size), the average bit rate at which data is generated (Mean Data Rate), the maximum delay allowed for queuing and transport of frames across the channel (Delay Bound), the maximum time allowed between consecutive service periods granted to the station (Maximum Service Interval) for the traffic stream, and the minimum physical bit rate to be assumed in establishing a schedule (Minimum PHY Rate). Each stream may have a different polling schedule. Alternatively, a station may request a single aggregate polling schedule for all admitted traffic streams. It does so by setting the Aggregation subfield in the TS Info Field of the TSPEC element equal to 1.

If the AP can accommodate the stream specified in the ADDTS request, it will indicate so in an ADDTS response that includes the Schedule element, specifying the schedule of the delivery of data and polls. If an ADDTS request is declined, the station may employ contention-based access for the traffic stream. A traffic stream is deleted when a station sends a DELTS frame to the AP. The negotiation between the station and the AP in establishing a polling schedule for each traffic stream, through the submission of ADDTS frames, provides a stand-alone admission control mechanism. As explained above, polled access has priority over contention-based access. It is not necessary, therefore, to restrict access of coexisting contention-based stations through admission control in order to enable polled stations to enjoy guaranteed delay/jitter performance.

The enhanced polled access mechanism of the 802.11e standard may operate during both the contention and the contention-free periods into which the channel time is typically partitioned. The AP can access the channel during the contention period by using PIFS, a shorter waiting requirement than that for stations, to initiate service periods for the stations with admitted traffic streams [13 – 15]. As a consequence, it is expected that, in practice, 802.11e APs will allocate most of the channel time to contention periods.

Compared to the legacy PCF mechanism, the 802.11e polled access mechanism results in a polling schedule that better matches the generation of frames in a periodic traffic stream. This results in superior delay/jitter performance and better channel use efficiency. The transmission of multiple uplink frames per poll also increases channel use efficiency.

Relative to contention-based access, scheduled polled access leads to better channel use efficiency because stations in the same WLAN (that is, stations served by the same AP) do not contend for the channel, thus eliminating the possibility of collision among them. The superior delay/jitter performance of polled access in 802.11e makes it the ideal choice for voice and streaming multimedia applications.

3.2.4 Illustrative Examples

Time-sensitive traffic occurs in diverse environments, with a different mix of traffic priorities. The prioritization capability of EDCA has been demonstrated in several performance studies [5, 9, 12]. Figure 3.3 illustrates the impact of AIFS differentiation on the average over-the-air delay experienced by nine high-priority voice streams using an 802.11b channel in the presence of lower priority data traffic, considered in [5]. Figure 3.3(a) shows the average delay experienced by the voice streams if the legacy DCF access mechanism was used and Figure 3.3(b) shows the delay experienced with EDCA.

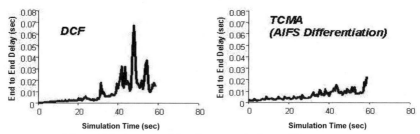

Figure 3.3: Average delay for top priority traffic category.

Prioritized access is useful if both low and high priority traffic are present in the same WLAN. The question thus arises whether EDCA would be of value in WLANs carrying mostly traffic of the same priority, such as call centers. The value of EDCA in such environments stems from its flexibility and the efficiency of channel utilization it introduces. For instance, EDCA can be of benefit because it allows the AP to use different access parameters than the stations.

Identical EDCA access parameters across all entities contending in a given priority class lead to consistent performance for all the traffic in that priority class only if these entities have comparable traffic loads. There is a pronounced load-induced inequity in the case of the AP. The AP has more traffic to transmit than any individual station since the uplink traffic is distributed among multiple stations and, in general, the downlink traffic in a WLAN is heavier than the uplink traffic. In the case of voice calls, the AP must transmit multiple voice streams, one for each station engaged in a voice call, while the stations transmit one voice stream each. By allowing the AP to contend for the channel with higher-priority EDCA parameters, downlink delays are shortened and become comparable to those of uplink voice streams.

Allowing the AP to access the channel with a shorter AIFS duration than the stations and no backoff requirement increases the voice capacity of a WLAN by as much as 38 per cent [16]. The voice capacity of a WLAN is the number of simultaneous voice calls that result in bounded over-the-air delays and no buffer overflow. Assuming an error-free channel, 46 voice calls with 20 milli-second frame interarrival time can be carried in an 802.11a WLAN when the AP uses the same access parameters as the stations. The WLAN voice capacity becomes 58 when the AP is allowed to transmit with AIFS equal to PIFS and a contention window of size zero.[1]

---

[1] Since the conference proceedings where the results in reference 16 of this chapter are not readily available, a synopsis is included as an endnote.

Using the HCCA polled access mechanism of 802.11e can increase the voice call capacity of a WLAN further, as it provides collision-free transmission. For comparable conditions, the voice capacity of an 802.11a WLAN is 65 voice calls [17].

The use of HCCA introduces a 12 per cent increase relative to the capacity achieved with optimized AP access parameters. Such a gain may seem insufficient to justify the complexity of implementing the scheduling algorithm required for HCCA. Considering EDCA as the alternative, some algorithmic complexity is also needed in order to achieve high capacity consistently. It relates to the choice of the access parameter values for different traffic conditions. The 802.11e standard does not specify how these parameters must be set; a task left to the user. The appropriate choice of a certain EDCA parameter value – namely, the contention window size – depends on the traffic conditions. The wrong choice could result in capacity loss, because of aggressive behavior and a high collision rate. This point is illustrated in [17], where choosing the standard default contention window value leads to a capacity of only 35 voice calls for an 802.11a WLAN employing EDCA. In the absence of special optimization algorithms for adaptation of the contention window size to traffic intensity, this EDCA parameter should be assigned a large fixed value.

Naturally, assigning large fixed values to the EDCA contention window size removes its effect on prioritization, leaving the AIFS size as the main priority differentiator between traffic classes. While this works for exclusively 802.11e WLANs, mixed systems are problematic. The range of ten time slots provided in the 802.11e standard for the AIFS duration is sufficiently wide to enable adequate priority differentiation [11, 12]. The entire range is not available, however, when 802.11e-compliant stations must co-exist with legacy stations. Legacy stations, which employ an AIFS interval of length DIFS+1, must be treated as having low priority traffic. The effective AIFS range is thus reduced to a single time slot, which may not be sufficient for differentiation among multiple classes. Additional differentiation would thus be of value. Hence the need to differentiate based on contention window size in this special case, in spite of the caveats.

3.3 Admission Control

Admission control provides bandwidth management to ensure that QoS-sensitive applications, such as voice and video, will be afforded a satisfactory quality of service. Overloading the WLAN with an excessive number of users entitled to high-priority access would make it hard to provide consistent QoS. Therefore, requests are submitted by stations for the admission of specific traffic streams to the AP, which keeps track of the traffic on the channel and accepts or declines the request. The information contained in this exchange will depend on the channel access method involved; it will be different for contention-based access and for polled access.

Admission control is an intrinsic part of polled access, and thus comes automatically with the decision to use this access method. Admission control is an option that is available for stations using contention-based access. It is important to note that admission control becomes imperative for contention-based stations with QoS traffic in a WLAN that supports polled access, unless polled access is limited to just top priority traffic. Stations using contention-based access will access the channel with lower priority than any station that uses polled access, regardless of their respective traffic priorities. Because the AP can transmit before any station, it can give a polled station an opportunity to transmit before any contention-based station.

3.3.1 Admission Control for Contention-Based Channel Access

Admission control for contention-based access is an optional feature for a station and an AP. It involves the decision at the AP to allow stations that employ contention-based access in the WLAN to transmit traffic using the parameters of an access category. This enables the AP to track and manage bandwidth use. It is not necessary to impose admission control on all access categories. The 802.11e contention-based access mechanism shields the admitted traffic from contention by lower-priority transmissions. It is important, however, to require admission control in all access categories of higher priority than the access category of the traffic of interest. The contention to be experienced by traffic in a given access category cannot be bounded if traffic in access categories of higher priority is unrestricted.

The basic procedure of admission control for distributed access is the following. In its beacons, the AP advertises to the WLAN the access categories that are protected by admission control. A station that has traffic to transmit or receive in a protected access category must request permission from the AP before it is allowed to do so. The signaling is similar to that used for the admission of a traffic stream for polled access [18]. A station's request, submitted in an ADDTS (add Traffic Specification) frame, describes the 'traffic stream' to be admitted. The description includes the data frame size, the mean data rate, and the minimum physical transmission rate for each of the directions on which the channel would be accessed with the parameters of the access category in question. If an ADDTS frame indicates a bi-directional traffic stream, traffic is specified for one of the two directions; the other is assumed to be the same.

The response to the ADDTS request, if affirmative, furnishes in the Medium Time field the 'channel time' the station is allotted for uplink transmissions using the parameters of the access category specified in the request. The allotted channel time is expressed as the number of time units the channel may be used by the station for its transmissions over a fixed known time interval. If the AP declines an ADDTS request, the station may still transmit, but with parameters of a lower-priority access category that requires no admission control. There should be at least one access category without the admission control requirement. Stations that do not support admission control may transmit only with parameters of access categories of equal or lower priority, and for which admission control is not mandatory.

Once a station receives its allotted channel time for a particular access category, it keeps track of the portion that has been used up for its transmissions, and for any retransmissions. The station may request additional channel time for an admitted traffic stream if its allocation is being used up too fast, or if a new data flow is added to the same traffic stream. A single admitted traffic stream could be specified per access category, which would be the aggregate of several data flows. The station updates the combined requirements of all data flows in the access category in question and sends a new ADDTS request for an updated allocation. To give up all of its allotted channel time for a particular access category, a station submits a DELTS (delete Traffic Specification) frame. The channel time allotted to a station for an access category is released if no transmissions in that access category to/from the station have occurred for a specified time period, the Inactivity Interval, which is indicated on the ADDTS frame.

3.3.2 Admission Control for Polled Channel Access

Admission control is exercised automatically when using polled channel access. The AP will reject an ADDTS request if it cannot meet the requirements for a service period schedule requested by a station for a traffic stream. If the requested requirements can be met, the AP responds with a service period schedule. Unlike in the case of contention-based access, a station using polled access may have several admitted traffic streams of the same priority.

During the negotiation, a minimum set of parameters must be specified in the ADDTS request so that the AP can schedule time on a service period for the traffic stream that is to be admitted. These parameters include mean data rate, frame size, minimum transmission rate, and either the maximum service interval or a delay bound. If a traffic stream is admitted, the ADDTS response will include non-zero values for mean data rate, frame size, minimum transmission rate, and the maximum service interval. The ADDTS response will include a Schedule element, which provides the schedule of the delivery of data and polls. The minimum transmission rate will be used in determining the length of TXOPs and service periods.

The priority of a traffic stream may be considered in admission control. An admission control request from traffic stream with a higher priority may cause an admitted stream to be dropped. The AP sends a DELTS frame to notify a station that a traffic stream is dropped. Admission of a traffic stream may therefore not be guaranteed.

3.4 Power Management

Several of the QoS-sensitive applications will involve multimedia traffic over battery-powered handheld devices, such as a PDA or a wireless VoIP phone. In crafting a standard of good QoS performance, it was thus considered important to prolong the battery life of such devices. The 802.11e standard amendment offers several new mechanisms to help battery-powered devices conserve power by enabling them to power down their receivers and transmitters intermittently without losing connectivity or data. The new power management mechanisms apply to WLANs served by an AP – such WLANs are known as 'infrastructure' WLANs, and for this reason, the discussion in this section will focus on power saving methods for infrastructure WLANs.

A station informs the AP of its operating power-management mode, 'power saving' versus 'active', when it associates with the WLAN. The mode can be changed during the association period by changing the Power Management bit, a bit in the frame control field of the frames transmitted by the station. The AP will not send transmissions to a station that has declared itself to be in 'power save' mode, unless it knows that the station has its receiver fully powered, i.e., it is in the 'awake' state, and ready to receive. Otherwise, the AP will assume that the station's receiver is powered down, i.e., it is in the 'doze' state, and for this reason, any incoming frames addressed to a power-saving station will be buffered for later transmission.

A simple, but not efficient, way for a power-saving station to retrieve multiple buffered frames at once is to switch its power management mode to 'active'. A data frame, or a Null frame sent by station to the AP with the Power Management bit set to 0 will enable the AP to transmit the buffered data. The station may subsequently return to power-save mode using another frame with the Power Management bit set to 1. The inefficiency in this approach stems from the fact that it requires extra frames to be transmitted for signaling purposes. 802.11e introduces delivery methods with reduced signaling.

The AP may deliver buffered frames to their destination power-saving stations either on a previously negotiated schedule or in response to transmissions from the respective stations that initiate such delivery. In order to initiate delivery in the latter case, a station should know that there are frames for it buffered at the AP. Notification of the presence of buffered frames at the AP typically comes through a special station-specific field contained in the beacon frames broadcast by the AP, or in reserved fields of downlink frames directed to the individual stations. In some situations, as we will see, notification is not provided by the power-saving mechanism, and thus it must be furnished either by the application running on the station, or by transitioning to a different power saving mechanism that provides such notification.

The station chooses the delivery and notification mechanisms and communicates it to the AP either upon association or re-association of the station with the WLAN or through explicit signaling using an ADDTS frame. The various mechanisms available in an infrastructure WLAN will be described in the following section. They include (1) the 'legacy' power save mechanism, which was available pre-802.11e. APSD (automatic power save delivery) was introduced by the 802.11e standard to reduce the signaling that would otherwise be needed for delivery of buffered frames to power-saving devices by an AP. APSD provides two ways to start delivery: (2) 'scheduled APSD' (S-APSD) and 'unscheduled APSD' (U-APSD). Unscheduled APSD can take (3) a 'full' U-APSD or (4) 'hybrid' U-APSD form. With full U-APSD, all types of frames use U-APSD independently of their priority. Hybrid U-APSD employs a combination of U-APSD and the legacy power save mechanism.

3.4.1 Legacy Power-Save Mechanism

The legacy power-save mechanism applies to both infrastructure WLANs and WLANs without an AP. We describe here how it works with the former since the new power save mechanisms deal only in WLANS served by an AP. For information on the latter, the reader is referred to the 802.11 standard.

Frames buffered at the AP for a power-saving station employing contention-based access are delivered when the station sends a special control frame, the power save poll (PS-Poll). The AP sends a single buffered frame to a station after receiving a PS-Poll, either immediately or soon thereafter. More PS-Polls are required in order to retrieve additional buffered frames. The presence of further frames remaining at the AP is indicated by the More Data bit of the control frames of the transmitted frame, which is set to 1.

A station using legacy power save can rely on the traffic indication map (TIM) to learn if the AP holds buffered data for it. The TIM is a bit map containing the buffer status per destination station. It is sent regularly on beacon frames broadcast by the AP at known times. If the station is in the 'doze' state, it will wake up at the beacon times to receive and interpret the TIM. Alternatively, a station can ascertain the presence of additional frames buffered for it at the AP while receiving a buffered frame. The More Data bit in the control field of that frame would have been set to 1 if additional frames remained buffered for the station.

A power-saving station that supports legacy polled access need not send PS-polls in order to receive its buffered frames. The station receives its buffered frames at the start of the contention-free period, when it awakens to listen to the TIM and learn of its buffer status at the AP. Such a station would probably not request to be on the polling list because that would require staying awake for the entire contention-free period. Uplink frames are sent by contention in that case.

3.4.2 Automatic Power Save Delivery

APSD is a mechanism for the delivery of unicast frames from the AP to a power-saving station. This mechanism was introduced by 802.11e in order to reduce the signaling traffic caused by PS-Polls and their acknowledgements. A station may use both APSD and legacy PS-Polls at the same time to retrieve buffered frames from the AP. Certain restrictions apply, however, which are discussed below. To use APSD, stations must have the Power Management subfield in the control field of all transmitted frames set to 1.

The AP may deliver buffered frames to their destination power-saving stations either on a previously negotiated schedule or in response to receiving transmissions from the respective stations that trigger such delivery. The two APSD approaches are thus known as 'scheduled' and 'unscheduled'. A station may use both approaches at the same time, provided that only one is used for a given access category.

3.4.2.1 Scheduled APSD

This mechanism is well suited for periodic traffic streams, such as voice and audio/video, and is especially good for unidirectional downlink periodic streams. With scheduled APSD, downlink transmissions to power-saving devices will occur at a schedule that is known in advance, obviating the need for special signaling between the station and the AP.

The AP and the station negotiate in advance a time schedule by which the station will power its receiver fully to receive any frames that are buffered for it at the AP. A station that wishes to use S-APSD must send an ADDTS request with the APSD subfield in TS Info field of the TSPEC element set to 1. The TSPEC element contains the time of the first downlink transmission (Start Service Time) as well as the time interval at which downlink transmission will be repeated (Service Interval), as in the case of polled access. The Start Service Time is expressed in terms of the time shared in the WLAN, known as the TSF timer [19]. While the Start Service Time field is used optionally with polled access, this field must be specified when using Scheduled APSD, as knowledge of the time of downlink frame delivery affords a station the longest stay in the doze state.

The AP is given the last say in setting the start time of the periodic transmissions to the station so that its transmissions to different power-saving stations are staggered in a way that minimizes the time the power-saving stations are awake. If the request is accepted, the AP will return an ADDTS response containing a Schedule element, which, among other, includes the Start Service Time selected by the AP. The station will wake up to receive its buffered frames at the times indicated by the returned schedule.

Either channel access method, polled or contention-based access can be used with Scheduled APSD. Scheduled APSD fits naturally with polled access. To indicate polled access, the Access Policy subfield of the ADDTS TS Info field would be set to (0, 1), and the Start Service Time field in the TSPEC element must have a nonzero value. When the station plans to use contention-based access with Scheduled APSD, the Schedule subfield of the ADDTS TS Info field must be set to 1, and the Access Policy subfield must be set to (1,0).

For stations using Scheduled APSD in conjunction with contention-based access, the uplink transmissions do not require polling. A power-saving device that uses contention-based access can transmit to the AP at any time.

3.4.2.2 Full Unscheduled APSD

Unscheduled APSD was introduced for stations accessing the channel by contention, in order to enhance the efficiency of legacy power save. A power-saving station may use not just a PS-Poll, but also any data or Null frame – referred to as a 'trigger' frame – in order to notify the AP that its receiver is fully powered and ready to receive transmissions [20, 21]. Using a data frame that is pending transmission at the station, instead of a PS-Poll, to initiate downlink transmission clearly reduces the traffic generated by the station and increases battery life and channel use efficiency.

Additional gains are achieved from relaxing the number of frames the AP is allowed to transmit to a power saving station when it receives notice to do so. While receiving a PS-Poll from a station allows the AP to transmit a single downlink frame -- of the highest priority access category buffered -- receiving a trigger frame will start an APSD service period for that station. During a service period, the AP may send multiple frames, subject to a limit specified by the station. Eliminating the extra signaling that would otherwise be necessary under legacy power save also increases the efficiency of channel use and conserves battery life.

Naturally, since the station does not know in advance the number of frames sent by the AP in a service period, it must be notified when the last frame has been transmitted for a given service period so it may transition to the doze state. The control subfield EOSP in the last delivered frame marks the end of a service period.

The AP need not deliver all frames buffered for a station in a single service period. As in the case of legacy power save, the More Data control subfield in a last frame transmitted in a service period indicates whether there are frames remaining buffered at the AP. Knowing its AP buffer status enables the station to send another trigger frame or PS-Poll to retrieve more of its buffered frames.

As with the legacy power-save mechanism, a station can learn about its buffer status by listening to the beacons for its TIM [21]. This is needed only while not receiving frames from the AP, as the More Data control subfield in downlink frames to the station conveys the same information.

To use full U-APSD, a station sets the first four bits of the QoS Info subfield of the QoS Capability element in the (re-) association request all to 1. The Max SP Length subfield of the QoS Info field is used to place a limit on the maximum number of frames to be delivered during a service period. For unrestricted delivery, this subfield should be set to 0.

3.4.2.3 Hybrid Unscheduled APSD

The hybrid U-APSD mechanism allows a station to choose between legacy power save delivery and APSD based on access category [22]. Trigger frames are used to initiate the delivery of buffered frames associated with access categories that have been designated as 'delivery enabled'. Buffered frames of access categories not so designated, can be retrieved with PS-Polls only. The station also designates in advance the access categories of the frames that may serve as trigger frames. An AP receiving frames in categories other than those designated by a station as trigger-enabled will not transmit buffered frames to the station.

The end of a station's service period and the presence of further frames remaining buffered at the AP are indicated by the control subfields EOSP and More Data of frames received by the station. However, unlike in full U-APSD where the More Data bit indicates the presence of buffered frames remaining at the AP, the same bit in hybrid U-APSD would indicate only whether frames of similar characterization (e.g., delivery enabled versus non-delivery enabled) as the received downlink frame remain buffered.

One can visualize the hybrid U-APSD mechanism as partitioning the incoming traffic of the different access categories into two sets, each directed to a different power-save buffer for a station, the legacy and APSD buffer. The notification and retrieval mechanisms work independently of one another, with PS-Polls used to retrieve frames from the legacy buffer, and a trigger frame causing frames to be delivered from the APSD buffer. The More Data bit on a downlink frame shows the status of the buffer in which the frame was held.

While stations receiving buffered frames know whether additional frames remain buffered from the More Data control subfield, stations not receiving any frames may have no way to knowing that frames are waiting at AP. The TIM in the beacons is used differently in hybrid U-APSD mechanism than the other power save mechanisms. It shows whether the AP has buffered for the station frames of non-delivery-enabled access categories only. Since the TIM does not account for frames of delivery-enabled access categories, a more pro-active way is needed for a station to figure out whether it has buffered frames of such access categories and must therefore initiate frame retrieval.

A station that has no uplink traffic may send a Null frame of a trigger-enabled access category uplink in order to both check buffer status and, if frames are buffered, initiate their retrieval. The AP responds with a Null frame if there are no frames buffered in the delivery-enabled access categories. The spacing of the uplink Null frames cannot be too long if the traffic in the delivery-enabled access categories is delay sensitive. The frequency of such Null frames cannot be high either, as that would increase channel load and battery drain unnecessarily.

Sending Null frames in order to retrieve buffered frames is inefficient for a power-saving station if it does not generate regular uplink traffic in a trigger-enabled access category and does not expect regular downlink traffic in a delivery-enabled access category. A more efficient way to retrieve buffered traffic in such conditions is to alter the characterization of the delivery-enabled access categories so that the status of the corresponding buffers will be included in the station's TIM. There are two options. One is to disable automatic delivery in all access categories, and then retrieve frames one at a time with PS-Polls. The second option is to enable automatic delivery for all access categories, and retrieve frames using full U-APSD. Changing the characterization of an access category requires further signaling – that is, re-association or the submission of TSPEC requests, one for each affected access category.

The designation of access categories as delivery- or trigger-enabled occurs through (re) association frames, by setting the corresponding subfields in the QoS Info subfield of the QoS Capability element. These designations may be also set or altered for an access category by submitting ADDTS frames for that access category, one for downlink and another for the uplink direction, indicating the new delivery and triggering capabilities, respectively.

To unify signaling for full and hybrid U-APSD, the convention was adopted to have a station employing full U-APSD designate all its access categories as delivery enabled.

3.4.3 Illustrative Examples

Examples of the use of APSD in various applications are given in this section. Aside from technical restrictions, some application can benefit more from a specific choice of a power-save mechanism.

Examples of an application for which Scheduled APSD provides an ideal power save mechanism are one-way periodic streams, like Internet Radio. In such applications, with acknowledgements suppressed, the traffic load consists primarily of periodic unidirectional streams from the AP with occasional uplink frames. Scheduled APSD enables the transmission of the periodic stream without the need for redundant uplink transmissions.

Scheduled APSD used in conjunction with contention-based access enables any frames generated by the station to be transmitted immediately without waiting for the station to be polled. In a congested WLAN, experiencing a lot of uplink delay jitter, Scheduled APSD prevents downlink traffic from assuming this delay jitter from uplink frames, as would be the case with U-APSD. Hence, Scheduled APSD would result in lower delay.

Finally, Scheduled APSD used in conjunction with contention-based access is ideally suited for wireless ad-hoc and mesh networks. In such networks, the power-saving device wakes up to receive frames that have been stored by a neighboring device according to a pre-negotiated schedule.

Full U-APSD is a simple, efficient, power-save mechanism appropriate for any mix of traffic, uplink and downlink, periodic and non-periodic. With bi-directional periodic streams, as for example a wireless phone application involved in a call, traffic flows back and forth between the station and the AP at regular intervals. Frames on the uplink stream cause the delivery of the downlink buffered frames. The More Data control subfield of the downlink frames notify the station as to whether more frames remain buffered at the AP. After receiving a portion of its buffered traffic, the station thus knows it must pursue further retrieval of the remaining frames. It is sufficient, therefore, for a voice-enabled device to listen for its TIM only while on standby in order to receive notice of the presence of any buffered frames, from any application, including signaling for incoming calls. Not having to listen for the TIM preserves the battery of the device.

Hybrid U-APSD offers a power saving device the ability to control the time used for the delivery of various types of traffic. This is useful when the device must tend to other time-critical activities that require postponement of the delivery of some types of traffic. Such activities would include a wireless phone engaged in channel scanning other channels for roaming possibilities. By restricting APSD delivery and triggering only to the top-priority access category during a call, a station engaged in call will receive from the AP only its buffered voice frames when it transmits voice frames uplink. However, the duration of power-save delivery can also be controlled directly by setting the Max SP Length field in the station's association request to the desired duration for a service period.

A VoIP-enabled station using hybrid U-APSD must take measures to receive notification of the arrival at the AP of any out-of-band signaling, or other traffic, regardless of whether this traffic is associated with a delivery-enabled access category or not. The station awakens to listen to the TIM in order to receive notice of buffered traffic not associated with a delivery-enabled access category. As for buffered voice frames at the AP, notification is received with the delivery of downlink voice frames in the frame's control field. When the call ends, however, there are no voice frames to convey the station's buffer status for the delivery-enabled categories. So, when the station goes on standby, it changes the delivery characterization of its access categories to non-delivery enabled in order to receive notice of their buffering through the TIM.

3.5 QoS in Wireless Mesh Networks

The concept of a wireless mesh takes on several forms, the most common being a collection of nodes that form an ad hoc network and are capable of serving as WLAN APs. Using the wireless channel, these nodes, which are called 'mesh points', can forward traffic received from 802.11 stations to other mesh points with ultimate destinations that include WLAN stations attached to other mesh points or points somewhere on the wired network. In addition to the 802.11 standard, protocols for forwarding, routing and channel access must be specified for the mesh points.[2] This requires an ad hoc networking standard with multi-hop capability.

A wireless mesh network can be used to enable WLAN service when wiring for APs is not readily available in an enterprise, or for a temporary network that can be easily set up and torn down. Mesh networks have found applications in public safety, disaster control, surveillance, and connectivity for municipal services. Municipalities and service providers are interested in wireless mesh for providing public access, an alternative to expensive home broadband in dense urban areas, or to offer inexpensive WiFi service to rural communities. In some applications, mesh points may be simply stations communicating wirelessly on a multi-hop ad hoc or infrastructure network.

The wireless mesh presents multiple challenges, including challenges in routing and security, especially when mobility is contemplated. While it is clear that tools exist to handle these issues, it is not clear what choices will ultimately be made for the IEEE 802.11 mesh standard. Channel selection, channel access, and meeting QoS requirements also present challenges in a wireless mesh network.

---

[2] The IEEE 802.11 Task Group s is developing a standard for 802.11 mesh networks.

The wireless medium providing forwarding and backhaul service for the mesh points may use either 802.11 connections, in the unlicensed spectrum used by WLANs, or some other wireless technology operating in different RF bands. Using 802.11 technology, while keeping costs low and connection simple because of the unlicensed RF spectrum, presents challenges arising from the competition between WLAN and mesh traffic. Dedicating different radios to the two types of traffic, each operating in one of the two unlicensed RF bands or on non-overlapping channels of the same RF band, eliminates the competition between WLAN and mesh traffic. By allowing these two types of traffic to be served by the same radio(s), however, and by properly managing the competition between them, one can also reduce hardware costs. RF management should be done in a distributed manner, as the requirement for controllers for this purpose might prevent hardware of different vendors from being interoperable.

Part of distributed RF management involves channel selection and channel access. Any group of mesh points that can hear one another must be able to operate on multiple channels in order to increase the mesh traffic carrying capacity. The value of single channel meshes is mostly in the home or small office where the total traffic does not exceed the traffic that can be carried by a WLAN. Single channel mesh points are useful also when there is need to set up communication quickly over a large area without wiring, as for instance in disaster control. The purpose of a single-channel mesh is primarily to extend coverage range of a WLAN through multiple-hop transmissions. As the number of hops increases, however, there would be a decrease in the mesh goodput (that is, the amount of successfully transmitted traffic that originates or terminates in the mesh), because more channel time is taken up to transmit frames end-to-end.

Another challenge is managing latency in order to meet QoS requirements. The latency experienced by frames traversing a wireless mesh increases fast when traffic bottlenecks arise as a result of traffic concentration in parts of the mesh that lack the necessary throughput capacity. For instance, this occurs in access mesh networks, where traffic concentrates at nodes near the point of attachment of the mesh to the wired network. If these nodes are not equipped with multiple radios in order to handle the higher throughput, they will become bottlenecks, contributing to frame delays and frame loss. Single radio mesh points are inadequate for a wireless mesh used to connect multiple APs. Use of two radios per node for the sole purpose of separating WLAN traffic from mesh traffic, though helpful in reducing the competition between them, is inadequate when the mesh comprises multiple APs.

As traffic is forwarded from node to node on a multiple-hop path of a wireless mesh, the latency experienced at each node adds. The 802.11e standard is useful in the wireless mesh where prioritization in a single hop can put time-sensitive traffic on the air before other traffic. However, 802.11e prioritization alone will not address the challenges arising in a multi-hop network due to accumulating latency. Methods for managing this latency are needed in order to meet the QoS requirements of real-time streaming applications such as VoIP. An example of such a mechanism is given below.

'Express' forwarding is a mechanism that illustrates how the 802.11e could be augmented in order to limit the latency accumulated over a multiple-hop path on a wireless mesh backbone. The transmitting node adds a set amount to the duration field of a QoS-sensitive transmission, while specifying that the transmission is using the express-forwarding mechanism. The immediate destination node knows that it can transmit that set amount sooner than the end of the requested duration, while neighboring nodes respect the full request. Thus, subsequent hops are spared contention delays.

The objective of express forwarding is to reduce access delay experienced by forwarded traffic after the first hop. The first hop experiences an access delay similar to any single-hop transmission using the same EDCA access parameters. A mesh point that is forwarding voice traffic beyond a single hop would not experience additional access delay, as it is allowed immediate contention-free access to the channel. Having the Duration field of the frame transmitted on the first hop set to a longer value than is necessary to complete transmission silences neighbor nodes. The node to which the frame is forwarded is permitted to transmit sooner. Without contention from neighbor nodes, the receiving node can access the channel immediately and forward the frame on to the next hop without further access delays. The Duration field value is shortened only for frames sent to intermediate nodes of a multi-hop path, and not on the final hop.

3.6 Summary

The IEEE 802.11e standard offers QoS functionality at the MAC Layer. Several new features are introduced for channel access, admission control, and power save. With the new mechanisms, a WLAN will be able to differentiate between traffic of different priorities and provide faster channel access for higher-priority traffic. At the same time, the new standard also pursues more efficient ways of utilizing the channel and better power management techniques for battery-based stations.

As in the case of the 802.11 standard, the 802.11e amendment offers two forms of channel access, polled access and contention-based access. Admission control is an intrinsic part of the polled access mechanism (HCCA) but it is an available option for the contention-based access mechanism (EDCA). Power save is an independent capability from channel access. However, when a station pursuing scheduled APSD uses polled access, it behaves very much as if it uses HCCA. Scheduled APSD can also be used with EDCA.

3.7 Endnote

The results in [16] were derived from a performance study where two access categories were used, Voice (VO) and Best Effort (BE), for voice and data respectively. Two scenarios were considered. In the first scenario, referred to as 'EDCA', the AP and the stations access the channel with the same access parameters for both traffic priorities. In the second scenario, referred to as 'PIFS Access', the AP uses different access parameters to transmit voice. With a shorter AIFS and no backoff, the AP can access the channel faster. Table 3.1 summarizes the differences in access parameters for the AP under the two scenarios. All other parameters were set at their default values indicated in the 802.11e standard [1].

Table 3.1: AP Access Parameter Values.

Parameters*	'EDCA'	'PIFS'
AIFSN[VO]	1	0
CWMin[VO]	15	0
AIFSN[BE]	2	2
CWMin[BE]	31	31

*AIFSN is the priority-dependent number of time slots that determines the AIFS length.

Tables 3.2 and 3.3 show the effect of 'PIFS' access on call capacity of a WLAN, how the capacity is impacted by traffic contending at a lower priority for the same channel, and the influence of voice packet aggregation prior to delivery to the MAC layer. For the latter, two RTP frame payload sizes are considered: one with 10 milliseconds and another with 20 milliseconds of audio data. These results are presented for the 802.11b WLANs (for a transmission rate of 11 Mbps) and 802.11a WLANs (for a transmission rate of 54 Mbps), respectively.

The presence of BE data traffic in the WLAN reduces call capacity. For the voice stream generating frames every 10 milliseconds, a capacity of 10 calls is possible in the 802.11b WLAN without any other traffic, as seen in Table 3.2. The capacity reduces to 6 calls when a station transmits data at 12.3 Mbps. For the 802.11a WLAN, the call capacity drops from 24 to 20 calls when introducing a data load of 12.3 Mbps, as seen in Table 3.3. Clearly, although EDCA expedites the transmission of higher-priority frames, it does not totally eliminate the competition for the channel from lower-priority frames.

Table 3.2: Call capacity of 802.11b WLAN.

Data traffic (Mbps)	10 ms audio		20 ms audio	
	EDCF	PIFS	EDCF	PIFS
0	10	11	18	19
0.5	9	10	17	17
2	7	10	12	17
12.3	6	10	11	17

Table 3.3: Call capacity of 802.11a WLAN.

Data traffic (Mbps)	10 ms audio		20 ms audio	
	EDCF	PIFS	EDCF	PIFS
0	24	33	46	58
0.5	23	33	44	58
2	23	33	43	58
12.3	20	33	41	58

'PIFS' access at the AP causes call capacity to increase. The benefit is greater in the presence of heavier data load, when the increased competition for channel by data transmissions leaves the AP at a greater disadvantage in accessing the channel than the voice stations. Specifically, the call capacity of the 802.11b WLAN increases from 6 calls to 10 calls (with a 10 millisecond voice interframe spacing) with 'PIFS' access when the data load is 12.3 Mbps, as seen in Table 3.2. The capacity goes from 18 to 19 calls with 'PIFS' access when no data traffic is present in the 802.11b WLAN. For the 802.11a WLAN, capacity goes from 22 to 33 calls without data traffic, and from 20 to 33 calls with a 12.3 Mbps data load (and the same audio payload), as seen in Table 3.3. 'PIFS' access at the AP brings robustness to competition from lower-priority traffic.

Finally, the effect of frame aggregation of voice traffic was considered. Increasing the payload size of RTP packets of a VoIP stream (thus reducing their arrival rate) causes the call capacity to increase, as both per call overhead and contention are reduced. In a voice-only 802.11b WLAN, the capacity goes from 10 to 18 calls when the RTP packet interarrival time increases from 10 to 20 milliseconds. A similar gain is experienced with 'PIFS' access, going from 11 to 19 in a voice-only 802.11b WLAN. The highest call capacity for an 802.11a WLAN is 58 calls under 'PIFS' access with a 20 millisecond RTP payload. Larger RTP payloads would not be advisable, as they add longer delay and jitter, affecting voice quality adversely.

3.8 References

[1]   Amendment to Wireless LAN Medium Access Control (MAC) and Physical (PHY) Layer Specification: Medium Access Control Quality of Service Enhancements, IEEE Std. 802.11e, November 2005.

[2]   IEEE Standard for Wireless LAN Medium Access Control (MAC) and Physical (PHY) Layer Specifications, ANSI/IEEE Std 802.11, 1999 Edition.

[3]   L. Kleinrock and F. Tobagi, "Packet Switching in Radio Channels: Part I – Carrier Sense Multiple Access Models and their Throughput Delay Characteristics", *IEEE Transactions on Communications*, Vol. 23, No. 12, pp. 1400 – 1416, 1975.

[4]   IEEE Standard 802.1D, "Media Access Control (MAC) Bridges," 2004.

[5]   M. Benveniste, "Tiered Contention Multiple Access (TCMA), a QoS-Based Distributed MAC Protocol", *Proceedings of PIMRC*, Lisboa, Portugal, September 2002.

[6]   M. Benveniste, "An Enhanced-DCF Proposal Based on 'Tiered Contention' Multiple Access (TCMA)", Doc IEEE 802.11-00/457, October 2000.

[7]   M. Hoeben and M. Wentink, "Enhance D-QoS through Virtual DCF", Doc IEEE 802.11-00/361, October 2000.

[8]   W. Diepstraten, "Distributed QoS resolution", Doc IEEE 802.11-00/361, October 2000.

[9]   S. Mangold, et al, "Analysis of IEEE 802.11e for QoS Support in Wireless LANs", *IEEE Wireless Communications*, pp. 40 – 50, December 2003.

[10]   S. Choi, et al, "Multiple Frame Exchanges during EDCF TXOP", Doc IEEE 802.11-01/566, January 2002.

[11]   J. W. Robinson, and T. S. Randhawa, "Saturation Throughput Analysis of IEEE 802.11e Enhanced Distributed Coordination Function", *IEEE Journal on Selected Areas in Communications*, Vol. 22, No. 5, pp. 917 – 928, June 2004.

[12]   Z. Tao and S. Panwar, "Throughput and delay analysis for the IEEE 802.11e Enhanced Distributed Channel Access", *IEEE Transactions on Communications*, Vol. 54, No. 4, pp. 596 – 603, April 2006.

[13]   M. Hoeben and M. Wentink, "Suggested 802.11 PCF Enhancements and Contention Free Bursts", Doc IEEE 802.11-00/113, May 2000.

[14] R. Meier, "An Integrated 802.11 QoS Model with Point-controlled Contention Arbitration", Doc IEEE 802.11-00/448, November 2000.

[15] M. Fischer, "A Hybrid Coordination Function for QoS", Doc IEEE 802.11-00/452, January 2001.

[16] M. Benveniste and V. Dham, "VoIP Call Capacity of a 802.11 Wireless LAN Using EDCF Access", *Wi-Fi Voice Conference 2004*, Paris - France 25-28 May, 2004, http://pubs.research.avayalabs.com/pdfs/WiFiVoice2004.pdf.

[17] S. Shankar, et al, "Optimal packing of VoIP calls in an IEEE 802.11 a/e WLAN in the presence of QoS constraints and channel errors", *IEEE GLOBECOM*, Vol. 5, pp 2974 – 2980, November 2004.

[18] R. Meier, et al, "Uniform 802.11e Admissions Control Signaling for HCF and EDCF", Doc IEEE 802.11-02/678, November 2002.

[19] M. Benveniste, et al, "Proposed Normative Text for Simplified APSD", Doc IEEE 802.11-03/107, January 2003.

[20] Y. Chen, et al, "Power management for VoIP over IEEE 802.11 WLAN", *IEEE Wireless Communications and Networking Conference*, Vol 3, pp. 1648 – 1653, March 2004.

[21] M. Benveniste, et al, "Some Power-save changes in 802.11e Draft", Doc IEEE 802.11-04/584r2, May 2004.

[22] S. Wang, et al, "Simplifications and Enhancements to Unscheduled APSD (U-APSD)", Doc IEEE 802.11-04/694, July 2004.

Performance Study of IEEE 802.11 DCF and IEEE 802.11e EDCA

Giuseppe Bianchi[a], Sunghyun Choi[b] and Ilenia Tinnirello[c]

4.1 Introduction

One of the key factors for the wide acceptance and deployment of IEEE 802.11 Wireless Local Area Networks (WLANs) is the simplicity and robustness of the Medium Access Control (MAC) protocol. Based on the well-known carrier sense paradigm, with an exponential backoff mechanism devised to minimize the probability of simultaneous transmission attempts by multiple stations, the protocol is able to work in presence of interference, which is very critical for networks operating in unlicensed spectrum. In fact, interfering sources are simply revealed by the carrier sense mechanism in terms of channel occupancy times, or by the acknowledgement mechanism in terms of collisions. However, the simplicity and the robustness have often been traded off with the efficiency of the access protocol, in terms of radio resources which are wasted or underutilized.

In this chapter, we provide a detailed analysis of the 802.11 distributed access protocol, by examining the protocol parameters which most critically affect the protocol efficiency. We quantify the protocol overheads due to control information (i.e., physical headers, frame headers, acknowledgement and other control frames) and to the distributed management of the channel grants (i.e., collisions and idle backoff slots). Then, we consider the distributed channel access extensions, defined in the recently-ratified 802.11e standard in order to support service differentiation among stations with different Quality-of-Service (QoS) requirements. Finally, we attempt to show how these parameters affect the resource repartitioning among the stations and how they can coexist with legacy DCF stations.

The IEEE 802.11 defines a basic service set (BSS) as the number of stations controlled by a single coordination function, where a coordination function is the 802.11 terminology for medium access control (MAC). There are two types of BSSs. One is the infrastructure BSS, and the other is the independent basic service set (IBSS). An infrastructure BSS is composed of a single access point (AP), (a bridge between an infrastructure, i.e., a wireline network typically, and the wireless link), and a number of

[a] *University of Roma Tor Vergata, Italy*
[b] *Seoul National University, Korea*
[c] *University of Palermo, Italy*

stations associated with the AP. Within an infrastructure BSS, a station communicate only with its AP, i.e., there is no direct transmission between two stations belonging to the same BSS. On the other hand, in an IBSS, there is no AP, and hence all the transmissions are between stations. We are primarily concerned with the IBSS in this chapter. The term "ad-hoc" is often used as an alternative term to refer to an IBSS in an IEEE 802.11 WLAN, and we use these terms equivalently throughout this chapter.

4.2 IEEE 802.11 MAC Protocol

The IEEE 802.11 legacy MAC [1] is based on logical functions, called the coordination functions, which determine when a station operating in a given 802.11 network is permitted to transmit and may be able to receive frames via the wireless medium. A data unit arriving from the higher layer to the MAC is referred to as a MAC Service Data Unit (MSDU), and the frame, which conveys the MSDU or its fragment along with the MAC header and Frame Check Sequence (FCS) based on CRC-32, is referred to as MAC Protocol Data Unit (MPDU). The MPDU is the frame that is transferred between stations from the MAC's perspective.

Two coordination functions are defined. The mandatory Distributed Coordination Function (DCF) allows distributed contention-based channel access based on carrier-sense multiple access with collision avoidance (CSMA/CA). The optional Point Coordination Function (PCF) provides centralized contention-free channel access based on a poll-and-response mechanism. Most of today's 802.11 devices operate only in the DCF mode. Accordingly, we limit ourselves to the DCF operation in this chapter.

The 802.11 DCF works with a single first-in-first-out (FIFO) transmission queue. The CSMA/CA mechanism is a distributed MAC protocol based on a local assessment of the channel status, i.e., whether the channel is busy (i.e., a station is transmitting a frame) or idle (i.e., no transmission). Basically, the CSMA/CA of the DCF works as follows.

When a frame arrives at the head of the transmission queue, if the channel is busy, the station waits until the medium becomes idle, and then defers for an extra time interval, called the DCF Interframe Space (DIFS). If the channel stays idle during the DIFS deference, the station then starts the backoff process by selecting a random backoff count. For each slot time interval, during which the medium stays idle, the random backoff counter is decremented. When the counter reaches zero, the frame is transmitted. On the other hand, when a frame arrives at the head of the queue, if the station is in either the DIFS deference or the random backoff process, the processes described above are applied again. That is, the frame is transmitted only when the random backoff has finished successfully. When a frame arrives at an empty queue and the medium has been idle longer than the DIFS time interval, the frame is transmitted immediately.

Each station maintains a contention window (CW), which is used to select the random backoff count. The backoff count is determined as a pseudo-random integer drawn from a uniform distribution over the interval [0,CW]. How to determine the CW value is further detailed as follows. If the channel becomes busy during a backoff process, the backoff is suspended. When the channel becomes idle again, and stays idle for an extra DIFS time interval, the backoff process resumes with the latest backoff counter value. The timing of DCF channel access is illustrated in Figure 4.1.

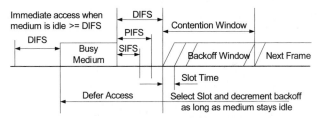

Figure 4.1: IEEE 802.11 DCF channel access.

For each successful reception of a frame, the receiving station immediately acknowledges the frame reception by sending an acknowledgement (ACK) frame. The ACK frame is transmitted after a short IFS (SIFS), which is shorter than the DIFS. Other stations resume the backoff process after the DIFS idle time. Thanks to the SIFS interval between the data and ACK frames, the ACK frame transmission is protected from other stations' contention. If an ACK frame is not received after the data transmission within the ACK Timeout[1], the frame is retransmitted after another random backoff.

The CW size is initially assigned CW_{min}, and increases when a transmission fails, i.e., the transmitted data frame has not been acknowledged. After an unsuccessful transmission attempt, another backoff is performed using a new CW value updated by

$$CW := 2(CW + 1) - 1, \tag{1}$$

with an upper bound of CW_{max}. This reduces the collision probability in case when there are multiple stations attempting to access the channel. After each successful transmission, the CW value is reset to CW_{min}, and the transmission-completing station performs the DIFS deference and a random backoff even if there is no other pending frame in the queue. This is often referred to as "post" backoff, as this backoff is done after, not before, a transmission. This post backoff ensures there is at least one backoff interval between two consecutive MPDU transmissions.

In the WLAN environment, there may be hidden stations. Two stations, which can transmit to and receive from a common station while they are out of range from each other, are known as hidden stations. Since the DCF operates based on carrier sensing, the existence of such hidden stations can severely degrade the network performance. To reduce the hidden station problem, 802.11 defines a Request-to-Send/Clear-to-Send (RTS/CTS) mechanism. If the transmitting station opts to use the RTS/CTS mechanism, then before transmitting a data frame, the station transmits a short RTS frame, followed by a CTS frame transmitted by the receiving station. The RTS and CTS frames include information of how long it takes to transmit the subsequent data frame and the corresponding ACK response. Thus, other stations hearing the transmitting station and hidden stations close to the receiving station will not start any transmissions; their timer called Network Allocation Vector (NAV) is set, and as long as the NAV value is non-zero, a station does not contend for the medium. Between two consecutive frames in the sequence of RTS, CTS, data, and ACK frames, a SIFS is used. Figure 4.2 shows the timing diagram involved with the RTS/CTS frame exchange. It should be noted that the RTS/CTS exchange can be very

---

[1] According to Annex C of [1], ACK Timeout is defined as SIFS + ACK transmission duration + SlotTime.

useful even if there is no hidden stations. For instance, when there are many contending stations, the bandwidth loss due to RTS collisions can be smaller than that due to longer frames [16].

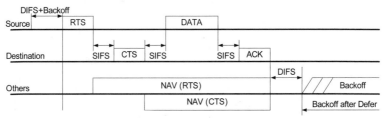

Figure 4.2: RTS/CTS frame exchange.

As explained thus far, the DCF normally waits for a DIFS interval before a backoff countdown. However, an EIFS interval shall be used instead for the contention immediately after an unsuccessful frame reception. Beginning any successful reception of a frame, the station starts using the DIFS instead of EIFS again. There are basically two different cases, which result in an unsuccessful frame reception: (1) the PHY has indicated the erroneous reception to the MAC, e.g., carrier lost; or (2) the error is detected by the MAC via an incorrect FCS value. The EIFS is defined to provide enough time for other stations to wait for the ACK frame of an incorrectly received frame before these stations start their frame transmission. Accordingly, the EIFS value is determined by the sum of one SIFS, one DIFS, and the time needed to transmit an ACK frame at the underlying PHY's lowest mandatory rate, which is 1 Mbps in the case of 802.11b PHY, i.e.,

$$EIFS = SIFS + ACK\_Tx\_Time_{@1Mbps} + DIFS \tag{2}$$

Figure 4.3 illustrates that stations receiving the data frame incorrectly defer for an EIFS period before starting a backoff procedure.

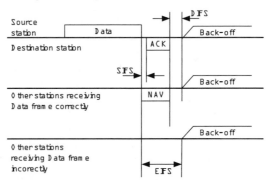

Figure 4.3: DCF access operation, where the ACK frame is transmitted at 1 Mbps.

All of the MAC parameters including SIFS, DIFS, Slot Time, CW_{min}, and CW_{max} are dependent on the underlying PHY. Table 4.1 shows these values for the popular 802.11b PHY [6]. Irrespective of the PHY, DIFS is determined by SIFS+2•SlotTime, and another important IFS, called PCF IFS (PIFS), is determined by SIFS+SlotTime.

Table 4.1: MAC parameters for the 802.11b PHY.

Parameters	SIFS (μsec)	DIFS (μsec)	Slot Time (μsec)	CW_{min}	CW_{max}
802.11b PHY	10	50	20	31	1023

4.2.1 DCF Overhead

It is very instructive to understand the protocol overhead introduced by the DCF operation. To this purpose, consider a scenario characterized by just a single transmitting station. For simplicity, neglect all the protocol overheads introduced by the upper layers (e.g., IP header, TCP/UDP header, etc.) as well as the interaction of the upper layers with the MAC operation (e.g., such as the TCP congestion control).

For a single transmitting station under the assumption that the frames are never corrupted by the channel noise, the maximum throughput can be immediately expressed as

$$S_{station} = \frac{E[payload]}{E[T_{Frame\_Tx}] + DIFS + SlotTime \cdot CW_{min}/2}, \tag{3}$$

where $SlotTime \cdot CW_{min}/2$ is the average time spent for the backoff between two consecutive data frame transmissions. Now, the time T_{Frame_Tx} spent to complete a frame transfer successfully depends on the considered handshake as well as the PHY employed. In the case of the basic access without an RTS/CTS exchange, this time duration is given by:

$$T_{Frame\_Tx} = T_{MPDU} + SIFS + T_{ACK}, \tag{4}$$

where T_{MPDU} and T_{ACK} represent the transmission times of an MPDU (or a data frame) and an ACK, respectively. These are dependent on the employed transmission rate as well as the frame size. Note that PHYs of the 802.11 provide a set of transmission rates. For example, the 802.11b PHY [6] provides four different transmission rates, namely, 1, 2, 5.5, and 11 Mbps. Which transmission rate to use for a particular frame transmission is not defined in the standard, and is left implementation-dependent. Some rate adaptation algorithms can be found in [12 - 15].

Now, for the transmission rate R_{MPDU_Tx}, the transmission time of a MPDU conveying a payload of L bytes is determined by:

$$T_{MPDU} = T_{PLCP} + 8 \bullet (28 + L)/R_{MPDU\_Tx}, \tag{5}$$

where T_{PLCP} is the overhead due to the PHY operation, e.g., preamble and header, and is given by 192 μsec for the case when the 802.11b PHY employs the long-preamble option. The number 28 in Eq. (5) represents the overhead of the MAC header plus the FCS field in number of bytes. Similarly, the transmission time of an ACK frame, which is 14 bytes long, for the transmission rate R_{ACK_Tx} can be represented by:

$$T_{ACK} = T_{PLCP} + 8 \bullet 14/R_{ACK\_Tx} \tag{6}$$

With RTS/CTS exchange, the time to complete a frame transfer is given by:

$$T_{Frame\_Tx} = T_{RTS} + SIFS + T_{CTS} + SIFS + T_{MPDU} + SIFS + T_{ACK}, \quad\quad (7)$$

where the RTS (of 20 bytes) and CTS (of 14 bytes) transmission times are given, respectively, by

$$T_{RTS} = T_{PLCP} + 8 \bullet 20/R_{RTS\_Tx},$$
$$T_{CTS} = T_{PLCP} + 8 \bullet 14/R_{CTS\_Tx}, \quad\quad (8)$$

Figure 4.4 presents the analysis of different DCF overheads for two different 802.11b 802.11b transmission rates for the MPDU when the payload is fixed at 1500 bytes. Irrespective of the MPDU transmission rate, 1 Mbps (minimum rate) was assumed for the transmission of the ACK, RTS, and CTS frames. We observe that the protocol overheads due to the backoff, RTS, CTS, and ACK are relatively large when the MPDU transmission rate is high since the time corresponding to the payload transmission is relatively short for a high transmission rate. We can also easily envision that the overheads will be relatively large as the payload size is reduced since the overheads are relatively fixed in time.

This is more clearly presented in Figure 4.5, which illustrates the normalized throughput, defined by the throughput divided by the MPDU transmission rate. We first observe that the normalized throughputs increase as the payload size increases. Second, we observe that the normalized throughputs are larger for a lower transmission rate due to the relatively smaller protocol overhead as observed in Figure 4.4. Finally, in the considered simple situation, i.e., a single transmitting station, the RTS/CTS exchange obviously results in a lower throughput performance.

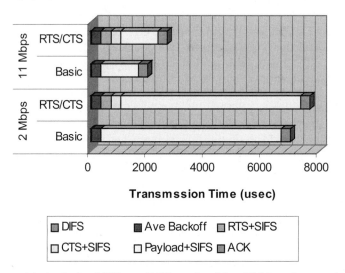

Figure 4.4: Analysis of different DCF overhead for 1500 byte-long payload.

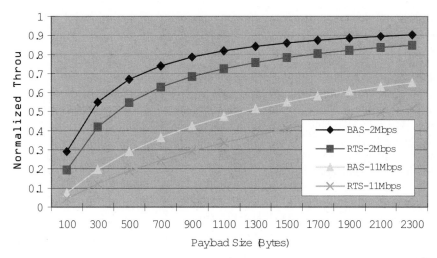

Figure 4.5: Normalized throughput versus payload size (bytes) for different MPDU transmission rates.

4.3 Performance Evaluation of the IEEE 802.11 DCF

Under suitable simplifying assumptions, it has been proven in [16] that the accurate performance evaluation of the DCF employed in the IEEE 802.11 legacy MAC can be carried out using elementary analytical techniques. The availability of such techniques allows us to easily obtain quantitative insights on the effectiveness of the DCF mechanism and the related parameter settings. The goal of this section is to provide the reader with a comprehensive overview of the modelling techniques suitable for evaluating the DCF performance. The reader interested in additional modelling details and extensions may refer to [16 - 25].

4.3.1 The Concept of Saturation Throughput

In Section 4.2.1, we have derived the maximum throughput that a single transmitting station can achieve, and we have quantified the DCF protocol overhead. From a practical point of view, this implies that all the traffic arriving at a long-term rate lower than the maximum throughput value will be delivered to the destination. Conversely, as long as the traffic arrival rate persistently grows above the maximum throughput threshold, the transmission buffer will build up until saturation, and the carried load will remain bounded to the maximum throughput value.

In the case of several competing stations, the situation is different. It is well known[2] that several random access schemes exhibit an unstable behaviour. In particular, as the offered load increases, the throughput grows up to a maximum value, referred to as

---

[2] See for example the well-known textbook by Dimitri Bertsekas and Robert Gallager, *Data Networks*, 2nd edition, Prentice-Hall, Inc., Englewood Cliffs, NJ, 1992 – this book provides a comprehensive and thorough discussion of the instability problems arising in random access protocols.

"Maximum Throughput." However, further increases of the offered load lead to a significant decrease in the system throughput eventually (which typically converges to zero in the case of infinite users).

Mild forms of instability arise also in the case of DCF's Binary Exponential Backoff operation, and for a scenario characterized by a finite number of competing stations. This can be illustrated by means of a simple simulation experiment. The plots shown in Figure 4.6 (taken from [16] – the simulation details and the throughput/load scales are not essential for the following discussion, and hence are omitted) have been obtained considering a finite number of stations. Each station has been loaded with a variable amount of traffic, linearly increasing with the simulation time (straight line). Fixed-size MPDUs have been considered.

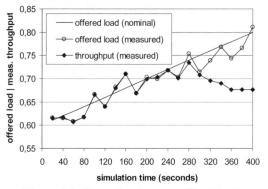

Figure 4.6: Throughput versus offered load.

The figure reports two additional plots: the offered load and the throughput, measured in time intervals lasting 20 seconds. Clearly, due to statistical fluctuations of the packet arrival process (Poisson in the specific case), the measured offered load in general differs from the nominal load. However, in normal (stable) conditions, the entire offered load will be delivered to the destination, and hence the measured throughput will be equal to the measured offered load. From the figure, we see that this happens only in the first part of the simulation run, specifically the first 260 seconds of the simulation time. After this time, the throughput measured in each 20-second time interval becomes smaller than the measured offered load (and excessive frames accumulate in the transmission buffer).

From the figure, we also see that the measured throughput appears to asymptotically converge to a constant value (0.68 in this specific example), regardless of the offered load. We define "saturation throughput" as the limit reached by the system throughput as the offered load increases.

The importance of the saturation throughput concept stays in the fact that it represents the maximum load that the system can carry in *stable conditions*, i.e., in practical operation. In fact, refer again to Figure 4.6, and consider, for an example, the load scenario encountered at simulation time 280s. Here, the nominal offered load is approximately 0.74, while the load measured in the 20-second time interval is approximately 0.75. If we now freeze the offered load to a constant value, i.e., 0.74, and we take measurements on a longer time interval, we would see that the longer the measurement time, the lower the measured throughput (ultimately, for an infinite measurement period, the measured throughput will result in a value equal to the saturation value).

Actually, the same situation will occur for *any* load greater than the saturation throughput bound, provided that a sufficiently long measurement time is considered. This results in the practical impossibility to maintain a sustained operation of the random access scheme at any load greater than the saturation value[3].

4.3.2 Maximum Saturation Throughput

Having defined the concept of saturation throughput, we are now ready to describe an analytical approach which allows us to derive this performance figure. In this section, we derive performance bounds on the throughput achievable by DCF [16, 26, 27]. Consider an 802.11 ad-hoc network scenario in which a finite and fixed number N of stations contend for the channel access. Assume that each station is in a saturation condition, meaning that its buffer is always non-empty and, at any time, a frame is always available for transmission. Moreover, assume that a frame transmission is never corrupted (either by noise or by interference due to hidden terminals), and that a transmission fails only when a collision with another frame occurs on the channel (this assumption of ideal channel conditions can be removed, as shown in Section 4.3.4 – see also [18, 21]).

The first important observation is that the 802.11 DCF rules allow us to introduce a discrete-time integer time-scale, which is the key to enable the model described below. As described in Section 4.2, a station can transmit only when it senses the channel idle for a DIFS. Moreover, after this time, it can schedule transmission only in discrete slot-time intervals, whose size is hereafter indicated as σ. Moreover, let us focus on the events occurring on the channel:

- When only one station has a scheduled transmission in a slot-time, neglecting wireless channel impairments, the transmission will be successful; all the other stations will freeze their backoff counters, until a DIFS time elapses after the end of the ACK.
- When two or more stations schedule transmission into a slot-time, a collision will occur, and the channel will be available for access only a DIFS or an EIFS[4] after the end of the longest transmitted frame.
- Finally, if no stations transmit in the given slot-time, the next transmission opportunity will be the following slot-time.

---

[3] Though traditional performance evaluation models for random access schemes frequently derive a theoretical maximum throughput value, this value is meaningless from a practical standpoint. Actually, this value is not even measurable from a simulation experiment. For example, in the case of Figure 4.6, we obtain a maximum measured throughput equal to about 0.74, but this measurement is only a rough estimate, as it is affected by the unreliability of a fairly short measurement time. If we rerun the same simulation experiment using measurement times longer than 20 seconds (and for consistency, we consider a slower increase of the nominal offered load), the measurement will be more reliable, but the maximum value will be lower.

[4] It is not trivial to determine whether a specific listening station will use a DIFS or an EIFS after a frame collision. If the listening station is able to synchronize with one of the colliding frames, and thus initiate a receiving process, an EIFS will be used (see the reason in Section 4.2). On the other hand, if this is not the case (e.g., comparable received power level for the two colliding frame preambles), the listening station will just see a busy channel: without initializing the reception of any frame, and a DIFS will be used. In all the simulation (as well as analytical) models we are aware of do not enter into this technical issue for simplicity, but consistently use either DIFS or EIFS for all collisions (for example, the 802.11 ns-2 implementation uses EIFS, and this leads to different results with respect to other models that use DIFS after a collision).

Hence, a discrete and integer time scale can be defined. Note that this discrete time scale does not directly relate to the system time, being a "slot" on the channel either an empty slot (in which case, the slot will last exactly one slot-time σ), or a busy slot, in which case the slot duration will depend on the events occurring (a transmission or a collision). In what follows, unless ambiguity occurs, with the term slot time, we will refer to either the (constant) value σ, representing the system slot-time, or the (variable) time interval representing the model slot time.

To determine the maximum throughput achievable in an 802.11 ad-hoc network composed of N saturated stations, let us now assume that each station randomly and independently accesses a slot time with probability τ. This is equivalent to assuming that every station follows a p-persistent backoff strategy, where the probability to access a random slot is constant and set to the value τ. In this assumption, the probability P_{idle} that no station accesses a given slot is readily given by:

$$P_{idle} = (1 - \tau)^N \tag{9}$$

Similarly, the probability $P_{success}$ that just one station accesses a given slot is expressed as:

$$P_{success} = N\tau (1 - \tau)^N \tag{10}$$

Let us now define T_s to be the duration of a period in which no other stations can access the channel because a successful transmission is occurring. This period not only includes the MPDU transmission time as well as the relevant ACK transmission time, but also includes a DIFS after the end of the ACK transmission, since in this period of time no other station can access the channel. Similarly, let T_c be the duration of a period in which other stations cannot access the channel because a collision is occurring. In what follows, we will refer to these two values as transmission slot and collision slot durations, respectively, and these values will be expressed in Section 4.3.2.1. Since a system slot-time σ elapses during an idle slot, we can derive the average slot duration by weighting in probability the three values T_s, T_c and σ:

$$E[slot] = P_{idle}\sigma + P_{success}T_s + (1 - P_{idle} - P_{success}) T_c \tag{11}$$

We are finally ready to define the system throughput S as the average amount of information transmitted into a slot. Given that $E[P]$ is the average MPDU payload size,

$$S = \frac{P_{success}E[P]}{E[slot]} = \frac{P_{success}E[P]}{P_{idle}\sigma + P_{success}T_s + (1 - P_{idle} - P_{success})T_c}$$
$$= \frac{E[P]}{T_s + \sigma\dfrac{P_{idle} + T_c^*(1 - P_{idle} - P_{success})}{P_{success}}}, \tag{12}$$

where $T_c^* = T_c/\sigma$ is the average collision time measured in slot-time units. Now, $E[P]$, σ, and T_s are constant values. Hence, the throughput above is maximized as long as we minimize the expression:

$$\frac{P_{idle} + T_c^* \left(1 - P_{idle} - P_{success}\right)}{P_{success}} = \left(1 - T_c^*\right)\frac{(1-\tau)}{N\tau} + \frac{T_c^*}{N\tau(1-\tau)^{N-1}} - T_c^* \tag{13}$$

If we approximate the average value T_c^* to be a constant value[5], independent on τ, we finally conclude that the optimal value τ_{max} that maximizes the system throughput is given by the solution of the equality:

$$\left(1 - \tau_{max}\right)^N - T_c^* \left\{N\tau_{max} - \left(1 - \left(1 - \tau_{max}\right)^N\right)\right\} = 0 \tag{14}$$

Under the condition $\tau_{max} \ll 1$, the approximation

$$\left(1 - \tau_{max}\right)^N \approx 1 - N\tau_{max} + \frac{N(N-1)}{2}\tau_{max}^2, \tag{15}$$

holds, and hence an explicit expression for the equality in Eq. (14) can be found:

$$\tau_{max} = \frac{\sqrt{1 + 2\left(T_c^* - 1\right)\frac{(N-1)}{N}} - 1}{(N-1)\left(T_c^* - 1\right)} \approx \frac{1}{N\sqrt{T_c^*/2}}, \tag{16}$$

where the last approximation holds for large values of N and T_c^*. An interesting alternative way to express Eq. (16) is to determine the optimal value of the contention window, CW_{opt}, which allows us to maximize the throughput. Specifically, consider a station which, instead of using the rules of exponential backoff, always extracts the backoff counter from the uniform range $[0, CW_{opt}]$. Under the assumption that the backoff counter is decremented at each slot (see a related discussion in Section 4.3.3.1), a transmission occurs every $(1+CW_{opt}/2)$ slots. Hence, the probability that a station transmits in a randomly chosen slot can be related to the Contention Window as:

$$\tau = \frac{1}{1 + CW_{opt}/2} \tag{17}$$

Substituting Eq. (17) into the τ_{max} expression given in Eq. (16), we conclude that CW_{opt} is, in the first approximation, proportional to the number N of competing stations, i.e.,

$$CW_{opt} \approx 2N\sqrt{T_c^*/2} - 2 \approx N\sqrt{2T_c^*} \tag{18}$$

---

[5] This is strictly true only if the frames have fixed-size payloads. In fact, with variable payload size, the length of a collision interval is bounded by the length of the longest packet involved in the collision. This gets longer when the number of packets simultaneously colliding gets higher. As the probability that a given number of packets are involved in a collision depends on τ, the value T_c^* is actually a function of τ.

4.3.2.1 Performance Bounds for 802.11b DCF

To compute the performance bounds for the DCF, we need to quantify[6] the parameters σ, T_s, and T_c. Similar to what have been done in Section 4.2.1, we refer to the 802.11b case; accordingly, the slot-time σ is set to 20 µsec. T_s is given by the time to transmit a frame (T_{Frame_Tx} is computed in Section 4.2.1) plus a DIFS interval. For the case of the basic access:

$$T_s = T_{MPDU} + SIFS + T_{ACK} + DIFS, \tag{19}$$

where T_{MPDU} and T_{ACK} have been computed in Eqs. (5) and (6), respectively (there, instead of $E[P]$, we have referred to the MPDU payload size as L). Similarly, for the case of RTS/CTS exchange,

$$T_s = T_{RTS} + SIFS + T_{CTS} + SIFS + T_{MPDU} + SIFS + T_{ACK} + DIFS \tag{20}$$

Refer to Eq. (8) for T_{RTS} and T_{CTS} expressions. The computation of T_c differs depending on whether we assume that a DIFS or an EIFS is used after a collision (see detailed discussion in footnote 4). Though it is perhaps more realistic, for a small scale single hop ad-hoc network, to assume destructive collisions, and thus that a DIFS elapses after a collision, in what follows, we will use an EIFS after a collision since such a setting requires more discussion on the station timing (in other words, the extension of what follows to the DIFS case is just a simplification). If all the frames have the same size, for the basic access case,

$$T_c = T_{MPDU} + EIFS \tag{21}$$

For the general case of frames having different sizes, then the duration of a collision depends also on the number of colliding frames, and thus on the transmission probability τ (refer to Eq. (15) in [16] for details). Instead, in the RTS/CTS case, collisions occur only for the RTS frames. Therefore, the duration of a collision is constant, and is given by

$$T_c = T_{RTS} + EIFS \tag{22}$$

Figure 4.7 presents the saturation throughput performance for 10 stations and 802.11b parameters (see Section 4.2.1). The case of 2 and 11 Mbps data rates are plotted, for both basic and RTS/CTS access cases. The rate for control frames (i.e., ACK, RTS, and CTS) is set to the minimum rate (i.e., 1 Mbps). The circle represents the maximum throughput computed by means of the approximate value τ given in Eq. (16). We first observe that the approximation in Eq. (16) is accurate. We find that, in the basic access case, the maximum throughput performance is more sensitive to the value τ compared with the RTS/CTS case. This implies that the RTS/CTS access mechanism is less sensitive to backoff parameters (as we will show in Section 4.3.3, DCF operates on a value τ which actually depends on the

---

[6] The discussion in this section will be further extended in Section 4.3.3.1, where additional considerations on the implication of the backoff countdown rules on T_s and T_c settings will be provided.

backoff settings, i.e., CW_{min}, CW_{max}, etc). Moreover, Figure 4.7 shows that in the case of low data rates (2 Mbps), the maximum throughput achievable by basic access and RTS/CTS exchange are very close. This is not true any more when higher data rates (11 Mbps) are considered, and the overhead due to the RTS/CTS exchange becomes a limiting factor in terms of achievable throughput.

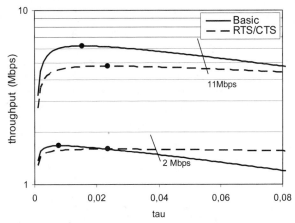

Figure 4.7: The maximum throughput with the number of stations, $N=10$.

Finally, let us evaluate the effect of the number of stations on the maximum achievable throughput. For convenience of notation, let

$$K = \sqrt{T_c^* / 2} \tag{23}$$

By using the approximation for the optimal τ given by Eq. (16), recalling that $T_c^* = T_c/\sigma$, and taking the limit for $N\rightarrow\infty$,

$$
\begin{aligned}
\lim_{N\to\infty} S_{max} &= \lim_{N\to\infty} \frac{E[P]}{T_s + \sigma \dfrac{P_{idle} + T_c^*\left(1 - P_{idle} - P_{success}\right)}{P_{success}}} \\
&= \lim_{N\to\infty} \frac{E[P]}{T_s + \sigma \dfrac{(1-\tau_{max})}{N\tau_{max}} - T_c + T_c \dfrac{\left(1 - (1-\tau_{max})^N\right)}{N\tau_{max}(1-\tau_{max})^{N-1}}} \\
&= \frac{E[P]}{T_s + \sigma K - T_c(1 + K - Ke^{1/K})}
\end{aligned}
\tag{24}
$$

We thus conclude that, even for a large number of stations, the maximum throughput tends to a constant finite value, which is a function of only the average transmission and collision times, T_s and T_c, and the slot size. It is interesting to note, via direct computation, that such an asymptotic maximum throughput is very close to that achievable in the case of $N=10$ given in Figure 4.7. In fact, through Eq. (24), we can show that in the 2 Mbps data rate case, the asymptotic throughput is 1.669 and 1.596 Mbps for the basic and RTS/CTS cases respectively, while it results in 6.210 (basic) and 4.763 (RTS/CTS) Mbps for the 11 Mbps data rate scenario respectively.

4.3.3 Saturation Throughput Analysis

Having derived the capacity limits of the IEEE 802.11 DCF, we now carry out an analysis devised to understand how far from its performance limits DCF operates. Such an analysis appears more complex: since each station accesses the channel according to Binary Exponential Backoff rules, the space state required to thoroughly model each individual station (e.g., the number of retransmission suffered by each station, and the backoff counter value) rapidly diverges, even in the presence of a small number of competing stations.

However, let us focus on a specific station, hereafter referred to as "tagged" station. This station will access the channel according to the Binary Exponential Backoff mechanism specified for DCF, and specifically, as described in Section 4.2, it will double the range in which the Contention Window is chosen every time a collision is encountered. Hence, the tagged station will access the channel with a "frequency" (measured in terms of number of accesses per channel slot) which depends on the number of retransmissions already suffered by the considered frame: a high frequency when the CW value is small, a small frequency conversely. Each of the remaining competing stations will, in turn, be characterized by complex exponential backoff rules and, thus, very different Contention Window (CW) values, depending on the specific history of each access attempt (e.g., the number of retransmissions suffered by the actual head-of-line MPDU). However, in stationary conditions, we argue that it is reasonable to consider their "aggregate" contribution as being, statistically speaking, invariant, and specifically to consider their effect as the result of individual stations accessing the channel via a suitable (i.e., to be determined) but constant permission probability.

Such an intuitive statement can be formally reworded by means of the two key assumptions:

1. Regardless of the history of the head-of-line (HOL) frame in terms of the number of retransmissions and accumulated backoff stage, we assume that each frame transmission suffers from a constant and independent collision probability;
2. If p is the collision probability and N is the number of competing stations, we assume that p is computed as the contribution of N-1 remaining stations, each independently accessing a channel slot with a constant permission probability τ.

As shown in what follows, these assumptions enable a very simple, though accurate, analytical modelling of the DCF.

For the sake of generality, it is useful to develop the model considering more general backoff rules than the exponential backoff specified in the DCF standard. To this end, let us define the term "Backoff Stage" as the number of retransmissions suffered by a HOL frame. A station in backoff stage 0, i.e., willing to transmit a new MPDU, will select[7] an integer random backoff value drawn from a general probability distribution B_0. If the

---

[7] Saturation conditions imply that a packet in backoff stage 0 immediately follows a previously transmitted one. Hence, consistent with the DCF specifications (see Clause 9.1.1 of the standard), a random backoff interval shall always be selected for the first packet transmission attempt.

transmitted frame collides, we say that the station enters backoff stage 1. The next backoff value will be drawn from a second probability distribution B_1, and so forth. In general, a station entering backoff stage i will extract a backoff value from a distribution B_i.

In the particular case of the DCF Binary Exponential Backoff, B_0 is a uniform distribution in the range $[0, CW_{min}]$, B_1 is a uniform distribution in the range $[0, 2(CW_{min}+1)-1]$ and in general, B_i is a uniform distribution in the range $[0, CW_i]$ where $CW_i = 2^i(CW_{min}+1)-1$. In addition, the IEEE 802.11 DCF specifies

1. the maximum Contention Window value as $CW_{max} = 2^m(CW_{min}+1)-1$ where m is a parameter that depends on the physical layer considered, and
2. a finite number of retries R, meaning that a frame whose first transmission has failed, will be retransmitted for at most R times, and then it will be dropped from the transmission queue.

In what follows, we will show that the performance do not depend on the probability distributions B_i, but only on their mean values $\beta_i = E[B_i]$. Moreover, we will show that the performance depends on the retry limit R, where R becomes eventually infinite in the analytical model.

Let us denote with (TX) the event that a station is transmitting a frame into a time slot, and denote with ($s=i$) the event that the station is found in backoff stage i.

We are ultimately interested in the unconditional probability $\tau = P\{TX\}$ that the station transmits in a randomly chosen slot. Thanks to Bayes' Theorem, for $i \in (0, ..., R)$,

$$P\{s = i \mid TX\} = \frac{P\{TX \mid s = i\}P\{s = i\}}{P\{TX\}} \tag{25}$$

which in turn can be rewritten as:

$$P\{TX\}\frac{P\{s = i \mid TX\}}{P\{TX \mid s = i\}} = P\{s = i\} \tag{26}$$

Since this equality holds for all $i \in (0, ..., R)$, it also holds for the summation:

$$\sum_{i=0}^{R} P\{TX\}\frac{P\{s = i \mid TX\}}{P\{TX \mid s = i\}} = \sum_{i=0}^{R} P\{s = i\} \tag{27}$$

However, the rightmost term in the equation is a probability distribution (namely, the probability that a station is in backoff stage i). Hence, the sum over all $i \in (0, ..., R)$ equals 1. We can thus derive an expression for τ:

$$\tau = P\{TX\} = \frac{1}{\displaystyle\sum_{i=0}^{R} \frac{P\{s = i \mid TX\}}{P\{TX \mid s = i\}}} \tag{28}$$

The value τ is thus known, as long as we find an expression for $P\{s=i|TX\}$ and $P\{TX|s=i\}$. Let us first focus on the conditional probability $P\{s=i|TX\}$ that a transmitting station is found in backoff stage i. Since, for $i>0$, this probability is given by the probability that the station, in the previous transmission event, was found in stage i-1 and that the transmission failed (by assumption, this occurs with constant probability p), it follows that $P\{s=i|TX\}$ is a geometric distribution[8] (truncated, in the case of finite value R), i.e.:

$$P\{s = i \,|\, TX\} = \frac{(1-p)p^i}{1-p^{R+1}} \qquad i \in (0,...,R)$$ (29)

Let us now find an explicit expression for $P\{TX|s=i\}$. This represents the transmission probability of a station in backoff stage i, or, in other words, the frequency of transmission (the number of transmission slots per channel slot) for a station assumed to always remain in backoff stage i. Under very general conditions[9], this probability can be computed by dividing the average number of slots spent in the transmission state while in stage i (owing to the time scale adopted, exactly 1 slot), with the average number of total slots spent by the station in stage i (i.e., the average number of backoff slots, plus the single transmission slot). According to the notation given above:

$$P\{TX \,|\, s = i\} = \frac{1}{1 + E[B_i]} = \frac{1}{1 + \beta_i} \qquad i \in (0,...,R)$$ (30)

In the special case of DCF, a station entering backoff stage i uniformly selects a backoff value in the range $[0,CW_i]$. Following [16], it is convenient to adopt the notation $W_i=CW_i+1$. Hence,

$$P\{TX \,|\, s = i\} = \frac{1}{1 + E[\,uniform(0,CW_i)\,]} = \frac{1}{1 + \dfrac{W_i - 1}{2}} = \frac{2}{W_i + 1}$$ (31)

---

[8] A more formal way to derive Eq. (29) is to envision $P\{s=i|TX\}$ as the steady-state probability distribution of a discrete-time mono-dimensional Markov chain describing the backoff stage evolution. One time step in this chain represents a backoff stage transition, driven by the success/failure of the packet transmission. At stage $0 \leq i < R$, the chain will evolve in the next time step in stage $i+1$ with probability p, and will return (or stay) in stage 0 with probability $1-p$; at stage R the chain will in any case return to stage 0 with probability 1. This interpretation allows us to simply extend the described analysis to more general backoff processes with memory, i.e., whose backoff evolution is regulated by a Markov chain. It suffices to substitute Eq. (29) with the steady-state distribution of the considered Markov Chain. A few proposals of backoff models with memory (for example, the slow CW decrease approach considered in [25]) has been reported.

[9] Since we are conditioning on the backoff stage i, we can envision the event of transmitting into a slot as the recurrence of transmission events separated by the time spent while in backoff stage i, assumed independent among transmission events. Hence, this computation can be interpreted as an application of the Long-Run Renewal rate theorem (see, e.g., William Feller, *An Introduction to Probability Theory and Its Applications*, Vol. II, Wiley, Cap. XI - pp. 368-380) and is shown to depend only on the average time spent while in the backoff stage i, and not on its distribution. As a side comment, there are some proposals that draw backoff counters from distributions different from the usual uniform one. As should be clear now, such a generalization does not appear to have practical significance, as the performance depend only on the mean value, and thus there is no reason in using backoff distributions more complex than the uniform one.

By substituting Eqs. (29) and (30) into Eq. (28) we can finally derive an explicit expression for τ.

$$\tau = \frac{1}{\sum_{i=0}^{R} \frac{(1-p)p^i}{1-p^{R+1}}(1+\beta_i)} = \frac{1}{1 + \frac{(1-p)}{1-p^{R+1}}\sum_{i=0}^{R}p^i\beta_i}, \tag{32}$$

where we have made use of the fact that $P\{s=i|TX\}$ is a distribution which, in the considered range $(0,\dots,R)$, sums to 1.

This expression depends on the values R (retry limit), and the sequence of β_i values (the mean per-stage backoff values), which are specified by the employed backoff model. Moreover, it depends on the conditional collision probability p which is still unknown. To find the value of p, it is sufficient to note that the probability p that a transmitted frame encounters a collision, is the probability that, in a time slot, at least one of the N-1 remaining stations transmits. The fundamental independence assumption #1 given at the beginning of this section implies that each transmission "sees" the system in the same state, i.e., in steady state. At steady state, according to assumption #2, each remaining station transmits a frame with constant permission probability τ. This yields:

$$p = 1 - (1-\tau)^{N-1} \tag{33}$$

Eqs. (32) and (32) represent a non linear system in the two unknowns τ and p, which can be solved using numerical techniques.

The above analysis was carried out for general backoff models. By properly choosing the sequence β_i and the value R, it can be immediately adapted to more a specific backoff model, e.g., the binary exponential backoff model adopted in DCF. For example, by setting:

$$R = \infty$$
$$W = CW_{min} + 1$$
$$m = log_2\left(\frac{CW_{max}+1}{CW_{min}+1}\right) \tag{34}$$
$$\beta_i = \begin{cases} \dfrac{2^i W - 1}{2} & 0 \leq i \leq m \\ \dfrac{2^m W - 1}{2} & i \geq m \end{cases}$$

We model a Binary Exponential Backoff scheme with no retry limit as an upper bound on the Contention Window (summarized by the value m). Eq. (32) becomes [16]:

$$\tau = \frac{2}{1+\left(1-p\right)\left\{\sum_{i=0}^{m-1} p^i 2^i W + \sum_{i=m}^{\infty} p^i 2^m W\right\}}$$

$$= \frac{2(1-2p)}{(1-2p)(W+1)+pW(1-(2p)^m)} \tag{35}$$

The same DCF Binary Exponential Backoff model, but with a finite retry limit R (for simplicity of computation, lower than or equal to the parameter m: for the general case refer to [17]) yields:

$$\tau = \frac{2(1-2p)(1-p^{R+1})}{(1-2p)(1-p^{R+1})+W(1-p)(1-(2p)^{R+1})} \tag{36}$$

4.3.3.1 Throughput Performance

Once the value τ is known, the throughput can be computed via Eq. (12), here reported for the convenience of the reader:

$$S = \frac{P_{success}E[P]}{P_{idle}\sigma + P_{success}T_s + \left(1 - P_{idle} - P_{success}\right)T_c} \tag{37}$$

where

$$P_{idle} = \left(1-\tau\right)^N$$
$$P_{success} = N\tau\left(1-\tau\right)^{N-1} \tag{38}$$

We recall that the denominator in Eq. (37) represents the average slot size $E[slot]$. If $E[P]$, instead of bits, is expressed in the same time unit of the parameters at the denominator (e.g., seconds), then Eq. (37) gives the "normalized" saturation throughput, defined as the fraction of channel time used to send the successful payload information.

The fundamental difference with respect to the treatment suggested in Section 4.3.2 is that, now, τ is no more a variable (i.e., a generic permission probability), but it is a numerical value function of the considered backoff model parameters, namely, the retry limit R and the sequence of mean per-stage backoff values β_i, for all i in $(0,\dots,R)$.

The values T_s and T_c have been computed in Section 4.3.2.1. However, some further remarks are needed when the analysis is applied to the DCF. The IEEE 802.11 standard discusses, in Clause 9.2.5.2, how the backoff counter is decremented. Here, it specifies that if the medium is determined to be busy at any time during a backoff slot, then the backoff procedure is suspended (meaning that the backoff timer shall not decrement for that slot). Let us assume that a station has a backoff counter equal to a value b at the beginning of a slot-time. If the current slot-time is idle, at the end of the slot-time, the backoff counter is duly decremented and the station will start the next slot-time with backoff value b-1. Conversely, if the current slot-time is busy (because another station starts transmitting in the considered slot), the station freezes the backoff counter to the value b. This implies that the station starts the slot immediately following a busy one with the same backoff value b. In other words, the backoff counter is decremented only during idle slots.

The implications of this operation is not immediately evident on the modelling framework described until now. Figure 4.8 illustrates what happens when two stations access the channel with different backoff values. In the example, at slot t, stations A and B start with a backoff counter equal to 2 and 3 respectively. Hence, we might expect them to transmit in consecutive slots, namely slot t+2 and slot t+3.

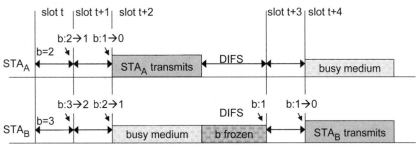

Figure 4.8: The slot immediately following a transmission can be accessed only by the transmitting station.

As expected, station A decrements the backoff counter to 0 at the end of slot t+1, thus transmitting a frame in slot t+2. It can then schedule the transmission for the next frame. With probability $1/(CW_{min}+1)$, it will extract 0 as the new backoff counter, and hence it will immediately transmit the next frame in the first slot available, incurring a DIFS interval after the end of the transmission (slot t+3, in the figure). Let us now focus on station B. In slot t+2, it will see station A's transmission on the channel and will freeze the backoff counter. It will thus start slot t+3 with a backoff counter value equal to 1, and (assuming slot t+3 to be empty) it will ultimately transmit only in slot t+4. We conclude that *a slot immediately following a successful transmission cannot be used for transmissions by any other station, except the transmitting one*. Hence, in ideal channel conditions, such a transmission is granted to be successful as no collision may occur.

The described effect can be accounted in the model by redefining the notion of successful transmission slot, and specifically by including either i) the extra slot-time at the end of a transmission, as well as ii) the possible extra frames transmitted in the "reserved" slot. In formulae:

$$\overline{T_s} = T_s + \sum_{k=1}^{\infty}\left(\frac{1}{CW_{min}+1}\right)^k T_s + \sigma = T_s\frac{CW_{min}+1}{CW_{min}} + \sigma, \qquad (39)$$

where T_s is the successful slot size given in Eqs. (19) and (20) for the Basic and RTS/CTS cases respectively, which accounts for a single frame transmission plus a DIFS. To be consistent, in the throughput computation, the amount of information transmitted into a successful slot shall also include the MPDU payload due to extra frames transmitted in such "reserved" slots:

$$\overline{E[P]} = E[P] \frac{CW_{min} + 1}{CW_{min}} \tag{40}$$

Since a successful "transmission slot" now includes an extra idle slot, a further model detail is that, in the slot immediately after such a transmission slot, the backoff counter will be found in the range $(0, CW_{min}-1)$ and not in the range $(0, CW_{min})$. This can be accounted in the DCF model by simply setting, in Eq. (31), $W_0 = CW_{min}$ instead of $W_0 = CW_{min} + 1$. Finally, since, according to the standard specification, a contending station will wait for an ACK_Timeout greater than an EIFS, before reattempting to transmit[10], then

$$\overline{T_c} = T_c + \sigma \tag{41}$$

The throughput shall be computed as in the usual case, but with the new values defined in Eqs. (39), (40) and (41):

$$S = \frac{P_{success} \overline{E[P]}}{P_{idle}\sigma + P_{success}\overline{T_s} + (1 - P_{idle} - P_{success})\overline{T_c}} \tag{42}$$

We point out that, in practical cases (e.g., the 802.11b parameters – we recall that $CW_{min}=31$ in this case), the difference between the results computed via Eq. (42) and that computed via the more approximate expression in Eq. (37) are negligible.

4.3.3.2 Delay Performance

In saturation conditions, the total delay experienced by a frame is not meaningful. The time elapsed from the time instant when the frame is inserted in the transmission buffer to the time instant when it is successfully transmitted depends on how long the system has remained in saturation conditions, and consequently how congested the transmission buffer has become (this in turn depends on how much greater is the offered load with respect to the saturation throughput bound).

Nevertheless, it is instructive to quantify the average *access* delay D, defined as the time elapsed between the time instant when the frame is put into service - i.e., it becomes head-of-line (HOL) - and the instant of time the frame terminates a successful delivery. Under the assumption of no retry limits, i.e., that all the HOL frames are ultimately delivered, this computation is straightforward. In fact, we may rely on the well-known Little's Result, which states that for any queueing system, the average number of customers in the system is equal to the average experienced delay multiplied by the average customer departure rate. The application of Little's result to our case yields:

---

[10] The Ack_Timeout is specified in the Annex C (For a formal description of MAC operation, see details of the Trsp timer setting on page 346) as:

Ack_Timeout = CTS_Timeout = aSIFS + Duration(Ack) + PLCPHeader + PLCPPreamble + aSlotTime.

According to this value, a station involved in a collision will be able to access the channel only a DIFS after the Ack_Timeout, and thus (in the assumption of 1 Mbps control rate) only a slot-time after the end of an EIFS for monitoring stations. Therefore, we conclude that the extra slot after the end of an EIFS will not be used by any station (either involved in a collision as well as other stations monitoring the channel). Of course, in the assumption that a DIFS is employed after a collision, then the fact that an Ack Timeout is greater than a DIFS is even more evident.

$$D = \frac{N}{S/E[P]} \qquad (43)$$

In fact, under the assumption that no frames are lost because of the retry limit (i.e., R=infinity), each of the N stations is contending with a HOL frame. Moreover, $S/E[P]$ represents the throughput S measured in frames per seconds, and thus represents the frame departure rate from the system.

The delay computation is more elaborate when a frame is discarded after reaching a predetermined maximum number of retries R. In fact, in such a case, a correct delay computation should take into account only the frames successfully delivered at the destination, while should exclude the contribution of frames dropped because of the frame retry limit (indeed, the delay experienced by dropped frames would have no practical significance).

To determine the average delay in the finite retry case, we can still start from Little's Result, but we need to replace N in Eq. (43) with the average number of HOL frames that will be successfully delivered. This value is lower than the number of competing stations, as some of the competing frames will ultimately be dropped. Thus, Eq. (43) can be rewritten as follows:

$$D = \frac{N(1 - P\{LOSS\})}{S/E[P]} \qquad (44)$$

where $P\{LOSS\}$ represents the probability that a *randomly* chosen HOL frame will ultimately be dropped. Let us now randomly pick an HOL frame among the N contending ones. Such an HOL frame can be found in any of the $i=0,...,R$ possible backoff stages. The probability that a random frame is found in backoff stage i has been expressed in Eq. (26), and can be rewritten in terms of known values p, $\tau = P\{TX\}$ and β_i by means of the equalities in Eqs. (29) and (30):

$$P\{s = i\} = P\{TX\} \frac{P\{s = i \mid TX\}}{P\{TX \mid s = i\}} = \tau \frac{(1-p)p^i}{1-p^{R+1}}(1+\beta_i) \qquad (45)$$

By conditioning on the backoff stage i, $P\{LOSS\}$ can be now computed as:

$$\begin{aligned}
P\{LOSS\} &= \sum_{i=0}^{R} P\{LOSS \mid s = i\} \cdot P\{s = i\} = \\
&= \sum_{i=0}^{R} p^{R+1-i} \cdot \tau \frac{(1-p)p^i}{1-p^{R+1}}(1+\beta_i) \\
&= \frac{p^{R+1}}{1-p^{R+1}} \tau(1-p) \sum_{i=0}^{R}(1+\beta_i)
\end{aligned} \qquad (46)$$

The average access delay expression is now found by substituting Eq. (46) into Eq. (44). In the derivation of Eq. (46), we have made use of the fact that the probability that a frame found in backoff stage i is ultimately dropped, is given by the probability that it first reaches the backoff stage R (i.e., it collides for R-i times), and then it also collides during the last transmission attempt. Hence $P\{LOSS|s{=}i\}{=}p^{R+1-i}$.

The average delay expression can be further simplified. From Eq. (33), the probability of a successful transmission, expressed in Eq. (10), can be rewritten as:

$$P_{success} = N\tau(1-p) \tag{47}$$

Hence, recalling that the throughput S was computed in Eq. (12) as the probability of successful transmission multiplied by the ratio $E[P]/E[slot]$,

$$
\begin{aligned}
D &= \frac{N(1 - P\{LOSS\})}{S/E[P]} \\
&= \frac{N - \dfrac{p^{R+1}}{1-p^{R+1}} N\tau(1-p)\sum_{i=0}^{R}(1+\beta_i)}{S/E[P]} \\
&= \frac{N}{S/E[P]} - E[slot]\frac{p^{R+1}}{1-p^{R+1}}\sum_{i=0}^{R}(1+\beta_i)
\end{aligned}
\tag{48}
$$

This final expression has an elegant intuitive interpretation[11]. From Little's Result, the first term represents the average inter-departure time between two *successfully* delivered frames from the same station. This differs from the average access delay as, between two successful transmissions, a number of dropped frames may occur. Now, p^{R+1} represents the probability that a new frame entering the system (i.e., placed in HOL position) will be dropped (note the difference with $P\{LOSS\}$, which instead, represents the probability that a randomly chosen frame, among the contending ones, is lost). Assuming independent frame dropping, the *average* number of dropped frames between two successful deliveries is thus given by the ratio $p^{R+1}/(1-p^{R+1})$. A dropped frame will be forced to cross all the backoff stages, from stage 0 to stage R, and in each stage i it will spend, in average, $(1+\beta_i)$ slots. Hence the average delay for a successfully transmitted frame is given by the average inter-departure time between successful frames, namely, the first in Eq. (48), minus the time spent by dropped frames. Hence, we might have directly written Eq. (48) from this intuitive reasoning, with no need to provide any formal derivation at all!

4.3.4 Non-Ideal Channel Conditions

The performance analyses described in the previous sections were based on the assumption of ideal channel conditions. In this section, we show how, with suitable simplifying assumptions, it can be extended to account for frame corruption.

Let us first comment that the thorough evaluation of the error probability encountered by a frame transmission would require a detailed investigation dealing with physical layer transmission details, fading channel modelling, and interference/capture issues. Such an

---

[11] This neat interpretation was suggested by A.C. Boucouvalas, P. Chatzimisios, and V. Vitsas in a private communication to one of the authors of this chapter. The delay expression given in Eq. (48) was first derived in their recent work with a technical approach different from that presented here [23].

investigation is out of the scope of the present section. However, with the simplifying assumption that all frames are subject to the same, known, corruption probability, the analysis becomes straightforward.

Let ζ be the probability that a transmitted frame is corrupted because of noisy channel conditions, and assume, for convenience of presentation, that errors may occur only on the transmitted MPDU (and not on control frames such as ACK, RTS and CTS – the extension of the analysis to account for control frame errors is immediate). Since the transmitting station will not receive an explicit acknowledgement, it will increment its backoff stage regardless of the fact that a collision occurred on the channel, or the frame was simply corrupted by channel noise. Hence, in case of transmission impairments, the conditional collision probability p, defined in the previous section as the probability that a transmitted frame collides, now represent the union of the events i) the frame collided, and ii) the frame was corrupted. In formulae:

$$p = 1 - (1 - \zeta)(1 - \tau)^{N-1} \tag{49}$$

As usual, τ represents the probability that a station transmits in a randomly chosen slot. With this new definition of p, it becomes clear that the computation of τ is not affected. Thus, Eq. (32) still holds and can be jointly solved with Eq. (49) to obtain the numerical expressions for p and τ.

Some additional care is required to compute the saturation throughput. In fact, it is necessary to determine the proper probabilities of the various events that may occur on the channel, events which now include the case of frame corruption. We can express the throughput S as:

$$S = \frac{(1 - \zeta)P_{success}E[P]}{P_{idle}\sigma + (1 - \zeta)P_{success}T_s + \zeta P_{success}T_e + (1 - P_{idle} - P_{success})T_c} \tag{50}$$

where the probability $P_{success}$ is still given by Eq. (10), but this time it represents the probability that a single frame is transmitted in a slot, i.e., it does not contend with other frames. Of course the frame will be successfully delivered only if it is transmitted alone in a slot and it is not corrupted, i.e., with joint probability $(1-\zeta)$ $P_{success}$.

In Eq. (50), a new value T_e is introduced to account for the duration of a period in which no other stations can access the channel because a corrupted transmission is occurring in the channel. A listening station which detects the transmitted frame as corrupted will wait for an EIFS time interval. Thus:

$$T_e = T_{MPDU} + EIFS \tag{51}$$

Note that some of the surrounding stations may correctly detect the frame. Hence, they will be able to read the duration field in the MAC header of the transmitted frame, and set the NAV accordingly. In other words, under the assumption that the ACK is transmitted at 1 Mbps, T_e will result in the same value regardless of the fact that a listening station sees the transmitted frame as a correct or corrupted one.

4.4 MAC Enhancements for QoS Support

In this section, we present the 802.11e MAC [3] for QoS provisioning. The IEEE 802.11e defines a single coordination function, called the hybrid coordination function (HCF). The HCF combines functions from the DCF and PCF with some enhanced QoS-specific mechanisms and QoS data frames. Note that the 802.11e MAC is backward compatible with the legacy MAC, and hence it is a superset of the legacy MAC. The HCF is composed of two channel access mechanisms: (1) a contention-based channel access referred to as the *enhanced distributed channel access* (EDCA), and (2) a controlled channel access referred to as the *HCF controlled channel access* (HCCA). The HCF sits on top of the DCF in the sense that the HCF utilizes and honors the CSMA/CA operation of the DCF. In a *QoS-enabled IBSS* (QIBSS), only the EDCA can be used since the HCCA requires an AP for channel control. Since we are considering the IBSS operation here, we will limit ourselves to the EDCA operation in the following sections.

4.4.1 IEEE 802.11e EDCA

The EDCA is designed to provide differentiated and distributed channel access for frames with 8 different user priorities (UPs) (from 0 to 7) by enhancing the DCF. Each MSDU from the higher layer arrives at the MAC along with a specific user priority value. Each QoS data frame also carries its user priority value in the MAC frame header. An 802.11e station shall implement four channel access functions, where a channel access function is an enhanced variant of the DCF, as shown in Figure 4.9. Each frame arriving at the MAC with a user priority is mapped into an access category (AC) as shown in Table 4.2, where one of the four channel access functions is used for each AC. Note the relative priority of UP 0 is placed between 2 and 3. This relative priority is obtained from the IEEE 802.1d bridge specification [4].

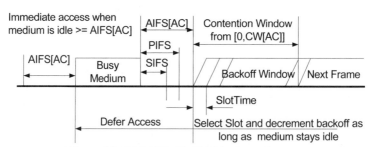

Figure 4.9: IEEE 802.11e EDCA channel access.

Basically, a channel access function uses AIFS[12][AC], CW_{min}[AC], and CW_{max}[AC] (instead of DIFS, CW_{min}, and CW_{max} of the DCF respectively) when contending to transmit a frame belonging to access category AC. AIFS[AC] is determined by

$$AIFS[AC] = SIFS + AIFSN[AC] \bullet SlotTime, \qquad (52)$$

---

[12] AIFS: Arbitration Interframe Space, referring to IEEE 802.11e MAC [3].

where AIFSN[AC] is an integer greater than one. Figure 4.9 shows the timing diagram of the EDCA channel access. One big difference between the DCF and EDCA in terms of the backoff countdown rule is as follows: the first countdown occurs at the end of the AIFS[AC] interval. Moreover, at the end of each idle slot interval, either a backoff countdown or a frame transmission occurs, but not both. Note that according to the legacy DCF, the first countdown occurs at the end of the first slot after the DIFS interval, and if the counter becomes zero during a backoff process, it transmits a frame at that moment.

Figure 4.10 shows the 802.11e MAC with four channel access functions, where each function behaves as a single enhanced DCF contending entity. Each channel access function has its own AIFS and maintains its own backoff counter. Accordingly, these four channel access functions contend for the medium in parallel independently. The channel access function completing the backoff the earliest transmits its pending frame into the medium, and the rest suspend their backoff process until the medium becomes idle again. However, when there is more than one channel access function completing the backoff at the same time, the collision is handled in a virtual manner. That is, the highest priority frame among the to-be colliding frames is chosen and transmitted, and the others perform a backoff with increased CW values.

Apparently, the values of AIFS[AC], CW_{min}[AC], and CW_{max}[AC], referred to as the EDCA parameters, play a key role for differentiated channel access among different user priority (or AC, more accurately speaking) frames. Basically, the smaller AIFS[AC], CW_{min}[AC], and CW_{max}[AC], the shorter the channel access delay for access category AC, and hence the more bandwidth share for a given traffic condition. In the infrastructure mode, these EDCA parameters can be determined and announced by the AP via beacon frames. However, there is no AP in an IBSS, and hence the default parameters as shown in Table 4.3 are used in an IBSS [3]. The parameters aCWmin and aCWmax in the table refer to the CW_{min} and CW_{max} values for different PHYs respectively, e.g., the values found in Table 4.1.

One distinctive feature of the 802.11e is the concept of transmission opportunity (TXOP), which is an interval of time when a particular station has the right to initiate transmissions. During a TXOP, there can be multiple frame exchange sequences, separated by SIFS, initiated by a single station. A TXOP can be obtained by a successful EDCA contention, and it is referred to as an EDCA TXOP. The duration of a TXOP is determined by another EDCA parameter, called TXOP limit. This value is determined for each AC, and hence is represented as TXOPLimit[AC]. The default values of the TXOP limits are also shown in Table 4.3.

Table 4.2: User priority to access category mappings.

Priority	User Priority (UP)	Access Category (AC)	Designation (Informative)
Lowest	1	AC_BK	Background
	2	AC_BK	Background
	0	AC_BE	Best Effort
	3	AC_BE	Best Effort
	4	AC_VI	Video
	5	AC_VI	Video
	6	AC_VO	Voice
Highest	7	AC_VO	Voice

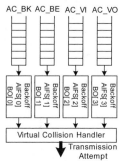

Figure 4.10: Four channel access functions for EDCA.

Table 4.3: Default EDCA Parameter Set.

AC	CW$_{min}$	CW$_{max}$	AIFSN	DS-CCK[13]	TXOP Limit Extended Rate/OFDM[14]	Other PHYs
AC_BK	aCWmin	aCWmax	7	0	0	0
AC_BE	aCWmin	aCWmax	3	0	0	0
AC_VI	(aCWmin+1)/2-1	aCWmin	2	6.016 ms	3.008 ms	0
AC_VO	(aCWmin+1)/4-1	(aCWmin+1)/2-1	2	3.264 ms	1.504 ms	0

During an EDCA TXOP, a station is allowed to transmit multiple MSDUs of the same AC with a SIFS time gap between an ACK and the subsequent frame transmission. Figure 4.11 shows the transmission of two QoS data frames of user priority UP during an EDCA TXOP, where the entire transmission time for two data and ACK frames is less than the EDCA TXOP limit. Multiple consecutive frame transmissions during a TXOP can enhance the communication efficiency by reducing unnecessary backoff procedures.

4.4.2 Further QoS Enhancement for Ad-Hoc Networks

As explained above, the current 802.11e EDCA for the ad-hoc mode can provide differentiated channel access for different user priority frames. However, its capability as defined currently seems to be limited since it relies on the fixed default EDCA parameters. Without a centralized decision-making entity (e.g., an AP), this seems a reasonable choice. However, one can invent better ways for enhanced QoS provisioning in ad-hoc networks.

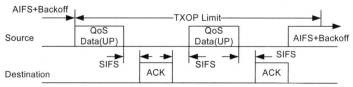

Figure 4.11: EDCA TXOP operation timing structure.

---

[13] Referring to IEEE 802.11b PHY [6].
[14] Referring to IEEE 802.11a [5] and 802.11g PHYs [7].

First, we expect that the 802.11e EDCA in the ad-hoc mode can be further enhanced by incorporating a dynamic EDCA parameter adaptation through negotiation among the participating stations. Along with a distributed admission control, e.g., the one discussed in [8], such a distributed dynamic EDCA parameter adaptation is a possibility.

Second, we can also develop a distributed version of a centralized QoS-supporting MAC, e.g., IEEE 802.11e HCCA [3]. By a distributed version, we mean a dynamic and distributed election of a centralized controller. Since the centralized controller does not need to be located in a fixed access point, this kind of centralized MAC can be easily applicable in the ad-hoc network as well. Note that Bluetooth, a wireless personal area network (WPAN) technology, defines this type of MAC [9], and HIPERLAN/2, another WLAN technology, supports this kind of dynamic centralized controller election optionally [10]. As a matter of fact, a similar concept was once discussed for the standardization as part of IEEE 802.11e MAC as well. However, it was not included in the 802.11e specification eventually due to many different reasons including its immaturity and involving complexity [11].

4.5 Performance Understanding of IEEE 802.11e EDCA

In order to understand the EDCA prioritization mechanisms, it is useful to describe the channel access in terms of low-level channel access operation.

As shown in Section 4.3.2, whenever all the stations operate under saturated conditions, the DCF channel access can be considered as slotted, since packet transmissions start only in discrete time instants. These instants correspond to an integer number of backoff slots which follow the previous channel activity period plus the DIFS time. By looking only at the time instants in which a packet transmission can be originated, the granted channel resources can be represented in terms of a sequence of idle slots, corresponding to the backoff slots in which no station accesses the channel, and busy slots, corresponding to the time interval required for packet transmission (which includes the corresponding acknowledgment when the packet transmissions is a success) plus the DIFS. Given a channel slot, the DCF fairness property implies that each station has the same probability to start a transmission and to experience a success.

The same slotted channel operation can be assumed for describing the channel access occurrence in EDCA. However, the major difference is that the time instants in which the packet transmissions can be originated, which delimit the channel slots, now depend on the minimum AIFS employed by the contending traffic classes. Moreover, because of the different AIFS values, some slots can be accessed only by a subset of the competing traffic classes.

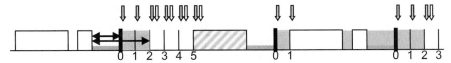

Figure 4.12: Protected slots in EDCA, in the case of AIFSN[AC$_1$]=AIFSN[AC$_2$]+2.

Figure 4.12 shows an example of slotted EDCA channel. The discrete time instants in which channel access can be granted are indicated by arrows, and numbered according to the time elapsed from the last channel activity period. A transmission originated after the minimum AIFS employed in the network[15] belongs to the transmission slot 0, while a transmission originated after x idle backoff slots belongs to the transmission slot x. Each arrow represents the probability that a station belonging to a given priority class transmits on the channel. Only two classes are considered in the figure. Since each class employs a different AIFS value (in the example, the difference between the two values is equal to two backoff slots), some slots can be accessed by only one class (in the example, slots labelled with index 0 and 1). We define these slots, which are shaded and pointed by a single arrow, as *protected*. Note that protected slots occur after each busy slot, and therefore the percentage of protected slots grows as the network congestion increases (this is an immediate consequence of the fact that, with a high number of competing stations, the average number of idle backoff slots, between two consecutive frames transmitted on the channel clearly reduces, and as such the relative amount of protected slots increases).

At the end of each channel access, the stations contend for acquiring the right of the next transmission grant. The contention is based on the comparison of the backoff counter values of each contending station, since the station with the lowest backoff expiration time acquires the right to initiate the next transmission.

The backoff expiration time does not depend only on the backoff counter value, but also on the specific AIFS setting, since the resumption of the backoff counters after each channel activity is not synchronous among the stations. In other words, there are two complementary factors which jointly affect the channel access contention. In fact, consider a number of competing stations, and let b_i be the backoff counter value for each station i at the end of a generic channel activity period. If all stations use the same inter-frame space, then the station that first transmits, and thus "win" the current ongoing contention, will be the one with smaller b_i value. The first way to differentiate performance would then be to configure the backoff operation so that, on average, a group of stations extracts a smaller backoff counter value with respect to the remaining ones (and this can be accomplished by setting different CW_{min} values for different classes of stations – see analysis in the next section 4.5.1). However, when different AIFS intervals are employed, if we define δ_i as the number of extra slots spent while waiting for the station's AIFS to elapse with respect to the minimum possible AIFS value, at the end of a generic channel activity period, the "winning" station would be that one with smaller value $b_i + \delta_i$. The impact of such AIFS differentiation on the channel access contention performance will be tackled in section 4.5.2. In the following, we refer to this slotted contention resolution model in order to investigate on the effects of the CW_{min} and AIFS differentiation.

4.5.1 CW_{min} Differentiation

The generalization of the analysis presented in Section 4.3 to multiple traffic classes is straightforward if we only consider the differentiation of the contention windows. In this

---

[15] As described in section 4.5.3.1, thanks to a new specification of the backoff counter decrement, unlike DCF (see section 4.3.3.1), in EDCA the first slot time immediately following a channel activity period is no more implicitly reserved to the station that has just transmitted. This difference will be extensively discussed later when dealing with EDCA/DCF coexistence.

case, the hypothesis of the model presented in Section 4.3 it is still valid. According to such hypothesis, the behavior of each station can be summarized by a unique parameter, the channel access probability τ, which is uniform slot by slot. However, the probability τ is now different depending on the service class the station belongs to, because the different backoff extraction ranges have the effect of increasing/reducing the probability that some stations win the contention against the others. In order to compute the access probability τ_k for a target station belonging to class k, we can still use Eq. (36), by partitioning the specific backoff extraction ranges (given by $W_k = CW_{min}[k]$ and R_k):

$$\tau_k = \frac{2(1-2p_k)(1-p_k^{R_k+1})}{(1-2p_k)(1-p_k^{R_k+1}) + W_k(1-p_k)(1-(2p_k)^{R_k+1})} \tag{53}$$

Note that, due to the access probability differentiation, the collision probability p_k experienced by each station depends on its service class. Specifically, C is the number of different classes, $k \in (1, C)$ is the class index, n_k is the number of terminals per class, and τ_k is the per-class transmission probability, the per-class conditional collision probability results:

$$p_k = 1 - \frac{\prod_{r=1}^{C}(1-\tau_k)^{n_r}}{(1-\tau_k)} \tag{54}$$

which simply states that the considered station competes with n_r stations of class $r \neq k$ and with n_k-1 stations of the same class (excluding the considered one).

The per-class successful access probability is then given by:

$$P_{success}(k) = n_k \tau_k (1-\tau_k)^{n_k-1} \prod_{r=1, r \neq k}^{C}(1-\tau_r)^{n_r} = n_k \tau_k (1-p_k) \tag{55}$$

Thus, the per-class throughput S_k can be computed as:

$$S_k = \frac{P_{success}(k)E[P]}{\prod_{r=1}^{C}(1-\tau_r)^{n_r}\sigma + \sum_{r=1}^{C}P_{success}(k)T_s + \left(1 - \prod_{r=1}^{C}(1-\tau_r)^{n_r} - \sum_{r=1}^{C}P_{success}(k)\right)T_c} \tag{56}$$

where the numerator represents the average number of payload bits transmitted by stations belonging to class k in each slot, and the denominator represents the average slot duration, which is common for all the classes. The ratio between the aggregated throughput perceived by different service classes, in the case of fixed payload size, is simply expressed by the ratio of the successful access probabilities:

$$\frac{S_j}{S_k} = \frac{P_{success}(j)}{P_{success}(k)} \tag{57}$$

If the number of contending stations is low and the per-class collision probability is negligible, the successful access probability can be approximated by $P_{success}(k)=n_k\tau_k$. In the same hypothesis of negligible collision probability p_k, the per-class access probability τ_k is:

$$\tau_k \approx \frac{2}{1+W_k}\qquad\qquad(58)$$

In this case the throughput ratio among the classes can be immediately related to the minimum contention window settings:

$$\frac{S_j}{S_k} \approx \frac{n_j\tau_j}{n_k\tau_k} = \frac{n_j(1+W_k)}{n_k(1+W_j)} \approx \frac{n_j}{W_j}\cdot\frac{W_k}{n_k}\qquad\qquad(59)$$

where it is evident that the throughput repartition among the stations is proportional to the inverse of the minimum contention window.

Figure 4.13 compares our approximated evaluation of the throughput repartition (points) with some results obtained via simulation (lines). We used the ns-2 simulator with custom-made 802.11e extensions. We assume that an equal number of $n_k=N$ high priority stations share the channel with $n_j=N$ low priority stations. In this case, the throughput repartition S_j/S_k is approximated by W_k/W_j, which does not depend on the network congestion status. In fact, from the figure we observe that the points match very well for both the $N=2$ and $N=10$ simulation curves, which are very close each other. We can conclude that *the throughput repartition due to the CW$_{min}$ differentiation is almost independent on the network load.* Thus, as the network congestion increases, high priority and low priority stations suffer proportionate throughput degradation as the collision probability increases.

4.5.2 AIFS Differentiation

AIFS differentiation is motivated by a completely different (and somewhat more complex) physical rationale. Rather than differentiating the performance by changing the backoff structure (through different settings of the CW$_{min}$ and CW$_{max}$ parameters), the idea is to reserve channel slots for the access of higher priority stations. In fact, as in the case considered in Figure 4.12, when some stations employ different AIFS values, there exists a period of time in which the stations with shorter AIFS value (namely, the higher priority stations) may access the channel, while the stations with longer AIFS (lower priority stations) are prevented from accessing the channel.

A fundamental issue of the AIFS differentiation is that protected slots occur after every busy channel period. This implies that the percentage of protected slots significantly increases as long as the network congestion increases. In fact, a greater number of competing stations implies that the average number of slots between consecutive busy channel periods reduces, and thus the fraction of protected slots over the total number of idle slots gets larger.

We already observed that the number of stations which can access each channel slot is not constant slot by slot, but depends on the time elapsed from the previous transmission. This means that in presence of AIFS differentiation the hypothesis about uniform per-slot collision probability is no more valid. Thus, the protocol analysis cannot be a simple generalization of the DCF one, as in the case of CW_{min} differentiation.

Indeed, several approaches have been recently proposed in literature to solve this issue [24, 28, 29, 30], with different levels of complexity and accuracy. Since the analytical approaches employed to model AIFS differentiation are somewhat complex and formally cumbersome, we refer the reader interested in the AIFS modeling details (e.g., in the derivation of absolute throughput values) to specialized literature works. In this book chapter, we propose a much simpler and intuitive modeling approach targeted to derive relative throughput figures (i.e., the ratio between the throughput figures expected by different classes). Although limited to operate under the assumption of limited collision probability, nevertheless the proposed model has the advantage of being straightforward and providing an immediate physical understanding of why AIFS differentiation is effective in differentiating performance between distinct traffic classes.

For convenience of presentation, let us restrict to the case of two service classes j and k which employ the same contention window parameters, but differentiate each other in terms of AIFS values. More specifically, let the AIFS setting of stations belonging to class j be greater than the AIFS setting of class k stations of an integer number δ_j of backoff slots. Obviously, class j stations will experience a lower channel access priority than class k ones, due to the longest AIFS setting.

Let us now assume that the stations belonging to both classes experience a marginal collision probability. Owing to this approximation[16], we can conclude that the transmission probabilities τ_j and τ_k of stations belonging to these two classes are the same, and that such a transmission probability is given by

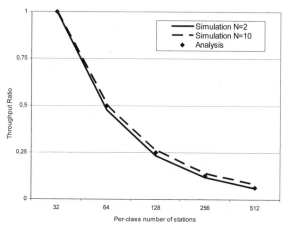

Figure 4.13: Throughput ratio among two service classes with CW_{min} equal to 32 and CW_{min} equal to the x-axis.

---

[16] Different traffic classes will in fact experience different transmission probability values, if the number of collisions is not negligible. In fact, low priority stations (class j) will experience collisions more frequently, when compared to high priority stations (class k), and hence the exponential backoff operation will increase the average contention window value employed throughout the contention process. This will in turn reduce the transmission probability τ_j when compared with τ_k.

$$\tau_k \approx \tau_j \approx \frac{2}{1+W} = \tau \tag{60}$$

Even if the different AIFS settings are assumed not to change the channel access probability τ, stations belonging to different classes will not access the same set of slots (as depicted by the arrows in Figure 4.12), thereby yielding prioritization to class k. In fact, at the end of a generic transmission occurring on the channel, class j stations will be able to access the channel only in the non-protected slots, i.e., when at least δ_j empty backoff slots have elapsed. Now let n_k and n_j be the number of competing high priority and low priority stations respectively. The probability that no high priority station accesses the channel during the δ_j protected slots is given by the probability that neither of the n_k high priority class k stations transmit in any of the δ_j protected slots, i.e.,

$$(1-\tau)^{n_k \delta_j} \tag{61}$$

In the non-protected slots, since all stations are assumed to employ the same transmission probability, all the stations will have the same probability to win the contention, regardless of the service class they belong to. Therefore, the probability that the next successful transmission will be generated from a station in class j is readily given by

$$\frac{n_j}{n_k + n_j} \tag{62}$$

Neglecting collisions, we can trivially approximate the throughput repartition among the two priority classes in terms of the ratio between the contention winning probabilities:

$$\frac{S_j}{S_k} \approx \frac{\dfrac{n_j}{n_j + n_k}(1-\tau)^{n_k.\delta_j}}{1 - \dfrac{n_j}{n_j + n_k}(1-\tau)^{n_k \cdot \delta_j}} \tag{63}$$

Figure 4.14 compares our approximated evaluation of the throughput repartition (points) with some results obtained via simulation (lines). Also in this case, we assume that an equal number of $n_k = N$ high priority and $n_j = N$ low priority stations share the same channel. From the figure we observe that, despite its simplicity, our approximation is quite accurate in capturing the throughput ratio. As expected, the accuracy degrades slightly as N increases because of the emerging occurrence of non negligible collision probabilities, although the difference with simulation results remain fairly limited even when as many as 10 stations ($N=5$) compete against each other.

The figure clearly highlights that the effectiveness of AIFS differentiation significantly depends on the number of competing stations (i.e., the network congestion status). For example, two protected slots correspond to a throughput ratio of about 65% in the case of $N=2$ and to a throughput ratio of about 37% (i.e., even if the same number of stations per priority class are competing, high priority stations share almost ¾ of the total available channel capacity) in the case of $N=5$. We can conclude that *the throughput repartition due to the AIFS differentiation is strongly affected by the high priority load: the higher the load, the greater the effectiveness of AIFS differentiation in protecting the high priority class.* This results in a complementary behavior when compared with the CW_{min} differentiation effect.

4.5.3 Coexistence of EDCA AC_BE and Legacy DCF Stations

In this section we try to clarify the rationale of the AC_BE default settings suggested in the standard through intuitive insights rather than through formal derivations [31]. Since EDCA is backward compatible with standard DCF, we expect that the best effort traffic category is somehow equivalent to the legacy DCF traffic. However, from Table 4.3, we see that the access parameters have some differences. Despite the same minimum and maximum contention window value, the inter-frame time value for the AC_BE is higher than a DIFS (we recall that a DIFS is equal to an AIFS with AIFSN=2).

Figure 4.15 shows the throughput results in a scenario in which N legacy DCF stations share the channel with the same number of EDCA stations (i.e., N). Curves with the same symbol refer to the same simulation. The bold lines represent the aggregate EDCA throughput, while the thin lines represent the aggregate DCF throughput. EDCA stations have been configured with the standard DCF backoff parameters (CW_{min}=31 and CW_{max}=1023). The packet size has been fixed to 1500 bytes (Ethernet MTU) and the retransmission limit is set to 7 for all the stations. Control frames are transmitted at a basic rate equal to 1 Mbps, while the MPDU is transmitted at 11 Mbps. Unless otherwise specified, these settings have been maintained in all the simulations. We measured performance in saturated conditions. Although this assumption is not realistic for real-time applications, it represents a very good representation of elastic data traffic. It is interesting to derive the limit performance, i.e., the maximum amount of bandwidth that AC_BE can obtain sharing the channel with best effort DCF stations. From the figure, we see that in the case AIFSN=2, the EDCA stations receive much more resources than DCF stations while in the case AIFSN=3, they achieve a performance close to that of legacy DCF stations.

This counter-intuitive result confirms that the default settings have been chosen in order to guarantee backward compatibility with the DCF. However, we need a detailed analysis of the channel access operations in DCF and EDCA to fully understand how this compatibility is provided.

4.5.3.1 Backoff Counter Decrement Rules

EDCA differs slightly from DCF in terms of how the backoff counter is managed (decremented, frozen, resumed). Such a minor difference (which may perhaps appear as a technicality) has some important consequences on the performance of EDCA access categories, especially when they compete with legacy DCF stations.

In standard DCF, the backoff counter is decremented at each idle slot-time, frozen during channel activity periods, and resumed after the medium is sensed idle again for a DIFS interval. This implies that a legacy DCF station, after a DIFS, resumes the backoff counter to the discrete value the station had at the instant of time the busy channel period started. An illustrative example is shown in Figure 4.16. Here, a busy channel period (i.e., a transmission from one or more other stations) starts while the backoff counter of the considered DCF station is equal to 4. This value will be frozen during the busy channel period, and will be resumed, again to the value 4, only a DIFS after the end of the busy period. As a consequence, it will be decremented to the value 3 only a slot after the DIFS. In EDCA, the backoff counter is also decremented at every idle slot-time and frozen during channel activity periods, but it is resumed one slot-time before the AIFS expiration. This means that when the AIFS timer elapses, the backoff counter will be decremented by one unit.

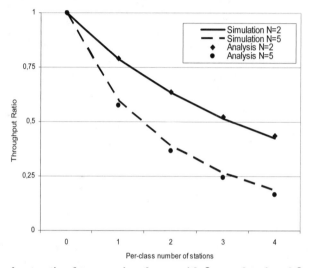

Figure 4.14: Throughput ratio of two service classes with δ_k equal to 0 and δ_j equal to the x-axis.

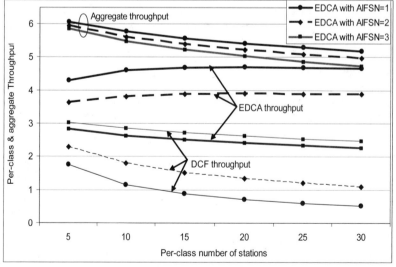

Figure 4.15: DCF versus EDCA throughput with AIFS Differentiation.

Moreover, since a single MAC operation per-slot is permitted (backoff decrement or packet transmission, see [2], Clause 9.9.1.3), when the counter decrements to 0, the station cannot transmit immediately, but has to wait for an additional backoff slot if the medium is idle, or for a further AIFS expiration if the medium is busy. Figure 4.16 shows how these different rules affect the channel access probability.

Let us first focus on the case AIFSN=2 (top figure), which corresponds to using an AIFS equal to the DCF DIFS. In the example, two stations encounter a busy channel period with the same backoff counter value. However, at the end of the channel activity, we see that the DCF station resumes its counter to a value equal to the frozen value (4 in the example), while the EDCA station resumes and decrements its counter. In the case of a single busy channel period encountered during the backoff decrement process, this difference will be compensated by the fact that the EDCA station will have to wait for an extra slot, i.e., unlike the DCF station, it will transmit in the slot following the one in which the backoff counter is decremented to 0 (as illustrated in Figure 4.16 for the top EDCA station). However, in the presence of several busy channel periods encountered during the backoff decrement process (which is very likely to happen in the presence of several competing stations), the EDCA station will gain a backoff counter decrement advantage for every encountered busy period with respect to the DCF station. This implies that, for an AIFSN equal to a DIFS, the EDCA station has an advantage over DCF, as we observed in Figure 4.15, and this advantage grows as the number of contending stations increases.

Figure 4.16 shows that there is a second reason why, with same inter-frame space AIFSN=2, the EDCA station gains priority over DCF stations. In fact, as shown in the figure, an EDCA station may actually transmit in the slot immediately following a busy channel period (it is sufficient that the busy channel period was encountered while the backoff counter was equal to 0 – last case in Figure 4.16 with AIFSN=2). Conversely, a DCF station cannot reset a backoff counter value to zero. Thus, the only case in which it can access the slot immediately following a busy period is when it extracts a new backoff counter, after a successful transmission, exactly equal to 0.

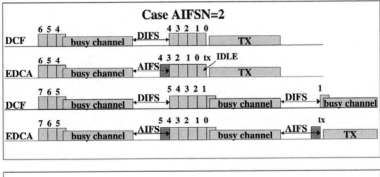

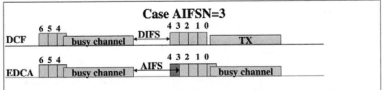

Figure 4.16: Backoff counter management in EDCA and DCF.

In order to synchronize the EDCA and DCF backoff decrements, it may be appropriate to set AIFSN=3. In this case, as we can see in the bottom part of Figure 4.16, although the EDCA station has a higher interframe space, after each busy slot the backoff evolution of the two target stations is the same. However, since the EDCA station has to wait for a further channel slot after the counter expiration, the access probabilities of the two stations does not coincide, since for a given extraction, the EDCA station must wait for an additional slot than the DCF station. However, this results in only a slight increase in access probability for the DCF station, which justifies the slightly higher throughput performance observed in Figure 4.15 in the comparison with DCF stations and EDCA stations with AIFSN=3.

4.5.3.2 Analysis of AC_BE Default Settings

The throughput results shown in Figure 4.15 show that, for the same contention window parameters, EDCA throughput performance are similar to that of legacy stations with AIFSN=3 (i.e., the EDCA AC_BE Access Category, see Table 4.3) rather than to a legacy DIFS (i.e., AIFSN=2). The discussion carried out in the previous section has provided a qualitative justification.

The goal of Figure 4.17 is to back-up the previous qualitative explanation with quantitative results. To this purpose, we have numbered slots according to our previous description of the channel access operations. The slot immediately following a DIFS is indexed as slot 0. In the assumption of ideal channel conditions, a successful transmission occurs if, in a transmission slot, only one station transmits; otherwise a collision occurs.

Figure 4.17 shows the probability distribution that a transmission occurs at a given slot, for two different load scenarios: N=5 (i.e., 5 EDCA stations competing with 5 DCF stations) and N=30. Only the first 10 slots are plotted, since most transmissions are originated after very few idle backoff slots. In addition, the figure further details, in different shades, the probability that a transmission occurring at a given slot results in a collision, in a success for an EDCA station, or in a success for a DCF station.

Figure 4.17 shows that DCF stations are the only ones that can transmit in the slot immediately following the last busy period. It also confirms that a transmission in slot 0 is always successful (as it is originated by a station that has just terminated a successful transmission). Indeed, a transmission in the slot immediately following a busy period is a rare event, since it requires that the station that has just experienced a successful transmission extracts a new backoff counter exactly equal to 0. Thus, the slot 0 is a *protected slot* for the DCF stations, but it is rarely[17] granted. The figure also shows that, in the slots with index greater than 0, DCF and EDCA stations experience almost the same success probability, with a negligible advantage for DCF. For example, in the case N=5 a DCF success occurs, almost constantly through the various slot indexes, in about 42.5% of the cases versus 41% for EDCA, while for N=30, these numbers reduce to about 32.5% and 31.3% respectively due to the increased probability of collision. The fundamental conclusion is that *by using AIFSN=3, an EDCA station can be set to operate like a legacy*

---

[17] Quantification is easy: after a successful transmission, a DCF station transmits in the slot 0 only if it extracts a backoff counter equal to 0. This occurs with probability $1/(1+CW_{min}) \approx 3.1\%$. This conditional probability is consistent with the absolute probability value reported in Figure 4.17 (about a half of this), since about half of the busy periods are successful DCF transmissions.

DCF station. With reference to the proposed EDCA parameter settings reported in Table 4.3, we thus conclude that an EDCA station belonging to the Access Category AC_BE will experience similar performance than a legacy DCF station. The above quantitative analysis also justifies why DCF shows a slightly superior throughput performance over EDCA AC_BE, as depicted in Figure 4.15 under the case of AIFSN=3.

4.5.3.3 AIFSN=2 and Legacy DCF Stations

As shown in Table 4.3, AIFSN=2 is the minimal setting allowed for an EDCA station. The rationale is that both AIFSN=0 and AIFSN=1 are already reserved in the 802.11 standard for SIFS and PIFS respectively. However, as discussed above, the different mechanisms employed in EDCA for decrementing the backoff counter suggests that by using AIFSN=2 (i.e., AIFS=DIFS), an EDCA station is expected to gain priority over a legacy DCF station.

This was shown in Figure 4.15 and confirmed by Figure 4.18. Similar to Figure 4.17, Figure 4.18 reports the probability distribution of a transmission occurring at a given slot in the scenario where N DCF stations compete with N EDCA stations that are configured with AIFSN=2 and standard contention window parameters (i.e., CW_{min}=31 and CW_{max}=1023).

Figure 4.18 shows, for two different load conditions (N=5 and N=30), how the channel slots are occupied by the contending stations. From the figure, we see that slot 0, as shown before, is rarely used by DCF stations, resulting in almost full protection for EDCA stations. Instead, channel slots with index higher than 0 are accessed by both classes with comparable probability. Figure 4.18 allows us to draw a number of interesting observations. First, the probability of collision in the protected slots (specifically, slot 0) is lower than the other slots (e.g., for the case N=5, a collision in slot 0 occurs only in about 8.5% of the cases, versus an average of 17% in the remaining slots. For the case N=30, these numbers become 24.5% versus 38.5%) due to the reduced number of competing stations. Second, and most interesting, as the network load increases, the probability of accessing low-indexed slots increases significantly. The reason is that the number of slots between two consecutive busy channel periods significantly reduces in high load. But this implies that much channel access occurs in slot 0 (more than 40% in the case of N=30 as shown in Figure 4.18), and is almost exclusively dedicated to EDCA stations, with a definite gain in terms of service differentiation effectiveness (as shown earlier in Figure 4.15). As a conclusion, *the use of AIFSN=2 in EDCA (i.e., AIFS=DIFS) provides a significant priority of EDCA stations over legacy DCF stations.* This is an extremely important fact, as it permits the effective deployment of AIFS differentiation even when DCF stations share the same channel, and thus, there seems to be no room for AIFS levels intermediate between the interframe spaces reserved by the standard (SIFS and PIFS) and the legacy DIFS[18].

---

[18] We in fact recall that a legacy DIFS is defined as SIFS + 2·slot, while a PIFS is defined as SIFS + 1·slot. Since SIFS and PIFS are parameters reserved for other purposes, the minimum deployable AIFS value coincides with that of a legacy DIFS, and this argument was indeed considered as a limiting factor for EDCA, when competing with legacy 802.11 stations. However, the above discussion show that this is not the case, and that the legacy DIFS is somewhat equivalent – in practice – to an AIFS setting equal to SIFS + 3·slot.

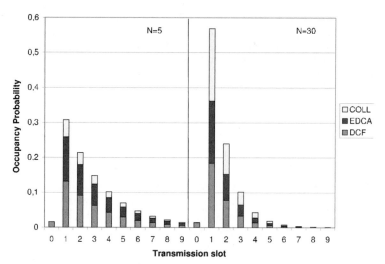

Figure 4.17: Per-slot occupancy probability - AC_BE versus DCF.

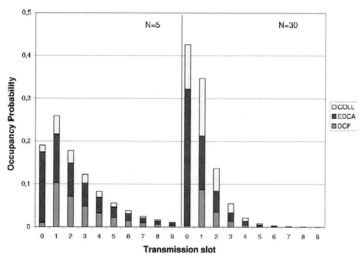

Figure 4.18: Per-slot occupancy probability - AIFSN=2 versus DCF.

4.6 Conclusions

In this chapter, we have presented a performance study, based on some common modeling approaches, on the IEEE 802.11 DCF and the recent IEEE 802.11e EDCA extensions. Regarding the standard DCF, we have first quantified the header and protocol overheads for a single transmitting station. Then we introduce the concept of saturation throughput, when two or more stations contend greedily for channel access. We have shown that in such conditions, the protocol behavior can be summarized into a single channel access probability τ, and we have derived the system throughput and delay as a function of the number of contending stations and τ. We have illustrated the DCF performance bounds for the cases of 2-way and 4-way handshake, by showing how the system throughput can be maximized for a given τ value, which depends on the number of contending stations and on the frame payload. Finally, we concluded the DCF analysis by deriving τ as a function of the contention window ranges, with elementary conditional probability arguments.

Regarding the EDCA extensions, we have discussed some simple model generalizations to differentiate the per-station contention window ranges. Then we focused on the problems arising from modeling the AIFS time differentiation. We proposed an intuitive approximation to overcome such modeling complications and to allow an easy tuning of the contention parameters, targeted to guarantee a desired throughput repartition among the service classes. Finally, we presented the system performance for the case of coexistence between EDCA and legacy DCF stations, and justified the equivalence between DCF stations and EDCA best effort class.

4.7 References

[1]   IEEE 802.11-1999 (R2003), Part 11: Wireless LAN Medium Access Control (MAC) and Physical Layer (PHY) specifications, ANSI/IEEE Std 802.11, 1999 Edition (Reaffirmed 2003), June 2003.

[2]   IEEE 802-2001, IEEE Standard for Local and Metropolitan Area Networks: Overview and Architecture, March 2002.

[3]   IEEE 802.11e-2005, Draft Supplement to Part 11: Wireless Medium Access Control (MAC) and physical layer (PHY) specifications: Medium Access Control (MAC) Enhancements for Quality of Service (QoS), Sept. 2005.

[4]   IEEE, Part 3: Media Access Control (MAC) bridges, ANSI/IEEE Std. 802.1D, IEEE 802.1d-1998, 1998 edition, 1998.

[5]   IEEE 802.11a-1999, Supplement to Part 11: Wireless LAN Medium Access Control (MAC) and Physical Layer (PHY) specifications: High-speed Physical Layer in the 5 GHZ Band, 1999.

[6]   IEEE 802.11b-1999, Supplement to Part 11: Wireless LAN Medium Access Control (MAC) and Physical Layer (PHY) specifications: Higher-speed Physical Layer Extension in the 2.4 GHz Band, 1999.

[7]   IEEE 802.11g-2003, Supplement to Part 11: Wireless LAN Medium Access Control (MAC) and Physical Layer (PHY) specifications: Further Higher-Speed Physical Layer Extension in the 2.4 GHz Band, 2003.

[8]   Yang Xiao, Haizhon Li, and Sunghyun Choi, "Protection and Guarantee for Voice and Video Traffic in IEEE 802.11e Wireless LANs," in *Proc. IEEE INFOCOM'04*, Hong Kong, March 2004.

[9]   Brent A. Miller and Chatschik Bisdikian, *Bluetooth Revealed: The Insider's Guide to an Open Specification for Global Wireless Communications*, Prentice Hall, September 2000.

[10] ETSI TS 101 761-4, Broadband Radio Access Networks (BRAN); HIPERLAN Type 2; Data Link Control (DLC) Layer; Part 4: Extension for Home Environment, V1.1.1, June 2000.

[11] Adrian P. Stephens, "AP Mobility Mechanism," IEEE 802.11-02/066r9, March 2002.

[12] Javier del Prado and Sunghyun Choi, "Link Adaptation Strategy for IEEE 802.11 WLAN via Received Signal Strength Measurement," in *Proc. IEEE ICC'03*, Anchorage, Alaska, USA, May 2003.

[13] Daji Qiao, Sunghyun Choi, and Kang G. Shin, "Goodput Analysis and Link Adaptation for IEEE 802.11a Wireless LANs," *IEEE Trans. on Mobile Computing (TMC)*, vol. 1, no. 4, pp. 278-292, October-December 2002.

[14] Jean-Lien C. Wu, Hunh-Huan Liu, and Yi-Jen Lung, "An Adaptive Multirate IEEE 802.11 Wireless LAN," in *Proc. 15th International Conference on Information Networking*, 2001, pp. 411-418.

[15] Ad Kamerman and Leo Monteban, "WaveLAN-II: A High-Performance Wireless LAN for the Unlicensed Band," *Bell Labs Technical Journal*, Summer 1997, pp. 118–133.

[16] Giuseppe Bianchi, "Performance Analysis of the IEEE 802.11 Distributed Coordination Function," *IEEE Journal on Selected Areas in Communications*, vol. 18, no. 3, March 2000.

[17] H. Wu, Y. Peng, K. Long, S. Cheng, and J. Ma, "Performance of Reliable Transport Protocol over IEEE 802.11 Wireless LANs: Analysis and Enhancement", in *Proc. IEEE INFOCOM'02*, 2002.

[18] V. M. Vishnevsky and A. I. Lyakhov, "802.11 LANs: Saturation Throughput in the Presence of Noise", in *Proc. IFIP Networking'02*, Pisa, Italy, 2002.

[19] P. Chatimisios, V. Vitsas, and A. C. Boucouvalas, "Throughput and Delay Analysis of IEEE 802.11 Protocol," in *Proc. the 5th IEEE International Workshop on Network Appliances (IWNA)*, Liverpool, UK, Oct. 2002.

[20] Y. Xiao, "A Simple and Effective Priority Scheme for IEEE 802.11," *IEEE Communications Letters*, vol. 7, no. 2, Feb. 2003, pp. 70-72.

[21] Z. Hadzi-Velkov and B. Spasenovski, "Saturation Throughput – Delay Analysis of IEEE 802.11 DCF in Fading Channels," in *Proc. IEEE ICC'03,* Anchorage, Alaska, May 2003, vol. 1, pp. 121-126.

[22] T.-C. Hou, L.-F. Tsao, and H.-C. Liu, "Throughput Analysis of the IEEE 802.11 DCF Scheme in Multi-Hop Ad Hoc Networks," in *Proc. ICWN'03*, Jun. 2003.

[23] P. Chatzimisios, A. C. Boucouvalas, and V. Vitsas, "IEEE 802.11 Packet Delay – A Finite Retry Limit Analysis," in *Proc. IEEE Globecom'03*, Dec. 2003.

[24] G. Bianchi and I. Tinirello, "Analysis of Priority Mechanisms Based on Differentiated Inter Frame Spacing in CSMA-CA," in *Proc. IEEE VTC'03-Fall*, October 2003.

[25] Q. Ni, I. Aad, C. Barakat, and T. Turletti, "Modeling and Analysis of Slow CW Decrease for IEEE 802.11 WLAN," in *Proc. IEEE PIMRC'03*, Sept. 2003.

[26] G. Bianchi, L. Fratta, and M. Oliveri, "Performance Evaluation and Enhancement of the CSMA/CA MAC Protocol for 802.11 Wireless LANs," in *Proc. IEEE PIMRC'96*, Taipei, Taiwan, October 1996, pp. 392-396.

[27] F. Calì, M. Conti, and E. Gregori, "Dynamic Tuning of the IEEE 802.11 Protocol to Achieve a Theoretical Throughput Limit," *IEEE/ACM Trans. Networking*, vol. 8, no. 6, Dec. 2000, pp. 785-790.

[28] J. Zhao, Z. Guo, Q. Zhang, W. Zhu, "Performance Study of MAC for service differentiation in IEEE 802.11", in *Proc. IEEE Globecom'02*, Taipei, 2002, pp. 778-782.

[29] J. W. Robinson and T. S. Randhawa, "Saturation Throughput Analysis of IEEE 802.11e Enhanced Distributed Coordination Function," *IEEE Journal on Selected Areas in Communications*, Vol.2 N.5, June 2004, pp. 917-928.

[30] Y. Xiao, "Performance Analysis of IEEE 802.11e EDCF under Saturation Conditions," in *Proc. IEEE ICC'04*, Paris, 2004, pp. 170-174.

[31] G. Bianchi, I. Tinnirello, and L. Scalia, "Understanding 802.11e contention-based prioritization mechanisms and their coexistence with legacy 802.11 stations," *IEEE Network*, Vol. 19, Issue 4, July-Aug. 2005, pp. 28-34.

5

Cross-layer Optimized Video Streaming over Wireless Multi-hop Mesh Networks

Yiannis Andreopoulos[a], Nicholas Mastronarde[b], Mihaela van der Schaar[b]

The proliferation of wireless multi-hop communication infrastructures in office or residential environments depends on their ability to support a variety of emerging applications requiring real-time video transmission between stations located across the network. We propose an integrated cross-layer optimization algorithm aimed at maximizing the decoded video quality of delay-constrained streaming in a multi-hop wireless mesh network that supports quality-of-service (QoS). The key principle of our algorithm lays in the synergistic optimization of different control parameters at each node of the multi-hop network, across the protocol layers - application, network, medium access control (MAC) and physical (PHY) layers, as well as end-to-end, across the various nodes. To drive this optimization, we assume an overlay network infrastructure, which is able to convey information on the conditions of each link. Various scenarios that perform the integrated optimization using different levels ("horizons") of information about the network status are examined. The differences between several optimization scenarios in terms of decoded video quality and required streaming complexity are quantified. Our results demonstrate the merits and the need for cross-layer optimization in order to provide an efficient solution for real-time video transmission using existing protocols and infrastructures. In addition, they provide important insights for future protocol and system design targeted at enhanced video streaming support across wireless mesh networks.

5.1 Introduction

Wireless mesh networks are built based on a mixture of fixed and mobile nodes interconnected via wireless links to form a multi-hop ad-hoc network. The use of existing protocols for the interconnection of the various nodes (hops) is typically desired as it reduces deployment costs and also increases interoperability [1]. However, due to the network and channel dynamics, there are significant challenges in the design and joint optimization of application, routing, MAC, and PHY adaptation strategies for efficient video transmission across such mesh networks.

---

[a] *Queen Mary University of London*
[b] *University of California, Los Angeles*

In this work, we are addressing some of these challenges by developing an integrated video streaming paradigm enabling cross-layer interaction across the protocol stack and across the multiple hops. The problem of multi-hop video streaming has recently been studied under a variety of scenarios [2] [3] [4]. However, the majority of this research does not consider the protection techniques available at the lower layers of the protocol stack and/or optimizes the video transport using purely end-to-end metrics, thereby excluding a significant amount of improvement that can occur by cross-layer design [5] [6] [7]. Consequently, the inherent network dynamics occurring in a multi-hop wireless mesh network as well as the interaction among the various layers of the protocol stack are not fully considered in the existing video streaming literature. Indeed, recent results concerning the practical throughput and packet loss analysis of multi-hop wireless networks [8] [9] have shown that the incorporation of appropriate utility functions that take into account specific parameters of the protocol layers such as the expected retransmissions, the loss rate and bandwidth of each link [8], as well as expected transmission time [9] or fairness issues .[10], can significantly impact the actual end-to-end network throughput. Motivated by this work, we show that, for delay-constrained video streaming over multi-hop wireless mesh networks, including the lower layer network information and adaptation parameters in the cross-layer design can provide significant improvements in the decoded video quality.

We focus on the problem of real-time transmission of an individual video bitstream across a multi-hop 802.11a/e wireless network and investigate i) what is the video quality improvement that can be obtained if an integrated cross-layer strategy involving the various layers of the protocol stack is performed and ii) what is the performance and complexity impact if the optimized streaming solution is performed using only limited, localized information about the network status, as opposed to global, complete information.

We assume that the mesh network topology is fixed over the duration of the video session and that, prior to the transmission, each application (video flow) reserves a predetermined transmission opportunity interval, where contention-free access to the medium is provided[1]. This reservation can be performed following the principles of the HCCA[2] protocol of IEEE 802.11e [12] and can be determined based on the amount of flows sharing the network. Although the design of such a reservation system is an important problem and it affects our results, recent work showed that scheduling of multiple flows in the context of a mesh topology can be done such that the average rate for every flow is satisfied and the interference to neighboring nodes is minimized [13]. Hence a similar solution can be applied for our case and the available nodes and links within the entire mesh topology can be pre-established by a central coordinator prior to the video streaming session initiation. This minimizes the probability of additional delays and link failures due to routing reconfigurations during the video streaming and also decouples the problem of optimized media streaming and optimized route and link-reservation establishment within the wireless multi-hop network. Once the available network infrastructure to the video streaming session has been established, we assume that an overlay network topology can convey (in frequent intervals) information about the expected bit error rate (BER), the queuing delay for each link, as well as the guaranteed

---

[1] Existing IEEE standards [12] already support such QoS mechanisms, which, barring interference and environment noise, provide guaranteed transmission time for each admitted application (video flow).
[2] HCCA: HCF Controlled Channel Access, where HCF stands for Hybrid Coordinator Function [12].

bandwidth under the dynamically-changing modulation at the PHY. Several examples of such application-layer overlay networks have been proposed in the literature [19] [20].

Under the above assumptions, we make the following contributions. For video packets of each hop in the mesh network, we propose an optimization framework that jointly determines per packet: (a) the optimal modulation at the PHY, (b) the optimal retry limit at the MAC, (c) the optimal path (route) to the receiver in the remaining part of the mesh network and (d) the application-layer optimized packet scheduling, given a predetermined topology and time reservation per link using the concepts of IEEE 802.11e HCCA. This chapter is an extension of our recent work [21] presenting our proposals in more detail and containing new experiments with a variety of video content.

The presentation of the material is organized as follows. Section 5.2 defines the scenarios examined in this work and provides the necessary definitions and formulations for the expected bandwidth, transmission error rate as well as the expected delay for streaming under various network paths. Section 5.3 presents the cross-layer optimization problem. The proposed solutions are presented in Section 5.4. Section 5.5 analyzes the complexity and feedback requirements of the proposed approaches. Section 5.6 presents indicative results, including comparisons with other well known approaches from the literature. Our conclusions are presented in Section 5.7.

5.2 Proposed Integrated Cross-Layer Video Streaming

Consider that N nodes (hops) of a wireless multi-hop mesh network decide to participate in a video streaming session. Example topologies with $N = 3$ and $N = 7$ are shown in Figure 5.1 and Figure 5.2. Node h_1 represents the original video source, while node h_N is the destination node (video client). Each link ij is associated with the corresponding allocated bandwidth for the video traffic (g_{ii}), the error rate observed on the link (e_{ii}), as well as the corresponding delay due to the video queue (d_{ij}^{queue}). Within the reserved time for the video traffic, each link exhibits a certain throughput given the chosen modulation strategy. Video packets are lost due to the experienced BER. This error is due to noise and interference in the wireless medium stemming from background noise, node mobility or simultaneous link transmissions. In addition to this error, under delay-constrained video streaming, packets are discarded due to delays incurred in the transmission, e.g., the queuing delay of each link. Notice that Figure 5.2 displays different connectivity structures for the network topology, as specified by the indicated links. Obviously, the tightly-connected multi-hop mesh topologies T1 and T2 of Figure 5.2 offer more alternative paths for the video traffic that topology T3; however, the overall reserved time across the various nodes of the network is also increased. In general, the decision on the connectivity as well as the number of nodes participating in the video streaming session depends largely on a number of system-related factors that transcend the video streaming problem (e.g., node cooperation strategy/incentives and network coordination and routing policies imposed by the utilized protocols). Hence, in this work we investigate cross-layer optimization for video over multi-hop wireless mesh networks given the network specification (participating nodes and connectivity) as well as the available reservation time on each link for the video traffic.

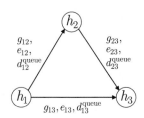

Figure 5.1: A simple topology with three hops.

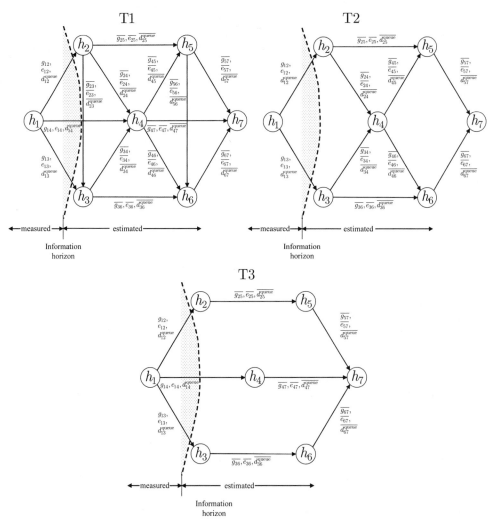

Figure 5.2: More complex topologies with seven hops. In these cases, the link information up to hop h_4 is directly conveyed by the overlay network infrastructure, while other link information is inferred based on theoretical estimates using average or past information (where $\overline{m_{ij}}$ indicates the estimated or average value for the metric m_{ij}, m = {g,e,d}).

Under the existence of feedback from an overlay network infrastructure, the BER and queuing delay per link can be disseminated to the remaining network hops at frequent intervals (via a hop-to-hop feedback mechanism[3]), or when the incurred change in network parameters is larger than a preset threshold. Thus, they can be considered to be known (Figure 5.1). However, in certain cases, feedback from remote hops may arrive with an intolerable delay, or, alternatively, it can be deemed unreliable due to the rapidly-changing network conditions. As a result, a certain "horizon" of information retrieval can be envisaged for each hop (Figure 5.2), where network information within the horizon is deemed reliable and can be received in a timely manner, while information beyond the horizon can only be theoretically estimated based on average or previous measurements.

5.2.1 Wireless Multi-hop Mesh Topology Specification

For a generic multi-hop wireless mesh network, we consider the *connectivity structure* **P**:

$$\mathcal{P} = \{\mathbf{p}_1,...,\mathbf{p}_M\} \tag{1}$$

where each element $\mathbf{p}_i$, $1 \leq i \leq M$ is the *connectivity vector* (end-to-end network path) given by:

$$\mathbf{p}_i = \begin{bmatrix} l_{i,1} & l_{i,2} & \cdots & l_{i,\rho_i^{\text{total}}-2} & l_{i,\rho_i^{\text{total}}-1} \end{bmatrix} \tag{2}$$

where each component $l_{i,j}$ ($1 \leq j < \rho_i^{\text{total}}$) indicates a particular wireless link (the j-th link of path i), and $\rho_i^{\text{total}} - 1$ is the total number of links participating in the network path $\mathbf{p}_i$. For example, for the topology of Figure 5.1 with $N = 3$ we have:

$$\mathcal{P} = \left\{ \begin{bmatrix} l_{1,1} & l_{1,2} \end{bmatrix}, l_{2,1} \right\} \tag{3}$$

with $\rho_1^{\text{total}} = 3$, $\rho_2^{\text{total}} = 2$, and:

$$\begin{aligned} l_{1,1} &= (h_1 \rightarrow h_2); \ l_{1,2} = (h_2 \rightarrow h_3) \\ l_{2,1} &= (h_1 \rightarrow h_3) \end{aligned} \tag{4}$$

Notice that (1) and (2) apply both for the end-to-end topology of interest but also for the topology between any intermediate node and the terminal (client) node in the mesh network utilized for video transmission. For example, if we consider the sub-network of topology T2 of Figure 5.2 consisting of nodes h_4, h_5, h_6 and h_7, there are two paths from h_4 to h_7, and the equivalent definitions apply locally. Hence, the subsequent problem specification and analysis is inherently scalable and can be applied in a similar fashion to either the entire end-to-end topology or only part of the topology (sub-network). Finally, it is important to mention that all the proposed algorithms in this chapter assume the non-

---

[3] For example, in order to utilize the medium more efficiently, it is possible to piggyback feedback about the link status information onto the acknowledgment packets.

existence of routing loops, i.e., the mesh network between the current hop and the destination hop can be represented by a tree graph.

5.2.2 Link and Path Parameter Specification

For each link $l_{i,j}$, given a certain modulation $m(l_{i,j})$ at the physical layer, we denote the expected bit error rate as $e(l_{i,j})$. Notice that this error is usually estimated based on channel modeling as well as experimental studies in the network which analyze the effects of interference [11]. As a result, the higher layers of the protocol stack can assume $e(l_{i,j})$ to be independent and randomly distributed [22]. Under a predetermined negotiation of traffic specification parameters for each link in the mesh network (e.g., following the HCCA protocol [12]), each link can provide a guaranteed bandwidth $g(l_{i,j})$ at the application layer. Following the HCCA specification [12], this bandwidth is linked with the traffic specification parameters by [14]:

$$g(l_{i,j}) = \frac{t_{\text{TXOP}}(l_{i,j}) \cdot \overline{L}}{\left[\overline{L} \cdot \left(R_{\text{phy}}(l_{i,j}) \right)^{-1} + T_{\text{overhead}} \right] \cdot t_{\text{SI}}(l_{i,j})} \tag{5}$$

where $t_{\text{TXOP}}(l_{i,j})$ is the transmission opportunity duration provided by the HCCA admission control for the video flow traffic of link $l_{i,j}$, $\overline{L}$ is the nominal MAC service data unit (MSDU) size[4], $t_{\text{SI}}(l_{i,j})$ is the specified duration of the service interval [12] for the video flow traffic at link $l_{i,j}$, $R_{\text{phy}}(l_{i,j})$ is the physical layer rate and T_{overhead} represents the duration of the required overheads corresponding to polling and acknowledgment policies. As demonstrated by (5), even though the negotiated transmission opportunity duration is constant per link, the guaranteed bandwidth depends on the provided physical layer rate $R_{\text{phy}}(l_{i,j})$, which in turn makes it dependent on the chosen modulation[5] $m(l_{i,j})$. Finally, depending on the chosen modulation, $R_{\text{phy}}(l_{i,j})$ may change for each MSDU. Hence, the guaranteed bandwidth $g(l_{i,j})$ of (5) can be determined for each MSDU.

Under the aforementioned assumptions for the error model of each link, the probability of error for the transmission of MSDU v of size L_v bits is:

$$e_{l_{i,j}}(L_v) = 1 - \left(1 - e(l_{i,j}) \right)^{L_v} . \tag{6}$$

Consequently, the probability of error for the packet transmission in path $\mathbf{p}_i$ is:

$$e_{\mathbf{p}_i}(L_v) = 1 - \prod_{j=1}^{p_i^{\text{total}}-1} \left(1 - e_{l_{i,j}}(L_v) \right) = 1 - \prod_{j=1}^{p_i^{\text{total}}-1} \left[\left(1 - e(l_{i,j}) \right)^{L_v} \right] . \tag{7}$$

Notice that the derivation of (7) computes the end-to-end probability of error under the assumption that the packet is sequentially transmitted from each link of path i.

---

[4] We assume that one video packet is encapsulated in one MSDU and the two terms are used interchangeably.

[5] For notational simplicity we do not particularly indicate the dependence of $e(l_{i,j})$ and $g(l_{i,j})$ on modulation $m(l_{i,j})$.

Under a single (successful) MSDU transmission via each link $l_{i,j}$, the transmission delay for path $\mathbf{p}_i$ can be calculated as:

$$d_{\mathbf{p}_i}(L_v, 1) = \sum_{j=1}^{\rho_i^{\text{total}}-1} \left(\frac{L_v}{g(l_{i,j})} + d_{\text{queue}}(l_{i,j}) + T_{\text{overhead}} \right) \tag{8}$$

where $d_{\text{queue}}(l_{i,j})$ depends on the transmitting-link queue length and will be discussed in the next subsection.

Considering an end-to-end scenario, if we denote the retransmission limit for each MSDU v (transmitted via path $\mathbf{p}_i$) as $T_{\mathbf{p}_i}^{\text{max}}$, the average number of transmissions over path $\mathbf{p}_i$ until the packet is successfully transmitted, or the retransmission limit is reached, can be calculated as:

$$
\begin{aligned}
t_{\mathbf{p}_i}^{\text{mean}}(T_{\mathbf{p}_i}^{\text{max}}) &= \sum_{j=1}^{T_{\mathbf{p}_i}^{\text{max}}+1} \left(j \cdot [e_{\mathbf{p}_i}(L_v)]^{j-1} [1 - e_{\mathbf{p}_i}(L_v)] \right) + [T_{\mathbf{p}_i}^{\text{max}} + 1] \cdot [e_{\mathbf{p}_i}(L_v)]^{T_{\mathbf{p}_i}^{\text{max}}+1} \\
&= \frac{1 - [e_{\mathbf{p}_i}(L_v)]^{T_{\mathbf{p}_i}^{\text{max}}+1}}{1 - e_{\mathbf{p}_i}(L_v)}
\end{aligned} \tag{9}
$$

The derivation of (9) is detailed in the appendix. The (end-to-end) expected delay for the transmission of an MSDU of size L_v through $\mathbf{p}_i$ under $t_{\mathbf{p}_i}^{\text{mean}}(T_{\mathbf{p}_i}^{\text{max}})$ transmissions can be approximated by:

$$d_{\mathbf{p}_i}\left(L_v, t_{\mathbf{p}_i}^{\text{mean}}(T_{\mathbf{p}_i}^{\text{max}})\right) = t_{\mathbf{p}_i}^{\text{mean}}(T_{\mathbf{p}_i}^{\text{max}}) \cdot \sum_{j=1}^{\rho_i^{\text{total}}-1} \left(\frac{L_v}{g(l_{i,j})} + T_{\text{overhead}} \right) + \sum_{j=1}^{\rho_i^{\text{total}}-1} d_{\text{queue}}(l_{i,j}). \tag{10}$$

The last equation derives the end-to-end delay estimate by joining all links of path $\mathbf{p}_i$ via the summation terms, thereby forming a "virtual" link from the sender to the receiver node in the multi-hop network. We follow this approach since the maximum number of retransmissions $T_{\mathbf{p}_i}^{\text{max}}$ required on each path $\mathbf{p}_i$ can only be defined end-to-end, based on the maximum permissible delay from the sender node to the receiver. We remark that in our experiments, the retransmission limit for any part of a path or even for one link $l_{i,j}$ is set equal to $T_{\mathbf{p}_i}^{\text{max}}$, since, in principle, all possible retransmissions (until the MSDU expires due to delay violation) could occur at an individual link. Following the analysis of (7)–(10), it is straightforward to define the average MSDU transmissions and the expected delay for sub-paths that include only a subset of links, or even for an individual link. This will be proven to be very useful for some of the derivations.

5.2.3 Application and Network-layer Parameter Specification

Since we are considering real-time video streaming through the multi-hop wireless network, each MSDU v is associated with a corresponding delay limit d_v^{deadline}, before which the video data encapsulated in the MSDU should arrive at the destination node h_n. In addition, decoding the video data at the video receiver incurs a reduction in the perceived distortion, which is represented by D_v. Several models exist for the definition of D_v (e.g., see [15]). Recent results [16] [17] demonstrated that acknowledgment-based transmission of scalable video under a strict distortion-reduction prioritization of the video packets leads to an additive distortion-reduction model at the receiver side under packet losses in the multi-hop network. This additive model is codec-specific and typically expresses the expected mean square error (MSE) reduction at the video decoder instead of the visual distortion reduction, since the latter is harder to quantify. See [15] for an example and [18] for further details on linking the distortion-reduction estimates (D_v) with the packetization process at the application layer. We assume that an optimized packet scheduling is performed at the application layer, where all packets v with the same delay deadline d_v^{deadline} are ordered at the encoder (sender) side according to their expected distortion reduction D_v [15] [3]. The delay deadline d_v^{deadline} will also be considered as a parameter in the proposed optimization strategy and it will be defined based on the application requirements.

At each link in the mesh wireless network, each video flow is subjected to a queuing delay[6], which depends on: (a) the MSDUs from a particular flow (user) that are scheduled for transmission via the link of interest at the moment when MSDU v arrives; (b) the queue output rate. The queue output rate depends on the quality of the link (error probability given in (6)) and the average number of retransmissions for each MSDU in the particular link (given by (9) with the replacement of $\mathbf{p}_i$ by $l_{i,j}$). If we assume that link $l_{i,j}$ is shared among multiple paths, then at the arrival of MSDU v at the queue of link $l_{i,j}$, another u MSDUs (where, typically, $0 \leq u \leq v - 1$) will be in the same queue. For each $l_{i,j}$, by indicating the group of u MSDUs by vector $\mathbf{v}_{\text{queue}}(l_{i,j})$, the queuing delay can be estimated as:

$$d_{\text{queue}}(l_{i,j}) = \sum_{\forall w \in \mathbf{v}_{\text{queue}}(l_{i,j})} d_{l_{i,j}} \left(L_w, t_{l_{i,j}}^{\text{mean}} \left(\mathrm{T}_{\mathbf{p}_i}^{\max} \right) \right). \tag{11}$$

For the optimization of the routing strategy of each MSDU v (presented in the next section), the determination of (11) can be performed dynamically during the path estimation, under the knowledge of the previous decisions for the MSDUs that were transmitted by the current node. Alternatively, each node can independently calculate (11) based on the queue contents of the particular link and disseminate the result at frequent intervals in the mesh network via the overlay infrastructure.

5.3 Problem Formulation

Assume a set of N wireless hops (nodes), with h_1 being the video encoder (server) and h_N the video decoder (client), and a connectivity structure $\mathbf{P}$ with M paths, where each path i, $1 \leq i \leq M$ consists of ρ_i^{total} hops. In addition, assume a predefined HCCA transmission opportunity duration $t_{\text{TXOP}}(l_{i,j})$ for each link $l_{i,j}$, with $1 \leq j < \rho_i^{\text{total}}$, and a link adaptation mechanism at the physical layer that can operate at an MSDU granularity. The end-to-end

---

[6] We assume that the MSDUs of each flow are accommodated with an independent queue at each link.

cross-layer optimization which determines the chosen path (routing), the maximum MAC retry limit, and the chosen modulation (at the PHY layer) for the transmission of each MSDU is:

$$\forall v : \left\{ \mathbf{p}_i^*, T_{\mathbf{p}_i}^{\max*}, m(l_{i,j}) \right\} = \arg \max_{\forall \mathbf{p}_i \in \mathcal{P}, \forall l_{i,j} \in \mathbf{p}_i} \left[\min_{1 \leq j < \rho_i^{\text{total}}} \left\{ c(l_{i,j}) \right\} \cdot \Delta_{v,\text{expected}} \right] \quad (12)$$

where:

$$\Delta_{v,\text{expected}} = \left[1 - \left[e_{\mathbf{p}_i}(L_v) \right]^{T_{\mathbf{p}_i}^{\max}} \right] \cdot \Delta_v \quad (13)$$

with $e_{\mathbf{p}_i}(L_v)$ given by (7), $T_{\mathbf{p}_i}^{\max}$ the maximum number of retransmissions for MSDU v if scheduled via path $\mathbf{p}_i$, and $c(l_{i,j})$ corresponding to the remaining time interval for which link $l_{i,j}$ can support the video-flow traffic under HCCA. For the transmission opportunity intervals belonging to the current service interval t_{SI}, $c(l_{i,j})$ can be calculated as:

$$c(l_{i,j}) = \max \left\{ \sum_{t_{\text{SI}}} t_{\text{TXOP}}(l_{i,j}) - d_{\text{queue}}(l_{i,j}), 0 \right\}. \quad (14)$$

Under the constraints set by the video codec and the mesh wireless network infrastructure, the optimization of (12) attempts to find the cross-layer parameters that maximize a capacity-distortion utility function. This function is formulated as the product of the minimum path capacity (expressed by the remaining reserved time within the current service interval at the most congested link) and the expected source distortion-reduction of (13). In this way, we minimize congestion across the various links (since the path whose worst link is having the highest capacity is selected under $D_{v,\text{expected}}$ given by (13)), and concurrently maximize the expected distortion reduction (under the current path's minimum link capacity $\min_{1 \leq j < \rho_i^{\text{total}}} \left\{ c(l_{i,j}) \right\}$). The granularity of this optimization is one MSDU. However, coarser granularities could also be considered, in order to reduce complexity. The problem constraints can be expressed for each MSDU v as:

$$\forall v, \text{ and } \forall \mathbf{p}_i \in \mathcal{P} : d_{\mathbf{p}_i} \left(L_v, T_{\mathbf{p}_i}^{\max} \right) \leq d_v^{\text{deadline}} \quad (15)$$

i.e., the maximum transmission delay through each possible path must be below or equal to the MSDU deadline (d_v^{deadline}) in order for the video data to be useful to the decoder. Moreover, the timing constraint set from HCCA scheduling is:

$$\forall v, \text{ and } \forall \mathbf{p}_i \in \mathcal{P} : \frac{\left[T_{\mathbf{p}_i}^{\max} + 1 \right] \cdot L_v}{g(l_{i,j})} \leq \min_{\forall l_{i,j} \in \mathbf{p}_i} \left\{ c(l_{i,j}) \right\}. \quad (16)$$

The two constraints of (15) and (16) can provide two bounds for the maximum number of retries for each MSDU v for each link $l_{i,j}$. Since both the MAC-layer scheduling and the application-layer deadline constraints are concurrently imposed, if $\min_{\forall l_{i,j} \in \mathbf{p}_i} \left\{ c(l_{i,j}) \right\} \frac{g(l_{i,j})}{L_v} - 1 > 0$ we set the tightest bound for the maximum retry limit:

$$T_{\mathbf{p}_i}^{\max} = \min \left\{ \frac{d_v^{\text{deadline}} - \sum_{j=1}^{\rho_i^{\text{total}}-1} d_{\text{queue}}(l_{i,j})}{\sum_{j=1}^{\rho_i^{\text{total}}-1}\left(\dfrac{L_v}{g(l_{i,j})} + T_{\text{overhead}}\right)}, \frac{g(l_{i,j}) \cdot \min_{\forall l_{i,j} \in \mathbf{p}_i}\{c(l_{i,j})\}}{L_v} - 1 \right\}. \quad (17)$$

Obviously, if there is no path $\mathbf{p}_i$ for which $d_v^{\text{deadline}} > \sum_{j=1}^{\rho_i^{\text{total}}-1} d_{\text{queue}}(l_{i,j})$ then the current MSDU may be dropped.

5.4 Video Streaming Optimization in the Multi-hop Mesh Network

In this section we derive an algorithm that determines the optimal parameters for (12) under a predetermined deadline for each MSDU v (given by d_v^{deadline}) and a predetermined transmission opportunity duration per link, given by $t_{\text{TXOP}}(l_{i,j})$, which is set by the HCCA admission control once the video flow is scheduled for transmission. Moreover, although the conditions of the various links vary over time, we assume the network topology to be fixed for the duration of the video transmission.

5.4.1 End-to-End Optimization

The optimization of (12) can be performed for each node of the mesh wireless network under the assumption that, for every link $l_{i,j}$, the parameters $g(l_{i,j})$, $e(l_{i,j})$ are determined based on the chosen modulation $m(l_{i,j})$ and the experienced signal to interference plus noise ratio (SINR) [22]. In addition, we assume that $d_{\text{queue}}(l_{i,j})$ is communicated to the sender node via frequent feedback using an overlay network infrastructure [19] [20] that uses real time protocols for conveying information from different layers.

The proposed optimization algorithm is given in Figure 5.3. Notice that, although an entire path is selected at the sender node, the algorithm is executed for each node in the network independently by assuming each node is the sender and considering only the network (and MSDU) subset corresponding to the node of interest. This ensures that the algorithm can scale well under a variety of topologies. In addition, in this way, potential network variations that invalidate the error, bandwidth or queuing-delay assumptions used when scheduling at the sender node, can be incorporated/corrected during the scheduling of a subsequent node. Finally, the independent algorithm execution at each node ensures that expired MSDUs will not propagate through the entire network unnecessarily. This facilitates the conservation of network resources in the mesh topology and reduces link congestion.

The algorithm of Figure 5.3 searches through all the possible routing configurations (line 4) that emerge under varying modulation strategies (line 6) and determines the retransmission limit for each case (line 8). The utility function is evaluated (line 9) and the overall maximum is retained. Although this is a greedy approach, it is guaranteed to obtain the maximum under dynamic feedback from the overlay network (parameters calculated in line 7).

1.	For each node that has non-expired MSDUs in its queue
2.	Extract the network connectivity structure $\mathbf{P}$ (eq. (1), (2))
3.	For the MSDU v existing at output of the queue of the sender node
4.	For each path $\mathbf{p}_i$ (network topology emanating from the sender node)
5.	For each link $l_{i,j}$ of path $\mathbf{p}_i$
6.	For each modulation strategy $m(l_{i,j})$
7.	Calculate $e_{l_{i,j}}(L_v)$, $e_{\mathbf{p}_i}(L_v)$, $d_{\text{queue}}(l_{i,j})$ from eq. (6),(7),(11)
8.	Calculate $T_{\mathbf{p}_i}^{\max}$ from (15)-(17)
9.	Under the calculated $T_{\mathbf{p}_i}^{\max}$, evaluate eq. (12)
10.	Compare with previous best, retaining the maximum (eq.(12))
11.	Schedule the MSDU according to the established $\left\{\mathbf{p}_i^*, T_{\mathbf{p}_i}^{\max *}, m(l_{i,j})\right\}$

Figure 5.3: Exhaustive algorithm for the determination of the cross-layer optimized mesh-network path selection, MAC retry limit and physical-layer modulation. The algorithm is applied for each MSDU existing in the queue of each node in the multi-hop wireless network.

5.4.2 Optimization under a certain Horizon of Network Information

In this case we are only considering the part of the mesh network topology that immediately connects to the node of interest. This may be advantageous in comparison to the previous case, since a limited set of network parameters needs to be communicated to the sender node.

For analytical purposes, this can be considered as the previous case with $2 \leq \rho_i^{\text{sub}} < \rho_i^{\text{total}}$, where ρ_i^{total} is the total length of the path that was used in the end-to-end optimization of the previous section. In this case, every path i originating from the current node consists of one or more links, but we do not consider the entire path to the destination. The advantage offered by this scenario is that the required information for the MSDU scheduling is localized (limited).

For each path $\mathbf{p}_i$, we assume that the information for the optimization process is known only for links $l_{i,1}, \ldots, l_{i,\rho_i^{\text{sub}}-1}$. For the remaining links of each path ($l_{i,\rho_i^{\text{sub}}}, \ldots, l_{i,\rho_i^{\text{total}}-1}$), we assume that the allocated transmission opportunity duration available for the MSDUs of each link is known, as well as the limits for the SINR experienced by each link. Our goal is to establish d_v^{deadline} for the video transmission up to links $l_{i,\rho_i^{\text{sub}}-1}$, i.e., the known network "horizon", in order to perform the optimization of Figure 5.3 locally. With respect to IEEE 802.11a networks, it can be shown [22] [23] that the physical-layer throughput of each link $l_{i,j}$ can be approximated by:

$$R_{\text{phy}}(l_{i,j}) = \frac{R_{\text{phy}}^{\max}(l_{i,j})}{1 + e^{-\mu \cdot (s(l_{i,j}) - \delta)}} \qquad (18)$$

where $R_{\text{phy}}^{\max}(l_{i,j})$ is the maximum achievable data rate for each modulation $m(l_{i,j})$, $s(l_{i,j})$ is the observed SINR, and m, d are constants whose values for each modulation $m(l_{i,j})$ can be extracted based on the observation for s and predetermined experimental points [22].

Assuming that, for every link $l_{i,j}$, the SINR s is a random variable following a certain probability distribution $F(s)$ we have:

$$\mathbf{E}\{R_{\text{phy}}(l_{i,j})\} = R_{\text{phy}}^{\max}(l_{i,j}) \int_{s_{\min}}^{s_{\max}} \frac{1}{1 + e^{-\mu \cdot (s-\delta)}} \cdot \mathcal{F}(s)ds \qquad (19)$$

where $s_{\max}$, $s_{\min}$ are the bounds of the observable values for the SINR for each link $l_{i,j}$. In addition, under a given or estimated probability distribution $F(s)$, (5) can be used in order to derive the expectation for the guaranteed bandwidth of each link, which, after a few straightforward manipulations, is:

$$\mathbf{E}\{g(l_{i,j})\} = \frac{t_{\text{TXOP}}(l_{i,j})}{t_{\text{SI}}(l_{i,j})} \cdot \int_{s_{\min}}^{s_{\max}} \frac{\mathcal{F}(s)}{[R_{\text{phy}}^{\max}(l_{i,j})]^{-1} \cdot (1 + e^{-\mu \cdot (s-\delta)}) + (\overline{L})^{-1} \cdot T_{\text{overhead}}} ds \qquad (20)$$

since the remaining parameters of (5) are constants (in our analysis we consider a nominal MSDU size $\overline{L}$). In a similar manner, the expected error of the sub-path $l_{i,\rho_i^{\text{sub}}}, \ldots, l_{i,\rho_i^{\text{total}}-1}$ within each path $\mathbf{p}_i$ is:

$$\mathbf{E}\left\{e_{\left[l_{i,\rho_i^{\text{sub}}} \cdots l_{i,\rho_i^{\text{total}}-1}\right]}(\overline{L})\right\} = \int_{s_{\min}}^{s_{\max}} \left(1 - \prod_{j=\rho_i^{\text{sub}}}^{\rho_i^{\text{total}}-1}\left[\left(1 - \frac{\kappa}{1+e^{\mu \cdot [s-\delta]}}\right)^{\overline{L}}\right]\right) \cdot \mathcal{F}(s)\ ds\ . \qquad (21)$$

The last equation was derived based on (7) under the assumption that the bit error probability can be approximated by [22] [23]:

$$e(l_{i,j}) = \frac{\kappa}{1 + e^{\mu \cdot [s(l_{i,j})-\delta]}} \qquad (22)$$

where k, m, d are derived experimentally depending on the observation for $s(l_{i,j})$ and the chosen modulation $m(l_{i,j})$ [22], with $\rho_i^{\text{sub}} \le j < \rho_i^{\text{total}}$.

Having the expected values for the full path's guaranteed bandwidth, the maximum transmission delay for an MSDU L_v transmitted through links $l_{i,\rho_i^{\text{sub}}}, \ldots, l_{i,\rho_i^{\text{total}}-1}$ can be derived based on (10) as:

$$d_{\left[l_{i,\rho_i^{\text{sub}}} \cdots l_{i,\rho_i^{\text{total}}-1}\right]}(L_v, \mathrm{T}_{\mathbf{p}_i}^{\max}) = \mathrm{T}_{\mathbf{p}_i}^{\max} \sum_{j=\rho_i^{\text{sub}}}^{\rho_i^{\text{total}}-1}\left(\frac{L_v}{\mathbf{E}\{g(l_{i,j})\}} + T_{\text{overhead}}\right) + \sum_{j=\rho_i^{\text{sub}}}^{\rho_i^{\text{total}}-1} d_{\text{queue}}(l_{i,j}) \cdot \quad (23)$$

Notice that (23) involves also the knowledge of the queuing delay $d_{\text{queue}}(l_{i,j})$ of each link $l_{i,j}$, $\rho_i^{\text{sub}} \le j < \rho_i^{\text{total}}$, i.e., for the subsequent links after the "horizon". For each link $l_{i,j}$, the expected $d_{\text{queue}}(l_{i,j})$ can be updated within intervals of $d_{\left[l_{i,\rho_i^{\text{sub}}} \cdots l_{i,j-1}\right]}\left(L_v, t_{\left[l_{i,\rho_i^{\text{sub}}} \cdots l_{i,j-1}\right]}^{\text{mean}}\left(\mathrm{T}_{\mathbf{p}_i}^{\max}\right)\right)$

(that correspond to the expected time required for MSDU v to reach link $l_{i,j}$, after it passes the link $l_{i,\rho_i^{\mathrm{sub}}-1}$ which is at the "horizon") as:

$$d_{\mathrm{queue}}(l_{i,j}) \leftarrow d_{\mathrm{queue}}(l_{i,j}) + t_{l_{i,j}}^{\mathrm{mean}}\left(\mathrm{T}_{l_{i,j}}^{\mathrm{max}}\right) \cdot \left(\frac{L_v}{\mathbf{E}\{g(l_{i,j})\}} + T_{\mathrm{overhead}}\right) \cdot \varphi_{\mathrm{enter}}$$

$$-d_{\left[l_{i,\rho_i^{\mathrm{sub}}} \cdots l_{i,j-1}\right]}\left(L_v, t_{\left[l_{i,\rho_i^{\mathrm{sub}}} \cdots l_{i,j-1}\right]}^{\mathrm{mean}}\left(\mathrm{T}_{\mathbf{p}_i}^{\mathrm{max}}\right)\right) \cdot \varphi_{\mathrm{exit}} \tag{24}$$

where operator $q_0 \leftarrow q_0 + u_0$ indicates the update of quantity q_0 by u_0 and:

$$\varphi_{\mathrm{enter}} = 1 - \mathbf{E}\left\{e_{\left[l_{i,\rho_i^{\mathrm{sub}}} \cdots l_{i,j-1}\right]}(L_v)\right\}^{t_{\left[l_{i,\rho_i^{\mathrm{sub}}} \cdots l_{i,j-1}\right]}^{\mathrm{mean}}\left(\mathrm{T}_{\mathbf{p}_i}^{\mathrm{max}}\right)} \tag{25}$$

$$\varphi_{\mathrm{exit}} = 1 - \mathbf{E}\left\{e_{l_{i,j}}(L_v)\right\}^{t_{l_{i,j}}^{\mathrm{mean}}\left(\mathrm{T}_{\mathbf{p}_i}^{\mathrm{max}}\right)} \tag{26}$$

The derivation of (24) is performed as follows. The queuing delay of the previous iteration is incremented by the product of the factor which indicates the expected delay due to retries for the new MSDU in link $l_{i,j}$ with the probability that the MSDU will reach link $l_{i,j}$ successfully (after maximally $\mathrm{T}_{\mathbf{p}_i}^{\mathrm{max}}$ retries are performed at links $l_{i,\rho_i^{\mathrm{sub}}},\dots,l_{i,j-1}$), which is given by j_{enter}, defined in (25). At the same time, the queuing delay is decremented by the product of the factor indicating the time duration for the possible successful MSDU transmissions with the probability of a successful MSDU transmission, which is given by j_{exit}, defined in (26).

Assuming that the value for $\sum_{j=\rho_i^{\mathrm{sub}}}^{\rho_i^{\mathrm{total}}-1} d_{\mathrm{queue}}(l_{i,j})$ is provided based on (24), (23) can be used in the constraint of (15) by updating the delay deadline:

$$d_v^{\mathrm{deadline}} \leftarrow d_v^{\mathrm{deadline}} - d_{\left[l_{i,\rho_i^{\mathrm{sub}}} \cdots l_{i,\rho_i^{\mathrm{total}}-1}\right]}\left(L_v, \mathrm{T}_{\mathbf{p}_i}^{\mathrm{max}}\right) \tag{27}$$

and the optimization process follows, as explained in the beginning of this section.

The analytical formulation of this section is also useful in defining low-complexity scheduling algorithms at each node without the need for real-time network feedback. For example, if we assume that, due to the random interference caused by the simultaneous operation of the wireless nodes in the mesh network, the probability distribution $F(s)$ for each link $l_{i,j}$ is uniform within an interval of $[s_{\mathrm{min}}, s_{\mathrm{max}}]$ we have:

$$\mathcal{F}(s) = \frac{1}{s_{\mathrm{max}} - s_{\mathrm{min}}}. \tag{28}$$

With the explicit expression of $F(s)$ from (28), we can derive the expected physical layer rate of each link $l_{i,j}$ from (19) as:

$$\mathbf{E}\big\{R_{\mathrm{phy}}(l_{i,j})\big\} = \frac{R_{\mathrm{phy}}^{\max}(l_{i,j})}{\mu(l_{i,j})\cdot\big(s_{\max}(l_{i,j})-s_{\min}(l_{i,j})\big)}\left[1+\ln\!\left(\frac{1+e^{-\mu(l_{i,j})\cdot(s_{\max}(l_{i,j})-\delta(l_{i,j}))}}{1+e^{-\mu(l_{i,j})\cdot(s_{\min}(l_{i,j})-\delta(l_{i,j}))}}\right)\right] \quad (29)$$

where $m(l_{i,j})$, $d(l_{i,j})$ indicate the particular modulation choices for each link. Similarly, the integration of (20) provides the following expectation for the link bandwidth:

$$\mathbf{E}\big\{g(l_{i,j})\big\} = \frac{R_{\mathrm{phy}}^{\max}(l_{i,j})\cdot t_{\mathrm{TXOP}}(l_{i,j})}{\left(1+\big(\overline{L}\big)^{-1}\cdot R_{\mathrm{phy}}^{\max}(l_{i,j})\cdot T_{\mathrm{overhead}}\right)\cdot t_{\mathrm{SI}}(l_{i,j})\cdot\big(s_{\max}(l_{i,j})-s_{\min}(l_{i,j})\big)}$$
$$\cdot\left[1+\ln\!\left(\frac{1+\big(\overline{L}\big)^{-1}\cdot R_{\mathrm{phy}}^{\max}(l_{i,j})\cdot T_{\mathrm{overhead}}+e^{-\mu(l_{i,j})\cdot(s_{\max}(l_{i,j})-\delta(l_{i,j}))}}{1+\big(\overline{L}\big)^{-1}\cdot R_{\mathrm{phy}}^{\max}(l_{i,j})\cdot T_{\mathrm{overhead}}+e^{-\mu(l_{i,j})\cdot(s_{\min}(l_{i,j})-\delta(l_{i,j}))}}\right)\right] \quad (30)$$

and from (21) we can derive the expected path error rate per packet:

$$\mathbf{E}\big\{e_{\mathbf{p}_i}\big(\overline{L}\big)\big\} = 1-\prod_{j=1}^{\rho_i^{\mathrm{total}}-1}\left[\left(1-\frac{\kappa(l_{i,j})}{\mu(l_{i,j})\cdot\big(s_{\max}(l_{i,j})-s_{\min}(l_{i,j})\big)}\left[1+\ln\!\left(\frac{1+e^{-\mu(l_{i,j})\cdot(s_{\max}(l_{i,j})-\delta(l_{i,j}))}}{1+e^{-\mu(l_{i,j})\cdot(s_{\min}(l_{i,j})-\delta(l_{i,j}))}}\right)\right]\right)^{\overline{L}}\right]$$
$$(31)$$

where $k(l_{i,j})$ is chosen based on the modulation.

Based on the explicit expressions of (29)–(31), we derive a less complex solution for the scheduling of each group of MSDUs corresponding to a video GOP. The algorithm is given in Figure 5.4. Based on this algorithm, for every new MSDU, all the cross-layer parameters are established analytically for each path (lines 1-3 of Figure 5.4) and only the search through all the possible paths (i.e., line 4 of Figure 5.4) is required in order to derive the optimal solution. Consequently, this optimization has minimal complexity. In the following section we formulate the complexity requirements of the three different optimization solutions, while Section 5.6 presents comparative experimental results.

```
1.   The optimal modulation parameters  κ*(l_{i,j}),  μ*(l_{i,j}),  δ*(l_{i,j})  are estimated
     only once per link during the optimization of the first MSDU and they
     remain constant throughout the remaining MSDUs until an update is received
     for the interval  [s_min,s_max]  or for the values of  R_phy^max(l_{i,j}),  t_TXOP(l_{i,j}),
     t_SI(l_{i,j}) .
2.   Per MSDU, the expected physical layer rate and guaranteed bandwidth per
     link are estimated from (19), (20), and the error for each path is
     estimated from (21).
3.   Having the calculated queuing delay for each link from (23), as well as the
     available time interval from (27), and the estimated MSDU deadline from
     (27), the maximum retry limit is established from (17) and the average
     number of retries, t_{p_i}^mean(T_{p_i}^max) from (9).
4.   For each node, the link that maximizes  min_{1≤j<ρ_i^total}{c(l_{i,j})}·[1 - E{e_{p_i}(L_v)}]^{T_{p_i}^max}  is
     selected.
```

Figure 5.4: Algorithm for cross-layer optimization under an estimation-based framework.

5.5 Complexity and Information Requirements of the Different Alternatives

Each proposed cross-layer optimization approach explores a different search space in order to determine the optimal parameters and also requires a varying amount of feedback on the conditions of the various links in the multi-hop mesh network. This results in varying computational and communication requirements for the presented algorithms.

Consider the case of a mesh network consisting of N nodes. Each node h_n, $1 \leq n < N$, is the origin of M_n paths. Each path $\mathbf{p}_i$ stemming from node h_n consists of $\rho_{M_n}^{\text{total}}$ nodes, with $1 \leq \rho_{M_n}^{\text{total}} \leq N$. For each link $l_{i,j}$ of these paths, with $1 \leq i \leq M_n$ and $1 \leq j \leq \rho_{M_n,i}^{\text{total}} - 1$, there are S_{mod} possible modulations at the physical layer, which result in a different error rate and different guaranteed bandwidth at the MAC layer. For the end-to-end cross-layer optimization with network feedback from each node (Section 5.4.1), the overall complexity for the scheduling of an MSDU at node h_n is:

$$C_{\text{exhaustive}}^{\text{full}}(h_n) = \sum_{i=1}^{M_n} \sum_{j=1}^{\rho_{M_n,i}^{\text{total}}-1} \sum_{m=1}^{S_{\text{mod}}} C_{\text{exhaustive}}(n) . \tag{32}$$

where $C_{\text{exhaustive}}(n)$ represents the complexity for the dissemination of the necessary network information from node h_n, as well as the execution of the algorithm of Figure 5.3. In order to derive $C_{\text{exhaustive}}^{\text{full}}(h_n)$ in (32) we are calculating the iterations that need to be performed for the subset of paths emanating from node n (M_n), each link within these paths (i.e., links $j = 1, ..., \rho_{M_n,i}^{\text{total}} - 1$), and all possible modulation strategies for each link (S_{mod}).

Similarly, considering a scenario with partial network information, i.e., when the overlay network provides feedback only until node r_i (with $1 \leq \rho_i < \rho_{M_n,i}^{\text{total}}$), we have:

$$C_{\text{exhaustive}}^{\text{partial}}(h_n) = \sum_{i=1}^{M_n'} \sum_{j=1}^{\rho_i} \sum_{m=1}^{S_{\text{mod}}} (C_{\text{exhaustive}}(n) + C_{\text{estimation}}(n)) . \tag{33}$$

where $C_{\text{estimation}}(n)$ represents the complexity for the estimation of the various parameters based on the analysis of the previous section and M_n' is the number of different paths (within the partial network topology under examination) originating from node h_n, with $1 \leq M_n' \leq M_n$. Similar to (32), (33) is estimated by summing all iterations for each path, each link up to r_i (network horizon for node n) and all modulation strategies.

Lastly, for the optimization of Figure 5.4 where the best modulation strategy is a-priori determined:

$$C_{\text{estimation}}(h_n) = \sum_{i=1}^{M_n''} C_{\text{estimation}}(n) . \tag{34}$$

where M_n'' is the number of links that are directly connected to node h_n. As an indication of the different complexity requirements as well as the different information requirements of each case, Table 5.1 presents numerical results for the three mesh network topologies of

Figure 5.2 based on (32)–(34) and we set S_{mod} = 8 [22]. The normalized information requirements are expressed in terms of the number of links in all possible paths (whose error, guaranteed bandwidth and queuing delay is conveyed by the overlay network) multiplied by the total number of times this information is updated by the overlay network per MSDU ($I_{refresh}$ with $0 < I_{refresh} \le 1$). First we considered the case of the first node (h_1, n = 1 in (32)–(34)) since this includes all the possible paths and all the links in the mesh topology (top of the table). Hence, the results of the top part of Table 5.1 show the worst-case complexity and information requirements from the viewpoint of an individual node.

Notice that the information cost depends on the frequency of updates received by the overlay network per MSDU, denoted by $I_{refresh}$. Given $I_{refresh}$ and the required bytes for conveying the status of each link via hop-to-hop feedback, it is straightforward to convert the provided information cost for each case into actual bandwidth overhead for the overlay infrastructure in the multi-hop wireless mesh network. Since $I_{refresh} \cong 0$ for the estimation-based case, the information cost of this case is practically negligible.

Table 5.1: Complexity comparison and the associated information requirements of different alternatives for cross layer optimization for: (top) the first node of each of the three topologies of Figure 5.2; (bottom) all nodes in the topology. The basic complexity unit for (32)–(34) is set to $C_{exhaustive}(1) \equiv 1$ and we additionally set $C_{estimation}(1) = 0.2 \times C_{exhaustive}(1)$ based on experimental observations.

Method/Topology Node h_1	Normalized Complexity			Normalized Information requirements		
	T1	T2	T3	T1	T2	T3
End-to-end	624	176.0	64	$14\,I_{refresh}$	$10 I_{refresh}$	$8 I_{refresh}$
Localized $\forall_i : \rho_i = 1$	28.8	19.2	28.8	$3 I_{refresh}$	$2 I_{refresh}$	$3 I_{refresh}$
Estimation based	0.6	0.4	0.6	0	0	0
Method/Topology All nodes in each topology	**Normalized Complexity**			**Normalized Information requirements**		
	T1	T2	T3	T1	T2	T3
End-to-end	1120	352.0	120	$40 I_{refresh}$	$28 I_{refresh}$	$15 I_{refresh}$
Localized $\forall_i : \rho_i = 1$	134.4	96	76.8	$14 I_{refresh}$	$10 I_{refresh}$	$8 I_{refresh}$
Estimation based	2.8	2	1.6	0	0	0

As a second step, we considered the cumulative complexity and information requirements for all the nodes in the multi-hop mesh network in order to estimate the streaming complexity and information overhead at the system level; the results are presented in the bottom of Table 5.1. We remark that, depending on the topology specification (i.e., average node connectivity) and the chosen method, the estimated complexity scales up to three orders of magnitude. Similarly, there is a large gap between the lowest and highest information requirements for the various approaches among the different topologies. As expected, the more complex the mesh topology, the higher the rate of increase of complexity and information requirements.

5.6 Experimental Results

We experimented with a variety of typical video content (Common-Interchange Format – CIF sequences "Foreman", "Silent", "Hall Monitor", and "Stefan", each consisting of 300 frames with 30 Hz replay rate). We used a fully-scalable codec [24] and the produced bitstream was extracted at an average bitrate of 2 Mbps and packetized into MSDUs of data payload not larger than 1000 bytes. The end-to-end delay for the MSDUs of each GOP was set to 0.54 sec, which corresponds to the replay duration of one GOP. We remark that although the utilized video coder is not a member of the MPEG family of coders, the assumptions made in Subsection 5.2.3 for the distortion-reduction estimation and the application-layer packet scheduling are also valid for the scalable coder currently standardized by the JSVM group of MPEG/VCEG [25] since it is based on open-loop motion-compensated prediction and update steps followed by embedded quantization and context-based entropy coding. Hence our methods and experiments are relevant to future systems that will utilize such scalable video coding technology in the context of mesh networks.

We simulated the cases of the multi-hop mesh network topologies of Figure 5.2, labeled T1-T3, under predetermined transmission intervals for each link. Our simulation took into account the different parameters for the various layers, such as varying SINR, transmission overheads at the MAC layer due to MSDU acknowledgements and polling overheads, as well as queuing and propagation delays in the various links of the mesh network. In order to incorporate the effect of noise and interference, we performed a number of simulations using random values for the SINR of each link, chosen between 15 and 25 dB. Network feedback via the overlay network was conveyed to each hop whenever a significant change in the experienced channel condition occurred. For the end-to-end optimization with network feedback (termed "End-to-end" in our results) this includes the information conveyed from all hops. However, we also considered a localized case where the information horizon was set to the direct neighborhood of each hop (termed "Localized" in our results – this information horizon is shown pictorially in Figure 5.2) and the remaining network parameters were estimated as explained in Subsection 5.4.2. In addition, a purely estimation-based case was also considered with no "horizon", where the only available information is the channel SINR range (s_{min}, s_{max} of (28)) for each link, communicated by the overlay network infrastructure whenever the channel variation exceeded 2 dB (termed "Estimation based") from the estimated value given by (28). This ensured that the information cost for the dissemination of the network information is minimal compared to the other alternatives, as indicated in Table 5.1. Notice that, both for the "Localized" case, as and the "Estimation based" case, the theoretical framework of (18) –(28) was used.

Apart from the various alternatives of the proposed optimization, we also derive results with streaming under two other optimization algorithms. The first case is optimization based on the expected transmission count (ETX) [8], where the utility function is chosen such that the retransmission limit of each MSDU is set based on the effective network bandwidth and the expected error rate. This case considers the MSDU delay deadline from a purely network-centric approach [8], i.e., it does not use the constraints set in (15), (16), but rather restricts the MSDU delay deadline based on link loss ratios and the available throughput [8]. It was termed "ETX optimized" in our results. Secondly, the case of selecting the link with the highest effective bandwidth was realized for the routing of each MSDU, since it corresponds to the popular solution for optimized routing [26] (termed as the "Highest Bandwidth" solution). Notice that, in both cases, the best modulation was established as in the "end-to-end" case, and each link's status information was also used for these cases, as conveyed by the overlay network infrastructure. As a result, the differences in performance stem purely from the different performance utilities that were chosen during the MSDU routing and path selection. Effectively, this separates the fully network-aware methods (proposed "End-to-End", "Highest Bandwidth" [26], and "ETX optimized" [8]) from the partial network-aware approaches (proposed "Localized" and "Estimation based"). In addition, within the fully network-aware methods, the difference in the performance utilities means that only the "End-to-End" approach fully utilizes application-layer, MAC, and PHY parameters via the optimization framework of (12)-(16).

Indicative results for the obtained average PSNR of each method are given in Table 5.2 (25 runs per sequence/method/topology). Two representative cases of medium and low average transmission bandwidth were chosen.

Table 5.2: Average PSNR results (Y-channel – 25 runs with 300 video frames per run) for video streaming in the multi-hop networks of Figure 5.2.

Method/Topology	Medium bandwidth case PSNR (dB)			Low bandwidth case PSNR (dB)		
	T1	T2	T3	T1	T2	T3
End-to-end	35.42	34.28	32.89	32.11	30.56	31.54
ETX optimization	34.15	31.89	31.58	30.33	29.74	29.55
Highest Bandwidth	33.18	30.51	30.45	28.66	27.22	27.00
Localized	34.08	32.48	30.86	29.67	28.55	28.19
Estimation based	33.21	30.11	29.81	29.31	27.45	27.12

In order to understand better the relationship between the obtained PLR for each case and the derived PSNR, the percentage of losses for the video packets when clustered into eight distinct distortion categories is presented in Figure 5 (example for the sequence "Foreman"). The second topology of Figure 5.2 was used for these results; similar results have been obtained for the remaining topologies and the remaining video sequences. Notice that our choice of eight distinct categories is only performed for illustration purposes, since each packet is associated with its own distortion-reduction. In our simulations, the packet losses were mainly due to deadline violation, since each hop drops the packets which have already expired. The results of Figure 5 indicate, for all the scenarios under consideration, that scheduling at the application layer by expected distortion-reduction leads to reduced losses for the most significant classes of packets. This justifies our use of a scalable video coder that permits such a scheduling. However, each method achieves different PSNR performance and PLRs depending on its chosen utility and the presence of network feedback.

As shown in the results of Table 5.2, the "End-to-end" case outperforms all other methods by a significant margin. The "ETX optimization" appears to perform relatively well, even though it is outperformed by approximately 1.5 dB by the "End-to-end" case for the medium-bandwidth case, and by approximately 1.7 dB for the low-bandwidth case. The "Localized" case outperforms the popular "Highest Bandwidth" case by approximately 1.1 dB for both medium-bandwidth and low-bandwidth cases, even though the "Highest Bandwidth" case uses full feedback for the status of all the links in each multi-hop topology. In fact, the "Highest Bandwidth" performs virtually on par to the "Estimation based" case for all our experiments, even though the latter requires almost no network feedback and, as shown in Table 5.1, has the lowest complexity. This is expected since the "Highest Bandwidth" approach provides less intelligent decisions when most of the links have low effective throughput. Finally, a comparison of the results for the different topologies reveals that, as expected, the higher the connectivity, the better the average performance of all methods. Nevertheless, this comes at a higher allocation of resources in the multi-hop mesh topology and it additionally has higher complexity and requires more feedback for the condition of all the links, as demonstrated in Table 5.1.

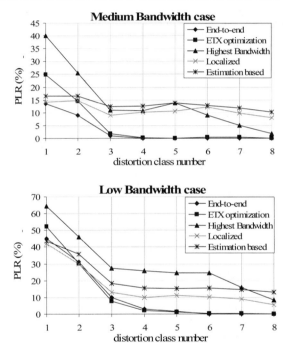

Figure 5.5: Percentage of losses for each packet distortion-reduction class (Cat.1=least significant packets; Cat.8=most significant); (a) Medium-bandwidth case; (b) Low-bandwidth case. The results correspond to topology T2 of Figure 5.2 and to the video sequence "Foreman".

Our results highlight several important issues in network design and infrastructure. Firstly, it was shown that having frequent feedback via an overlay network about the link conditions and performing end-to-end optimization with the appropriate utility function offers significant improvements in the achievable video quality. Indeed, the "End-to-end" and "ETX optimized" cases outperform the remaining algorithms by 3-5 dB, in all cases (Table 5.2). Secondly, the importance of choosing a cross-layer distortion-capacity utility function is highlighted by the fact that both methods outperform the conventional "Highest

Bandwidth" scenario. Moreover, the proposed utility of (12) and the derivation of the MSDU retransmission limit based on the delay limit for the video transmission ((15)-(17)) appear to be the best choice for video streaming applications. Thirdly, higher connectivity in the multi-hop mesh topology leads to better video streaming performance, at the expense of complexity and network feedback requirements.

Fourthly, the study of the PLRs reported in Figure 5 in conjunction with the PSNR results of Table 5.2 reveals that prioritization of video packets with respect to distortion-reduction incurred in the decoded video is extremely important. For example, even though the "Estimation based" case has lower average PLR from the "Highest Bandwidth" case, it performs worse in terms of PSNR since the latter achieves lower PLR for the most significant classes of packets. This result emphasizes the fact that, in the case of analysis of multimedia transmission over wireless, average PLRs that do not consider the significance of the various packets for the application are not always relevant metrics for the system performance.

Finally, it appears that even a limited horizon of information in the network infrastructure can be extremely beneficial. We believe that the determination of an appropriate "horizon" of information that provides the optimal trade off between the overhead at the overlay network versus the improvement offered by utilizing dynamic network feedback is an interesting research direction. Moreover, the dynamic adaptation of such a "horizon" in function of the network variations or the mesh topology specification (i.e., simple vs. complex mesh networks and static vs. dynamic scenarios) could be examined.

5.7 Further Reading

In this chapter, we explored the problem of real-time transmission of an individual video bitstream across a multi-hop 802.11a/e wireless network. Several key assumptions were made, however, to limit the scope of the problem. For instance, we assumed that each application (video flow) reserves a predetermined transmission opportunity interval, where contention-free access to the medium is provided, thereby neglecting any resource management and resource allocation issues. Moreover, we limited our study to a single video stream when, in actuality, there will often be several video streams sharing the same multi-hop wireless infrastructure.

In the literature, different centralized and distributed approaches have been adopted to solve the resource management problem for wireless networks. Centralized approaches solve the end-to-end routing and path selection problem as a combined optimization using multi-commodity flow [27] algorithms, as this ensures that the end-to-end throughput is maximized while constraints on individual link capacities are satisfied. In contrast, distributed approaches use fairness or incentive policies to resolve resource allocation issues in a scalable manner [28]-[31]. For instance, in [28], a new solution to the problem of engineering non-monetary incentives for edge based wireless access services was proposed which offers both higher throughput for bursty data and more stable allocation for real-time applications.

In our recent work, we consider the benefits of dynamic resource and information exchanges among multiple wireless users [39] when provisioning paths (and allocating resources) for different users' video flows. Additionally, in [32], we explore a multi-user dynamic, decentralized (i.e., at each network node), route adaptation scheme that is driven by localized network information within each node's information horizon.

In this chapter, we also assumed that the delay and resource overheads associated with information feedback over different length horizons are negligible. These overheads, however, can have significant impact on delay-sensitive multimedia applications. To enable efficient distributed multi-user video streaming over a wireless multi-hop infrastructure, nodes need to collect and disseminate timely network information based on which the various nodes can collaboratively adapt their cross-layer transmission strategies. For instance, based on the available information feedback, a network node can select an alternate (less congested) route for streaming video packets that have a higher contribution to the overall distortion or a more imminent deadline. These issues are investigated in more detail in [33].

To enable the information feedback, we assumed that a directed acyclic overlay network topology can be superimposed over any wireless multi-hop network to convey (in frequent intervals) information about the expected BER, the queuing delay for each link, as well as the guaranteed bandwidth under the dynamically-changing modulation at the PHY. Methods for constructing such overlay structures given a specific multi-hop network and a set of transmitting-receiving pairs can be found in [34] [35] [36]. Additionally, several examples of such application-layer overlay networks have been proposed in [19] [20].

Although we consider the SINR experienced at each link, the cross-layer streaming solution proposed in this chapter ignores the effects of interference that arise due to the broadcast nature of the wireless medium (e.g., transmitter-receiver, transmitter-transmitter, or receiver-receiver interference). In other words, our analysis assumes the existence of multiple orthogonal channels for transmission. In practice, however, interference effects in wireless networks can severely degrade the throughput. Consequently, particularly in a multi-user scenario, resource management and routing become of paramount importance for minimizing interference. In [37], a rate matrix was introduced to describe the state of the network at a given time. In [7], an elementary capacity graph was used to represent the physical layer state of the various links. In [38], a node-link incidence matrix was used. More details about wireless video transmission with interference consideration can be found in [32].

5.8 Conclusions

Delay-constrained video streaming over multi-hop wireless mesh networks is an application that deserves considerable attention due to the research challenges imposed by such a service, as well as due to the important role that robust and efficient multimedia services have when it comes to commercial deployment of such networks in office and residential areas. We investigated a framework where QoS guarantees are provided for video transmission over a variety of links in a multi-hop network using IEEE 802.11a/e. The integrated cross-layer solution that maximizes the product of the expected video quality with the link utilization appears to provide significant improvement over other optimized solutions. Moreover, the utilization of network information (for the dynamically changing conditions of the various hops) gathered via overlay-network feedback, appears to be of paramount importance for the overall video quality at the receiver hop.

Although the proposed algorithm operates per video packet and can potentially incur significant complexity and communication overhead for the overlay network infrastructure, there is a significant potential for improved video streaming performance. This motivates us to investigate the problem further and attempt to explore the best granularity for the optimization as well as the network feedback that provides optimal quality/complexity/robustness in a distributed video streaming scenario over the hops of the mesh network. Finally, under the proposed paradigm, the issues of collaborative streaming of multiple flows and fairness deserve significant attention in future research.

5.9 Appendix – Derivation of (9)

We want to derive the result for:

$$m = \sum_{j=1}^{N} j \cdot p^{j-1} \cdot (1 - p) + N \cdot p^{N} . \tag{35}$$

with $p = e_{\mathbf{p}_i}(L_v)$ and $N = \mathrm{T}_{\mathbf{p}_i}^{\max} + 1$.

Consider:

$$\sum_{j=1}^{N} p^{j} = p(1 + p + \ldots + p^{N-1}) = \frac{p(1 - p^{N})}{(1 - p)} . \tag{36}$$

Taking the derivative of both sides with respect to p, we have:

$$\sum_{j=1}^{N} j \cdot p^{j-1} = \frac{d}{dp} \left[\frac{p(1 - p^{N})}{(1 - p)} \right] = \frac{(1 - p^{N})}{(1 - p)} - \frac{p(N \cdot p^{N-1})}{(1 - p)} + \frac{p(1 - p^{N})}{(1 - p)^{2}} . \tag{37}$$

or:

$$\sum_{j=1}^{N} j \cdot p^{j-1} = \frac{(1 - p^{N})}{(1 - p)^{2}} - \frac{N \cdot p^{N}}{(1 - p)} . \tag{38}$$

Hence:

$$\sum_{j=1}^{N} j \cdot p^{j-1}(1 - p) + N \cdot p^{N} = \frac{(1 - p^{N})}{(1 - p)} . \tag{39}$$

5.10 References

[1] K. Holt, "Wireless LAN: past, present, and future," *Proc. IEEE Design, Automation, and Test in Europe*, DATE, vol. 3, pp. 92-93, 2005.

[2] W. Wei, and A. Zakhor, "Multipath unicast and multicast video communication over wireless ad hoc networks," *Proc. Internat. Conf. on Broadband Networks*, Broadnets, pp. 496-505, 2002.

[3] P. A. Chou and Z. Miao, "Rate-distortion optimized streaming of packetized media," *IEEE Trans. on Multimedia*, vol. 8, no. 2, pp. 390-404, April 2006.

[4] E. Setton, T. Yoo, X. Zhu, A. Goldsmith, and B. Girod, "Cross-layer design of ad hoc networks for real-time video streaming," *IEEE Wireless Communications Mag.*, vol. 12, no. 4, pp. 59-65, Aug. 2005.

[5] M. van der Schaar, S. Krishnamachari, S. Choi, and X. Xu, "Adaptive cross-layer protection strategies for robust scalable video transmission over 802.11 WLANs," *IEEE J. on Select. Areas in Comm.*, vol. 21, no. 10, pp. 1752-1763, Dec. 2003.

[6] A. Butala, and L. Tong, "Cross-layer design for medium access control in CDMA ad-hoc networks," *EURASIP J. on Applied Signal Processing*, to appear.

[7] Y. Wu, P. A. Chou, Q. Zhang, K. Jain, W. Zhu, and S.-Y. Kung, "Network planning in wireless ad hoc networks: a cross-layer approach," IEEE *J. on Select. Areas in Comm.*, vol. 23, no. 1, pp. 136-150, Jan. 2005.

[8] D. S. J. De Couto, D. Aguayo, J. Bicket, and R. Morris, "A high throughput path metric for multi-hop wireless routing," *Proc. ACM Internat. Conf. on Mob. Computing and Networking*, MOBICOM, pp. 134-146, 2003.

[9] R. Draves, J. Padhye, and B. Zill, "Routing in multi-radio, multi-hop wireless mesh networks," *Proc. ACM Internat. Conf. on Mob. Computing and Networking*, MOBICOM, pp. 114-128, 2004.

[10] V. Gambiroza, B. Sadeghi, and E. W. Knightly, "End-to-end performance and fairness in multihop wireless backhaul networks," *Proc. ACM Internat. Conf. on Mob. Computing and Networking*, MOBICOM, pp. 287-301, 2004.

[11] R. L. Cruz, and A. V. Santhanam, "Optimal routing, link scheduling and power control in multihop wireless networks," *Proc. Joint Conf. of IEEE Comput. and Commun. Societ.*, INFOCOM, vol. 1, pp. 702-711, 2003.

[12] *IEEE 802.11e/D5.0, Draft Supplement to Part 11*: Wireless Medium Access Control (MAC) and physical layer (PHY) specifications: Medium Access Control (MAC) Enhancements for Quality of Service (QoS), June 2003.

[13] M. Kodialam, and T. Nandagopal, "Characterizing achievable rates in multi-hop wireless networks: The joint routing and scheduling problem," *Proc. ACM Internat. Conf. on Mobile Computing and Networking*, pp. 42-54, 2003.

[14] P. Ansel, Q. Ni, and T. Turletti. "An efficient scheduling scheme for IEEE 802.11e," *Proc. IEEE Workshop on Model. and Opt. in Mob., Ad-Hoc and Wireless Net.*, WiOpt 2004, Cambridge, UK, March 2004.

[15] M. Wang and M. van der Schaar, "Operational rate-distortion modeling for wavelet video coders", *IEEE Trans. on Signal Processing*, to appear.

[16] D. Taubman, and J. Thie, "Optimal erasure protection for scalably compressed video streams with limited retransmission," *IEEE Trans. on Image Processing*, vol. 14, no. 8, pp. 1006-1019, Aug. 2005.

[17] M. van der Schaar, and D. Turaga, "Cross-layer packetization and retransmission strategies for delay-sensitive wireless multimedia transmission", *IEEE Trans. On Multimedia*, to appear.

[18] Y. Andreopoulos, R. Keralapura, M. van der Schaar and C.-N. Chuah, "Failure-aware, open-loop, adaptive video streaming with packet-level optimized redundancy," *IEEE Trans. on Multimedia*, to appear.

[19] D. Krishnaswamy, and J. Vicente, "Scalable adaptive wireless networks for multimedia in the proactive enterprise." *Intel Technology Journal*, vol. 8, no. 4, Nov. 2004.

[20] S. Roy, M. Covell, J. Ankcorn, S. Wee, and T. Yoshimura, "A system architecture for managing mobile streaming media services," *Proc. IEEE Distrib. Computing Syst. Workshop*, pp. 408-413, 2003.

[21] Y. Andreopoulos, N. Mastronarde and M. van der Schaar, "Cross-layer optimized video streaming over wireless multi-hop mesh networks," *IEEE Journal on Select. Areas in Communications*, vol. 24, no. 11, pp. 2104-1215, Nov. 2006.

[22] D. Krishnaswamy, "Network-assisted link adaptation with power control and channel reassignment in wireless networks," *Proc. 3G Wireless Conference*, pp. 165-170, 2002.

[23] K.-B. Song, and S. A. Mujtaba, "On the code-diversity performance of bit-interleaved coded OFDM in frequency-selective fading channels," *Proc. IEEE Vehicular Technol. Conf.*, VTC, vol. 1, pp. 572-576, 2003.

[24] Y. Andreopoulos, A. Munteanu, J. Barbarien, M. van der Schaar, J. Cornelis and P. Schelkens, "In-band motion compensated temporal filtering," *Signal Processing: Image Communication* (special issue on "Subband/Wavelet Interframe Video Coding"), vol. 19, no. 7, pp. 653-673, Aug. 2004.

[25] Ohm, "Advances in scalable video coding," *Proc. 13th European Signal Process. Conf.* EURASIP, Sept. 2005.

[26] D. B. Johnson, and D. A. Maltz, "Dynamic source routing in ad hoc wireless networks," Chapter in *Mobile Computing* (eds. T. Imielinski and H. Korth), Kluwer Acad. Pub., 1996.

[27] B. Awerbuch and T. Leighton, "Improved approximation algorithms for the multicommodity flow problem and local competitive routing in dynamic networks," in *Proc. 26th ACM Symposium on Theory of Computing*, May 1994.

[28] R.-F. Liao, R. Wouhaybi, and A. Campbell, "Wireless incentive engineering," *IEEE J. Select. Areas Commun., Special Issue on Recent Advances in Multimedia Wireless*, 4th quarter, 2003.

[29] S. Shenker, "Efficient network allocations with selfish users," in *Proc. Perform.*, Sep. 1990, pp. 279-285.

[30] J. MacKie-Mason and H. Varian, "Pricing congestable network resources," *IEEE J. Select. Areas Commun.*, vol. 13, no.7, pp. 1141-1149, Sep. 1995.

[31] R. La and V. Anantharam, "Optimal routing control: Repeated game approach," *IEEE Trans. Automat. Contr.*, vol. 47, no. 3, pp. 437-450, Mar. 2002.

[32] H. Shiang and M. van der Schaar, "Multi-user video streaming over multi-hop wireless networks: A distributed, cross-layer approach based on priority queuing," *IEEE J. Select. Areas Commun.*, to appear.

[33] H. Shiang and M. van der Schaar, "Informationally decentralized video streaming

over multi-hop wireless networks," *IEEE Trans. on Multimedia*, to appear.

[34] J. R. Evans and E. Minieka, *Optimization Algorithms for Networks and Graphs*, NY: Marcel Dekker, 1993.

[35] M. Waldvogel and R. Rinaldi. "Efficient Topology-Aware Overlay Network, " *ACM SIGCOMM Computer Comm. Review*, vol. 33, no. 1, pp. 101-106, Jan 2003.

[36] J. Jannotti, "Network-layer support for overlay networks," in *Proc. IEEE Conf. Open Architectures and Network Programming*, NY, June 2002.

[37] S. Toumpis, A. J. Goldsmith, "Capacity Regions for wireless Ad Hoc Network", *IEEE Transactions on Wireless Communications*, vol. 2, no. 4, pp. 736-748, July 2003.

[38] L. Xiao, M. Johansson, S. P. Boyd, "Simultaneous Routing and Resource Allocation Via Dual Decomposition," *IEEE Transactions on Communications*, vol. 52, no. 7, pp. 1136-1144, July 2004.

[39] N. Mastronarde, D. Turaga, and M. van der Schaar, "Collaborative resource exchanges for peer-to-peer video streaming over wireless mesh networks," *IEEE J. Select. Areas Commun.*, vol. 25, no. 1, pp. 108-118, Jan. 2007.

6

Understanding and Achieving Next-Generation Wireless Security[a]

Wireless networking is quickly becoming a defacto standard in the enterprise, streamlining business processes to deliver increased productivity, reduced costs and increased profitability. Security has remained one of the largest issues as companies struggle with how to ensure that data is protected during transmission and the network itself is secure. Wi-Fi Protected Access (WPA) offered an interim security solution, but was not without constraints that resulted in increased security risks. The new WPA2 (802.11i) standards eliminate these vulnerabilities and offer truly robust security for wireless networks. As a global leader in wireless networking, Motorola, through the acquisition of the former Symbol Technologies, not only offers this next-generation of wireless security - but also builds on the new standard with value-added features that further increase performance and the mobility experience for all users.

6.1 Overview

Corporations are increasingly being asked to allow wireless network access to increase business productivity, and corporate security officers must provide assurance that corporate data is protected, security risks are mitigated and regulatory compliance is achieved. This chapter will discuss:

- The risks of wireless insecurity;
- The progression of security standards and capabilities pertaining to Wi-Fi security;
- How the 802.11i standard provides robust security for demanding wireless environments;
- How Motorola incorporates 802.11i in its wireless switching products in a way that optimizes scalability, performance and investment protection.

---

[a] *Motorola, Inc*

6.2 Risks of Wireless Insecurity

The advent of wireless computing and the massive processing power available within portable devices provides organizations with an unprecedented ability to provide flexible computing services on-demand to enable business initiatives. While functionality is being rapidly adopted, the process complicates the ability of IT departments to control their own Intranets and enforce their own standards. Where the high performance computers of yesterday required a dedicated room and special environmental controls, today's computers arrive via a visitor's pocket or traveling employee's notebook, and require no hard-wired connections to reach the corporate Intranet.

With this great mobile computing power and flexibility comes major risk. War driving can enable hackers to obtain unauthorized access to corporate resources and proprietary intellectual property. The login credentials of legitimate wireless users can be sniffed or cracked. Malicious insiders can move throughout an enterprise network with impunity via sessions with insecure wireless access points.

The consequences of these risks are significant. We have seen spammers and phishers leverage open access points to send unsolicited and malicious electronic mail in stealth mode. Worms are introduced through a new infection vector. Customer lists and account numbers are routinely downloaded to portable devices. Enterprise databases are accessed and modified by unauthorized users.

The bottom line is that wireless insecurity, as long as it is unaddressed, enables the theft of data, lowers productivity, and causes quantifiable financial losses.

6.3 Understanding Wi-Fi Protected Access (WPA)

Created by Motorola and other Wi-Fi Alliance vendors, WPA was based on an early draft of IEEE 802.11i to address critical flaws in WEP. These security shortcomings required an interim solution that would not require hardware upgrades or replacements for existing consumer devices. A series of compromises was made in order to "fix" WEP through software-based firmware upgrades. The majority of existing WEP devices had extremely minimal CPU resources, often based on sub-40 MHz chips based on older hardware such as the 80486. As these devices are typically incapable of encryption work, the implementation of RC4 for WEP was often offloaded onto secondary chips. This is a primary consideration. The replacement for WEP must still use RC4 and RC4 primitives for any and all encryption. The main problems with WEP are:

- WEP does not prevent forgery of packets.
- WEP does not prevent replay attacks. An attacker can simply record and replay packets as desired and they will be accepted as legitimate.
- WEP uses RC4 improperly. The keys used are very weak, and can be brute-forced on standard computers in hours to minutes, using freely available software.
- WEP reuses initialization vectors. A variety of available cryptanalytic methods can decrypt data without knowing the encryption key.
- WEP allows an attacker to undetectably modify a message without knowing the encryption key.

6.3.1 WPA TKIP

The IEEE's primary response to the problems of WEP was Temporal Key Integrity Protocol (TKIP). TKIP acts as a wrapper for WEP, adding a layer of security around WEP's otherwise weak encryption.

One of the first problems TKIP solves is that of key length. WEP uses small keys, and their effective length is shorter due to several design flaws. TKIP uniformly uses a 128-bit encryption key, and while WEP can support 128-bit encryption keys, maintaining compatibility with older WEP devices inevitable leads to standardization upon 64-bit encryption keys within a wireless network. However, TKIP still makes use of RC4, a relatively weak encryption algorithm that was used due to hardware constraints on most of the devices originally designed to provide WEP.

TKIP also reduces the chance of replay attackers. TKIP expands the initialization vector (IV) to 48 bits from 24 bits, and combines this IV with the fixed key in a more cryptographically secure manner. Using a 48-bit IV means that any particular value of the IV cannot be duplicated with a particular key. Thus, packets cannot be replayed. Guaranteeing that a particular key-IV pair is never reused also denies an attacker the ability to capture multiple packets that are identically encrypted, which would lead to the ability to extract the plain text messages.

Further, TKIP addresses WEP's use of a single key by all clients. To create a base key, TKIP uses either a passphrase or a master key derived from the authentication process, and several other pieces of information, such as a client's MAC address. This base key in turn is used with the IV to create per-packet keys. So in theory, every packet sent over WPA is encrypted by a separate and unique key.

Finally, TKIP takes on weaknesses in key deployment by creating a base key that is different for each client. A client provides a shared secret for authentication and various other pieces of information. On wireless networks secured using WEP, all clients constantly use the same key, providing a large amount of cipher text for attackers to analyze. This also increases the probability of reuse of the 24-bit IV, exposing encrypted messages to attackers.

One fundamental problem continues for networks that have switched from WEP to WPA, or deployed WPA directly, yet do not use authentication. The initial passphrase or secret deployed on clients and access points is often weaker than needed, since it usually must be human-readable and entered by a human. This immediately limits the passphrase or secret to a subset of readable characters that can easily be entered from the keyboard. Furthermore, the length is often limited to 20 characters or less due to the difficulties associated with remembering or entering long strings of seemingly random text.

It is important to note that if robust authentication methods are not used with WPA, it must rely upon Pre-Shared Keys (PSK). The same secret phrase must be entered on all clients and all access points. This carries forward the key management issues inherent in WEP. In addition, it is virtually impossible to securely distribute the key or passphrase, as the secret information must be provided to all clients. A single malicious client can use this data to compromise other client sessions. Unfortunately, WPAPSK is relatively common due to the lack of a need for a separate authentication system.

6.3.2 802.1X - User Authentication and Network Access

In an attempt to address the lack of user authentication in WEP, support for the 802.1X protocol was added to WPA. The 802.1X protocol was originally designed for wired networks and only facilitates authentication, therefore it cannot guarantee secure authentication on wireless networks.

The first problem with 802.1X in a wireless network is that an attacker has access to the authentication packets sent and received by clients. If weak authentication methods are used (several are supported) or weak encryption is used (such as RC4), it may be possible for an attacker to discover the authentication credentials.

The second problem is that an attacker can execute a man-in-the-middle attack on the 802.1X authentication sequence. On a wired network this attack would be far more difficult, as an attacker would need physical access to the cable in between the client and the switch being accessed. On a wireless network, anyone within broadcast range has the ability to access. An attacker could be several hundred feet away with directional antennas. We will discuss in the later section "WPA2: Under the Covers" how implementation of Extensible Authentication Protocol (EAP) methods such as TLS can mitigate against possible man-in-the-middle attacks.

The third problem is that an attacker can execute denial-of-service attacks against clients by sending packets to the wireless access point, telling it to drop the client connection. On a wired network, this would again require access to the physical cable between the client and the switch.

The fourth problem is that with hard-wired devices, 802.1X will drop the port if the interface goes down - that is, if the cable is unplugged, or the device at the endpoint is not responsive. However, on wireless networks the status of the physical link condition cannot be trusted. An attacker can access the physical medium used to transmit the signal - the air, for example. Thus anyone within broadcast range could execute a denial-of-service attack against a client system and then take the client's place before the wireless access point notices.

Additionally, attackers can send disassociation messages to wireless clients, preventing them from disconnecting properly from the access point by sending an 802.1X EAPOL Logoff message.

6.3.3 WPA Cracking Tools

There are a number of WPA cracking tools which attempt to determine the initial shared secret when WPA-PSK is used. Once this secret is known, the base key and session keys can be recreated, and traffic to and from clients and the access point can be decrypted on the fly. Alternatively, attackers can record traffic and then mount an offline attack at a later time, allowing use of greater computational resources.

For the majority of these tools to work, the attacker must be able to monitor the entire initial key exchange. An attacker who starts monitoring wireless traffic while clients are already connected will not be able to gather the proper data to crack the WPA encryption. However, it is relatively simple for an attacker to create a denial-of-service condition by sending disassociation packets to the clients. The clients then disconnect and reconnect, re-authenticating in the process and enabling attackers to view the needed data.

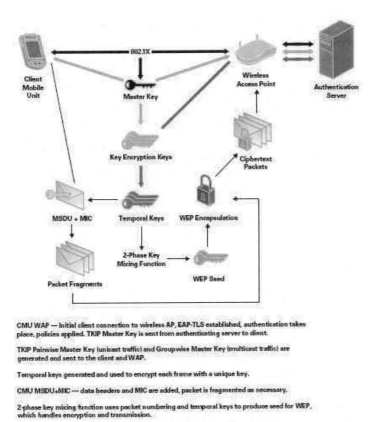

CMU WAP — Initial client connection to wireless AP, EAP-TLS established, authentication takes place, policies applied. TKIP Master Key is sent from authenticating server to client.

TKIP Pairwise Master Key (unicast traffic) and Groupwise Master Key (multicast traffic) are generated and sent to the client and WAP.

Temporal keys generated and used to encrypt each frame with a unique key.

CMU MSDU+MIC — data headers and MIC are added, packet is fragmented as necessary.

2-phase key mixing function uses packet numbering and temporal keys to produce seed for WEP, which handles encryption and transmission.

Figure 6.1: WEP, TKIP, and 802.1X.

6.3.4 WPA Summary

WPA is generally accepted as an interim step to provide incrementally improved security until WPA2 is available. Most devices that were upgraded to WPA capability are not capable of further upgrades. These devices are generally hardware-constrained, with minimal processing power and with RC4 as the only onboard encryption option.

6.4 The Way Forward: Wi-Fi Protected Access 2 (WPA2) and 802.11i

The 802.11i standard is virtually identical to WPA2, and the terms are often used interchangeably. 802.11i and WPA2 are not just the future of wireless access authentication - they are the future of wireless access. Wireless access is still in its infancy, in spite of the purchase and deployment of several million access points and wireless clients. The majority of these access points and clients are relatively immobile. Users sit down with their laptops at a conference table and connect, or a clerk stays within a relatively small area such as a warehouse, using wireless equipment to track inventory.

6.4.1 Increased Density of Access Points

Wireless access in the future will feature increased density of access points. There are several reasons for this, including a greater need for bandwidth. An area covered by a single access point will not be able to provide as much bandwidth to clients as two access points. Also, in office buildings and other areas with stores located near each other, access points are typically not shared, but deployed separately within each location. Some residential areas already have multiple access points on each block. Finally, increased density benefits availability. If two or more access points cover an area and one of the access points fail, the area retains some degree of coverage.

Considering these factors, organizations will likely use more than one access point to cover a given area. But increasing the number of access points without a strategy for centralized management creates additional security risks. Robust management of access points is a key design requirement for Motorola in the implementation of 802.11i-based products.

6.4.2 Roaming Wireless Clients

Critical to 802.11i is the addition of fast secure roaming support for clients. This assists Voice-over-IP (VoIP) and other mobile applications that require continuous access. While some wireless 802.11 equipment currently supports fast secure roaming, it is usually vendor-specific, as no official standard existed prior to 802.11i that supported this function.

Currently, most wireless clients are relatively immobile, as few truly portable devices have come into use. Laptop users generally sit down at fixed locations to use their systems. However, in the future, more wireless devices (such as phones and PDAs) will support 802.11 - and these devices must roam. Also, due to the data types such portable roaming devices will likely carry, such as live voice and video, users will immediately notice any interruptions in service. And if such interruptions are common, the viability and use of live voice and video services becomes untenable. For network providers, access must be smooth and seamless while clients are roaming.

6.4.3 Failover Requirements

Robustness and availability are commonly forgotten yet important aspects of wireless networks. The majority of wireless networks have no failover or redundancy capabilities other than manual connection to a new access point when the one in use fails. As more wireless networks are deployed to carry critical traffic such as phone calls via the VoIP protocols, reliability and robustness will become more important. One benefit of 802.11i roaming support is that a client has de facto support to connect seamlessly to a new access point should the one in use fail. Of course, this will require service coverage of areas by one or more access points, but with costs falling, this is not a serious issue.

6.5 WPA2: Under the Covers

WPA was provided as an interim solution, and it had a number of major constraints. WPA2 was designed as a future-proof solution based on lessons learned by WEP implementers. Motorola is a key contributor and proponent of the WPA2 standard, and provides next-generation products based on this standard.

WPA2 will be a durable standard for many reasons. One of the most important choices was that of the encryption algorithm. In October 2000, the National Institute of Standards and Technology (NIST) designated the Advanced Encryption Standard (AES) as a robust successor to the aging Data Encryption Standard. AES is an extremely well-documented international encryption algorithm free of royalty or patent, with extensive public review.

WPA2, like WPA, supports two modes of security, sometimes referred to as "home user" and "corporate." In "home user" mode a pre-shared secret is used, much like WEP or WAP. Access points and clients are all manually configured to use the same secret of up to 64 ASCII characters, such as "this_is_our_secret_password." An actual 256-bit randomly generated number may also be used, but this is difficult to enter manually into client configurations.

The "corporate" security is based on 802.1X, the EAP authentication framework (including RADIUS), one of several EAP types (such as EAP-TLS, which provides a much stronger authentication system), and secure key distribution. This paper discusses "corporate" security. "Home user" security introduces the same security problems present in WEP and WPA-PSK.

6.5.1 WPA2 and 802.1X

While 802.1X as a standard preceded 802.11i, it is proving to be a key enabler for secure and flexible wireless networks, allowing for client authentication, wireless network authentication, key distribution and the pre-authentication necessary for roaming. In using 802.1X in conjunction with 802.11i, it is strongly suggested to use EAP as a framework for authentication, and use an EAP type for the actual authentication that provides the optimal balance between cost, manageability and risk mitigation. Most often an 802.1X setup uses EAP-TLS for authentication between the wireless client (supplicant) and the access point (authenticator). In theory, several options may replace EAP-TLS, but in practice this is rare.

The 802.1X authentication protocol as deployed with 802.11i provides a number of services:

- Capabilities negotiation between the client and wireless network provider.
- Client authentication to the wireless network provider.
- Authentication of the wireless network provider to the client.
- A key distribution mechanism for encryption of wireless traffic.
- Pre-authentication for roaming clients.

In wired 802.1X, the network port is in a controlled state prior to authentication. But on wireless networks, no such port exists until the client connects and associates to the wireless access point. This immediately poses a problem, since beacon packets and probe request/response packets cannot be protected or authenticated. Fortunately, access to this data is not very useful for attackers, other than for potentially causing denial-of-service attacks, and for identifying wireless clients and access points by their hardware MAC addresses.

An 802.1X wireless setup consists of three main components:

- Supplicant (the wireless client).
- Authenticator (the access point).
- Authentication server (usually a RADIUS server).

The supplicant initially connects to the authenticator, as it would to a WEP- or WPA-protected network. Once this connection is established, the supplicant has in effect a network link to the authenticator (access point). The supplicant can then use this link to authenticate and gain further network access. The supplicant and authenticator first negotiate capabilities. These consist of three items:

- The pairwise cipher suite, used to encrypt unicast (point-to-point) traffic.
- The group cipher suite, used to encrypt multicast and broadcast (point-to-multiple-points) traffic.
- The use of either a pre-shared key (PSK, or "home user" security, using a shared secret) or 802.1X authentication.

Figure 6.2: WPA2 and 802.1X.

Once a common set of capabilities is agreed upon - and assuming the network uses 802.1X - the supplicant and the authenticator begin the authentication process. At this point, wireless encryption keys have not been exchanged and the exchange is in the clear. It is EAP-TLS that comes into play to protect this data, providing the essential SSL encryption. Signable X.509 certificates add the benefit of allowing the supplicant to prove its identity to the authenticator, and vice versa.

Several problems arise from the constraints of wireless networking. First, the supplicant must have local copies of the root certificates used by the certificate authority to sign the authenticator's certificate. Because the authenticator (the wireless access point) fully controls the supplicant's access to the network, a hostile authenticator can modify or redirect traffic in any way, and could point the user to fake certificate authority sites. Even if the supplicant has a local copy of the root certificate used to sign the authenticator's certificate, a compromised certificate placed in a certificate revocation list (CRL) may not be detected if the authenticator provides the supplicant with false or old CRL data. Therefore, any compromised authenticator certificates pose a significant risk to wireless network clients, especially since many will not check for certificate revocation.

The key exchange consists of a Master Key (MK) generated on the authentication server and in the supplicant. The MK is sent to the authenticator. The Pairwise Master Key (PMK) is generated from the MK and the Group Master Key (GMK) is generated by the authenticator. The PMK and GMK keys are then used as needed to generate temporal keys, used to encrypt individual frames sent on the wireless network. These keys are known as Pairwise Transient Keys (PTK) and Groupwise Transient Keys (GTK).

The PTK is used to encrypt traffic to and from the supplicant and the authenticator. The GTK is used to encrypt broadcast or multicast traffic sent to all hosts on a particular wireless network.

6.5.2 WPA2 and TKIP

WPA2 supports the use of the TKIP encryption scheme to provide backward compatibility with WPA equipment. As 802.11i equipment becomes ubiquitous, networks will drop support for TKIP and WPA, removing a number of potential security vulnerabilities.

TKIP uses a new key for each frame that is encrypted; the keys used to encrypt these frames are called either the Pairwise Temporal Key (for unicast traffic) or Groupwise Temporal Key (for multicast and broadcast traffic). These keys are generated from the Pairwise Master Key (PMK) and Groupwise Master Key (GMK).

The majority of weaknesses in TKIP under WPA are due to a weak encryption algorithm. This problem is securely addressed with TKIP under WPA2. By using 802.1X and EAP-TLS to handle key distribution, keys are transferred securely and are not as prone to attack. The use of an extremely strong cipher, AES, addresses the weaknesses of RC4. Finally, the use of strong key lengths, 128 bits, significantly reduces the chance of a successful brute force attack against AES-encrypted wireless traffic.

6.5.3 WPA2 and CCMP

Moving forward, the 802.11i (and by extension WPA2) standards call for the use of Counter Mode with Cipher Block Chaining Message Authentication Code Protocol (CCMP), which specifies use of CCM with the AES block cipher. CCM is a general-purpose cipher mode that does not specify the block cipher to use. The CBC-MAC portion provides data integrity and authentication while the counter mode does the actual encryption, protecting the data from eavesdroppers. It is expected that TKIP (using RC4) will be phased out in favor of CCMP as networks transition to pure WPA2 configurations, removing security risks present in support for WPA.

When a packet is encrypted using CCMP, a number of data fields are added. The first field is the message integrity code (MIC), which is appended to the data. The MIC includes the hardware MAC addresses of the source and destination; this data essentially acts as a very strong cryptographically secure hashing function, which prevents man-in-the-middle attacks and other risks. The data and the MIC are then encrypted using the appropriate encryption key.

The packet is then modified with a data header. The first portion of data contained in the packet is the IV and key ID (4 bytes), which is needed to identify the encryption key used to encrypt the packet. At this point an extended IV (4 bytes) is attached to the packet. This field and the IV with key ID field are not encrypted, as the remote end must identify which key was used to encrypt a packet and the packet's sequence number. The first IV ensures that data is ordered properly. The rest of the packet contains the encrypted payload of data and MIC. The resulting packet is shown below.

As with TKIP, the addition of MIC to packet data does not prevent replay attacks. MIC only ensures the data is not tampered or modified in transit. In order to prevent replay attacks, the IV included with the packet is referenced to ensure that sequential packets have increasing IV numbers; if an out-of-order IV is received, the client knows something is not right.

The data encrypted in the data payload and MIC field use the temporal key. This key changes for each frame and is generated from the master key, which is in turn generated from the 802.1X authentication performed by the user.

The MIC calculation and encryption of the data payload are done at the same time. This greatly speeds up encryption of packets and reduces the latency introduced by encryption.

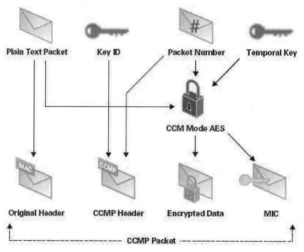

Figure 6.3: CCMP Packet.

6.5.4 WPA2 and Fast Roaming

WPA2 neatly solves the problem of roaming (and failover) in two ways: through the use of Pre-Authentication and PMK Caching.

6.5.4.1 PMK Caching

When a client re-associates with an access point, it uses a PMK from an older 802.1X authentication executed on the same access point. On this new association, no 802.1X exchange happens; the client immediately carries out the 802.11i handshake and is ready to send/receive data.

When the client loses the connection with the first access point - or otherwise decides to move to the second access point (because of signal strength, for example), it must only change radio frequencies and establish a base 802.11 connection with a second access point that it associated with previously. Once this is completed, the client only needs to perform the 802.11i handshakes to establish the PTK and the Groupwise Master Key before beginning communication, since authentication has already taken place.

6.5.4.2 Pre-Authentication

When a client is associated with an access point and hears a beacon from another access point with the same SSID and security policy, it carries out an 802.1X authentication with that access point over the wire. The client and access point derive the PMK and keep it cached. Now if the client roams over to the new access point, it already has a PMK - the 802.1X authentication phase is skipped.

6.6 Opportunistic PMK Caching: Fast Roaming at Its Fastest

As described in an earlier section, roaming is a key technical advantage of WPA2. However, even the "fast roaming" options included in the standard cause the client a brief disruption of service, which is too lengthy for some time-sensitive applications.

Currently, establishing a connection to an 802.11 access point, authenticating to it, and establishing encryption keys can take anywhere from 150 to 350 milliseconds - and in extreme cases, 800 milliseconds or more. Long delays in establishing connections for clients occur when, due to the need for access points to communicate with back-end authentication servers, network equipment is spread out over a wide area.

While PMK Caching and Pre-Authentication within WPA2 help reduce this latency by reducing redundant instances of 802.1X authentication, experts recognize that this does not close the "disruption gap" that impacts quality of service. Not only does the standard not address intensive applications, but several implementation and architecture specific factors can exacerbate the problem in wireless networks. In many standard fast roaming scenarios, establishing communications with the network can take 100 to 150 milliseconds, which is an acceptable delay for some activities such as web browsing, but which can result in a very noticeable interruption in service during a Wi-Fi VoIP phone call or video conference.

In July 2004, the IEEE formed a Fast Roaming Task Force to begin work on the 802.11r fast roaming standard for wireless networking. The goal of 802.11r is to improve handoff times between access points. The final product will be known as Fast BSS Transition. Motorola will extend its support to include this new standard, as it does for all other key standards. However, optimized roaming must be enabled today.

To meet this need now, Motorola has employed a unique fast roaming capability in its WPA2-compatible products that improves the roaming latency of WPA2. This feature, Opportunistic PMK Caching, has gained the support of such leading supplicant providers as Microsoft and Funk Software. While PMK Caching and pre-authentication enable fast roaming within the WPA2 standard and are supported by Motorola, Opportunistic PMK Caching takes a big step beyond these techniques. Opportunistic PMK Caching improves fast roaming in order to create a transparent environment for users of latency-sensitive wireless applications.

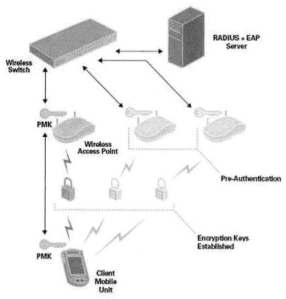

Figure 6.4: Opportunistic PMK Caching.

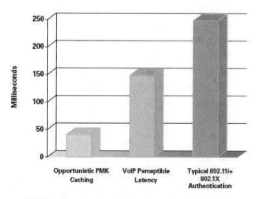

Figure 6.5: Performance of Opportunistic PMK Caching.

In a wireless switch environment, the switch has access to all PMKs from all connected access points. Depending upon the switch policy configuration, it is possible for a PMK from any one of the access points to be used for all connected access points. Therefore, a client may perform PMK caching with any other PMK that is available, bypassing the 4-way CCMP authentication handshake. This can greatly speed up the roaming process between access points, and in some cases will lead to virtually no disruption in service - critical to wireless VoIP and other devices.

Motorola's own testing has shown that the access point handoff can occur in less than 40 milliseconds - far below the 150 milliseconds time considered the threshold of VoIP latency as perceptible to humans. Motorola achieved this key breakthrough with its switch-based architecture, making possible seamless interoperation with several third-party clients and transparent wireless roaming for demanding user applications.

6.7 Summary

There is no disputing the benefits that wireless local area networks provide to enterprises. Information technology departments have a mandate to provide wireless services in support of business initiatives, as well as a responsibility to provide these services securely. The history of 802.11-based WLANs has been a legacy of insecurity and significant risk to early adopters. Both WEP and its interim successor WPA have provided only minor obstacles to determined hackers and should be deployed with caution.

The ratification of IEEE 802.11i laid the foundation for drastic improvements in wireless security. WPA2 offers more formidable encryption, better key management, and robust authentication, as well as access point roaming.

Motorola surpasses a key shortcoming within WPA2 - insufficient access point roaming speeds. The current specification permits a credentials negotiation process, which can cause lengthy disruptions in the connections, a fatal problem for time-sensitive applications such as wireless-based VoIP. Motorola developed Opportunistic PMK Caching to overcome this issue, a technique that leverages the centralized wireless switch architecture and provides the highest-speed Wi-Fi roaming available on the market.

7

Wireless Local Area Network Security

Dorothy Stanley, Joshua Wright[a]

The specification and broad adoption of strong AES-based encryption and data authentication and strong end user authentication in IEEE 802.11 Wireless Local Area Network (WLAN) systems provide strong link layer security. Since the wireless link for data traffic is secure, standards work now turns to the protection of management frames and implementers look to deploy intrusion detection tools, while attackers look for implementation-flaw based attacks, such as "fuzzing". This chapter discusses the topics of WLAN link security, key management, end user authentication, standards, wireless driver vulnerability attacks and wireless intrusion detection techniques.

7.1 Introduction

The level of required security in a system changes over time, as technology and export regulations change and as the processing capabilities of both valid users and potential attackers increase. One static aspect, however, is the need for end users to adhere to recommended security practices, such as keeping up-to-date virus software and intrusion detection software on their laptops or client devices. There are conflicting requirements of security and convenience. End users desire a simple, quick logon using stored passwords on client devices; however, for stronger authentication, particularly in enterprise networks, two separate credentials from the user, a password and a time-changing code are typically required. This is similar to the credentials required to withdraw cash from an ATM, you must present both a password (something you know) and the appropriate ATM card (something you have).

The level of privacy and authentication required may also be a function of the application or location in which the WLAN is deployed. Enterprise applications may have different needs than public space applications. A given residential application may need the security level of an enterprise. Some government applications can use commercially available WLAN security technologies. The security solutions must be broad enough to support a variety of application spaces. The solutions must be easy to configure and use, as

---

[a] All authors currently affiliated with *Aruba Networks, Inc*

the same laptops and client devices may be used for Internet access in more than one location.

7.2 Current Application Solutions

Strong link-layer security is now available in WLAN systems, as a result of the broad implementation and deployment of IEEE 802.11 Advanced Encryption Standard – Counter mode with Cipher-block-chaining Message authentication code Protocol (AES-CCMP) based systems. In what now seems to be almost ancient history, Wired Equivalent Privacy (WEP) encryption defined by IEEE 802.11 was identified as not providing "industrial-strength" link security. Papers by Borisov [1] and Walker [2] discussed the vulnerabilities of WEP. The results of Fluhrer et al. [3] enabled easy-to-mount passive attacks [4] on WEP, which have been commoditized in attack tools.

In response to the identified flaws on WEP, customers deployed overlay VPN solutions, while the IEEE 802.11 Working Group completed work on secure link layer protocols. The Wi-Fi Alliance [5] provided Wi-Fi[1] Protected Access (WPA) interoperability certification for the Temporal Key Encryption Protocol (TKIP), which was deployable on legacy WEP hardware. This was followed in 2004 with interoperability certification of AES-CCMP based link security, termed WPA2. Both TKIP and AES-CCMP based solutions incorporated use of IEEE 802.1X [6] and Extensible Authentication Protocol [7] based end user authentication, significantly raising the level of secure authentication provided in WLAN systems.

Wi-Fi Alliance has continued to promote deployment of secure WLAN systems, including Wi-Fi Protected Set-up (WPS). Developing the security technology is not sufficient; the technology must be used, and made easy to use and configure, especially in non-managed (consumer) environments. WPS provides a means for consumers to simply configure a security-enabled WLAN system, thus increasing the number of deployed secure systems.

In addition to link layer solutions, security of end user data can be provided above the MAC layer. Virtual Private Network (VPN) overlays can be costly however, requiring additional equipment (VPN servers) and maintenance expenses. Most WLAN infrastructure equipment, including Aruba WLAN Access Point (AP) and Mobility Controller (MC) products is designed to work transparently with VPN solutions.

For some customers, the deployment of low-cost access points coupled with application-level security solutions is desirable. In public space deployments, for example, many operators' primary concern is with economically deploying wireless access networks. Service providers offer Wi-Fi interoperable access to the greatest number of customers and provide easy-to-use Web portal interfaces for customer registration. Security is provided at the network level, via VPN access to corporate networks, and secure web server access.

Attacks targeting wireless driver software vulnerabilities (see Section 7.9) and a proliferation of exploit tools have appeared, in a continuing effort to compromise the integrity, availability and confidentiality of wireless LAN environments. As a response to these attacks, many vendors have produced wireless LAN intrusion detection systems

---

[1] Wi-Fi is a trademark of the Wi-Fi Alliance (WFA).

(WIDS), also known as wireless LAN intrusion prevention and detection systems (WIPDS). These systems can provide significant value to organizations, accommodating an additional layer of security through detection and attack countermeasures techniques (see Section 7.10).

7.3 MAC-Level Encryption Enhancements

IEEE 802.11 MAC-level enhancements have been specified to provide standard, strong encryption and data authentication at the wireless MAC level and to enable use of upper-layer authentication. The IEEE 802.11i MAC Security Enhancements amendment, now incorporated into the IEEE 802.11-2007 standard [8] defined standards for

- TKIP, a strengthened version of the RC-4/per-frame IV encryption protocol;
- CCMP, a 128-bit AES encryption and data authentication protocol.

TKIP was intended to provide a backwards-compatible solution for WEP-capable devices, incorporating improvements and enhancements to address the shortcomings of WEP. These enhancements include:

- The addition of a per-frame hash function and IV sequencing rules [9, 10];
- The addition of temporal key derivation algorithms [11];
- The addition of a message authentication code, termed message integrity code [12].

Taken together, TKIP addresses the flaws identified in the WEP algorithm that were identified by the cryptographic community. A critical constraint placed on TKIP was that it be able to be implemented and deployed via software upgrade to the then existing base of millions of 802.11 devices, to avoid requiring a total deployed-base upgrade to provide a secure system.

TKIP was designed to have a lifetime of about 5 years, and was intended to provide a secure mechanism that could be deployed on WEP capable hardware. TKIP has met its design goals. Since 2002, no practical attacks have been mounted against TKIP, and one theoretical attack [13] has been identified. In 2007, virtually all new WLAN products support AES-CCMP, and WFA support of WPA2 is required for WFA interoperability certification. New amendments to the IEEE 802.11 standard are likely to use only CCMP security (see 7.3.5).

7.3.1 The TKIP Per-Packet Hash Function

The RC4 key used to encrypt a given data frame in WEP is a combination of an initialization vector (IV) and the secret key. Unfortunately, in the key-scheduling algorithm of RC4, the first bytes of the key stream are predictable for certain known IV values [4]. Because the IV used to encrypt a given frame is sent in the clear, a passive observer can easily identify the frames to target for attack. The TKIP per-frame hash function is introduced primarily to eliminate this flaw in WEP. The hash function is also defined to include the MAC address of the transmitting station. This enables each transmitting station to generate a unique IV stream and thus prevents the reuse of IV values among stations using a shared secret key. IV values must not be reused, to prevent the reuse of RC4 key streams and subsequent data recovery attacks.

A simplified description of the TKIP per-frame hash algorithm is shown below. The details of the hash function are provided in [8]. The algorithm is described in two phases, both of which use S-boxes to mix and substitute 16-bit values. In phase 1, the 128-bit temporal key, the high 32 bits of the transmitting station's MAC address and the Sequence Counter (IV) are hashed into an 80-bit value, composed of 5-16-bit values, as illustrated in Figure 7.1.

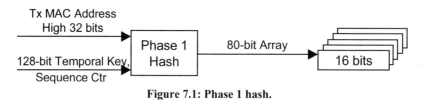

Figure 7.1: Phase 1 hash.

Phase 2 of the temporal key hash function takes the 80-bit array from phase 1, together with the Temporal Key and Sequence Counter (IV), and generates a 128-bit per-frame key. As the name implies, the key that is generated will be used for one frame only; the phase 2 hash is calculated for each frame that is encrypted (Figure 7.2). The per-frame key is subsequently used as a WEP key, with the first 24 bits transmitted in the clear.

The phase 2 hash uses an S-box mixing function that operates on 16-bit values of the array, a mixing function that uses rotate and addition operations, and an algorithm to calculate the 48-bit IV value. The phase 2 hash eliminates the effects of the WEP/RC4 key scheduling algorithm flaw. Use of the extended 48-bit IV eliminates the need to re-key due to exhaustion of the IV space and eliminates the issue of IV reuse seen in WEP, as quadrillions of frames can be sent before the TKIP IV space (2^{48}) is exhausted.

7.3.2 TKIP Temporal Key Derivation

TKIP temporal key derivation defines a method whereby the "secret key" or master key is not used to encrypt data packets but rather is the basis from which temporal or transient encryption keys are derived (Figure 7.3). These temporal keys may then be used as input to the per-frame hash function described above. This approach is very different from the WEP definition and implementations, in which the provisioned key is used directly as the secret portion of the encryption key.

TKIP uses a pseudorandom function (PRF), operating on the secret key, a text string, the MAC addresses of the station and the authenticator, and nonce values, to generate a temporal key. The temporal key is then used for the encryption and MIC temporal keys (described in Section 7.3.3). The transient key provides the key material for the TKIP per-packet RC4 encryption and the TKIP MIC function.

In TKIP, it is critical that for any encryption key, a given IV be used to encrypt one and only one frame. Proper use of IVs is ensured by the application of IV sequencing rules. First, the notion of a sequence counter is introduced. The sequence counter is incremented by the transmitter on a per-frame basis. As part of the per-packet mixing function, the sequence counter is mapped to the WEP IV. Then, the receiver must verify that the received frames increment sequentially, per Quality of Service traffic class. If the IV of a frame is less than or equal to that of a previously received frame, it is discarded by the receiver.

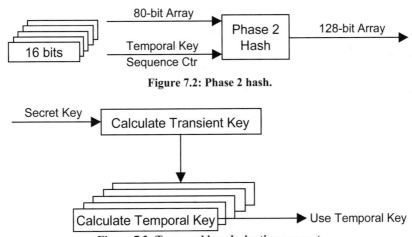

Figure 7.2: Phase 2 hash.

Figure 7.3: Temporal key derivation concept.

7.3.3 TKIP Message Integrity Code

A Message Integrity Code (MIC)[2] is needed to verify the authenticity of a transferred data packet. Use of the MIC verifies that the packet was not modified in transit and that the source and destination addresses were not changed. The ability to verify message integrity is viewed by cryptographers to be as important as, if not more important than, the privacy provided by encryption. The MIC is required to prevent the "bit-flipping" attacks identified in [2]. A MIC algorithm known as "Michael" is the TKIP MIC [12].

Because of the design constraint that the TKIP MIC be implementable on legacy WEP devices, Michael is a relatively weak MIC algorithm. Countermeasures are introduced, which log MIC-error events and rate-limit the number of MIC failures. This prevents an attacker from generating a large number of forgery attempts within a short period of time. For example, countermeasures require that if 2 MIC-error frames are received within 60 seconds at an AP, that the AP disassociate all TKIP stations, and not accept any new stations using TKIP for 60 seconds.

7.3.4 AES Based Encryption and Data Authentication

The Advanced Encryption Standard (AES) Rijndael algorithm [14] was selected by NIST [15] as the next-generation encryption algorithm, to replace DES and 3DES. Several modes or ways of using the AES algorithm have been defined. AES-CCMP [16] is used in IEEE 802.11 to provide strong link layer encryption and data authentication. AES-CCMP

---

[2] Here the term "MIC" is used, as "MAC" is already used for Medium Access Control. Message Authentication Code (MAC) is the standard cryptographic term.

combines Counter mode encryption with Cipher Block Chaining message integrity/authentication.

Figure 7.4 below [17] shows the AES-CCMP encapsulation processing for an 802.11 frame; that is, the encryption and authentication using the AES-CCMP algorithm. The AES cipher is used to calculate the MIC value (top half of Figure 7.4) and to encrypt the data payload (lower half of Figure 7.4). The IV and counter contents and frame field definitions are specified in [8].

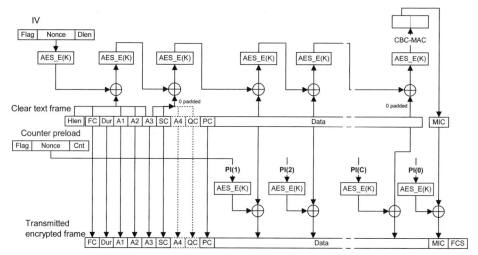

Figure 7.4: CCMP Encapsulation.

Use of CCMP enables deployment of 802.11 systems in Federal Information Processing Standard (FIPS) applications, as AES in CCM mode is a FIPS approved mode of operation. CCMP supports, but does not guarantee government use of commercially available WLAN equipment. The CCMP decapsulation operation [17] is shown in Figure 7.5.

7.4 Secret Key Generation and Distribution

In the 802.11 specification, key distribution mechanisms are not defined. Keys are obtained via an upper layer EAP method protocol exchange or via a manual (Pre-shared Key) configuration. Upper-Layer Authentication Messages, specifically the 802.1X EAPOL-Key message are used in the 4-Way Handshake [8] to exchange information needed for the supplicant (client) and authenticator (network entity) to generate encryption and authentication keys from a Master Key and to derive a new transient key if needed.

7.5 Authentication

Authentication of end users or end systems is needed to control access to the WLAN. In enterprise applications, only authorized users must be allowed to access the corporate intranet. In public space applications, user identification is needed by the service provider to accurately bill the end user. This section gives a brief overview of the EAP-TLS [18], EAP-MD5 [19], and EAP-TTLS [20] EAP methods, together with 802.11 authentication (used with the EAP methods) and RADIUS MAC-based authentication.

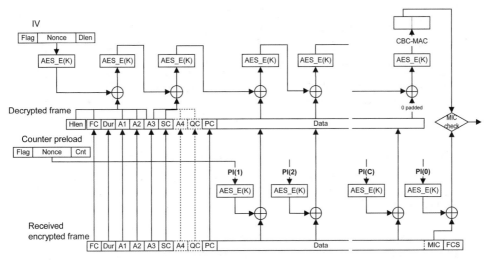

Figure 7.5: CCMP Decapsulation.

Digital certificates and shared secrets (passwords) are common credentials used to authenticate an end user or device. A standard, common certificate-based authentication method is EAP-TLS. Multiple EAP methods have been defined, and each authentication method has advantages and disadvantages [21]. The needs of individual deployments may require use of a method supporting a specific type of user credential. IEEE 802.11 EAP method requirements are defined in [22].

The benefit of the using EAP for authentication is that additional EAP types can be easily defined and added to a system. Additional EAP types include EAP-SIM [23], which reuses the mobile GSM authentication credentials, EAP-SRP [24], a secure password-based method, and EAP-AKA [25], which uses symmetric key credentials. EAP-PEAP [26] is similar to EAP-TTLS in concept, but tunnels only EAP-based authentication methods.

7.5.1 802.1X EAP Authentication

The EAP-TLS [18] protocol provides a mechanism for certificate-based mutual authentication. Upon completion of successful EAP-TLS authentication, a secret master key is known at the station and the RADIUS server. This key is subsequently delivered to the authenticator (AP or MC) by the RADIUS server. EAP-TLS requires prior distribution of client side and server side certificates via a secure connection. RADIUS Authentication servers supporting EAP-TLS and certificate management capabilities are also required. A simplified message diagram for EAP-TLS is shown in Figure 7.6. EAP authentication messages sent to/from the station to the RADIUS Server transit the AP/MC.

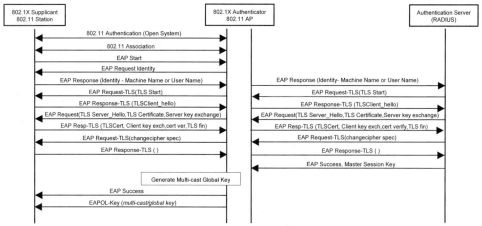

Figure 7.6: Simplified EAP-TLS message flow.

7.5.2 EAP-MD5

The EAP-MD5 [19] authentication algorithm provides one-way password-based network authentication of the client. It is expected to be used in 802.1X wired Ethernet switch deployments. This algorithm can be used for wireless applications with no WLAN security requirements. The impediment to using EAP-MD5 in wireless LAN applications is that no encryption keys are generated. Also, although the protocol can be used by the client to authenticate the network, it is typically used only for the network to authenticate the client. Finally, as the Disassociation message is not currently authenticated, a valid established session can be hijacked by an attacker [27]. The message flow is shown in Figure 7.7.

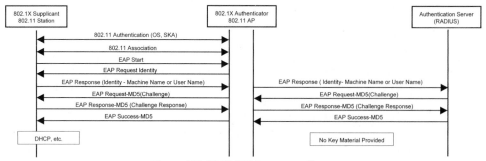

Figure 7.7: EAP-MD5 message flow.

7.5.3 EAP-TTLS

EAP-TTLS [20] can be viewed as an interesting combination of both EAP-TLS and traditional password-based methods such as Challenge Handshake Authentication Protocol (CHAP) [28], and One Time Password (OTP). In this method, a TLS tunnel is first established between the station Supplicant and the Authentication Server. The client authenticates the network to which it is connecting by authenticating the digital certificate provided by the TTLS server. This is exactly analogous to the techniques used to connect to a secure web server. Once an authenticated "tunnel" is established, the authentication of the end user occurs. EAP-TTLS has the added benefit of protecting the identity of the end user from view over the wireless medium, providing anonymity of the end user, a desirable attribute. EAP-TTLS also enables existing end user authentication systems to be reused. The simplified message protocol exchange for EAP-TTLS is shown in Figure 7.8.

7.5.4 IEEE 802.11 and RADIUS MAC Authentication

The IEEE 802.11 standard [8] supports two subtypes of MAC layer authentication services: open system and shared key. Open system authentication is the default authentication service and is used by a station to indicate its intent to associate with an Access Point. 802.11 open system authentication at the MAC level is used with upper layer 802.1X EAP authentication. 802.11 shared key authentication provides the ability to verify that the AP and the station share the same WEP key before 802.11 association. A challenge-response protocol is used, and vulnerabilities have been identified. Shared key authentication is not included in the Wi-Fi interoperability requirements and is not recommended for use.

RADIUS-based MAC authentication is a technique supported by most infrastructure equipment. The MAC addresses of valid 802.11 devices are provisioned into the AP or MC, and only traffic from these MAC addresses is allowed through the AP or MC. Authentication is tied to the hardware that is used and not to the identity of the user. Software does exist to change the MAC address of a wireless device, and thus MAC-based authentication provides a only a very minimal level of access control to wireless networks.

7.6 Evolution, Standards, and Industry Efforts

The future growth of 802.11 WLANs requires interoperable, standards-based, evolvable solutions that extend the features and security capabilities of 802.11 systems and also support the bandwidth needs of 802.11n High Throughput systems. This section describes security-related extensions under development at the time this chapter was written.

7.6.1 Security-related changes in the TGn High Throughput Amendment

The security-related changes in High Throughput systems are very limited. Pre-standard High Throughput systems are being produced and deployed, and those that are WFA certified require support for AES-CCMP based encryption only. The TKIP algorithm requires computation of the MIC on the MAC Service Data Unit (MSDU), and encryption of the MAC Packet Data Unit (MPDU), which is difficult to implement at extremely high data rates. In addition, the design life-time of TKIP has been passed. Thus support of TKIP at the High Throughput data rates is not required.

The TGn Draft 2.0 amendment [29] defines MSDU aggregation, that is, construction of a frame payload with multiple concatenated MSDUs. One bit of the QOS field is used to indicate the presence of the aggregated MSDU frame when Quality of Service (QoS) mechanisms are used, and that bit must be included in the CCMP MIC calculation to guarantee correct interpretation of a received frame.

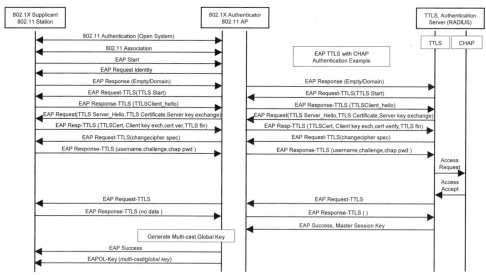

Figure 7.8: Simplified EAP-TTLS message flow.

7.6.2 Security-related changes in the TGr Fast BSS Transition Amendment

The TGr amendment adds use of AES-128-CMAC [30] as a MIC algorithm, used to provide data authenticity of the data exchanged to establish Fast Transition security associations. In the TGr Fast BSS Transition amendment, the contents of the 4-Way Handshake messages, with additional optional Quality of Service information are essentially overloaded into the 802.11 Authentication and Association frames. A new IEEE 802.11 Authentication type is defined for the Fast Transition protocol. The contents of information elements included in the Authentication and Association frames are protected using AES-128-CMAC. Also, the allowable MIC algorithms used with 4-Way Handshake messages are extended to include AES-128-CMAC.

7.6.3 Security in the TGs Mesh Amendment

The TGs amendment will define mesh operation for IEEE 802.11 WLAN systems (also see Section 16.4.5). New security mechanisms must be defined which authenticate Mesh Point to Mesh Point links. Multiple credential types should be supported for authentication, including pre-shared keys and digital certificates. While the first application of mesh networks that typically comes to mind is to provide wireless connectivity between Access Point devices, mesh functionality can be used to interconnect any set of WLAN devices. Mesh applications for simultaneous peer-to-peer connections, particularly in the consumer market, will require a simple pre-shared key authentication mechanism.

As of this writing, the proposed mesh authentication approach is to extend the IEEE 802.11 EAP based authentication and CCMP protection mechanisms to mesh applications [31], as shown in Figure 7.9. While this approach has the benefit of using a familiar authentication technology, it does require both authenticator and supplicant functionality to be present at each Mesh Point, requires a known Authenticator to be accessible to all Mesh Points, and requires a new key hierarchy to be defined. It also requires a very orderly progression of authentication, beginning with a designated Mesh Key Distributor Mesh Point.

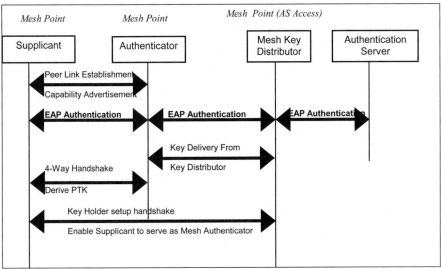

Figure 7.9: EAP Authentication and Mesh Points.

In this illustration, the Mesh Key Distributor (MKD), (the device with access to the Authentication Server), and the Authentication Server must be configured and the MKD authenticated to the AS. Once the MKD is authenticated, a second device (the Authenticator in Figure 7.9) initially takes on the role of the Supplicant, and authenticates to the AS via the Mesh Key Distributor, which serves as its Authenticator. This device is then authorized to function as an authenticator. Each subsequent Mesh Point that joins the mesh then follows the scenario shown in Figure 7.9. The Supplicant Mesh Point first establishes a wireless link to the Authenticator. The Authenticator Mesh Point facilitates the EAP exchange between the Supplicant and itself, to authenticate the supplicant device to the AS. The key derived from the Supplicant-AS authentication is delivered to the Mesh Point Authenticator from the Mesh Key Distributor.

Management action frames are defined that are used by the Mesh Key Distributor and each Authenticator Mesh Point, to transport the EAP messages across (potentially multiple) mesh points. After a Mesh Point establishes a security association with an MKD, it can receive key material and serve as a Mesh Authenticator for other Mesh Points.

A new key hierarchy is defined to support Mesh Point authentication, which includes new keys used for distribution of keys from the MKD to new Mesh Points (see Figure 7.10).

A proposed alternative approach to continuing use of EAP authentication is termed "comminus" and uses a Diffie-Hellman based authentication protocol, based on SKEME [32]. The comminus authentication protocol provides a mutual authentication of equals, in that there is no concept of Authenticator and Supplicant. It can be used with either digital certificates or pre-shared keys, and consists of a 4-message protocol exchange, proposed to be instantiated as a new 802.11 authentication type. A reference implementation of comminus is available at http://www.lounge.org/comminus.tgz.

Figure 7.11 uses the following definitions:

- *Na* is a pseudo-random number chosen by the Initiator
- *Nb* is a pseudo-random number chosen by the Responder
- *K* is a secret derived from *gab modp,* the Diffie-Hellman shared secret
- *p* is an authenticating key derived from *K*
- *w* is a key-encrypting key derived from *K*
- *A to B is PRF(Nb, p|ga|gb|BSSIDa|BSSIDb,* and
- *B to A is PRF(Na, p|gb|ga|BSSIDb|BSSIDa)*

While comminus appears to be a simple and elegant approach, it has not been adopted into the TGs draft amendment to date.

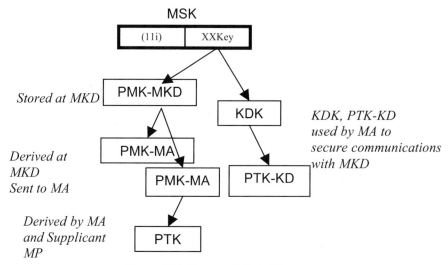

Figure 7.10: Proposed Mesh Hierarchy.

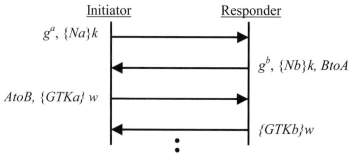

Figure 7.11: Comminus pre-shared key example.

7.6.4 Security in the TGw Protected Management Frames Amendment

The IEEE 802.11i MAC Security Enhancements amendment defined security mechanisms for IEEE 802.11 data frames, that is, frames carrying user data payloads. Subsequent amendments, particularly the Radio Resource Management and Network Management amendments under development define management action frames, which carry user-related and system-related management information.

The TGw amendment was initiated to define mechanisms to protect selected management frames. To date, the means proposed in the IEEE 802.11w amendment is to extend AES-CCMP to protect directed (unicast) management action frames. A new Integrity group key is defined, which provides AES-128-CMAC based data authenticity for (a) broadcast and multicast management action frames, (b) broadcast Disassociation and (c) broadcast Deauthentication management frames. Data authenticity is used, rather than encryption to allow for "mixed" TGw and non-TGw station deployments. If the broadcast and multicast management frames were encrypted, then only TGw-enabled client devices could receive the secured frames.

Note that in the IEEE 802.11 security work to protect the link layer, frames are not protected until keys and corresponding security associations are established. Thus the contents of some frames, such as Beacon and Probe Response frames are not protected.

7.7 Wireless and Software Vulnerabilities

With the availability of strong link layer encryption and data authentication support with the TKIP and AES-CCMP protocols, and support for strong EAP-based authentication mechanisms, attackers are seeking alternative mechanisms that can be used to exploit wireless networks. One such mechanism is to take advantage of software implementation flaws in wireless devices.

7.7.1 Exploiting Wireless Device Drivers

As the attack community has discovered, some wireless device driver software has been designed and implemented to comply with the requirements of the IEEE 802.11 specification, but without sufficient handling for malformed frames, or inappropriate frames that violate the IEEE 802.11 state machine. For example, wireless station drivers such as those written for Microsoft Windows workstations must process the data payload portion of Beacon frames to extract information about available networks. The payload of Beacon frames includes both fixed parameters which are strictly ordered and always present, and tagged parameters or Information Elements (IEs) which can be in any order. The data in tagged parameters uses the familiar type/length/value encoding format, as shown in Figure 7.12.

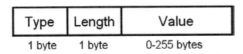

Figure 7.12: Tagged Management Parameters Format.

The SSID IE is always present in Beacon frames, formatted as "The length of the SSID information field is between 0 and 32 octets. A 0 length information field indicates the broadcast SSID" [8].

From reading this description in the 802.11 specification, a wireless device driver developer could assume that the SSID information will never be longer than 32 bytes, since 32 bytes is the maximum described value. However, 32 bytes do not represent the potential maximum size of the SSID. Since the length field of a tagged management parameter is an 8-bit unsigned value (as shown in Figure 7.10), the SSID length can range from 0 to 255 bytes. One attack against wireless drivers takes advantage of this assumption, as shown in Figure 7.13.

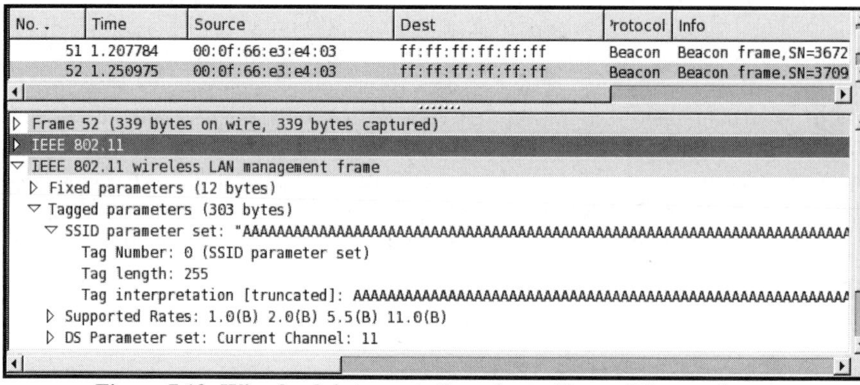

```
No. .    Time        Source             Dest               'rotocol  Info
      51 1.207784    00:0f:66:e3:e4:03  ff:ff:ff:ff:ff:ff  Beacon    Beacon frame,SN=3672
      52 1.250975    00:0f:66:e3:e4:03  ff:ff:ff:ff:ff:ff  Beacon    Beacon frame,SN=3709

▷ Frame 52 (339 bytes on wire, 339 bytes captured)
▷ IEEE 802.11
▽ IEEE 802.11 wireless LAN management frame
  ▷ Fixed parameters (12 bytes)
  ▽ Tagged parameters (303 bytes)
    ▽ SSID parameter set: "AAAAAAAAAAAAAAAAAAAAAAAAAAAAAAAAAAAAAAAAAAAAAAAAAAAAAAAAAAAAAAAAAAAAAAAAAAA
        Tag Number: 0 (SSID parameter set)
        Tag length: 255
        Tag interpretation [truncated]: AAAAAAAAAAAAAAAAAAAAAAAAAAAAAAAAAAAAAAAAAAAAAAAAAAAAAAAAAAAAAAAAA
    ▷ Supported Rates: 1.0(B) 2.0(B) 5.5(B) 11.0(B)
    ▷ DS Parameter set: Current Channel: 11
```

Figure 7.13: Wireshark interpretation of a malformed Beacon frame.

In Figure 7.13, the Wireshark [33] traffic analyzer software displays the contents of a Beacon frame, where the SSID IE length is set to the maximum potential value of 255 bytes. In one Windows XP wireless driver implementation, the driver software will attempt to copy all 255 bytes of the SSID into a static 32-byte memory buffer, allowing the attacker to overwrite other memory addresses with arbitrary data (e.g., the contents of the SSID), and making it possible to execute arbitrary code of the attacker's choosing on the target wireless device.

The implications of such an attack against wireless networks is significant, since it represents an opportunity for the attacker to compromise wireless networks, even though they may be using strong encryption and authentication mechanisms. In the specific case of the malformed SSID IE in a Beacon frame, an attacker does not require any authentication credentials to the wireless network nor does he need to manipulate clients into a state where they will be expecting the delivery of Beacon frames. All the attacker needs is physical proximity to the target wireless network to attempt to exploit vulnerable systems. Further, since Beacon frames are sent to the broadcast address, an attacker can potentially compromise multiple victims simultaneously with one transmitted frame.

Exploiting wireless driver vulnerabilities is an attractive attack technique for an adversary for several reasons;

- No network access is required: The attacker does not require authenticated access to the wireless network to exploit many of the reported driver vulnerabilities. Physical proximity to a vulnerable station is the only requirement for the attacker to take advantage of these flaws.

- Applicable regardless of encryption or authentication selection: As of the time of this writing, all of the reported wireless driver vulnerabilities take advantage of unauthenticated management frames, which are present in all IEEE 802.11 wireless networks, independent of a selected encryption mechanism or EAP type. With the ratification of the IEEE 802.11w amendment, Deauthentication and Disassociation frames will be protected, mitigating driver vulnerabilities that take advantage of these frames. Unprotected management frames, including the Beacon, and Probe Response frames, and control frames will still be potential targets for an attacker.
- Maximum privileges on victim systems: In the Intel processor architecture model used by modern Microsoft Windows operating systems and Mac OS X, the operating system uses split privilege levels for core OS components (the kernel and hardware drivers) and for user space applications (office productivity applications, web browsers, etc.) This architecture is commonly known as the Intel processor ring architecture, where core OS components run with the greatest privilege level (ring 0) and userspace applications run in the least privilege level (ring 3). Most operating systems do not make use of the varied privilege ring 1 or ring 2 levels.

This split privilege architecture is advantageous to users, since user space applications only have a limited set of privileges when interacting with the system. A user space application crash only crashes the application itself and does not affect other operating system components. Core operating system components run with the greatest system privilege levels in order to have access to the system hardware. Because wireless drivers run in ring 0, they have complete access to the operating system components and all hardware devices. This privilege level exceeds even that of the OS "administrator" account, allowing any code execution supplied by the attacker to circumvent popular client security mechanisms including firewalls, anti-virus scanners and host-based intrusion prevention systems.

In order to take advantage of software flaws, an attacker must first discover a vulnerability to exploit, or take advantage of a known vulnerability. The next section describes how attackers identify wireless driver vulnerabilities.

7.7.2 Discovering Driver Vulnerabilities

Before being able to exploit a client driver, the attacker must first identify a vulnerability. Typically, attackers and researchers will develop their own lab environments, potentially mirroring the network they are targeting and wish to compromise, and will leverage their own clients to discover and exploit vulnerabilities before attempting to exploit their desired target. One vulnerability discovery technique is known as protocol fuzzing (fuzzing).

Fuzzing is a well-known technique in the research and attack communities, where malformed input is created and sent to a target software process. The researcher looks for and may intend for the process to fail by crashing, noting the data that was used to cause the crash. Often, the ability to crash a software process gives the attacker the ability to execute arbitrary code on the vulnerable system.

Fuzzing begins with no preconceived notions on how the target software should operate. By contrast, an engineer developing software to implement a protocol designs the software to comply with the guidelines set forth (often in a standards document). Any assumptions by the developer represent opportunities for an attacker to take advantage of the system.

The use of fuzzing tools has been a popular technique in the discovery of vulnerabilities in software, targeting many different protocols such as ISAKMP, Windows RPC, SNMP, FTP, image file formats and many others. Fuzzing tools have recently started to gain commercial appeal for software vendors who wish to test the robustness of their software products with offerings from companies such as Mu Security [34] and Codenomicon [35].

When using fuzzing techniques, the attacker or researcher selects the target software to evaluate and uses targeted or focused fuzzing with selected aspects of the protocol. In IEEE 802.11 wireless driver fuzzing, the researcher may use a focused approach and limit fuzzing tests to a specific frame type, or even a specific element of a selected frame type. Alternatively, the researcher may adopt a broad fuzzing approach, supplying random information in the payload and even the header of an IEEE 802.11 frame before sending it to the target system. The application of focused or broad fuzzing techniques is examined later in this section.

The tools for wireless device driver fuzzing are currently a mix of public and privately-developed tools. One example of a public 802.11 fuzzing tool is fuzz-e developed by Jon Ellch [36]. The fuzz-e tool is an example of a broad 802.11 fuzzing tool, supplying random information in the payload and header of IEEE 802.11 management frames. Fuzz-e also includes a mechanism to assert the availability of the target system to identify when one or more malformed frames cause the target system to crash, allowing the researcher to run this tool in an unattended fashion.

With fuzz-e, the adversary does not need to have a detailed understanding of the IEEE 802.11 MAC layer. With a target device and a fuzzing client, the researcher runs fuzz-e with the desired parameters and waits until fuzz-e has caused the wireless driver software to fail.

Another tool that can be used for fuzzing is file2air [37]. File2air takes an input binary file representing the contents of an IEEE 802.11 frame and sends it to the identified target. While not written as a fuzzing tool, file2air transmits any data that is identified, so a researcher could create several binary file "test cases" representing focused fuzzing targets.

Figure 7.15 illustrates an example of an IEEE 802.11 Probe Response frame that has been modified with an unusually long supported data rates field, where the data in the supported rates tagged parameter is a series of 0x41 ("A") values. Using file2air, this frame can be transmitted to a specified destination as shown in Figure 7.16.

A third tool used for wireless driver fuzzing is Scapy [38], an extension to the Python scripting language. As an option to bridge the gap between the overly broad targeting used by the fuzz-e tool and the very narrow targeting used in file2air, a researcher can quickly develop Python scripts to create and transmit IEEE 802.11 wireless frames with arbitrary content, exploring how a target station reacts to malformed input. A sample Python/Scapy 802.11 fuzzing script is shown in Figure 7.17.

```
# ./fuzz-e -P rausb0 -A -T 0 -S 5 -i wifi0 -f pcap-out.dump -c
0 -R -E logging.txt -D dest-addys.txt
fuzz-e <johnycsh@gmail.com>
Reading in destination addys.
00:13:CE:55:98:EF
----fuzz-e-cfg summary----
Autonomous mode:  1
type value:       0
subtype value:    5
random times:     1
DestFilename      dest-addys.txt
Event Log         logging.txt
Num Hosts         1
                  00:13:CE:55:98:EF
-------------------------
00:13:CE:55:98:EF maps to 172.16.0.108
PING 172.16.0.108 (172.16.0.108) 56(84) bytes of data.
From 172.16.0.110 icmp_seq=1 Destination Host Unreachable

--- 172.16.0.108 ping statistics ---
1 packets transmitted, 0 received, +1 errors, 100% packet loss,

ping returned 256
Host [00:13:CE:55:98:EF / 172.16.0.108] is Down.
```

Figure 7.14: Probe Response fuzzing with fuzz-e.

```
$ xxd proberesp.bin
0000000: 5000 3a01 0020 a64f 0140 0011 926e cf00  P.:.. .O.@...n..
0000010: 0011 926e cf00 2004 495f 5159 3200 0000  ...n.. .I_QY2...
0000020: 6400 2104 0007 4c45 4150 4e45 5401 0841  d.!...LEAPNET..A
0000030: 4141 4141 4141 4141 4141 4141 4141 4141  AAAAAAAAAAAAAAAA
0000040: 4141 4141 4141 4141 4141 4141 4141 4141  AAAAAAAAAAAAAAAA
0000050: 4141 4141 4141 4141 4141 4141 4141 4141  AAAAAAAAAAAAAAAA
0000060: 4141 4141 4141 4141 4141 4141 4141 4141  AAAAAAAAAAAAAAAA
0000070: 4141 4103 010b                           AAA...
$ ▌
```

Figure 7.15: Hex interpretation of a malformed Probe Response frame.

```
# file2air -i ath6 -t -n 1000 -r madwifing -f proberesp.bin -d 00:13:ce:55:98:ef
file2air v1.0RC4 - inject 802.11 packets from binary files <jwright@hasborg.com>
Transmitting packets ... Done
# []
```

Figure 7.16: Using file2air to transmit the proberesp.bin file contents.

In this Python script, the Scapy fuzz() function supplies random information for the input values that are not specified. Used within the while() loop, the script will use the well-formed packet options specified in the "basep" packet object (consisting of the IEEE 802.11 header, the Probe Response frame fixed payload, the SSID IE and the DS parameter set IE), appending randomized information selected in the fuzz() function for the supported data rates IE. Looping after 20 packets (after a delay or .1 seconds for each packet), this script will continue to supply random information in the supported data rates IE, sending the Probe Response frames to the specified destination MAC address.

```
#!/usr/bin/python
# Import the sys and scapy modules needed for various functions
import sys
from scapy import *

target = "00:09:5B:64:6F:23"
ap = "00:40:96:01:02:03"
conf.iface = "wlan0"

# Base packet object "basep". Creating a standard 802.11 frame
# and specify the content
basep = Dot11(
    proto=0, type=0, subtype=5,        # Probe response frame
    addr1=target, addr2=ap, addr3=ap,  # sent to target from AP
    FCfield=0, SC=0, ID=0)             # other fields set to 0

# Append the probe response fixed payload contents next
basep /= Dot11ProbeResp(
    timestamp = random.getrandbits(64),    # Random BSS timestamp
    beacon_interval = socket.ntohs(0x64),  # byte-swap BI, ~.10 sec
    cap = socket.ntohs(0x31))             # AP/WEP/Short Preamble

# Next, add an Information Element, fixed SSID conent
ssid = "fuzzproberesp"
basep /= Dot11Elt(ID=0, len=len(ssid), info=ssid)

# Append the DS parameter set IE, channel 1
basep /= Dot11Elt(ID=3, len=1, info="\x01")

while 1:
    # Fuzzing on the supported rates IE. Use the basep packet
    # repeatedly, changing the supported rates IE after 20 packets
    tmpp = basep
    # fuzz() fills in missing arguments with random data
    tmpp /= fuzz(Dot11Elt(ID=1))
    # Send a packet every 1/10th of a second, 20 times
    sendp(p, count=20, inter=.1)
```

Figure 7.17: Sample Python script using fuzzing on the data rates parameter.

Using Python and Scapy for 802.11 driver fuzzing requires some knowledge of the Python scripting language, but allows the researcher to quickly develop scripts that can be broad or narrow in fuzzing scope. Due to the unpredictable nature of the random information that is generated and sent to the target station, the fuzzing process will typically be accompanied by a packet capture process that is stopped when the fuzzing process produces a target crash. After the crash, the researcher can review the packet capture logs to identify the exact content that caused the crash.

Another mechanism used in identifying driver vulnerabilities is to apply reverse-engineering techniques to the driver code. Reverse-engineering tools examine the binary driver and represent it in the native assembly language instructions that represent the driver functionality. This process gives the researcher the ability to inspect the functionality of the driver to identify potential bugs or vulnerable function calls used throughout the code.

Several popular tools are available to aid the researcher in analysis of driver software instructions, including the IDA Pro disassembler [39]. Using IDA Pro, an attacker can automate some analysis activities to look for vulnerable driver functions. Figure 7.18 is an example of the IDA Pro disassembler used for the analysis of a buffer overflow when handling a SSID IE with a SSID longer than 32 bytes.

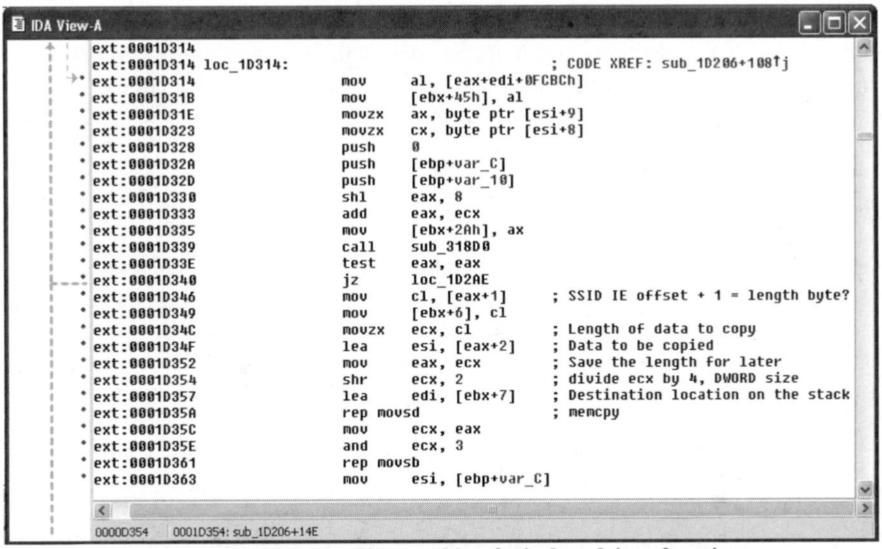

Figure 7.18: IDA Pro disassembly of wireless driver function.

Once a crash is identified, the crash result information can be investigated to identify if it is possible to turn the crash into a more useful exploit for the attacker. This is not a trivial process, and often requires many hours of experimentation and analysis before developing an exploit that can be used against a vulnerable station. Once the exploit has been developed, however, it can be used by any adversary and integrated into generic exploit frameworks.

7.7.3 Exploiting Driver Vulnerabilities

As of this writing, the majority of wireless driver exploits have been integrated into the Metasploit [40] framework. Metasploit is a framework to leverage software exploits with a variety of payloads and encoding options to aid in system penetration testing. While it takes significant skill to develop exploits discovered through driver fuzzing or reverse engineering analysis, the integration of the exploit into Metasploit makes it very straightforward to compromise vulnerable systems.

Figure 7.19 demonstrates an example of Metasploit using a Windows driver exploit targeting a popular driver. Once an exploit is selected with the Metasploit "use" command, the attacker can select from a variety of payloads supplied by Metasploit. The payload represents the code that will be executed on the target system once it is exploited. Metasploit includes over 100 different payloads that can be used with almost any exploit, including the ability to open a listening port returning a Windows CMD shell on the target host, to uploading and starting a remote console management interface using the Virtual Network Computing (VNC) protocol. In Figure 7.18, Metasploit is configured to use the Windows/Adduser payload, which creates a new user with Administrator privileges on the target system. The INTERFACE and DRIVER parameters are used to identify the wireless card on the attacker's system that will deliver the exploit; PASS is used to specify the password for the attacker's account.

Using Metasploit, an attacker can take advantage of vulnerable systems using the exploits and payloads available in this framework. This increases the risk to organizations that have not updated vulnerable drivers, since it allows a significantly larger population to take advantage of these vulnerabilities, as opposed to the population who is capable of developing the exploit.

7.7.4 Mitigating Driver Vulnerabilities

While the remedy for addressing driver exploits (namely, updating vulnerable drivers to fixed versions) is simple, implementing it is often difficult. Organizations with a mixed environment of hardware may have to address driver vulnerabilities for multiple wireless device providers. In some cases, even the same hardware devices may have different internal wireless cards, making it difficult to assess the vulnerability posture of the network.

Another concern for sufficiently mitigating driver vulnerabilities is the need to stay actively informed as to the status of vulnerable driver vendors and driver versions. In some cases, wireless card manufacturers are reluctant to widely disclose serious vulnerabilities in a product, which may make it difficult for an organization to assess vulnerabilities. One resource that attempts to make this information openly and widely available is the Wireless Vulnerabilities and Exploits (WVE) [41] project.

WVE is a vendor-neutral project aiming to clearly identify and classify wireless-related vulnerabilities, including software flaws. By monitoring the resources at www.wve.org, organizations can stay abreast of new vulnerabilities in wireless network as they are discovered.

```
 --------------
< metasploit >
 ------------
        \   ,__,
         \  (oo)____
            (__)    )\
               ||--|| *


        =[ msf v3.0-beta-dev
+ -- --=[ 178 exploits - 104 payloads
+ -- --=[ 17 encoders - 5 nops
        =[ 30 aux

msf > use windows/driver/b_____wifi_ssid
msf exploit(_____wifi_ssid) > set PAYLOAD windows/adduser
PAYLOAD => windows/adduser
msf exploit(_____wifi_ssid) > set INTERFACE wifi0
INTERFACE => wifi0
msf exploit(_____wifi_ssid) > set DRIVER madwifing
DRIVER => madwifing
msf exploit(_____wifi_ssid) > set PASS moo
PASS => moo
msf exploit(_____wifi_ssid) > exploit
[*] Sending beacons and responses for 60 seconds...
```

Figure 7.19: Metasploit example using a wireless driver exploit.

Once a vulnerability in a wireless driver is known, an organization must enumerate all the wireless drivers in use throughout their network to identify vulnerable stations. A free tool designed to help organizations in this process is the WiFi Driver Enumerator (WiFiDEnum) [42].

WiFiDEnum accepts a range of IP address targets, and connects to Windows hosts with administrator credentials to access the system registry. By examining registry information, WiFiDEnum identifies all the wireless drivers installed on the host, and correlates the version information to a database of known driver vulnerabilities, reporting drivers that are vulnerable to known exploits. Using WiFiDEnum, an authorized administrator can scan all Windows hosts on the network to identify vulnerable systems. The results of a single-host scan are shown in Figure 7.20.

Once a scan is complete, WiFiDEnum can generate a report identifying all the vulnerable drivers with links to information on how to address the vulnerabilities, as shown in Figure 7.21. WiFiDEnum is available at http://labs.arubanetworks.com.

Identifying wireless driver vulnerabilities on wireless client devices is currently a fruitful avenue for attackers who wish to take advantage of wireless LAN environments. These techniques are not only applicable to IEEE 802.11 wireless LAN environments, and several exploits have been identified that take advantage of Bluetooth wireless environments [43]. It is expected that this trend will continue into other wireless environments such as WiMAX and WUSB.

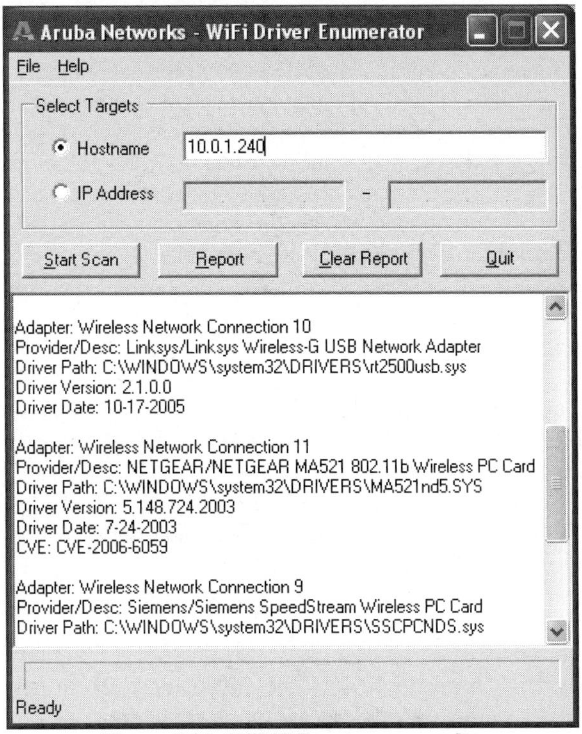

Figure 7.20: WiFiDEnum scan results.

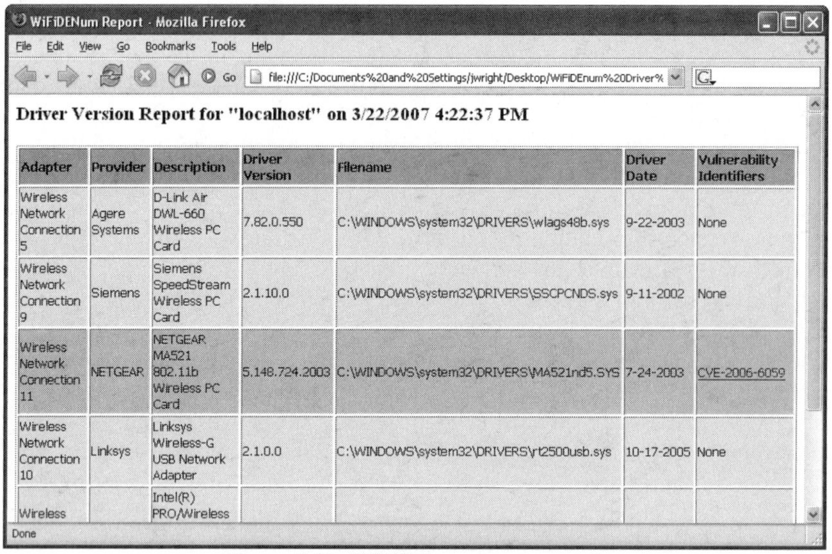

Driver Version Report for "localhost" on 3/22/2007 4:22:37 PM

Adapter	Provider	Description	Driver Version	Filename	Driver Date	Vulnerability Identifiers
Wireless Network Connection 5	Agere Systems	D-Link Air DWL-660 Wireless PC Card	7.82.0.550	C:\WINDOWS\system32\DRIVERS\wlags48b.sys	9-22-2003	None
Wireless Network Connection 9	Siemens	Siemens SpeedStream Wireless PC Card	2.1.10.0	C:\WINDOWS\system32\DRIVERS\SSCPCNDS.sys	9-11-2002	None
Wireless Network Connection 11	NETGEAR	NETGEAR MA521 802.11b Wireless PC Card	5.148.724.2003	C:\WINDOWS\system32\DRIVERS\MA521nd5.SYS	7-24-2003	CVE-2006-6059
Wireless Network Connection 10	Linksys	Linksys Wireless-G USB Network Adapter	2.1.0.0	C:\WINDOWS\system32\DRIVERS\rt2500usb.sys	10-17-2005	None
Wireless		Intel(R) PRO/Wireless				

Figure 7.21: WiFiDEnum scan report.

While attackers are currently focused on exploiting wireless client vulnerabilities as long as these activities continue to identify vulnerable and exploitable client systems, the techniques are not limited to identifying client vulnerabilities. It is assumed that attackers will also use these same techniques to identify and exploit vulnerabilities on access points as well. Further, it is conceivable for an attacker to exploit other devices on the network, such as RADIUS servers providing EAP authentication, by investigating and targeting vulnerabilities that may exist in the handling of malformed EAP frames. Many of the advantages of attacking client systems still apply to attacks against the EAP exchange, including the ability to communicate with an organization's RADIUS server with no authentication credentials for IEEE 802.1X environments.

7.8 Wireless Intrusion Detection

In reaction to the stream of vulnerabilities and exploit tools designed to compromise the integrity, availability and confidentiality of wireless LAN environments, many vendors have produced wireless LAN intrusion detection systems (WIDS), also known as wireless LAN intrusion prevention and detection systems (WIPDS). These systems can provide significant value to organizations, accommodating an additional layer of security through detection and attack countermeasures techniques.

7.8.1 Deployment Models

Several approaches have been applied to the deployment of WIDS systems, offering organizations flexibility in the selection and implementation of WIDS solutions. Two models have emerged as the dominant deployment models as a network overlay, and as an integrated solution.

7.8.1.1 WIDS Overlay Deployment Model

In an overlay deployment model, the organization deploys dedicated "air sensors" that passively collect traffic from one or more wireless interfaces. This approach gives organizations visibility into the events on the wireless spectrum within range of any deployed wireless sensor. In this model, the vendor supplying the wireless sensors and supporting architecture is not an integral part of the wireless network and the sensors do not service any legitimate client devices; rather, the sensors passively listen to the events on the wireless network and use various analysis techniques to identify attacks.

The overlay deployment model has several advantages for an organization, including the ability to deploy a monitoring solution that is separate from the wireless transport provider. In this model, if a vulnerability is discovered in the transport network implementation equipment, the organization can leverage the overlay WIDS product to monitor for the vulnerability. Overlay deployment models are also useful in the enforcement of "no-wireless" environments, allowing organizations to monitor for the presence of unauthorized wireless devices. Often, overlay solutions offer the most comprehensive analysis capabilities, since the vendors offering these systems focus solely on WIDS features without the need to also offer IEEE 802.11 transport capabilities.

A disadvantage of the overlay deployment model can be cost; organizations must deploy wireless sensor devices in all areas that require WIDS monitoring, which can be costly in terms of hardware and supporting infrastructure (cabling, power over Ethernet adapters and wired switch ports). The lack of integration with the transport network can constrain the analysis capabilities of the overlay model, since the overlay WIDS system can only assess layer 1 and layer 2 traffic characteristics without knowledge of dynamic encryption keys.

7.8.1.2 WIDS Integrated Deployment Model

In an integrated deployment model, the organization relies on the same devices that are supplying wireless access (access points or APs) to also perform intrusion detection services. In this model, the access point operates as both a transport provider, and as a WIDS sensor, identifying and reporting attacks.

The integrated deployment model is often a more cost-effective solution for organizations that are already deploying WLAN environments, since existing devices can also perform WIDS analysis. As an integrated part of the IEEE 802.11 network, this deployment model has knowledge of dynamic encryption keys, allowing it to analyze not only layer 1 and layer 2 traffic characteristics that are unencrypted, but also to analyze upper-layer protocol characteristics as well.

A disadvantage of the integrated deployment model is the lack of resources the integrated device has to dedicate to WIDS analysis. Unlike the overlay model, the integrated model must provide service and analysis capabilities, restricting the AP's ability to devote available CPU and memory resources to monitoring tasks. An AP servicing users is also constrained to monitoring the frequency for which it is servicing users. This prevents the integrated model AP from scanning other frequencies as freely as the overlay device that does not provide AP service to users.

It should also be noted that some vendors implement hybrid deployment models, leveraging deployed AP's for monitoring in the integrated model, while offering dedicated sensors that can be deployed in the overlay model.

7.8.2 Analysis Techniques

Regardless of the deployment model, WIDS systems have several mechanisms, including signature analysis, trend analysis and anomaly analysis that can be used to identify attacks against the wireless network.

7.8.2.1 Signature Analysis

Signature analysis is a common technique used in nearly all intrusion detection systems where the analysis units identify predefined patterns or signatures in wireless traffic that indicates the presence of an attack.

For example, consider the attack tool ChopChop [44]. Designed to exploit weaknesses in the WEP Integrity Check Validation (ICV) mechanism, ChopChop allows an attacker to decrypt WEP frames one byte at a time by using the AP as a decoder, repeatedly transmitting malformed frames until the AP responds with a valid frame. In order to manipulate the AP into transmitting a frame, ChopChop transmits the manipulated frame with a multicast destination address, as shown in Figure 7.22.

In frame 869, we see the attacker impersonating the legitimate station at 00:04:23:63:88:d7, using the destination MAC address ff:2d:8d:24:bc:15. While this is a legal multicast MAC address, it is unusual for multicast addresses to begin with 0xff. In frames 817 and 873, the attacker modifies the 6th octet of the destination address with the values 0x0a and 0x0b. In frames 875 and 877, the 5th octet of the MAC address as well as the 6th octet of the MAC address in what appears to be a sequential pattern.

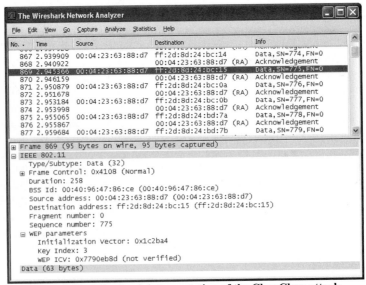

Figure 7.22: Wireshark interpretation of the ChopChop attack.

Using signature analysis techniques, a WIDS system could generate an alert with the following conditions:

- First byte of the destination MAC address is 0xff;
- Frame type is 1 (data) and subtype is 0 (data);
- Flags byte is 0x41 (protected, To DS bits set);
- WEP initialization vector is present.

This signature can be expressed in a Wireshark display filter, as shown in Figure 7.23.

```
wlan.da[0:1] eq ff && wlan.fc.tods eq 1 && wlan.fc.fromds eq 0 &&
wlan.fc.type_subtype eq 32 && wlan.wep.iv
```
Figure 7.23: Wireshark display filter to identify ChopChop activity.

Using this method of signature analysis, a WIDS system could easily identify this attack, while likely mitigating false-positive events. This highlights the major benefit of signature analysis; that signature analysis requires few resources to identify an attack, and signatures can be developed rapidly.

Unfortunately, while signature analysis is a useful feature to identify attacks where the adversary is using an unmodified attack tool, a cautious attacker may be successful in evading the attack with simple modifications to the attack tool. Figures 7.24 and 7.25 show the original source and a minor modification to ChopChop, respectively.

```
481     // prepare the dmac
482     randomMAC(dmac);
483     if (smac) dmac[0]=0;              // unicast
484     else dmac[0] = 0xff;             // multicast
485     dmac[1] = dmac[1] & 0x7f;        // not ff:ff:ff:ff:ff:ff
```
Figure 7.24: Source code excerpt from chopchop.c, original.

```
481     // prepare the dmac
482     randomMAC(dmac);
483     if (smac) dmac[0]=0;              // unicast
484     else dmac[0] |= 0x10;            // multicast
485     dmac[1] = dmac[1] & 0x7f;        // not ff:ff:ff:ff:ff:ff
```
Figure 7.25: Source code excerpt from chopchop.c, modified.

In Figure 7.24, line 484 sets the first octet of the destination MAC address to 0xff, causing the frame to be recognizes as multicast traffic, as we observed in the Wireshark capture displayed in Figure 7.22. Figure 7.25 changes this line to retain the leading random MAC address byte generated in line 482, but sets the most-significant bit which also marks this frame as multicast data. Using this trivial modification to the attack tool, an adversary could evade simple signature analysis mechanisms.

7.8.2.2 Trend Analysis

Another analysis mechanism commonly used by WIDS systems is trend analysis, or analyzing events on the wireless network over a period of time. Using trend analysis, a WLAN IDS analyst can apply signature-based analysis mechanisms to each event that is observed by the monitoring agent, raising an alert when a trend of events is characterized. For example, consider the packet trace information presented in Figure 7.26 generated with the fakeap tool [45].

```
    1         3.118933    00:90:96:8d:af:8a    ->    ff:ff:ff:ff:ff:ff    Beacon
frame,SN=2046,FN=0,   BI=100,   SSID:   "trude"    "Current   Channel:   5"
wlan_mgt.fixed.timestamp == "0x0000000000019276"
    2         5.219032    00:40:ae:76:4d:bc    ->    ff:ff:ff:ff:ff:ff    Beacon
frame,SN=2071,FN=0,   BI=100,   SSID:   "Lilaea"    "Current   Channel:   4"
wlan_mgt.fixed.timestamp == "0x00000000000192E4"
    3         7.411628    00:04:76:ed:64:91    ->    ff:ff:ff:ff:ff:ff    Beacon
frame,SN=2091,FN=0,   BI=100,   SSID:   "urim"    "Current   Channel:   2"
wlan_mgt.fixed.timestamp == "0x00000000000323BD"
    4        11.466788    00:80:0f:bc:f6:fc    ->    ff:ff:ff:ff:ff:ff    Beacon
frame,SN=2128,FN=0,   BI=100,   SSID:   "costly"    "Current   Channel:   4"
wlan_mgt.fixed.timestamp == "0x00000000000192D3"
    5        13.535755    00:40:96:6d:1b:f3    ->    ff:ff:ff:ff:ff:ff    Beacon
frame,SN=2147,FN=0,   BI=100,   SSID:   "prometheus"    "Current   Channel:   1"
wlan_mgt.fixed.timestamp == "0x00000000000192B5"
```

Figure 7.26: TShark interpretation of fakeap attack.

This trace displayed with the TShark tool indicates the presence of several unique BSSs, each with a unique network SSID. Careful inspection of this trace reveals that the even though each of the beacon frames are reportedly from unique source MAC addresses, the BSS timestamp is consistently a very small value. Since the BSS timestamp is a 1 microsecond counter that indicates how long the AP has been online, the only legitimate case for this packet capture is for each of these AP's to have been online less than .2 seconds (frame 3).

This behavior is the result of the fakeap tool, designed to impersonate a list SSID names with random MAC addresses, signal levels and channels. With signature analysis, it would be possible to write a rule to flag any AP's with a very small BSS timestamp, but this technique would generate multiple false-positives each time an AP rebooted and naturally reset the BSS timestamp to zero. Through using trend analysis however, the WIDS analyst can raise an alert when multiple unique BSSs are identified with very small BSS timestamp values, preventing excessive false-positives.

While trend analysis is useful for WLAN attack analysis, it must be implemented carefully to avoid attacks that target the WIDS system itself. Consider the packet trace in Figure 7.27.

In this example, we see a steady stream of Beacon frames from the same source address (with the exception of frame 21, which is advertising an alternate SSID). We can further determine that the standard Beacon interval has been applied (BI=100) indicating that only 10 beacons should be transmitted per second. Matching the Beacon interval advertisement to the time distribution pattern for these frames indicates that more beacons are being transmitted than is otherwise appropriate for this network, which would be grounds to raise an alert.

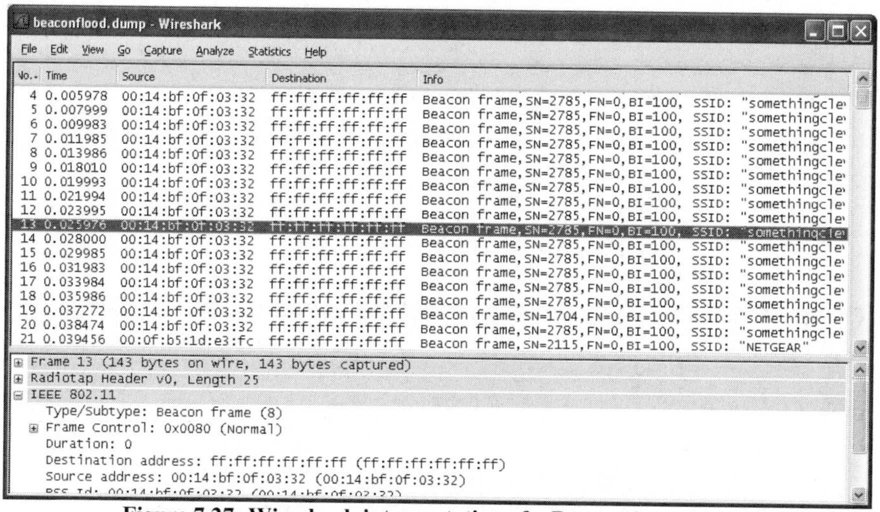

Figure 7.27: Wireshark interpretation of a Beacon frame flood.

When implementing an alert of this type, the analyst may choose to allocate a block of memory for each unique BSSID to track the advertised Beacon interval versus the number of actual frames observed. An attacker could choose to try to exhaust memory resources on the WLAN monitoring agent by flooding the network with Beacon frames, each having a different BSSID address and requiring an individual portion of memory for trend analysis. WLAN IDS vendors must use caution not to expose themselves to resource exhaustion attacks in these cases that could give the adversary the ability to render the monitoring system ineffective.

7.8.2.3 Anomaly Analysis

A third technique that is valuable for WIDS systems is the use of IEEE 802.11 anomaly analysis. Using a baseline of known behavioral characteristics for wireless devices, anomaly analysis allows organizations to obtain one method of 0-day or "unknown attack" detection capabilities by identifying traffic characteristics that deviate from normal operating behavior.

For example, several popular consumer wireless cards have been show to be flawed in the handling of the Supported Rates information element. The format of this information element is described in [8] section 7.3.2.2, where the information field is between 1 and 8 bytes in length. Consider the packet capture shown in Figure 7.28, generated with the Metasploit 3.0 Framework.

Using this technique, researchers have discovered that is it possible to compromise vulnerable devices with a long supported rates IE field. Through IEEE 802.11 anomaly analysis techniques, it was possible to identify this attack before this vulnerability was discovered by monitoring for inappropriate use of this information element.

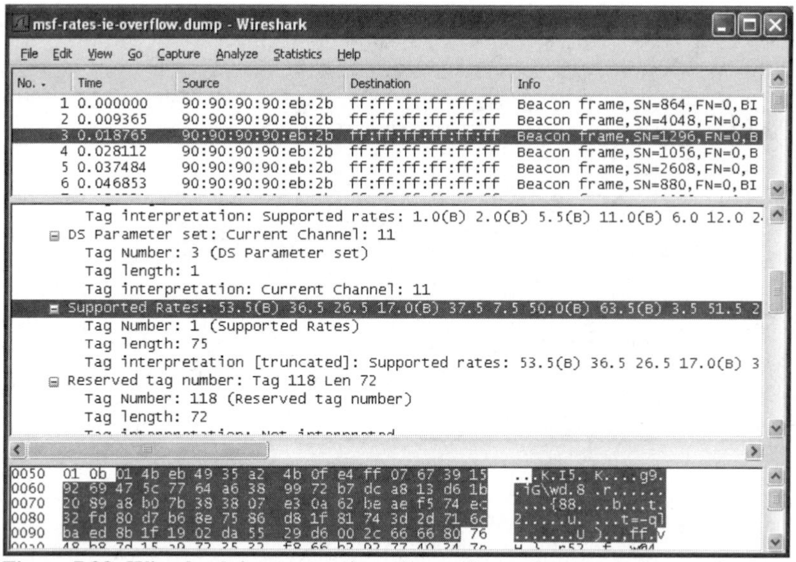

Figure 7.28: Wireshark interpretation of a malformed supported rates element.

While anomaly analysis can be beneficial for unknown attack detection, it does little to help the organization reacting to the alert to assign a criticality index to the event. Since the WIDS system does not recognize the event as a particular kind of attack, it can do little to describe the potential impact to the organization with any degree of confidence. WIDS analysts must perform their own analysis with whatever information if available on the targeted device and the information provided by the WIDS system to evaluate the criticality of the event.

7.8.3 Upper-Layer Analysis Mechanisms

An emerging technique in improving the quality of WIDS analysis mechanisms is the integration of network-based IDS (NIDS) systems WIDS capabilities. Historically, WIDS systems have predominately focused on analyzing wireless events from users outside of the organization, leaving insider attack analysis to upstream NIDS devices such as Snort [46]. While NIDS systems provide tremendous analysis capabilities to identify attacks on the wire, they are seldom positioned to be able to characterize attacks that happen between users on the wireless network.

Figure 7.29 illustrates a common deployment example of using a NIDS system to monitor activity on WLAN. By configuring the network switch to forward a copy of all network activity to the NIDS system, all traffic going to the upstream network through the network switch will be mirrored and assessed by the NIDS. However, this monitoring mechanism does not accommodate monitoring traffic that remains within the AP that is shared by the attacker and other users. By directing upper-layer attacks to other users on the same AP, the attacker can evade the NIDS system and preserve a significant degree of stealth in the attack.

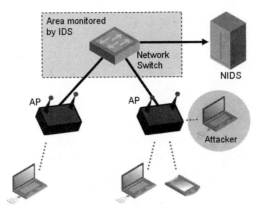

Figure 7.29: NIDS Monitoring Deployment Detection Example.

The need for integrated WIDS and NIDS monitoring systems is becoming more apparent as wireless attacks begin to cross functional boundaries between layer 2 and layer 3. For example, the KARMA attack [47] manipulates client systems by responding to all Probe Request frames regardless of the desired SSID, allowing an attacker to impersonate all access points within range of the selected channel. Once the victim station roams to the KARMA attacker, the attacker establishes a "fake" network environment, responding on behalf of any requested server for any service in an attempt to collect sensitive authentication credentials. KARMA is also deployed with the Bring Your Own Exploits (BYOX) model, where an attacker can automate the process of targeting various exploits against victim systems.

In the KARMA attack, the victim station roams from a legitimate AP to the KARMA attacker. This process may appear to a WIDS system a simple network roaming event, since the WIDS system is not inspecting upper-layer traffic patterns that characterize the true nature of the network impersonation and BYOX attacks launched by KARMA. A NIDS system would also be unaware of this attack, since the traffic remains on a wireless environment and does not traverse a boundary monitored by a NIDS sensor.

In order to adequately address insider wireless attacks and attacks that cross functional boundaries between layer 2 and layer 3 environments, some vendors are deploying integrated monitoring solutions where the WIDS system is able to communicate with NIDS systems for comprehensive monitoring capabilities. One example of this model is the integration of the Snort NIDS system with the Aruba Networks Mobility Controller (MC) architecture [48]. In this deployment model, the MC communicates with a designated Snort sensor after inspecting and decrypting all wireless traffic through the knowledge of dynamic encryption keys used on the network. Using the Snort alert_aruba_action output plugin [49], the NIDS administrator can designate events that characterize unauthorized activity using the Snort rules language, and communicate back to the MC to manipulate wireless user network privileges accordingly.

7.8.4 Wireless Countermeasures

Many WIDS providers also provide one or more mechanisms to mitigate the effectiveness of attacks on the WLAN. Commonly referred to as wireless countermeasures, or wireless intrusion prevention services, these features are often attractive to organizations since they can automate the process of responding to attacks to minimize the impact to organizations.

While each vendor will characterize their wireless countermeasure techniques as unique in their own perspective, common implementations include adversary denial of service attacks and role-based access control measures.

7.8.4.1 Adversary Denial of Service

Early in the history of attacks against 802.11 wireless networks, attackers identified weaknesses in the handling and verification of 802.11 management and control frames. This allows attackers to implement numerous denial of service (DoS) attacks against wireless networks, often impersonating as a legitimate station or access point by transmitting spoofed frames. A brief summary of IEEE 802.11 DoS attacks is summarized in Table 7.1.

Table 7.1: Summary of common IEEE 802.11 DoS attacks.

Name	Description
RF Jamming	A basic technique using commodity RF jammers or even popular IEEE 802.11 wireless cards to interfere with the legitimate use of shared spectrum space.
Deauthenticate, Disassociate Flood	An early technique where the attacker sends spoofed Deauthenticate or Disassociation frames on behalf of the AP or a victim station, causing the recipient to believe the source has disconnected from the wireless network.
Associate Flood	A technique where the adversary spoofs association request frames from random station addresses, attempting to force the AP to run out of association identifiers, forcing it to stop servicing new clients.
Network Allocation Vector Reservation Flood	Leverages request-to-send or fragmented data frames to reserve the wireless medium for the maximum duration value (32,767 μsec), forcing other stations to wait for the reservation period to end before transmitting. An attacker who transmits these frames at a rate greater than 31 frames per second can sustain the attack and prevent all users from accessing the medium.

While these attacks are well-known to the IEEE 802.11 Working Group, and will be partially addressed with the IEEE 802.11w amendment, WIDS vendors have been able to detect these attacks and prevent adversaries from using unauthorized network resources. The deauthenticate and/or disassociate flood DoS attack is commonly implemented as a rogue AP countermeasure technique. In this model, a WIDS sensor that characterizes an unauthorized AP that is connected to the organization's LAN can mark the device as a rogue AP. When an unauthorized user attempts to access the rogue AP, the WIDS system impersonates the rogue AP and possibly the unauthorized station and sends deauthenticate messages to both parties, forcing them to disconnect from the network. Sustaining this flood of deauthenticate frames prevents the unauthorized station from accessing the network through the rogue device.

This mechanism can be a valuable feature for organizations; however, it can also disclose sensitive information about the nature of the WIDS system.. Once a passive analysis mechanism, the transmission of deauthenticate frames allows an attacker to characterize the vendor selected for WLAN IDS monitoring through the use passive device fingerprinting techniques [50]. If an attacker is able to identify the vendor who supplies

WIDS services, the attacker may be able to modify their attacks in such a way to avoid detection, or to attack the WIDS service itself.

7.8.4.2 Role-Based Access Control Measures

A distinct advantage for integrated deployment model vendors is the ability to implement role-based access control mechanisms for wireless users. Since the integrated WIDS vendor is also servicing the client users, it is in a position to grant or deny any network resources as designated by the network administrator, or to apply policies such that access is revoked if unauthorized activity is detected. This can be powerful WIPS mechanism for organizations, allowing an enterprise network to design roles that grant only the necessary access privileges for the applications in use.

When deploying role-based access control measures, however, organizations should be cautious not to expose themselves to DoS attacks from malicious activity. If an attacker is able to identify the conditions that trigger a privilege revocation event (such as using an unauthorized file-sharing application), they may leverage this condition to sustain a DoS attack against multiple users. This is mostly a concern for open wireless networks, where any unauthorized attacker can impersonate legitimate stations without knowledge of encryption keys used on the network.

WIDS systems can provide a tremendous degree of value to organizations for monitoring and reacting to attacks against IEEE 802.11 WLANs. By themselves, WIDS systems cannot successfully mitigate deficiencies in open networks or weak encryption mechanisms, but they can provide a well-rounded security foundation to a strong wireless authentication and encryption architecture.

7.9 References

[1] N. Borisov, I. Goldberg, and D. Wagner, "802.11 Security," http://www.isaac.cs.berkeley.edu/isaac/wep-faq.html
[2] J. Walker, "Unsafe at Any Key Size: An Analysis of the WEP Encapsulation," November 2000.
[3] S. Fluhrer, S. Mantin, and A. Shamir, "Weaknesses in the Key Scheduling Algorithm of RC4," Eighth Annual Workshop on Selected Areas in Cryptography (August 2001).
[4] http://aircrack-ng.org
[5] http://www.wi-fi.org
[6] Standards for Local and Metropolitan Area Networks: Port-Based Network Access Control, International Standard ISO/IEC, IEEE P802.1X, April 2001.
[7] Aboba, B. et al, "Extensible Authentication Protocol (EAP)", RFC 3748, June 2004.
[8] Standard for Information Technology – Telecommunications and information exchange between systems – Local and metropolitan are networks – Specific Requirements. Part 11: Wireless LAN Medium Access Control (MAC) and Physical Layer (PHY) specifications, IEEE Std 802.11-2007.
[9] R. Housley and D. Whiting, "Temporal Key Hash," IEEE 802.11-01/550.

[10] R. Housley, D. Whiting, and N. Ferguson, "Alternative Temporal Key Hash," IEEE 802.11i, 11-02-282r0.

[11] T. Moore and C. Chaplin, "TGi Security Overview," IEEE 802.11i, 11-02-114r1.

[12] N. Ferguson, "Michael-an-improved-MIC-for-802.11-WEP," IEEE 802.11i, 11-02-020r0.

[13] V. Moen, H. Raddum, and K.J. Hole, "Weakness in the Temporal Key Hash of WPA," ACM SIGMOBILE Computing and Communications Review, Vol 8, Issue 2, ACM Press, pp. 76-83, 2004.

[14] http://www.esat.kuleuven.ac.be/~rijmen/rijndael

[15] http://www.nist.gov

[16] R. Housley, D. Whiting, and N. Ferguson, "AES-CTR-Mode-with-CBC-MAC," 80211-02-001r1.

[17] O. Letanche and D. Stanley, "Proposed_TGi_Clause_8_AES-CTR_CBC-MAC_(CCM)_text", 80211-02-144r4.

[18] B. Aboba and D. Simon, "PPP EAP TLS Authentication Protocol," IETF RFC 2716, http://www.ietf.org/rfc/rfc2716.txt

[19] R. Rivest, "The MD5 Message-Digest Algorithm," http://www.ietf.org/rfc/rfc1321.txt

[20] See work in progress - P. Funk and S. Blake-Wilson, "EAP Tunneled TLS Authentication Protocol (EAP-TTLS)," http://tools.ietf.org/html/draft-ietf-pppext-eap-ttls-01

[21] C. Ellison and B. Schneier, "Ten Risks of PKI: What You're Not Being Told About Public Key Infrastructure," http://www.counterpane.com/pki-risks.html

[22] Stanley, D., Walker, J. and Aboba, B., "Extensible Authentication Protocol (EAP) Method Requirements for Wireless LANs", RFC4017, March 2005.

[23] Haverinen, H. Ed. And J. Salowey, Ed., "Extensible Authentication Protocol Method for Global System for Mobile Communications (GSM) Subscriber Identity Modules (EAP-SIM)", RFC 4186, January 2006.

[24] See work in progress—J. Carlson, et al., "EAP SRP-SHA1 Authentication Protocol," http://tools.ietf.org/html/draft-ietf-pppext-eap-srp-03

[25] Arkko, J. and H. Haverinen, "Extensible Authentication Protocol Method for 3rd Generation Authentication and Key Agreement (EAP-AKA)", RFC4187, January 2006.

[26] See work in progress - http://www.drizzle.com/~aboba/IEEE/draft-josefsson-pppext-eap-tls-eap-02.txt

[27] See http://www.cs.umd.edu/~waa/1x.pdf

[28] W. Simpson, "PPP Challenge Handshake Authentication Protocol (CHAP)," http://www.ietf.org/rfc/rfc1994.txt

[29] P802.11N (D2) Draft STANDARD for Information Technology Telecommunications and information exchange between systems - Local and metropolitan area networks-Specific requirements- Part 11: Wireless LAN Medium Access Control (MAC) and Physical Layer (PHY) specifications: Amendment: Enhancements for Higher Throughput.

[30] FIPS SP800-38B – Dworkin, M., "Recommendation for Block Cipher Modes of Operation: The CMAC Mode for Authentication", May 2005, http://csrc.nist.gov/CryptoToolkit/modes/800-38_Series_Pubications/SP800-38B.pdf.

[31] Braskich, T. et al., "Efficient Mesh Security and Link Establishment", 11-06-1470-03-000s, November 2006.

[32] Harkins, D. and Kuhtz, C. "Secure Mesh Formation" 11-06-1092-02-000s-comminus-preso, November 2006.

[33] http://www.wireshark.org

[34] http://www.musecurity.com

[35] http://www.codenomicon.com

[36] http://802.11mercenary.net

[37] http://wve.org/entries/show/WVE-2005-0059

[38] http://www.secdev.org/projects/scapy

[39] http://www.datarescue.com/idabase

[40] http://metasploit.com

[41] http://www.wirelessve.org

[42] http://labs.arubanetworks.com

[43] http://wve.org/entries/show/WVE-2006-0018,   http://wve.org/entries/show/WVE-2006-0054, http://wve.org/entries/show/WVE-2006-0069

[44] www.wve.org/entries/show/WVE-2006-0038

[45] www.wve.org/entries/show/WVE-2005-0056

[46] http://www.snort.org

[47] KARMA attack, www.wve.org/entries/show/WVE-2006-0032

[48] Aruba Networks, www.arubanetworks.com

[49] alert_aruba_action, Snort output plugin, http://snort.org/docs/snort_htmanuals/htmanual_261/node142.html

[50] "Weaknesses in WLAN Session Containment", Joshua Wright, http://802.11ninja.net/~jwright/802/papers/wlan-sess-cont.pdf

Acknowledgments

The authors thank Benny Bing for his suggestions on additional content to include in the chapter.

8

The 802.11n MIMO-OFDM Standard

Richard van Nee[a]

The IEEE 802.11n standard is the first wireless LAN standard based on MIMO-OFDM, a technique that significant range and rate relative to conventional wireless LAN. This chapter describes the main features of the 802.11n standard including packet structures, preamble formats, and coding aspects. Performance results show that net user throughputs over 100 Mbps are achievable, which is about four times larger than the maximum achievable throughput using IEEE 802.11a/g. For the same throughput, MIMO-OFDM achieves a range that is about 3 times larger than non-MIMO systems.

8.1 Introduction

The appetite for higher data rate continues as consumer demand for bandwidth hungry applications like gaming, streaming audio and video grows. Advancement in handset processors and further integration of technologies like higher mega-pixel cameras into handsets, create a never ending need for more bandwidth consuming applications at longer ranges and more efficient utilization of the limited spectrum available to Network Operators. 3G technology falls short in meeting this demand, while coverage is often worse than what customers are used to from 2.5G networks.

On the other hand, wireless LAN, the technology initially expected to provide only limited range and bandwidth has come a long way. Since the introduction of proprietary WLAN products in 1990 and the adoption of the first IEEE 802.11 standard in 1997, maximum data rates have made an impressive growth that is depicted in Figure 8.1. Till 2004, the growth in data rate was achieved by going from single carrier direct-sequence spread-spectrum to OFDM using higher order constellation sizes up to 64-QAM [1]. Unfortunately, this increase in rate came at the expense of a loss in range. The use of highly spectral efficient higher order modulations requires a significant larger SNR than the simple BPSK modulation used for the lowest 1 Mbps rate, resulting in a loss of range. In addition, the link becomes more vulnerable to co-channel interference, which reduces the total system capacity.

---

[a] *Qualcomm, Inc*

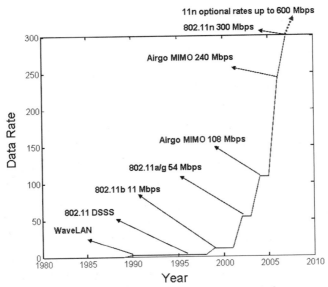

Figure 8.1: Wireless LAN data rate growth.

The solution to obtain significant higher data rates and increase range performance at the same time is MIMO-OFDM (Multiple Input Multiple Output Orthogonal Frequency Division multiplexing) [2]-[3]. MIMO-OFDM increases the link capacity by simultaneously transmitting multiple data streams using multiple transmit and receive antennas. It makes it possible to reach data rates that are several times larger than the current highest 802.11a/g rate of 54 Mbps without having to employ a larger bandwidth or a less robust QAM constellation [4]. With the introduction of MIMO-OFDM wireless LAN products in 2004 by Airgo Networks and the advent of the MIMO-OFDM based 802.11n standard in 2007, the performance of wireless LAN in terms of throughput and range is brought to a significantly higher level, enabling new applications outside the traditional wireless LAN area. The one time vision to replace wires in home entertainment applications, like TV cable replacement, has become a reality.

8.2 IEEE 802.11n

In July 2003, the 802.11n task group was formed to create a new wireless LAN standard. The main goal of this new standard is to give a throughput of at least 100 Mbps at the MAC data service access point [6]. A number of proposals were made that all share three common elements: the use of MIMO-OFDM, 20 and 40 MHz channels, and packet aggregation techniques. Based on this common ground, in July 2005 a joint proposal group was formed to create the first draft 802.11n standard [5].

The 802.11n standard defines a range of mandatory and optional data rates in both 20 and 40 MHz channels. Table 8.1 lists the Modulation and Coding Schemes (MCS) and their corresponding data rates for the cases of 1 and 2 spatial streams. For every MCS, 4 data rates are shown, as every MCS can be used in either a 20 MHz channel or a 40 MHz channel, using either a normal 800 ns guard interval or an optional 400 ns short Guard Interval. The use of 2 spatial streams with a short guard interval in a 40 MHz channel gives a highest possible data rate of 300 Mbps. Even higher data rates are possible by using the optional MCS listed in Table 8.2 that uses 3 and 4 spatial streams. For 4 spatial streams, the highest possible data rate becomes 600 Mbps.

In addition to the MCS sets listed below, an 802.11n device also needs to support all mandatory 802.11g rates if it operates in the 2.4 GHz band, or all 802.11a rates if it operates in the 5 GHz band. This ensures full interoperability with legacy WiFi equipment.

Table 8.1: Modulation and Coding Schemes (MCS) for 1 and 2 spatial streams.

MCS	Code Rate	Modulation	Number of Spatial Streams	Data Rate in 20 MHz, 800 ns GI	Data Rate in 20 MHz, 400 ns GI	Data Rate in 40 MHz, 800 ns GI	Data Rate in 40 MHz, 400 ns GI
0	½	BPSK	1	6.5	7.2	13.5	15
1	½	QPSK	1	13	14.4	27	30
2	¾	QPSK	1	19.5	21.7	40.5	45
3	½	16-QAM	1	26	28.9	54	60
4	¾	16-QAM	1	39	43.3	81	90
5	2/3	64-QAM	1	52	57.8	108	120
6	¾	64-QAM	1	58.5	65	121.5	135
7	5/6	64-QAM	1	65	72.2	135	150
8	½	BPSK	2	13	14.4	27	30
9	½	QPSK	2	26	28.9	54	60
10	¾	QPSK	2	39	43.3	81	90
11	½	16-QAM	2	52	57.8	108	120
12	¾	16-QAM	2	78	86.7	162	180
13	2/3	64-QAM	2	104	115.6	216	240
14	¾	64-QAM	2	117	130	243	270
15	5/6	64-QAM	2	130	144.4	270	300
32	½	BPSK	1	N/A	N/A	6	6.7

8.3 Preambles

Figure 8.2 shows the packet structure of IEEE802.11a. One of the most important criteria for the choice of the new preambles for IEEE802.11n is compatibility with IEEE802.11a/g. To achieve this, a mixed-mode preamble is constructed as depicted in Figure 8.3. The mixed-mode preamble starts with an 802.11a preamble with the only difference that multiple transmitters transmit cyclically delayed copies of the preamble. A legacy 802.11a receiver is able to receive this preamble up to the legacy signal field, which guarantees a proper defer behavior of legacy devices for 802.11n packets.

Table 8.2: Optional MCS for 3 and 4 spatial streams.

MCS	Code Rate	Modulation	Number of Spatial Streams	Data Rate in 20 MHz, 800 ns GI	Data Rate in 20 MHz, 400 ns GI	Data Rate in 40 MHz, 800 ns GI	Data Rate in 40 MHz, 400 ns GI
16	½	BPSK	3	19.5	21.7	40.5	45
17	½	QPSK	3	39	43.3	81	90
18	¾	QPSK	3	58.5	65	121.5	135
19	½	16-QAM	3	78	86.7	162	180
20	¾	16-QAM	3	117	130	243	270
21	2/3	64-QAM	3	156	173.3	324	360
22	¾	64-QAM	3	175.5	195	364.5	405
23	5/6	64-QAM	3	195	216.7	405	450
24	½	BPSK	4	26	28.9	54	60
25	½	QPSK	4	52	57.8	108	120
26	¾	QPSK	4	78	86.7	162	180
27	½	16-QAM	4	104	115.6	216	240
28	¾	16-QAM	4	156	173.3	324	360
29	2/3	64-QAM	4	208	231.1	432	480
30	¾	64-QAM	4	234	260	486	540
31	5/6	64-QAM	4	260	288.9	540	600

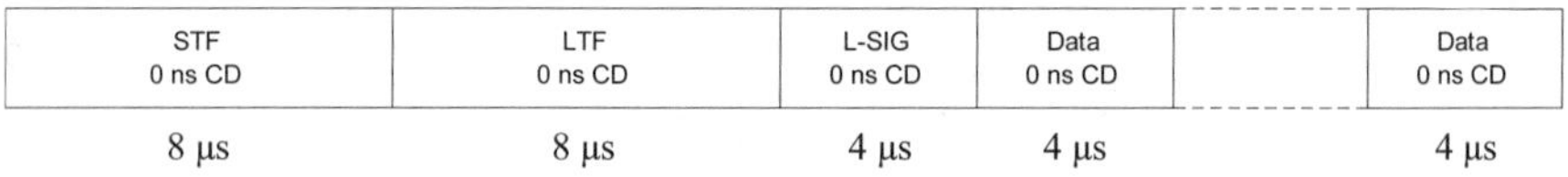

STF 0 ns CD	LTF 0 ns CD	L-SIG 0 ns CD	Data 0 ns CD		Data 0 ns CD
8 µs	8 µs	4 µs	4 µs		4 µs

Figure 8.2: IEEE802.11a/g packet structure.

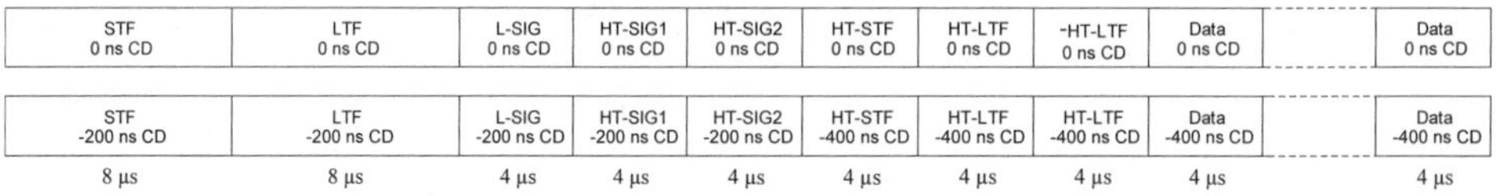

STF 0 ns CD	LTF 0 ns CD	L-SIG 0 ns CD	HT-SIG1 0 ns CD	HT-SIG2 0 ns CD	HT-STF 0 ns CD	HT-LTF 0 ns CD	-HT-LTF 0 ns CD	Data 0 ns CD		Data 0 ns CD
STF -200 ns CD	LTF -200 ns CD	L-SIG -200 ns CD	HT-SIG1 -200 ns CD	HT-SIG2 -200 ns CD	HT-STF -400 ns CD	HT-LTF -400 ns CD	HT-LTF -400 ns CD	Data -400 ns CD		Data -400 ns CD
8 µs	8 µs	4 µs	4 µs	4 µs	4 µs	4 µs	4 µs	4 µs		4 µs

Figure 8.3: IEEE802.11n mixed-mode packet with two spatial streams.

The short training field (STF) of 802.11n is the same as for 802.11a/g, except that different transmitters use different cyclic delays. The latter is done to avoid undesired beamforming effects and to get accurate power estimates that can be used to set the receive gain. For a proper AGC setting, it is important that the received power during the short training field is the same as the power during the rest of the packet. To achieve this, the short training fields from different transmitters must have a low cross-correlation, also after being convolved with the wireless channel impulse response which has a typical rms delay spread in the order of 50 to 100 ns. It can be seen from the autocorrelation function in Figure 8.4 that applying a cyclic delay of -400 ns (or a cyclic advance of 400 ns) minimizes the correlation between two different transmitted short symbols, which is the reason that this value was selected for the greenfield short training field and for the second short training field in the mixed-mode preamble. The short training field of the mixed-mode preamble uses a cyclic delay of only -200 ns for the case of two transmitters, because of a fear that legacy 802.11a/g receivers might not be able to deal with larger cyclic delay values. Legacy receivers are also the reason that only negative cyclic delays are used. To a receiver the presence of a second transmit signal with a positive cyclic delay appears like a multipath signal that arrives later than the signal of the first transmitter. If the receiver uses a correlation approach to set its symbol timing [7], then it will set the symbol time too late, which can result in inter-symbol interference. When the second transmitter uses a negative cyclic delay, the receiver will set its symbol timing too early, which will eat into the OFDM guard time without causing inter-symbol interference.

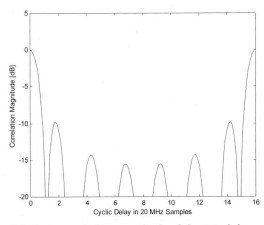

Figure 8.4: Autocorrelation magnitude of short training symbol.

After the legacy signal field (L-SIG), a new high throughput signal field (HT-SIG) is transmitted that contains 48 bits with information including a 16-bit length field, a 7-bit field for the modulation and coding scheme (MCS), bits to indicate various options like LDPC coding, and an 8-bits CRC. To enable detection of the presence of a high throughput signal field, it uses a BPSK constellation that is rotated by 90 degrees.

After the high throughput signal field, a second short training field is transmitted. This high throughput short training field (HT-STF) can be used to retrain the AGC, which may be needed for two reasons; first, the transmitter may employ beamforming for the high throughput part of the packet only, such that there may be a large power difference between the received signal before and after the start of the high throughput short training field. Second, there may also be a power difference because of non-zero cross-correlations between the cyclically shifted short training fields of the legacy part of the mixed mode preamble. This effect is small when using a cyclic delay of -200 or -400 ns like explained earlier. For the case of 4 transmitters, however, cyclic delays as small as 50 ns are used, which can result in the difference of a few dB between the received power before and after the high throughput short training field.

The high throughput short training field is followed by one or more high throughput long training fields (HT-LTF) that are used for channel estimation. The number of HT-LTF symbols is equal to the number of spatial streams. For the case of 2 spatial streams, the second HT-LTF of the first spatial stream is inverted to create an orthogonal space-time pattern. The receiver can obtain channel estimates for both spatial streams by adding and subtracting the first and second HT-LTF, respectively. The channel estimates can then be used to process the MIMO-OFDM data symbols that follow the HT-LTF. The only remaining training task after channel estimation is pilot processing. Every data symbol has a few pilot subcarriers - 4 in 20 MHz modes, 6 for 40 MHz modes - that can be used to track any residual carrier frequency offset.

In addition to the mixed-mode preamble, the 802.11n standard also defines a greenfield preamble. This preamble that is shown in Figure 8.5 is 8 microseconds shorter, resulting in a larger net throughput. It is not compatible with legacy 802.11a or 802.11g devices as such devices will not be able to decode the signal field of a greenfield preamble. Because of this, the greenfield preamble is useful in 2 situations; first, in networks without any legacy devices, and second, in pieces of reserved time, also referred to as "green time". Green time can be reserved for instance by an RTS/CTS (Request-to-Send/Clear-to-Send). In the reserved time, the 11n standard allows a burst of packets to be sent using a RIFS (Reduced Interframe Spacing) of 2 microseconds only. Using a greenfield preamble instead of the longer mixed-mode preamble minimizes the training overhead for such packet bursts, while a mixed-mode preamble can be used for the RTS/CTS to make sure that both 802.11n and legacy devices will properly defer.

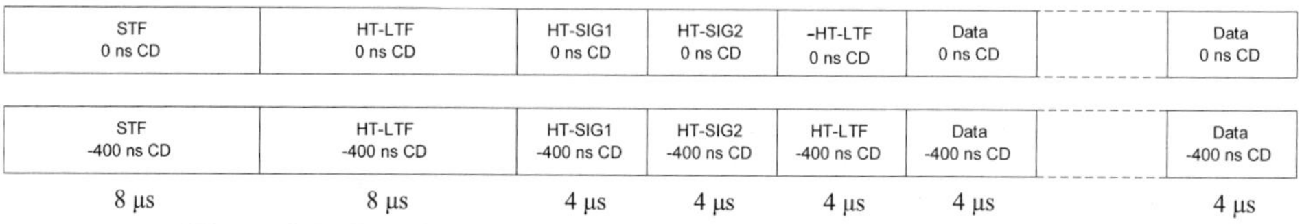

STF 0 ns CD	HT-LTF 0 ns CD	HT-SIG1 0 ns CD	HT-SIG2 0 ns CD	-HT-LTF 0 ns CD	Data 0 ns CD		Data 0 ns CD
STF -400 ns CD	HT-LTF -400 ns CD	HT-SIG1 -400 ns CD	HT-SIG2 -400 ns CD	HT-LTF -400 ns CD	Data -400 ns CD		Data -400 ns CD
8 µs	8 µs	4 µs	4 µs	4 µs	4 µs		4 µs

Figure 8.5: IEEE802.11n greenfield packet with two spatial streams.

It is mandatory for an 802.11n device to transmit or receive 2 spatial streams. In addition to this, optional modes are defined for 3 and 4 spatial streams. Figure 8.6 shows the structure of the optional mixed-mode preamble for the case of 4 spatial streams. This preamble has 4 high throughput long training symbols that are encoded with an orthogonal pattern such that the receiver is able to obtain channel estimates for all 4 spatial streams. Together with the optional short guard interval option and the use of a 40 MHz channel, the 4 spatial stream mode gives a highest possible raw data rate of 600 Mbps.

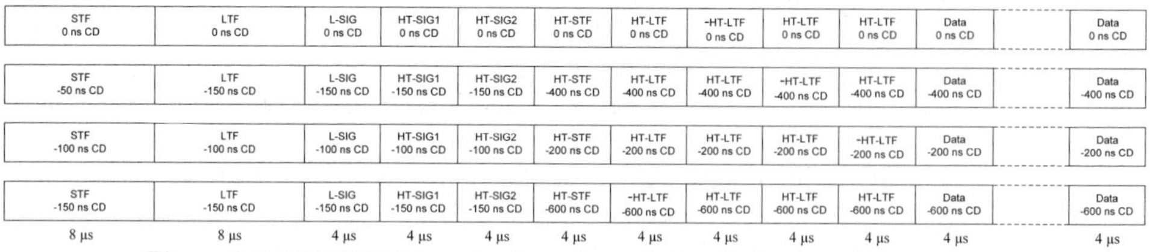

STF 0 ns CD	LTF 0 ns CD	L-SIG 0 ns CD	HT-SIG1 0 ns CD	HT-SIG2 0 ns CD	HT-STF 0 ns CD	HT-LTF 0 ns CD	-HT-LTF 0 ns CD	HT-LTF 0 ns CD	HT-LTF 0 ns CD	Data 0 ns CD		Data 0 ns CD
STF -50 ns CD	LTF -150 ns CD	L-SIG -150 ns CD	HT-SIG1 -150 ns CD	HT-SIG2 -150 ns CD	HT-STF -400 ns CD	HT-LTF -400 ns CD	HT-LTF -400 ns CD	-HT-LTF -400 ns CD	HT-LTF -400 ns CD	Data -400 ns CD		Data -400 ns CD
STF -100 ns CD	LTF -100 ns CD	L-SIG -100 ns CD	HT-SIG1 -100 ns CD	HT-SIG2 -100 ns CD	HT-STF -200 ns CD	HT-LTF -200 ns CD	HT-LTF -200 ns CD	HT-LTF -200 ns CD	-HT-LTF -200 ns CD	Data -200 ns CD		Data -200 ns CD
STF -150 ns CD	LTF -150 ns CD	L-SIG -150 ns CD	HT-SIG1 -150 ns CD	HT-SIG2 -150 ns CD	HT-STF -600 ns CD	-HT-LTF -600 ns CD	HT-LTF -600 ns CD	HT-LTF -600 ns CD	HT-LTF -600 ns CD	Data -600 ns CD		Data -600 ns CD
8 µs	8 µs	4 µs	4 µs	4 µs	4 µs	4 µs	4 µs	4 µs	4 µs	4 µs		4 µs

Figure 8.6: IEEE802.11n mixed-mode packet with four spatial streams.

8.4 802.11n Transmitter

Figure 8.7 shows the block diagram of an IEEE802.11n transmitter. Input data is first scrambled using the same length-127 pseudo-noise scrambler that is used in IEEE802.11a. The convolutional encoder is also the same as IEEE802.11a, with the only difference that for 3 and 4 spatial streams, odd and even bits are separately encoded by two different encoders which is done to limit the maximum decoding rate at the receive side.

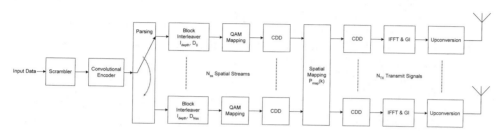

Figure 8.7: Block diagram of an IEEE802.11n transmitter.

After encoding, a parser sends consecutive blocks of $s = \max(N_{bpsc}/2,1)$ bits to different spatial streams, with N_{bpsc} being the number of bits per subcarrier. The bits are then interleaved by a block interleaver with a block size equal to the number of bits in a single OFDM symbol of the nth spatial stream, $N_{CBPS,n}$. By interleaving the bits across both spatial streams and subcarriers, the link performance benefits from both spatial diversity and frequency diversity. The interleaver for spatial stream n within its block of $N_{CBPS,n}$ bits is defined by the following relations, where k_n is the input bit index for spatial stream n and j_n is the output bit index.

$$k_n = 0, 1 \ldots N_{CBPS,n}-1$$
$$s_n = \max(N_{BPSC,n}/2,1)$$
$$i = (N_{CBPS,n}/I_{DEPTH}) (k_n \bmod I_{DEPTH}) + \text{floor}(k_n/I_{DEPTH})$$
$$j = s_n \times \text{floor}(i/s_n) + (i + N_{CBPS,n} - \text{floor}(I_{DEPTH} \times i / N_{CBPS,n})) \bmod s_n$$
$$j_n = (j + N_{CBPS,n} - N_{BPSC,n}D_n) \bmod (N_{CBPS,n})$$

The interleaving depth I_{DEPTH} and the subcarrier rotation D_n are defined in Table 8.3.

Table 8.3: Interleaving parameters.

N_{SS}	N_{SD}	I_{DEPTH}	D_0	D_1	D_2	D_3
1, 2, 3, 4	52	13	0	22	11	33
1, 2, 3, 4	108	18	0	58	29	87

After interleaving, bits are mapped onto QAM symbols. Then, a spatial stream dependent cyclic delay (CD) is applied in the frequency domain. More details about this cyclic delay can be found in section 8.3. At this point, a spatial mapping matrix is applied to each subcarrier to convert N_{ss} spatial stream inputs into N_{tx} transmitter outputs. If the number of transmitters is identical to the number of spatial streams, the spatial mapping matrix can simply be the identity matrix. To transmit legacy 802.11a/g rates that have only one spatial stream, the spatial mapping matrix reduces to a column of ones. After the spatial mapping matrix, an additional cyclic delay can be applied per transmitter to provide transmit cyclic delay diversity (CDD) and prevent undesired beamforming effects. Each transmitter subsequently applies an IFFT, inserts a guard interval, upconverts and transmits the signal.

The subcarrier mapping for 20 MHz and 40 MHz channels is depicted in Figures 8.8 and 9, respectively. The 20 MHz mode uses 56 subcarriers for the high throughput data symbols, which is 4 more than the number used by 802.11a. The extra tones increase the throughput at the cost of some extra transmitter complexity to keep the transmitted spectrum within the spectral mask. The 802.11n spectral mask for 20 MHz mode is actually more tight than the 802.11a mask, so the use of more tones does not decrease the adjacent channel performance relative to 802.11a. The legacy part of the mixed-mode preamble and the high throughput signal field use the same subcarriers as 802.11a, which are shown as grey blocks in the figures. The 20 MHz 11n modes use 4 pilots just like 802.11a, while the 40 MHz 11n mode uses 6 pilots. A difference with 802.11a is that 802.11n uses a space time mapping for the pilots when transmitting from multiple antennas. In this way some extra transmit diversity is obtained on the pilots, and undesired beamforming effects are prevented.

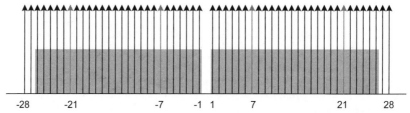

Figure 8.8: Subcarrier allocations for a 20 MHz channel. Data tones are black. Pilots are dark grey. Light grey blocks are the subcarriers used for 802.11a, for the legacy part of the mixed-mode preamble, and for the HT-SIG field.

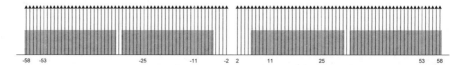

Figure 8.9: Subcarrier allocations for a 40 MHz channel. Pilots are dark grey. Light grey blocks are the subcarriers used for 802.11a, for the legacy part of the mixed-mode preamble, and for the HT-SIG field.

8.5 LDPC Coding

The mandatory code used in 802.11n is the same binary convolutional code (BCC) that is used by 802.11a/g. An optional LDPC code is specified to get some extra gain over the mandatory BCC. The LDPC code is a systematic block code with possible block lengths of 648, 1296, and 1944. The same coding rates as for BCC are provided, which are ½, 2/3, ¾, and 5/6. The parity check matrices are sparse and highly structured, which facilitates both encoding and decoding. Figure 8.10 shows the gain of LDPC over BCC for the case of 2 spatial streams using 2 transmit antennas and 2 receive antennas, assuming perfect training and an MMSE receiver for MIMO detection. The channel model used in the simulation is the non-line-of-sight channel D, which is a typical indoor wireless channel with a delay spread of 50 ns [8]. The LDPC decoder used layered belief propagation [9] with 10, 20 and 50 iterations. It can be seen that a gain of about 2 dB over BCC can be achieved when doing 50 ·

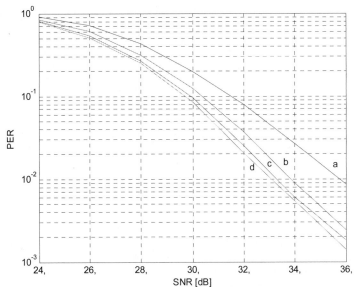

Figure 8.10: LDPC versus BCC PER curves for 1000B packets in 802.11n channel D NLOS with ideal training, using a rate of 270 Mbps in 40 MHz with 2 spatial streams, 2 transmitters, and 2 receivers. a) BCC, b) LDPC with 10 iterations, c) LDPC with 20 iterations, d) LDPC with 50 iterations.

8.6 Space Time Block Coding

Space Time Block Coding (STBC) is an optional feature in the 11n standard to provide extra diversity gain in cases where the number of available transmitters is larger than the number of spatial streams. STBC operates on groups of 2 symbols, mapping N_{ss} spatial stream inputs onto N_{sts} space time stream outputs. For the case of 1 spatial stream and 2 space time streams, for instance, STBC mapping is done as follows: if $\{d_{ke}, d_{ko}\}$ are QAM symbols for an even and odd symbol for subcarrier k, respectively, then STBC encoding maps this single spatial stream input onto 2 space time stream outputs $\{d_{ke}, d_{ko}\}$ and $\{-d^*_{ko}, d^*_{ke}\}$. Hence, the even symbol contains d_{ke} on the first space time stream and $-d^*_{ko}$ on the second space time stream; the odd symbol contains d_{ko} on the first space time stream and d^*_{ke} on the second space time stream. The benefit of this type of STBC is that it doubles the diversity order of the link. In addition to the single spatial stream STBC mode described above, the 802.11n standard also specifies STBC modes with 2 and 3 spatial streams.

8.7 Beamforming

Beamforming is a way to provide extra performance gain in cases where the number of transmit antennas is larger than the number of spatial streams, and where the transmitter has knowledge about the channel. Beamforming is done by multiplying the N_{ss} spatial stream inputs for every subcarrier by an N_{tx} by N_{ss} beamforming matrix. The 802.11n standard specifies a number of optional methods that can be used to support beamforming. Two categories of beamforming exist, implicit beamforming and explicit beamforming. When using explicit beamforming, the beamformee – i.e., the device that is beamformed to – provides channel information to the beamformer, which is the device that is actually doing the beamforming. It can do this by sending a packet containing the received channel values for all subcarriers, receivers, and spatial streams. It can also calculate the beamforming matrix coefficients and send these to the beamformer. The standard specifies two types of beamforming weight formats, an uncompressed format and a compressed format to limit the amount of overhead. To minimize the amount of overhead, implicit beamforming can be used where the beamformer uses received preambles from the beamformee to calculate the beamforming weights, assuming a reciprocal channel. Implicit beamforming does have more stringent calibration constraints than explicit beamforming.

8.8 MAC Enhancements

The 802.11n standard includes several enhancements to the Medium Access Control (MAC) layer that help to increase the net throughput, especially when using the newly defined high data rates. One important new feature is aggregation; by making the packets as large as possible, the relative throughput impact of preamble overhead is minimized. This is very important as a typical Ethernet packet with a length of 1500 bytes takes only 40 microseconds of transmission time for the data part at a rate of 300 Mbps, which is the same duration as the mixed-mode preamble for a 2-spatial stream packet. Hence, the net throughput is reduced by 50% just by the preamble overhead. To minimize this throughput hit, the 802.11n standard specifies two type of aggregation. Several MAC Service Data Units (MSDU) can be aggregated into one A-MSDU up to length of 7935 bytes. It is also possible to aggregate MAC Protocol data Units (MPDU) into one A-MPDU with a maximum aggregated length of 65535 bytes. A limitation of A-MSDU compared to A-MPDU is that for A-MSDU, all MSDUs need to be targeted to the same destination address.

Another new mechanism that can be used to increase throughput is the Reduced Inter Frame Spacing (RIFS). An 11n device may transmit a burst of packets separated by a RIFS of 2 microseconds, thereby minimizing the amount of protocol overhead duration.

With the introduction of many new data rates in 802.11n, link adaptation becomes more problematic. Rather than searching through a limited set of data rates, the sender now has to decide how many spatial streams and what channel width to use in order to maximize the link throughput. To facilitate link adaptation, the 802.11n standard introduces a way to provide MCS feedback, whereby a receiver can inform the sender what MCS it could use. The receiver can deduce this MCS recommendation from the received channel estimates and the received signal-to-noise ratio.

8.9 Use of 40 MHz Channels

The 802.11n standard allows the use of 40 MHz channels, while legacy 802.11a/g devices only use 20 MHz channels. In the 5 GHz band, the use of a 40 MHz mixed-mode preamble or the use of duplicate 11a RTS/CTS in both primary and secondary 20 MHz channels ensures that all legacy 802.11a devices can correctly defer for each 40 MHz transmission. At the same time, 40 MHz 802.11n devices correctly defer for legacy devices because the standard requires a 40 MHz device to do a Clear Channel Assessment based on activity in both primary and secondary channel.

In the 2.4 GHz band, the situation for using 40 MHz transmissions is much more complicated than it is for the 5 GHz band. First, there are less channels available, only three 20 MHz channels. Second, the channels can be partially overlapping as the center frequencies are specified on a 5 MHz grid rather than on a 20 MHz grid like in the 5 GHz band. Figure 8.11 shows measured percentages of channel occupancy in the 2.4 GHz band. Channels 1, 6, and 11 are used most frequently, but there is also a significant percentage of other channels. The disadvantage of these channel spacings is that is not possible to transmit a mixed-mode preamble that can be correctly received by all legacy devices. For instance, if an 802.11n device would use primary channel 1 and secondary channel 5 to do a 40 MHz transmission, then only legacy devices centered on channels 1 and 5 could receive the legacy portion of a 40 MHz mixed-mode preamble, while devices on channels {2,3,4,6,7} would not be able to receive the legacy portion, while they would be interfered by a partial overlap with the 40 MHz packet. Because of this, WiFi is discouraging the use of 40 MHz channels by requiring devices to have the 40 MHz capability turned off by default in the 2.4 GHz band.

8.10 MIMO-OFDM Performance Results

In 2004, Airgo Networks (acquired by Qualcomm in 2006) launched the first wireless LAN chipset based on MIMO-OFDM. This first generation MIMO-OFDM system uses a 20 MHz channel to transmit at either standard 802.11a/g data rates with a large range increase compared to conventional wireless LAN, or at significantly higher data rates up to 108 Mbps. In 2005, a second generation MIMO-OFDM product was introduced that uses Adaptive Channel Expansion to transmit either in a 20 or 40 MHz channel, increasing the top data rate to 240 Mbps.

Figure 8.12 demonstrates the performance impact of a wireless LAN using MIMO-OFDM versus a conventional wireless LAN in a 20 MHz channel. The plot shows the cumulative distribution function of the measured TCP/IP throughput, where the client device has been put on a slowly rotating turntable to get throughput results for all possible orientations. From Figure 8.12 it can be seen that in 10% of all possible orientations, the MIMO wireless LAN has a throughput less than about 33 Mbps, so for 90% of all orientations the throughput exceeds 33 Mbps. For the non-MIMO wireless LAN, this 10% number is only 4 Mbps. Hence, for a 10% outage probability, the MIMO throughput is more than 8 times better in this particular case. For the 1% outage probability, the performance difference is even more pronounced.

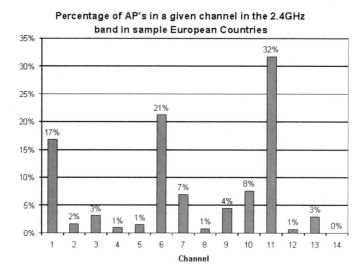

Figure 8.11: Percentage of Access points seen in a particular 2.4 GHz channel, based on 1722 measured Access Points in cities in the Netherlands, Belgium, and Italy.

Figure 8.13 shows measured TCP/IP throughput results for both 20 and 40 MHz channel width. Each throughput curve consists of 8 points that correspond to different locations of the client inside a house, with an increased range towards the access point, but also with an increasing number of walls between the client and the access point. For the first test point, the client device is in the same room as the access point at a distance of 17 feet, while at the last test point the distance is 102 feet including 5 walls in between client and access point. The results show that for any range, the MIMO-OFDM throughput is 2.5 to 5 times larger than the throughput of non-MIMO products. Notice that several of these other products use channelbonding to provide a proprietary maximum raw data rate of 108 Mbps in a 40 MHz channel. This explains why these products are able to achieve maximum TCP/IP throughputs over 40 Mbps, while conventional 802.11a/g products have a maximum TCP/IP throughput of about 25 Mbps. The maximum throughput of MIMO-OFDM in a 40 MHz channel exceeds 100 Mbps, which meets the throughput goal set by 802.11n [6]. Notice that the MIMO-OFDM device used for Figure 8.13 did not use all 802.11n MAC throughput enhancements yet and also had a top rate of 240 Mbps that is lower than the highest 802.11n data rate of 300 Mbps for 2 spatial streams in 40 MHz. An 802.11n device implementing all optional throughput enhancements can be expected to achieve a maximum net throughput around 150 Mbps.

Another way to look at the results of Figure 8.13 is in terms of range increase for a given throughput. For instance, for a required throughput of at least 40 Mbps, the best non-MIMO-OFDM product has a maximum range of about 25 feet including 1 wall. For the same 40 Mbps throughput, MIMO-OFDM has a range of more than 80 feet including 4 walls. This range increase of more than a factor of 3 makes it possible to guarantee a high throughput throughout an entire house, which opens the way to new throughput-demanding applications such as wireless video distribution.

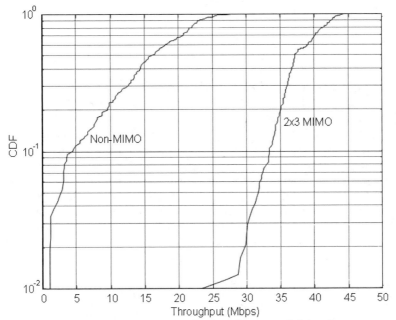

Figure 8.12: Cumulative distribution of measured throughput.

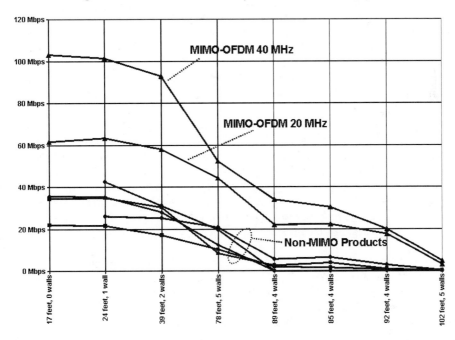

Figure 8.13: TCP throughput measured at various distances.

8.11 Conclusions

The performance of wireless LAN in terms of range and throughput is increased significantly by the use of MIMO-OFDM, which is the basis of the new IEEE 802.11n standard. Performance results show that net user throughputs over 100 Mbps are achievable with just 2 spatial streams, which is about four times larger than the maximum achievable throughput using IEEE 802.11a/g. For the same throughput, MIMO-OFDM achieves a range increase of about a factor of 3 compared to conventional wireless LAN. This performance boost makes MIMO-OFDM the ideal successor to the current OFDM-only wireless LAN. Also, it enables new throughput-demanding applications such as wireless video distribution. Seeing the effectiveness and superior capability of MIMO-OFDM in enhancing data rate and extending range, other standards organizations have realized that it can do wonders for other technologies, both fixed, mobile and cellular. Standard bodies like 3GPP, WiBro, WiMax and the 4G Mobile Forum have started exploring the use of MIMO-OFDM in their respective technology areas, making it the technology of choice for future wireless networks.

8.12 References

[1] R. van Nee, G. Awater, M. Morikura, H. Takanashi, M. Webster, and K. Halford, "New High Rate Wireless LAN Standards," *IEEE Communications Magazine*, Vol. 37, No. 12, pp. 82-88, Dec. 1999.

[2] G. G. Raleigh and J. M. Cioffi, "Spatio-Temporal Coding for Wireless Communications," *Proc. 1996 Global Telecommunications Conf.*, Nov. 1996, pp. 1809-1814.

[3] G. G. Raleigh and V. K. Jones, "Multivariate Modulation and Coding for Wireless Communication," *IEEE Journal on Selected Areas in Communications*, Vol. 17, No. 5, May 1999, pp. 851-866.

[4] R. van Nee, A. van Zelst, and G. Awater, "Maximum Likelihood Decoding in a Space Division Multiplexing System," *IEEE VTC 2000*, Tokyo, Japan, May 2000.

[5] IEEE 802.11 Working Group, *IEEE P802.11n/D2.0 Draft Amendment to Standard for Information Technology-Telecommunications and information exchange between systems-Local and Metropolitan networks-Specific requirements-Part 11: Wireless LAN Medium Access Control (MAC) and Physical Layer (PHY) specifications: Enhancements for Higher Throughput*, January 2007.

[6] A. Stephens, "802.11n Functional Requirements, " IEEE document 802.11-02/813r12, March 2004.

[7] R. van Nee and R. Prasad, *OFDM for Mobile Multimedia Communications*, Boston: Artech House, Dec. 1999.

[8] V. Erceg, et al., "TGn Channel Models," IEEE document 802.11-03/940r4, May 2004.

[9] J. Chen, et al., "Reduced-Complexity Decoding of LDPC Codes", *IEEE Transactions on Communications*, Vol.53, No.8, August 2005, pp. 1288-1299.

9

MIMO Spatial Processing for 802.11n WLAN

Bjørn Bjerke, Irina Medvedev, John Ketchum, Rod Walton,
Steven Howard, Mark Wallace and Sanjiv Nanda[a]

9.1 Introduction

Tremendous consumer interest in multimedia applications is driving the need for successively higher data rates in wireless networks. The IEEE 802.11n standard for high throughput Wireless Local Area Networks (WLANs) improves significantly upon the data rates experienced by end users of current WLAN systems, e.g., 802.11a, b, and g.

The soon-to-be ratified 802.11n standard specifies a high data rate multiple-input, multiple-output (MIMO) based physical layer which employs orthogonal frequency division multiplexing (OFDM) and up to four spatial streams [1]. Both high data rate and long-range coverage are achieved by employing spatial signal processing techniques such as spatial spreading and transmit beamforming [2], among others. 802.11n introduces a range of MAC-layer enhancements also, but these are beyond the scope of this chapter.

In this chapter, we give an overview of two spatial processing alternatives available to implementers of 802.11n. We examine spatial spreading and transmit beamforming schemes, as well as possible receiver structures. Comparisons in terms of performance and complexity are also given.

The chapter is organized as follows. Section 9.2 gives a brief overview of MIMO OFDM, as well as the relevant system aspects of the 802.11n physical layer (PHY). Section 9.3 describes spatial spreading. Section 9.4 describes eigenvector-based transmit beamforming and schemes for channel sounding and calibration. Section 9.5 describes receiver structure alternatives for use with the above mentioned techniques. A comparison of the schemes, including simulation results illustrating the performance of the various receivers, is provided in Section 9.6, and a complexity analysis is given in Section 9.7. Conclusions are drawn in Section 9.8.

9.2 MIMO OFDM System Overview

In a MIMO communication system, the transmitter and receiver are both equipped with multiple antennas, thus allowing multiple data streams to be transmitted over parallel

---

[a] All authors currently affiliated with *Qualcomm, Inc*

spatial channels [3]. In a MIMO OFDM system with N_T transmit antennas and N_R receive antennas, the wideband channel can be characterized at discrete frequencies $l\Delta_f$, $l_1 \leq l \leq l_2$ by a set of $N_R \times N_T$ channel matrices, $\mathbf{H}(l)$. Here we address the 802.11n 20 MHz baseband channel that is divided into 64 subcarriers, with $\Delta_f = 312.5$ kHz and $-32 \leq l \leq 31$. Up to N_M = min$\{N_T, N_R\}$ parallel channels may be synthesized in the MIMO system, with the number of parallel spatial streams transmitted, N_S, upper-bounded by N_M.

Figure 9.1 shows a simplified system diagram for the 802.11n PHY, including transmitter and receiver blocks. Scrambled data bits are encoded using the rate-1/2, 64-state convolutional encoder that was originally introduced to the 802.11 standard by the 802.11a amendment, and later also employed by 802.11g [4]. The coded bits are punctured to achieve the desired code rate, which must belong to the set {1/2, 2/3, 3/4, 5/6}. A stream parser distributes the coded and punctured bits to the N_S spatial streams in a round robin fashion (see [1] for details), and each stream is independently interleaved and mapped to a complex constellation. The interleavers are based on the interleaver employed by 802.11a/g, but with one extra, stream-dependent permutation. 802.11n allows the following modulations to be used: BPSK, QPSK, 16-QAM, and 64-QAM. Table 9.1 shows the mandatory data rates resulting from the various modulation and code rate schemes (MCSs). These rates are achieved using $N_S = 1$ and $N_S = 2$ spatial streams. Extending the code rate and modulation combinations in Table 9.1 to $N_S = 3$ and $N_S = 4$ spatial streams is optional. In addition, 802.11n allows MCSs using unequal modulations on a per-stream basis. The latter is particularly useful in combination with transmit beamforming schemes, as we shall see in the subsequent sections. The code rate is always uniform across the streams. The total number of code rate and modulation schemes specified in 802.11 is 77 [1].

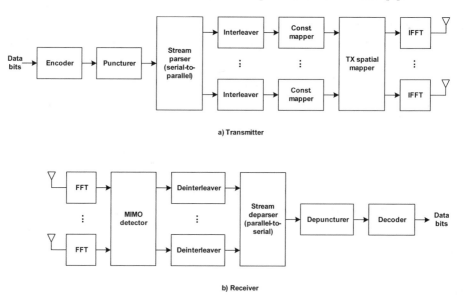

a) Transmitter

b) Receiver

Figure 9.1: Simplified system diagram.

The transmitter may choose to apply spatial processing that rotates and/or scales the constellation mapper output vector. This amounts to a matrix mapping operation and is useful in systems where there are more transmit chains than spatial streams. Several alternatives for such transmit spatial processing exist. In subsequent sections, we consider two distinct approaches, referred to as spatial spreading and transmit beamforming, respectively.

Table 9.1: 802.11n mandatory modulation and coding schemes,
1 and 2 streams (20 MHz, 800 ns guard interval)

MCS index	Code rate	Stream 0 modulation	Stream 1 modulation	Data rate (Mbps)
0	1/2	BPSK	-	6.5
1	1/2	QPSK	-	13.0
2	3/4	QPSK	-	19.5
3	1/2	16-QAM	-	26.0
4	3/4	16-QAM	-	39.0
5	2/3	64-QAM	-	52.0
6	3/4	64-QAM	-	58.5
7	5/6	64-QAM	-	65.0
8	1/2	BPSK	BPSK	13.0
9	1/2	QPSK	QPSK	26.0
10	3/4	QPSK	QPSK	39.0
11	1/2	16-QAM	16-QAM	52.0
12	3/4	16-QAM	16-QAM	78.0
13	2/3	64-QAM	64-QAM	104.0
14	3/4	64-QAM	64-QAM	117.0
15	5/6	64-QAM	64-QAM	130.0

Before we discuss the spatial processing in more detail, we develop a mathematical model for the MIMO OFDM system. In particular, the received waveform in subcarrier l may be expressed as

$$\mathbf{y}(l) = \mathbf{H}(l)\mathbf{x}(l) + \mathbf{n}(l) \tag{1}$$

where $\mathbf{y}(l)$ is the N_R-element vector of received symbols, $\mathbf{x}(l)$ is the N_T-element transmit vector, $\mathbf{H}(l)$ is the $N_R \times N_T$ channel matrix whose elements represent the complex gains of the channel coupling between individual transmit and receive antennas, and $\mathbf{n}(l)$ is the $N_R \times 1$ additive white Gaussian noise (AWGN) vector. Letting $\mathbf{s}(l)$ be the N_S-element vector of modulation symbols to be transmitted, one element for each spatial stream, the N_T-element transmit vector may be expressed by

$$\mathbf{x}(l) = T\big[\mathbf{s}(l)\big] \tag{2}$$

where $T[\bullet]$ is a transformation of $\mathbf{s}(l)$ that is dependent on the transmit spatial processing employed. In the case where the number of spatial streams equals the number of transmit chains and no transmit spatial processing is applied, the transformation simply reduces to matrix multiplication by the identity matrix, i.e.,

$$T\big[\mathbf{s}(l)\big] = \mathbf{I} \cdot \mathbf{s}(l). \tag{3}$$

9.3 Spatial Spreading

Spatial spreading (SS) is a transmit processing scheme that may be used when the transmitter has no or limited channel state information (CSI) available to it. In this case, the receiver spatial processing is solely responsible for separating and demodulating the parallel data streams. The spatial spreading technique described here is an extension of the work found in [5], which is the original idea as applied to multiple-input, single-output (MISO) systems.

With spatial spreading, the transmitted vector, $\mathbf{x}(l)$, is formed by multiplying the vector of modulation symbols to be transmitted, $\mathbf{s}(l)$, by the $N_T \times N_S$ matrix $\mathbf{W}(l)$, whose columns are orthonormal. Thus, the transmitted vector for the OFDM subcarrier with frequency $l\Delta_f$ may be expressed as

$$\mathbf{x}(l) = T\left[\mathbf{s}(l)\right] = \mathbf{W}(l)\mathbf{s}(l) \tag{4}$$

where $\mathbf{W}(l)$ consists of the first N_S columns of the orthonormal spatial spreading matrix, and the N_S-element vector $\mathbf{s}(l)$ consists of the modulation symbols to be transmitted on each of the N_S spatial channels. The received signal is then given by

$$\mathbf{y}(l) = \mathbf{H}_e(l)\mathbf{s}(l) + \mathbf{n}(l) \tag{5}$$

where $\mathbf{H}_e(l) = \mathbf{H}(l)\mathbf{W}(l)$ represents the effective channel as observed by the receiver. As before, $\mathbf{n}(l)$ is a column vector of complex Gaussian noise elements each with zero mean and variance N_0.

The spatial spreading matrix, $\mathbf{W}(l)$, varies with subcarrier frequency, $l\Delta_f$, in order to maximize the transmit diversity order and to provide many independent "looks" at the channel over the set of OFDM subcarriers. A very simple and effective construction uses a single fixed unitary spreading matrix in combination with a linear phase shift across the OFDM subcarriers per transmit chain. The resulting spatial spreading matrix may be expressed as

$$\mathbf{W}(l) = \mathbf{C}(l)\hat{\mathbf{W}} \tag{6}$$

where $\hat{\mathbf{W}}$ consists of the first N_S columns of a unitary matrix such as a Hadamard matrix or a Fourier matrix, and $\mathbf{C}(l)$ is a diagonal matrix representing the per-transmit chain phase shifts. The linear phase shifts may be implemented as cyclic transmit diversity (CTD) by introducing a different fixed cyclic time shift per transmit chain, which is represented in the frequency domain by the $N_T \times N_T$ matrix

$$\mathbf{C}(l) = diag\left\{1, e^{-j2\pi\delta_0 l\Delta_f}, \ldots, e^{-j2\pi\delta_{N_T-1} l\Delta_f}\right\} \tag{7}$$

where δ_i is the time shift on antenna i, $0 \le i \le N_T - 1$.

9.4 Transmit Beamforming

Transmit beamforming techniques may be employed when the transmitter has sufficient channel state information available to compute transmit beamforming vectors. The transmitter may acquire sufficient CSI by estimating the channel matrices using known sounding sequences sent to it by the intended receiver. This method assumes a time division duplex (TDD) system in which the downlink and uplink channels are reciprocal. Alternatively, the intended receiver may explicitly feed back either CSI or beamforming vectors to the transmitter.

9.4.1 Eigenvector Beamforming

Eigenvector beamforming is a technique in which the transmitter employs optimum transmit steering using the eigenvectors of the MIMO channel. The MIMO channel associated with a single OFDM subcarrier is decomposed into orthogonal spatial channels commonly referred to as eigenmodes [6]. The channel matrix for each subcarrier can be diagonalized by means of the singular value decomposition (SVD), as follows:

$$\mathbf{H}(l) = \mathbf{U}(l)\mathbf{D}(l)\mathbf{V}^{H}(l) \tag{8}$$

where $\mathbf{U}(l)$ ($N_R \times N_R$) and $\mathbf{U}(l)$ ($N_T \times N_T$) are matrices consisting of the left and right singular vectors, respectively, of the channel at subcarrier frequency $l\Delta_f$, and $\mathbf{D}(l)$ is a diagonal matrix of dimension $N_R \times N_T$ whose diagonal elements are the ordered singular values $\sqrt{\lambda_0(l)}, \sqrt{\lambda_1(l)}, \ldots, \sqrt{\lambda_{N_M-1}(l)}$ of the channel at that subcarrier. $\lambda_r(l)$ is an eigenvalue and $N_M = \min\{N_T, N_R\}$. The notation $\mathbf{A}^{H}$ denotes the complex conjugate transpose of the matrix $\mathbf{A}$.

The largest eigenvalue is sometimes referred to as the principal eigenvalue, and the associated eigenmode is referred to as the principal eigenmode. We can synthesize a set of *wideband eigenmodes* consisting of the eigenmodes associated with an eigenvalue of a specific rank across the entire set of subcarrier frequencies, $l\Delta_f$, $l_1 \leq l \leq l_2$. Thus, the principal wideband eigenmode consists of the collection of principal eigenmodes at each frequency $l\Delta_f$.

The resulting wideband eigenmodes exhibit interesting properties that make them particularly suitable for communication over frequency selective channels and that reflect the underlying statistics of the individual single-frequency eigenmodes. The most important of these is that the largest wideband eigenmodes exhibit relatively little frequency selectivity, while the smallest tend to reflect the frequency selectivity of the underlying single-input, single-output (SISO) channel.

The optimum transmit and receive steering vectors may be obtained from the SVD of the channel, and, in particular, when the columns of $\mathbf{V}(l)$ are used as transmit steering vectors and the rows of $\mathbf{U}^{H}(l)$ are used as receive steering vectors, up to N_M parallel channels can be synthesized [6].

The transmitted signal vector is formed by multiplying the vector of modulation symbols to be transmitted by the matrix of right singular vectors, as follows:

$$\mathbf{x}(l) = T[\mathbf{s}(l)] = \mathbf{V}(l)\mathbf{s}(l) \qquad (9)$$

The steering matrix $\mathbf{V}(l)$ maximizes the coupling of the transmitted signal into the natural modes of the channel. The resulting received signal vector is given by

$$\mathbf{y}(l) = \mathbf{H}(l)\mathbf{x}(l) + \mathbf{n}(l) = \mathbf{U}(l)\mathbf{D}(l)\mathbf{s}(l) + \mathbf{n}(l) \qquad (10)$$

where $\mathbf{n}(l)$ is a column vector of complex Gaussian noise elements each with zero mean and variance N_0.

At the receiver, processing the received vector with the matrix of left singular vectors results in an estimate of the transmitted modulation symbol vector:

$$\tilde{\mathbf{s}}(l) = \mathbf{U}^H(l)\mathbf{y}(l) = \mathbf{D}(l)\mathbf{s}(l) + \tilde{\mathbf{n}}(l) \text{ where } \tilde{\mathbf{n}}(l) = \mathbf{U}^H(l)\mathbf{n}(l). \qquad (11)$$

Alternatively, a matched filter or minimum mean-squared error (MMSE) receiver may be employed in place of the matrix of left singular vectors. We shall explore this further in the subsequent sections.

With noiseless channel state information at the transmitter and receiver, there is no crosstalk between the symbols in the estimate at the receiver as $\mathbf{D}(l)$ is a diagonal matrix. Therefore the elements of $\mathbf{s}(l)$ are received as though they were transmitted in N_S parallel channels. Up to N_M spatial streams may be created and each spatial stream may be assigned an individual rate that is kept fixed across the subcarriers.

Assuming a TDD communication link between two stations where the uplink channel is expressed as

$$\mathbf{H}_U(l) = \mathbf{U}(l)\mathbf{D}(l)\mathbf{V}^H(l), \qquad (12)$$

the downlink may be expressed as

$$\mathbf{H}_D(l) = \mathbf{H}_U^T(l) = \mathbf{V}^*(l)\mathbf{D}(l)\mathbf{U}^T(l) \qquad (13)$$

as a result of the channel reciprocity inherent in TDD links. $\mathbf{A}^T$ and $\mathbf{A}^*$ denote the transpose and the complex conjugate of the matrix $\mathbf{A}$, respectively.

Assume Station A transmits on the uplink and receives on the downlink. Then its optimum transmit and receive steering vectors are $\mathbf{V}(l)$ and $\mathbf{V}^T(l)$, respectively. Likewise, if Station B transmits on the downlink and receives on the uplink, its optimum transmit and receive steering vectors are $\mathbf{U}^*(l)$ and $\mathbf{U}^H(l)$, respectively. Thus, once a station has obtained its receive vectors, it may derive the transmit steering vectors by conjugating the receive vectors.

9.4.2 Channel Sounding and Calibration

Before transmit beamforming can commence, the transmitter must acquire or compute an estimate of the MIMO channel over which it is about to transmit. This may be accomplished efficiently by having the intended receiver of the beamformed transmission send MIMO channel sounding sequences back to the transmitter. The transmitter then estimates the channel and derives its transmit steering vectors by means of the SVD, as discussed above.

The beamforming channel model is shown in Figure 9.2. Beamforming transmissions from Station A to Station B are enabled when Station B sends Station A a sounding sequence, allowing Station A to form an estimate of the MIMO channel from Station B to Station A, for all subcarriers. In a TDD channel in which the downlink and uplink channels are reciprocal, the channel from Station A to Station B is the matrix transpose of the channel from Station B to Station A, to within a complex scaling factor, i.e.,

$$\mathbf{H}_{AB}(l) = \rho(l) \cdot \mathbf{H}_{BA}^{T}(l) \tag{14}$$

Station A uses this relationship to compute transmit steering vectors that are suitable for transmitting to Station B over $\mathbf{H}_{AB}(l)$.

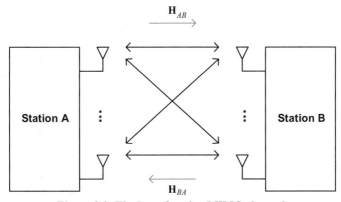

Figure 9.2: The beamforming MIMO channel.

While the over-the-air channel between the antennas at Station A and the antennas at Station B is reciprocal, the observed baseband-to-baseband channel used for communication may not be, as it includes the transmit and receive chains of the stations. Differences in the amplitude and phase characteristics of the transmit and receive chains associated with individual antennas degrade the reciprocity of the over-the-air channel, and cause performance degradation. An over-the-air calibration procedure may be used to restore reciprocity. The procedure provides the means for calculating a set of correction matrices that can be applied at the transmit side of a station to correct the amplitude and phase differences between the transmit and receive chains in the station. If this is done at both ends of the link, reciprocity is restored in the baseband-to-baseband response of the downlink and uplink channels. Note that it is also possible to compute and apply correction matrices for the receive side.

Figure 9.3 illustrates the baseband-to-baseband channel, including reciprocity correction, as seen by Stations A and B. The amplitude and phase responses of the transmit and receive chains can be expressed as diagonal matrices with complex valued diagonal entries, of the form $\mathbf{A}_{TX}(l)$ (or $\mathbf{A}_{RX}(l)$). The relationship between the baseband-to-baseband channel, $\tilde{\mathbf{H}}_{AB}(l)$, and the over-the-air channel $\mathbf{H}_{AB}(l)$ is

$$\tilde{\mathbf{H}}_{AB}(l) = \mathbf{B}_{RX}(l)\mathbf{H}_{AB}(l)\mathbf{A}_{TX}(l) \tag{15}$$

and, similarly, the relationship between $\tilde{\mathbf{H}}_{BA}(l)$ and $\mathbf{H}_{BA}(l)$ is

$$\tilde{\mathbf{H}}_{BA}(l) = \mathbf{A}_{RX}(l)\mathbf{H}_{BA}(l)\mathbf{B}_{TX}(l). \tag{16}$$

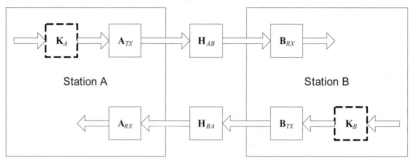

Figure 9.3: The baseband-to-baseband channel.

As an example, consider the case where calibration is performed at both Station A and Station B [1]. The objective is to compute transmit side correction matrices $\mathbf{K}_A(l)$ and $\mathbf{K}_B(l)$ that restore reciprocity such that

$$\tilde{\mathbf{H}}_{AB}(l)\mathbf{K}_A(l) = \eta(l) \cdot \left[\tilde{\mathbf{H}}_{BA}(l)\mathbf{K}_B(l)\right]^T \tag{17}$$

where $\eta(l)$ is a complex scaling factor. The correction matrices are diagonal matrices with complex valued diagonal entries. The reciprocity condition above is enforced when

$$\mathbf{K}_A(l) = \alpha_A(l)\mathbf{A}_{TX}^{-1}(l)\mathbf{A}_{RX}(l) \tag{18}$$

and

$$\mathbf{K}_B(l) = \alpha_B(l)\mathbf{B}_{TX}^{-1}(l)\mathbf{B}_{RX}(l) \tag{19}$$

where $\alpha_A(l)$ and $\alpha_B(l)$ are complex valued scaling factors. Using these expressions for the correction matrices, the calibrated baseband-to-baseband channel between Station A and Station B is expressed as

$$\hat{\mathbf{H}}_{AB}(l) = \tilde{\mathbf{H}}_{AB}(l)\mathbf{K}_A(l) = \alpha_A(l)\mathbf{B}_{RX}(l)\mathbf{H}_{AB}(l)\mathbf{A}_{RX}(l) \tag{20}$$

and if both sides apply the correction matrices, then we have that

$$\hat{\mathbf{H}}_{BA}(l) = \alpha_B(l)\mathbf{A}_{RX}(l)\mathbf{H}_{BA}(l)\mathbf{B}_{RX}(l) = \frac{\alpha_B(l)}{\alpha_A(l)} \cdot \hat{\mathbf{H}}_{AB}^T(l) \tag{21}$$

Focusing on Station A, a possible procedure for estimating $\mathbf{K}_A(l)$ would be as follows:

1. Station A sends Station B a sounding sequence, allowing Station B to estimate the channel matrices $\tilde{\mathbf{H}}_{AB}(l)$.

2. Station B sends Station A a sounding sequence, allowing Station A to estimate the channel matrices $\tilde{\mathbf{H}}_{BA}(l)$.

3. Station B sends its estimates of $\tilde{\mathbf{H}}_{AB}(l)$ to Station A.

4. Station A uses its local estimates of $\tilde{\mathbf{H}}_{BA}(l)$ and the estimates $\tilde{\mathbf{H}}_{AB}(l)$ received from Station B to compute the correction matrices $\mathbf{K}_A(l)$.

Steps 1. and 2. must occur over a short time interval to ensure that the channel changes as little as possible between measurements. A similar procedure may be used to estimate $\mathbf{K}_B(l)$ at Station B.

Transmitting a MIMO sounding sequence using transmit steering vectors allows a receiving station to directly estimate receive steering vectors without the intermediate steps of estimating the channel and performing an SVD calculation. Due to the reciprocity of the TDD channel, the receiving station can then calculate transmit steering vectors from the estimate of the receive vectors. The steering vectors must be updated sufficiently often to reflect changes due to the time-varying nature of the channel.

The MIMO sounding sequences may also be used by the receiving station to determine the number of active spatial streams and data rates that may reliably be supported. The rate recommendations can then be fed back to the transmitting station.

9.5 Receiver Structures

The theoretically optimal receiver for the transmission schemes described above is a maximum likelihood (ML) sequence receiver that is capable of making joint decisions on all the information bits using knowledge of the correlation introduced by the channel code across blocks, data subcarriers and all the OFDM symbols in a packet. Such a receiver would be prohibitively complex, as it would need to perform an exhaustive search over all combinations of bits for the whole sequence of information bits transmitted in a packet. In this section, we shall instead investigate some more practical receiver structures.

9.5.1 Near-Optimal Iterative Receiver

Near-optimal performance can be achieved with a receiver that performs iterative joint detection and decoding, as shown in Figure 9.4. This receiver consists of a MIMO detector and a channel decoder, both of which compute soft decisions on the coded bits. The soft information is exchanged between the two units in an iterative manner, increasing the reliability of the decisions with each iteration performed. There are several choices for both the detector and the decoder.

The decoder must be capable of generating soft outputs from soft inputs, and may, for example, be a soft-output Viterbi (SOVA) decoder [7] or a maximum a posteriori probability (MAP) decoder, e.g., the BCJR [8].

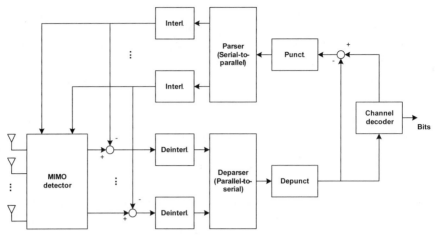

Figure 9.4: Iterative MIMO receiver.

The optimal MIMO detector is a MAP detector that minimizes the error probability for each received bit [9]. Soft bit decisions in the form of a posteriori probabilities (APPs) are usually expressed as log-likelihood ratios (LLRs) for each coded bit b_k in subcarrier l (for ease of notation, the subcarrier index is suppressed in the following):

$$L(b_k \mid \mathbf{y}) = \ln \frac{P\{b_k = +1 \mid \mathbf{y}\}}{P\{b_k = -1 \mid \mathbf{y}\}}, \; k = 1, 2, \ldots, N_S \cdot K \qquad (22)$$

where K denotes the number of bits per QAM symbol. The log-MAP LLRs may be separated into two parts, namely a priori LLRs $L_a(b_k)$ provided to the detector by the decoder, and extrinsic ("new") LLRs computed by the detector. That is,

$$L(b_k \mid \mathbf{y}) = L_a(b_k) + L_{ext}(b_k) \text{ where } L_a(b_k) = \ln \frac{P\{b_k = +1\}}{P\{b_k = -1\}}. \qquad (23)$$

A priori LLRs provided by the decoder are subtracted from the output of the MIMO detector, and the resulting extrinsic information is deinterleaved to become a priori LLRs for the decoder. Likewise, this a priori information is subtracted from the output LLRs computed by the decoder before being fed back to the detector in the subsequent iteration, as shown in Figure 9.4.

A commonly used approximation of the log-MAP LLR is known as max-log-MAP, and results in the following expression for the LLR for bit b_k:

$$L_{ext}(b_k \mid \mathbf{y}) \approx \frac{1}{2} \max_{\mathbf{b}: \, b_k = +1} \left\{ -\frac{1}{\sigma^2} \|\mathbf{y} - \mathbf{H}_e \mathbf{s}\|^2 + \mathbf{b}_{[k]}^T \mathbf{L}_{a,[k]} \right\} - \frac{1}{2} \max_{\mathbf{b}: \, b_k = -1} \left\{ -\frac{1}{\sigma^2} \|\mathbf{y} - \mathbf{H}_e \mathbf{s}\|^2 + \mathbf{b}_{[k]}^T \mathbf{L}_{a,[k]} \right\} \qquad (24)$$

where $\mathbf{b}_{[k]}$ denotes the subvector of $\mathbf{b}$ excluding the kth element, $\mathbf{L}_{a,[k]}$ denotes the vector of a priori LLRs also excluding the kth element, and $\|\mathbf{y}-\mathbf{H}_e\mathbf{s}\|^2$ represents the Euclidean distance cost function. The summations are performed over all vectors $\mathbf{b}$ having $b_k = +1$ and $b_k = -1$, respectively, associated with subcarrier l.

9.5.2 List Sphere Decoding

List sphere decoding (LSD) is a method for reducing the complexity of the log-MAP or max-log-MAP detectors known from the recent literature [9]. LSD is based on sphere decoding principles and seeks to reduce the search space of the log-MAP or max-log-MAP by considering only those hypotheses for which the cost function

$$\|\mathbf{y}-\mathbf{H}_e\mathbf{s}\|^2 \tag{25}$$

in the expression for the bit LLR is small, i.e., less than or equal to the parameter r^2, which is referred to as the sphere radius.

The cost function may be rewritten as follows:

$$J(\mathbf{s}) = \|\mathbf{y}-\mathbf{H}_e\mathbf{s}\|^2 = (\mathbf{s}-\hat{\mathbf{s}})^H \mathbf{U}^H \mathbf{U}(\mathbf{s}-\hat{\mathbf{s}}) + C \tag{26}$$

where $\hat{\mathbf{s}}$ represents the center point for the search (e.g., the zero-forcing (ZF) or minimum mean-squared error (MMSE) solution) and $\mathbf{U}$ is the upper-triangular result of a QR or Cholesky decomposition of $\mathbf{H}_e$, $\mathbf{U}^H\mathbf{U} = \mathbf{H}_e^H\mathbf{H}_e$. The constant C is omitted since it is not a function of $\mathbf{s}$.

By reformulating the cost function into a summation over the individual transmit stream candidates, the LSD can incrementally compute the cost of each transmit stream symbol, and thereby prune the symbol combinations for which J is large (greater than r^2):

$$J = \sum_{i=N_S}^{1} T_i \text{ where } T_i = \left| u_{ii}(s_i - \hat{s}_i) + \sum_{j=i+1}^{N_S} u_{ij}(s_j - \hat{s}_j) \right|^2 . \tag{27}$$

By using the upper-triangular matrix $\mathbf{U}$ instead of $\mathbf{H}_e$ for the cost calculations, each term in the summation depends only on the current symbol decision, s_i, as well as the decisions made on the symbols in the data streams considered thus far, i.e., $s_{N_S}, \ldots, s_{i+1}$.

9.5.3 Linear Receivers

When operating in eigenvector steering mode, the channel is effectively orthogonalized by the application of steering vectors at the transmitter. Inter-stream cross talk is minimized, and, consequently, the need for complicated receiver processing to separate the streams is greatly reduced. A simple suboptimal receiver may be sufficient in this case. Such a receiver may employ ZF or MMSE processing at the front-end, followed by per-data stream bit LLR computation. Figure 9.5 illustrates an MMSE-based receiver structure.

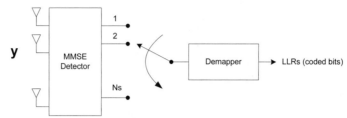

Figure 9.5: Linear receiver structure.

As an example, MMSE processing involves applying a spatial filter $\mathbf{M}$ to the received signal vector:

$$\tilde{\mathbf{s}} = \mathbf{M}\mathbf{y} \text{ where } \mathbf{M} = \left(\sigma^2\mathbf{I} + \mathbf{H}_e^H\mathbf{H}_e\right)^{-1}\mathbf{H}_e^H \tag{28}$$

and $\mathbf{I}$ denotes the identity matrix. An unbiasing operation then follows:

$$\hat{\mathbf{s}} = \mathbf{\Gamma}^{-1}\tilde{\mathbf{s}} \text{ where } \mathbf{\Gamma} = \mathrm{diag}\left(\mathbf{M}\mathbf{H}_e\right). \tag{29}$$

The per-stream max-log-MAP LLR is given by

$$L_{ext}(b_k \mid \hat{s}_i) \approx \frac{1}{2}\max_{\mathbf{b}_i:\, b_k=+1}\left\{-\frac{\gamma_i^2}{N_0}\left|\hat{s}_i - s_i\right|^2 + \mathbf{b}_{i,[k]}^T\mathbf{L}_{a,i,[k]}\right\} - \frac{1}{2}\max_{\mathbf{b}_i:\, b_k=-1}\left\{-\frac{\gamma_i^2}{N_0}\left|\hat{s}_i - s_i\right|^2 + \mathbf{b}_{i,[k]}^T\mathbf{L}_{a,i,[k]}\right\} \tag{30}$$

for $k = 1, 2, \ldots, K_i$ and $i = 1, 2, \ldots, N_S$, where γ_i is the ith diagonal element of $\mathbf{\Gamma}$. As unequal modulations per stream may be used, K_i denotes the number of bits per QAM symbol on stream i. The maximizations are performed over all vectors $\mathbf{b}_i$ having $b_k = +1$ and $b_k = -1$, respectively, for a single stream at a time.

9.6 Comparison of Spatial Spreading and Transmit Beamforming

One significant difference between spatial spreading and eigenvector-based transmit beamforming is how the signal-to-noise ratio (SNR) is distributed per spatial stream at the receiver. With eigenvector beamforming, the SNR per spatial stream in a given subcarrier is directly proportional to the eigenvalue in that subcarrier. As a result, the larger eigenmodes contribute a larger fraction of the total capacity. Furthermore, the lower SNR variation exhibited by the larger eigenmodes permits the use of error correcting codes with lower redundancy. For example, punctured convolutional codes can be used without significant loss of performance. With stronger codes, such as turbo codes or low density parity check (LDPC) codes, the performance is much less affected by SNR variations.

With spatial spreading, the received SNR is determined in part by the cross talk among the symbol streams transmitted over the spatial channels. As a result, all spatial channels have statistically similar SNR distributions, with a variance that can be significantly greater than that of the larger eigenmodes created using eigenvector beamforming. Spatial spreading creates diversity across the band by exploiting the spatial dimensionality of the channel. It can increase the SNR variation across the band, which may degrade the average throughput when using weak error correcting codes. However, in return spatial spreading reduces the outage probability.

Figure 9.6 shows the cumulative probability distributions of the power gain obtained by using various transmission schemes in a 2×2 channel whose elements are complex Gaussian random variables with unity average power. For eigenvector beamforming, the distributions of the maximum eigenvalue (λ_{max}) and the minimum eigenvalue (λ_{max}), are shown. The maximum eigenvalue follows the fourth-order maximum ratio diversity combining distribution, while the minimum eigenvalue is Rayleigh distributed. Also shown is the power gain distribution associated with transmit beamforming where the steering vectors are fixed across subcarriers ("Fixed Steering"). Fixed steering is better than Rayleigh, but not quite as good as fourth-order diversity. Finally, we observe that the distribution of power gains achieved with spatial spreading is closer to that of the principal eigenvalue, but off by about 3 dB.

In the following sections, we compare eigenvector beamforming and spatial spreading in terms of simulated average physical layer data rate vs range performance, and packet error rate versus SNR performance.

9.6.1 Simulation Setup

The simulation results presented here were obtained with an 802.11n baseband link simulator using the 20 MHz mode of operation. This mode uses 64 subcarriers, of which 52 are used for data and four as pilot. The DC subcarrier is unused, and the remaining subcarriers are used as guard subcarriers. Each 3.2 μs OFDM symbol is preceded by a guard interval of duration 0.8 μs, resulting in a total OFDM symbol duration of 4.0 μs. The packet size is 1000 bytes, and the simulation results shown have been averaged over thousands of channel realizations. The SNR is given in terms of total E_s/N_0 per receiver, where E_s reflects the captured energy in the 3.2 μs OFDM symbol, i.e., exclusive of the guard interval. The simulation results were obtained with IEEE 802.11n channel model B, which is representative of an indoor wireless environment. This model has an rms delay spread of 15 ns, a total delay spread of 80 ns, two clusters of scatterers, and a 6 Hz Doppler component at the carrier frequency [10].

Where noted, the simulation results also reflect RF impairments modeled according to the requirements specified in the 802.11n technology selection criteria document [11]. These impairments include 1) phase noise at both transmitter and receiver, 2) power amplifier non-linearity, 3) carrier frequency offset, and 4) symbol timing clock offset.

MIMO training sequences are used to perform channel estimation and compute coefficients for the spatial processing at the receiver. Acquisition and MIMO channel estimation are performed on a per-packet basis. In the case of eigenvector beamforming, a multi-packet sequence is simulated in order to let the transmitter acquire the necessary CSI to compute its transmit steering vectors. Several MIMO configurations and receiver alternatives were evaluated.

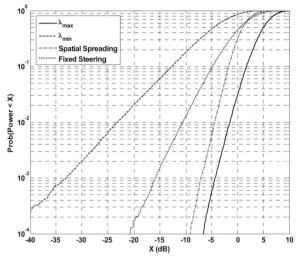

Figure 9.6: Gain distributions for 2 × 2 system.

9.6.2 Data rate vs Range Performance

Figures 9.7 to 9.9 show the average physical layer data rate as a function of distance for eigenvector beamforming and spatial spreading with 2×2, 4×2, and 4×4 configurations. The rates were achieved using a practical rate control algorithm that seeks to choose the highest data rate that can be sustained under the current channel conditions while maintaining operation with a target packet error rate of 1%. We assume a total transmit power of 17 dBm, a noise figure of 10 dB, and operation at 2.4 GHz and 5.25 GHz, respectively. The path loss model consists of free space loss with a slope of 2 up to a breakpoint distance of 5 m and a slope of 3.5 beyond the breakpoint distance. A linear MMSE receiver was used in all cases. For these simulations, a single packet was transmitted per channel realization, and 2000 channel realizations were generated per range point.

We observe that eigenvector beamforming with link adaptation can achieve the same data rate as spatial spreading at 50-100% greater range, depending on configuration and frequency band.

9.6.3 Packet Error Rate Performance

In this section, we focus on comparing the performance of the various receiver structures discussed previously in Section 9.5. In particular, Figure 9.10 shows the performance improvement achieved by iterative LSD and soft-output Viterbi decoding over the linear MMSE and ZF receivers for spatial spreading in a 2 × 2 configuration with data rates of 39 Mbps and 52 Mbps [12]. For QPSK, rate-3/4 (39 Mbps), a single iteration of the LSD results in an 8 dB gain over the MMSE receiver at 1% PER. The second iteration provides an additional 1 dB gain. For 16-QAM, rate-1/2 (52 Mbps), the first iteration of the LSD results in a gain of 5 dB at 1% PER, while the second iteration improves the performance by another 1.5 dB.

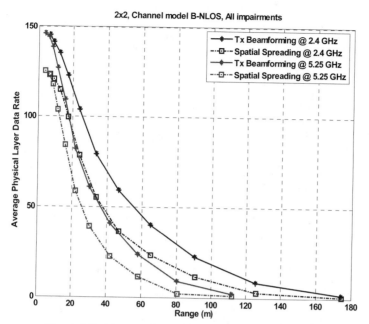

Figure 9.7: Average data rate vs range for 2 × 2 systems.

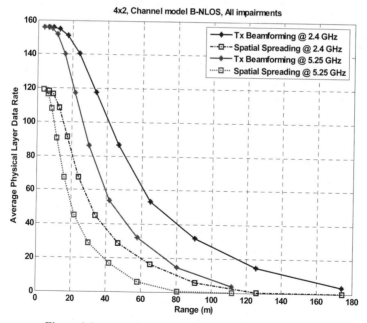

Figure 9.8: Average data rate vs range for 4 × 2 systems.

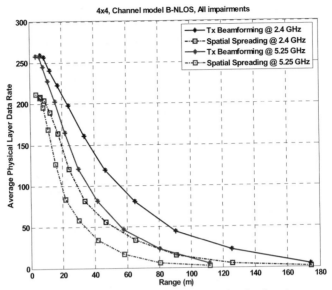

Figure 9.9: Average data rate vs range for 4 × 4 systems.

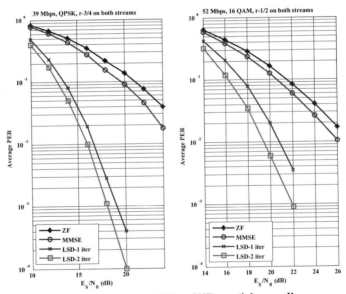

Figure 9.10: Average PER vs SNR, spatial spreading.

Figure 9.11 compares the performance of spatial spreading and eigenvector beamforming in a 2×2 configuration (beamforming results are labeled "ES"). With eigenvector beamforming, the SNR on a given subcarrier of a spatial stream is directly proportional to the eigenvalue in that subcarrier. As mentioned previously, this property allows independent modulations to be used on each spatial stream. In principle, independent code rates could also be applied, but this has not been adopted by 802.11n.

The beamforming performance is shown for six different code rate/modulation combinations that achieve total rates of 39 Mbps and 52 Mbps, respectively, all using the MMSE receiver. With spatial spreading, the same modulations are used on both streams. As shown in Figure 9.11, the best choice of the beamforming rate with the linear MMSE receiver outperforms spatial spreading with two iterations of the LSD by 2 dB, in both data rate cases. In this case, the best choice is transmitting on a single stream with high-order modulation.

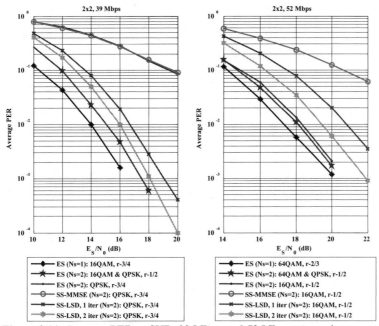

Figure 9.11: Average PER vs SNR, 39 Mbps and 52 Mbps comparison.

With noiseless CSI at the transmitter and receiver, eigenvector beamforming results in complete channel diagonalization and a set of orthogonal spatial channels. Loss of orthogonality due to channel estimation errors and other mismatches is mitigated by the MMSE receiver. With spatial spreading, cross talk between spatial streams results in a lower SNR; furthermore, the received SNR variance across frequency can be significantly greater than that on the larger eigenmodes, resulting in a degradation in performance of the convolutional code.

9.7 Complexity Analysis

In this section, we compare the computational complexity of iterative LSD and decoding with that of the simple linear receiver that we use in conjunction with eigenvector steering. The complexity is stated in terms of the equivalent number of real multiplies (MUL) and additions (ADD) required per decoded (i.e., information) bit. Table 9.2 shows the assumed equivalent complexity of the various mathematical operations used.

Table 9.2: Equivalent complexity of mathematical operations.

	Real MUL	Real ADD
Complex ADD; max; min	-	2
Complex MUL	4	2
Square root	5	6
Mul. by -1; compare; div. by 2	-	1
Real DIV	1	-

The LSD consists of three functional blocks: a precomputer, a search unit, and an LLR computer [13]. The precomputer calculates the center point for the search, and performs a triangular decomposition of the channel matrix for use by the search unit. The search unit traverses a search tree of possible symbol candidates, and computes incremental costs for each node at each level in the tree. When the search has been completed, it provides a pruned list of symbol candidates and their associated cost. Finally, the LLR computer calculates the extrinsic LLRs, but based on the pruned rather than the exhaustive list of candidates.

For this analysis, we assume that the MMSE solution is used as the center point for the search, and that Cholesky decomposition is used to triangularize the channel matrix. The SOVA is used for soft-output decoding in all iterations except the last one, in which the conventional Viterbi algorithm (VA) is used. We ignore the complexity of interleaving/deinterleaving and puncturing/depuncturing. Consequently, the overall complexity of the iterative receiver may be expressed as

$$
\begin{aligned}
C = {} & C(\text{MMSE}) + C(\text{Chol}) + C(\text{LSD search}) \\
& + C(\text{LSD-MLM})|_{full} + (N_{it} - 1) \cdot C(\text{LSD-MLM})|_{it} \\
& + (N_{it} - 1) \cdot C(\text{SOVA}) + C(\text{VA})
\end{aligned}
\tag{31}
$$

where N_{it} denotes the number of iterations, $C(\text{LSD-MLM})|_{full}$ denotes the complexity of the operations needed for max-log-MAP LLR computations performed only during the first iteration, and $C(\text{LSD-MLM})|_{it}$ denotes the complexity of the operations performed only during the subsequent iterations. The complexity of the LSD tree search is dependent on the number of tree nodes visited, which, in turn, depends on the channel, the SNR, the sphere radius and the number of symbol candidates allowed in the list. For this reason, we derive an expression for the *average* complexity, based on the average number of nodes visited at each level in the tree. The complexity of the simple non-iterative MMSE-based receiver may be expressed as

$$
C = C(\text{MMSE}) + C(\text{MMSE-MLM}) + C(\text{VA})
\tag{32}
$$

where $C(\text{MMSE})$ denotes the complexity of MMSE processing, $C(\text{MMSE-MLM})$ denotes the complexity of per-stream max-log-MAP LLR computation, and $C(\text{VA})$ denotes the complexity of the Viterbi algorithm.

The complexity of the various sub-blocks is analyzed in the following sections.

9.7.1 MMSE Processing

Define $\mathbf{P} = (\sigma^2 \mathbf{I} + \mathbf{H}_e^H \mathbf{H}_e)^{-1}$. The MMSE filter is then given by $\mathbf{M} = \mathbf{P}\mathbf{H}_e^H$. To avoid the matrix inversion, the matrix $\mathbf{P}$ may be computed iteratively using the recursion

$$\mathbf{P}_i = \mathbf{P}_{i-1} + \frac{\mathbf{P}_{i-1} \mathbf{H}_e^H(i,:) \mathbf{H}_e(i,:) \mathbf{P}_{i-1}}{\sigma_n^2 + \mathbf{H}_e(i,:) \mathbf{H}_e^H(i,:)} \tag{33}$$

for $i = 1, 2, \ldots, N_R$, where $\mathbf{H}_e(i,:)$ denotes the ith row of $\mathbf{H}_e$.

The matrices $\mathbf{P}_1, \ldots, \mathbf{P}_{N_R}$ are hermitian, so only the upper triangles need to be computed. The MMSE solution is computed each time the channel estimate is updated, typically once per packet. The number of real MUL and ADD operations per data bit for the MMSE processing are summarized in Table 9.3. R is the code rate which is common across the streams, $N_{SD} = 52$ is the number of data subcarriers, and N_O is the number of OFDM symbols given by

$$N_O = \left\lceil \frac{N_b}{\sum_{i=1}^{N_S} K_i \cdot R \cdot N_{SD}} \right\rceil \text{ where } N_b \text{ is the number of information bits.} \tag{34}$$

9.7.2 Cholesky Decomposition

The Cholesky decomposition is computed each time the channel estimate is updated, which we assume is once per packet. We also assume a scaled and decoupled decomposition such as the one described in [14]. The number of real MUL and ADD operations are summarized in Table 9.3.

9.7.3 LSD Search

The LSD tree search has been described in [9] and [13], and involves computing the incremental costs for 2^{K_i} branches for each surviving node in the search tree. For the purpose of this analysis, we assume that the average number of candidates that satisfy the sphere radius criterion at level i in the tree is given by $N_{ave}(i)$, where, typically, $N_{ave}(i) < 2^{K_i}$. The tree search produces a list of candidates that all satisfy the radius criterion, and whose cost functions are stored and ready to be used in the subsequent max-log-MAP LLR computations. The number of real MUL and ADD operations as a function of $N_{ave}(i)$, $i = 1, 2, \ldots, N_S$, are summarized in Table 9.3. The number of surviving nodes per level may be limited to N_{cand}, which can be made dependent on the constellation size used. Consequently, the upper bound on the number of symbol candidates in the list is given by $N_{cand}^{N_S}$, and this quantity determines the hardware complexity.

9.7.4 LSD Max-log-MAP

On the average, only $\sum_{i=1}^{N_S} N_{ave}(i)$ hypotheses for **s** are considered when computing the bit LLRs, and only $\sum_{i=1}^{N_S} N_{ave}(i) - 2 \max\{\cdot\}$ operations are performed. The corresponding upper bounds on these quantities are given by $N_{cand}^{N_S}$ and $N_{cand}^{N_S} - 2$, respectively. The resulting average complexity, expressed as the equivalent number of real MUL and ADD operations, is summarized in Table 9.3, where the indicator function $\mathbf{1}_{full}$ is 1 only in the first iteration (otherwise 0), and $\mathbf{1}_{it}$ is 1 if a-priori LLRs are available, as they would normally be in the second and subsequent iterations.

9.7.5 Per-Stream LLR Computation

When performing per-stream LLR computation as in the suboptimal MMSE-based receiver, the number of hypotheses that need to be considered is reduced from $2^{\sum_{i=1}^{N_S} K_i}$ (for full-complexity max-log-MAP) to $\sum_{i=1}^{N_S} 2^{K_i}$. Further complexity reduction is made possible by the fact that the cost functions take scalar inputs rather than vectors and matrices. The complexity of per-stream max-log-MAP LLR computation is summarized in Table 9.3.

9.7.6 Viterbi Decoding

The complexities of the conventional Viterbi algorithm as well as the soft-output Viterbi algorithm were analyzed in [15]. Following a similar approach, the complexity in terms of number of real ADD operations per decoded bit is summarized in Table 9.3. The constraint length is L, and the truncation length is $D = 5(L - 1)$.

Table 9.3: Complexity of LSD sub-blocks.

$C(\text{MMSE})$	Real MUL	$\dfrac{11N_R N_S^2 + 7N_R N_S - 4N_S^2 - 2N_S + \left(4N_R N_S + 3N_S\right)N_O}{\sum_{i=1}^{N_S} K_i \cdot R \cdot N_O}$
	Real ADD	$\dfrac{10N_R N_S^2 + 2N_R N_S - 5N_S^2 + \left(4N_R N_S - 2N_S\right)N_O}{\sum_{i=1}^{N_S} K_i \cdot R \cdot N_O}$
$C(\text{Chol})$	Real MUL	$\dfrac{24N_R^3 + 4N_S^3 - 3N_S^2 + 29N_S}{6 \cdot \sum_{i=1}^{N_S} K_i \cdot R \cdot N_O}$
	Real ADD	$\dfrac{12N_R^3 - 6N_R^2 + 2N_S^3 - 3N_S^2 + 19N_S}{3 \cdot \sum_{i=1}^{N_S} K_i \cdot R \cdot N_O}$

$C(\text{LSD search})$	Real MUL	$\dfrac{\displaystyle\sum_{i=1}^{N_S-1}\left[4\left(N_S-i\right)+4\right]\cdot 2^{K_i}\cdot\prod_{j=i+1}^{N_S}N_{ave}(j)+4\cdot 2^{K_{N_S}}}{\displaystyle\sum_{i=1}^{N_S}K_i\cdot R}$
	Real ADD	$\dfrac{\displaystyle\sum_{i=1}^{N_S-1}\left[4\left(N_S-i\right)+5\right]\cdot 2^{K_i}\cdot\prod_{j=i+1}^{N_S}N_{ave}(j)+4\cdot 2^{K_{N_S}}}{\displaystyle\sum_{i=1}^{N_S}K_i\cdot R}$
$C(\text{LSD-MLM})$	Real MUL	$\dfrac{\displaystyle\prod_{i=1}^{N_S}N_{ave}(i)\left[\mathbf{1}_{full}+\mathbf{1}_{it}\cdot\left(\sum_{j=1}^{N_S}K_j-1\right)\right]}{\displaystyle\sum_{i=1}^{N_S}K_i\cdot R}$
	Real ADD	$\dfrac{\displaystyle\prod_{i=1}^{N_S}N_{ave}(i)\left[\mathbf{1}_{full}+\mathbf{1}_{it}\cdot\left(\sum_{j=1}^{N_S}K_j-1\right)+2\sum_{l=1}^{N_S}K_l\right]-2\sum_{i=1}^{N_S}K_i}{\displaystyle\sum_{i=1}^{N_S}K_i\cdot R}$
$C(\text{MMSE-MLM})$	Real MUL	$\dfrac{3\displaystyle\sum_{i=1}^{N_S}2^{K_i}}{\displaystyle\sum_{i=1}^{N_S}K_i\cdot R}$
	Real ADD	$\dfrac{4\displaystyle\sum_{i=1}^{N_S}2^{K_i}+2\sum_{i=1}^{N_S}K_i\cdot 2^{K_i}-2\sum_{i=1}^{N_S}K_i}{\displaystyle\sum_{i=1}^{N_S}K_i\cdot R}$
$C(\text{VA})$	Real ADD	$2+3\cdot 2^L+\left(2^L-2\right)\big/D$
$C(\text{SOVA})$	Real ADD	$4\cdot 2^L+2^{L-1}+3D-1$

9.7.7 Examples

Previously, we saw that eigenvector beamforming coupled with a simple MMSE-based receiver can outperform spatial spreading even when an iterative receiver is employed together with the latter. In this section, we quantify the computational complexity associated with these two approaches [12]. Specifically, we consider the 2×2 system operating at 39 Mbps introduced in Section 9.6.3. The number of information bits is $N_b =$

8000. Table 9.4 shows the equivalent number of real MUL and ADD operations for the following four configurations:

1. Spatial spreading; 2 streams of QPSK, $R = 3/4$; $N_O = 52$; iterative LSD-SOVA receiver with $N_{it} = 1$ iteration ($N_{ave}(1) = 2.20$; $N_{ave}(2) = 2.33$).
2. Same as 1., but with $N_{it} = 2$ iterations.
3. Eigenvector beamforming; 1 stream of 16-QAM, $R = 3/4$; $N_O = 52$; MMSE receiver.
4. Eigenvector beamforming; 2 streams, 16-QAM, QPSK, $R = 1/2$; $N_O = 52$; MMSE receiver.

Table 9.4: Complexity comparison of iterative LSD and MMSE-based receivers.

	Real MUL	Real ADD
1.	40.12	440.90
2.	45.25	1122.02
3.	19.86	^53.66
4.	27.95	465.30

Note that the fourth configuration employs different modulations on the two streams. We note that, on average, the iterative receiver with more than one iteration incurs a complexity penalty over the MMSE-based receiver. Specifically, for the same 39 Mbps data rate, the iterative receiver with two iterations requires about twice as many MUL operations and more than twice as many ADD operations as the MMSE-based receiver for a single stream. Although not analyzed here, the complexity advantage of the MMSE receiver is even more pronounced for systems with higher dimensionality and modulation orders. It is also worth noting that iterative processing introduces additional delays proportional to the number of iterations.

9.8 Conclusions

In this chapter, we have discussed two alternatives for spatial signal processing applicable to 802.11n high throughput MIMO WLAN systems, known as spatial spreading and eigenvector transmit beamforming. We have also examined possible receiver structures for use with these transmission schemes, including near-optimal iterative LSD and SOVA, and linear MMSE-based approaches. We have shown that with an MMSE-based receiver, eigenvector beamforming achieves a substantial performance gain over spatial spreading. Applying two iterations of LSD and SOVA decoding to spatial spreading results in as much as 9 dB improvement in performance at 1% PER over the simple MMSE receiver. However, the best choice of code rate and modulation with eigenvector beamforming and an MMSE receiver still outperforms spatial spreading with two iterations of the LSD.

An extensive complexity analysis comparing the receiver structures has been presented. Illustrative examples show that on the average, the iterative receiver may require up to twice the number of MUL and ADD operations, compared to the simple MMSE-based receiver, for the same data rate.

9.9 References

[1] IEEE 802.11 WG, "IEEE P802.11n D2.0 Draft Amendment to Standard for Information Technology - Telecommunications and information exchange between systems - Local and metropolitan networks - Specific requirements. Part 11: Wireless LAN Medium Access Control (MAC) and Physical Layer (PHY) specifications: Enhancements for Higher Throughput," February 2007.

[2] I. Medvedev, et al., "Transmission Strategies for High Throughput MIMO OFDM Communication," *IEEE International Conference on Communications*, pp. 2621-2625, May 2005.

[3] G. J. Foschini and M. J. Gans, "On Limits of Wireless Communications in a Fading Environment When Using Multiple Antennas," *Wireless Personal Communications*, pp. 311-335, June 1998.

[4] IEEE 802.11 WG, "IEEE Std 802.11-2003 Standard for Information Technology - Telecommunications and information exchange between systems - Local and metropolitan area networks - Specific requirements. Part 11: Wireless LAN Medium Access Control (MAC) and Physical Layer (PHY) specifications," 2003.

[5] A. Narula, M. D. Trott, and G. W. Wornell, "Performance Limits of Coded Diversity Methods for Transmitter Antenna Arrays," *IEEE Trans. Inform. Theory*, Vol. 45, No. 7, pp. 2418-2433, Nov. 1999.

[6] J. B. Andersen, "Array Gain and Capacity for Known Random Channels with Multiple Element Arrays at Both Ends," *IEEE Journal on Selected Areas in Communications*, pp. 2172-2178, Nov. 2000.

[7] J. Hagenauer and P. Hoeher, "A Viterbi Algorithm with Soft-Decision Outputs and its Applications," *Proc. IEEE Globecom*, pp. 47.1.1-47.1.7, 1989.

[8] L. R. Bahl, J. Cocke, F. Jelinek, and J. Raviv, Optimal Decoding of Linear Codes for Minimizing Symbol Error Rate," *IEEE Trans. Inform. Theory*, Vol. 20, pp. 284-287, Mar. 1974.

[9] B. M. Hochwald and S. ten Brink, "Achieving Near-Capacity on a Multiple-Antenna Channel," *IEEE Trans. on Commun.*, vol. 51, pp. 389-399, Mar. 2003.

[10] IEEE 802.11 WG, "TGn Channel Models," IEEE 802.11-03-940r4, May 2004.

[11] IEEE 802.11 WG, "TGn Comparison Criteria," IEEE 802.11-03-814r30, May 2004.

[12] I. Medvedev, et al., "A Comparison of MIMO Receiver Structures for 802.11n WLAN – Performance and Complexity," *IEEE International Symposium on Personal, Indoor and Mobile Radio Communications*, pp. 2621-2625, Sept. 2006.

[13] D. Garret, L. Davis, S. ten Brink, B. Hochwald, and G. Knagge, "Silicon Complexity for Maximum Likelihood MIMO Detection Using Spherical Decoding," *IEEE Journal of Solid-State Circuits*, Vol. 39, pp. 1544-1552, Sept. 2004.

[14] L. M. Davis, "Scaled and Decoupled Cholesky and QR Decompositions with Application to Spherical MIMO Detection," *IEEE Wireless Commun. and Networking Conference*, pp. 326-331, 2003.

[15] P. H.-Y. Wu, "On the Complexity of Turbo Decoding Algorithms," *IEEE Vehicular Tech. Conference*, pp. 1439-1443, 2001.

10

Capacity of Wireless Mesh Networks: Comparing Single-Radio, Dual-Radio and Multi-Radio Networks

Stephen Rayment[a]

This chapter focuses on wireless mesh infrastructure systems used for creating large scale Wi-Fi based infrastructure networks, and examines three different approaches currently available for implementing them. It examines the strengths and weaknesses of each approach with particular focus on an analysis of the capacity that is available to users.

10.1 Introduction

Mesh is a type of network architecture. Other common network architectures have included Ethernet, originally a shared bus topology for local area networks (LANs) in which every node taps into a common cable that carries all transmissions from all nodes to an egress point. In bus networks, any node on the network senses all transmissions from every other node in the network. Today, most LANs use a star architecture in which every node is connected using a dedicated link to a central switch connected to an egress point (switches can be interconnected to form larger networks).

Mesh networks are different – physical layer connectivity from every node to the egress is not required. As long as a node is connected to at least one other node in the mesh network, it will have full connectivity to the entire network because each mesh node forwards packets to other nodes in the network as required. Mesh protocols automatically determine the best route through the network and can dynamically reconfigure the network if a link becomes unusable.

There are many different types of mesh networks. Mesh networks can be wired or wireless. For wireless networks there are ad-hoc mobile mesh networks and permanent infrastructure mesh networks. There are shared and switched mesh networks, single-radio, dual-radio and multi-radio mesh networks. All of these approaches have their strengths and weaknesses. They can be targeted at different applications and used to address different stages in the evolution and growth of the network.

The first wireless mesh networks were mobile ad hoc networks – with wireless stations dynamically participating in a peer-to-peer network. Mesh was an attractive approach for this form of wireless networking since wireless nodes may be mobile and a

[a] *BelAir Networks*

mesh approach allows a wireless node to participate in a network without needing to communicate with all of the other nodes in the network. Mobile peer-to- peer networks benefit from this sparse connectivity requirement. The combination of wireless stations and mesh architecture can provide a reliable network with a great deal of flexibility.

The popularity of Wi-Fi has generated a lot of interest in developing wireless networks that support Wi-Fi access across very large areas. Large coverage access points (AP) are available for these scenarios, but the cost of deploying these wide area Wi-Fi systems is dominated by the cost of the network required to interconnect the APs and connect them to the Internet egress point - the backhaul network.

Even with fewer APs, it is very expensive to provide T1, DSL or Ethernet backhaul for each access point. For these deployments, wireless backhaul is an attractive alternative and a good application for mesh networking. If wireless mesh links can be used between most of the APs then just a few wired egress connections at mesh portals back to the Internet are required to support the entire network.

Wireless links work better when there is clear line of sight between the communicating stations. Permanent wireless infrastructure mesh systems deployed over large areas can use the forwarding capabilities of the mesh architecture to go around physical obstacles such as buildings. Rather than blasting through a building with high power, as would be the case with a traditional point to multipoint system, a wireless mesh system will forward packets through intermediate nodes that are within line of sight and go around the obstruction with robust wireless links operating at much lower power. This approach works very well in dense urban areas with many obstructions.

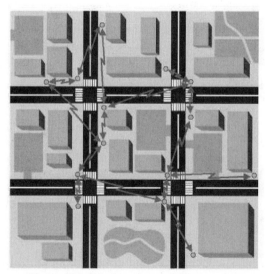

Figure 10.1: Meshing Around Obstructions.

Because of the existence of multiple redundant paths from an AP to the egress portal, reliability of mesh architectures is higher than that of centralized point to multipoint systems with one point of aggregation. If one link fails in the mesh, traffic can be forwarded via an alternate route. Finally, particularly in dense urban areas with obstructions between nodes, high levels of frequency re-use can be achieved between mesh links. This evenly distributes and increases overall system capacity compared to point to multipoint systems where traffic congests at the central egress point

There are many different types of mesh systems and they often get lumped together. Since early wireless mesh systems were focused on mobile ad-hoc networks, it had been assumed that wireless mesh systems were low bandwidth or temporary systems that could not scale up to deliver the capacity and quality of service required for enterprise, service provider and public safety networks. That is not the case. Engineered, planned and deployed effectively, wireless mesh networks can scale very well in performance for a variety of permanent network infrastructure requirements.

10.2 Terminology

A shared mesh network is a wireless mesh network that uses a single-radio to communicate wirelessly via mesh links to all the neighbouring nodes in the mesh. As we shall see, the total available bandwidth of the radio channel is thus 'shared' between all the neighbouring nodes in the mesh.

Wireless mesh nodes or mesh points (MPs) typically include both mesh interconnection links and client access, in which case the node is referred to as a mesh AP (MAP). MPs, that is mesh nodes used just for forwarding, are possible but are not included in the analysis that follows – all nodes are assumed to provide both mesh links and client access.

A dual-radio shared mesh AP uses separate access and mesh link radios. Only the mesh link radio is shared. In a single-radio mesh AP, access and mesh links are collapsed onto a single-radio. Now the available bandwidth is shared between both the mesh links and client access.

A switched mesh network is a wireless mesh network using multiple radios to communicate via dedicated mesh links to each neighbouring node in the mesh. Here all of the available bandwidth of each separate radio channel is dedicated to the link to the neighbouring node. A switched mesh node always uses separate access and multiple mesh link radios.

The collection of mesh APs that "home" to a particular wired egress point a referred to as a mesh cluster. The mesh point located at that egress connection is referred to as a mesh portal. In fact multiple mesh portals can be used for a single cluster to increase system reliability and capacity.

10.3 Single-radio Shared Wireless Mesh

In a single-radio mesh, each mesh AP node acts as a regular AP that supports local Wi-Fi client access as well as forwarding traffic wirelessly to other mesh points. The same radio is used for access and wireless mesh links. This option has the advantage of providing the lowest cost deployment of a wireless mesh network infrastructure. However, each mesh AP typically uses an omni-directional antenna to allow it to communicate with all of its neighbour mesh APs, where all the mesh APs share the same channel for their mesh links.

Further, in a single-radio shared mesh every packet generated by local clients must be repeated on the same channel to send it to at least one neighbouring mesh AP. The packet is thus forwarded to successive mesh nodes and ultimately to the mesh portal that is connected to a wired network.

This packet forwarding generates excessive traffic on the channel shared by all the mesh links and clients. As more mesh APs are added, a higher percentage of the wireless traffic in any cell is dedicated to mesh link forwarding. Very little of the channel capacity is available to support users.

The impact of mesh forwarding is that capacity varies with between $1/N$ times the channel capacity and $(1/2)^N$ times the channel capacity where N is the number of mesh link hops in the longest path between a client and the wired infrastructure mesh portal.

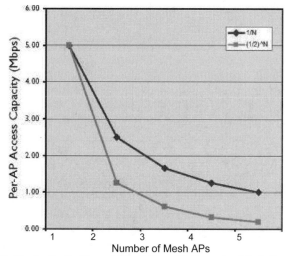

Figure 10.2: Single-Radio Shared Mesh Per-AP Capacity, Mesh Portal at End.

Figure 10.2 shows mesh AP capacity estimates for a single-radio Wi-Fi mesh network using these equations. Shared radio meshes always display the undesirable characteristic that user capacity available at each mesh AP declines as you add more mesh APs to the network and increase the number of wireless links.

The starting capacity of 5 Mbps assumes a single channel of 802.11b, which has a raw data rate of 11 Mbps and useful throughput measured at the TCP/IP layer of about 5 Mbps. This throughput is shared between the access traffic and the mesh link traffic in a single-radio mesh. This is the maximum throughput available in an 802.11b system. As distances from the AP increase, throughput will of course decrease with the varying modulation schemes used by 802.11. The results presented here will still be valid for an entire cell, but should be scaled accordingly.

The choice of model to use, $1/N$ or $(1/2)^N$, will vary with the topology of the mesh, the location of the mesh portal and the extent of the "interference domain" between mesh APs. The interference domain describes the number of nodes in the mesh whose transmissions will be sensed by and hence block the transmission of other nodes. More details on this can be found in the Appendix. The $1/N$ model is obviously the most optimistic. In all cases, capacity available in each mesh cluster declines rapidly as more mesh APs are added.

There are mesh routing protocols that can optimize the forwarding behaviour and eliminate unnecessary transmissions. But the best these optimizations can do is to bring the network closer to 1/N performance.

It should be noted that these analysis assume perfect mesh forwarding, no interference and perfect coordination of the Wi-Fi channel access. Actual delivered throughput and capacity will usually be lower.

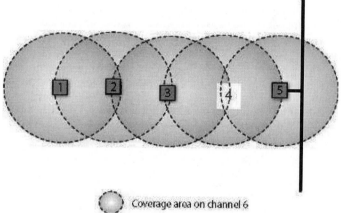

Coverage area on channel 6

Figure 10.3: Single-Radio Mesh Architecture, String of Mesh APs.

Consider a linear string of mesh APs arranged so that each one can sense only one adjacent neighbour on either side (Figure 10.3), that is, with an interference domain of one node. This is not a likely real world deployment, but it simplifies the analysis and we will use this example to compare each of the wireless infrastructure mesh approaches. Throughout this chapter we will also assume that client access load is evenly distributed across the mesh APs. In this string of mesh APs with the mesh portal on the end, N the number of hops from Figure 10.4, is same as the number of mesh APs.

The total channel capacity is 5 Mbps. It can be seen that in this topology the best case 1/N performance is not achievable. N=5, so each AP should have 1 Mbps of capacity. All of the traffic from the entire mesh cluster will have to flow through AP5 to get to the wired portal. If each mesh AP accepts a load of exactly 1 Mbps of traffic from its clients, then AP5 will have to forward 4 Mbps of traffic from APs 1, 2, 3 and 4; and has exactly 1 Mbps of capacity left for its local clients. This analysis assumes perfect contention and collision management. If that is not the case, then more collisions and re-transmissions will result in further congestion and still lower capacity than shown by the simplified analysis.

In a single-radio Wi-Fi mesh network, all clients and mesh APs must operate on the same channel and use the 802.11 Media Access Control (MAC) protocol to control contention for the physical medium. As a result, the entire mesh ends up acting like a single access point - all of the mesh APs and all of the clients must contend for a single channel. As we have seen, this shared network contention and blocking reduces capacity. It also introduces unpredictable delays in the system as forwarded packets from mesh APs and new packets from clients contend for the same channel.

The configuration in Figure 10.3 has the minimum connectivity required to complete the mesh and minimum interaction between adjacent mesh APs for a 5-node mesh AP network. Mesh APs 2, 3, and 4 can hear two other mesh APs; and mesh AP1 and AP5 hear one other mesh AP each. Each time mesh AP3 transmits, mesh AP2 and AP4 must defer and hold off their transmissions since they are using the 802.11 MAC protocol, which uses a "listen before talk" random exponential back-off algorithm.

As can be seen, a capacity analysis of mesh systems should include the effects of the network topology on mesh forwarding, which can be significant. It is difficult to accurately predict mesh capacity without knowing mesh AP placements.

Consider the string of mesh APs in Figure 10.3. If we move the wired backhaul from AP5 to AP3, what happens to the capacity? N, the number of forwarding hops, is reduced from 5 to 3, so we might expect the capacity to be higher than the N=5 capacity shown in Figure 10.4. However, due to the shared network behaviour and the fact that AP3 can hear more mesh AP neighbours than AP5, the capacity is actually lower as shown in Figure 10.4. (Note: The x axis in Figure 10.4 is the number of mesh APs in the cluster, not the number of mesh link hops in the longest path through the cluster.)

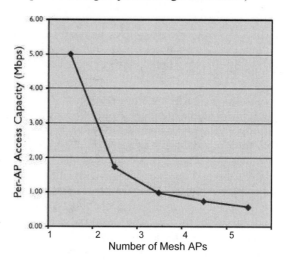

Figure 10.4: Single-Radio Shared Mesh Per-AP Capacity, Mesh Portal in Middle.

The 1/N equation used earlier predicts that per-AP capacity will be 1.67 Mbps when N=3. However, when we factor in the effects of contention and blocking when the wired portal is in the middle of a string of 5 APs (Figure 10.3 with the wired connection at mesh AP3), the estimated capacity is 0.58 Mbps. This matches the $(1/2)^N$ prediction of .56 Mbps when N=5.

The string of mesh APs that we have described so far is not a typical mesh configuration. The cluster of mesh APs shown in Figure 10.5 is a more common example of a small mesh network. In this case, contention reduces the capacity available for client access beyond what we have described in the string of mesh APs examples previously discussed.

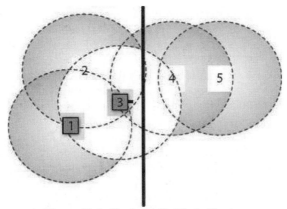

Figure 10.5: Single-Radio Mesh Cluster.

Large coverage mesh APs in these systems have high power radios and high gain antennas. The mesh APs can sense each other at a much greater range than they can sense the clients they support, because most Wi-Fi client devices are low power with low gain antennas.

In this cluster, mesh AP3 can sense all the other mesh APs except for mesh AP5. All traffic for the entire cluster flows through mesh AP3 so it will frequently hold off the other mesh APs, limiting their ability to handle traffic from their local clients. A more complicated formula is required to characterize the impact of neighbouring mesh APs in a shared backhaul network as well as the mesh forwarding.

The capacity in a single-radio mesh is limited by both access and mesh link traffic. Optimizing the mesh forwarding protocol will have limited impact on capacity. The physical layer capacity is the limiting factor. Adding more mesh nodes to a cluster will always reduce capacity per node.

Single-radio meshes offer the lowest cost of deployment. In an infrastructure network, single-radio mesh systems are best used for small mesh clusters of a few nodes. Larger systems may be created by providing many wired or wireless backhaul connections to the mesh portal in each cluster. This of course will increase the operating cost associated with these networks and will complicate the logistics of physical deployment, where for example, rooftop locations may be required for wireless backhaul.

Single-radio mesh solutions are also appropriate for mobile, ad hoc peer-to-peer wireless networks where the emphasis is on dynamic connectivity or they can be used for large sensor networks (eg. for meter reading) where the data rate is low.

10.4 Dual-radio Shared Wireless Mesh

The capacity and scaling ability of wireless mesh infrastructure networks can be improved by using mesh APs that have separate radios for client access and the mesh links.

In a dual-radio shared mesh, the mesh APs have two radios operating on different frequencies. One radio is used for client access and the other radio provides the shared wireless mesh links. Because the radios operate in different frequency bands, they can operate in parallel with no interference. A typical configuration is 2.4 GHz Wi-Fi for client access and 5 GHz wireless for the mesh links. Since the mesh interconnection is performed by a separate radio operating on a different channel, client access is not affected by mesh forwarding and can run at full speed.

In a dual-radio mesh, the mesh links are still a shared network, subject to the same network contention issues that hamper the single-radio mesh. The contention for the mesh links still limits capacity, but as we shall see, to a lesser extent than is the case with single-radio shared mesh.

The mesh links in dual-radio mesh architectures is again running the 802.11 MAC protocol. With one radio dedicated to mesh links at each node, all of the mesh APs use the same channel for connectivity. Parallel operation of mesh links is not possible, as most of the mesh APs will sense multiple other mesh APs. They must contend for the channel and at the same time will block each other. The result again is reduced system capacity as the network grows.

As noted earlier, the useful capacity of each Wi-Fi access coverage cell is 5 Mbps. In dual-radio mesh systems, the access radios of adjacent cells can use different channels. There are three non-overlapping channels in the 2.4 GHz band, so they will be able to operate independently in most cases (Figure 10.6).

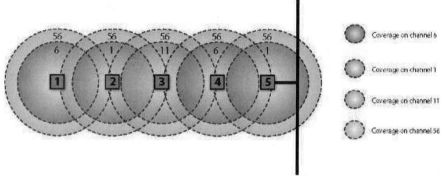

Figure 10.6: Dual-Radio Shared Wireless Mesh, String of Mesh APs.

A commonly used shared mesh protocol is 802.11a, which has a raw date rate of 54 Mbps and useful TCP/IP throughput of approximately 20 Mbps in this type of network.

Again, capacity is limited because of the behaviour of the shared network used for the mesh links. And again, contention and blocking vary depending on the placement and hence the interference domain of the mesh APs.

Figure 10.7 shows the capacity for the minimal overlap string of mesh APs shown in Figure 10.6. The backhaul network offers 20 Mbps of capacity, so per-AP capacity is only limited by AP client capacity for a few node mesh. After three or four nodes, the per-AP capacity drops off because of the shared mesh link effects. In a more typical mesh cluster with more overlap between mesh APs, useful access capacity could be worse than shown here.

Dual-radio systems are a significant improvement over single-radio mesh designs and provide for more potential growth of a mesh cluster.

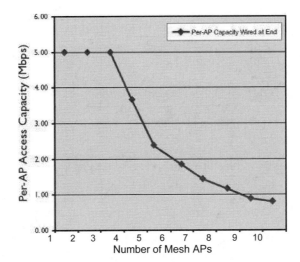

Figure 10.7: Dual-Radio Shared Mesh Per-AP Capacity, Mesh Portal at End.

10.5 Multi-Radio Switched Wireless Mesh

Like a dual-radio shared wireless mesh, a Multi-Radio switched wireless mesh separates access and mesh links. It goes a step further, however, to provide increased capacity by overcoming the shared mesh limitations inherent to single and dual-radio mesh architectures.

In a Multi-Radio switched mesh, multiple radios in each mesh node are dedicated to the mesh links. The mesh is no longer a shared network, since it is built from multiple mesh links where each of the mesh links operates on different independent channels. Traffic is switched in each mesh point from one channel to another – hence the name.

Typically, but not necessarily, each switched mesh link is formed using directional antennas. In this way each of the mesh links is a dedicated link between mesh points. This is sometimes referred to as a multiple point to point switched mesh. It is possible to create very rich mesh topologies with this Multi-Radio approach and just a few mesh link radios at each node. (Figure 10.8)

The performance of a Multi-Radio switched mesh is similar to switched, wired connections. The mesh link radios operate independently on different channels. There are only two nodes per mesh link, so contention is very low. In fact, it is possible to run a customized point-to-point protocol that optimizes throughput in this two-node contention-free environment. These dedicated mesh links typically operate in the unlicensed 5 GHz band and are based on "pre-WiMAX" 802.11a chipsets. These pre-WiMAX wireless links have a potential throughput of approximately 25 Mbps in point to point mode. Point to point links are also a good application for 802.16d WiMAX.

Performance in a Multi-Radio mesh is much higher than dual-radio or single-radio mesh approaches. The mesh delivers more capacity and exhibits the desirable attribute of continuing to scale as the size of the network is increased - as more nodes are added to the system, overall system capacity grows.

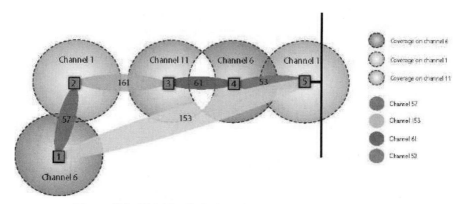

Figure 10.8: Multi-Radio Switched Wireless Mesh, String of Mesh APs.

Figure 10.9 shows the capacity per mesh AP for the Multi-Radio configuration shown in Figure 10.8. We assume a channel capacity of 23 Mbps for each of the point-to-point wireless mesh links. In this string of mesh APs, without the direct link between mesh AP1 and AP5, total system capacity is limited to the capacity of the single link connecting mesh AP4 to AP5, the mesh portal. This delivers maximum capacity for up to five mesh APs, limited only by the AP client capacities, and only then declines with each additional mesh AP.

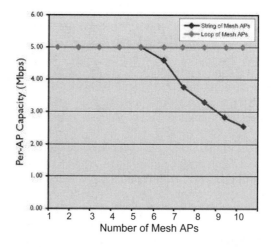

Figure 10.9: Multi-Radio Switched Mesh Per-AP Capacity, Mesh Portal at End or Loop.

Creating a loop by adding a mesh link between mesh AP1 and AP5 doubles system capacity and delivers maximum per-AP capacity through 10 APs, as shown. So the bottleneck in a multi-radio architecture is not in the wireless mesh. System capacity in this architecture is limited by the wired backhaul connection. System capacity will increase and per-AP capacity will remain stable as more mesh APs are added to the network - as long as there is enough wired backhaul capacity. Capacity increases beyond that shown in Figure 10.9 are possible if there are multiple mesh portals supplying the cluster.

A more typical multi-radio switched mesh configuration is shown in Figure 10.10. In this design there are multiple paths through the network, and a mesh protocol would eliminate the forwarding loops and minimize the number of links to the mesh portal. Larger networks would typically have additional wired portals to increase capacity and offer more redundancy in the system.

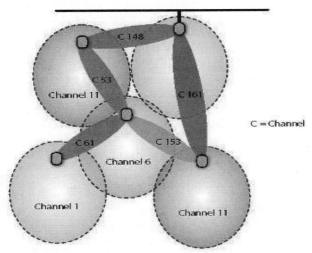

Figure 10.10: Multi-Radio Wireless Mesh, Typical Cluster.

The estimated capacity of switched multi-radio mesh is compared to shared single-radio and dual-radio designs in Figure 10.11. This graph shows the capacity of the different approaches when deployed in a rectilinear grid around a wired connection in the middle.

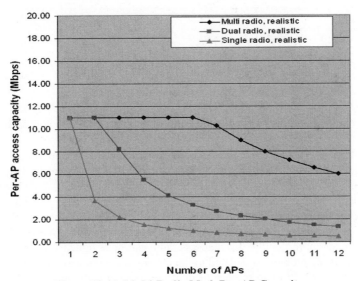

Figure 10.11: Multi-Radio Mesh Per-AP Capacity.

Note that, unlike the previous graphs, the capacity of each node in these graphs has been scaled to reflect 802.11 'b/g' compatibility mode operating at an over the air rate of 54Mbps. The delivered peak TCP/IP rate in this mode is 22-25Mbps. The normal operating scenario of 'b/g' compatibility mode delivers approximately 11Mbps TCP/IP capacity as extra packets are transmitted to enable 'b' client radios to determine the presence of 'g' packets on air. 802.11g only mode, while capable of delivering about 22Mbps of TCP/IP capacity is not typically deployed in public environments due to backwards compatibility with the large pool of 802.11b devices.

Figure 10.12 plots the performance of the different mesh architectures in terms of the overall system capacity, for a cluster of nodes connected to a single mesh portal. It is interesting to see that the single and dual-radio mesh systems asymptote to a system capacity close to the capacity of the medium (the air) whereas the capacity of the multi-radio system rises until limited by the capacity of the node at the mesh portal. In practice this means that in single and dual-radio systems, adding more nodes to a cluster does not increase the overall system capacity.

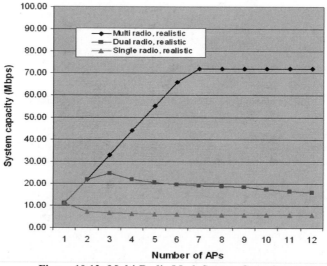

Figure 10.12: Multi-Radio Mesh System Capacity.

There are also other advantages of the multi-radio switched mesh approach.

- Co-existence – most large wireless meshes are designed to support Wi-Fi clients in the 2.4 GHz unlicensed band. There are many other Wi-Fi devices in that band. It is important for large infrastructure to fit into the RF environment. A single-radio mesh must use the same channel throughout the system. (Similarly, the backhaul mesh in a dual-radio system uses the same 5 GHz channel for the whole system.) It is unlikely that this channel will be the best at each location in a large network. A multi-radio mesh is much more flexible. Each access radio can be assigned a different channel, so the co-existence problem is isolated to the coverage area of a single mesh AP - not the whole system. Multi-radio meshes fit into their environment and share the unlicensed spectrum better.

- Interference – multi-radio meshes are very flexible in terms of channel assignment on the access or mesh link radios and so can be deployed to minimize interference with other networks in the same area. The backhaul network consists of point-to-point links. They use directional antennas that have high gain, but they project their signals in a narrow pattern in a specific direction. This minimizes the impact of the multi-radio backhaul mesh on other systems in the area. In addition, interference from external networks in one location will disrupt service across an entire dual or single-radio mesh network. Multi-Radio meshes have very little self-interference, or blocking, because of their flexible channel assignment and multiple radios operating on different channels at the same time. Both dual-radio and single-radio meshes cause self-interference, since all the nodes in the mesh must share a common channel for the mesh links.
- Latency – the dedicated point-to-point links in the multi-radio mesh keep backhaul latency low and predictable. Single-radio and dual-radio mesh approaches have a shared backhaul network using a contention-based protocol with unpredictable latency.
- Mesh link range – because the point to point links typically used with switched mesh architectures are typically formed using directional antennas, high gains can be used to increase inter node spacing.
- Egress portals – as we have shown, capacity increases with the number of nodes in a switched mesh architecture. Hence mesh clusters of much larger node counts are feasible. Indeed, now latency becomes a principal determinant of cluster size. Still, mesh clusters of 50 to 100 nodes are readily implemented. This greatly facilitates and reduces the operational cost of deployment as far fewer backhaul connections to egress portals are needed to deliver bandwidth to the cluster. This reduces both the direct operating costs of backhaul connections as well as more indirect costs such as rooftop access for providing wireless backhaul.

10.6 Conclusion

Capacity in a wireless mesh infrastructure is principally affected by the shared network contention of the mesh links between mesh point used o forward packets from one mesh point to another. These physical layer limitations cannot be addresses through mesh forwarding protocols. Improper design of mesh forwarding protocols will further reduce the performance from that shown here

Single-radio wireless mesh, representing the lowest cost mesh network, is low capacity and will not effectively scale to implement a complete large network. Single-radio mesh is best used in small mesh clusters at the edge of a network. The dual-radio mesh architecture represents evolution in the growth of a mesh network. Multi-Radio mesh systems separate wireless access and mesh links, and use dedicated typically point-to-point connections to form those mesh links. This eliminates both in-channel mesh forwarding and shared mesh link contention. The result is a high capacity system that can scale to support large networks with broadband service for many users.

In the real world, large wireless networks require an integrated combination of the three mesh approaches described. It is possible to deploy a very low cost, low capacity network based mostly on single-radio mesh with some Multi-Radio mesh nodes acting as aggregators for single-radio mesh clusters. Over time, the network can be upgraded to more capacity by replacing single-radio mesh nodes at the edge with multi-radio nodes. Network design can be customized to meet the application requirements and budget by using the appropriate mix of the different wireless mesh approaches.

10.7 Appendix: Capacity Analysis for Single, Dual, Multi-Radio Meshes

Let h be the number of mesh links, and let n be the number of mesh APs. Let k be the resulting available per-AP access bandwidth. Let C_a be the channel capacity for the access radios, let C_m be the mesh link channel capacity, and let C_p be the point-to-point link channel capacity.

Consider various cases:

1. String of pearls (i.e., mesh APs connected in a chain), mesh portal at one end.
2. String of pearls, mesh portal in the middle.
3. Full mesh on rectangular grid.

10.7.1 String of pearls, mesh portal on one end

For this string of pearls case with the mesh portal on one end of the chain, $h = n - 1$.

10.7.1.1 Single-radio

The channel capacity is C_a and is shared by all mesh APs and all client devices.

10.7.1.1.1 Lower bound

In this case, we assume that every mesh AP sees all traffic. This is realistic for a small sized network. Then the channel capacity is used by each of the n mesh AP's access traffic, plus the backhaul traffic. The farthest hop from the mesh portal carries k units of traffic; the next carries $2k$, the one after that $3k$, and so on, and the last hop carries $(n - 1)k$ units of backhaul traffic.

$$
\begin{aligned}
C_a &= kn + \sum_{i=1}^{n-1} ki \\
&= kn + k \sum_{i=1}^{n-1} i \\
&= kn + \frac{kn(n-1)}{2} \\
&= \frac{kn(n+1)}{2}.
\end{aligned}
$$

The available per-AP access capacity is then

$$k = \frac{2C_a}{n(n+1)}.$$

10.7.1.1.2 Upper bound

Assume that every mesh AP sees only the traffic with the adjacent mesh APs, plus the traffic between those APs and their access clients. When there are two mesh APs, the channel carries the access traffic for two APs plus the mesh link of one AP's access traffic. When there are three or more mesh APs, the throughput is constrained by the mesh AP next to the one at the mesh portal. That mesh AP sees access traffic for itself and its two neighbours (for a total of $3k$ units of traffic), plus it sees the backhaul traffic on either side: $(n - 1)k$ units on the side towards the mesh portal, and $(n - 2)k$ units on the other side. Thus,

$$C_a = \begin{cases} k & n = 1 \\ 3k & n = 2 \\ 2nk & n \geq 3 \end{cases}$$

The available per-AP access capacity is then

$$k = \begin{cases} C_a & n = 1 \\ \dfrac{C_a}{3} & n = 2 \\ \dfrac{C_a}{2n} & n \geq 3 \end{cases}$$

10.7.1.2 Dual-radio

The access channel capacity is C_a and the mesh link channel capacity is C_m. The mesh link farthest from the wired link carries the access traffic of a single AP, k units of traffic. The next hop carries $2k$ units, etc., and the last hop carries $(n - 1)k$ units of traffic.

10.7.1.2.1 Lower bound

Assuming that at least one of the mesh link radios sees all other mesh link transmissions,

$$C_m = k \sum_{i=1}^{n-1} i$$
$$= \frac{kn(n-1)}{2}.$$

The available per-AP access capacity is then (taking into account the fact that k is also constrained by the access capacity, C_a)

$$k = min\left(C_a, \frac{2C_m}{n(n-1)}\right).$$

10.7.1.2.2 Upper bound

Assuming that each mesh link radio can only see its immediate neighbours, the mesh AP adjacent to the mesh AP at the mesh portal is the bottleneck (because it sees mesh link traffic on either side). That radio sees a total of $(n-2)k + (n-1)k = (2n-3)k$ units of traffic. That analysis assumes that there are at least three mesh APs, and hence two hops. If there are only two mesh APs then the traffic is just k units. But since $(2n-3)k = k$ when $n = 2$, the available per-AP access capacity is

$$C_m = (2n-3)k, \qquad n \geq 2.$$

$$k = \begin{cases} C_m & n = 1 \\ \dfrac{C_m}{2n-3} & n \geq 2 \end{cases}.$$

10.7.1.3 Multi-Radio

This option differs from the dual-radio case because the mesh link radios are point-to-point links so the bandwidth is not shared. The mesh link with the most traffic is the one immediately before the mesh portal, hence

$$C_p = (n-1)k.$$

The available per-AP access capacity is then

$$k = min\left(C_a, \frac{C_p}{n-1}\right).$$

10.7.2 String of pearls, mesh portal in middle

In these scenarios, the mesh portal is located in the middle of the chain of mesh APs. The number of mesh links, h, is given by

$$h = floor\left(\frac{n}{2}\right).$$

10.7.2.1 Single-radio

In the case where n is odd we have the same number of links on either side of the central mesh AP, but when n is even we have one less link on one side of the central mesh AP. Note that $n - 2h$ is 1 when n is odd and 0 when it is even.

10.7.2.1.1 Lower bound

Assume that every mesh AP can see all traffic. Note that $n - 2h$ is 1 when n is odd and 0 when it is even. Using this fact we can write

$$
\begin{aligned}
C_a &= nk + \sum_{i=1}^{h} ik + \sum_{i=1}^{h-1} ik + (n - 2h)hk \\
&= nk + 2k\sum_{i=1}^{h-1} i + hk + (n - 2h)hk \\
&= nk + \frac{2kh(h-1)}{2} + hk + (n - 2h)hk \\
&= k\left(n(1 + h) - h^2\right)
\end{aligned}
$$

The available per-AP access capacity is

$$
k = \frac{C_a}{n(1+h) - h^2}.
$$

10.7.2.1.2 Upper bound

Assume that each mesh AP only sees traffic of immediately adjacent mesh APs, plus those APs' access clients.

$$
\begin{aligned}
C_a &= 3k + hk + (h - 1)k + (n - 2h)k \\
&= k(2 + n).
\end{aligned}
$$

Then the available per-AP access capacity is

$$
k = \frac{C_a}{2 + n}.
$$

10.7.2.2 Dual-radio

10.7.2.2.1 Lower bound

Using the $n-2h$ factor as in the single-radio case, and assuming that the central mesh AP can see all other mesh traffic,

$$
\begin{aligned}
C_m &= \sum_{i=1}^{h} ik + \sum_{i=1}^{h-1} ik + (n - 2h)hk \\
&= k\left(h(h - 1) + h(1 + n - 2h)\right) \\
&= kh(h - 1 + 1 + n - 2h) \\
&= kh(n - h).
\end{aligned}
$$

The available per-AP access capacity is

$$k = min\left(C_a, \frac{C_m}{h(n-h)}\right).$$

10.7.2.2.2 Upper bound

Assume that the central mesh AP can see only its traffic to and from its immediate neighbours:

$$C_m = hk + (h-1)k + k(n-2h)$$
$$= k(n-1).$$

The available per-AP access capacity is

$$k = min\left(C_a, \frac{C_m}{n-1}\right).$$

10.7.2.3 Multi-Radios

This case is straightforward. If the maximum number of mesh link hops is h then the highest amount of traffic on any one link is hk,

$$C_p = hk.$$

The available per-link access capacity is

$$k = min\left(C_a, \frac{C_p}{h}\right).$$

10.7.3 Full mesh on rectilinear grid

Assume that the mesh APs are positioned on a rectilinear grid, as shown below. The central node is assumed to be the mesh portal, and all traffic is destined to or from that mesh portal.

Figure 10.13 shows a full mesh with 9 nodes (solid circles) and the location of subsequent nodes as they are added, for a total of up to 25 nodes. The links shown are applicable to the single-radio and dual-radio cases.

Different link configurations may be used in the Multi-Radio depending on the number of mesh link radios present in each of the mesh APs. One feasible point-to-point mesh configuration using 3 mesh links per mesh AP is shown in Figure 10.14.

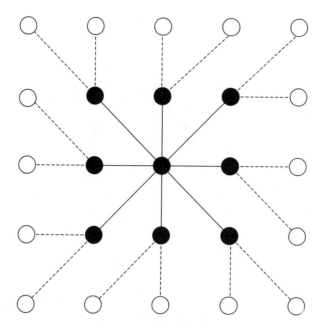

Figure 10.13: Full meshes with 9 nodes. Open circles show locations of subsequent nodes as they are added. Solid lines show data transmissions with the 9 nodes, and the dotted lines show data transmissions that will appear as new nodes are added.

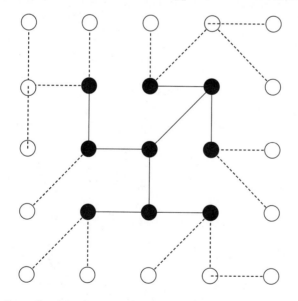

Figure 10.14: Full mesh with nine nodes (solid circles) with point-to-point links using Multi-Radio mesh APs. The open circles and dotted lines represent additional mesh APs and their point-to-point backhaul links.

10.7.3.1 Single-radio

10.7.3.1.1 Lower bound

To determine a lower bound of capacity, assume that the central mesh AP can sense all traffic to all other mesh APs. With n mesh APs there are nk units of access traffic. All of the first nine mesh APs except the central mesh AP have one mesh link to get to the mesh portal. The next 16 mesh APs each have an additional link to get to the mesh portal. Thus, for n at most 25,

$$C_a = nk + (n-1)k + k\,max(0, n-9)$$
$$= k[2n - 1 + max(0, n - 9)].$$

The per-AP access capacity (assuming that access traffic is uniformly distributed across the APs) is then

$$k = \frac{C_a}{2n - 1 + max(0, n - 9)}.$$

10.7.3.1.2 Upper bound

To determine an upper bound of capacity, assume that the central mesh AP can sense all traffic to or from its immediate neighbours, but cannot sense any other traffic. Then

$$C_a = k(min(n,9) + n - 1).$$

The per-AP access capacity is then bounded by

$$k = \frac{C_a}{n + min(n - 1, 8)}.$$

10.7.3.2 Dual-radio

The analysis for the dual-radio case is similar to the single-radio case, except that (1) the access radio traffic does not interfere with backhaul traffic, and (2) the channel capacity is C_m rather than C_a. The available per-AP access bandwidth, k, remains bounded by C_a.

10.7.3.2.1 Lower bound

As for the single-radio case, we restrict our attention to the case that $n \leq 25$. Then

$$C_m = k(n - 1 + max(0, n - 1)).$$

Hence

$$k = min\left(C_a, \frac{C_m}{n-1+max(0,n-9)} \right).$$

10.7.3.2.2 Upper bound

For $n \le 25$

$$C_m = (n-1)k.$$

Hence

$$k = \left(min\, C_a, \frac{C_m}{n-1} \right).$$

10.7.3.3 Multi-Radio

The analysis is complicated by the constraint that each AP can have at most three point-to-point backhaul links, as shown in Figure 10.15. To understand the pattern of backhaul aggregation, it is helpful to redraw Figure 10.13 by putting the mesh portal at the top of a diagram with its mesh links joining it to its immediate neighbours (where "neighbour" is used in the sense of mesh connectivity rather than geography), and so on, as shown in Figure 10.15.

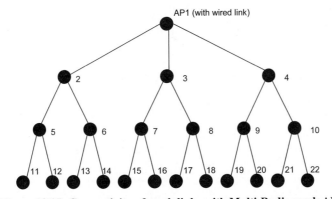

Figure 10.15: Connectivity of mesh links with Multi-Radio mesh APs

It is easy to verify that the traffic on the most congested mesh link is given by

$$C_p = k\ floor\left(\frac{n+1}{3} \right).$$

Solving for k, and remembering that the per-AP access capacity is also bounded by the access channel capacity, we get

$$k = min\left(C_a, \frac{C_p}{floor\left(\dfrac{n+1}{3}\right)} \right).$$

11

Autonomous Mobile Mesh Networks and their Design Challenges

Ambatipudi Sastry[a]

This chapter presents an overview of design challenges involved in mobile mesh networks that support multimedia applications. First, various types wireless mesh networks are enumerated and the historical developments in the area of mobile mesh networks are briefly reviewed. The need for and value of autonomous mobile mesh networks for broadband applications is described later. This is followed by an overview of technical challenges that need to be addressed in designing autonomous mobile mesh networks and for providing useful multimedia peer-to-peer services over such networks. Emphasis is placed on describing generic system level challenges rather than on specific solutions for component subsystems, some of which are only beginning to evolve.

11.1 Introduction

Ad hoc wireless networks are interconnected sets of mobile nodes that are self-organizing, self-healing, survivable, and instantaneously available, without any need for prior infrastructure. Since Internet Protocol (IP) suite is now recognized as the universal interface or "glue" for interconnecting dissimilar networks, an IP-based ad hoc network has the potential to solve the interoperability problems faced by various conventional stovepipe networks that are designed for specific usage cases.

A multi-hop mesh network can be defined as a communications network that has two or more paths to any node, providing multiple ways to route data and control information between nodes by "hopping" from node to node until a connection can be established. Mobile mesh networks enable continuous efficient updates of connections to reconfigure around blocked or changed paths. However, not all mesh networks are similar or capable of delivering the same types of services.

Table 11.1 shows a way to categorize various types of mesh networks based on their characteristics. The focus of this chapter is on *autonomous, multi-hop, highly dynamic, human-centric, broadband mesh networks* that are capable of supporting decentralized applications without requiring a connection to fixed infrastructure and external servers. We refer to such mobile ad hoc networks as "mobile mesh networks" throughout rest of this

[a] *PacketHop, Inc*

chapter. The design criteria for such mobile mesh networks differ significantly in a number of ways from other types of mesh networks, such as, for example, sensor networks, if they are only required to support transport of infrequent low rate data with emphasis on preservation of battery life.

Table 11.1: Categories of meshes and their characteristics.

Categories of mesh	Characteristics
Human centric Vs Sensor data	Presence of a human user at each node requires user interfaces and access to applications
Fixed Vs Mobile	Mobility can be either nomadic or highly dynamic
Radio-specific	Specific to certain types of radios (Ex: IEEE standards such as 802.11, 802.15, 802.16-2005, etc), some times referred to as layer-2 mesh (as in OSI model)
Radio-agnostic	Mesh protocol structure independent of specific radios
Overlay mesh	Connecting dissimilar systems with different types of wireless and/or wired links (layer-3 or higher)
Peer-to-peer mesh	All nodes have identical networking functions
Bandwidth	Broadband, Wideband, or Narrowband (these definitions are somewhat arbitrary – one common usage is narrowband for <100 Kb/s, broadband for >1 Mb/s, and wideband in between)
Services	Centralized Vs Distributed (it is necessary to distinguish between peer-to-peer networks Vs peer-to-peer applications)

11.2 Evolution of mobile mesh networks

The growth in multimedia mobile services is being driven by rapid changes in the capabilities of wireless devices. The increasing computing capabilities of these devices together with built-in cameras for images and video and recording capabilities are enabling the latest mobile devices to usher in a rapid transition from what were essentially voice traffic capabilities to delivering rich multimedia communications. Built-in location methods such as Global Positioning Satellite (GPS) and other advanced features will also enable location-based information services. These developments are leading to a fundamentally new user paradigm from a centralized server-based set of user applications to pervasive distributed information creation and dissemination as evidenced through the rise of social networking. Such pervasive mobile peer-to-peer services will be a natural fit over autonomous mobile ad hoc peer-to-peer networks.

An autonomous mobile mesh is distinguished from traditional wireless networks by its ability to operate in a self-organizing and self-managing way without requiring any pre-installed fixed network infrastructure. The network itself is a peer-to-peer network, which is ideal for supporting distributed applications. A mobile mesh can establish and update connectivity automatically and instantaneously as users move. Its reliability is enhanced as it updates the mesh connectivity in real-time to changes due to mobility and provides alternate paths when any particular link is lost. This also addresses a key need for survivability and availability in reasonably dense networks. Although no infrastructure is required, a mobile mesh should be able to utilize fixed infrastructure where available to permit communications with a command center, a remote agency, or resources on the Internet. A significant difference between Internet routing protocols and mobile mesh protocols is that the latter require the network topology to be updated more frequently due to mobility and hence the update overhead should be minimized. Also, service quality of user applications should be carefully designed to take into account for the changes in connectivity and for countering propagation anomalies. As the nodes in a mobile mesh are not tethered to a particular secured location, distributed authentication for nodes and users is another important element.

Mobile mesh networking is a technology that possesses a long lineage of research, development, and testing spanning nearly three decades. Early research activity in this area drew support from Department of Defense and various other government agencies such as the National Science Foundation within the United States and by several organizations worldwide. The initial impetus for this technology was provided by the need for communications-on-the-move in a battlefield, which has long been a significant concern for the military, especially in highly mobile environments. The advent of packet switched data communications and the Internet were considered by the Department of Defense as a model to be extended to mobile communications. This is reflected in the technology research led by the Defense Advanced Research Projects Agency, which spanned nearly three decades starting in the 1970s, with several programs in the areas such as Packet Radio [1], Global Mobile Information Systems (GloMo) [2], Small Unit Operations and Situational Awareness (SUO-SAS) [3], and Future Combat Systems [4]. The main goal of the research and development in this key technology area is to develop efficient mobile packet-switched networking techniques that support mobility as well as methods for creating and maintaining interconnections with a fixed network such as the Internet.

The Mobile Ad Hoc Networks (MANET) Working Group of the Internet Engineering Task Force (IETF) initiated and is continuing a standardization process for routing protocols and has developed several Experimental RFCs (Request for Comments) [5]. The set of Experimental RFCs on routing protocols include Ad hoc On-Demand Distance Vector (AODV) Routing [6], Optimized Link State Routing Protocol (OSPF) [7], and Topology Broadcast with Reverse Path Forwarding (TBRPF) [8].

Recent years saw a rapid growth in mobile communications, particularly in multimedia services in addition to conventional voice service. These trends are expected to continue and gain momentum, resulting in integrated multimedia services that include voice within the Internet Protocol (IP) paradigm. Mobile mesh networks supporting multimedia services can be connected to the Internet either through mobile wireless point-to-point access, such as cellular, or the emerging fixed wireless local network access such as wireless local area networks (WLANs), as shown in Figure 11.1.

An autonomous mobile mesh can leverage commercial off-the-shelf (COTS) mobile computing platforms and industry-standard wireless devices as cost-effective building blocks. Considerable attention is currently focused on leveraging the IEEE 802 standards-based broadband radio technology both for point-to-point and local area networks. The feasibility of deploying mobile mesh networks got a significant boost due to the confluence of the availability of standards-based commercial broadband wireless devices, improved security mechanisms, affordable and powerful computing platforms, multi-hop mobile mesh networking technology, and IP-based multimedia applications. This allows a mobile mesh to take full advantage of the latest computing and WLAN technologies and lower costs due to worldwide markets.

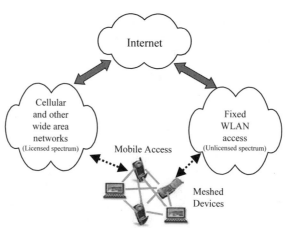

Figure 11.1: Elements of the evolving wireless network architecture.

11.3 Usage Scenarios for Mobile Mesh Networks

Most of today's mobile applications require central servers and directories to communicate even though some of these applications may appear to be operating in a peer-to-peer environment. The proliferation of broadband radios in mobile devices combined with pervasive information such as video, photos and files created by individuals is leading to a new use case for mobile broadband communications. People want to be able to easily communicate with their friends and co-workers, sharing rich content anywhere, anytime. The emerging mobile mesh networking technologies together with distributed multimedia user applications are changing the mobile services paradigm, especially for communications within groups with common interests. Autonomous mobile mesh networks and multimedia services need not require a connection to infrastructure and central application servers. However, the new paradigm should allow optional use of infrastructure where it is available.

The usage cases for mobile mesh applications in the real world can be varied and the challenges in design of systems for such usage cases, though some of them are common, may need specific adaptations. Some of the potential usage cases are briefly described below.

11.3.1 Mobile Mesh Networks for Public Safety Services

Public safety services require mobile, instantaneous, infrastructure-free, self-organizing broadband networks to effectively and safely respond to incident management, emergency management, and surveillance. In addition to the traditional voice-based applications, broadband services such as video, map-based situation awareness, web and database access, and high resolution image transfer are needed for effective incident response. In most situations, first responders will not have access to broadband data network infrastructure and in a large number of cases such infrastructure is not feasible. Mobile mesh networks provide a means of rapidly establishing survivable instantaneous stand-alone communications with ability connect to the infrastructure when available. The need for multimedia communications has been well recognized by the first responders for effective situation awareness and incident management and IP-based solutions would also solve critical interoperability issues.

Several organizations are contributing to the understanding and documenting of the technical and operational requirements of public safety services, including Project SAFECOM [9], Project MESA (Mobility for Emergency and Safety Applications) [10], National Public Safety Telecommunications Council (NPSTC) [11], Association of Public-Safety Communications Officials (APCO) International [12], and various Regional Planning Committees (RPCs). SAFECOM's Statement of Requirements (SoR) [9] emphasizes the information aspects of communications - specifically, the need for wireless exchange of data, video, and other types of media beyond voice. Many of these organizations identified the need for broadband ad hoc networks to serve the needs of first responders at incidents. Guidelines for management of various incidents and their communications needs have also been established in the National Incident Management System (NIMS) [13].

In May 2003, the Federal Communications Commission (FCC) allocated spectrum in the 4.940-4.990 GHz band for the exclusive use of public safety [14]. Numerous chipset developers, device manufacturers, RF system developers, and standards development organizations are currently providing 4.9 GHz band-based chipsets and systems with 802.11 PHY and MAC for use in public safety markets that can be used to build mobile mesh networks for public safety. Such availability is facilitated and accelerated by the IEEE 802.11j specification, which defines an Amendment to 802.11 for use in the unlicensed 4.9 GHz band (4.9-5.0 GHz) in Japan and the related commercial activities. 4.9 GHz band allows higher power than the unlicensed UNII band and the ambient noise can be expected to be much lower for 4.9GHz as there will not be any unlicensed users in that band.

Utilization plans for the 4.9 GHz public safety spectrum are still evolving within the constraints of FCC allocation rules. Key issues in determining the use of the 4.9 GHz band will be the choice of MAC/PHY layer protocols and QoS requirements to meet mission-critical user needs. The Telecommunications Industry Association (TIA) standards group TR8.8 and APCO Project 25 Interface Committee Broadband Task Group are currently working to define the broadband standards for public safety.

11.3.2 Disaster Relief Operations

The most crippling element in relief operations in the aftermath of a major disaster is the loss of fixed communications infrastructure of all types, as happened during the hurricane Katrina in 2005 in Louisiana and Mississippi in the United States [15]. Practically all landlines, cellular, and land mobile radio systems were knocked out either due to the hurricane winds and rain or subsequent flooding. Those systems that were hastily put together lacked interoperability and thus were unable to communicate directly with each other. It was also clear that for disasters of such magnitude as hurricanes and earthquakes, voice communications alone would not be adequate for damage assessment and location determination without good situation awareness tools through video, images, and position information.

The commercial availability of broadband wireless technologies such as WLANs and the feasibility to extend their range through ad hoc mesh formations greatly enhance their ability to support mobile communications at a disaster site. By pre-positioning such facilities ahead of time, significant enhancements can be made towards rapidly improving recovery coordination efforts. Provisioning of multimedia services in a standards-based IP-compatible environment assures interoperability. Also, with broadband multimedia incident communications applications, need for voice communications can be dramatically reduced, ensuring that critical voice traffic has the best chance of getting through. Software-defined (or cognitive) radios can manage multi-band multi-mode communications effectively, leading to efficient combined use of both the licensed and unlicensed spectrum by dynamically sharing based on need, availability, cost, and performance.

11.3.3 Defense Network-centric Operations

Self-organizing mobile mesh networks provide excellent support for on-the-move communications for forward echelons in military tactical communications, working in conjunction with the hierarchy of systems of networks such as satellite systems, airborne nodes, and high speed backbone networks. This is the reason for early support for research in mobile mesh networks in the form of packet radio networks [1]. Rapidly reconfigurable and reliable tactical communications are critical for achieving information superiority and supporting efficient mission executions. Highly dynamic battlefield environments impose stringent communications requirements such as survivable and mobile on-the-move support, scalability, fault-tolerance and ease of deployment. In this context, many commercially available WLAN products and solutions, mesh networks, and IP-based multimedia applications can be leveraged for tactical communications.

11.3.4 Enterprise Applications

Current WLANs deployments within enterprises involves use of a large number of access points (APs) due to their short range and the APs need to be tethered to wired infrastructure through interfaces such as Ethernet. Autonomous mobile mesh networks of user devices such as laptops or handheld devices would enable affinity groups to network among themselves on an ad hoc basis with or without connections to the infrastructure and would allow distributed applications to be shared among users without a central server. Examples of such distributed applications include collaborative planning, distributed designs and documentation, and distributed information sharing at meetings and conference if external connection is not required. Also, use of multi-hop mesh networks would greatly reduce the number of APs needed to provide connectivity to an enterprise's intranet and to the Internet and allow access to hard-to-reach locations by APs alone. A number of enterprises such as office complexes, hospitals, and academic institutions can benefit from mesh networks to facilitate mobile users.

11.3.5 Logistics

Mobile mesh networks would be very valuable for providing operational support in situations where a network infrastructure neither exists nor is feasible. Examples include rapid deployment of temporary facilities such as construction or repair sites, operations in isolated facilities such as oil rigs, tracking of frequently moved objects as in large warehouses and ports. The information requirements in these applications can vary over a wide range, from low rate identification to high resolution surveillance video and would frequently require remote control and positioning of resources such as cameras and robots.

11.3.6 Consumer/Home Networking

The availability of broadband WLANs opens up a number of opportunities to interconnect various devices and services such as audio and video entertainment, high speed Internet access, and security and surveillance within home. The rise of capability for pervasive content creation and distribution by consumers creates new opportunities for affinity groups to interact with each other while being mobile, such as social networking, distributed multi-party games, and information sharing. Connecting the devices that support such services through commercially available WLANs requires clear line of sight connections. Use of multi-hop mesh networks will alleviate this restriction through range extension and allow flexible location of devices and mobility where needed. Such networking and applications also facilitate location-dependent generation of information and services.

11.3.7 Transportation Applications

Several applications such as vehicle safety and traveler information services are emerging as part of the transportation information systems and tests and experimental installations are already under way. Mobile mesh networks can play a key role in providing rapid multi-hop connectivity to mobile vehicles and facilitate broadband information transfer in city streets and over freeways and can leverage any existing broadband communications infrastructure. The Dedicated Short Range Communications (DSRC) [16] standard, which utilizes 802.11a for the medium access physical layers, can be used to build a mobile mesh. Except for a portion of the band dedicated for public safety, DSRC is designed for sharing between public safety and private users. The DSRC frequency allocation involves 75 MHz contiguous band, 5.850-5.925 GHz with 10 MHz channels with option to combine two channels up to 20 MHz bandwidth and with dedicated common control channel for announcements and warnings. IEEE 802 Task Group P (TGp) is in the process of developing amendments to the 802.11 standard for wireless access in vehicular environments. When 2 adjacent 10 MHz channels are combined, the 20 MHz spectrum is similar to the maximum allowed in the 4.9 GHz band for public safety, which creates a potential synergy between the two bands through a common MAC protocol and possible leverage in developing commercial hardware.

11.3.8 Video Surveillance

The growing need for security is causing significant increase in use of video surveillance in private and public places. The traditional analog video cameras and recording devices and human scanning with multiple displays are giving way to video surveillance with IP-ready digital cameras, digital recording, real-time scene analysis, and event-triggered surveillance. However, extending the physical areas of surveillance through wired connections and optical fiber is costly, especially for outdoor surveillance. Wireless mesh connecting deployed cameras can play an important cost-effective role in surveillance of large areas such as ports, parking lots, warehouses, and for border patrol. These applications may involve both fixed and mobile mesh nodes and the installed network can be used for other applications as well with appropriate traffic priorities in place.

11.4. Performance Requirements for Mobile Mesh and Applications

Understanding of performance of integrated mobile mesh networks and multimedia applications in the literature is still evolving through simulations and limited field tests. The overall performance perceived by users is influenced by a number of factors at various layers, such as physical channel characteristics based on the radios and frequency bands used, medium access techniques, network protocols, security mechanisms used, types of applications and supporting quality of service mechanisms, and usage scenarios. Since the entire stack of protocols determines the overall performance of a mobile mesh system, it is difficult to characterize the performance based on any one component protocol alone, though each component is to be designed to be efficient by itself and with flexibility to interoperate with other components for cross-layer optimizations. For example, the issue of scalability cannot be addressed by any one component alone, but should be addressed at the total system level in terms of its ability to provide user services at a specified quality of service.

While the understanding and specifications for performance criteria for multimedia information over mobile mesh is beginning to emerge, more detailed information is available on performance requirements at the routing layer from IETF and for WLANs through IEEE 802.11 standards and field experience.

11.4.1 General Performance Metrics for the Internet

For example, RFC 2330 [17] describes a framework for IP performance metrics for Internet in general, aimed at providing users and service providers a common understanding of the performance and reliability. It describes a set of criteria and elements to be considered in defining metrics and measurement methodologies. The measurement methodology may involve direct measurements of a metric, projections from lower-level measurements, estimations from a set of more aggregated measurements of constituent, or correlated measurements over time. The most significant parameters appear to be (i) queuing delays, both short term and long term, and (ii) loss rates, short term and long term. Also, service differentiation over a time scale should be independent of traffic loads. Performance parameters may also be expressed under specific conditions such as bandwidth reservation or with statistical assurances that are defined appropriately.

11.4.2 Performance Metrics for Mobile Ad hoc Networks

RFC 2501 [18] identifies several qualitative performance requirements for ad-hoc network routing protocols such as (i) distributed operation, (ii) freedom from loops, (iii) demand-based operation (within latency limits), (iv) security, prevention of snooping and manipulation, (v) sleep mode of operation for energy conservation and emission control, and (vi) unidirectional link support. Several quantitative metrics for performance assessment are also identified such as (i) statistical measures for end-to-end data throughput and delay, (ii) route acquisition time (particularly relevant to on-demand protocols), (iii) percentage of packets delivered out-of-order (important for transport layer protocols), and (iv) efficiency (reflecting the amount of overhead that can impact protocol performance). Efficiency may be measured in terms of average number of data bits transmitted/data bits delivered over all hops, average number of control bits transmitted/data bits delivered (includes control bits in routing and data packets), and average number of control and data packets transmitted/data packets delivered (this metric captures a protocol's channel access efficiency). The network "context" in which a protocol's performance is measured should be considered, essential parameters that should be varied include, (i) network size (number of nodes), (ii) network connectivity (average number of neighbors of a node), (iii) rate of change in topology (speed of change), link capacity (bits/s), (iv) fraction of unidirectional links, (v) traffic patterns (effectiveness in non-uniform and bursty traffic), (vi) mobility patterns, and (vii) fraction and frequency of sleeping nodes. In general, quality of service (QoS) parameters of interest are end-to-end delay, delay jitter, loss probability, and stability and/or reliability.

Several wireless radio technologies may be used to build mobile mesh networks and/or backhaul links to connect to Internet such as IEEE 802.11, 802.15.3 and 4, and 802.16-2005 - 16 and 16e (also referred to as WiMax), and high speed cellular links (such as Evolutionary Data Only (EvDO), High Speed Downlink Packet Access (HSDPA), MediaFlo). Many of these systems have their own performance specifications either announced by their vendors or others based on available standards specifications.

In general, desirable features for a mobile mesh system can include (i) autonomous operation (independent of infrastructure), (ii) ability to use infrastructure simultaneously where available, (iii) instantaneous (rapidly deployable), (iv) interoperable (standards-based), (v) hardware independent, where feasible, and (vi) distributed services (instead of centralized, which might need infrastructure connectivity).

11.5. Design Challenges for Mobile Mesh Networks

There are significant differences in the design issues and requirements between a dynamic mobile mesh and an infrastructure fixed mesh, some of which are listed in Table 11.2. A fixed wireless mesh network is a network of Access Points (APs) mounted on towers, utility poles, etc., that extend the reach of the wireless network to clients. Fixed networks extend the range of a network by enabling data to hop from AP to AP till it reaches an Internet service point. This has applicability for providing permanently installed outdoor broadband access in places such as cities or other large areas that have large user populations widely spread out. Fixed mesh networks can be accessed through the nearest AP, which usually connects the user to an Internet service point. Currently, such fixed mesh infrastructures are mostly using IEEE802.11 based WLAN radio devices but other radios may also be used as well. While a user can access any AP within the network, access of the APs while the user (or station) actually in motion is more complex due to the need for re-authentication with different APs and for real-time application handoffs without interrupting ongoing sessions. IEEE 802.11 Task Group R is currently formulating

procedures   for   such   fast   handoff   techniques   (see
http://grouper.ieee.org/groups/802/11/index.html).

Table 11.2: Differences between mobile mesh and fixed infrastructure mesh.

Mobile Mesh	Fixed Infrastructure Mesh
Mobile autonomous network	Static and preplanned
A network of user clients	A network of Access Points
Easy migration to other frequencies	Frequency-specific hardware
Distributed authentication	Centralized authentication
Self-organizing routing for mobility	Routing between fixed nodes
Need to support application QoS	Application QoS handled by service providers
Need to support Distributed server-less applications in isolated mesh	No applications – acts as transport

Autonomous mobile mesh networks, by contrast, do not require access points. Using mobile mesh technology, ad-hoc networks can be created among entities (stations or clients) instantaneously without requiring infrastructure, as shown in Figure 11.2. Some of the principal characteristics and benefits of mobile mesh networks are listed in Table 11.3. An infrastructure can be used where available to reach external sites, such as Internet. Mobile nodes need to carry out authentication in both fixed and mobile meshes - when nodes see a new neighbor in a mobile mesh or when they access a new AP in a fixed infrastructure. However, in a mobile mesh, authentication may need to be carried out in a distributed manner within the mesh itself as no infrastructure connection can be assumed to be present. In a fixed infrastructure, mobile nodes may be able to connect to a central authenticating server, which might cause more handoff delay.

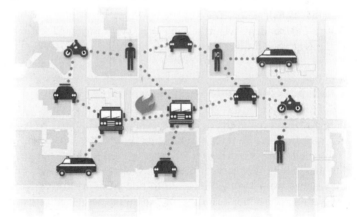

Figure 11.2: An autonomous mobile mesh network forming instantly at an incident scene.

Table 11.3: Characteristics and benefits of autonomous mobile mesh networks.

Characteristics	Benefits
Self-forming	Infrastructure is optional
Self-healing	Alternate paths automatically found
Multi-hop routing	Bandwidth reuse
Distributed	Survivable

There are thus significant distinctions between mobile mesh and fixed infrastructure systems. Mobile mesh provides instantaneous and autonomous peer-to-peer communications and server-less distributed applications with ability to function without connection to infrastructure, providing self-organizing ability and flexibility. Because of mobility, topology changes can frequently occur in a mobile mesh, unlike in a fixed mesh. These differences between fixed and mobile mesh networks create significant additional challenges in designing highly dynamic mobile mesh networks. These design challenges are briefly discussed below.

11.5.1 Physical Radio Channels

Mobile mesh networks using commercial-off-the-shelf (COTS) WLAN devices that conform to standards such as IEEE 802.11 will need to build the higher layers on top of the physical and medium access (MAC) layers defined within the standards. Because of their widespread availability, unlicensed frequency spectrum, and low cost, WLAN devices have become popular in both commercial mesh offerings and laboratory type networks. Broadband radios based on standards such as IEEE 802.11 allow access interoperability across multiple vendors. Depending on the type of client devices such as laptops, personal digital assistants (PDAs), cellular phones, and home appliances, the radios (also referred to as network interface cards) may be used in several form factors, such as Peripheral Component Interconnect (PCI) cards, Personal Computer Memory Card International Association (PCMCIA) or PC cards, Mini-PCI cards, CompactFlash (CF), Universal Serial Bus (USB) adapters, and PCMCIA ExpressCard. Some of them have internal antennas and some may also have the capability to connect to an external antenna.

11.5.2 Medium and Mesh Network Access

The 802.11 MAC is basically designed to operate either in the infrastructure mode (or basic service set, BSS) in which stations or clients connect to access points (APs) based on established mutual handshake procedures for association and authentication or in independent BSS (or IBSS) mode in which client devices directly (one hop) connect with each other without requiring an AP. However, to create a multi-hop mobile mesh, additional mechanisms need to be created for node discovery and authentication on a stand alone basis, without requiring connections to infrastructure. While several proprietary solutions for a multi-hop mesh are currently available, Task Group S (TGs) of the IEEE 802.11 Working Group (802.11WG) is currently developing mechanisms to create a OSI layer 2 mesh called 'Extended Service Set (ESS) Mesh Networking' (see http://grouper.ieee.org/groups/802/11/index.html), some of which would be useful in creating a basic mobile mesh network, though it may not be adequate to cover all types of usage cases. An autonomous mobile mesh network can work in a stand alone mode, allowing devices to freely join and leave. This requires suitable discovery methods to recognize the presence or absence of devices rapidly and to complete handshake procedures without significant latency. Devices in a mesh may need simultaneous connectivity beyond the mesh through infrastructure access points to reach external applications such as corporate databases or the Internet. Thus, it is desirable to use techniques for autonomous mobile mesh networks that also allow integration with commercial wireless infrastructure as is, without costly upgrades or replacements to wireless access points.

ESS Mesh Amendment being developed by TGs is expected to have a core component to assure interoperability and an extensible framework to enable use of extended routing protocols, QoS mechanisms, and management mechanisms. Since mesh networking requirements for various usage scenarios are widely different, the core mechanisms to be specified by the 802.11s standard may need to be extended for different usage scenarios. It is important to highlight that 802.11 based TGs standardization efforts are relevant only to mesh networks that use devices based on 802.11 technology.

11.5.3 Routing and Multicasting

Peer-to-peer autonomous multi-hop mobile mesh networks form instantly in an unplanned manner as users congregate in an area. The network begins to form when two or more first responders come within range of one another. As others arrive on the scene, they join the network automatically and immediately. As the network grows, more links are formed and it becomes more resilient from survivability point of view. It also adapts itself as users move about and others arrive on the scene or depart the area. When any of the users acquires access to a fixed infrastructure network, other users can then access the infrastructure through that "gateway" user. A user thus may traverse through multiple nodes (multiple hops) to reach the destination user if that user is not directly connected and thus requires establishment of routes.

In general, there can be three principal components in establishing dynamic routing in a mobile mesh

- a node discovery mechanism,
- a topology update methodology, and
- a routing mechanism, that can employ different criteria for route selection such as minimize number of hops, maximize end-to-end bandwidth, minimize total path delay, or minimize power consumption, etc.

The design goal of the protocols is to reduce dramatically overhead for discovery and topology updates compared to simple flooding, which involves notification by all nodes to each of its neighboring nodes regarding any changes in connectivity.

Ad hoc network protocol mechanisms for autonomous mobile mesh networks can be broadly divided into two categories:

- On-demand (or Reactive): Routes established whenever a packet is to be sent by a source; and
- Proactive: Routes set up and updated continuously.

On-demand protocols generally exhibit higher latencies as a network becomes larger which could inhibit their scalability, while proactive protocols have slightly higher overhead. Specific characteristics of both depend on the number and density of the network nodes, rate of change of topology due to mobility, and traffic load.

Every node in an autonomous mobile mesh network maintains a routing table and forwards incoming packets toward the destination using criteria such as number of hops, network congestion, link bandwidth, or energy. Figure 11.3 illustrates how an autonomous mobile mesh network functions in a multi-hop mode in which a message from source device A is routed to the destination device B through a series of "hops." Each node maintains a routing table and forwards the incoming packet towards the destination node. Alternate routing paths are automatically established if changes in topology occur due to mobility or a node in the network cloud goes down.

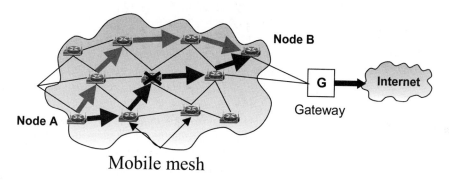

Figure 11.3: Route reestablishment in an autonomous mobile mesh network when a node or link goes down.

Major challenges in the design and implementation of routing protocols include achieving low overhead, low delay, fast updates to the routing table (convergence), scalability, and reliability. In addition, multicasting plays a key role in an autonomous mobile mesh network for efficient bandwidth utilization to accommodate large volumes of multimedia traffic. Such multimedia broadband applications include video, situational awareness, Web access, and image transfer, among others. The relative scarcity of bandwidth in wireless networks makes multicasting particularly attractive.

Routing at OSI layer 3 allows mesh nodes to use different physical and medium access layers, communicating through IP packets and interfaces. This will facilitate an "overlay mesh," fully exploiting the IP paradigm's ability for interoperability. In such overlay networks, node discovery and topology updates need to be carried out in layer 3, independent of any such functions in the layer 2. This is in contrast to mesh formed and restricted to using homogeneous radio medium only such as 802.11, as in a layer 2 mesh currently being defined in IEEE 802.11 TGs. When overlay mesh routing is used, different parts of the mesh using a specific medium may use different backhaul channels to connect to the Internet.

11.5.4 Security

Autonomous mobile mesh networks face severe challenges for security, including wireless eavesdropping, denial of service, and routing attacks. Centralized network security architectures break down in these scenarios because connectivity between a node and the central authority cannot be guaranteed. Figure 11.4 shows the security requirements at various layers in an autonomous mobile mesh network. Many existing standard algorithms for encryption, key management, and authentication can be utilized when designing the security methodology for an autonomous mobile mesh network.

| **System** |
| Data privacy and security |
| Intrusion detection and response |
| **Network** |
| Router authentication |
| Prevention of traffic analysis |
| Hardening against denial of service attacks |
| **Medium Access** |
| Authentication and Authorization |

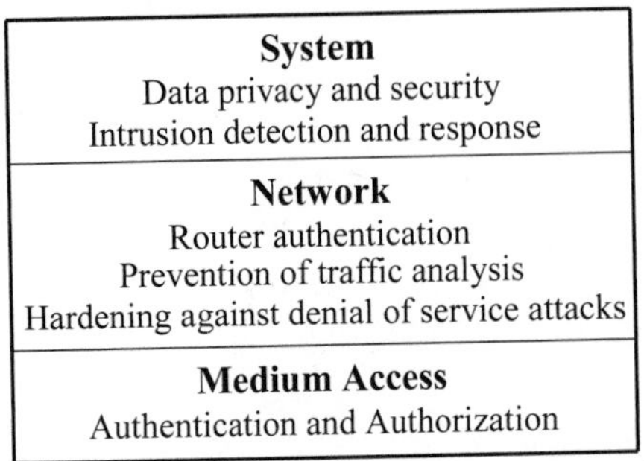

Figure 11.4: Security requirements for an autonomous mobile mesh network.

Mesh communications are inherently peer-to-peer at the applications, network, and link levels. The need for interoperability across administrative domains creates special security issues at all levels. Mesh communication networks need to integrate authentication and authorization mechanisms at various layers. For example, for systems based on IEEE802.11, mesh authentication in layer 2 can be accomplished by leveraging the IEEE 802.11i and 802.1x protocols [19]. 802.11i was defined for infrastructure mode and one-hop IBSS modes operation and thus needs to be extended for use in authentication for multi-hop mobile mesh networks that do not involve access points and connections to external infrastructure.

A mobile mesh node must be capable of detecting other peer nodes and establishing a security context between the peers. Dynamic authentication among mesh nodes must be achieved without a central authentication server as the mobile mesh may not have access to or may get disconnected from any fixed network infrastructure. Also, compromised nodes must be detected and refused communication with other nodes as quickly as possible to protect the network. Maintaining privacy and data integrity in the mesh network requires a strong link level encryption algorithm at each hop. In addition, end-to-end path security may also be created if necessary. Also, whenever a path traverses external link layer elements such as an access point that is not part of the secure mesh network, a virtual private network connection might be necessary to reach the secure end-node in the mesh. Data privacy at the application level can be assured by using end-to-end application level encryption.

11.5.5 IP addressing

In the traditional method of IP address allocation, called 'stateful,' a central entity assigns addresses to network nodes either permanently or dynamically, through mechanisms such as Dynamic Host Configuration Protocol (DHCP) [20]. This methodology creates a set of nodes with known addresses without duplicates. In a stateless approach, nodes attempt to construct their own addresses in a distributed manner, but this can cause duplicate addresses. As a mobile mesh may not have access to an infrastructure, a distributed stateless approach looks desirable. A principal problem with the approach is the need to carry out duplicate addresses detection (DAD). Several solutions are proposed in the literature and research is still continuing [21]-[23]. Zero configuration approaches depend upon broadcast mode (one-hop) access among nodes so that DAD operation can be efficiently done. However, extending this process to multi-hop networks requires flooding of the DAD check information and may not scale well because of the overhead, particularly in large mesh networks. Stateful addressing can be used treating a network as a stub network but this can cause a problem when two or more meshes merge as the address sets may not be distinct.

Stateless auto configuration needs a large set of address space to avoid frequent duplicate address occurrences but is difficult within the limited IPv4 addressing system. With IPv6 addressing, which allows a much larger address set, problems arise since IPv6 networks need to interoperate with the currently predominantly IPv4 networks surrounding it [24] and also requires applications to be IPv6 compliant. Even with a large set of IPv6 address set, auto address configuration may still require DAD since devices may not come with globally unique addresses binding to hardware. Even for MAC addresses, such binding has not become universal as in some devices they are changeable and for others there are no such addresses, especially for some sensors. Thus, addressing in mobile mesh networks continues to be an important issue. In view of this, selection of an addressing solution for a specific mesh seems to be highly influenced by specific usage case considerations, such as the need for globally routable addresses, operation of a network as a stub network with interaction with the Internet initiated from within the mesh rather than from outside, need for frequent mesh merging, size of the network, etc.

11.5.6 Roaming

Nodes in an autonomous mobile mesh network may roam in and out of coverage area and between networks, either among homogeneous or different types of networks. It is highly desirable that mobile devices should be able to join and leave a mesh and/or connect to public or private fixed infrastructure, while retaining real-time seamless connectivity to critical applications. Such real-time roaming usually involves transparent handoffs at both in medium access layer and IP layer. When roaming is within a homogeneous network (such as say within a network of 802.11 APs or even within a mobile client mesh acquiring new neighbors), it might involve re-association and re-authentication. When the roaming is between non-homogeneous physical media, different association and authentication mechanisms may have to be invoked. Depending on how the IP addresses are administered, it might require that a node acquire a new IP address which then needs to be communicated to the external corresponding nodes to preserve the ongoing session. These handoffs steps at both layers together can accumulate considerable latency, sometimes long enough to disrupt the session or cause information to be lost. For roaming within 802.11 APs in Layer 2, IEEE 802.11 WG's TGr is developing mechanisms for Fast BSS transition. At the IP layer, IETF developed MobileIP for transparent IP address handoffs for individual nodes in IPv4 [25], and in IPv6 [26] and their extensions to mobile network as a whole entity (multi-homing) [27] for Internet connections. When a distant node is a server supporting the ongoing session, several support steps are needed to maintain the continuation of the client-server handshake process through the underlying transport layer connection and for meeting the real-time requirements of the transport connection itself.

11.5.7 Data Transfer Reliability

An autonomous mobile mesh network must provide reliable transport of data between source and destination. However, conventional networking technologies rely on classic timeout-based transport protocols not designed for wireless networks. For example, it is well known that the widely used Transmission Control Protocol (TCP) presents significant problems on wireless links. TCP interprets temporary loss of quality in a wireless link as congestion and invokes retransmission procedures which could terminate the TCP connection leading to the loss of a session that will not be automatically restored when the link comes back up. Several mechanisms are suggested in the literature to improve TCP response or to mimic TCP exchanges so as to prevent TCP connection termination for wireless links. Assuring data reliability in an autonomous mobile mesh network environment is a harder problem since an end-to-end path may have multiple hops of different and time-varying levels of quality and latency. Protocols relying on User Datagram Protocol (UDP) together with various acknowledgment strategies are an interesting avenue to pursue for developing "mesh-aware" applications that are tolerant of intermittent connectivity and might be more appropriate to handle real-time multimedia traffic.

11.5.8 Quality of Service (QoS)

Statistical sharing of the Internet infrastructure between real- time and non-real-time services in an efficient and fair manner has become an important issue with the rise of multimedia applications with simultaneous flows of different data having different real-time requirements. Several working groups of the IETF developed solutions such as Differentiated Services (DiffServ) for connectionless handling of quality of service (QoS) [28]-[32]. The approach consists of defining a variety of aggregate behaviors for traffic and using corresponding bit patterns marked in each packet for it to receive a particular forwarding treatment, or Per-Hop Behavior (PHB), at each network node. A PHB is a description of the externally observable forwarding behavior of a DiffServ node applied to a particular DiffServ Behavior Aggregate (BA). A BA is a collection of packets with the same forwarding treatment, when crossing a link in a particular direction. The DiffServ Working Group has standardized a common layout for a six-bit field called the DiffServ field (RFC 2474) [28] and (RFC 2475) [29]. The working group defined an assured Forwarding (AF) PHB group (RFC 2597) [30], suggested as a means for a service provider to offer different levels of forwarding assurances for IP packets. The DiffServ Working Group also defined the Expedited Forwarding (EF) PHB (RFC 2598) [31], in which the packet departure rate for each BA must equal or exceed a configurable rate. Several mechanisms such as priority queuing, class-based queuing, and using preemption can be invoked to implement the PHBs.

DiffServ mechanisms are primarily oriented to fixed Internet in which some level of stability in network operation is generally assumed. DiffServ methodologies as they are defined currently would not work well in multi-hop mobile mesh networks because of their peculiar characteristics such as (i) unpredictable bandwidth conditions due to changing propagation conditions and ambient noise, (ii) frequent changes in topologies due to mobility, (iii) energy constraints due to battery power limitations, and (iv) vulnerability to intrusion and denial-of-service attacks than that of wired networks. The DiffServ mechanisms need to be augmented through enhancements to permit management of critical data flows in a mobile mesh through such measures as dynamic use of radio link quality metrics and bandwidth reallocation, optimization of multicast protocols, and application-level control to provide enhanced QoS. Since the wireless medium is a particularly scarce resource, admission control mechanisms should be implemented to manage the QoS and to enforce graceful degradation where feasible.

The provision of real-time QoS services over mobile mesh networks requires a combination of techniques involving adaptation of applications and network resource management. Also, dynamic allocations of bandwidth are needed rather than guaranteed services and fixed resource reservations, which are infeasible in a dynamically changing bandwidth environment. QoS management needs to be carried out in a coordinated manner at various layers, such as admission control and adaptation at the applications layer, QoS-based routing and multicasting at the routing layer, and access and radio resource management at the medium access and physical layers. It is highly desirable that such QoS methodologies build on and be compatible with the standards-based approaches at various layers such as IETF methods mentioned above and the IEEE 802.11e MAC amendment for WLANs.

Scalability should be specified in terms of the total system performance including user applications, rather than any one particular component. For example, a routing protocol may be able to support a large number of nodes but the rest of the elements of the network should be able to assure the over all system performance in terms of QoS for the set of applications supported. For example, number of cameras that can be installed at each node should be consistent with the overall network capacity. Or, the size of the network should be such that the accumulated end-to-end latency from multiple hops should be less than the allowed limit for interactive real-time applications such as voice or video conferencing.

11.5.9 Network Management

Each node in an autonomous mobile mesh network is a critical resource not just for the individual user but for the network as a whole. Since a connection to a centralized management entity from the autonomous mobile mesh network at an incident site cannot be assured, mobile mesh management methodology should provide for a combination of distributed and centralized approaches. For example, a secure centralized management resource may provide long term configuration, security and policy support, while all dynamic functions of a mesh may be facilitated through distributed management. Simple Network Management Protocol (SNMP) [33], which consists of SNMP entities (or agents) in NETWORK nodes and at least one SNMP entity containing command generator and/or notification receiver applications (or a manager), and a management protocol, may be a better fit as an interface between mesh and its external infrastructure rather than within the mesh, because of its client-server structure.

11.5.10 Distributed Services in a Mobile Mesh

An autonomous mobile mesh network carries a variety of types of traffic, ranging from control messages to multicast video or voice streams. To support distributed user applications in a server-less environment, several distributed support services will be needed such as network addressing, timing reference, domain name service, security management and policy management. Such distributed services include messaging services to distribute presence information, service priorities, and adaptation between network and applications. Distributed storage services support disconnected computing and distributed address allocation methods facilitate dynamic addressing for nodes as they join and leave. Distributed network timing functions can provide timing references within a mesh in the absence of access to Internet network timing reference, and domain name caching needs to be in place to provide distributed domain name services within the mesh. Management functions can include both distributed aspects for autonomous management and centralized facilities for configuration and policy management.

11.5.11 Applications

Mobile network applications face several constraints compared to applications on fixed networks, such as unpredictable behavior of the underlying wireless network because of propagation conditions, mobility, and changing congestion levels, radio interference, wide variety in mobile devices with varying capabilities to create and display content, limited power, and limited local storage. Also, the applications need to be agile and adaptive to dynamically respond to changing network conditions. While all desktop applications may not be suited to wireless environment, there can be scenarios and applications unique to the mobile wireless environment. It is also important to tailor the user interfaces for ease of use consistent with device constraints.

Some examples of broadband mobile wireless applications are:

- Real-time video: Depending on the usage cases, both full motion and slow scan video may need to be supported together with video source adaptation as applicable. To meet the demand for high video bandwidth, a mobile mesh system can benefit from multicast protocols. Multicasting can provide efficient use of bandwidth by reducing the number of transmissions compared to separate unicast transmissions to each recipient in the mesh.
- Voice over multi-hop mesh: IP-based voice has now been established well over fixed Internet and is also is finding its way over WLANs. The primary issues concerning real-time voice are end-to-end delay and delay jitter. In a multi-hop mobile mesh, the delay jitter might vary significantly and requires appropriate buffering and replay mechanisms.
- Instant multimedia messaging: Distributed instant messaging application for peer-to-peer or peer-to-group sessions will be a natural fit in a mobile mesh as presence information becomes available as soon as a node joins the network. Thus, it is not necessary to have a central server for authentication or presence indication. In addition to text-based instant messages, users can have the ability to share files in a variety of

formats, including documents, spreadsheets, diagrams, still digital photographs, and selected video frames.

- Resource tracking: The presence information, together with location data using for example GPS data, can be used to for resource tracking using standard maps indicating where the nodes are at any time. White boarding tools can be used on maps, images, and video clips for interactive planning.

- Affinity group communications: Stand-alone affinity groups can use distributed collaborative group applications such as real-time multimedia conferencing and multi-party games and can include other members over the Internet if any node in the mesh has the capability to connect to infrastructure.

11.6 Conclusions

The availability of affordable commercial broadband radios and mobile computers creates an excellent backdrop for the evolution of broadband mobile mesh networks. The IEEE 802.11 set of standards for WLANs ushered in a new era of broadband wireless access and applications. Widespread WLAN markets helped lower the costs of WLAN devices dramatically and also are acting as catalysts for continued improvements in bandwidth, energy consumption, security and quality of service. The increasing processing capabilities of mobile computing devices and their integration with WLAN technologies not only support multimedia services but make broadband mobile mesh technology affordable and widely deployable. The integration of WLANs with cellular handsets and integrated PDAs opens up additional opportunities. Such standards-based commercial developments provide an excellent framework and potential to deliver networking solutions through autonomous mobile mesh networks that are interoperable, cost-effective, secure, and reliable.

11.7 References

[1] Jubin, J. and Tornow, J. D.; "The DARPA Packet Radio Network Protocols", *Proceedings of the IEEE* (Special issue on Packet Radio Networks), January 1987, Vol. 75, No. 1, pages 21-32.

[2] B. M. Leiner, R. J. Ruth, and A. R. Sastry, "Goals and Challenges of the DARPA GloMo Program," *IEEE Personal Communications*, Vol. 3, No. 6, pp. 34-43, Dec. 1996.

[3] M.A. McHenry, J.G. Allen, "Small Unit Operations Situation Awareness System," *Proc. SPIE* Vol. 3709, p. 27-34, Digitization of the Battlespace IV, Raja Suresh; (Ed.), July 1999.

[4] W.W. Brown, V. Marano, IV, W.H. MacCorkell, and T. Krout, "Future Combat System-Scalable Mobile Network Demonstration Performance and Validation Results," Conf. Proceedings, *IEEE Military Communications Conference (MILCOM)*, Vol.2, 1286- 1291, October 13-16, 2003.

[5] Mobile Ad hoc Networks (MANET) Working Group, http://www.ietf.org/html.charters/manet-charter.html, Internet Engineering Task Force (IETF).

[6] C. Perkins, E. Belding-Royer, and S. Das, "Ad hoc On-Demand Distance Vector (AODV) Routing," IETF RFC 3561, July 2003 http://www.ietf.org/rfc/rfc3561.txt

[7] T. Clausen and P. Jacquet (Eds), "Optimized Link State Routing Protocol (OLSR)", IETF RFC 3626, October 2003, http://www.ietf.org/rfc/rfc3626.txt

[8] R. Ogier, F. Templin, M. Lewis, "Topology Dissemination Based on Reverse Path Forwarding (TBRPF)," IETF RFC 3684, February 2004. http://www.ietf.org/rfc/rfc3684.txt

[9] "Statement of Requirements for Public Safety Wireless Communications & Interoperability," The SAFECOM program, Department of Homeland Security, Version 1.0, March 10, 2004, http://www.safecomprogram.gov/SAFECOM/library/technology/1200_statementof.htm

[10] Project MESA (Mobility for Emergency and Safety Applications) – Statement of Requirements (SoR), http://www.projectmesa.org/ftp/Specifications/MESA_70.001_V3.1.2_SoR.doc

[11] National Public Safety Telecommunications Council (NPSTC), http://www.npstc.org

[12] Association of Public-Safety Communications Officials (APCO), International, http://www.apcointl.org

[13] National Incident Management System (NIMS) document, Department of Homeland Security, March 1, 2004, http://www.fema.gov/nims/nims_compliance.shtm#nimsdocument

[14] *FCC Memorandum Opinion and Order and Third Report and Order* (*MO&O & Third R&O*), FCC03-99, WT Docket No. 00-32, May 02, 2003, and *FCC Memorandum Opinion and Order (MO&O) on Petition for Reconsideration*, FCC04-265, November 12, 2004.

[15] "Report and Recommendations to the Federal Communications Commission", Independent Panel Reviewing the Impact of Hurricane Katrina on Communications Networks, June 12, 2006 (included as Appendix B in FCC 06-83 regarding EB Docket No. 06-119).

[16] ASTM International E2213-03 Standard Specification for Telecommunications and Information Exchange Between Roadside and Vehicle Systems - 5 GHz Band Dedicated Short Range Communications (DSRC) Medium Access Control (MAC) and Physical Layer (PHY) Specifications.

[17] V. Paxson, G. Almes, J. Mahdavi, and M. Mathis, "Framework for IP Performance Metrics," IETF RFC 2330, May 1998. http://www.ietf.org/rfc/rfc2330.txt

[18] S. Corson and J. Macker, "Mobile Ad hoc Networking (MANET): Routing Protocol Performance Issues and Evaluation Considerations, IETF RFC 2501, January 1999. http://www.ietf.org/rfc/rfc2501.txt

[19] IEEE 802.11i-2004, Amendment to IEEE Std 802.11, 1999, IEEE Standard for Information technology--Telecommunications and information exchange between system--Local and metropolitan area networks Specific requirements--Part 11: Wireless LAN Medium Access Control (MAC) and Physical Layer (PHY) specifications--Amendment 6: Medium Access Control (MAC) Security Enhancements.

[20] T. Lemon and B. Sommerfield, "Node-specific Client Identifiers for Dynamic Host Configuration Protocol Version Four (DHCPv4)," IETF RFC 4361, February 2006. http://www.ietf.org/rfc/rfc4361.txt

[21] M. Mohsin and R. Prakash, "IP Address Assignment In A Mobile Ad Hoc Network," *Proc. IEEE MILCOM*, Anaheim, USA, pp. 856-861, 2002.

[22] K. Weniger, "Passive Duplicate Address Detection in Mobile Ad Hoc Networks," *Proc. IEEE WCNC 2003*, New Orleans, LA, pp. 1504–1509, Mar. 2003.

[23] K. Weniger, "PACMAN: Passive Autoconfiguration for Mobile Ad hoc Networks," *IEEE Journal on Selected Areas in Communications (JSAC)* Special Issue "Wireless Ad hoc Networks", Volume: 23, pp. 507- 519, March 2005.

[24] E. Nordmark and R. Gilligan, "Basic Transition Mechanisms for IPv6 Hosts and Routers," IETF RFC 4213, October 2005. http://www.ietf.org/rfc/rfc4213.txt

[25] C. Perkins, (Ed), "IP Mobility Support for IPv4," IETF RFC 3344, Aug 2002. http://www.ietf.org/rfc/rfc3344.txt?number=3344

[26] D. Johnson, C. Perkins, and J. Arkko, "Mobility Support in IPv6," IETF RFC 3775, June 2004. http://www.ietf.org/rfc/rfc3775.txt

[27] V. Devarapalli, R. Wakikawa, A. Petrescu, and P. Thubert, "Network Mobility (NEMO) Basic Support Protocol," IETF RFC 3963, January 2005. http://www.ietf.org/rfc/rfc3963.txt

[28] K. Nichols, S. Blake, F. Baker, and D. Black, "Definition of the Differentiated Services Field (DS Field) in the IPv4 and IPv6 Headers", IETF RFC 2474, December 1998. http://www.ietf.org/rfc/rfc2474.txt?number=2474

[29] S. Blake, D. Black, M. Carlson, E. Davies, Z. Wang, and W. Weiss, "An Architecture for Differentiated Services", IETF RFC 2475, December 1998. http://www.ietf.org/rfc/rfc2475.txt?number=2475

[30] J. Heinanen, F. Baker, W. Weiss, and J. Wrocklawski, "Assured Forwarding PHB Group", IETF RFC 2597, June 1999. http://www.ietf.org/rfc/rfc2597.txt?number=2597

[31] B. Davie, A. Charny, F. Baker, J.C.R. Bennett, K. Benson, J. Le Boudec, A. Chiu, W. Courtney, S. Cavari, V. Firoiu, C. Kalmanek, K. Ramakrishnam, and D. Stiliadis, "An Expedited Forwarding PHB (Per-Hop Behavior)", IETF RFC 3246, March 2002. http://www.ietf.org/rfc/rfc3246.txt?number=3246

[32] D. Grossman, "New Terminology and Clarifications for Diffserv," IETF RFC 3260, April 2002. http://www.ietf.org/rfc/rfc3260.txt

[33] D. Harrington, R. Presuhn, and B. Wijnen, "An Architecture for Describing Simple Network Management Protocol (SNMP) Management Frameworks," IETF RFC 3411, December 2002, http://tools.ietf.org/html/rfc3411

12

Service Provisioning for Wireless Mesh Networks

John Macchione[a]

What you can learn from this chapter is what services are being commonly deployed in Municipal Wireless networks, for what type of customers, some of the networking considerations that each service may drive, and some high-level architectural diagrams. As you will see, one of the key issues is where and how much network control should be implemented. One of the fundamental decisions that a network operator has to determine for an IP network is whether to centralize or distribute the network control. In this context, network control is based on the control of data flow associated with each user. There are advantages and disadvantages to both approaches. Note that this chapter does not intend to make any recommendations regarding this design issue. The high-level diagrams shown throughout this chapter convey the design concepts that network operators will encounter as they build their network infrastructures.

12.1 Introduction

Municipal Wireless networks are a hot new topic that is changing the face of telecom today. With the ability to offer broadband speeds over the airwaves, governments and service providers have all looked at this network approach as a way to enhance their services to the community. Over 300 governments have created Municipal Wireless networks, ranging in size up to 2 square miles. Many more governments are planning deployments with the world's largest cities planning deployments of over 100 square miles.

The drivers for the creation of these networks are varied. In some cases, municipalities seek to create business for downtown districts, some seek to bridge the digital divide, some seek to create combined networks for government services and public access, and some seek to improve their public safety communications. Companies that operate for-profit, such as telcos and ISPs, and cable MSOs, are looking at municipality-wide deployments in order to be an additional broadband carrier or an alternative telephony provider.

---

[a] *Juniper Networks*

In the end, these networks will be owned and managed by different types of entities. In some cases, it will be the city government, in other cases it will be a non-profit, and in some places it will be a more traditional service provider such as an MSO, Wireless ISP, or phone company.

City governments tend to build one of three types of networks. Many networks are dedicated to Public Safety, although some will add on some municipal government functions. Many networks are dedicated to Public Access, although some will add some government functions. The final type is a network dedicated to government functions. It is very rare when both Public Safety and Public Access are built into the same network.

While the "how's" of the deployments are still unknown, it is clear that the proliferation of WiFi devices, whether laptop, PDA, or the coming dual-mode phones, will push network deployments to occur.

Even the phrase, "Municipal Wireless" is a catch all. While the early deployments were owned by Municipalities, the coming networks may be built or owned by a range of entities. However, for the sake of this chapter, we will use the catch-all phrase, "Municipal Wireless".

The final decision for building a Municipal Wireless network, or for that matter any type of network should be based on a financial analysis. This analysis should consider the revenue that the network operator will bring in or the cost savings that the network will enable. It must also capture the costs of building the network. This chapter covers many of the service offerings that can generate the revenue or reduce the costs and can be used as a guide to making decisions on offerings.

12.2 Wireless Mesh Networks

The basic building block of these networks is the wireless mesh. Today, these networks are based on 802.11 WiFi - every laptop that ships today has this capability. Many PDAs and a growing number of cell phones also contain WiFi capability. In the future, this may migrate to WiMAX, but for now, WiFi is the norm.

By deploying a number of access points across the municipality, the network operator can establish connectivity to users throughout the area. Depending on the wireless vendor and the locations that can be used to mount the access point, it can range from 8 to 30 access points to cover a square mile.

At a high level, the Wireless Mesh Network appears similar no matter what the service set happens to be. Figure 12.1 below shows a series of eight access points, each covering a portion of the geography. These access points use their mesh capabilities to bring all network traffic back to a single access point. This access point provides the backhaul for all 8 access points. The backhaul can be either a wired network connection or a wireless point-to-point connection. Common backhaul connections would be DSL, Cable, T1, point-to-point wireless, or WiMAX.

The circles represent the area covered by an access point. The circle with the asterisk (*) is the access point that serves as the collection point for all traffic moving in and out of this cluster of access points. The dotted line represents the backhaul of network traffic from this cluster.

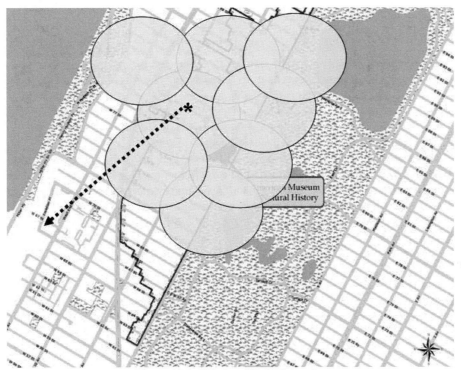

Figure 12.1: Wireless Mesh Networks.

Admittedly, this diagram oversimplifies the access point coverage. In the real world, the Radio Frequency (RF) coverage would not be circular due to differing amounts of buildings, trees, windows, brick, wood, and vehicles through which it travels.

The backhaul through a "foreign" IP network mandates a tunneling solution be put in place. Typically, this is a basic GRE tunnel. As the industry gains more experience with these types of networks, it is possible that Mobile IP or L2TP tunnels will come into favor.

12.3 Service Offerings

Municipal WiFi networks range from simple to high-end services. These service offerings differ depending on whether the network operator is the municipality itself or a for-profit service provider. Some of these services are based on the network access, others are applications that can be offered on the network, either by the network operator or a 3rd party, and others are government specific functions.

This section also discusses some network-based services that would be used by the network operator to control the network, rather than as a fee-based service to the customer. For example:

- *Free Public Access (unregistered)* – To promote usage of their downtown districts; cities provide free unlimited Internet access to anyone with a WiFi device. In some cases, the service is provided by a non-profit, such as a Chamber of Commerce.
- *Free Public Access (registered)* – Free Internet access is available to any registered user. This allows the city either to restrict usage to its citizens or to collect information regarding visitors to the city.
- *Flat-Rate Fee-based Public Access* – This allows Internet access for a fee. The fee can be based on a daily, monthly or longer period. Different fees can be charged for different types of users (e.g., citizens vs. visitors).
- *Differentiated-Rate Fee-based Public Access* – This allows the network operator to charge different fees for users who received different network behavior. For example, different fees can be charged based on peak transfer rate, total bytes downloaded, QoS guarantees or other rate-limiting or metering functions.
- *Web (URL) Filtering* – This feature (or service) limits the web sites that can be visited from the network. Network operators may want to limit access to web sites containing pornography or hate group materials. Conversely, they may want to offer a listing of sites approved for usage.
- *Spectrum Preservation* – The wireless network and backhaul network have limited capacity. As the network operator adds more users to the network, this spectrum and bandwidth may become saturated. Several different approaches and service options can help keep the network running in accordance with users' expectations.
- *Public Safety* - Police and fire department networks based on IP can share the same network infrastructure as Public Access networks. Public Safety networks have special network requirements for security and reliability.
- *Video Surveillance* – The ability to place cameras with wireless networks allows for quick setup and installation. This can provide video coverage for traffic, high-crime areas or for special events.
- *Mobile Government Users* - City officials can stay connected while moving throughout the city. Costs are contained by staying on a city-owned network. If the network is not city- owned, the presence of an additional broadband wireless network will give the city more options to acquire an affordable solution.
- *Virtual Private Network (VPN)* – A service for small or medium businesses that finds the Municipal WiFi network more cost effective than traditional services for connecting between sites. Government buildings are also a candidate for this service.
- *Voice over IP* (VoIP) – Users of broadband networks will begin making VoIP phone calls over the network as soon as it is set up. The network operator can provide improved bandwidth and QoS to these users as an incremental service.
- *Meter Reading* – A wireless network that stretches across the municipality allows for collecting data from many types of devices. These devices can include water meters, electric meters, traffic meters, handheld devices for ticketing and other transactions, and any other type of device that can provide a data link.
- *Government as the Anchor Tenant* – Municipalities often have many smaller buildings where it is difficult or expensive to receive network services. As an anchor tenant on a municipal Wi-Fi network, a cost savings can be achieved if the network is

municipally owned. Additionally, the anchor tenant would reduce financial risks for a service provider that is considering building a Municipal Wireless network.

- *Dedicated Internet Access* – A service for small or medium businesses to have broadband always-on internet access.

Table 12.1 provides a quick reference guide for Services and the types of operators that are likely to offer them. The column *Network Operator* indicates the type of network operator that is most likely to offer this type of service. These are not bounded conditions, but are based on current market general practices.

Every network service places incremental requirements on the network. By building the network up from the simplest service to a fully featured network, the requirements at each layer can be reviewed, thereby determining the types of equipment required to support the services. Note that the layering methodology used in this chapter is illustrative in nature. Since network operators will be choosing a unique set of services to serve their set of customers, the inclusion or exclusion of certain solution components will differ.

12.3.1 Free Internet Access (Unregistered)

Providing free Internet access is a relatively simple task. However, the roaming capability present in a Municipal WiFi mesh network makes it unique relative to other network types. Users on laptops or PDAs can simply move between access points, requiring that the wireless and router network understand how to keep the session active despite changes to the physical and logical network connections. Besides actual roaming, a stationary device may transfer between access points due to changes in the signal strength from a given access point (perhaps due to a truck blocking the straight line of the RF signal path).

The network needs to be able to accommodate these transfers very quickly. The edge routers therefore must support large numbers of host routes, BGP route advertisement, no re-authentication when roaming, and GRE to cross IP backhaul networks.

As depicted in Figure 12.2 below, the end user is using a laptop or PDA to connect to an access point on the wireless network. That access point connects to another access point that is the backhaul device for a cluster of access points (the circle with the *). The dotted line represents the backhaul (either wired or wireless). Several dotted lines are shown to represent that multiple backhaul devices may be active. The depicted Juniper Networks Services router provides the tunneling to the end users and serves as the gateway to the Internet.

12.3.2 Free Internet Access (Registered)

As depicted in Figure 12.3, by registering users, the network will now require a RADIUS server to authenticate the user. Standards-based Authentication, Authorization, and Accounting (AAA) functionality now must be deployed. The Juniper Networks Steel Belted RADIUS server is deployed to ensure that only registered users are allowed onto the network.

Table 12.1: Network Services.

Network Services	Network Operator*	Customers	Description
Free Public Access (unregistered)	City or non-profit	City Residents, those who work in the city, and visitors	Free internet access to anyone within the network's range.
Free Public Access (registered)	City or non-profit	City Residents, those who work in the city, and visitors	Free internet access to registered users.
Flat-Rate Fee-based Public Access	City, non-profit, or Service Provider	City Residents and those who work in the city	Internet Access for a fee to registered users.
Differentiated-Rate Fee-based Public Access	City, non-profit, or Service Provider	City Residents and those who work in the city	Internet Access for a fee to registered users. Each user receives different network behavior based on their QoS requirements.
Web Filtering	City or non-profit	City employees, SMB, and schools	Black-listing or white-listing of web sites to restrict users
Preservation of Wireless Spectrum	City, non-profit, or service provider	City Residents, those who work in the city, visitors, city employees, SMB, and schools	Prevention of abusive user from using up the wireless spectrum preventing other users from getting their share
Public Safety	City	Fire, Police, EMS	IP-based Wireless services to Fire, Police, and Rescue
Video Surveillance	City	Fire, Police, EMS, Schools	Video surveillance cameras send video over wireless network.
Mobile Government Users	City, non-profit, or Service Provider	City employees such as health inspectors, building inspectors, Police, Fire, EMS, and School employees	Mobile City officials conducting city business around town.
Meter Reading	City, non-profit, or Service Provider	Water department, electric department, parking enforcement, traffic signals	Meter reading of common data devices such as water and electric meters
Government as Anchor Tenant	City, Non-profit, or Service Provider	City owned buildings	Network Services to Government Buildings and Mobile Government Users
Virtual Private Networks	City, non-profit, or Service Provider	City owned buildings and SMB	VPN Services to connect multiple buildings of SMB or government locations
Voice over IP	City, non-profit, or Service Provider	All users	VoIP calls over the broadband network, whether made with the network operators knowledge or not

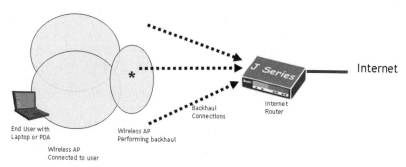

Figure 12.2: Free Internet Access Network Diagram.

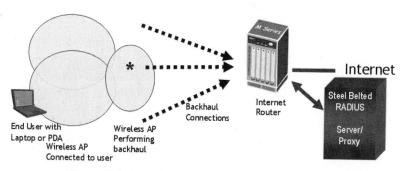

Figure 12.3: Free Internet Access to Registered Users.

12.3.3 Flat-Rate Fee-Based Public Access

When fee-based services are layered into the network, new requirements emerge. These new requirements include the need to have billing systems that interact with the network and that customer service representatives can access. The network now needs to be able to limit users' bandwidth to ensure that each user gets the bandwidth for which they paid. The network would need to be aware if a customer is active and has paid for their access.

In Figure 12.4, the Billing and Accounting system works with Steel Belted RADIUS using industry standards. Steel Belted RADIUS will have all the information required to setup the Multiservice Edge Router to grant access for the user.

12.3.4 Differentiated-Rate Fee-based Public Access

Customers may receive different charges for different levels of service. In this scenario, the network operators need to ensure that users get the level of service for which they have paid. The network now needs to be able to limit user's bandwidth usage in a variety of ways, for example:

- Users may be paying for unlimited usage;
- Users may be limited to a peak information rate;

- Users may be limited to a certain number of bytes of download per day (or week or month);
- Users may be guaranteed a minimum (committed) information rate or any of a number of other QoS parameters.

It is also possible that a user's payments are not current and need to be restricted from the network.

In Figure 12.5, an Edge Router has been introduced into the network configuration. The Juniper Networks Edge Router can handle each user's flow with a different set of rules, enabling the network operator to sell different packages with a variety of bandwidth, QoS, and access capabilities. The router itself needs to be capable of handling a large number of flows to support the network.

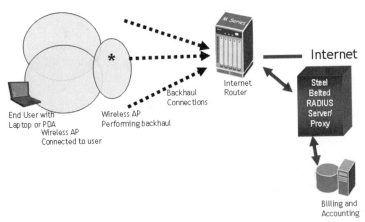

Figure 12.4: Flat-Rate Fee-based Internet Access Network Diagram.

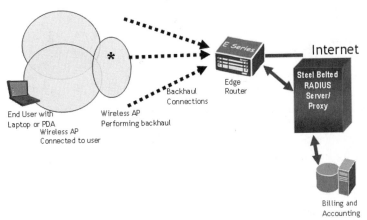

Figure 12.5: Differentiated-Rate Fee-based Internet Access Network Diagram.

12.4 Web Filtering

Non-profits and municipalities frequently want to restrict users from being able to reach certain web sites (e.g., pornography). The introduction of a firewall that can blacklist (a listing of sites that are banned) or white-list (the rarer case of a listing of sites that are allowed) sites may become important.

Figure 12.6 introduces a firewall that can look at site requests and then either allow or deny those requests. In addition, the firewall works with a URL database server to access permission lists. These listings indicate sites that users *are* or *are not* allowed to visit. Permissions can be granted on a system-wide basis, a user community (e.g., per school) basis, or on an individual user basis.

The network operator can either create or maintain these lists or they can be provided as a service by a Juniper Networks partner. The lists often are grouped into useful categories such as "children safe" or "sports."

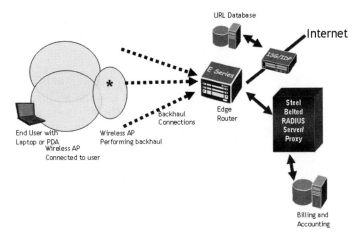

Figure 12.6: Web Filtering Network Diagram.

12.5 Wireless Spectrum Preservation

In a Municipal WiFi network, the bandwidth available within each access point and within each backhaul cluster is limited. For initial deployments, with a small number of users, this bandwidth will seem infinite. However, with the success of adding users - both in tiny increments and in big blocks - the network will become more constrained. Planning for success is critical.

Network operators obviously will want to place control on several types of users. *Abusive users* are those who spread viruses, worms, or flood the network with Denial of Service (DoS) requests. Abusive users therefore must be prevented from putting malicious traffic into the network as it uses up the wireless spectrum and network bandwidth (in addition to a host of other issues). While the network can be protected from these types of activities, the amount of wireless and backhaul bandwidth that can be used up by a spreading worm or DoS attack is relatively large. A solution to the problem is the implementation of Intrusion Detection and Prevention (IDP) firewalls that can inspect every packet to ensure that these users are isolated and shut off.

The network operator also may want to limit access to at-risk users. These users are those whose devices are not up to current standards for threat protection. For example, users who have not updated their virus protection software in the past month could be denied access until they are compliant. By limiting the users before they become infected, the network operator protects the user from potentially crippling attacks and prevents the users from spreading the attack--even when they are attached to other networks.

The network operator can choose to restrict certain applications on the network. These restrictions can be based on network conditions (e.g., limiting real-time applications such as video or on-line gaming) or based on services that are allowed (e.g., Skype). An IDP-capable firewall can detect and restrict these activities (see Figure 12.7).

The Juniper Networks ISG/IDP appliance provides the functionality to place restrictions on users. As shown in Figure 12.7, issuers cannot run on the network unless their device meets the operator requirements for preparedness, applications in use, and security standards.

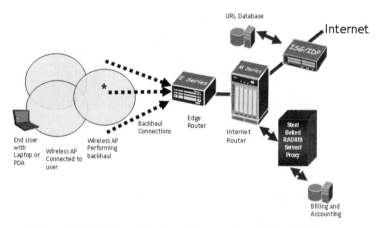

Figure 12.7: Wireless Spectrum Preservation Network Diagram.

The Juniper Networks ISG/IDP appliance detects over 3600 attacks across 60+ protocols using multiple attack detection mechanisms including stateful signatures and protocol anomalies to deliver Zero-Day coverage against existing and emerging threats. It extends the application level attack protection that Deep Inspection provides by detecting and preventing attacks that target the internal network protocols and applications. Operating as an inline device identifies and stop attacks such as worms, viruses, and Trojans from propagating across the network, thereby minimizing the time and costs associated with intrusions.

12.6 Public Safety

Public Safety communications obviously have high standards to meet. The standards have risen over the past five years due to well-known failures of emergency networks to work together. This drive to implement converged emergency networks comes at the same time when the networking world is converging on IP as a communications protocol.

This leads to quandaries as to how to deploy these networks in a solid, secure environment. One of the first decisions is whether to deploy the network for multi-use, e.g.:

- Can other government agencies use the same network?
- Can the network also serve for Public Access?

Municipalities throughout the world are making different decisions. Not only does the network need to be protected from malicious users (who, in fact, may be trusted government employees), but the network traffic from police, fire, and rescue teams needs to be given higher priority to assure it gets through the multi-use network in a timely fashion. Various schemes may be used, ranging from packet prioritization to bandwidth reservation. At times of unusual activity, the network can be tuned to restrict non-critical traffic even further.

Operators must be able to control the access rate that devices send traffic onto the network. This has to occur from the central management location for the network. There can be restrictions (such as Public Access) during an emergency. This can be a bandwidth enhancement, such as guaranteeing a police officer higher bandwidth when accessing certain functions (e.g., an onsite video camera).

Operators also need the ability to set QoS parameters throughout the network. A packet may traverse through several access points, the backhaul, and the IP network before reaching its destination. The more ability to control the path, the more likely it will reach its destination at the time when expected.

Public Safety requirements mandate that the network be highly reliable. The devices involved need to have redundant capabilities to maximize uptime and throughput. The network shown in Figure 12.8 contains a number of features that meet Public Safety requirements. The wireless network must support automatic re-routing in the event that an access point fails or an RF link degrades.

The Juniper Networks Service Deployment System (SDX) meets many of these requirements. The SDX can communicate to a wireless system's EMS of over standard interfaces. This allows the wireless system to set up QoS and CAC functionality in the access points.

The SDX communicates directly with the Juniper Networks routers to provide controls in the IP network. The SDX also communicates with common applications to understand their network requirements and instructs the network devices to raise or lower bandwidth and QoS to ensure proper handling of the traffic. To assure that the network will be available, Juniper Networks recommends that the Steel Belted RADIUS HA version be deployed in a redundant configuration.

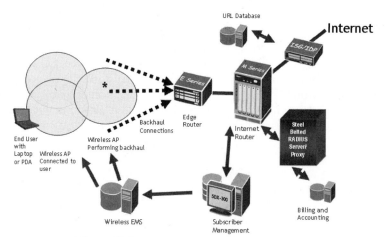

Figure 12.8: Public Safety Network Diagram.

12.7 Video Surveillance

Video Surveillance is becoming a common tool for government agencies. In fact, crime prevention is the leading application for video surveillance. The presence of cameras can be a crime deterrent and the storage of video helps ensure convictions. For example, New Orleans reported a 57% decrease in violent crimes after wireless cameras were deployed in some neighborhoods.

Video cameras also can be used to monitor traffic. Video at special events can be arranged quickly, just by adding a camera with a WiFi connection. The need for wiring, with its extensive leadtime, would no longer exist. High quality cameras and camera phones can be deployed in real-time to emergency situations.

Video is bandwidth intensive since it provides a steady flow of video traffic through the network. This is particularly true across a wireless network. The video itself will only make sense at the reception point if a sufficient number of packets make it all the way through the network. In an uncontrolled network, packets will be dropped or delayed due to contention or bottlenecks unless the right QoS parameters are set to give video priority.

In addition, the ability of an operator to "turn up the bandwidth" for a specific video camera when an event is in progress would be desirable to gain a better quality picture. When events are unfolding at a crime or accident scene, an operator should be able to increase the bandwidth allocated to a camera to improve the quality of the captured video.

Success in building a video network requires all of the same components outlined above in the *Public Safety* section of this chapter. Please note that, although beyond the scope of this paper, video control, video server, and storage equipment would be required for full-scale video surveillance.

12.8 Mobile Government Users

Mobile government users are government employees who spend a significant amount of time traveling to locations throughout the municipality. Examples of these types of employees include health inspectors, building inspectors, and DPW employees. These employees are most effective when they are out in the city, rather than driving back to the office to fill out paperwork and read email. Enabling these workers to stay in contact while deployed will make them more productive and potentially will save taxpayers' money.

Mobile government users continually work with data that is part of a public trust. This data needs to be treated with the utmost security when it is transmitted over a frequency that is available for public access. The requirements for this form of data transmission include wireless encryption with flexible private/public keys.

To illustrate, in Figure 12.9, the Odyssey Client on the user laptop integrates into the WiFi network by enabling public/private keys working with both the wireless network and Steel-Belted RADIUS. Odyssey Client is an enterprise-class 802.1x access client that is deployed easily across all desktop and handheld devices for the lowest overhead. It provides a fine level of control over how users can connect to the network – ensuring compliance with a company's security policies.

Based on the IEEE security standard 802.1x and with full support for advanced WLAN security protocols, the Odyssey Client provides the strong security required for wireless access to the LAN. Together with an 802.1x-compatible RADIUS server, the Odyssey Client secures the authentication and connection of WLAN users, ensuring that only authorized users can connect, that login credentials will not be compromised, and that data privacy will be maintained over the wireless link.

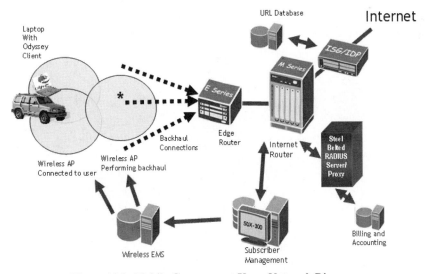

Figure 12.9: Mobile Government Users Network Diagram.

12.9 Virtual Private Networks (VPNs)

The business case for a successful municipal wireless deployment often depends upon reaching small to medium sized businesses (SMB). One of the key offerings that can be made to these businesses is a VPN connecting multiple locations. By creating a VPN between locations, businesses know that their data is safe in transit.

Note that this is only one type of VPN. It is among the simplest to sell and provision to new customers. The network operator would (most likely) offer other types of VPNs with different designs and components, e.g., VPNs connecting to corporate locations via other network operators. Network operators will be attempting to satisfy customers while using the minimal amount of their wireless spectrum. The use of WAN acceleration tools can accomplish both goals.

As shown in Figure 12.10, the Juniper Networks Secure Service Gateway (SSG 520) provides the business with a secure VPN for this point-to-point connection. The WXC WAN Accelerator provides a significant improvement in the utilization of the WAN connection and improvement in application speed. Improvement in application speed and reductions in network usage will range from 30-200%. Applications that have compressible, redundant data as well as applications that use chatty protocols (such as Microsoft Exchange), will benefit most significantly from this acceleration.

The CPE devices shown in the diagram above can be purchased retail. However, many network operators have found it better to provide the equipment themselves (either as a sale or as part of a monthly fee). This approach has several advantages, for instance it:

- Helps customers to start using the network quickly.
- Reduces calls to the customer service center
- Improves the response capability of the customer service center since the CPE and its setup is standardized.

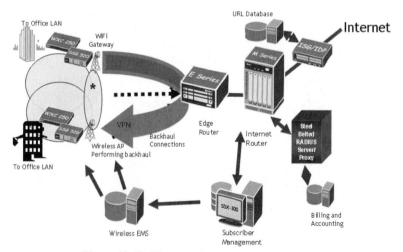

Figure 12.10: Virtual Private Network Diagram.

A *WiFi gateway* is a device that is required to connect the business to the Municipal Wireless network. These devices connect to the business on a wire and to the municipal network wirelessly. The municipal wireless equipment vendor typically certifies these devices.

12.10 Voice over IP (VoIP)

The very nature of a broadband wireless network means that it will handle VoIP calls. This will break down into two categories:

- Calls conducted *with* the network and network operator's knowledge.
- Calls conducted *without* the network and network operator's knowledge.

In the first case, users can use tools such as Skype, Vonage, or other voice-knowledgeable IM packages. Since the network generally will supply enough bandwidth to users to make these calls, the users will connect using the same tools they use on DSL and cable networks today.

This expands incrementally to SMB or small government offices that may have an IP PBX or VoIP Gateway and place their calls through the data network over a VPN. This approach simply puts voice calls on the network with their data traffic through the mechanisms that have already been discussed in this paper.

That simple expansion can cause two difficulties. The SMB will find that call quality varies throughout the day, as it shares the network equally with users on laptops. The network operator will find that the SMB puts a relatively steady load on one portion of the network, so it will not perform for other users as anticipated.

Use of the tools that have been previously introduced (e.g., the Juniper Networks SDX) allows network operators to monitor these users and offer sales packages that will fit their specific requirements. Once an SMB or individual user either has self-identified or has been identified by the network as a VoIP user, they can be sold packages that give them more dedicated bandwidth, prioritization over other (data) users, and improved network QoS. In extreme cases, the identification of these users can help the network operator to add another access point to their network.

12.11 Meter Reading

Currently, meter reading is a labor- intensive function. Since the Municipal Wireless network will blanket the area with WiFi, any device that can connect to the network by WiFi (or by Ethernet to a WiFi bridge) can send its information directly across the network.

This capability has many uses. Water and electric meter reading occurs at every residence or building. Traffic measuring can occur at many points on the roadways, with control exercised remotely at each traffic light. Handheld devices can automatically transmit traffic and parking tickets, public safety inspection reports, and other government forms.

Automation of the meter reading function benefits the municipality in many ways. The most obvious is the reduction in labor cost by not having meter readers walking the streets every day. The challenges and liabilities presented by dogs and locked gates disappear. Meters that malfunction or report some anomalous event (burst pipe) can be dealt with the day it occurs rather than weeks later. When linked with time-of-day rates, meters that can report usage around the clock can be used to encourage the saving of electricity during peak usage periods.

All of the networks discussed above support these functions.

12.12 Government as Anchor Tenant

Having the Government as an anchor tenant is often a key to the overall success of a Municipal Wireless Network. From the network operator's point of view, the Government is a very large enterprise with significant telecom needs. Having the Government as a customer provides fiscal security. In addition, the Government often either can hinder or encourage the deployment. This can be established by how it handles permitting, road closures, access to light poles and other access point locations.

The network requirements of a government are varied, ranging from mobile users to municipal offices to schools. In each of these cases, the security aspects are paramount, with cost savings a close second. Schools present a unique challenge in that students are inclined to break the security models in an inquisitive manner.

12.13 Dedicated Internet Access

Bridging the digital divide is often a stated goal for Municipal Networks. Another common goal is to create economic growth and development by serving the underserved SMB market. The municipalities are looking to ensure the every resident can be reached by a broadband connection and that additional competition lowers the cost of broadband services.

At the same time, MSOs, telcos, and wireless ISPs have the obvious goal of expanding their customer bases. In many cases, municipalities will find ways to work with these service providers to encourage Municipal Wireless networks to achieve these goals. Providing Internet Access to this market is simple; making it work effectively is the challenge. One of the fundamental roadblocks to bridging and serving is overcoming the fact that small businesses do not have an IT department. In many cases, they are struggling to keep their computers running either standalone or with dialup Internet access.

Having an always-on broadband Internet connection requires a different level of IT support. The challenges that a small business therefore would include:

- Setting up the router and firewalls;
- Creating a LAN, wiring the location;

- Keeping the PC operating system up to date;
- Keeping application software up to date;
- Maintaining a security environment for antivirus and spyware.

SMBs also need to be compliant with legal issues relative to web sites visited, specifically, they need to keep employees working as opposed to web surfing. All of these issues have to be accomplished without hiring an IT department.

Network operators can simplify all of these challenges by offering a firewall and wireless LAN as options for their services. These options could be offered either as a resale or as part of the monthly service fee thereby allowing for quick access to the network. Since the network operators would choose the equipment and recommend a default configuration, it is likely that this would cut down on expensive calls to the customer support center.

In addition, offering network-based antivirus services in addition to URL Web filtering can provide many of the IT functions that these small businesses require. The Juniper Networks NetScreen-5GT provides all the security functions of larger firewalls for a small office. Additionally, it can provide a wireless LAN inside the building. This approach eliminates the SMB from having to worry about running cables and minimizes the overall number of devices requiring monitoring.

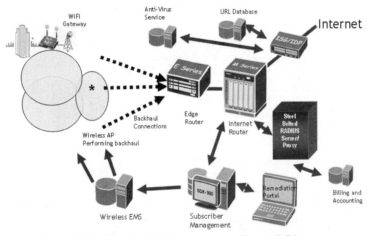

Figure 12.11: Dedicated Internet Access Network Diagram.

The Juniper Networks NetScreen-5GT can be run with a service for antivirus protection so that the SMB does not have to be concerned with learning the nuance of viruses. The NetScreen-5GT can perform URL filtering and Deep Packet Inspections (discussed previously). In addition, the NetScreen-5GT Wireless can provide embedded antivirus protection that complements the user's desktop antivirus software.

Further, network operators can use the IDP/SDX threat mitigation solution to send the user to a remediation portal to recommend upgrades to antivirus software, operating system software, and other steps to ensure a positive experience for the end user.

12.14 Advanced Network Services

Network operators will want to introduce new, advanced services after their initial deployment. In addition, they will want to up sell what they have. The use of a branded captive portal can serve both purposes. In the case of a new service offering, users can be directed to a portal to ensure they are aware of the service. The network can be setup to direct users each time they access the network or only after a new service is introduced.

As shown in Figure 12.12, Juniper Networks SDX/captive portal process also can help in upselling to a customer. For example, users may have contracted for a 512 Kb/sec rate with a cap of 4 Gb/month. When the total of 4 Gb/month is reached, they are rate limited to 64 kb/sec. This is an opportunity to send them to a captive portal to inform them about their current restriction and to sell them a larger per month limit. This mechanism is in use in many DSL networks throughout the world.

Also depicted in Figure 12.12, the Juniper Networks SDX and E-Series combination provides the rate limiting functions and can extend to the Wireless EMS. The branded portal can be pushed from the SDX.

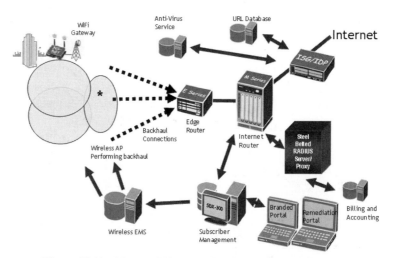

Figure 12.12: Advanced Network Services Network Diagram.

12.15 Conclusion

The Municipal Wireless Network faces three distinct challenges. The enabler that allows these networks to be built is the emergence of new types of wireless technologies. However, successful deployments will have taken into account their competition and the services to be offered on these networks.

The technology of the wireless networks will continue to evolve as mature telecom vendors and startups push the envelope with both proprietary and standards-based solutions. This will enable the building of newer, faster, and more fully featured wireless.

The competition comes from a number of sources. Existing networks built on DSL, cable modems and 3G Wireless networks provide bandwidth to large number of users. Other users are used to traditional dial-up, T1 services or Public Safety networks. The Municipal Network will need to compete with these existing networks. In general, the business plans for these networks rely on both "new users" and taking users from existing networks.

This chapter has focused on the services that will be offered on the Municipal Wireless Network. Network operators will need to determine which set of services is right for their specific network. These services will have to be structured to create success -- a success measured by a positive ROI.

The network also must have built-in flexibility as the services offered are likely to change over time. Juniper Networks provides solutions that deliver those requisite levels of flexibility. We have built IP networks for the world's leading carriers and governments to meet the strictest of standards. As such, we anticipate supporting Municipal Wireless Networks by building secure and assured networks.

13

Metro-Scale Wi-Fi Networks

Josef Kriegl and William Merrill[a]

In this chapter we present the requirements, architecture and design challenges for municipal wireless broadband access networks. We first provide an overview of the requirements and services anticipated for municipal Wi-Fi networks based on five recent municipal RFPs. Next, we present design guidelines for a layered network using access, mesh, injection and backhaul tiers to create a cost-effective municipal broadband network utilizing equipment conforming to the 802.11 family of standards in the mesh and access tiers. To illustrate the challenges involved in creating mesh networks we explore the costs and benefits of building single versus multi-radio mesh networks and of creating a network capable of supporting session-persistence roaming across a municipality. We provide design equations supported by measured data for multi-hop mesh networks using a single mesh radio configuration. Our roaming solution is designed to operate in the context of a routed mesh network and offers seamless integration with network access controllers, and in conjunction with dedicated network branch routers, session-persistent roaming within metro-scale Wi-Fi networks.

13.1 Introduction

Envisioned as a replacement for traditional Ethernet Local Area Networks in home and office environments, 802.11-based broadband access networks have seen an explosive growth in the public space over the last five years. Beginning with small scale hotspot deployments in high use public areas such as airports, cafes and businesses, cities have begun to embrace Wi-Fi technology as a catalyst for attracting customers and revitalizing downtown areas. While traditional hot-spot and hot-zone deployments usually rely on the availability of a wired backhaul network to route client traffic to an access controller and further on to an Internet gateway, the emergence of low cost mesh networking equipment has spurred the interest of using 802.11 wireless networks. This technology has the potential to provide Internet access for large portions of the population.

Over the last two years municipalities across the U.S. and in other nations have considered public wireless access networks to bridge the digital divide and provide free or

[a] All authors currently affiliated with *Tranzeo Wireless Technologies USA Inc* (*formerly Sensoria Corporation*)
Wireless Fabric is a trademark of Tranzeo Wireless Technologies. All other trademarks are the property of their respective owners.

low cost access. Many municipalities have come forward with requests for proposals to design and build metro-scale Wi-Fi networks. The financial viability of such networks depends to a large extent on the cost of the network's distribution system, i.e. the links, switches and routers required to connect all wireless 802.11 access points to a network operations center and further on to Internet peering routers. One common approach is to partition the distribution network into a hierarchical topology using different networking technologies at each level or tier. These will subsequently be referred to as the access, mesh, injection and backhaul tiers. The access tier consists of the collection of spatially distributed 802.11 access points, which are co-located with mesh bridges or routers. These mesh nodes, collectively called the mesh tier, aggregate user traffic and forward it across one or more wireless hops to a gateway. Each gateway is in turn connected to one endpoint of a system of point-to-point and point-to-multi point wireless links called the injection tier. The upstream endpoints of these links are connected to a wired or wireless distribution system called the backhaul tier. In the context of this chapter, the transition point between the backhaul and the injection tier will be called the backhaul point of presence, while the mesh node connecting the mesh tier with the injection tier will be referred to as the mesh gateway.

The following sections begin with an overview of municipal broadband initiatives and the typical set of functional and performance requirements defined in the respective RFP documents. For this purpose, the network specifications of five municipal RFPs have been surveyed [1], [2], [3], [4], [5]. Subsequently, a tiered approach to designing a metro-scale network is presented, including possible technologies that can be used to realize the individual tiers. This section also provides a discussion of Quality of Service (QoS) considerations, network capacity planning, and network-wide roaming and concludes with an overview of Network Operations Center (NOC) functions. A detailed discussion of design issues access, mesh and injection tier design follows. The chapter concludes with the description of a network-wide seamless roaming architecture based on the Sensoria ER500 mesh router and supporting infrastructure.

13.2 Wireless Broadband Initiatives

Early adopters of metro-scale Wi-Fi networks in the U.S. include the cities of Portland, OR [1], Philadelphia, PA [3], San Francisco, CA [2], Thornton, CO [4], Rockville, MA [5], New Orleans, LA and Anaheim, CA, just to name a few. In their mission statements, these municipalities emphasize the need for a city-wide Wi-Fi network as an alternate means to provide broadband Internet access to all inhabitants, businesses, schools, community organizations, and visitors. The anticipated benefits of a city-wide broadband access network include affordable Internet access for economically disadvantaged citizens, reduction of operational cost for the city's mobile workforce, facilitation of community access to government services as well as the streamlining of citizen–government interaction. In the majority of cases, the city also requires the successful proposer to own, operate and maintain the network, with the mandate to continuously upgrade the network as relevant network technologies mature or emerge [1], [2], [3]. Furthermore, the municipalities also intend to stimulate competition in the broadband Internet access market while providing low cost or free of charge access to low-income households, facilitating the creation of home-based businesses, increasing telecommuting opportunities and providing a telecommunications infrastructure that attracts visitors, conventions and new businesses [1], [2], [3].

The municipalities frequently assert themselves as anchor tenants, with the intention to use the network for their own mobile workforce and also provide broadband fixed wireless access to distributed city assets such as utility equipment, parking meters and surveillance cameras [1]. In all surveyed RFPs some form of dedicated network access for police and emergency services is envisioned, featuring secured over-the-air transmissions and the ability to reserve a portion of the total available bandwidth on demand. In some cases, dedicated access in the 4.9GHz band is either required or needs to be on the provider's roadmap for public safety applications [2], [4]. Cities typically ask for a staged built-out plan of the network. First, an initial proof of concept covering a small area is requested to allow the successful bidder to demonstrate the viability of their technical solution. By the end of the second phase the network is expected to cover major portions of a city's downtown area as well as underserved neighborhoods, thereby providing feedback on overall network scalability, market acceptance and profitability of the underlying business model. The final build-out phase is expected to cover all initially identified service areas, covering most of the city's neighborhoods, public spaces and possibly transit corridors.

13.3 Network Use Cases and Performance Requirements

All surveyed municipalities desire to establish a wireless broadband network infrastructure to meet the needs of multiple user groups, each associated with different usage models and preferred applications. Concurrent access by multiple users associated with distinct groups must be supported for client devices that are based on the IEEE 802.11b/g standard. Wi-Fi compliance, which certifies compatibility between 802.11-based devices, is considered the least common denominator for ensuring interoperability and hence unconditionally required. Given the various project stakeholders and anticipated user groups, a metro-scale Wi-Fi network must be able to provide logical partitioning of the network at the link layer (i.e. support for multiple network identifiers or ESSIDs) and provide configurable levels of security for each of the logical networks. The available security mechanisms must be compatible with the 802.11i standard, with support for WPA/2 authentication mechanisms using various Extensible Authentication Protocol (EAP) modes [37]. Furthermore, mutual client isolation is required and subscribers must be able to use Virtual Private Network (VPN) technology providing end-to-end transport layer security. Additional pro-active and retro-active security mechanisms must include link layer traffic encryption, firewall support on each 802.11 Access Point (AP), an intrusion detection mechanism, rogue AP and interference detection as well as the ability to exclude or disable individual users by MAC address and user account.

Recognizing that a wireless access network infrastructure inherently offers the opportunity for people and devices to connect anytime, anywhere, in a stationary manner as well as in-motion, cities ask for different levels of mobility support. Mobile access scenarios include nomadic, portable (at speeds less than 30mph) and mobile (greater than 30mph). RFPs typically acknowledge the limited support of fast 802.11 handoff available today and hence require seamless mobility (i.e. the ability to maintain TCP and UDP sessions across AP handoffs) only for low speed mobility scenarios (less than 30mph). For each of the different roaming scenarios authentication level persistence is required (i.e. once logged-on, a user is not required to re-authenticate when moving across the network within the time limits of an authentication session, which will usually be significantly longer than the roaming time).

The network also needs to provide QoS capabilities, including packet characterization and prioritization based on the Internet Protocol Type of Service (ToS) field as well as higher layer protocol information (port numbers and application signatures) in order to support latency-sensitive applications such as VoIP and streaming video. This requirement applies to both the access tier (via 802.11e) and the distribution system. The QoS mechanism should also be capable of controlling bandwidth allocations (service level agreements) on a per-user and per-ESSID basis as well as on-demand bandwidth re-configuration (e.g. all available bandwidth may be allocated to first responders during emergency situations). Individual service levels ranges anywhere from 256Kbps for free access to premium (1Mbps) and business access at 3Mbps per subscriber [1].

Service coverage areas are usually specified in detail in the RFP, including requirements for indoor coverage. Within the defined service area, standard Wi-Fi client devices should be able to connect to and use the network at the subscribed data rate from any perimeter room facing a public street [2]. Additional Customer Premise Equipment (CPE) such as wireless repeaters, bridges, etc. should be supported to allow for improved indoor coverage. All networking equipment must support high levels of availability (99% to 99.9% on the access tier and 99,9% to 99.999% on the distribution tier [1], [2]). Furthermore, system level fault mitigation mechanisms are encouraged, including the availability of backup power sources and automatic reconfiguration and route management in the wireless backhaul layer when a link or network element fails.

In many instances, the Wi-Fi network should be designed to allow multiple service providers (wholesale customers) to offer their respective service plans to retail customers, either by providing multiple ESSIDs or a single ESSID combined with a capture portal accommodating multiple providers. The municipality usually requires the network operator to offer a basic free service. In addition, the operator, its affiliates or a third-party wholesale customer (an independent service provider) are expected to offer multiple service levels. This implies that the network is to be designed to offer fixed and portable broadband access to multiple users associated with a given AP at the 1Mbps (premium service) and 3 Mbps (business service) level. Furthermore, bi-directional roaming support is required thereby enabling users with service accounts on a third-party network (such as an existing commercial hot-spot or Wi-Fi network in a different town) to seamlessly connect to the municipal wireless network. Customer support is mandated to a varying degree, with multiple support levels for premium service customers. These include self-service via web-based sign up, issue resolution and higher level interactive support options (call center, email, and chat, respectively). While level 1 support may be provided through third-party wholesale customers (service providers), all higher support levels are to be provided by the network operator, to both its retail and wholesale customers.

The Network Operations Center (NOC) may be centralized or distributed over multiple facilities. All user traffic is expected to traverse the NOC, thereby providing the basis for user Authentication, Authorization and Accounting (AAA) functions as well as third-party roaming support. The network's Operations Support System (OSS) is part of the NOC and includes network element and link monitoring functions as well as fault detection mechanisms. It offers a centralized interface to network element configuration and software upgrade. The OSS also implements the service ordering and fulfillment system by providing the necessary interfaces to operators, retail and wholesale customers to add, modify and cancel services. Furthermore, this system is responsible for collecting system usage statistics as a basis for network upgrade and expansion decisions. Given that a municipal Wi-Fi network requires major capital investment, future extensions need to be supported including capability and capacity upgrades. The network should be designed to allow for the incremental addition of backhaul, mesh and AP equipment to increase its capacity or coverage. Finally, the network elements should allow for modular hardware upgrades wherever possible (such as the addition or replacement of an AP or backhaul radio to cope with new access standards and/or the need for more capacity).

The following sections introduce a metro-scale Wi-Fi network architecture that is aimed at addressing the aforementioned network requirements. The focus of the discussion will be on the wireless tiers of the network. On order to simplify the discussion the scale of the network is assumed such that it can be served by a single centralized NOC.

13.4 Multi-Tier Network Design Overview

Users connect to a Wi-Fi network by means of the access tier. This portion of the network includes the 802.11 client devices (stations) and the collection of spatially distributed 802.11 Access Points (APs). In some instances where network access is to be further distributed within buildings, a fixed wireless bridge can be used to connect to the outdoor network. The most popular versions of the IEEE 802.11 standard are defined in the 802.11b and 802.11g amendments [38]. The popularity of 802.11b/g is evident from the plethora of available client devices and outdoor Wi-Fi APs such as [6], [7], [8], [9], [10] [42], [43]. The 802.11 standard defines two operation modes, *ad-hoc* and *infrastructure* [38]. In infrastructure mode, an AP coordinates one or more stations within a Basic Service Set (BSS) called an AP cell. Multiple AP cells can be deployed to extend a network's coverage area, resulting in a configuration called an Extended Service Set (ESS). The need for partitioning the coverage area into a series of cells arises from several factors, the most compelling of which include the operation of the 802.11 MAC, FCC limits on radiated power in the unlicensed 2.4GHz ISM band as well as the need to provide sufficient throughput to a large user base.

Given the segmentation of the coverage area into a collection of AP cells, a mobile user on foot or in a vehicle will typically traverse multiple cell boundaries during an active session, resulting in significant delays incurred by the link layer handoff process [32]. In order to maintain transport layer sessions, HTTP, FTP, SSH and streaming multimedia protocols rely on the network to 1) always offer a candidate AP for a potential handoff and 2) continuously update routing information to the client device's point of attachment in the network for successful packet delivery. If these requirements are met, then the network is said to provide seamless roaming capabilities. As noted in [32], the 802.11 standard does not provide for an efficient handoff mechanism, resulting in significant packet delivery delays as well as lost packets. Typical handoff times were measured between 300 and

500ms, with over 90% of the handoff time reportedly spent scanning for a suitable candidate AP in all eleven available channels. While several schemes are being proposed to reduce the handoff time [45], [46], [47] to meet the VoIP latency limit of 50ms [32], the suggested approaches require extensions to both the radio management software and card drivers and hence are not suitable for existing products. The 802.11r task group is developing a standardized way to reduce handoff latency in order to meet the needs of VoIP for seamless roaming scenarios [48], however, final approval of the amendment is not expected until early 2008 [36].

Hot-spot and hot-zone deployments in indoor environments have traditionally relied on a wired distribution network based on an Ethernet backbone, an approach that is impractical to replicate in an outdoor setting where the coverage area is expected to extend over multiple blocks in a downtown district. Using traditional wired broadband technologies such as DSL, cable or dedicated optical fiber at each AP location in a municipal deployment would be prohibitively expensive. Instead, a wireless backbone network consisting of a collection of layer 2 and layer 3 relay nodes collocated with APs can greatly reduce installation cost and time by aggregating and forwarding the traffic from multiple APs. The relay nodes are generally layer 3 wireless routers capable of forming a mesh-like ad-hoc wireless multi-hop network. These nodes are called mesh routers, mesh APs or simply mesh nodes. Dual-radio, dual-band solutions featuring omni-directional antennas for both the AP and mesh radio have emerged as the most common product type [6], [7], [8] and represent a tradeoff between low-cost, low performance single radio solutions [7], [44], [43] and multi-radio designs, which usually offer higher throughput over mesh networks with more wireless hops at the expense of a higher deployment and per-unit cost [42], [9]. The wireless technology used for mesh links may be of any type, although cost considerations have prompted most vendors to employ 802.11 radios operating in the 2.4GHz or 5GHz band or to consider non-standard 802.11 radios frequency shifted to operate in 5, 10 or 20MHz wide channels within the 900MHz ISM band [11], [12].

The required density of the mesh nodes is driven by a number of factors including seamless AP coverage, reliable connectivity and the level of desired path redundancy across the mesh network. Path redundancy has a direct effect on the fault tolerance of the mesh tier, and, in mesh networks that employ interference-aware routing algorithms, can help increase the aggregate throughput across the mesh by routing traffic around interference hot spots [20]. Secondary factors that influence the effective mesh density include the availability of power at mounting sites as well as the need to limit the self-interference range in a deployment that uses a single mesh radio. There is a practical limit on the number of wireless mesh hops that can be realized while maintaining a certain level of aggregate network throughput. This is particularly true for single and dual radio mesh nodes, as discussed and analyzed in detail in [20], [21], [23]. When multiple radios share overlapping spectrum, mesh nodes may interfere with each other's transmissions every time a packet is relayed, thereby reducing the bandwidth available to forward data at each hop. This self-interference is compounded when forwarding data along a multi-hop path. Packets may be dropped because the mesh nodes furthest away from the egress point may be injecting traffic at a higher rate than can be forwarded by nodes along the path [20]. One method to mitigate this multi-hop starvation is by enforcing bandwidth limitations at the network ingress points [34]. In addition to dividing spectrum among all channel users, throughput over the mesh is further reduced due to collisions caused by the hidden station phenomenon and the need for subsequent retransmission, particularly if the RTS/CTS scheme is not employed [49]. Another source of inefficiency in mesh throughput is introduced by the 802.11 CSMA/CA MAC protocol's well known degradation when operating close to network capacity with many users on the channel [31].

In single mesh radio networks, since each wireless hop represents an additional use of the shared spectrum, limiting the number of wireless hops in the mesh network can help reduce the detrimental effect of self interference. In addition, the practical throughput limits of aggregate traffic over the mesh network and into the backhaul tier limit the size of the mesh supported by each network egress point. As a result, the geographical area covered by a single mesh network (i.e. the collection of mesh nodes dynamically associated with a given network egress point) will also be limited. Since municipal wireless networks typically require coverage of 135 square miles [1] and larger, multiple meshes, each with an associated egress point (mesh gateway), are required to cover the entire service area.

Mesh topology layouts vary in shape, size and number of mesh nodes in order to accommodate terrain, achieve (seamless) AP coverage and provide adequate throughput for different traffic patterns. Two representative topologies include the lattice (or grid) layout and the linear topology [17]. The grid topology can be found in urban area deployments with a regular street layout. A linear topology typically results from deployment in suburban communities or along transportation corridors. Mesh nodes on a grid are best placed at intersections, providing unobstructed Wi-Fi coverage in four directions, and connecting the mesh with up to four of its neighbors thereby providing a high degree of fault tolerance. However, lattice networks also provide more potential for self interference than linear meshes, as discussed in [20]. The self-interference effect may be mitigated somewhat by either providing more mesh egress points per unit area, thereby significantly reducing the average path length for traffic traversing the mesh or using multi-radio mesh nodes to avoid mesh interference altogether.

Traffic originating from client devices and bound for the Internet needs to be routed through the Network Operations Center (NOC) for client configuration and access control purposes (dynamic IP configuration, registration, authentication, access control, accounting and downstream bandwidth control) and further on to a peering router connected to the wider Internet[1]. Both wired and wireless links as well as a combination thereof are conceivable for this purpose. As illustrated in Figure 13.1, the distribution network can be divided into an injection tier, aggregating traffic from multiple mesh networks and a backhaul tier, which aggregates traffic from all injection tiers and connects to the NOC. Using a combination of dedicated P2P and P2MP links for the injection tier as well as optical fiber or high-speed wireless links on the backhaul network, can help reduce the overall cost of deployment and ownership of a municipal Wi-Fi network.

---

[1] While intra-mesh traffic, e.g. first responder traffic, does not need to be routed to the NOC, authentication, and in some cases dynamic IP configuration still requires interaction with services available at the NOC (RADIUS server, DHCP server).

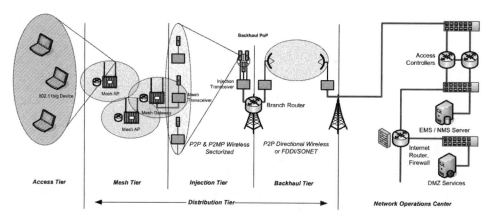

Figure 13.1: A multi-tier Wi-Fi network architecture. The figure shows the various tiers and interconnection points as well as the network operations center. Mesh to injection layer transition is accomplished by a mesh gateway connected via Ethernet to the mesh transceiver. The injection transceiver to backhaul transition point (labeled Backhaul PoP) includes a router with traffic shaping capabilities to satisfy network partitioning and QoS requirements discussed in subsequent sections. The figure does not illustrate how a typical municipal Wi-Fi network would scale from the network operations center across the various tiers toward the 802.11 access points.

The injection tier consists of multiple mid-range wireless P2P and P2MP links (on the order of kilometers) based on a variety of PHY and MAC layer technologies. These links can be used to extend an existing wired network infrastructure (such as a metropolitan fiber ring) at a much lower cost per mile (by a factor of 10 [35]), while dramatically reducing deployment time. P2P and P2MP links are generally set up to operate in a different (licensed or unlicensed) frequency band than the mesh networks they serve. As such, injection layer links are well suited to extend a grid mesh without the need to add mesh hops or to reach remote parts of the city. Common P2P and P2MP solutions rely on a time-division MAC access scheme [14], [12] or WiMax [13]. The collection of mesh networks associated with a P2MP link is called a branch of the city-wide network. The backhaul portion of the distribution network may be comprised of wired lines, high speed wireless P2P links or a combination thereof, reaching tens of kilometers. The interconnection point between the injection layer AP and the backhaul network called the Backhaul Point of Presence (Backhaul PoP). The use of a two-tiered distribution network strikes a balance between the number of required broadband connections (optical fiber, cable or DSL termination points) and the bandwidth available to the mesh networks it serves (if limited by the injection link capacity). A sectorized antenna configuration for the injection transceiver location (as indicated in Figure 13.1) can help increase the available bandwidth to the mesh networks in the served area by means of frequency re-use.

As is the case with wired infrastructure networks [15], a network-wide traffic characterization and prioritization framework is necessary to ensure that latency and loss-sensitive traffic flows such as video and VoIP streams are prioritized over background and best effort traffic. Depending on the technology and the number of attached mesh networks (and as a result the number of 802.11 APs and potential users), the injection layer capacity may represent a throughput bottleneck in the path from Wi-Fi user to the Internet PoP.

Proper capacity and topology planning as well as a QoS scheme are necessary to avoid this situation, which could result in random packet loss if the link becomes oversubscribed. The Backhaul PoP is one point in the network to implement this traffic shaping function for downstream traffic[2] as this allows for the traffic shaping parameters (reserved bandwidth for each traffic type, flow priorities) to be tuned to the available injection layer capacity. The available capacity can be expected to vary for each branch due to the employment of P2P and P2MP technologies that best meet the distance and foliage penetration requirements at each site. Besides traffic shaping, the Backhaul PoP also lends itself for network partitioning at layer 3 in order to manage the control traffic associated with roaming users. The entity responsible for traffic shaping and roaming is called the branch router.

The NOC shown in Figure 13.1 aggregates all user traffic and is responsible for traffic routing, subscriber-related administration functions as well as various network element management functions. An Access Controller (AC) acts as the layer 3 router for client IP traffic between the backhaul network and the NOC subnet. Outbound IP traffic is masqueraded by the AC and forwarded to the NOC's Internet router. The AC also manages the dynamic assignment of IP addresses to 802.11 client devices and handles session Authentication, Authorization and Accounting (AAA) functions, per-subscriber QoS policy enforcement (bandwidth limits, allowed service types, times of use, etc.) as well as subscriber usage statistics collection. The AAA functions, along with bi-directional roaming support may be implemented by an external AAA provider and delivered through standardized interfaces. The utilization of an external AAA provider reduces the burden of subscriber management by the network operator and simplifies the management of third-party roaming agreements for visiting users.

The network elements that require monitoring and management include the 802.11 APs, the injection layer equipment as well as the branch router at each Backhaul PoP location. Element management functions include the provisioning and activation of software upgrades, configuration changes to support new services and alter existing services as well as monitoring performance and state of the employed network elements and associated links. The network element management functions are usually implemented by the Element Management / Network Management (EMS/NMS) server part of the OSS. This entity streamlines business processes such as service ordering, service activation, billing and asset management with subscriber-related functions such as customer service, trouble ticket generation and tracking as well as roaming support for guest users in a metro-scale environment. Subscriber usage tracking as well as reporting of network-related issues experienced by individual subscribers as a means to assess network performance and plan for future upgrades are also functions supported by the OSS [25].

Estimating capacity needs for a Wi-Fi network is an important aspect of the design process as it influences technology choices at each tier and ultimately impacts the user's experience. A sound capacity plan that meets bandwidth and QoS needs of a variety of different user groups requires estimates on the number of initial and future subscribers, the number of concurrent users, the distribution of the users across the coverage area (user density on a per-AP, per-mesh and per-branch basis), communication patterns (intra- and

---

[2] Upstream traffic can be efficiently adjusted and controlled at each individual mesh AP such that the aggregate throughput of all mesh AP's in all mesh networks associated with a given injection layer AP does not exceed the injection capacity.

inter-mesh as well as Internet bound) and the envisioned application QoS requirements. These estimates are usually combined to establish a traffic model, which can then be used to describe the anticipated use of a network, both in terms of traffic (protocol) types and network utilization (consumed bandwidth) in the upstream and the downstream direction. Anticipated user densities on a per AP, per-mesh and per-branch basis can be estimated based on zoning information (residential vs. business district) and the number of potential subscribers in the area. This information can be combined with appropriate application data models describing the application types, per-flow bandwidth and associated network utilization statistics, resulting in average and peak bandwidth requirements. While the peak bandwidth must be met at any given location and time (in an instantaneous fashion), the average bandwidth and the number of anticipated users determine how much total network capacity must be available over each link.

When relying on traditional Internet traffic models, deviations can be expected based on a somewhat different application mix [16], the subscriber growth rate and alternate communication patterns. While the majority of the traffic can be expected to consist of Internet-bound flows (i.e. to and from the NOC), including all server-based VoIP applications, some network usage models call for intra-network communication capabilities to support emergency services. Typically, most intra-network traffic will be localized, either within a single AP cell or a single mesh network, although traffic exchange may happen across mesh and branch boundaries. Roaming users add some uncertainty to the bandwidth requirements for each network (AP) location; however, traffic characterization and prioritization at interconnection points can help maintain the QoS requirements for latency-sensitive applications such as VoIP. Traffic pertaining to periodic network monitoring tasks and route maintenance for roaming users also needs to be accounted for in the network capacity plan. Finally, the traffic associated with software and content provisioning to network elements represents a significant amount of data and needs to be considered in the network load analysis. However, since software upgrades and content provisioning may be scheduled at off-peak hours of the day or triggered by the available (averaged) bandwidth dropping below a pre-defined threshold, they usually don't need to be accounted for in the aggregate average bandwidth calculations. Network elements such as the branch router at the Backhaul PoP and the client Access Controller at the NOC are also potential bottlenecks in the Wi-Fi network. The presence of these devices may affect maximum achievable packet forwarding rate as well as the number of users and associated data that can be handled concurrently.

13.5 Wi-Fi Tier Design

The Wi-Fi tier offers an array of challenging design problems. Coverage, RF propagation and interference issues require detailed site surveying followed by appropriate AP placement, antenna selection and frequency reuse schemes. The resulting AP layout influences the scale and cost of higher layer network tiers. Furthermore, AP functions such as authentication, encryption, 802.11b/g compatibility, handoff and dynamic IP address assignment are expected to interoperate with a plethora of 802.11 client devices and web browsers without the need for hardware changes or software upgrades to the user's existing equipment. In addition, the AP layout has to allow for seamless roaming and the web-based service registration and authentication process must be suitable for users of different levels of expertise. Network management must incorporate the ability to dynamically reserve bandwidth for higher priority services (e.g. in the event of an emergency), respond to security breaches (e.g. by blacklisting individual MAC addresses) as well as monitor and mitigate unexpected local RF interference and rogue APs[3].

Outdoor AP products for the US market are generally certified as FCC class A digital devices with a maximum EIRP[4] of 36dBm [18]. In contrast, consumer 802.11b/g radios achieve much lower EIRP values, due to lower-gain antennas and lower card output power resulting in asymmetric link performance and throughput. The ambient noise floor and the receiver sensitivity in combination with the received signal strength of a transmission ultimately determine the ability of each endpoint to decode transmissions successfully. The ratio of signal strength to noise floor is defined as the Signal to Noise Ration (SNR). As described in detail in [50], the probability of wireless communication is described by a link power budget, which establishes the relationship between transmitter power, propagation loss, transmitter and receiver antenna gain, antenna polarization and cable and connector loss at each transceiver. The absolute difference between the received power and the receiver's sensitivity (in dBm) is called the link margin and should be high enough to account for variability in the received power due to short-term fading effects. A Wi-Fi link is generally considered reliable if the link margin is equal or greater 6-10dBm [26]. Given that outdoor Wi-Fi communication usually takes place in urban areas, the rules governing free space pass loss practically never apply. As a result, path loss models incorporating the effects of multi-path propagation (due to reflection), diffraction (due to stationary and moving objects), scattering (due to foliage) and absorption (due foliage and building materials) tend to be complex [50]. Commercially available modeling tools do exist to assist in the process of creating an RF propagation map, examples include [27], [28]. Even with a detailed RF map, transmit power, receiver sensitivity, cable loss and antenna gain are subject to the variations in user equipment. While outdoor AP vendors strive to maximize receiver sensitivity[5], maximum EIRP is limited by FCC regulations. Note, however, that maximizing the output power at every AP location is not always advisable as this may cause interference with neighboring APs operating on the same channel, especially under traffic burst conditions and mixed b/g operation.

The maximum achievable throughput over the Wi-Fi tier is limited by the following factors: the channel contention as a result of concurrent use of the wireless channel by multiple users and the propagation characteristics of each link. Furthermore, the concurrent use of 802.11b and 802.11g devices will result in degraded aggregated throughput, due to the use of RTS/CTS for backward compatibility and the longer transmission intervals used

---

[3] A rogue AP in the context of a municipal Wi-Fi network is an unauthorized AP that broadcasts the same ESSID as the legitimate network, thereby interfering with a subscriber's ability to connect to the legitimate network and use it for Internet access. Besides the potential for such a denial of service attack, the rogue AP operator could also gain access to the user's computer through social engineering techniques (e.g. give instructions on the login splash page asking the user to disable their firewall).

[4] EIRP (Effective or Equivalent Isotropically Radiated Power) is a measure of the peak power density radiated by a combined radio and antenna system. It is equivalent to the input power provided to a theoretical antenna that evenly distributed the power density across all three dimensions, so that the power in any direction matches the peak power density of the described system.

[5].[53] estimates that a 3dB increase in receiver sensitivity translates to a 30% reduction in AP density due to increased range of client device transmissions, while 6dB result in 50% fewer AP's. The study does mention that high data rates at these low signal levels are only possible if the noise floor is reduced by the same amount.

by of 802.11b devices for the same amount of data [29], [30]. The aggregate throughput from multiple stations is a function of packet size and the mode in which channel arbitration is performed. In DCF basic mode, channel contention occurs without the overhead of an additional Request-To-Send / Clear-To-Send (RTS/CTS) message exchange cycle, resulting in higher throughput while the number of contending stations is small. As the user count increases beyond approximately five, and each of the users send at the maximum possible rate, the throughput sharply decreases due to lost packets following collisions [31]. The level of channel contention (i.e. how many devices compete for access to the wireless channel) can be reduced through spatial and/or frequency re-use. While spatial reuse (by adding more APs) translates to higher equipment cost[6], frequency reuse not only reduces channel contention but also helps reduce co-channel interference from neighboring APs. For this purpose, 802.11b/g operating in the 2.4GHz ISM band offers three non-overlapping channels (1, 6 and 11), which may be assigned to AP cells as shown in Figure 13.2.

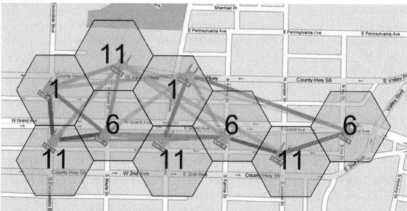

Figure 13.2: Frequency reuse and AP placement for continuous coverage. The actual coverage areas will deviate from this idealized layout due to variations in the RF propagation environment at each access point location. The numbers shown in the hexagons correspond to the three non-overlapping 802.11b/g channels. The bold lines between the mesh nodes represent the various one-hop and gateway routes in a nine node mesh deployment.

In the majority of outdoor Wi-Fi AP products, omni-directional antennas are used for the AP radio. Vertically mounted omni-directional antennas exhibit a torodial radiation pattern, although antenna variations and equipment mounted nearby can distort this pattern. Ideally, the designer would strive to place the antenna at a height that minimizes path loss (due to foliage and other obstacles) while providing sufficient signal strength in the area directly below the center of the antenna. However, antenna height is often dictated by available infrastructure (street lights, power poles, building roof tops). Sectorized antennas can help mitigate lack of coverage by directing the antenna beams appropriately. An additional benefit of using sectorized antennas is to increase the per-area channel capacity by reducing access contention, although care must be taken in system and antenna design

---

[6] An increased number of APs per unit area results in denser mesh networks, which in turn requires a larger number of injection layer links and/or points of presence. As will be argued in the mesh tier design section, the aggregate throughput of dual radio mesh nodes is limited by the number of hops within the mesh.

to prevent cross-sector interference. Multi-story buildings require AP cells that also extend into the third dimension. Alternatively, mesh networking technology as discussed in the next section may be used to accomplish this task.

13.6 Mesh Tier Design

Mesh networks in use today for municipal wireless networks have evolved from wireless ad-hoc networks conceived for both fixed and mobile wireless networking environments such as mobile communication and sensor networks [19]. RF range and data rates comparable to the access tier and the relatively low cost have been driving factors for mesh networking companies to use 802.11-based radios to close the links between mesh APs. Mesh radios are usually configured to operate in *ad-hoc mode*, allowing mesh nodes within a broadcast domain to discover each other and thus serve as mutual gateways for client traffic to reach destinations within and beyond the mesh. A variety of mesh routing protocols for path discovery and adaptation are available today. They differ in terms of delay incurred for route establishment (pro-active versus re-active) as well as the metrics used to find the optimal path [19]. The following sections begin with an overview of the traffic patterns encountered in a municipal mesh network, discuss the advantages and drawbacks of single, dual and multi-radio mesh nodes in the context of typical mesh topologies and give guidance on the design of a mesh network based on dual radio nodes, which represent a reasonable tradeoff between performance, cost, deployment time and configuration effort.

A municipal Wi-Fi network's main purpose is to provide wireless Internet access to a large number of users. In addition, albeit to a smaller extent, localized intra-mesh traffic can be expected to support emergency communication services and other collaboration efforts. The prevailing traffic pattern, however, is defined by traffic that flows from 802.11 client devices through the nearest mesh AP and its mesh gateway to a server or peer on the Internet. Both traffic patterns are illustrated in Figure 13.3. Various throughput models and predictions have been explored by the research community, suggesting lower and upper bounds for total network throughput under the assumption that each node is concurrently communicating with another node in the network [21], [22], [20]. For the purpose of deriving throughput for a municipal mesh network, these usage scenarios can be relaxed in that most flows will be directed from a mesh node towards the gateway, and, if intra-mesh communication is present, the traffic will be localized (i.e. emergency communication traffic will remain contained either within a single mesh AP cell or at most across a few mesh nodes).

In the context of this chapter a mesh network is defined as the collection of mesh nodes associated with a single egress point. The association of mesh nodes with an egress point – the mesh gateway - can be dynamic due to gateway re-selection in case of mesh gateway failure. Figure 13.3 depicts two co-located mesh networks, where Mesh AP 1 and 2 may be associated with either Mesh Gateway 1 or 2. If Mesh Gateway 1 fails, Mesh APs 1 and 2 will be able to re-select Mesh Gateway 2. Conversely, if Gateway 2 fails, Mesh AP 3 and 4 will become isolated as none of their transmission boundaries overlap with any of the remaining nodes. The reliability provided by mesh gateway re-selection is highly desirable from a network operator's point of view.

As indicated earlier, a path selection mechanism is required for mesh nodes to determine an efficient route towards the mesh gateway. Path selection may happen at either OSI layer 2[7] or layer 3. A path (or route) is considered efficient if it meets the following performance requirements:

- It maximizes the throughput for the current flow as well as the aggregate throughput of the mesh
- It minimizes the delay and jitter imposed on packets of the current flow as well as other flows in the mesh

The metrics used to determine an efficient route vary by vendor and are often proprietary. Common metrics include bit error rate, number of retransmission attempts, Signal to Noise Ratio (SNR), packet latency, and hop count. Advanced path selection algorithms also take into account local interference hot-spots created by the presence of multiple flows traversing a specific area, thereby improving throughput despite a longer path length [22].

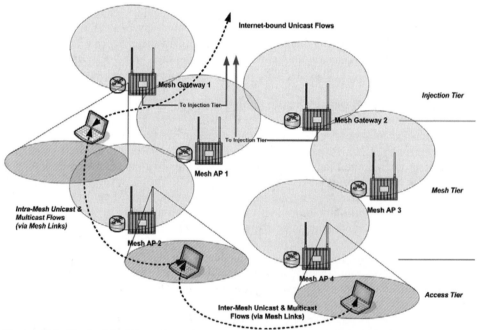

Figure 13.3: Typical traffic patterns in a municipal mesh network include intra- and inter-mesh unicast and multicast communication (emergency and municipal work force use cases) as well as Internet-bound unicast traffic flows (general Internet access use case). In this illustration two mesh networks are shown with the overlapping grey ovals representing the transmission range of each mesh node. Note that in this example all depicted client traffic must traverse mesh links in order to reach their destination.

---

[7] The 802.11s amendment proposes a mesh path selection algorithm implemented at layer 2. This standard is expected to be approved in mid-2008.

A fundamental issue with multi-hop wireless networks is that throughput decreases sharply with the number of hops traversed when mesh transmissions share the same spectrum. For example, with dual radio mesh nodes the throughput is cut in half when going from one to two mesh hops [22]. In the case of a single radio mesh node, the result is worse because the access tier represents an additional wireless hop in the network and the client devices have to contend for the same channel as the mesh transmissions. In addition to re-transmitting local client traffic, a mesh AP often acts as a relay node for other nodes' traffic, thereby further degrading the available throughput for users.

The aggregate throughput of a single radio mesh network is estimated to be in between 1/n (best case) and 0.5/n (worst case) [23], [24]. To better understand this analysis, consider the example in Figure 13.4. The total wireless channel capacity is assumed to be 24Mbps consistent with the default settings of many 802.11a radios operating at a signaling rate of 54Mbps. Assuming each mesh AP is loaded with 6Mbps, AP6, which also acts as the mesh gateway, will have to receive and forward 6Mbps from each of the other APs, in addition to the 6Mpbs generated by its local client(s). This is based on an interference range equal to the transmission range (i.e. AP3's transmissions only interfere with AP2's and AP4's transmissions, but not AP1's or AP5's), perfect channel access coordination and collision-free operation, none of which are realistic in the context of the 802.11 MAC protocol. The interference range typically extends beyond two wireless hops and is dependent on the communication signal rate's required SNR. The medium access algorithm involves random back off times likely resulting in unnecessary medium idle times [20]. Furthermore, the hidden station phenomenon may cause collisions if no RTS/CTS scheme is employed, resulting in additional transmission back offs and retransmissions. Two-radio mesh networks realize a somewhat improved aggregate throughput as they eliminate the interference caused by transmissions on the AP side, yet they still must manage contention on the mesh.

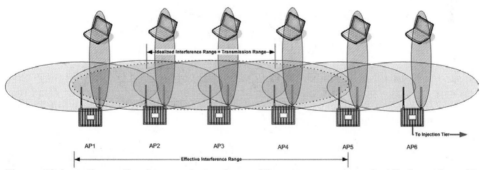

Figure 13.4: A linear five hop mesh topology with one user per mesh AP. In real-world deployments, the effective interference range, shown here for mesh AP3, will extend over two or more wireless mesh hops.

Figure 13.5 shows the measured throughput degradation over five mesh hops using the Sensoria dual radio ER500 mesh router. The results confirm the 1/n relationship between throughput and hop count cited by [23] and [24] for a single client. A latency test demonstrated 5ms over five hops on a lightly loaded network, increasing to 35ms when the network is fully loaded [34]. A more realistic load scenario involves traffic injection at each mesh AP in the chain. For this case, the traffic injected at each mesh router was limited to 1.5Mbps (the choice of 1.5Mbps will be clarified shortly). Figure 13.5 shows that the aggregate throughput for this case is approximately evenly distributed amongst all mesh APs, while the aggregate client throughput (7.7Mbps) is larger than the throughput achieved previously by a single client over the maximum number of hops.

Following the analysis and measurements described in [33], the total network throughput can be estimated by dividing the anticipated per-hop throughput by the sum of the hops for each traffic-sourcing mesh AP. For example, applying this formula to the five hop single client case results in an anticipated throughput of 24Mbps / 5 = 4.8Mbps, which matches the measured result. For the five-hop, five-client case applying the formula results in 24Mbps / (1 + 2 + 3 + 4 + 5) = 1.6Mbps, justifying the choice of 1.5Mbps bandwidth restriction in order to not overload the 802.11 MAC. The sum of the hops rule represents an estimate for the total mesh tier throughput of a dual radio mesh network. As noted in [33], the throughput rule assumes a collision-free MAC model, which is usually not the case given the impact of the hidden terminal phenomenon, mesh self interference and non-optimal use of the wireless channel. However, as shown in Figure 13.5, practical tests largely confirm the model and demonstrate that the rule represents a lower bound on the aggregate throughput when the separation between the first and last transmission exceeds the interference range.

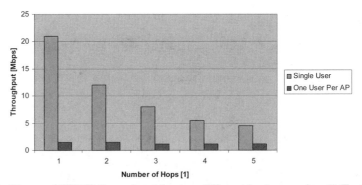

Figure 13.5: Measured ER500 throughput for two different load scenarios: 1) Traffic from a single user as a function of hop count 2) Traffic from one user per mesh AP with a QoS rate limit of 1.5Mbps. For the second case, the combined throughput for all users is higher than the throughput achieved for a single user over five mesh hops.

In a multi-radio mesh, more than two radios are dedicated to backhaul connectivity. The backhaul network no longer represents a shared network since each link operates as a dedicated P2P link. Contention is limited to the two communicating nodes, resulting in constant backhaul performance regardless of the scale of the mesh network. Multi-radio mesh nodes are also superior in dealing with local extraneous interference sources as they can be configured to operate in a frequency range that experiences the least amount of RF disruption in a given area. The obvious drawbacks are the higher cost of multi-radio mesh nodes and the added deployment and configuration expense, often complicated by the setup of directional antennas. Finally, automatic path re-selection as well as mesh gateway failover is difficult to achieve when mesh nodes do not share a common communication channel.

13.7 Injection Tier Design

The injection layer's purpose is to extend the reach of individual mesh networks as well as facilitate linking all meshes to a wired backhaul. Wireless injection technologies differ widely in their PHY and MAC layer, the supported topology and their bandwidth capabilities. As outlined in the introduction, P2P as well as P2MP topologies are conceivable. When planning injection layer links, the same principles regarding link budget, transmit power, receiver sensitivity and propagation loss discussed for the access and mesh tier apply. P2P and P2MP equipment such as [12] and [14] operates in the 900MHz, 2.4GHz and 5GHz range, offering different levels of throughput. 900MHz equipment is preferable in areas with high foliage density, long range and/or no LOS to the mesh gateway; however, it usually provides lower supported data rates than 2.4GHz or 5GHz equipment. On the other hand, 2.4GHz equipment will suffer from co-channel interference with the access tier, third-party 802.11b/g devices and non-802.11 interferers such as cordless phones, microwave ovens and therefore the utilization of this frequency band only makes sense when the following conditions exist:

- 5GHz range is insufficient due to propagation path loss,
- 900MHz and 5GHz transmission suffers from non-802.11 interference or
- AP frequency reuse patterns in the respective areas are such that self-interference does not occur.

While 5GHz equipment usually doubles or triples the available bandwidth compared to 2.4GHz, path loss through foliage and buildings is much higher, limiting the deployment scenarios. Nonetheless, the limited range can be enhanced by reducing the amount and size of obstructions present, such as elevating both transceivers, by increasing the angle at which the downlink enters the obstructed area or by using high gain antennas where allowed by the FCC (such as for 5.8GHz PTP links).

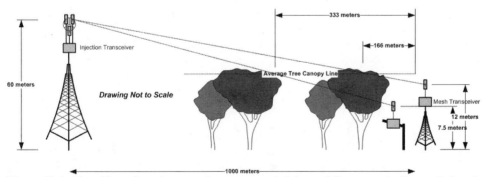

Figure 13.6: The effect of mesh transceiver antenna height on foliage penetration path length over an injection layer link. The cost of dedicated mesh transceiver towers has to be weighed against an increased number of Backhaul PoPs. Note that in this illustration the distance and height relationships are not shown to true scale.

As an example of the impact of obstructions on injection links designed to be significantly longer than mesh or AP tier links, Figure 13.6 illustrates the effect of antenna height on foliage penetration distance, and as a result on the achievable link distance. For the purpose of describing the characteristics of the injection tier, the transceiver closest to the Backhaul PoP will be called the injection transceiver, while the transceiver(s) associated with individual mesh networks will be called mesh transceivers. When applying simple trigonometric laws it becomes quickly evident that given the distance between injection and mesh transceiver is generally two orders of magnitude greater than the injection tower height (e.g. 1000 meters versus 30 meters), the foliage-penetrating portion of the link (often the dominant component of link loss) can be significantly reduced by elevating the mesh transceiver tower. For example, given an average foliage (tree) height of 15 meters, an injection tower height of 30 meters and a projected link distance of 1000 meters, the elevation of the mesh transceiver from an average street light height of 7.5 meters to 12 meters reduces the foliage penetration path length from 333 to 166 meters. Since signal attenuation through foliage exponentially increases based on the linear addition of foliage penetrated, injection link range will increase drastically using elevated mesh transceivers to service tree covered areas.

A key aspect of the injection tier design is the reliability and throughput available over each of its wired or wireless links. As a simple illustration, if a T1 wired connection is considered for an injection link, providing 1.544MBps concurrent uplink and downlink, then this limits the maximum mesh or access link throughput to 3Mbps (considering throughput as shared uplink and downlink). Similarly if P2P or P2MP wireless links are used, these must be considered in the context of the anticipated traffic model for each mesh network. The injection tier acts as the network trunk for a potentially dynamic mesh tree (changing due to gateway re-selection). The result of aggregating a larger user base within each injection link smoothes and increases the traffic load of each injection link as compared to the traffic dynamics at the individual AP and mesh layers as well as increasing each injection link's robustness requirements. However, this aggregation also means each injection link is a potential throughput bottleneck in the municipal network. A wireless injection link's robustness is compounded in comparison with lower layer links, since the range of a wireless injection link will most likely be much longer than the access or mesh layer links, and longer links provide more opportunities for interference and higher link margin variability. As a result wireless injection links may require dedicated towers, high gain antennas and labor-intensive antenna sighting to achieve system reliability and throughput requirements.

13.8 Network-wide Seamless Mobility Support

Network-wide roaming is considered an important enabler of value-added services within municipal wireless networks, for example to support mobile VoIP and other multimedia streaming applications. However, this presents a problem for 802.11 networks due to their long average link layer handoff times between 300ms and 500ms [32] and due to the added capabilities needed to support seamless network-wide roaming. The development of the 802.11r standard is designed to improve the existing handoff scheme, making 802.11r-compliant wireless access networks a viable alternative to traditional cellular networks for voice and other real-time communication. The following paragraphs describe the mobility challenges for municipal Wi-Fi networks based on the current 802.11 standard and provide an example mobility solution available on the Sensoria ER500 mesh router. This solution is first presented in the context of a single mesh network to illustrate intra-mesh routing and is subsequently applied to a metro-scale Wi-Fi network.

While 802.11 specifies in principle how Basic Service Sets (i.e. single 802.11 broadcast domains) can be combined to form Extended Service Sets (i.e. collections of 802.11 broadcast domains) covering large areas, it does not specifically address the issue of fast and seamless roaming, especially for a routed (layer 3) mesh network. Such routed networks as part of a metro-scale distribution system are now common among popular mesh vendors including [6], [7], [8], [9]. A mobility solution for this type of mesh network must thus provide layer 3 route tracking functions in addition to basic layer 2 handoff functions to ensure delivery of both Internet-bound and intra-network traffic to stations changing their point of attachment in the network with a low degree of signaling overhead, packet loss and latency. In other words, it must be capable of delivering IP traffic to a large number of clients in a micro-mobility environment with frequent handoffs without putting unnecessary traffic load on an already throughput-limited mesh layer.

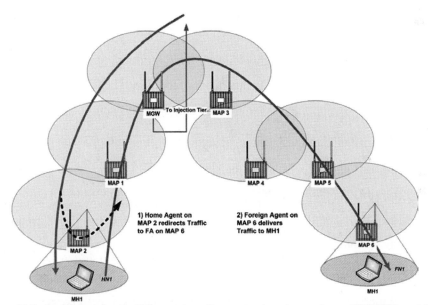

Figure 13.7: Inefficient bandwidth consumption scenario when using a RFC2002 mobile IP solution in a mesh network. HN1 denotes the home network of the mobile host MH1 (802.11-enabled laptop), while FN1 denotes the current foreign network. Traffic destined for MH1 first is relayed to the home network before being redirected over the same path over a dedicated IP tunnel towards FN1 and MH1's new point of network attachment.

While traditional mobile IP solutions following RFC 2002 [52] do provide seamless connectivity at the transport layer and are generally well suited for macro-mobility scenarios across heterogeneous networks [51], they can not be expected to perform well within a (homogeneous) municipal Wi-Fi network for several reasons. First, mobile IP relies on tracking mechanisms that employ custom protocols not available on standard 802.11 client devices such as laptops, PDAs and Wi-Fi phones. In order to employ mobile

IP in the network, a mobile device must be equipped with non-standard software components to allow it to establish and update relationships with the network's home and foreign agents. Second, the mobility architecture as described in RFC 2002 has been designed with low mobility rates in mind and thus suffers from relatively large latencies at the network layer, which would further increase the already excessive handoff times imposed by the 802.11 link layer [45]. Third, every mobile node away from its home network requires the setup and maintenance of a dedicated IP tunnel, resulting in a mobility solution which does not scale well with the number of mobile users envisioned in a municipal network[8]. Finally, the fact that IP packets destined for a mobile host must always be routed to the home agent first can result in an inefficient use of the available network bandwidth if the mobile device has changed its point of attachment in the network. Figure 13.7 illustrates such a case where the traffic destined for the mobile host MH1 attached to Foreign Network 1 (FN1) first is sent to its Home Network (HN1) and then forwarded across a dedicated IP tunnel to mesh AP 6. The path MGW – MAP 1 – MAP 2 is traversed twice in this case.

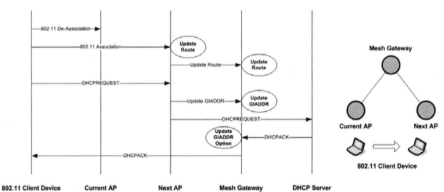

Figure 13.8: Sequence diagram illustrating the message exchange between an 802.11 client device, the roaming daemon on its current AP as well as the subsequent AP to facilitate intra-mesh roaming. The mesh gateway plays a key role in that 1) it updates the client's route and local gateway information following reception of an association event and 2) it intercepts the DHCPACK message from the server and updates the GIADDR option field to match the local gateway address of the new mesh AP before forwarding the DHCPACK to the client device.

The Sensoria ER500 seamless roaming solution addresses these concerns using an alternate approach to mobility management based on Dynamic Host Configuration Protocol (DHCP), 802.11 layer 2 triggers (association and de-association events) as well as a fixed number of static IP tunnels. Furthermore, it leverages the proxy ARP mechanism, the WirelessFabric[TM] mesh routing algorithm and a single, network-wide overlay network (IP subnet) facilitating Internet-bound communication between stations within a mesh as well as across mesh boundaries. The re-routing delay following an 802.11 handoff is minimized and bound by the transmission time of the 802.11 association event (from the new mesh AP to the mesh gateway). Each mesh node implements these functions by means of a user

---

[8] Assuming each AP cell represents its own network with associated home agent, the number of IP tunnels is bound by $n*(n-1)$ if a single 'Care of Address' is used, and potentially higher if each station uses its own Co-located Care of Address.

space roaming agent, which also acts as a DHCP relay agent and utilizes interfaces to the operating system for setup and maintenance of IP tunnels, L2 trigger processing as well as routing table manipulation. One IP tunnel per mesh AP is set up during initialization and is subsequently used to deliver inbound client traffic to the associated client devices. This tunneling scheme minimizes the number of points in the network required to keep client routes up to date and ensures the most efficient use of mesh bandwidth by minimizing the number of retransmissions across the mesh backhaul.

DHCP is a common mechanism to dynamically assign an IP address to a network device and is leveraged in the ER500 to establish the initial binding between network point of attachment (mesh AP) and associated IP tunnel at the mesh network egress point (mesh gateway). As a station changes its point of attachment or leaves the network, L2 triggers combined with internal signaling messages are used to update routing information on the mesh gateway. The subsequent DHCP message exchange between the client device and the DHCP server (DHCPREQUEST/DHCPACK) is used to update the client's local gateway address by modifying the respective option field in the DHCPACK message. Routes and state data structures for stations which leave the network without sending a de-association frame or DHCPRELEASE message are cleaned up on both the mesh AP and mesh gateway after a configurable idle time has lapsed. Figure 13.8 illustrates the process of intra-mesh roaming, which leverages the sequence of message exchanges between a mobile station as it de-associates from one AP and moves to a neighboring AP. Once a station is assigned an IP addresses it will keep this address for at least the duration of the authentication session with the Access Controller (AC), allowing the station to maintain transport layer sessions as it moves from AP cell to AP cell.

In an ER500-based mesh network both Internet-bound and intra- and inter-mesh traffic between client devices is routed along paths determined by the WirelessFabric[TM] mesh router using pro-active and reactive routing techniques. On the mesh gateway, return traffic from the Internet destined for a client device is either routed through the local AP interface or through the appropriate IP tunnel device to the respective mesh AP, where the tunnel header is removed and the native IP packets are forwarded to the local recipient. In order to support intra-mesh and inter-mesh unicast and multicast communication among client devices on the common overlay subnet, the MANET router receives notification of newly arrived stations from the roaming agent through a dedicated IPC mechanism, allowing it to respond to routing requests originating from other mesh nodes. The tunneling scheme is implemented to support mesh gateway re-selection due to equipment failure or detection of a more efficient path to the mesh gateway. Once an alternate gateway is identified, the roaming agents on all affected mesh APs initiate IP tunnel creation with the new gateway and transfer all current client device records to it. Subsequently, the roaming agent on the new gateway completes the IP tunnel setup to the new set of mesh APs and adds the new client routes, followed by the notification of the previous gateway of the failover event in order to have it remove its client routes and clean up related data structures. Finally, the upstream router is informed of the new gateway by means of gratuitous ARP messages.

In a small-scale municipal Wi-Fi network involving only one or a few mesh gateways directly connected to an AC (i.e. without a branch router), the proxy ARP mechanism can be used on the mesh gateway to ensure accurate routing of traffic inbound from the Internet. The choice of using an ARP subnet gateway rather than a dynamic routing protocol in the ER500 mobility solution is driven by the limitations imposed by the need to support commercially available ACs that do not allow the designation of a remote subnet [39], [40] and/or lack dynamic routing protocols facilitating inter-mesh roaming [41]. Proxy ARP enables a mesh gateway to respond to requests for a client device's link layer address on behalf of the client connected to a mesh AP. Upon reception of an IP packet destined to its link layer address, the mesh gateway router consults its local routing table to make the forwarding decision. If the device is attached locally, IP packets are sent through the appropriate AP interface; otherwise, the appropriate IP tunnel is used to forward the packet. Outbound traffic routing is accomplished in the usual manner using the native routes set up and maintained by the WirelessFabric^{TM} mesh router. The ER500 inter-mesh mobility solution utilizes a broadcast notification mechanism via the backhaul network segment to complete the handoff process. Reception of a broadcast or multicast-based notification message triggers route removal on the previous gateway.

In a multi-mesh network configuration, the distribution tier can be partitioned into two network segments using a branch router. The main benefit of an additional network segment is that this eliminates client-related ARP traffic from unaffected injection links and confines it to the high capacity backhaul tier. As an added benefit, the introduction of one additional IP subnet per branch offers the network operator the ability to logically group and subsequently address network elements such as mesh gateways and injection tier equipment by network branch. By running a DHCP relay agent on each branch router, the IP configuration of all network elements can be centrally managed at the NOC. Also, the branch router can be used to implement traffic shaping functions (traffic characterization and prioritization) for inbound client traffic, giving priority to latency and loss-sensitive multimedia flows and preventing random packet loss over bandwidth-constrained injection layer links.

In this configuration a branch router and its attached mesh gateways run a dynamic routing protocol (RIP or OSPF) allowing the mesh gateways to update routing information for roaming clients on the branch router, both following the initial association of client devices as well as a result of inter-mesh roaming events. The branch router in turn runs proxy ARP on its backhaul tier interface, allowing it to respond to ARP requests from the ACs and subsequently route traffic inbound towards clients. Branch routers also need to engage in the exchange of client routing information over the backhaul tier using dynamic routing protocols to facilitate inter-branch roaming. When a client device crosses a branch boundary while roaming, its new mesh gateway informs its branch router about the presence of a new client and the branch router in turn broadcasts a notification message on the backhaul network and sends a gratuitous ARP message to the ACs at the NOC. Upon reception of the broadcast message, the previous branch router removes the associated client route. Potential improvements of the broadcast mechanism include using positive acknowledgement as well as directly forwarding the de-association event from the previous mesh AP to the previous branch router.

13.9 Conclusion

This chapter provided a review of several municipal Wi-Fi RFPs followed by a discussion of common functional and performance requirements for city-wide Wi-Fi networks. Next, a multi-tiered network design approach for solving the various technical challenges of meeting the coverage, bandwidth, mobility, control, monitoring and cost objectives set forth in the RFPs was presented and complemented with an example network architecture. The design challenges of the access, mesh and injection tier were discussed and a multi-hop throughput calculation rule for single mesh radio networks was presented. The last section described a solution supporting mesh and network-wide roaming in the context of a routed mesh and backhaul network. The building blocks and mechanism for such a scalable network-wide roaming architecture include the use of DHCP, IP tunnels, network segmentation, proxy ARP and dynamic routing protocols.

13.10 References

[1] "Unwire Portland: A Citywide Broadband Wireless System", City of Portland, Oregon, September 16, 2005, RFP No. 104112.

[2] "TechConnect Community Wireless Broadband Network", City and County of San Francisco, December 22, 2005, RFP No. 2005-19.

[3] "Wireless Philadelphia", RFP for a Citywide Wireless Network, City of Philadelphia, Pennsylvania, April 7, 2005.

[4] "Colorado Wireless Communities Community Wireless Broadband Network", City of Thornton, Colorado, January 11, 2007.

[5] "WiFi Implementation in Rockville Townsquare", City of Rockville, Maryland, RFP No. 20-07.

[6] Sensoria ER500 Data sheet, http://www.sensoria.com/pdf/EnRoute500_Outdoor_Mesh_Network_Router.pdf

[7] Tropos MetroMesh5210 Data Sheet, http://tropos.com/pdf/5210_datasheet.pdf

[8] Mesh Dynamics MD4000 Data Sheet, http://www.meshdynamics.com/prod-md-4000.html

[9] Strix Systems OWS3600 Data Sheet, http://www.strixsystems.com/products/datasheets/Datasheet_OWS_3600.pdf

[10] Motorola MWR7300 Data Sheet, http://www.motorola.com/governmentandenterprise/northamerica/en-us/generic.aspx?FilePath=en_US/Solution/MWR7300_mesh_wireless_router.xml

[11] Ubiquiti Networks Super Range 900 MHz Mini PCI Card Data Sheet, http://www.ubnt.com/super_range9.php4

[12] Tranzeo Wireless Technologies Products Overview, http://tranzeo.com/products

[13] IEEE 802.16 Broadband Wireless Access Working Group, http://www.ieee802.org/16/docs/02/C80216-02_05.pdf

[14] Motorola Canopy Products Data Sheets, http://motorola.canopywireless.com/products/specshome.php

[15] Räisänen, V., *Implementing Service Quality in IP Networks*, John Wiley & Sons, New York 2003, ISBN 978-0470847930.

[16] Balachandran, A. et. al, "Characterizing User Behavior and Network Performance in a Public Wireless LAN", *Proceedings of ACM SIGEMTRICS*, June 2002.

[17] Camp, J. et. al, "Measurement Driven Deployment of a Two-Tier Urban Mesh Access Network", *Proceedings of the 4th International Conference on Mobile Systems*, Applications and Services. pp. 96-109, 2006.

[18] FCC Part 15 Regulations, http://www.fcc.gov/oet/info/rules/part15/part15-8-14-06.pdf

[19] Akyildiz, I. et. al, "Wireless Mesh Networks: A Survey", *Computer Networks and ISDN Systems*, Volume 47 Issue 4, pp. 445-487, March 2005.

[20] Arpacioglu, O., et. al, "On the Scalability and Capacity of Wireless Networks with Omnidirectional Antennas", *Proceedings of the Third International Symposium on Information Processing in Sensor Networks*, pp. 169 – 177, Berkeley, California, USA, 2004.

[21] Li, J. et. al, "Capacity of Ad Hoc Wireless Networks", *Proceedings of the 7th annual International Conference on Mobile Computing and Networking*, pp. 61–69, Rome, Italy, 2001.

[22] Jain, K. et. al, "Impact of Interference on Multi-hop Wireless Network Performance", *Proceedings of the International Conference on Mobile Computing and Networking*, pp. 66-80, San Diego, CA, Sept. 2003.

[23] BelAir Networks, Capacity of Wireless Mesh Networks, White Paper, 2006, http://www.belairnetworks.com/resources/pdfs/Mesh%5FCapacity%5FBDMC00040%2DC02%2Epdf

[24] Rongdi, C., "Performance Comparison of Two Wireless Mesh Networks", Network Research Center of Tsinghua University, Outubro 2005.

[25] IEC Online Tutorials, Operations Support System, http://www.iec.org/online/tutorials/acrobat/oss.pdf

[26] Link Planning for Wireless LAN, http://home.deds.nl/~pa0hoo/helix_wifi/linkbudgetcalc/wlan_budgetcalc.html

[27] WaveCall Radio Modeling Software, http://www.wavecall.com

[28] Softwright Radio Modeling Software, http://www.softwright.com

[29] Cisco Systems, Capacity Coverage and Deployment Considerations, http://www.cisco.com/application/pdf/en/us/guest/products/ps430/c1244/ccmigration_09186a00801d61a3.pdf

[30] Proxim Wireless Networks, Maximizing Your 802.11g Investment, http://www.proxim.com/learn/library/whitepapers/maximizing_80211g_investment.pdf

[31] Ferre, P. et. al, "Throughput Analysis of IEEE 802.11 and IEEE 802.11e MAC", *IEEE Wireless Communication and Networking Conference*, Volume 2, pp. 783-788, March 2004.

[32] Mishra, A. et. al, "An Empirical Analysis of the IEEE 802.11 MAC Layer Handoff Process", *University of Maryland Technical Report*, UMIACS-TR-2002-75, 2002.

[33] Merrill, W. M. "A Simple Model for Multi-Hop 802.11 Throughput", Sensoria Corporation White Paper, 2006.

[34] Merrill, W. M. "Sensoria's Enroute500 Performance Analysis and Prediction", Sensoria Corporation White Paper, 2006.

[35] "Broadband Wireless: Now Playing in Select Locations", *Data Communications Networking News*, http://www.rysavy.com/Articles/BroadbandWireless/BroadbandWireless.html

[36] Official IEEE 802.11 Working Group Project Timelines,
     http://grouper.ieee.org/groups/802/11/Reports/802.11_Timelines.htm
[37] IEEE 802.11i Standards Download, http://standards.ieee.org/getieee802
[38] IEEE 802.11b/g Standards Download, http://standards.ieee.org/getieee802
[39] IP3 Networks, NetAccess Access Controller, http://www.ip3.com/naxk.htm
[40] Nomadix, AG3000 Data Sheet,
     http://www.nomadix.com/Files/Downloads/Products/AG3000_Data_Sheet.pdf
[41] Bluesocket, Access Controller Family Overview,
     http://bluesocket.com/products/controllerfamily.html
[42] BelAir300 Converged Multi-service Wireless Node Data Sheet,
     http://www.belairnetworks.com/resources/pdfs/BelAir300_Data_BDMA30010-
     B01.pdf
[43] Nortel Wireless Access Point 7220 Data Sheet,
     http://products.nortel.com/go/product_content.jsp?segId=0&parId=0&prod_id=50180
     &locale=en-US
[44] BelAir50C Wireless Mesh Node Data Sheet,
     http://www.belairnetworks.com/resources/pdfs/BelAir50C_Data_BDMA05030-
     B01.pdf
[45] Shin, M. et. al, "Improving the Latency of 802.11 Hand-offs using Neighbor Graphs",
     *Proceedings of the 2nd international Conference on Mobile Systems, Applications,
     and Services*, pp. 70-83, 2004.
[46] Huang, P. et. al, "A Fast Handoff Mechanism for IEEE 802.11 and IAPP Networks",
     63rd IEEE Vehicular Technology Conference, pp. 966-970, Spring 2006.
[47] Pack, S. et. al, "Fast Handoff Scheme based on Mobility Prediction in Public Wireless
     LAN Systems", *IEE Proceedings of Communications*. Volume 151, No. 5, October
     2004.
[48] IEEE 802.11r Task Group,
     http://grouper.ieee.org/groups/802/11/Reports/tgr_update.htm
[49] Kim, Y. et. al, "A Novel Hidden Station Detection Mechanism in IEEE 802.11
     WLAN", *IEEE Communications Letters*, Volume 10, Issue 8, pp. 608-610, August
     2006.
[50] McLarnon, B., "VHF/UHF/Microwave Radio Propagation: A Primer for Digital
     Experimenters",
     http://www.ictp.trieste.it/~radionet/2000_school/lectures/carlo/linkloss/INDEX.HTM
[51] Blondia, C. et. al, "Performance Comparison of Low Latency Mobile IP Schemes",
     *Proceedings of the 7th ACM International Symposium on Modeling, Analysis and
     Simulation of Wireless and Mobile Systems*, pp. 297-300, 2004.
[52] IETF RFC 2002, "IP Mobility Support", http://www.faqs.org/rfcs/rfc2002.html
[53] Tropos Spectrum White Paper, http://www.tropos.com/pdf/Spectrum_Whitepaper.pdf

14

Usage and Performance Comparison of Mobile Metro Mesh Networks

Devabhaktuni Srikrishna[a]

The chapter details the operational performance characteristics of two Tropos® MetroMesh™ networks operating in two different U.S. cities. One of the networks offers access free-of-charge; the other requires a paid subscription for access. Both networks support a mix of consumer, business and city automation services. Cities are not named to maintain customer confidentiality.

14.1 MetroMesh Network Architecture

In MetroMesh networks, Tropos MetroMesh routers are placed throughout the city at densities ranging from 25 to 40 per square mile depending upon use case (e.g., DSL alternative, Mobile Internet), foliation and topography.

Figure 14.1: Tropos 5210 MetroMesh Router.

[a] *Tropos Networks*

A fraction of the MetroMesh routers, configured as gateways, are connected, using optical fiber or point-to-multipoint wireless links, to the wired Internet backbone. The remaining routers, referred to as nodes, are connected via mesh wireless links. The ratio of gateways to nodes is driven by capacity needs and is determined by typical over-subscription models for fixed and mobile broadband networks.

The remaining MetroMesh routers, configured as nodes, route traffic over one or more mesh hops back to the gateways. The Tropos MetroMesh OS selects the best path through the mesh for high throughput and low latency, provides reliable mobile client connectivity for a wide range of standard Wi-Fi clients and extracts the maximum packet-carrying capacity from available airtime across all available Wi-Fi channels.

Wi-Fi client devices in MetroMesh networks access the Internet by associating with Tropos MetroMesh routers. MetroMesh routers employ omni-directional antennas both to provide coverage to connect these Wi-Fi client devices and to route client traffic through the wireless mesh to/from the Internet.

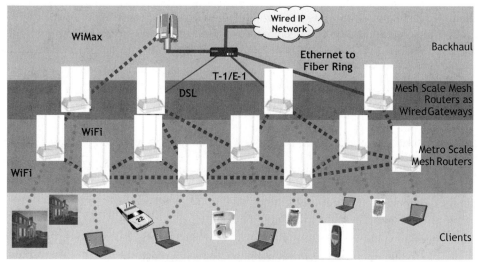

Figure 14.2: Tropos MetroMesh Architecture.

After providing a summary of the routing and RF management algorithms used in the mesh, we use intimate knowledge of two live networks including detailed measurements, to explore the unique benefits of MetroMesh based on the Predictive Wireless Routing Protocol (PWRP™). The data presented in this chapter was collected during a 24-hour period in October 2006 using standard Tropos analysis and control tools, Tropos Insight and Tropos Control. These tools provide a wide range of real-time, end-to-end information for operational management and business analytics. The data in this chapter represents a fraction of the data that is available from Tropos Insight analysis tools in any Tropos citywide installation.

14.2 Predictive Wireless Routing Protocol (PWRP)

The algorithms incorporated into PWRP have been developed and honed over six years and incorporate real-world experience from more than 500 successful deployments in 30 countries. Applications as diverse as mobile broadband for public safety, residential broadband, video surveillance, voice services and mobile public access operate on these networks.

14.2.1 Scalable routing

With PWRP, the network scales easily to deployments of thousands of nodes covering hundreds of square miles while keeping the routing overheads to less than 5% of airtime utilization. This is in contrast to distance vector and link state approaches where the overheads grow linearly or quadratically with the size of the network. The architecture is designed to accommodate incremental capacity addition as the usage grows and automated tools that Tropos has developed can be used to pinpoint nodes in the network that can be converted to gateways to further increase system capacity.

14.2.2 Throughput-optimized routing

PWRP selects the highest throughput and most reliable paths through the mesh by reducing the total transmission time (and link layer retransmissions) required per packet across multiple mesh links between the client and the gateway. In 802.11, all links are subject to stochastic physical effects in the environment such as RF fading and interference—these effects are the primary root causes of reduced link performance manifested as link-layer retransmissions. PWRP exploits dynamic path diversity to consistently select the highest performing end-to-end path through the wireless mesh, achieving more than two times higher usable throughput. This is in contrast to other approaches that just attempt to minimize hop count without factoring in the impact to throughput of selecting paths with high packet error rates that necessitate high numbers of retransmissions resulting in lower throughput. Even when there are only a handful of paths to choose between, the difference between choosing the best path and any other path can often result in dramatically different throughput and reliability. In deployments with 25+ nodes per square mile, the average node often has over 10 available paths to choose between. This is independent of the ratio of nodes to gateways. Therefore, selecting the best path is important even when the gateway-to-node ratio is small. MetroMesh consistently delivers 1+Mbps throughput with less than 50ms latency across the mesh using paths that vary from one to four hops.

14.2.3 RF spectrum management

PWRP continuously and dynamically optimizes the use of available spectrum through the use of patented algorithms for automatic channel selection, adaptive data rate selection, automatic transmit power adjustment and adaptive noise immunity. These algorithms have many benefits including:

- dynamic interference rejection by switching to unutilized or interference-free channels factoring in a number of metrics including airtime utilization, noise level, and other measurements.
- maximization of airlink capacity and efficient spatial-use by enabling concurrent operation in multiple frequency channels in a given area, selecting the optimum data rates for link operation, and tuning MAC level packet detection registers to adapt to the RF environment.
- contention management by automatically adjusting transmission power and data rates to ensure balanced link budgets on mesh and client links to ensure uniform and reliable connectivity to lower power client devices.

These algorithms have been extensively tested and verified in real-world deployments and are growing in importance as deployments become denser and coverage requirements increase.

14.2.4 Multi-mode routing

On multi-mode platforms, PWRP has the ability to automatically use 5 GHz to close mesh links when line-of-sight happens to be available to increase system capacity. However it also has the ability to operate those mesh links that cannot close over 5 GHz using 2.4 GHz. This enables the mesh to be deployed at node densities similar to those for 2.4 GHz-only networks while simultaneously increasing overall system capacity through the opportunistic use of 5 GHz. In addition, the use of the higher capacity multi-mode product enables larger clusters with higher overall hop counts without sacrificing throughput or capacity.

14.2.5 Seamless session-persistent mobility

PWRP ensures that client sessions are maintained as the client devices roam, even at vehicular speeds exceeding 70 mph. This is accomplished transparently to the end-user with no special client software required. PWRP enables fast handoffs even in cross-subnet roaming situations through the use of IP-in-IP tunnels and BGP. PWRP mobility is scalable to networks of thousands of nodes, hundreds of subnets and hundreds of thousands of client devices. A number of public safety agencies throughout the U.S. utilize PWRP mobility for mission-critical applications today.

14.2.6 Dynamic rate-limiting and traffic management

In networks with heavy usage, a small number of users often end up consuming a disproportionate share of the bandwidth through applications like peer-to-peer transfers. These high traffic streams can have a negative impact on the network performance of the remainder of the users, unless proactive traffic shaping policies are applied to these bandwidth-hungry applications. PWRP provides the ability to proactively identify these classes of users (based on traffic stream characteristics) and to apply selective rate-limiting policies so as to allow for a more equitable distribution of bandwidth and to apply more granular operator-specified policies on network usage. Tropos developed this feature-set based on operator feedback and an analysis of usage data gathered from some of our most heavily-used production networks.

14.2.7 Correlated Mesh Data Protocol (CMDP)

CMDP™ collects and correlates mesh-wide usage and performance data. CMDP uses a distributed data-collection paradigm. Mesh nodes leverage CMDP's extensive data and statistics collection during mesh operation and CMDP edge intelligence is used to filter summarize the collected data to feed back to Tropos Insight for analysis and post-processing. Using this protocol, the mesh provides precise information correlated across nodes on the network about network performance and end-user statistics. For example, data collected using CMDP was used recently to optimize a production network by generating automated recommendations on optimal nodes to convert to gateways in order to improve overall system capacity.

14.2.8 Patents

Many patents on PWRP are pending and five patents have been issued by the USPTO to date:

- "Selection of routing paths based upon path quality of a wireless mesh network", US Pat. 6965575;
- "Method for allowing a client to access a wireless system", US Pat. 7016328;
- "Method and system to provide increased data throughput in a wireless multi-hop network", US Pat. 7031293;
- "Method and apparatus to provide a routing protocol for wireless devices", US Pat. 6704301;
- "Selection of routing paths based upon routing packet success ratios of wireless routes within a wireless mesh network", US Pat. 7058021.

14.3 Overview of the Networks

The cities profiled in the paper are suburban, with populations ranging from 20,000 and 30,000, footprints of 9 to 13 square miles and housing densities in the neighborhood of 1,000 homes per square mile. The installations described here are for the most part dual-band designs using Tropos 5210 single-radio (2.4 GHz) MetroMesh routers for client access and inter-mesh links and 5.8 GHz point-to-multipoint wireless links for backhaul.

14.3.1 Client Usage

Based on statistics gathered using Tropos Insight, the for-fee network uses mostly higher power Wi-Fi modems (for DSL-alternative service) that are stationary and always on while the free network employs a mix of fixed and mobile client types. All networks are heavily-used based on traffic-per-client statistics. In addition, mobility is becoming important, with 5-6% of clients moving more than half a mile during the 24-hour period, though other networks display mobility patterns reflecting 10-15% of mobile devices roaming over distances of half a mile. A wide range of Wi-Fi device types (64-83 unique device types) connected to these open-standard networks during the day.

Table 14.1: Network Usage by End-Users.

	For-Fee Network	Free Network
Daily Routed Client Devices	2,242	1,861
7-day Routed Client Devices	3,735	3,377
Daily Active Clients > 1 MByte Downloaded	1,626	1,309
7-day Repeat Clients > 1 MByte Downloaded	1,528 (94% of daily)	595 (46% of daily)
7-day Active Clients > 1 MByte Downloaded	2,310	2,382
Total Download (24 Hours)	143 Gigabytes	71 Gigabytes
Average Download per Client (24 Hours)	64 Megabytes	38 Megabytes
Total Bytes Uploaded (24 Hours)	50 Gigabytes	22 Gigabytes
Average Upload per Client (24 Hours)	29 Megabytes	12 Megabytes
Total Number of Client Device Types	64	83
Number of Devices Roaming > 2500 Feet	104 (5%)	117 (6%)

14.3.2 Client Link Performance

The average packet error of the client link varies widely (3-22%) depending on the type of client device. Fixed, high-power bridges placed in the home or office such as the 200 mW PePWave and Senao bridges experience lower packet error rates and they transfer more data on average. In general, devices such as the PePWave bridge that are compliant with Tropos' TMCX (Tropos MetroMesh Compatible eXtensions) show superior performance compared with non-TMCX-compliant devices. Portable and handheld devices tend to have lower transmit power (battery-constrained) and also are highly mobile, resulting in higher packet error rates. Conditional measurements of packet error rate based on the manufacturer of client device below show that it is varies inversely with increases in the transmit power and directly with increases in the mobility of the client device. These trends are exhibited across both networks. Since TCP/IP protocol retransmits lost packets allowing the applications to recover from temporary outages, the client link packet error rates observed in these networks result in a satisfactory user experience as shown by the average hourly bytes transferred for devices from each manufacturer. Usage rates of 5 MB/hour (as shown below) correspond to steady usage average of 11 Kbps, consistent with usage patterns in consumer DSL networks.

14.3.3 Mesh Network Performance

Using the PWRP, the 2.4GHz mesh creates a network with a path length that is on average less than three hops, and 85-90% packet success probability (PSP - the probability of a packet traversing the mesh successfully) and, on average, delivers 3-4 Mbps and < 25 ms round-trip latency upstream and downstream. Using multiple 5 GHz links at the gateways to backhaul traffic, the mesh architecture is resilient to backhaul failures and provides reliable access for clients.

Table 14.2: For-Fee Network.

Most Common Client Devices (% of total)	Fixed Senao 200 mW modem 977 (44%)	Fixed PePWave 200 mW modem 694 (31%)	Mobile Intel Centrino 196 (9%)	Gemtek 111 (5%)
Average Uplink Signal Power	-64 dBm	-68 dBm	-80 dBm	-77 dBm
Average Packet Error Rate	4%	3%	12%	13%
Average Distance Roamed	44 feet	22 feet	148 feet	89 feet
Number of Devices Roaming > 2500 Feet (% of this device)	39 (4%)	12 (2%)	9 (5%)	3 (3%)
Average Hourly Bytes Transferred	3.6 Mbytes	5.5 Mbytes	0.8 Mbytes	1.0 Mbytes

Table 14.3: Free Network.

Most Common Client Devices (% of total)	Fixed PePWave 200 mW modem 477 (26%)	Mobile Intel Centrino 298 (16%)	Gemtek 223 (12%)	Brand X (12%)
Average Uplink Signal Power	-75 dBm	-82 dBm	-81 dBm	-78 dBm
Average Packet Error Rate	6%	22%	15%	11%
Average Distance Roamed	82 feet	1194 feet	359 feet	189 feet
Number of Devices Roaming > 2500 Feet (% of this device)	12 (3%)	40 (13%)	11 (5%)	4 (2%)
Average Hourly Bytes Transferred	2.3 Mbytes	1.4 Mbytes	2.1 Mbytes	2.6 Mbytes

Table 14.4: Mesh Network Performance.

	For-Fee Network	Free Network
Average Mesh Node Downstream	4.0 Mbps	3.1 Mbps
Average Mesh Node Upstream	3.8 Mbps	2.7 Mbps
Average Mesh Node Round-Trip Latency	20.2 ms	16.8 ms
Average Mesh Node Unicast Packet Loss	0.36 %	1.1%
Average Mesh Node Hop Count*	1.7	1.9
Average Mesh Node Broadcast PSP on Active Link	89%	88%
Average Gateway Backhaul Downstream	26.4 Mbps	7.2 Mbps
Average Gateway Backhaul Upstream Throughput	14.6 Mbps	2.7 Mbps
Average Gateway Round-Trip Latency	3.4 ms	13.1 ms
Average Gateway Unicast Packet Loss	0.02%	0.24 %

* add one additional hop for clients

14.4 Hourly Usage Patterns

14.4.1 For-Fee Network

The relatively low variability in network usage throughout the day suggests a good amount of business utilization ("Active Clients" in Figure 14.4). The constant background of upstream traffic ("Uploaded Mega Bytes By Clients") in Figure 14.3 across all hours is generally indicative of peer-to-peer applications.

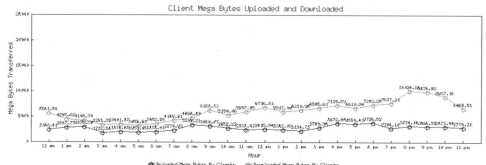

Figure 14.3: Total Client Bytes Transferred Per Hour (For-Fee Network).

Figure 14.4: Active Clients Per Hour (For-Fee Network).

14.4.2 Free Network

As shown in Figure 14,5, this network shows a strong pattern of consumer usage with the peak in users and traffic during prime-time "busy hours" between 5:00 P.M. and 10:00 P.M. This peak usage is also reflected in the three-fold increase in download traffic during "busy hours" seen in Figure 14.5 ("Download Mega Bytes By Clients").

14.5 Summary

Results from these Tropos MetroMesh networks suggest highly utilized, heavily used networks serving a wide variety of Wi-Fi devices that are providing fixed as well as mobile services. Throughput, latency, capacity and packet error rate statistics suggest utilization levels similar to U.S. DSL networks and performance significantly better than typical 3G cellular data offerings.

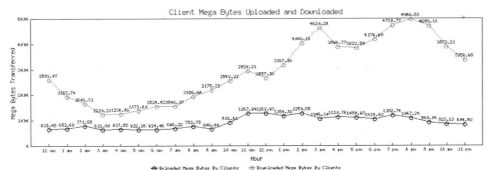

Figure 14.5: Total Client Bytes Transferred Per Hour (Free Network).

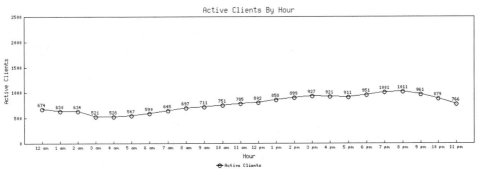

Figure 14.6: Active Clients Per Hour (Free Network).

14.6 References

[1] "Metro-Scale Mesh Networking with Tropos MetroMesh™ Architecture" http://www.tropos.com/pdf/tropos_metro-scale.pdf

[2] "Tropos MetroMesh Proven: Metro-Scale Wi-Fi in Chaska, MN" http://www.tropos.com/pdf/chaska_performance.pdf

[3] "Price-Performance Comparison: 3G and Tropos MetroMesh Architecture" http://www.tropos.com/pdf/price-performance.pdf

[4] For more information on MetroMesh and PWRP, please see http://www.tropos.com/technology/whitepapers.html

15

First, Second and Third Generation Mesh Architectures

Francis daCosta[a]

Evolving from ad hoc 802.11 networking, earlier generations of wireless mesh provided basic networking over extended outdoor areas. With the emergence of demanding data applications along with video and voice, single-radio "First Generation" single-radio wireless mesh solutions are proving unsatisfactory in many of these demanding environments. Third Generation wireless mesh solutions are based on multi-radio backhauls and deliver 50-1000 times better performance, but some custom hardware-oriented approaches limit flexibility and create deployment challenges. Software-oriented Third Generation wireless mesh based on distributed dynamic radio intelligence delivers the same high performance but with the additional benefits of easier installation, better avoidance of interference, and the added flexibility of easy mobility. These new capabilities are enabling many new types of applications beyond the traditional wireless mesh metro/muni environment.

15.1 Introduction

Mesh network requirements have evolved from their military origins as requirements have moved from the battlefield to the service provider, and residential networking environments. Today, to cover large areas with a single wired Internet link, more cost effective and efficient means of bandwidth distribution are needed. This implies more relay nodes (hops) than were needed before. Further, growing demands for Video and Voice-over-IP require packets to be moved over the mesh at high speeds with both low latency and low jitter. These new mesh requirements (more hops to cover large areas, more efficient bandwidth distribution and better latency and jitter for Video and VoIP) has given rise to the third-generation of mesh architectures.

15.2 Three Generations of Mesh Architectures

Three generations of evolving mesh architectures are depicted in Figure 15.1. They are (Left to Right):

---

[a] *MeshDynamics*

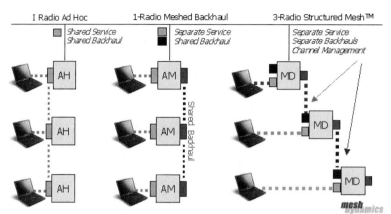

Figure 15.1: (L-R) - Ad Hoc, 1-Radio Meshed Backhaul, 3-Radio Structured Mesh.

- First Generation: 1-Radio Ad Hoc Mesh (left). This network uses one radio channel both to service clients and to provide the mesh backhaul. The ad hoc mesh radio, marked AH, provides both services – client access and backhaul. This architecture provides the worst services of all the options, as expected, since both backhaul and service compete for bandwidth.

- Second Generation: Dual-Radio with Single Radio Ad-Hoc meshed backhaul (center). This configuration can also be referred to as a "1+1" network, since each node contains two radios, one to provide service to the clients, and one to create the mesh network for backhaul. The "1+1" appellation indicates that these radios are separate from each other – the radio providing service does not participate in the backhaul, and the radio participating in the backhaul does not provide service to the clients. These two radios can operate in different bands. For example, a 2.4 GHz IEEE 802.11 b/g radio can be used for service and an IEEE 802.11a (5.8 GHz) radio can be used exclusively for backhaul. Though this configuration is sometimes called a Dual Radio Mesh, only one radio participates in the mesh. Performance analysis indicates that separating the service from the backhaul improves performance when compared with conventional ad hoc mesh networks. But since a single radio ad hoc mesh is still servicing the backhaul, packets traveling toward the Internet share bandwidth at each hop along the backhaul path with other interfering mesh backhaul nodes - all-operating on the same channel. This leads to throughput degradations which are not as severe as for the ad-hoc mesh, but which are sizeable nevertheless.

- Third Generation: 3-Radio Structured Mesh (right). The last architecture shown is one that provides separate backhaul and service functionality and dynamically manages channels of all of the radios so that all radios are on non-interfering channels. Performance analysis indicates that this provides the best performance of any of the methods considered here. Note that the two backhaul radios for the 3-radio configuration shown in Figure 15.1 are of the same type - not to be confused with 1+1 so-called dual radio meshes where one radio is for backhaul) and the other for service. In the 3-radio configuration, 2 radios are providing the up link and down link backhaul functionality, and the third radio is providing service to the clients.

15.3 Bandwidth degradation on Single Channel Backhauls

With one backhaul radio available for relaying packets, all nodes communicate with each other on one radio channel. For data to be relayed from mesh node to mesh node, that node must repeat it in a store-and-forward manner. A node first receives the data and then retransmits it. These operations cannot occur simultaneously because, with only a single radio channel, simultaneous transmission and reception would interfere with each other (Figure 15.2).

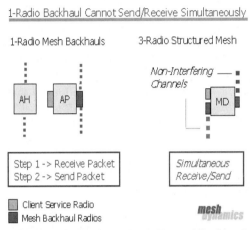

Figure 15.2: Single vs. Dual Channel Backhauls.

This inability - to simultaneously transmit and receive - is a serious disadvantage. If a node cannot send and receive at the same time, it loses ½ of its bandwidth as it attempts to relay packets up and down the backhaul path. A loss of ½ with each hop implies that after 4 hops, a user would be left with ($½×½×½×½$) = 1/16 of the bandwidth available at the Ethernet link. This is $1/(2^N)$ relationship defines the fraction of the bandwidth available to a user after N hops.

Third generation mesh products eliminate bandwidth degradation with a dual channel backhaul. There is been no measurable bandwidth degradation and this is the significant departure from both first and second-generation mesh architectures. Figure 15.3 shows live video from an IP camera part of a 9-hop mesh network at a ski resort. A single channel backhaul would be incapable of delivering this video feed beyond 1-2 hops: typical video bandwidth requirements would cripple the system. Latency and jitter would be unacceptable.

15.4 Latency/Jitter Degradation on Single Channel Backhauls

Latency is inversely related to available bandwidth: thin pipes can provide only so much flow. Single channel backhauls suffering from bandwidth degradation also suffer from poor latency and jitter over multiple hops. This is primarily due to the need for a single radio to serve both backhaul and client traffic. The result is that most First Generation single-channel backhaul networks provide reliable video- or voice service over only one- or two hops. As a result many more costly wired or fiber Internet or intranet drops are needed to deliver adequate service, increasing the ongoing total cost of ownership.

Third Generation mesh products do not suffer from bandwidth degradation – they use multiple backhaul radios to obviate radio channel interference. Field tests indicate a latency of less than 1 millisecond per hop even under heavy traffic. Since latency is thus not a factor of traffic or user density, jitter (variation in delay) is also very low. This makes Third Generation products suitable for networks serving large number of users, demanding applications, video, and voice - even simultaneously.

These capabilities of Third Generation wireless mesh networks make them especially useful where video surveillance is part of a metro/muni requirement, for expanding coverage into under-served areas with limited high speed wired or fiber infrastructure, and for border and perimeter networking, where bandwidth must be extended node-to-node as if in a long string of pearls.

Additionally, MeshDynamics multi-radio backhauls also incorporate VoIP concatenation for timely VoIP packet delivery. VoIP packets are small – typically less than 300 bytes but sent frequently, generally once every 20 milliseconds.

Networking protocols like CSMA/CA do not transport small and time sensitive packets well. The VoIP concatenation engine aggregates small VoIP packets into a larger packet for more efficient delivery. This aggregation takes place every 5-10 milliseconds. USAF tests (Figure 15.4) show overall latency is less than 10 ms + 1 ms per hop over a 4 hop network. Jitter is less than 1 ms per hop.

Figure 15.3: Live Feed from a multi-hop dual channel backhaul.

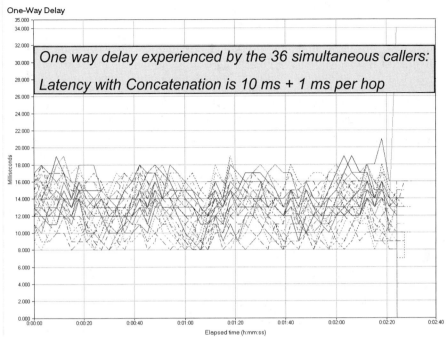

Figure 15.4a: Latency for 36 simultaneous VoIP calls over a 4 hop Mesh Network running VoIP Concatenation over the backhaul.

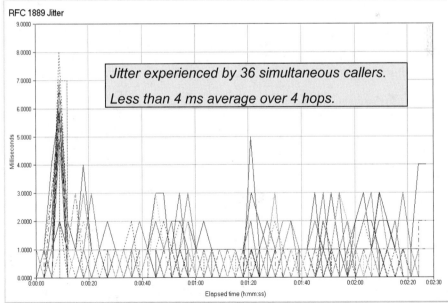

Figure 15.4b: Jitter for 36 simultaneous VoIP calls over a 4 hop Mesh Network running VoIP Concatenation over the backhaul.

USAF tests found comparable latency/jitter for single channel backhauls to be an order of magnitude higher. Some reasons for this:

- Bandwidth degrades with each hop. As an analogy to water pipes, the smaller the pipe, the slower the flow.
- With all radios on the same channel, there is compounded contention will packets fighting each other, all on the same channel.
- The overall efficiency of the CSMA/CA protocol used degrades exponentially as the number of clients on the same channel increase.

15.5 Frequency Agility through Distributed Intelligence

Wireless is a shared medium. Radios communicating on the same channel and within range of each other contend for available bandwidth. In single channel backhauls, there is one radio acting as the repeater between nodes: all backhaul radios must be on the same radio channel. The entire network is susceptible to channel interference/jamming. System performance is compromised. Figure 15.5 shows how the dual channel backhaul can switch channels to avoid debilitating external interference effects.

This first step of providing two-radio backhauls provides an obvious theoretical advantage over single-channel backhauls, but management of channel selection and interference avoidance becomes a challenge addressed in a number of different ways.

One of the first approaches applied to this problem was to segregate the backhaul links with hardware radio switching and sectored directional antennas. Since each sectored antenna "sees" only a narrow field of view, radio emissions from adjacent nodes do not create interference.

The limitations of this hardware-centric approach are the costs of sectored antennas and custom-developed radio hardware, the relative inflexibility of the system in dealing with perturbations and new external interference sources, and the complexity of site-surveying and installation. The alignment of the directional antennas must be precise and installation must include some manual determination of channel choices to maximize the efficient use (and re-use) of channels as well as a manual configuration of network topology.

A more-recent alternative approach utilized by MeshDynamics automates the channel-selection and topology-definition tasks by distributing dynamic radio intelligence in each node, in effect creating multiple "RF robots". Sophisticated algorithms allow each node to listen to its environment continuously to determine its relationship with neighboring nodes as well as extraneous and potentially interfering radio sources. Based on this analysis of the environment, an individual node selects the best channels to use to connect to the optimal nearby node for highest performance.

In this distributed dynamic radio intelligence approach, the network forms a tree-like structure emanating from one or more "root" nodes that have the wired or fiber connection to the Internet or intranet. As the branches of the "tree" radiate outward, eventually they become geographically distant enough from one another for nodes to begin re-using channels. This greatly increases the data-carrying capacity of the network, since it makes better use of the scarcest resource in an outdoor WiFi environment: the fixed number of unlicensed channels.

It may be obvious that this approach works with both sectored directional antennas and with omnidirectional antennas. The greatest flexibility comes with the use of omnidirectional antennas, since no pre-engineering of paths must be done. Each node ascertains the best connection and coordinates that connection with adjacent nodes through the periodic exchange of routing and other information.

Although the structure that results gives the appearance of a tree, the software intelligence in each node permits it to function exactly as the single-radio First Generation wireless mesh. A failure of any node prompts immediate coordinated reconnections around the network to bypass the failed node, as in the case of a traffic accident felling the light pole on which the node is mounted. When the missing node is returned to service, its neighbors recognize its presence and recalculate the best connections once again. This capability also makes additions and expansions to the network very straightforward, as new nodes may be simply configured with the proper security information, then powered-up. New nodes automatically are added to the network based on an exchange of information between the existing nodes, which are continually monitoring the environment.

The independent but coordinated "RF robots" in each node are also useful in detecting and avoiding external interference sources. Because 802.11 WiFi is an unlicensed medium, new independent and uncontrollable RF courses may appear in the network unpredictably and at any time. First-Generation wireless mesh networks are challenged by such sources since every device is sharing the same channel, but all may not be close enough to "hear" the offending point interference source. Hardware-oriented Third Generation networks may also be affected if an unexpected interference source appears in one of the sectored node-to-node paths.

Such an interference source is easily with managed distributed dynamic radio intelligence. Nodes close to the offending RF source may move in a coordinated fashion to a new and unused channel. By coordinating this movement, the impact to end users may be minimized. In extreme cases where a powerful interference source blankets an area, nodes cut off from communication with the rest of the network simply repeat the start-up process of listening for other nodes on non-interfered channels and the network rapidly reconverges to a stable state.

Quick and easy deployment without significant pre-engineering, automated and rapid avoidance of interference sources, and fast additions and reconfigurations of the network by the simple addition of nodes are all distinctive features of Third Generation architectures delivered through distributed dynamic radio intelligence as developed by MeshDynamics. An emerging application of this technology is in temporary and event-driven networking such as that needed for sporting events. An additional benefit of this software-based distributed intelligence is the ability to place a node in motion relative to other nodes in the network.

One use of mobility has been in security and other types of rapid-response applications. Nodes may be mounted on vehicles, even man-carried to new locations while remaining in communication with the overall wireless mesh network at all times. No management or other user interaction is needed, and with the use of omnidirectional antennas, the mobile nodes may move in any direction around the perimeter or through the middle of the geographic area supported by the fixed wireless mesh node. Mobile nodes have even been mounted in unmanned aerial vehicles or tethered balloons to provide coverage of an area.

Another mobile application supported by MeshDynamics' distributed dynamic radio intelligence is support for networking in rail corridors. Mobile nodes installed in commuter trains link with a series of fixed nodes installed along the rail line. As the train moves, the mobile node is constantly listening to the environment. Typically, the signal from the fixed node being approached will be increasing while the signal from the fixed node recently passed will be decreasing. In a coordinated fashion, the "RF robots" in all three nodes coordinate the hand-off from one fixed node to another, assuring seamless connectivity for commuters in the railcar, whether the train is in motion or halted at a station.

15.6 Radio Agnostic Mesh

Frequency agility is taken one step further in MeshDynamics Modular Mesh products. The mesh control "RF robot" software runs above the MAC layer of the radio: the same mesh control software supports radios operating on different frequency bands. Decoupling the logical channel-selection and topology-definition processes from the specific physical radio in this fashion delivers distributed dynamic radio intelligence benefits for current as well as emerging radio standards.

Figure 15.5 shows how this level of flexibility is supported. There are 4 mini-PCI slots on the board, two on the bottom and two on top. Each of the four slots can house a different frequency radio. This opens up some interesting possibilities including 2.4 GHz backhaul systems being part of a mesh with 5.8 GHz backhauls. Since the service and backhaul radios are distinct, it is possible to use a service radio to bridge over from a 5.8 GHz backhaul to 2.4 GHz backhaul – as shown in Figure 15.7. The 4325 Mobility Relay node on the bottom left has joined the mesh – even though the upper links are 5.8 GHz (blue) – through the service radio (pink).

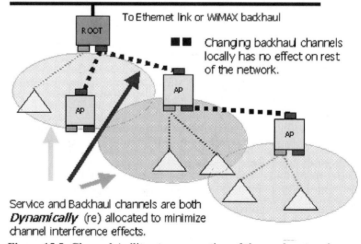

Figure 15.5: Channel Agility at every section of the mesh network.

Switching to another channel contains local interference at one section of the network. With one radio backhauls, this is not possible: the entire network is on the same channel and switching to another channel is simply not practical. The performance of single channel backhauls is therefore heavily compromised in RF polluted environments or under malicious attacks. Military field trials with dual channel backhaul have demonstrated that frequency agility ensures the mesh is running even with malicious RF interference- the backhaul radios (blue) simply switched to non-interfering channel.

One advantage of this level of flexibility includes supporting longer range and lower bandwidth 2.4 GHz 802.11b radios with shorter range but more bandwidth capable 5.8 GHz 802.11a radios, all part of the same mesh network. The longer range enables the edge of the network – where bandwidth requirements are low – to be serviced adequately by 2.4 GHz edge/mobility nodes, (Figure 15.7). Bandwidth is thus distributed more efficiently and cost effectively managed: the node spacing is adjusted for the subscriber density based on the range of the radios used.

In the future, this flexibility will also permit the incorporation of new radio types such as WiMAX (802.16) as an adjunct to the 802.11 WiFi mesh. These potentially higher-speed links may offer a flexible way to inject bandwidth into the mesh along with a cost-effective distribution strategy for high-speed WiMAX links.

One additional benefit of decoupling distributed the dynamic radio intelligence software controlling channel selection and topology configuration from the radio hardware is the potential to incorporate commercial-off-the-shelf (COTS) radio and other modules. This substantially decreases time-to-market and greatly increases manufacturing scale for some components, reducing both development- and unit cost over the use of custom hardware development.

Figure 15.6: 4 Slots (two on bottom) supports different radio types.

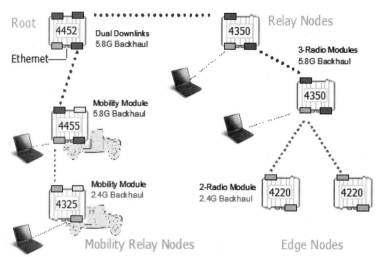

Figure 15.7: MeshDynamics 5.8 GHz and 2.4 GHz Backhaul Interoperability.

15.7 New Applications Enabled by Third Generation Wireless Mesh

Applications of earlier generations of wireless mesh technology were limited to a small number of node-to-node hops due to performance limitations, particularly the limited support for video- and voice-over-IP. This has created the perception that wireless mesh is suitable only for the delivery of basic networking such as casual web surfing. The emergence of Third Generation wireless mesh, with its inherent higher performance, is engendering many new and useful applications. Chief among these are applications requiring bandwidth to be extended across long distances by use of the wireless mesh backhauls alone. In these cases, it is impossible or impossibly costly to add additional wired or fiber network drops every two or three hops.

In one example, MeshDynamics wireless mesh nodes are being used along a national border to provide connectivity to distant locations in a long "string of pearls" configuration. In this case, there is no need to provide blanket WiFi coverage for client PCs or PDAs over the entire area, so the links between nodes are up to 14 miles in length. Mobile nodes mounted in security forces vehicles join the network dynamically and while in motion. Service radios in the vehicles provide connectivity for staff in the vehicles and operating nearby.

Similarly, a video surveillance application required extension of bandwidth into an undeveloped area with no installed high-speed infrastructure. Cameras are cabled directly to MeshDynamics nodes and linked back to a central site via many hops. Third Generation technology provides high performance and minimal delay and jitter to support the high fidelity video CODECs in use.

MeshDynamics' distributed dynamic radio intelligence technology, combined with Third Generation performance, is also key to a military application deploying sensors on moving combat vehicles with no fixed root wired or fiber connection. These vehicles are in constant motion in relation to one another, but information from sensors mounted on each must be brought together for threat analysis. The network topology must constantly and automatically adapt to the varying distances between vehicles while the mesh must deliver high performance with very low latency and jitter to permit the sensor data to arrive in a timely fashion. And as noted in section 15.5 above, this combination of high performance and automated topology flexibility is enabling many other mobility applications that are not possible with other hardware-oriented Third Generation wireless mesh solutions.

Even in more traditional metro/muni applications, Third Generation technology is being used where earlier generations of wireless mesh have failed. In some localities, high speed internet infrastructure is not yet available from cable or telco providers or is prohibitively expensive. A single high-speed connection at the root node must be extended over a broad area using only the node-to-node connections for a backbone. This requirement for many hops resulted in the removal of earlier-generation wireless mesh and replacement with MeshDynamics equipment. The ease of deployment that comes with distributed dynamic radio intelligence can also be a benefit in these underserved areas where skilled RF engineers may be in short supply. And the reduced number of wired or fiber drops contributes to a lower total cost of ownership, permitting these networks to be deployed more quickly and more broadly.

15.8 Conclusions

New mesh requirements (more hops to cover larger areas, more efficient bandwidth distribution, better latency and jitter for Video and VoIP) have given rise to the Third Generation of mesh architectures. Third Generation multi-radio backhaul architectures deliver the higher bandwidth and more-deterministic performance necessary to meet these new requirements. While all Third Generation solutions deliver demonstrably higher performance, some custom hardware-oriented solutions come with requirements for extensive pre-engineering and offer limited capabilities for mobility and avoidance of interference. By contrast, software-oriented approaches based on distributed dynamic radio intelligence support frequency agility, automated channel selection, dynamic topology configuration, and radio agnostic meshes, providing more effective single framework solutions for larger scale and diverse application environments. This combination of features delivers higher performance for traditional metro/muni applications, but is also opening many new applications for wireless mesh.

16

Wireless Mesh Networks

Jan Kruys, Luke Qian[a]

16.1 Introduction

Wireless mesh networking is rapidly gaining in popularity with a variety of users: from municipalities to enterprises, from telecom service providers to public safety and military organizations. This increasing popularity is based on two basic facts: ease of deployment and increase in network capacity expressed in bandwidth per footage.

So what is a mesh network? Simply put, it is a set of fully interconnected network nodes that support traffic flows between any two nodes over one or more paths or routes. Adding wireless to the above brings the additional ability to maintain connectivity while the network nodes are in motion. The Internet itself can be viewed as the largest scale mesh network formed by hundreds of thousands of nodes connected by fiber or other means, including, in some cases, wireless links.

In this chapter we will look more closely into wireless mesh networks.

16.1.1 History

Mesh networking goes back a long time; in fact tactical networks of the military have relied on stored and forward nodes with multiple interconnections since the early days of electronic communications. The advent of packet switching allowed the forwarding function of these networks to be buried in the lower layers of communication systems, which opened up many new possibilities of improving the capacity and redundancy of these networks. Attracted by the inherent survivability of mesh networks, the US Defense research agency DARPA has funded a number of projects aimed at creating a variety of high-speed mesh networking technologies that support troop deployment on the battlefield as well as low speed, high survival sensor networks. In parallel, and motivated by the above research programs, the IETF has investigated mesh networking technology and standards under the heading of MANET (Mobile Ad-hoc NETwork). More recently, the IEEE (its 802.11 Working Group) has taken up the development of a standard for wireless mesh networks. It has not taken long for the wireless industry to recognize that Wi-Fi technology provides a suitable basis for secure, cost-effective wireless mesh networking

---

[a] All authors currently affiliated with *Cisco Systems, Inc*

and today, we see such networks being deployed in enterprises, mining, transportation, public safety, and military applications throughout the world. In fact, mesh networking has entered outer space: some satellite based communication systems include IP routers aboard satellites.

16.1.2 The Benefits of Wireless Mesh Networking

The three characteristics of wireless mesh networks mentioned above (mobility, ease of deployment, and scalable capacity) derive from the two underlying facts: the unconstrained propagation of radio waves and the increased bandwidth made possible by reducing the distance between network nodes.

Unconstrained RF propagation allows wireless mesh nodes to maintain connectivity with more than one peer node. This gives resilience to the network by allowing traffic to follow multiple paths. Having potential connections with more than one peer node also makes it possible to let the network form itself: this facilitates deployment.

Increased bandwidth resulting from shorter links goes back to the fact that RF transmission power is limited by regulations as well as practical considerations. Since received signal strength drops exponentially with distance, shorter links mean much higher mesh point density and therefore much higher throughput (=capacity per area).

The following goes a long way towards explaining the commercial interest in wireless mesh networking:

- Fixed connections are expensive to put in place and therefore expensive to lease or rent. Extended connectivity allows many nodes to share access to the same fixed infrastructure point of presence. This reduces the operating cost of a large network by a large margin.
- Multiple connections in the mesh assure continuity of services in case of device failure or link blockage. Continuity of service is important for service providers because their users expect that in return for the fee they pay for using the network
- Scalability assures that a network can grow in capacity without massive investment in infrastructure. This reduces the business risks by allowing a network to grow with demand.

As we shall see below, these benefits do not come free. They depend on such basics as sufficient and suitable RF spectrum being available and on a broad palette of technical features, including suitable RF transceivers, cryptographic security functions, responsive routing and forwarding, and systems management functions etc.

16.1.3 Some Typical Deployment Scenarios

Mesh technology can be used to provide a variety of services and support a wide range of applications. These range from small in-home mesh networks that connect consumer devices like TVs and audio systems to large-scale municipal networks that cover entire cities. Another major type of mesh network is a fully mobile network for public safety applications: a mobile access point in the police car is a node of the wireless mesh that also supports fixed cameras and a portable, on-site network. Additional redundancy and coverage in such networks may be provided by other means such a WiMAX access link or a 3G access link.

The figure below sketches a municipal mesh network that serves not only the police and the cities public services but also businesses and consumer residences. Such a network combines high-speed broadband infrastructure or backhaul links as well as access links that operate over shorter distances. The access links may, in turn, serve mesh devices that link to other devices, possible mobile ones. Mesh technology makes it possible that these networks combine structured as well as unstructured elements and yet provide a unified architecture.

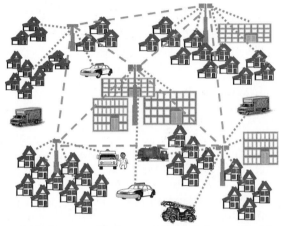

Figure 16.1: An example of a municipal wireless mesh network.

Figure 16.2 shows how mesh networking can be used to provide public safety staff the necessary communications to deal with major incidents: the municipal grid provides the backhaul for the mesh network at the site of the incident. Here, mesh technology allows the police, the firemen and the ambulance staff to communicate directly with each other rather than having to rely communications links to the respective headquarters.

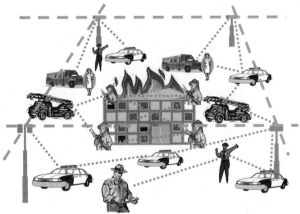

Figure 16.2: Public Safety Wireless Mesh Network.

16.1.4 Other Wireless Solutions

Cellular mobile phone networks and WiMAX are two examples for wireless networks that are able to provide services that overlap with the services that can be offered by a wireless mesh network.

The following example of service "profile" of a large scale mesh network makes this clear: the profile is composed of 6 services, some on demand, some always on. The latter are reserved for the "owner" of the network: the municipal government.

On-demand services
1 Mbps down, 0.5 Mbps up - free
2 Mbps down, 1 Mbps up – paid, consumer oriented
3 Mbps down, 2 Mbps up – paid, business oriented

Basic Services
160 kbps up/down – paid, public VoIP
80 kbps up/down – city services
60 kbps up/down – city emergency services

Although 3G and 4G networks as well as WiMAX can provide such services, in principle, the actual coverage area in which these service levels can be realized vary significantly with the level of the service: more bits/seconds means a shorter operating range. Although the typical wireless mesh based on Wi-Fi technology is subject to the same constraint, its effect is not necessarily as significant because the range of a Wi-Fi access point can be significantly shorter without affecting deployment cost or operating costs. Thus, the Wi-Fi mesh has a higher typical capacity (in terms of bits/second/area) than the typical service provider network.

16.2 Current Issues and Solutions

16.2.1 Network Structure

The structure of a mesh network plays a major role in the performance of voice services because it determines the delays encountered by the voice packets as they travel through the mesh network.

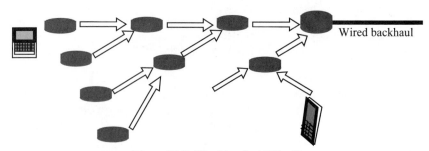

Figure 16.3: "Parking Lot Effect".

In Figure 16.3 above, the voice packets from the PDA have to contend for link access with packets coming from 5 other mesh nodes whereas the packets from the cell phone contend for link access only with packets coming from one other node. The build up of traffic load towards the portal node is sometimes called the "parking lot effect". The parking lot effect is a function of the number of hops in a mesh and the peering ratio. In the left hand part of the mesh above, the hop depth H is 3 and the peering ratio R_p is 2. In this case, the load on the portal link is equivalent to 7 peer links. In general, the load on a portal link is proportional to R_p^{H}.

Remedies for the parking lot effect include the proper dimensioning of the links of mesh network and/or limiting access to the mesh so as to keep the offered load within the limits imposed by the capacity of the available RF links.

16.2.2 Intra-mesh Channel Re-use

The links between mesh nodes may use one or more RF channels. One RF channel facilitates rapid peer acquisition and mesh formation but it also causes interference between links. In its most benign form, this re-use causes sharing of channel capacity over multiple links. This re-use effect exacerbates the parking lot effect described above: the further it originates from the portal, the more a packet has to compete for the channel. In some cases, nodes may not see all of the nodes whose transmission they could interfere with. This leads to destructive interference and packet loss.

Because of the exponential back-off mechanism built into the medium access of the IEEE 802.11 protocol, packet loss may lead to significant delays per packet per link. Again, if the packet has to traverse many hops, the effect is incremental and leads to significant impairment of performance, notably for voice services.

Remedies for medium re-use require increasing the isolation between links. Three dimensions are available: space (directional antennas), frequency (multiple RF channels), and coding (using orthogonal codes for mesh links). Which of these is the best choice for a given case depends on the requirements and conditions of a given network deployment.

16.2.3 Medium Access Contention

The basic IEEE 802.11 Medium Access Control protocol was designed for bursty data traffic in an unpredictable RF environment. Because channel state is re-established for every packet through listen-before-talk and collision avoidance through the use of a *contention window*, medium access is inherently subject to jitter although is typically much less than a millisecond in magnitude.

In order to facilitate voice and video service, QoS extensions have been added to the standard: the most frequently used mechanism of the standard is EDCA which is an enhancement of the basic distributed access control mechanism with a priority bias for certain types of traffic like voice, video, and best effort data. In addition to the minimum contention window size, the priority bias is also expressed as an additional back off parameter called the *Arbitration Inter Frame Space* or AIFS for each *traffic class*, which specifies a back off that is executed at the start of each contention window. The larger the AIFS, the better the separation of traffic classes but the lower the link's efficiency.

Although the EDCA mechanism does a fair job in improving the statistics of a given traffic class, it requires a certain operating margin to avoid the rapid build-up of packet delay that is typical of a heavily loaded 802.11 link.

16.2.4 Mesh Routing and Forwarding

In addition to mesh structure and RF link properties, the routing of packets plays an important role in the service level offered by a mesh network. In the simplest case, a mesh network is fixed and has a single portal that allows all users of the mesh to communicate with the outside world. If there is no user-to-user traffic, the routes in the mesh form a simple tree and routing is largely static. Route set-up is simple: finding the most efficient path to the portal. Once the links are formed, the best path between a given node and the portal does not change, except in case of node or link failure.

Mobile mesh networks do not have the benefit of static links. The useful life of a link between two peers typically exceeds the time required to establish the link but re-forming the routes in a mesh, as the nodes move and their distances and RF connectivity change, remains a costly and potentially disruptive event that must be avoided whenever feasible. Much research effort has been invested in finding an optimum solution to routing in fixed and mobile mesh networks but so far most of that work has focused on the ad-hoc, mobile mesh network. Examples include the following:

- Spanning tree routing
  Networks consisting of wireless bridges naturally form tree-like structures. Such a tree structured mesh is rooted in a portal node. Routes may be set up using a combination of parameters, e.g., hop count and link transit delay as metrics to determine the best path between a given node and the portal. Monitor functions which detect link failure and/or portal loss, may be used to trigger link or tree rebuilding. Because it focuses on solving a particular problem, the fixed and slowly changing mesh, such a tree-like mesh structure can be both simple and efficient.

- AODV: Ad-hoc, On-demand Distance Vector protocol (IETF RFC-3561)
  AODV is one of the many developments in the MANET space. Although basically fairly simple, it lacks the inherent efficiency of spanning tree routing, a price paid for being able to deal with a changing environment. Whenever a route is needed or needs repair, its originator floods the network with a request for a destination. The latter replies with a unicast that is forwarded back to the originator. Loss of a link triggers a route error message up and down the route. The actual cost of flooding in AODV depends on the rate of change of the environment: no changes means the routes do not change and there is no overhead expended on route maintenance.

- OLSR: Optimized Link State Routing protocol (IETF RFC 3621)
  OLSR is a pro-active routing protocol that uses the link state as a driving factor and includes a multicast capability: a subset of nodes, called multi-point relays, provides anchors for neighbor nodes. Link state information distribution can remain local and multicasting is supported "naturally". Although conceptually simple, the implementation of OLSR requires many different control messages and although it is

more efficient than its flooding based parents, overall efficiency is not an obvious property.

- HWMP: Hybrid Wireless Mesh Protocol (IEEE 802.11-TGs)
  In 2005, the IEEE 802.11 Working Group started the development of an extension of the wireless LAN standard with a layer 2 mesh protocol. The choice quickly converged on a combination of a simple tree building protocol to handle static meshes and AODV elements to support mobile mesh deployments. HWMP supports both modes of operation with a single set of primitives. This allows a mesh to use the most efficient routing protocol applicable for a given deployment or application.

The choice between a standard protocol and a proprietary protocol may be driven by technical and operational considerations as well as by commercial considerations. The former include how well a protocol copes with the primary mission of the mesh network, (e.g., mobile deployment or fixed deployment), and how well it handles specific service requirements like resilience or efficient multicasting.

Commercial considerations include the availability of hardware from multiple sources - this is particularly important for consumer type applications. Typically commercial considerations override technical considerations.

With routes in place, forwarding in the mesh network remains non-trivial. Regardless of the network structure, there will be alternative routes to a given destination and each node has to decide which next hop to send a given packet to.

Determining the best path to a given destination requires the use of a consistent set of metrics: all nodes involved must share the same meaning and measurement of the metrics of a given link or path, the latter being a function of a collection of the former. Path metrics may take many forms: from the number of hops to traverse or the airtime needed to reach the destination to complex values that bring together a variety of parameters such as hop count, link load, and SNR.

Ideally each node will keep track of the cost of reaching each of the destination node for which it has a route rather than the cost of reaching the next node on that route. Deriving route cost may not be straightforward: hop count and airtime have the advantage of being additive: the metric for a given route in the mesh is simply the addition of the metrics of each of the hops involved. Parameters like SNR do not have that property and require some form of normalization to yield an "addable" parameter.

Which type of metric is the best depends on the type of mesh network and on the operational conditions. In general, the complexity of metric depends on resource constraints: if link throughput typically exceeds traffic demand throughout the network, hop count is adequate. Conversely, if link throughput and data rate are limited or vary throughout the network, link parameters will be important, possibly more important than the hop count. The picture is further complicated by the fact that different types of service have different requirements: voice packets are short and must be delivered within certain time constraints and therefore link reliability – possibly obtained by lowering the transmission rate - and hop count are important routing metrics. At the other extreme is a background data service: packets may be large and timing is not an issue and therefore hop count is less important but because of the volume of this type of traffic a high data rate is more important than link reliability.

This problem of optimal forwarding is exacerbated by multicasting. Intra-mesh multicast requires some nodes to be multicast-aware so that they can detect and forward multicast packets. As noted above, OLSR has built-in features for efficient multicasting. The Practical General Multicast protocol [see RFC 3208] and the Protocol Independent Multicast Protocol [see RFC 4602] offer examples of how one might go about providing multicast as an overlay capability.

16.2.5 Mesh Security

The security of wireless systems is a well-known subject that receives much attention of the research community. The absence of a physically closed channel between sender and receiver is not only psychologically unsettling, but also does introduce many opportunities for an attacker, regardless of motivation, to gain information or to place or modify information. Protective measures are well-known and broadly applied: cryptographic data protection and identification and authentication of users, the latter also making use of cryptography.

In wireless mesh networks, security issues are complicated by the need to build the network by wireless means and by the absence of human users that can be used as "trusted" parties during network initialization. The need to keep the equipment cost low rules out effective physical protection; security functionality and the storage of cryptographic data (notably keys) cannot be fully relied upon. Mesh security is therefore very much an exercise in selecting the adequate measures – in the absence of perfect solutions.

Mesh nodes perform a number of functions, each of which has its own security concerns:

- Discovery
  This function serves to detect other mesh nodes that belong to the same owner or administrative domain. Here, security is limited to authentication of the information provided by other nodes. This can be provided by means of public key ciphers that allow the verification of digital signatures; with symmetric ciphers, this is not possible. Whether such signature verification during discovery is done depends on operational needs, the tolerance for overhead and the available budget. In commercial mesh networks, the discovery function is left unprotected. This exposes the network to spoofing and denial-of-service attacks.

- Peer Link Establishment
  This function creates secure links between mesh nodes and in the process it validates the non-protected data that nodes obtain during discovery. Depending on the type of mesh and the operational requirements, link establishment may precede setting up routes or it may be done prior to setting up routes. See below under *Routing*. Secure link establishment requires that nodes are able to identify themselves, that they can be authenticated and that they are able to set up a cryptographic session with each other that protects the flow of data and management information between nodes. Various means are available to secure link set-up: symmetric key ciphers are simple to apply but suffer from scalability issues (administering thousands of crypto keys is nightmare) whereas asymmetric ciphers scale very well but suffer from the "man-in-

the-middle" attack. In practice, combinations are used. The IEEE 802.1X protocol is a well-known example: the supplicant requests a connection to an authenticator who makes use of a security server to validate the supplicant's credentials and to generate cryptographic session keys. Its security relies on a secret shared between a given node and one (or more) security server(s). The latter validates the identity of a node and generates a set of cryptographic keys that are passed in two separate packages to the supplicant and the authenticator. This allows any two nodes to set up a secure link – assuming at least one of them has access to a security server. 802.1X requires a certain measure of physical security to store security information (e.g., cryptographic keys and certificates).

- Routing
  The security of routing is a much debated subject. Routing protocols form paths through a network that is used to forward data and management information. Interfering with route set-up allows an attacker to force changes in the routes so as to cause loss of connectivity among nodes as well as loss or hiding of data. These effects can be soft and temporary – and therefore not easily detected, or hard and permanent, in which case recovery is clearly required. In general we can say that attacks against the routing function of a mesh network tend to result in denial of service like effects. These attacks can be thwarted by expanding the routing protocols so as to make them harder to interfere with. One example is the Tesla approach: authentication information is broadcast in advance of the actual data and its associated authentication information. Since the attacker has no knowledge of the data at the time the advance authentication information is received, spoofing becomes more difficult. For further reading, see [1]. Another practical approach is to first secure the links between the nodes of a mesh before routing is initiated. This has the disadvantage of setting up links that may not be needed but it avoids to a large extend the complexities of securing the routing process itself.

- Forwarding
  The forwarding function delivers packets to their destinations, either directly or via intermediate nodes that lie on a path known to include the destination. The protective measures typically include data confidentiality and integrity, but in some application data origin authentication may be needed as well. For reasons of efficiency, symmetric key cryptography is the preferred mechanism. The protection of forwarded data may be either hop-by-hop or end-to-end between the originating mesh node and the destination mesh node. As described above under *Link Establishment*, communication between nodes takes place over secured links – which is adequate for securing hop-by-hop forwarding. Where operational considerations demand end-to-end protection between origin and destination, a second layer of cryptographic protection may be applied. This adds both overhead and (management) complexity. If the mesh is connected to other networks, end-to-end protection of layer 2 forwarding is impractical and higher layer security solutions like IPSEC must be used.

- Management
  Management of mesh networks provides the means to set parameters in nodes and to collect operational data from nodes. The parameters control or modulate the execution of the above functions. The security of management protocols is sometimes overlooked but it clearly is critical to the functioning of a wireless mesh network. One way to look at this problem is to consider "management" as an application that makes use of the mesh functionality in order to manage that mesh functionality.

16.2.6 Congestion Control

As described in *network structure* above, funneling traffic into one or a few portal links is a property of a typical mesh network. Clearly, such links are subject to the danger of congestion. In ad-hoc and mobile networks, local disturbances of the network's routes may cause congestion of the links that are left. Therefore, a good mesh network design includes means to predict and prevent congestion from occurring.

Congestion occurs when a source produces more than its sink can handle. In a mesh network, each link is a sink and all links that feed it are potential sources. This applies recursively throughout the mesh. Each node must consider each of its links as a potential sink for traffic coming from other links terminating at that node. Note that each link will carry traffic in both directions and the changes in the offered load in one direction may cause congestion to occur in the other direction. Congestion avoidance and control requires monitoring each "outgoing" link and sending flow control messages to the nodes feeding the incoming links. This feedback percolates back to the traffic source nodes and, ideally, to the applications creating the traffic so that lossless traffic reduction results.

Practical mesh networks tend to have a few links per node and this simplifies the picture considerably. The required "upstream" feedback can take many forms, ranging from specific capacity allocations to a simple "reduce" flag. What is best in a given system or application depends on the mission and operating conditions of the network.

Other forms of congestion control include rate limiting the traffic sources. This is a pro-active from of source flow control that has the advantage of avoiding the loss of packets that occurs during actual congestion. The downside is inefficiency: the further away from a congested link a source the lower its contribution to the actual congestion condition. As the "throttling" needed to keep the network from congestion gets more imprecise with increasing distance from the portal, more bandwidth is wasted.

A full service mesh network will support different classes of service. The QoS extensions of the IEEE 802.11 standard (formally known as Amendment 11e) recognize 4 different traffic classes: voice, video, best effort and background and it provides independent prioritization for each. This, by itself, does provide a form of flow control on a hop-by-hop basis that protects the higher priority services from being crowded out but it does not provide congestion control. However, regardless of which form of congestion control is chosen, it makes good sense to apply it to all traffic class-aware manner such that the premium services are least affected by the throttling - pro-active or otherwise.

16.2.7 Fairness

"Fairness" is probably one of the least understood properties of wireless networks, probably because a broadly agreed definition is not available. The underlying problem is that wireless networks are assumed, or known, to offer limited capacity that is easily exceeded by demand. In simple access networks (e.g., per IEEE 802.11), the available capacity is determined by the RF channel, its load and its impairments. All users are assumed to get a fair share of that. In a mesh network, this picture is complicated by the fact that the same RF channel may be used for access to the mesh nodes as well as for connections between mesh nodes. Further complications arise in the case of overlapping or adjacent mesh networks: one may use a given channel for access whereas the other may use it for a node-to-node link.

A simple working definition of fairness could be: "*No user or other entity is able to obtain an excessive share of network resources (links, capacity) at the expensive of other users or entities*". In other words, an excessive share of the resources is fine, provided that there are no other users.

Note that service differentiation does not conflict with the basic fairness principle: within a given service class, all users are treated according to the principle but it is up to the network manager to decide how the available capacity is allocated to the different service classes. As a consequence, users of a lower service class may see their share of the network capacity drop as traffic from higher service classes increases.

Applied to mesh networks, the fairness principle requires that such a resource sharing policy is applied to all nodes and links.

16.2.8 UDP and TCP Performance

The behavior of a mesh network affects the services that it carries. In case of a UDP service, packet loss, delays, and jitter caused by the mesh become directly visible in the UDP packet delivery to the destination.

In the case of TCP the picture is complicated by the TCP Ack, which is the feedback from the destination that controls the downstream packet rate of the TCP source. Interference, hidden nodes, and congestion effects can affect the TCP Ack on its way back to the source and cause it to be delayed or lost. In the first case, the TCP source will reduce its downstream packet rate; in the second case, the flow may stop altogether. This "starvation" effect has been observed in simulations as well as in practice. See also [2].

16.2.9 Voice over Mesh

Voice over mesh poses its particular challenges because the human ear is very adept in detecting minute differences in timing of the signals it receives. Figure 16.4 below shows an abstraction of an end-to-end network with mesh networks forming the access networks that support mobile voice over IP devices.

Signal timing is determined by the wireless access links, by the meshes and by the Internet. Thanks to the continuing growth of the Internet infrastructure, the latter is typically an unvarying factor and much of the voice quality impairment is caused by variations in the access links and the mesh networks.

Figure 16.4: A Voice over Mesh Network.

According to the ITU-T guidance, 99th percentile delay and jitter on a high quality voice connection must remain below 150 msec and 50 msec respectively. Assuming that jitter is basically noise like and caused by independent sources, we can put the jitter "tolerance" at 50 milliseconds for each of the segment of the connection. Delays however, do add up and that means that per access network, the allowed delay budget is only 25 msec. In a 4-hop deep mesh network, this equates to about 5 msec per link (1 access link, 4 mesh links). This presents a significant challenge for a Wi-Fi network because medium access is a statistical process with potentially very large delays for individual packets (due to exponential back-off in case of transmission failure).

Assuming proper dimensioning of the RF links, the main determinant of mesh link performance is medium re-use and its impact on access contention and packet error rate. Thus, if a number of child nodes share the same RF channel with a parent node, the offered load will increase medium contention, which in turn causes transmission failures and therefore exponential back-offs. This situation can be avoided by keeping the number of voice calls below a given level (to reduce medium contention to low levels) and by dropping packets rather than re-trying a failed transmission.

There are a number of factors that can affect the quality of voice over mesh. These include:

- Choice of codec.
  Voice codecs have different requirements with respect to jitter, packet loss, etc. Codecs like Skype's offer good tolerance for packet loss. Some systems are clever and they choose a codec dynamically – depending on the connection quality. This allows optimization for voice quality. Alternatively, knowing the codec type, the lower layers could attempt to tailor the link behavior (e.g., packet priority and packet delay/loss tolerance) so as to optimize voice performance.

- Handover
  As a user moves through the area covered by a mesh network, its connections change – both the client link and the intra-mesh links may change. The former is known as handover, which is equivalent to "roaming" in IEEE 802.11 parlance. The actual handover time depends on a number of factors, including the need to detect another access point and setting up a secure link to it. Due to the intermittent nature of the voice packet stream, clients have some time to look for other mesh nodes, assuming they know at which frequency to probe for another node. That information can be supplied by neighbor mesh nodes or by a controller. What a client handover may also result is a significant change in the path between the access point and the mesh portal node. Even if alternate paths are available, the signaling needed to enable an alternate path may cause a noticeable delay.

- Other causes of voice quality impairment
The following factors all affect the performance of a mesh network although in a less severe manner than high traffic loads:

- Radio measurement traffic;
- Location signaling;
- Legacy 802.11 devices that share the channel;
- Non-802.11 devices like microwave ovens and cordless phones, Bluetooth (some of which exist at 2.4 GHz and 5 GHz).

Assuming the use of multiple RF channels for the backhaul, most of these factors become less of an issue at 5 GHz because in this band, due to the large number of RF channels, channel loading will be significantly less than at 2.4 GHz.

16.2.10 Mesh Network Management

Mesh Network Management covers a wide range of subjects that are common to most wireless networks. The main difference is that in a wireless mesh network the components of the network interact more strongly and in more complex ways.

Dynamic system-wide RF management facilitates smooth operations, such as dynamic channel assignment, transmit power control, and load balancing. For outdoor mesh networks where RF interference issues can be significant, this is a crucial capability. RF management requires the collection of RF channel data from all nodes of the network so as to allow optimization of channel assignments and reducing interference between links.

Network wide policy management, including VLAN assignments and QoS settings, are required to support real-time, mobile applications such as voice over WLAN. QoS policies in a wireless mesh will be more complicated due to the interaction of many factors, such as RF channel assignment, network structure, traffic types, and service level agreements.

Security management that provides uniformly enforced security and usage policies that can address the particular capabilities of different classes of devices, such as Mesh Access Points and Mesh Portals and that assist in the discovery and mitigation of Denial of Service attacks, and denial of rogue access points

Mobility Management functions may be used to assist handover of user devices and the re-configuration of mesh links in a mobile mesh network which, if properly implemented can cellular-like fast handoffs and maintain QoS for such critical applications as voice over WLAN.

16.3 Mesh Deployment Issues

This chapter gives a short introduction to this important and interesting subject. All the major mesh technology vendors provide advice and indeed deployment services, in many cases together with companies specialized in this area.

Deployment of a mobile mesh network (e.g., for the Police or the Military) is in many ways a simple affair: provided the available spectrum and transmitter powers match the operational requirements, deployment is largely a matter of provisioning and putting the mesh nodes to work.

For fixed mesh networks, deployment can be a complex issue, even though the self-organizing capability makes life a bit easier.

The main factors that drive a fixed mesh deployment are:

- The service level(s) to be offered in terms of bits per second per user and on which basis, best effort or guaranteed.

- The user density – basically people per area.
  In combination with a) this determines the access capacity that must be installed to achieve the desired service levels.

- The available sites for mounting the access transceivers.
  In cities concerned with visual pollution, there will be stringent requirements for size, shape, and color.

- The available favorable sites for mounting intra-network transceivers.
  Here, high sites are an advantage since these offer the best range to a peer node. Therefore, good sites may be scarce and/or costly.

- The available RF frequency band and the number of channels, shared or otherwise.
  Given an adequate number of channels, the capacity of the network may not be an issue, provided node frequency planning takes into account the network structure and required link capacity levels. In some case, multi-tier intra-mesh links may be needed. For example, the access points of a mesh network may linked to each other and one or a few "concentration nodes" through 5 GHz links using the 802.11 protocol. Between these and the gateways to the internet, high speed point-to-point bridges may used, or, as has been done in a number of cases, a point-to-multipoint system may be used as the second tier of the intra-mesh links. See also section 16.1.3.

16.4 IEEE 802.11, Amendment "s"

The IEEE 802.11 standard – now available as IEEE 802.11/2007 – remains the reference for license-exempt networks for all kinds of applications. Since its inception in the early 1990's, it has grown in scope and, even more so, in the number of pages it takes to explain in great detail its features and functions, including advanced security capabilities, support for quality of service by service differentiation and management. The specification of wireless mesh networks is being done in Task Group TGs.

16.4.1 Overview

The scope of the TGs is to develop a protocol for mesh networking using the IEEE 802.11 MAC/PHY layers to support both broadcast/multicast and unicast delivery over self-configuring multi-hop topologies.

The purpose of the project is to provide an Amendment[1] to the standard that specifies a protocol for auto-configuring paths between stations over self-configuring multi-hop topologies in a Wireless Distribution System (WDS) to support both broadcast/multicast and unicast traffic in a Mesh.

The purported applications range from simple ad-hoc networks to large-scale municipal networks although the work focuses on mesh of 32 nodes or less.

Early on the group decided to make its work open-ended so that existing metrics and routing protocols could be supported by real systems that would otherwise use the baseline functions defined by the standard. This "*extensibility framework*" avoided potential conflicts of interest of members who have developed and installed proprietary technology.

In November 2006, the group issued its first draft for a so-called letter ballot. Although this latter ballot generated thousands of detailed comments, the key choices made in the preparations leading up to the draft were confirmed. These are discussed below. For additional details, please refer to [3].

16.4.2 The IEEE 802.11s Mesh Network Model

The following diagram illustrates the mesh network model adopted by the TGs group. It defines three abstract types of mesh node: the Mesh Point (MP), the Mesh Access Point (MAP), and the Mesh Point Portal (MPP).

The following sections discuss the TGs Amendment in terms of the basic mesh functions: Discovery, Link Establishment, Routing, and Forwarding.

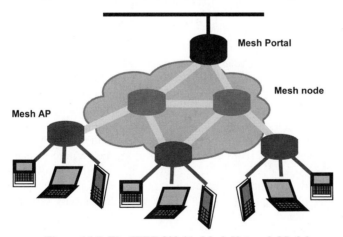

Figure 16.5: The IEEE 802.11s Mesh Network Model.

16.4.3 Mesh Discovery

Mesh discovery is facilitated by two mechanisms: Mesh beacons and Mesh probe/response. The beacon and probe response contain the same information, including: an identifier for

---

[1] In the following we refer to the IEEE 802.11 TGs "Amendment" in the understanding that it has not been completed at the time of writing. Eventually this Amendment will become part of the baseline standard.

the mesh network, the node's capabilities, the security scheme in use and a list of its neighbors.

These two mechanisms provide a node with sufficient information to select candidate peer with which to enter into secure peer link establishment.

Figure 16.6 below illustrates this. Node 7 is not a member of the mesh network but it sees Node 4 and 6(e.g., because Node 5 is hidden by a building). The link budget between 4 and 7 is less than between 6 and 7 and, since the number of hops to reach the Root is the same, Node 7 will establish a link with Node 4.

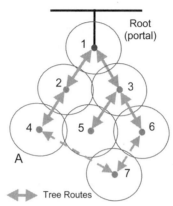

Figure 16.6: Mesh Discovery.

16.4.4 Peer Link Establishment

TGs has opted for setting up secure peer to peer links before performing any routing. Whereas in the baseline 802.11 standard, link establishment is an asymmetrical process in which control is exercised by the access point, in a mesh network, this is not the case: two nodes are peers and each has a "control" role. This difference has been taken into account by a true peer link set up mechanism that is completely symmetrical and robust. It uses bi-lateral open and bi-lateral confirm primitives and completes in two exchanges, provided all necessary security parameters (e.g., the right PMKs, see below) are available in the two nodes. If one of the two nodes is not able to authenticate the other, it reverts to the 802.1X supplicant role as defined in the baseline standard with the other node acting as authenticator. After obtaining cryptographic keys from the security server, the peer link establishment is completed.

16.4.5 Mesh Security

The security features of the 802.11 TGs Amendment correspond roughly to the description given above under *Peer Link Establishment* in section 16.2.5. IEEE 802.1X is used as the basic authentication and key generation mechanism.

In addition, the Amendment defines a Mesh Key Distributor (MKD) that is used to hold a copy of each pair-wise master key (PMK) resulting from a successful authentication. Mesh nodes can request downloads of these cached PMKs or the MKD may "push" these keys to the members of the mesh. By spreading the PMK cache to the nodes, the need for interaction with the security server or the MKD is greatly reduced and the speed of mesh re-forming is greatly increased. This is necessary for ad-hoc and mobile applications and it is a benefit to all mesh deployments.

In practice, the security server and MKD may be combined in a single device that could also act as portal.

16.4.6 Routing Metrics

The 802.11 TGs standard provides a default metric for routing and forwarding decisions: the airtime needed to transmit a default size packet. This airtime depends on the transmission speed and it has the advantage of being additive: the predicted airtime needed to reach a given destination is simply the sum of the airtimes of the hops to be traversed. Note that the airtime metric ignores the effect of traffic loading and therefore it says nothing about the throughput (i.e., the packet rate) of the path. The standard does allow for other metrics. See also section 16.4.1 above.

16.4.7 Routing and Metrics

As noted in section 16.2.4 above, there is a plethora of research on routing protocols and rather than add its own version, the TGs group decided to make a hybrid of two existing routing protocols: a pro-active tree based distance vector protocol and an ad-hoc, on-demand distance vector routing protocol known as AODV.

Because routing takes place after secure peer links have been established, many of the security risks – real or academic – associated with routing protocols, are avoided.

The hybrid protocol uses the protocol elements of AODV but infuses them with new meanings and parameters so as to support fast and efficient building of routing trees if and when appropriate. The rules for processing these primitives and parameters like Sequence Numbers and Time To Live in the primitives assure that indicate the freshness of the path information and that prevent forwarding loops and duplications.

- RANN – Root Announcement
  The Root Announcement is similar to the Portal Announcement: it distributes information about a Root node throughout the mesh. This allows nodes to set up on-demand paths to the Root node as and when required.

- PREQ – Path (Set-up) Request
  The Path Request can be used in two ways: to pro-actively establish a routing tree from the Root outwards and, secondly, to set up an on-demand route to a particular destination node. In the pro-active mode, the destination of the PREQ is left unspecified (wildcard value). The PREQ may request a PREP. In that case, every node that receives the PREQ responds with a PREP that establishes the reverse path to the Root and contains information about the peer nodes of the sender. A node may send a gratuitous PREP at any time (e.g., as the path to the Root changes or its peer links change). In the on-demand mode, the PREQ specifies a particular destination node. As in the pro-active case, the PREQ may specify that intermediate nodes may respond with a PREP message in case they have a path to the destination of the PREQ. The

propagation of the PREQ establishes a set of forward paths from the originator to – in principle – all other nodes in the mesh. Over time, nodes may receive multiple PREQs from the same originator; it will retain the most recent forward path information. Aging of the path data maintained by each node assures that unused forward paths are deleted.

- PREP – Path (Set-up) Reply
  Nodes that receive a PREQ addressed to themselves, respond with a PREP. Unlike the PREQ, the PREP is not flooded throughout the mesh but targeted at the originator of the PREQ. Its propagation establishes the reverse path from destination to originator. A flag in the PREQ allows intermediate nodes on forward path to respond to the PREQ if they have a path to the requested destination. In a mesh that is sufficiently connected, this mechanism considerably reduces the amount of PREQ flooding needed to set up a path to a given destination.

- PERR – Path Error
  The PERR is used for announcing a broken link to all traffic sources that have an active path over this broken link. A PERR may be either unicast or broadcast (if there are many nodes on the broken path). The PERR may contain a list of the destinations that have become unreachable.

The following figure shows how HWMP can be used for pro-active tree based routing as well as for on-demand routing.

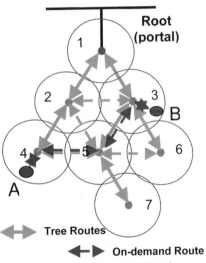

Figure 16.7: On-Demand Routing.

The numbered nodes are Mesh Access Points, the Root node is also a portal that ties the mesh to a wired network (e.g., another Metropolitan Ethernet).

Traffic to and from the portal typically follows the tree structure because that is the most economical. In case nodes communicate across the mesh rather than with external nodes, they can establish an on-demand route: MP4 will issue a PREQ for Address B. When this arrives at MP3, which is the proxy of B, the former will reply with PREP back to MP4 in order to complete the "direct" route.

16.4.8 Forwarding

Forwarding packets received from other nodes is an essential function of every mesh network node.

The selection of the next node to forward a given packet to, is determined by the number of routes available, the metrics associated with each route, and the service class of the packet to be forwarded. Unless the routing protocol allows setting up multiple routes, the forwarding node has little choice: there may be one only forward path to the destination node.

There are two types of forwarding in a mesh: intra-mesh forwarding and proxy forwarding. The former is concerned with packets originating in the mesh. The second is concerned with packets originating from or destined to nodes that are outside the mesh proper: attached layer 2 devices or entities on an external network.

Intra-mesh forwarding requires four addresses: receiver, transmitter, destination, and source. In order to avoid that all mesh nodes have to keep track of which entities they can reach (in other words, maintain a spanning tree) proxy forwarding requires 6 addresses: receiver, transmitter, destination and source, external destination and external source.

The additional external addresses are carried in a Mesh Header, which contains two additional parameters: the Time-to-Live and a Sequence Number. The latter facilitates in sequence delivery to the destination entity. See the figure below.

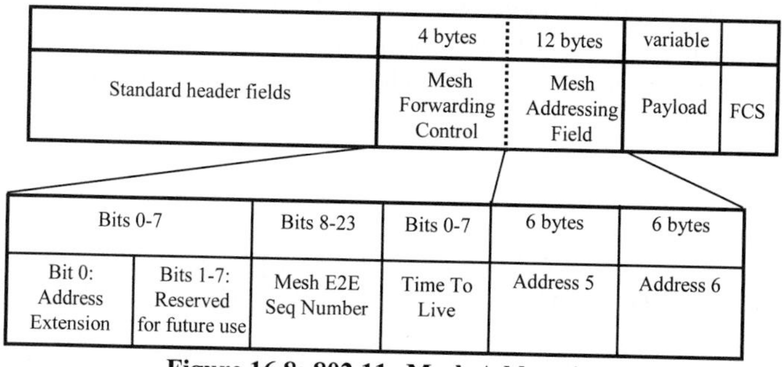

	4 bytes	12 bytes	variable	
Standard header fields	Mesh Forwarding Control	Mesh Addressing Field	Payload	FCS

Bits 0-7		Bits 8-23	Bits 0-7	6 bytes	6 bytes
Bit 0: Address Extension	Bits 1-7: Reserved for future use	Mesh E2E Seq Number	Time To Live	Address 5	Address 6

Figure 16.8: 802.11s Mesh Addressing.

16.4.9 Interworking

The 802.11 TGs Amendment defines a layer 2 mesh network as a conventional LAN segment that can be bridged with other LAN segments using standard IEEE 802.1D bridges. The combination of mesh functionality and the IEEE 802.1D bridging functionality is called a (Mesh) Portal.

A mesh network may have zero or more portals, which may be connected to one or more LAN segments. In case two or more portals connect the mesh to one external LAN segment, broadcast loops may occur and the IEEE 802.1D bridging protocol may cause the LAN ports of a portal to be closed.

The IEEE 802.11 TGs Amendment provides two mechanisms that facilitate interworking:

- Portal Announcements that are flooded throughout the collection of nodes that have the same Mesh ID – regardless whether they have formed a mesh network or not. These announcements help nodes decide which neighbor to link up with: they contain a hop count that is incremented as the announcement is forwarded.
- The Proxy Update protocol, which provides for the transfer of "proxy information" towards the portals. This allows the bridge of a portal to quickly learn the addresses of all the nodes of the mesh and their attached stations.

16.4.10 MAC Enhancements

In addition to the features above, the IEEE 802.11 TGs Amendment describes a number of optional enhancements of the CSMA/CA MAC: Congestion Control, Mesh Deterministic Access and Power Save mode.

The first feature attempts to reduce the incidence of congestion on mesh links by providing feedback to the sources feeding that link.

The second feature allows mesh nodes to "plan" their transmission events in time so as to avoid collisions for medium access as much as possible. This scheme requires a degree of synchronization between the participating nodes. In past work the 802.11 Working Group has found that a QoS approach based on synchronization of access points is less favorable because, in unlicensed spectrum, which is potentially an interference rich environment, synchronization is difficult or impossible to achieve and maintain.

The third feature is power saving or sleep mode. This requires synchronization of the nodes of the mesh. In an Access Point based wireless LAN all stations are in is proving a major challenge. At the time of writing the details of these features were not decided yet.

16.5 Conclusion

Both commercial and institutional users are increasingly showing interest in mesh networks, attracted by the potential for reduction of operating cost and/or the flexibility offered by these networks.

The design and deployment of wireless mesh networks present many interesting challenges, notably when such a network is mobile or when it has to use shared spectrum. Some of the main technical problem areas are outlined above; treating them in detail would fill a whole book.

A lot of research has gone into routing and related aspects of mesh network design, and that has facilitated the work of the IEEE 802.11 TGs group to develop mesh services and protocol that make use of the 802.11 PHY and MAC. Much less effort has been put into understanding how fixed rate services like voice and video can be deployed over networks that are subject to complex and somewhat unpredictable in their behavior.

Careful dimensioning and applying controls that keep the mesh links from operating at or near their critical throughput level will go long way towards satisfactory performance of mesh networks.

16.6 References

[1] http://www.monarch.cs.rice.edu/monarch-papers/ndss03rev.pdf

[2] http://networks.rice.edu/papers/cc.pdf

[3] http://www.ieee802.org/11

Acknowledgement

The authors would like to thank Benny Bing, the editor of this book, for his valuable suggestions for drafting this chapter.

Disclaimer

THE INFORMATION HEREIN IS PROVIDED "AS IS," WITHOUT ANY WARRANTIES OR REPRESENTATIONS, EXPRESS, IMPLIED OR STATUTORY, INCLUDING WITHOUT LIMITATION, WARRANTIES OF NONINFRINGEMENT, MERCHANTABILITY OR FITNESS FOR A PARTICULAR PURPOSE.

17

WLAN Interworking with 2G/3G Systems

Srinivasan Balasubramanian[a]

The wireline systems have moved to supporting richer voice over packet switched services with the introduction of Skype and Vonage type services. This has completely displaced the traditional circuit-switched services. This trend to move away from circuit-switched services is catching on in the cellular networks are moving towards supporting fundamental services like voice over the packet domain using the IMS over packet networks. In addition, there is widespread adoption of the wireless LAN systems at home, office and commercial environments. Cellular operators are seeing WLAN not as competing technology, but as something that complements it by allowing the offloading of MSs to the WLAN systems to increase the cellular system capacity and potentially extending coverage. Integrated WiFi and 2G/3G cellular is seen as having the best of two worlds, capitalizing on the strengths offered by each technology. Operators are equipping cellphones with WLAN capability and it was only natural to ask the question if the same applications supported over the packet domain in the wireline and cellular networks can also be supported over the WLAN systems.

This chapter discusses the topics of making WLAN systems interwork with the 2G/3G systems, service continuity between the two systems addressing domain registration and call continuity with emphasis on voice. This chapter is written covering the solutions as it applies to 3GPP2. Most of these solutions also apply to 3GPP and exceptions will be highlighted. This chapter is also written from a mobile device perspective since the device plays a central role in handling mobility between the different WLAN and 2G/3G technologies. This interworking between the WLAN and 2G/3G systems is also commonly referred to as IWLAN.

The interface to the 3GPP and 3GPP2 network is defined in a generic manner such that the same mechanism can be used for other technologies like I-Bluetooth, I-WiMax, I-UWB, I-Proprietary, etc apart from the I-WLAN.

Currently there are over 25 different cell phones that are WiFi enabled in the market with 802.11 a/b/g compatibility. The major players thus far have been Nokia and Samsung. Amongst the operators, T-Mobile with their numerous hotspots already deployed in the field, has WiFi services enabled in their phones. The year 2007 will likely bring in more phone vendors supporting WiFi on their phones and potentially with 802.11n capability. There is an increased demand for WiFi enabled phones, with operators starting to show

---

[a] *Qualcomm, Inc*

genuine interest towards supporting services over WLAN, sometimes retaining simultaneous operation over the cellular and WiFi networks.

17.1 Introduction

WLAN as a wireless technology has entered a significant number of office and households environments. In a typical urban environment, just by driving through the streets for 3 to 4 miles, one is likely to encounter over 100 to 200 access points.

Voice over IP (VoIP) in the wireline telephony is gaining widespread acceptance of the technology. VoIP is actively being worked on and is becoming a reality for the cellular world with the 3GPP2 deploying VoIP networks with HRPD systems. There is also significant effort being put toward making VoIP happen over HSPA in the 3GPP network. The main benefit seen in moving toward the packet-switched domain from the circuit-switched domain is that it provides flexibility in the realtime delivery of packets to applications with different QoS and avoids starvation of other applications with fixed channel allocation (e.g., circuit-switched channels). VoIP is no longer seen as an emerging technology, particularly with the IP phones taking hold in the enterprises and IP-PBXs quickly replacing traditional PBXs.

The WLAN standard is evolving to support QoS and increased network capacity. The WLAN technology is also making inter-AP handoff more seamless with pre-authentication and fast handoff procedures. These functions make supporting VoIP over WLAN system easier and WLAN IP phone capability is a natural technology extension. There are also forecasts on the increased number of mobile devices carrying WLAN technology being sold in 2007. With all the technology enhancements in WLAN, enterprise-wide mobility becomes easier, allowing for affordable mobile connectivity to all employees. With IP-PBX features on the mobile phone, using the VPN technique allows for a single mobile to be used to access the intra-office phone extensions as well as residential calling and serving as a cellular phone over the 2G/3G networks. A single device used for accessing both WLAN and cellular networks – gaining the best of two worlds, capitalizing on the services and the capabilities of each network, co-existing, and interplaying the two systems by using the mobile device to co-ordinate the activities between the two networks. This chapter addresses:

- The establishment/maintenance/releasing of the secure tunnel between the WLAN and the 2G/3G networks.
- Mobility support between the WLAN and 2G/3G systems without the support of the Voice Call Continuity (VCC) feature. Single and dual public and private identities are considered in describing the mechanisms.
- Voice call establishment procedures over the WLAN domain.
- Voice call continuity between 2G/3G and WLAN systems supporting the transfer of active calls in both directions.
- Mobility support between the 2G/3G and WLAN systems when idle and with traffic – addressing which systems the mobile device is associated with, whether it is CS/IMS

registered at a given time, etc.

Figure 17.1 shows the topics and the flow of information addressed in this chapter.

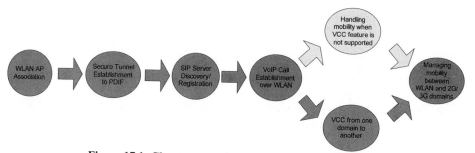

Figure 17.1: Chapter organization and the flow of topics.

17.2 Standards related activities

The standards activities can be broken into two main parts. The first part describes the procedures to establish the data pipes between the WLAN to the 2G/3G networks. The second part addresses the mobility of the idle and active services between the WLAN and 2G/3G domains.

This section identifies the relevant standards and provides the current status on the activities. Details providing association to 3GPP2, 3GPP and IETF related activities are identified.

17.2.1 3GPP2

The 3GPP2 standards can be downloaded from http://www.3gpp2.org/Public_html/specs/index.cfm.

WLAN Interworking

WLAN 3G Interworking; Version 1.0; Dated: TBD; 3GPP2 X.P0028-100 covers the details of:

- Scenario 1: Common billing and common customer care, which is restricted to the network implementation and does not impact the MS implementation directly.
- Scenario 2: Use of cellular authentication mechanisms over WLAN (AKA and TLS/PSK).

Access to Operator Service and Mobility for WLAN Interworking, Version 0.3, Dated: TBD; 3GPP2 X.P0028-200 covers the details of:

- Scenario 3: Mobile devices access to carrier's 3G packet data network via WLAN – essentially interaction with the PDIF node and

- Scenario 4: Addresses data session continuity – MIP and MOBIKE etc.

Voice Call Continuity

Voice Call Continuity between IMS and Circuit Switched Systems; Version 1.0; Dated TBD; 3GPP2 X.P0042 covers details of:

- This chapter defines an inter-technology handoff (HO) call model which supports procedures that allow a mobile subscriber to HO from an HRPD or WLAN VoIP multimedia session to a 1x circuit-switched voice session.
- It also specifies the interactions and signaling flows between a new functional entity node in the MMD network called the VCC Application Server (VCC AS) and the:
  - o Serving-Call/Session Control Function (S-CSCF)
  - o Home Location Register (HLR)
  - o High Rate Packet Data Access Terminal (HRPD AT)
  - o Wireless Local Area Network (WLAN) Station (STA)
  - o Media Gateway Control Function (MGCF)
- It is the goal of this specification to allow the core network to know as closely as possible the current accessibility of the MS and to deliver services efficiently across the appropriate access network while minimizing the impact on the legacy systems.
- The MSs in this specification include two kinds. One is HRPD/1x dual mode handset, and the other is WLAN/1x dual mode handset. HRPD/1x handsets are assumed to cannot be in simultaneous radio communications with both an HRPD radio access network and a 1x radio access network, While WLAN /1x handsets can be in simultaneous communication with both a WLAN access network and a 1x radio access network.

17.2.2 3GPP

WLAN Interworking

http://www.3gpp.org/ftp/Specs/html-info/23234.htm
3GPP TS 23.234: 3GPP system to Wireles Local Area Network (WLAN) interworking; System description
The current version is 7.4.0, published as of 2006-12-11.

http://www.3gpp.org/ftp/Specs/html-info/24234.htm
3GPP TS 24.234: 3GPP system to Wireless Local Area Network (WLAN) interworking; WLAN User Equipment (WLAN MS) to network protocols; Stage 3
The current version is 7.4.0, published as of 2006-12-15.

This chapter specifies system description for interworking between 3GPP systems and WLANs. This specification is not limited to WLAN technologies. It is also valid for other IP based Access Networks that support the same capabilities towards the interworking system as WLAN does. The intent of 3GPP-WLAN Interworking is to extend 3GPP services and functionality to the WLAN access environment. The 3GPP–WLAN

Interworking System provides bearer services allowing a 3GPP subscriber to use a WLAN to access 3GPP Packet Switched (PS) services.

This specification defines a 3GPP system architecture and procedures to do the following:

- Provide Accounting, Authentication and Authorization (AAA) services to the 3GPP-WLAN Interworking System based on subscription.
- Provide access to the locally connected IP network (e.g., the Internet) if allowed by subscription.
- Provide WLAN MSs with IP bearer capability to the operator's network and PS Services, if allowed by subscription.
- Provide WLAN MSs with IP bearer capability to access IMS Emergency calls for both UICC and UICC-less cases.

This specification defines two new procedures in the 3GPP System:

- WLAN Accounting, Authentication and Authorization, which provides for access to the WLAN and the locally connected IP network (e.g., the Internet) to be authenticated and authorized through the 3GPP System. Access to a locally connected IP network from the WLAN, is referred to as WLAN Direct IP Access.
- WLAN 3GPP IP Access, which allows WLAN MSs to establish connectivity with External IP networks, such as 3G operator networks, corporate Intranets or the Internet via the 3GPP system.

The Packet Data Gateway supports WLAN 3GPP IP Access to External IP networks. The WLAN includes WLAN access points and intermediate AAA elements. It may additionally include other devices such as routers. The WLAN User Equipment (WLAN MS) includes all equipment that is in possession of the end user, such as a computer, WLAN radio interface adapter etc. 3GPP also allows for the MS to be associated with multiple PDGs simultaneously based on the application needs and associated authentication credentials.

Voice Call Continuity

3GPP TS 23.206: Voice Call Continuity (VCC) between Circuit Switched (CS) and IP Multimedia Subsystem (IMS); Stage 2
http://www.3gpp.org/ftp/Specs/html-info/23206.htm
The current version is 7.1.0, published as of 2006-12-11. This stage 2 text covers the architecture details associated with the VCC feature.

3GPP TS 24.206: Voice call continuity between Circuit Switched (CS) and IP Multimedia Subsystem (IMS); Stage 3
http://www.3gpp.org/ftp/Specs/html-info/24206.htm
The current version is 7.0.0, published as of 2006-12-08. This stage 3 text covers the signaling details associated with the VCC feature.

The present chapter specifies the functional architecture and information flows of the Voice Call Continuity (VCC) feature which provides the capability to transfer the path of a voice call between a 3GPP CS system (GSM/UMTS) and IMS, and vice versa.

VCC is a home IMS application that provides capabilities to transfer voice calls between the CS domain and the IMS. VCC provides functions for voice call originations, voice call terminations and for Domain Transfers between the CS domain and the IMS.

The VCC application is implemented in the user's home network. Voice calls from and to a VCC MS are anchored at the VCC application in the home IMS to provide voice continuity for the user during transition between the CS domain and the IMS. VCC voice calls originated and terminated in the domain are anchored at the VCC application in home IMS using standard CS domain techniques available for redirecting calls at call establishment. A 3pcc (Third party call control) function is employed at the VCC application to facilitate inter domain mobility through the use of Domain Transfers between the CS domain and the IMS. Domain Transfers may be enabled in one direction (i.e., from the CS domain to the IMS or from the IMS to CS domain), or in both directions as per network configuration requirements. The VCC application has the capability to perform domain transfers for a VCC MS's voice session multiple times in both directions.

The capability of anchoring of Emergency calls and CS calls other than voice calls (e.g., CS Video, CS Data, CS Fax, SCUDIF calls...) is not specified in this chapter. Anchoring of voice calls originating from or terminating to a subscriber that does not have a VCC MS or an active VCC subscription is subject to operator policy. Calls that are not anchored are handled according to the normal CS call originating and terminating procedures. In 3GPP-CS (GSM/UMTS), calls are anchored by the MSC at VCC AS via CAMEL triggers. In 3GPP2-CS (1X) calls are anchored by the MSC at VCC AS via WIN triggers. 3GPP2 standard also allows for other non-WIN trigger options.

3GPP2 has solution for both single and dual radio, i.e., it allows for scenarios that make-before-break (dual-radio) and break-before-make (single-radio). 3GPP currently in Rel-7 there is support only for single radio (3GPP Rel-8 plans to work on dual radios).

In 3GPP2, during VCC handoffs, MSC can route the call to a local MGCF/MGW allowing for the bearer to in the local network without having to route to the home network always after VCC handoffs. Such an optimization is not currently available in 3GPP.

For both 3GPP and 3GPP2 supplementary services continuity is not addressed currently (e.g., Call Waiting, Call Transfer, 3 Way Calling, Call Hold) and is being actively worked on for the next release.

The standards bodies are also working towards making the mobility support between two packet networks like WLAN and HRPD systems a reality. Mobile-IP is being considered as the mechanism to move the call over to the new domain, but this does come with the impacts of triangulation problem and the delays involved in delivering the VoIP packets. One possible approach is to use Proxy-MIP in handling the mobility in the network itself without MS treating the assigned IP address as a Simple-IP address. If a network based Proxy-MIP is chosen then it will have to transition across domains which may not be always viable. A combination of MIP and Proxy-MIP will likely play a role in supporting mobility across packet domains.

The support for emergency calls over WLAN is currently being discussed in both the 3GPP and 3GPP2 standards.

17.3 WLAN Interworking Plumbing

Figure 17.2 shows the interacting blocks in making the WLAN systems interact with the 2G/3G networks. The newer nodes added to enable the interworking with the 2G/3G networks are the Packet Data Interworking Function (PDIF) and the DNS. All other components are in existence in the 2G/3G network or the WLAN network. The PDIF node provides the bridging between the external Internet and the 2G/3G Intranet, ensuring adequate security in exchanging traffic with the 2G/3G network. The DNS node deployed in the external Internet for the MS associated with the WLAN system to identify and locate the PDIF node. Typically a single PDIF node will support both IPv4 and IPv6 traffic.

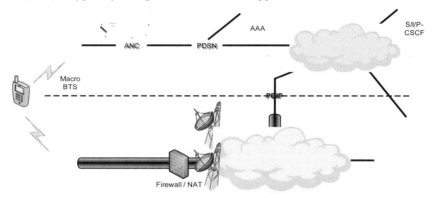

Figure 17.2: Architecture Diagram.

The interworking functions implemented at the MS handle NAT boxes deployed between the WLAN AP and the PDIF node. Any firewalls/NAT deployed beyond the PDIF node in the 2G/3G network needs to be explicitly handled by the network. When Mobile-IP is used, the PDIF node also acts as the foreign agent.

When IMS services are deployed, the nodes are deployed within the 2G/3G operator's network. There are a lot of nodes involved in deploying the IMS network. What is shown in the figure is the major node implementing the SIP service functionality – the Call Session Control Function (CSCF). The CSCF is broken into multiple pieces, the Proxy-CSCF (P-CSCF) is in the MS's visited network, the Interrogating-CSCF (I-CSCF) which is used by the P-CSCF to location the Serving-CSCF (S-CSCF) which is the home CSCF for the MS. The S-CSCF has the user's subscription information and the P-CSCF enforces the visited network's policies.

In the 3GPP networks the PDIF box is broken into Packet Data Gateway (PDG) and Wireless Access Gateway (WAG).

Figure 17.3 provides a very simplified architecture of the WLAN interworking with the 3GPP system. The shaded area refers to WLAN 3GPP IP Access functionality. The WAG is used to provide:

- Connectivity to the H-AAA server in the 3GPP network, used by the WLAN AP when performing WPA/WPA2 authentication.

- Cooperates with PDG to provide the required parameters like the DNS, DHCP address from the destination network to the MS.
- Identify the appropriate PDG to establish the IPSec tunnel with.
- Act as a router in sending the packets to the PDG and potentially act as a NAT/firewall for entry into the 3GPP network.
- Accounting information, particularly under roaming scenarios.

WAG is an optional node

- When WLAN authentication and the IKEv2 procedures to talk to PDG are not inter-connected. i.e., the MS does not use the pseudonym obtained from the EAP/AKA procedures over the WLAN to perform fast re-authentication with the PDG.
- When PDG is accessible from the public Internet.

The MS establishes the IPSec tunnel with the PDG to obtain packet data connectivity into the 3GPP network.

3GPP additionally also allows for NAT boxes to be deployed between the PDG and the P-CSCF with the MS requiring to handle the NAT traversal. The MS detects the presence of NAT boxes and uses IPSec tunnel mode when present and IPSec transport mode when there are no NAT boxes between the two nodes. In the 3GPP2, when NAT is deployed between the PDIF and the P-CSCF it is expected for the network to deploy ALG (Application Layer Gateway) to support IMS/SIP signaling for proper call routing.

Figure 17.4 shows the steps involved in establishing the interworking functions between the WLAN and 2G/3G network. A brief description of the overview steps involved in WLAN 3G interactions are provided below, followed by the detailed description.

1. *WLAN Association*: The MS finds a valid WLAN AP and decides to associate with the AP, having performed link level authentication and authorization. The WLAN AP may check with the H-AAA in executing the authentication and authorization procedures. The MS also determines other AP's qualities like QoS capabilities before it decides to associate with that AP.

Figure 17.3: WLAN Interworking architecture for 3GPP.

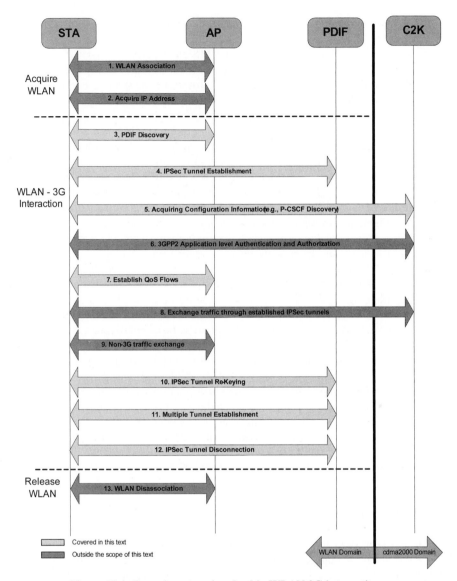

Figure 17.4: Overview steps involved in WLAN 3G interactions.

2. *Acquire IP Address*: MS acquires an IP address to use over the WLAN system before initiating access to cdma2000 packet data services.
3. *PDIF Discovery*: The PDIF is located in the cdma2000 network, either in the visited or the home network. The MS determines the address of the DNS server and performs a DSN query to obtain the address of the PDIF.
4. *IPSec Tunnel Establishment*: The IPSec tunnel is established between the MS and the PDIF using the IKEv2 procedures. IKEv2 initialization procedures are used to

discover the existence of NAT between the MS and the PDIF and the NAT traversed address of the MS.

5. *Acquiring Configuration Information*: The MS shall perform a DHCP query via the secure tunnel established with the PDIF to discover configuration parameters such as the address of the P-CSCF, DNS, etc. When provisioned with the address of the DHCP server, the MS shall perform a uni-cast DHCP query to obtain the address of the P-CSCF.

6. *3GPP2 Application level Authentication and Authorization*: The authentication and authorization to access cdma2000 application services are performed when the MS attempts to access the cdma2000 network e.g., IMS applications.

7. *Establish QoS Flows for cdma2000 applications*: Only after the successful establishment of the IPSec tunnel, the application is allowed to be activated and supported over the WLAN. When a given application is activated, it acquires the required radio QoS to support the QoS guarantees of the application.

8. *Exchange traffic through established IPSec tunnels*: The applications are free to exchange traffic via the appropriate security association.

9. *Non-3G traffic exchange*: Apart from supporting traffic via the IPSec tunnel for 3G applications, the MS also supports other applications that access the public Internet directly without using the PDIF node.

10. *IPSec Tunnel Re-Keying*: Initiated by the MS or network, an established security association will have a lifetime and will need to be re-keyed to retain the association.

11. *Multiple Tunnel Establishment*: When a given application requires a different level of encryption, different from the first Child SA created as part of the IKE AUTH exchange, the application requests to establish Child security associations as needed.

12. *IPSec Tunnel Disconnection*: The MS will have established the IKE SA and zero more Child SAs. The number of Child SAs created depends on the application needs of QoS and encryption support needed. The security association can be released, initiated either by the MS or the network once they are no longer required or the authorization has expired.

13. *WLAN Disassociation*: The MS disassociates itself explicitly as it transitions to the 3G network or just goes out of WLAN coverage.

17.3.1 WLAN Association

WLAN is not seen as a technology that provides ubiquitous coverage. With numerous WLAN AP that one encounters driving from one location to another and with even the basic security disabled in a lot of these APs, with a few minutes stop at a traffic it is possible to associate with an AP, download your email send a quick note to someone and have your engine revved up to take you to the next traffic light 300 yards away. With the each AP potentially associated with a different Internet service provider and the inter-AP HO taking a long time[1] to support any real-time services, HO from one AP to another maintaining continuity of service impossible to manage. With the coverage of the WLAN

---

[1] The fast handoff and pre-authentication feature has been added to the 802.11r, which have not yet been introduced into the market as yet. Also the assumption is that when the SSID changes there will likely be a change in the IP address assigned to the MS and the services cannot be moved over to another AP without disrupting the application.

AP being very small, it is also not possible provide effective rate adaptation to even support best-effort services under mobility scenarios.

Although WLAN APs are suitable for best effort data applications, WLANs are not suitable for QoS applications like VoIP. Such applications require extensive negotiation with the network prior to establishing and supporting the service over the WLAN domain. This implies care must be taken in selecting and associating with an AP in order to guarantee a stable WLAN environment prior to moving the services over to the WLAN domain. The following are seen as some typical requirements that need to be satisfied in association with an AP.

- The MS shall use the minimum time needed to acquire a WLAN system reliably.
- The MS shall not switch between registering in a WLAN based IMS network and a circuit-switched network no more than once every N (typical value of 60 secs) seconds.
- The MS after acquiring the WLAN system shall switch to another system, including other WLAN systems, when the RF performance measured based on current RSSI levels and/or packet erasure rates etc., drops below a configurable threshold.

The intent is to trade off these three requirements in making a decision for system selection.

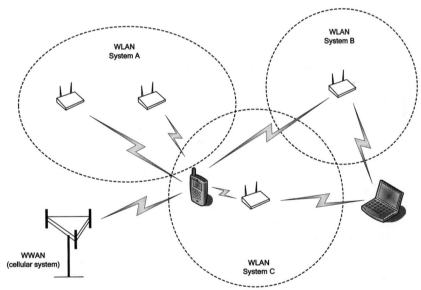

Figure 17.5: Typical RF environment impacting WLAN system selection.

17.3.1.1 Scanning Process

The scanning process can be initiated through a user-initiated event or automatically initiated by the system selection procedures of the MS. When a manual scan is initiated, the WLAN module scans for the available AP and presents the results to the user. The scan will be a passive scan of all the channels. If the MS knows the region of operation and regulatory restrictions allow for active scans to be supported, the MS uses active scan to determine the available APs instead of only using the passive scan. Passive scans are used to identify the list of available APs. Active scans are required to discover APs that do not include the SSID in the Beacon broadcast messages.

The following sections describe the algorithm that is run for manual/automatic scan procedures used in determining the AP for the MS to associate with.

17.3.1.2 Manual Scan Procedures

When the user initiates a manual scan, the MS scans for all available WLAN systems. Since this will not include APs do not transmit their SSIDs in the beacon, the MS also looks for APs identified in the preferred list with expect probe request/responses. The combined set of SSIDs found in the neighborhood along with the strengths is presented to the user.

17.3.1.3 Automatic Scan Procedures

The MS, based on system selection algorithms and triggers decides to power up the WLAN module and initiate an automatic scan of the available WLAN systems. The MS maintains a set of preferred APs. The system selection procedures with the intent to determine the best system available is triggered and the steps defined in the following section are executed. The best AP amongst the detected APs is selected by the MS to associate itself with.

Figure 17.6 describes the high level flow chart involved in making a WLAN system selection. Several iterations of the scan are executed and the measured RSSI levels are used to determining the CandidateSet. Different types of scans are used and the filtered RSSI levels are used to determine the ActiveSet for the MS.

The MS is provisioned with the preferred set of SSIDs that it prefers to associate it. The APs acquired as part of the automatic scan procedures will be restricted to the entries identified in this preferred list.

Each of the steps involved are detailed further in subsequent sections. These sections describe the algorithms to employ in making a system selection, in particular determining the suitability of supporting VoIP over the associated WLAN system.

17.3.1.4 Access Point Sets Definition

Access Point Sets provide a way to collectively refer to the APs that a given MS is seeing at a given point.

- *Preferred List*: The set of SSIDs that is provisioned by the operator and updated by the user identifying the preferred system that the MS automatically associates itself with when within the coverage area.
- *Scan Set*: The set of entries chosen for the current scan event. This is typically a subset of the Preferred Set, determined based on the current location of the MS using GPS or based on the location within the cellular network.

- *Detected Set*: All APs the MS is able to sense energy from, restricted to the entries in the WLAN preferred AP list.
- *Candidate Set*: A subset of the Detected Set that meets criteria, e.g., RSSI above the RSSIThreshold and is a member of the Scan Set, for being a potential active set member.
- *Active Set*: A member of the Candidate Set that the MS considers suitable for association.
- *Remaining Set*: A subset to the Detected Set APs that are not Candidate Set APs.

17.3.1.5 Iterations in making WLAN system selection

The MS is configured to go through a certain number of iterations before the WLAN system is considered stable for acquisition. The parameter regulating the number of iterations is defined below. This is the total number of scan iterations that are executed, after which the WLAN module either has a valid active set or gives up on the scan event.

17.3.1.6 Scan Types

The type of scan employed depends on the deployment scenarios. The likely choice will be to use a different scan type for each iteration in making a WLAN system selection. The type of scan to perform is identified for each iteration.

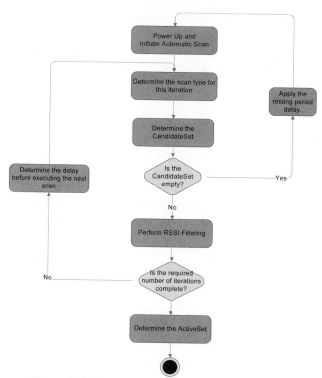

Figure 17.6: Automatic Scan - high level flow chart.

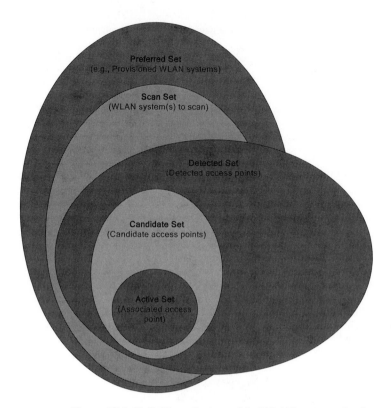

Figure 17.7: Definition of sets used in WLAN system selection.

Based on regulatory restrictions in different regions of the world, the MS is not free to transmit and perform an active scan without prior knowledge of the region the MS is operating in.

Listed below are mechanisms that the MS uses in determining if an active scan can be employed in acquiring WLAN AP information.

- The MS is provisioned by the operator with a parameter that identifies if the device is allowed to perform active scans or not. When the device is not associated with an AP that is 802.11d capable, then this parameter alone dictates if active scan can be employed.
- When the MS is associated with an 802.11d capable network, the AP announces the regions and the regulatory restrictions. The MS uses active scan only when not banned by the regulatory restrictions of the region and the operator's provisioned parameter allows for active scan.
- Maintain a table of country/region codes and the regulatory restrictions to use during roam. The MS uses active scans only in regions where active scans are allowed. The region of operation is determined by the association with the 3G network.
- When the MS is not able to determine the region and determine regulatory restrictions through the mechanisms described above, active scan is not enabled.

A potential list (but not exhaustive) list of possible scan types is provided below.

- Passive scan of non-overlapping channels;
- Passive scan of odd numbered channels;
- Passive scan of even numbered channels;
- Active scan of odd numbered channels;
- Active scan of even numbered channels;
- Active scan of all channels, restricted to the entries a subset of the entries in the preferred list.
- Active scan of non-overlapping channels, restricted to the entries a subset of the entries in the preferred list;
- A selected subset of the current CandidateSet, likely the top few entries.

The results will be saved for each scan iteration. Once the results are gathered for the specified number of iterations, processing will occur in order to determine the candidate/active set.

17.3.1.7 Candidate Set Selection

The MS retains the APs ordered based on the filtered RSSI values. Only APs identified by the SSIDs in the WLAN Preferred AP list are considered and all other APs ignored in this selection process. There are other criteria like requiring the RSSI to be above a certain minimum threshold to be retained part of the Candidate Set may also be employed.

17.3.1.8 RSSI Filtering

There are several ways to implement the RSSI filtering. Any chosen filter needs to prevent premature entry/exit to/from the AP and should trigger a timely exit both when idle and when in traffic from the WLAN AP. Entry into the AP can typically be delayed to determine reasonable stability as 2G/3G systems are expected to have ubiquitous coverage.

The RSSI measured is independent of the active or the passive scans performed. The measured RSSI values may fluctuate quite a bit and a suitable filtering mechanism is required to properly predict the suitability of a given AP. The RSSI is measured by measuring the strength of the beacons transmitted from the APs.

Figure 17.8 shows a plot of the RSSI and data rate experienced by the MS as the person walks away from the access point. The RSSI as seen from the card (can be converted to received power in dBm by subtracting with -95. e.g., an RSSI of 40 is −55 dBm). Some of the potential approaches are detailed below.

- Maintain a running average of the measured RSSI values:

$$RSSI_{Filtered} = (RSSI_1 + \ldots + RSSI_i)/i$$

where i is the current number of iterations that AP has remained in the CandidateSet since the last time it entered the CandidateSet.
- Discrete-time-filter:

$$\text{RSSI}_{\text{Filtered}} = \alpha_1 \text{RSSI}_{i-x} + \ldots + \alpha_x \text{RSSI}_{i-1}[1 - (\alpha_1 + \ldots + \alpha_x)]\text{RSSI}_i$$

where $\alpha_1, \ldots, \alpha_x$ are weights associated with the previous RSSI measurements; RSSI_0 and RSSI_{-1} are set to RSSI_1 upon the first measurement.

- Employs a siding window based exponential averaging mechanism to filter the measured RSSI values.

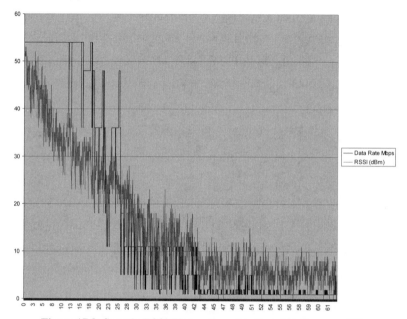

Figure 17.8: Sample RSSI Plot as a user walks away from an AP.

17.3.1.9 Time delay for subsequent scan event

A certain delay is required between two measurements to ensure that all the measurements are not done when a peak or a fast fade occurs. This is the time period the MS waits between two scan events. Some of the possible approaches include using a fixed interval or varying it using step sizes, being more aggressive when a very strong AP is found.

17.3.1.10 Active Set Selection

The APs are ordered based on the RSSI values. A fast rising Beacon from an AP may enter the CandidateSet and be the top entry without enough hysterisis. Additional quantities are considered in determining the ActiveSet. The thresholds for accepting a AP into the ActiveSet depends on choices made for the CandidateSet selection and the RSSI filtering process. Define $N_{\text{MinScansForActiveSet}}$ as the minimum number of scans for a AP to be considered for inclusion in the ActiveSet. Some of the possible approaches are defined below.

- Use a single absolute $RSSI_{ActiveSetThreshold}$ threshold and only allow APs that have a $RSSI_{Filtered}$ above this threshold to be part of the ActiveSet.

  * The AP with the strongest $RSSI_{Filtered}$ is a consideration for the MS to associate itself with. Or
  * The AP that has remained in the CandidateSet the longest and amongst those the AP with the strongest $RSSI_{Filtered}$ is a consideration for the MS associate itself with. Or
  * Some weighting between the two is used in determining the best ActiveSet AP - the number iterations the AP has remained in the CandidateSet and the $RSSI_{Filtered}$ value.

- Ramping down the required threshold based on the number of iterations the AP has remained in the CandidateSet.

  * $RSSI_{InitialThreshold}$: The RSSI level for an AP to be included in the Active Set based on the initial scan event.
  * $RSSI_{StepDownSize}$: The RSSI step down size that is used to reduce the current RSSI Threshold to use with each additional scan event performed in determining the Active Set subject to a lower bound of $RSSI_{MinimumThreshold}$.
  * $RSSI_{MinimumThreshold}$: The minimum RSSI threshold that is to be used in determining the members of the Active Set.

  Typically $RSSI_{IncludeThreshold} \leq RSSI_{MinimumThreshold} \leq RSSI_{InitialThreshold}$. Similar choices as in the previous mechanism need to be used in determining the best ActiveSet AP.

Parameter	Typical Values used for the filtering algorithm
Number of scan iterations	At least 10 iterations
Delay between two iterations	At least 5 seconds
$RSSI_{ActiveSetThreshold}$	At least -85 dBm

17.3.2 WLAN De-Selection

Much as it is important to ensure that an AP is capable of supporting the application traffic needs when associating with it, it is equally, if not more important, to make a determination of when to disassociate certain service and potentially the MS from the AP and the WLAN system as a whole. The MS when associated with an AP and supporting mobile-terminated services like VoIP, the MS needs to monitor the TIM bits when in power save mode or provide a transmit opportunity to the AP to send any queued up data when the EDCA/HCCA modes of 802.11e are enabled. This allows the AP to send the incoming page (e.g., SIP Invite message) to the MS, triggering the MS to go into active traffic in supporting a VoIP call.

The MS periodically monitors the beacons to measure the RSSI filters the values and determines if the currently associated AP is still suitable for supporting the best-effort and QoS applications. It also constantly monitors other APs with the same System ID and moves to a new AP when the current AP starts to under perform and the RSSI of the new AP with the same System ID goes above the active set threshold. When the current AP starts to under-perform and there are no APs with the same System ID to move to, the MS de-selects the WLAN system and moves to associate itself with the available 2G/3G system.

17.3.2.1 In Traffic Operation

"In traffic operation" refers to the duration of time when we are connected to a WLAN system. The assumption here is that we are operating within an ESS, and our goal is to remain in the ESS. The MS issues a periodic background scans and keeps track of the APs with the same SSID. In moving with APs, the MS uses the QoS compatibility, the past history, the measured/filtered RSSI value, and security compatibility in making a decision. To maintain the call across Inter-AP HOs, the MS needs to be able to use the same IP address. There are several ways to deal with this:

- The network guarantees that the Simple-IP address assignment remains the same across the APs with the same SSID.
- The MS uses MIP and the MS executes the procedures to point the MIP to the new AP.
- The network supports Proxy-MIP. The MS is assigned an IP address that looks like the Simple-IP address. When the MS moves to the new AP, the network handles the re-pointing of the IP address to the new AP using procedures like Network Localized Mobility Management (NetLMM). See [16].

The MS employs a similar mechanism to the one used for Idle mode operation, but with the measurements executed with increased frequency. Apart from measuring the RSSI of the current AP and the neighboring APs with the same System ID, the MS also monitors the packet performance of the both the downlink and uplink. The packet performance is measured at the MAC level and at the application level. When the performance of RSSI or the packet error rate cross a threshold, the active call is handed over to the 2G/3G system.

When active in a voice call over the 2G/3G system and when the MS enters the coverage of an AP, the MS executes the Idle Mode Operations in acquiring the AP. Once associated with the AP, the MS moves the active call to the WLAN system from the 2G/3G domain. Please see the Voice Call Continuity section for the actual procedure of moving the active call between WLAN and the 2G/3G systems.

17.3.2.2. IP Address Assignment

Once the MS has associated itself with the AP, it proceeds to get an IP Address assignment. The procedure depends on whether the MS is to be assigned an IPv4 or IPv6 address. For IPv4 assignment, the MS executes DHCPv4 procedures. For IPv6 assignment, the MS executed DHCPv6 procedures to get a Prefix assignment and uses privacy extension mechanisms to generate the Interface-Identifier (IID). The MS then executes DAD procedures to verify that the IPv6 address is uniquely assigned to this MS and does not clash with an assignment to another entity. For IPv6 the MS based on application needs, may generate multiple IIDs.

17.3.3 PDIF Discovery Mechanisms

This section describes the mechanism by which the MS identifies the IP address of the PDIF in associating with the 3G packet network via WLAN. The MS initiates this procedure after associating with the AP and deciding to move one or more 2G/3G services to the WLAN domain. Typically a single public identity (e.g., phone number) and private identity is used. Note that the IMS domain can be supported over only one of the packet domains when packet domains are available over both the WLAN and the 2G/3G domains. The MS is statically provisioned with the PDIF's IP address or MS is statically provisioned with the FQDN for the PDIF. Note for 3GPP, MS constructs an FQDN using the W-APN (WLAN Access Point Name) Network Identifier and VPLMN ID (Visited Public Land Mobile Network ID). For 3GPP2 the FQDN is provisioned. The MS may be provisioned with multiple PDIF IP Addresses or FQDN values and the MS chooses amongst them in associating with a PDIF. It is required that at least one of the PDIF addresses be routable to the WLAN AP the MS is associated with.

If the MS is provisioned with the IP address of the PDIF, it shall use that to establish communication with the 3G packet network.

If the MS is provisioned with the FQDN of the PDIF, then the MS executes the following steps. The MS determines the IP address of the DNS. The possible mechanisms of obtaining the IP address of the DNS server is by either static provisioning or use of DHCP over WLAN to obtain the IPv4 address of the DNS server. The MS then perform a DNS query with the FQDN of the PDIF. The DNS server provides the MS with the IP address of the PDIF. If more than one address is provided by the DNS server then the MS automatically selects an IP address amongst the ones provided, randomly. The MS could potentially execute "ping" procedures with the different PDIF nodes and select the nearest PDIF node to associate itself with. The nearest PDIF node will be the node that has the shortest round trip time, this value is selected based on multiple ping attempt. The ping procedure also verifies reachability before executing the IKE procedures, which is a slower way to detect accessibility.

17.3.4 Tunnel establishment procedures

This section describes the procedures involved in establishing the IPSec tunnel with the PDIF. This procedure is executed independently for IPv4 and IPv6, with each establishing it own IKE SA and IPSec tunnels. This procedure assumes that the MS has associated with a WLAN AP, has a valid IP address assignment and has acquired the address of the PDIF address. This procedure also assumes that the V-AAA shall act as a stateful proxy between the PDIF and the H-AAA. The establishment of the IPSec tunnels is initiated and the following procedure is executed.

With cellular networks, the network is typically a trusted entity and the MS does not need to authenticate the network it is associating with. Procedures like CHAP and PAP can be used by the 3GPP2/3GPP networks to authenticate the MS before allowing access to the MS. With the MS accessing the PDIF typically via the public Internet and going through one or more operators, a mutual authentication method needs to be used with both the MS and the PDIF verifying each other.

The standards support two mutual authentication schemes – IKEv2/EAP/AKA and

IKEv2/EAP/TLS. The more commonly used IKEv2/EAP/AKA procedures are described below. The RFC 4306 covers the IKEv2 procedures and the RFC 4718 describes the common practices used in implementing IKEv2. The information covered in this section is not intended to be completely representation, but more as a guideline/pointers in referring to the specifics in the RFCs.

For 3GPP, the WLAN AP authentication using WPA and WPA2 has been integrated with the establishment of the IPSec tunnel with the PDIF. The credentials in the H-AAA are used to authenticate the MS when it associates itself with the AP. The IKEv2 procedures use the established credentials from the AP authentication process. This does imply for the 3GPP the WLAN operator and the cellular operator need to have close ties and the MS cannot associate with any AP that the user chooses. Also the AP will have to be WPA/WPA2 capable for the interworking with the 3G domain to work. The pseudonym acquired during the WPA/WPA2 authentication process is used for the IKEv2 procedures. The IKEv2 instead of executing the full authentication only goes through fast re-authentication in establishing the IPSec tunnel with the PDG. 3GPP also supports independent authentication of the WLAN AP that includes WEP, WPA, and WPA2 authentication methods. When the MS does not receive the required credentials as part of the authentication procedures with the AP, it executes the regular IKEv2 authentication procedures in establishing the IPSec tunnel with the PDG.

3GPP, apart from the EAP-AKA and EAP-TLS also supports EAP-SIM and PAP and CHAP based authentication methods. 3GPP mandates that the CERT and AUTH payload included in Step 7 below, which allows for the MS to authenticate the PDIF/PDG it is associating itself with. This aspect is required as per RFC 4306 and is used to prevent the man in the middle attack. The MSs need to be provisioned with the required information for authenticating the CERT parameters received from the PDG/PDIF. It is likely 3GPP2 deployments also support this mechanism, although not directly mandated by the standards.

Described below is the procedure executed in the 3GPP2 domain.

Figure 17.9 shows the steps involved in establishing an IPSec tunnel with the PDIF node. There are two phases in this authentication process. Phase 1 consists of steps 1 to 11 in the call flow and Phase 2 consists of steps 12 and 13.

In phase 1 the IKE_INIT procedures are used to exchange the DH parameters to use. This DH parameter information is used to generate the key and encrypt the payloads exchanged in the first phase of the authentication procedure. This will encounter the man-in-the-middle attack, but should not be an issue as the tunnel establishment is completed only when phase 2 also completes. At the end of phase 1 the MSK received from the H-AAA is sent to the MS. Only a valid PDIF node can get the MSK from the H-AAA. The encrypted 64-byte MSK is sent to the MS.

In phase 2, the MS/PDIF uses this MSK to generate the payload for the mutual authentication as part of the second phase of the authentication. Both entities authenticate each other as part of this process. This step authenticates the IKE_INIT_REQUEST sent by the MS and at the end of Step 13 the IPSec tunnel is considered to be established.

The following steps provide the description for Figure 17.9:

- Step 1 – Determining that the application needs to move its 3G services over the WLAN system, requests to establish the IPSec tunnel with the PDIF.
- Step 2 – The MS initiates the IKEv2 exchange with the PDIF. The first set of messages is the IKE_SA_INIT exchange.

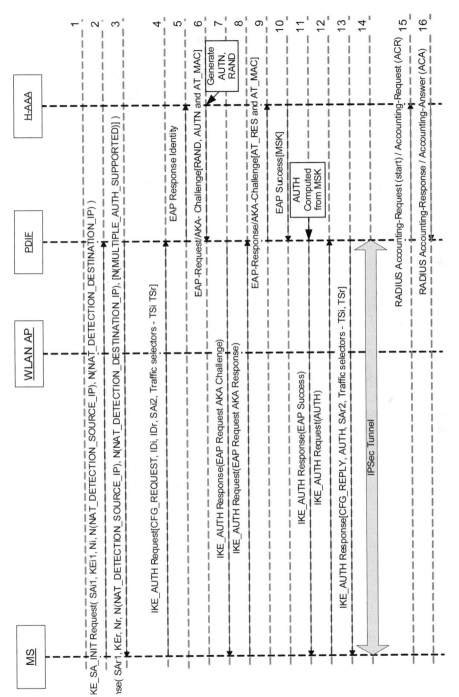

Figure 17.9: IPSec tunnel establishment procedures.

- The MS proposes all the supported ESP proposals as part of SA payload (SAi1). The security association specifies the encryption and pseudo random function to be used for the IKE procedure itself. The MS shall use Diffie-Hellman key exchange for negotiating keys for IKE SA.

  The MS detects the presence of NAT boxes between the WLAN AP and the PDIF node as part of the IKEv2 initialization procedures. Both the initiator and responder include   NAT_DETECTION_SOURCE_IP/Port   and NAT_DETECTION_DESTINATION_IP payloads that contains both the IP address and Port number information in the first round of messages and support the negotiation of UDP encapsulation. This procedure allows the MS to detect the presence of NAT boxes and adapt the packet transmission to the PDIF node based on the type of NAT between the AP and PDIF. Note that there may be one or more NAT boxes and what is of interest is the source and destination IP address/port used by the MS and source and destination IP address/port received by the PDIF for packets transmitted from the MS to the PDIF and similarly for the packets transmitted from the PDIF.

  NAT Traversal is required to handle address and port number translation while retaining the same level of security. Please refer to the appendix for details on the different types of NATs.

- Step 3 – The PDIF selects one proposal among the suggested proposals in SAi1 and responds back with that proposal as part of (SAr1). The PDIF shall send its Diffie-Hellman parameters as part of key exchange payload. The PDIF supports NAT Traversal as outlined in Section 2.23 of [8] and includes NAT_DETECTION_SOURCE_IP and NAT_DETECTION_DESTINATION_IP payloads in the response message.

- Step 4 – The MS initiates the IKE_AUTH exchange with the PDIF. These messages are encrypted and integrity protected with the keys established through the IKE_SA_INIT exchange. The MS omits the AUTH payload in order to indicate to the PDIF that it wants to use EAP over IKEv2. The MS includes its identity in the IDi payload of the IKE_AUTH request. The MS requests a Tunnel Inner IP Address (TIA), by setting the CONFIGURATION payload (CFG_REQ attribute value set to INTERNAL_IP4_ADDRESS or INTERNAL_IP6_ADDRESS depending on the IP version the MS supports) in the IKE_AUTH request. The MS includes its NAI in the IDi payload. When the MS wants to use EAP, it does not include the AUTH payload in the IKE_AUTH message. MS shall use source IP address as the identifier in IDi payload. As part of SAi2, MS proposes all supported ESP proposals that it supports for the packets carried in the IPSec tunnel. MS shall use global traffic selector in TSi and TSr payload. The MS also requests a DHCP or DHCPv6 server's IP address by including the INTERNAL_IP4_DHCP or INTERNAL_IP6_DHCP attribute in the CP(CFG_REQUEST) of the IKE_AUTH request. The traffic selectors identify the packet filters to use in routing the packets to the different IPSec tunnels. When all traffic is to flow through a given tunnel, the range is opened up to include all the IP address and port numbers.

- Step 5 – When the PDIF receives the IKE_AUTH request without the AUTH payload, it contacts the H-AAA to request service authorization and user authentication information by sending the EAP-Response/Identity message in the RADIUS Access-Request message, or Diameter-EAP-Request (DER) command. NOTE: Steps 4, 5, 6, and 7 can occur multiple times.
- Step 6 – EAP messages are exchanged between the MS and the H-AAA for mutual authentication. The H-AAA decides whether EAP-AKA authentication is suitable based on the user profile or on the indication from the MS on the preferred authentication mechanism. See [R5], Section 4.1.1.6. The PDIF generates a random value RAND and AUTN based on the shared WKEY and a sequence number. The H-AAA sends the EAP-Request/AKA Challenge to the PDIF, via the DIAMETER-EAP-Answer command or RADIUS Access-Challenge. The EAP-Request/AKA Challenge contains the AT_RAND, AT_AUTN and the AT_MAC attribute to protect the integrity of the EAP message.
- Step 7 – The PDIF sends an IKE_AUTH Response to the MS that contains the EAP-Request/AKA-Challenge message received from the H-AAA.
- Step 8 – The MS responds with the IKE_AUTH request message including the EAP-response message. The MS verifies the authentication parameters in the EAP-Request/AKA-Challenge message and if the verification is successful, it responds to the challenge with an IKE_AUTH Request message to the PDIF. The main payload of this message is the EAP-Response/AKA-Challenge message.
- Step 9 – The PDIF forwards the EAP-Response/AKA-Challenge message to the H-AAA via either RADIUS Access-Request message or Diameter DER command.
- Step 10 – In case of successful authentication the H-AAA sends a RADIUS Access-Accept message with EAP-Message attribute containing EAP Success when the RADIUS protocol is used or a Diameter-EAP-Answer command with a Result-Code AVP indicating success when Diameter is used. The H-AAA sends to the PDIF the EAP Success and the MSK generated during the EAP-AKA authentication process [R5]. In the case of RADIUS, as per EAP AKA [R5], the 64-byte MSK is split into two 32-byte parts with the first 32-bytes sent in the MS-MPPE-REC-KEY [R6] and the reaming 32-bytes sent in the MS-MPEE-SEND-KEY [R6]. Both of these attributes are needed to construct the 64-byte MSK at the PDIF. If any of those are missing, the PDIF rejects the session. In the case of Diameter, the entire MSK 64-byte key is transmitted in the EAP-Master-Session-Key AVP. Refer to RFC 4187 for a description of the EAP-AKA authentication procedures.
- Step 11 – Upon reception of a RADIUS Access-Accept message or a Diameter-EAP-Answer (DEA) Command with a Result-Code AVP indicating success, the PDIF sends an IKE_AUTH response message including the EAP Success.

- Step 12 – The MS sends the IKE_AUTH request message including the AUTH payload calculated from the MSK, which is generated upon successful EAP authentication. A description of how the AUTH payload is computed is provided. AUTH = PRF(PRF(Shared Secret,"Key Pad for IKEv2"), <message octets>), where the string "Key Pad for IKEv2" is 17 ASCII characters, and the Shared Secret is the MSK derived from the EAP authentication. The format of <message octets> is specified in section 2.15 of RFC 4306 for both initiator and responder. For the initiator, <message octets> is the IKE_SA_INIT Request message appended with Nr (responder's nonce) and PRF(SK_pi,IDi), where SK_pi is a key derived from the D-H exchange and IDi is the initiator's identity payload excluding the fixed header. For the responder, <message octets> is the IKE_SA_INIT Respond message appended with Ni (initiator's nonce) and PRF(SK_pr, IDr), where SK_pr is a key derived from the D-H exchange and IDr is the responder's identity payload excluding the fixed header.
- Step 13 – The PDIF replies with the IKE_AUTH response message including an assigned TIA, AUTH payload, SAs, etc. The PDIF uses the MSK to compute the AUTH payload. The MS calculates the MSK according to [R5] and uses it as an input to generate the AUTH payload to authenticate the first IKE_SA_INIT message. The MS sends to the PDIF the AUTH payload in an IKE_AUTH Request message. The PDIF obtains the MSK from H-AAA in Step 8.

  - When providing an IPv4 TIA, the PDIF provide the complete address to the MS.
  - When providing an IPv6 TIA, the PDIF provides the prefix and the IID generated by the MS. The PDIF assigns a unique prefix and the MS generates one or more IP addresses by appending a randomly generated IID.

  If the MS requests for a DHCP or DHCPv6 server's IP address via the IKE_AUTH request, the PDIF shall convey a DHCP or DHCPv6 server's IP address if available in the INTERNAL_IP4_DHCP or INTERNAL_IP6 DHCP attribute in the CP(CFG_REPLY) of the IKE_AUTH response.
- Step 14 – The PDIF uses the MSK to check the correctness of the AUTH payload received from the MS and calculates its own AUTH payload for the MS to verify. The PDIF sends the AUTH payload to the MS together with the configuration payload containing SAs and the rest of the IKEv2 parameters in the IKE_AUTH Response message, and the IKEv2 negotiation terminates. When the IKE_AUTH exchange completes, an IPsec tunnel is established between the MS and the PDIF. The MS and PDIF, also support UDP encapsulation of IPsec. AH is used to authenticate the packets exchanged over the established IPSec tunnel and ESP is used encrypt the packets when a NAT is detected as part of the IKEv2 Initialization procedures.
- Step 15 – If RADIUS is used, the PDIF sends RADIUS Accounting Request (Start) to the H-AAA; if Diameter is used, the PDIF sends ACR (Start) to the H-AAA.
- Step 16 – The RADIUS or Diameter H-AAA responds back to the PDIF with RADIUS Accounting Response or Accounting-Answer (ACA) respectively.

When Mobile-IP is used, the PDIF node provides the Agent advertisement as part of the IKE message exchange. The MS subsequently uses this to perform MIP registration.

17.3.4.1 Error Scenario 1: IPSec Tunnel Establishment Verification Failure

There are two levels of authentication that is enabled, one at the AP level and one at the PDIF level. When the user is not authenticated to access the 3G network, the IPSec tunnel establishment procedure fails blocking the user from enabling 3G services over the WLAN system. This section describes the scenario when the IPSec tunnel establishment encounters a verification failure.

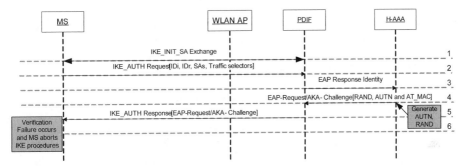

Figure 17.10: IPSec Tunnel Establishment Verification Failure.

- Steps 1-5 follow the description per the IKEv2 tunnel establishment procedures described before.
- If the verification is not successful, the MS aborts the IKEv2 exchange.

The MS subsequently abandons the IPSec tunnel establishment procedures. The IPSec tunnel establishment is attempted again when an application requests it.

17.3.4.2 Error Scenario 2: IPSec Tunnel Establishment H-AAA authentication failure

This section describes the alternative when the IPSec tunnel establishment fails authentication with the H-AAA.

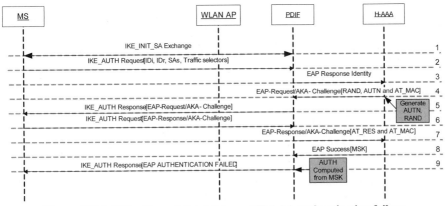

Figure 17.11: IPSec Tunnel Establishment H-AAA authentication failure.

- Steps 1-8 follow the description per the IKEv2 tunnel establishment procedures described before.
- Step 9: If the PDIF receives a RADIUS Access-Reject message or a Diameter EAP-Answer (DEA) Command with a Result-Code AVP indicating failure, the PDIF rejects the tunnel establishment towards the MS by sending an IKE_AUTH response message with the Notify payload set to 'AUTHENTICATION FAILED'. MS abandons the IPSec tunnel establishment procedures.

The MS subsequently abandons the IPSec tunnel establishment procedures. The IPSec tunnel establishment is attempted again when an application requests it.

17.3.5 UDP Encapsulation to Support NAT Traversal

See Appendix on NAT Types for a description of the types of NAT boxes and their behavior. Once the presence of NAT is detected through the IKEv2 procedures, the MS performs UDP encapsulation procedures as defined in RFC 3948 to carry the packets between the MS and the PDIF via the established IPSec tunnel.

The presence of NAT as described earlier is done using the IKE procedures. AH authentication data runs over source and destination IP Addresses: NAT and AH are hence inherently incompatible and AH cannot be used when NAT are present. Although NATs can correct TCP/UDP checksums, they will not be able to perform this function when the packets are encrypted using ESP/AH. NATs provide special treatment for port number 4500 and all traffic as soon as possible must be moved to this port.

When NATs are detected to allow for proper exchange of traffic two modes are defined when packet encryption is supported one for tunnel mode and the other for transport mode. The packet format of the original packet and the how the transport mode packets are generated for IPSec operation and transformation of these packets to include UDP encapsulation to handle NATs is show in Figure 17.12 and Figure 17.13.

To support IMS networks, there are three levels of IPSec tunnels that the MS needs to support when communicating with the 2G/3G network:

- Between the MS and the PDIF operating in the tunnel mode.
- Between the MS and the Proxy-Call Session Control Function (P-CSCF) or the SIP Server operating in the transport mode – for IMS based application.
- Between the MS and the Mobile-IP Home Agent (HA) operating in the tunnel mode – when MIP is used.

This section describes the mechanism of data encapsulation when MS detects the presence of NAT between MS and PDIF in sending data to the PDIF box via the IPSec tunnel. As per RFC 3948 the MS uses the UDP encapsulation method to send the NAT traversed packets to the PDIF. Once the MS has an IKE SA and hence an IPSec tunnel established with the PDIF, it can start exchanging packets via the IPSec tunnel. The MS enables the use of the UDP encapsulation method when during the IKEv2 procedure it detects the presence of one or more NAT boxes between MS and PDIF. In the 3GPP2 and 3GPP2 systems NATs are typically not deployed. When NATs are deployed in these networks it is left up to the network to build in Application Layer Gateways (ALG) to support these NAT boxes.

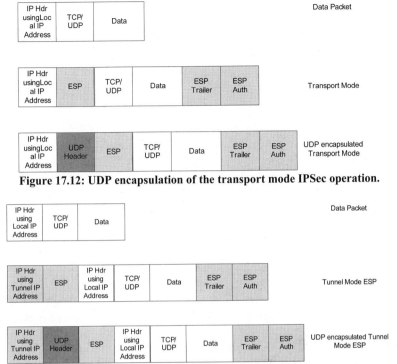

Figure 17.12: UDP encapsulation of the transport mode IPSec operation.

Figure 17.13: UDP encapsulation of the tunnel mode IPSec operation.

Given the overheads involved in sending packets through the IPSec tunnel, many deployments choose to send the media directly using the MS assigned IP address rather than using the IPSec assigned IP address.

In the description that follows it is assumes that both media signaling and RTP traffic are carried via the PDIF via the IPSec assigned address. It also assumes that the NAT boxes beyond the PDIF node are handled by the 2G/3G network. It will require the PDIF to behave as an ALG for SIP messaging, converting the required IP address and Port number fields for proper routing of the SIP Signaling and RTP packets.

Figure 17.14 shows the UDP packet encapsulation performed at the MS when it detects the presence of one or more NAT boxes between the MS and the PDIF node. The MS when preparing the media signaling (Session Initiation Protocol – SIP) message:

- During IKE negotiation with NAT traversal support in *Negotiation of NAT-Traversal in the IKE – RFC 3947* we can detect the IP address and port numbers that is carried in an independent payload as part of this exchange. The MS populates the Contact field with the IPSecGW IP Address and Port number that the IPSecGW uses to communicate within its subnet.
- Populates the Via field with the IPSecGW's (PDIF) IP address/Port Number.
- The Cline and Oline fields in the SDP as the IPSec Address assigned to the MS.

Signaling packet
encapsulation

IP Header (Src: Local IP address/port, Dest: PDIF IP address/ port)	UDP Hdr	ESP Hdr	Tunneled IP Header (Src: IPSec assigned IP address/port, Dest: IP address/port)	**Contact field in SIP**(IPSec Assigned IP address /port) **Via field in SIP** (IPSec Assigned IP Address , Port Number) Cline/OLine fields in the SDP (IPSec Assigned IP Address , Port Number)	ESP Auth

RTP packet
encapsulation

IP Header (Src: Local IP address/port, Dest: PDIF IP address/ port)	UDP Hdr	ESP Hdr	Tunneled IP Header (Src: IPSec assigned IP address/port, Dest: IP address/port)	RTP Payload

Figure 17.14: UDP Encapsulation to support carrying media signaling a RTP traffic via the PDIF.

The MS sends any Media Signaling and RTP traffic via the IPSec tunnel, wrapping the packets in the Tunneled IP address and the UDP IPSec header. This packet is sent to the PDIF wrapped with the IP header with the local IP address/Port number for source and PDIF address and port number for the destination as shown in Figure 17.14.

Some deployment may choose to handle the RTP packet directly to the WLAN AP instead of going via the PDIF node. The SIP signaling procedures will still use the IPSec tunnel with the PDIF. This can be achieved by populating the Cline and Oline fields appropriately in SDP parameters of the SIP message.

17.3.6 Acquiring configuration information

This procedure is used by the MS to discover the address of the P-CSCF (SIP Server), DNS, MIPv6 bootstrapping information, Broadcast Multicast Service (BCMCS) Server, when it is configured with the address of the DHCP server. This procedure is executed by the MS once it has established an IPSec tunnel with the PDIF. Using IKEv2 exchange the MS obtains the DHCP server address in the 2G/3G network.

DHCP query mechanisms are used to obtain the address of the different servers in the 2G/3G domain. The DHCP while operating within the 2G/3G domain, the MS typically issues a broadcast query. To optimize this behavior it is possible for the DHCP messages to be directed to the unicast IP address, which is the address acquired through IKEv2 negotiations.

17.3.7 Rekeying Procedures

Security associations once established have a lifetime associated with it. In order to maintain the security association, the context needs to be refreshed so that both ends understand that the other end is available and the association extended for a subsequent lifetime period. The IKE SA itself has a longer lifetime with the individual Child SA requiring more frequent refreshes with a shorter lifetime. Rekeying procedures are used to refresh the SA context.

Being in a wireless environment with short coverage and with the potential for the network to not be able to reach to MS, the rekeying procedures are always initiated by the MS. When the SA's lifetime expires at the network without the MS refreshing the context, the network quietly removes the SA, which includes the IKE SA itself or the individual Child SAs. When the IKE SA lifetime expires all the Child SA context is also removed. The MS is provisioned with the lifetime of the IKE SA and Child SAs based on operator needs.

17.3.7.1 Rekeying of IKE_SA

This section describes the procedures used by the MS to trigger and reestablish the IKE_SA through rekeying procedures. When the $(IKE\_SA\_Lifetime - T_{InitiateRe-Keying})^2$ timer (which was started when the SA was established/rekeyed) expires the MS initiates the rekeying procedures, if the MS is still under WLAN coverage. The MS may have one or more Child SAs established. When the MS is not under WLAN coverage when the timer expires, the IKE_SA is deprecated and a new association needs to be created once the MS reenters the WLAN coverage.

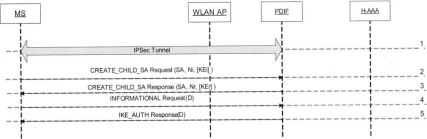

Figure 17.15: Rekeying of IKE_SA.

A description of the steps shown in Figure 17.15 is provided below.

- Step 1: The IPSec tunnel is established with the PDIF.
- Step 2: The $(IKE\_SA\_Lifetime - T_{InitiateRe-Keying})$ timer expires and rekeying is triggered. CREATE_CHILD_SA is sent identifying the IKE_SA being rekeyed, a nonce in the Ni payload and the Diffie-Hellman value in the KEi payload. The traffic selector is not included in this request.
- Step 3: The PDIF replies using the same Message ID with the accepted offer in an SA payload, and a Diffie-Hellman value in the KEr payload.
- Step 4: The INFORMATION Request is sent indicating the old IKE_SA to be deleted.

---

[2] Both the SA Lifetime, $T_{InitiateRe-Keying}$ is configurable. Typical values used for Lifetime is 24 hours and $T_{InitiateRe-Keying}$ is 30 min.

- Step 5: The INFORMATION Response is sent including the delete payloads for the deleted SAs. The delete payload needs to be included to avoid half-closed connections.

At the end of the procedure a new IKE_SA is created and the old IKE_SA is deleted. The Child_SAs associations are retained with the same configuration as with the old IKE_SA.

17.3.7.2 Rekeying of CHILD_SA

This section describes the procedures used by the MS to trigger and reestablish the IKE_SA through rekeying procedures. When the (Child_SA_Lifetime − $T_{\text{InitiateRe-Keying}}$)[3] timer (which was started when the Child SA was established/rekeyed) expires the MS initiates the rekeying procedures, if the MS is still under WLAN coverage.

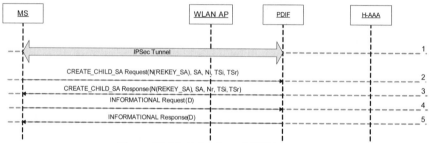

Figure 17.16: Rekeying of CHILD_SA.

A description of the steps shown in Figure 17.16 is provided below.

- Step 1: The IPSec tunnel is established with the PDIF.
- Step 2: The (Child_SA_Lifetime − $T_{\text{InitiateRe-Keying}}$) timer expires and the rekeying is triggered. CREATE_CHILD_SA is sent identifying the Child_SA being rekeyed, a nonce in the Ni payload and the Diffie-Hellman value in the KEi payload. The traffic selector may optionally be included in this request.
- Step 3: The PDIF replies using the same Message ID with the accepted offer in an SA payload, and a Diffie-Hellman value in the KEr payload.
- Step 4: The INFORMATION Request is sent indicating the old Child_SA to be deleted.
- Step 5: The INFORMATION Response is sent including the delete payloads for the deleted SAs. The delete payload needs to be included to avoid half-closed connections.

At the end of the procedure a new Child_SA is created and the old Child_SA is deleted. New traffic selectors if requested are put in place. The IKE_SA association is retained with the same configuration as before.

---

[2] Both the SA Lifetime, $T_{\text{InitiateRe-Keying}}$ is configurable. The default value for Lifetime is 24 hours and $T_{\text{InitiateRe-Keying}}$ is 30 min.

17.3.8 Tunnel disconnect procedures

The tunnel disconnect procedures can be initiated by the MS or by the network. The MS disconnects the SA, Child or IKE, based on the application requirements and the technology(ies) the MS is currently associated with. The network based on resource constraints and MS subscription/authentication expiration, initiates tunnel disconnect procedures.

17.3.8.1 MS-initiated tunnel disconnection

This section describes the procedures used by the MS to release the SA that it had established earlier. The MS closes all applications that require PDIF association or closes an application that results in closing of a Child SA.

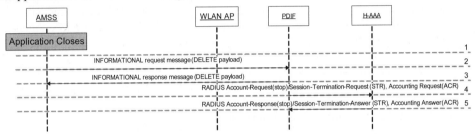

Figure 17.17: Mobile-initiated tunnel disconnect.

The following steps correspond to Figure 17.17:

- Step 1 – The closing of the application results in the filter entry being removed from the PS layer.
- Step 2 – This step describes the behavior for both IKE SA and Child SA.
  1. If this application was using a Child SA and if it is the last application using the Child SA, the MS decides to release the SA with the PDIF and sends the INFORMATIONAL request message identifying the Child SA to release.
  2. If all the applications that require association with the PDIF are released, the MS initiates releasing of the IKE SA itself.
  To disconnect one or more IPsec tunnels (SA(s)), the MS sends an IKEv2 INFORMATIONAL request message containing a "DELETE" payload including the SA(s) to be deleted (# of SPIs attribute set to the number of SAs to be deleted and SPI(s) attribute identifying the SA(s) to be deleted) if it wants to delete SA for a single protocol, e.g., ESP. If the MS wants to delete SAs for more than one protocol (e.g., IKE and ESP), the MS includes multiple delete payloads each containing SPI's corresponding to each of the protocols.

- Step 3 – The PDIF sends an IKEv2 INFORMATIONAL response message to the MS containing a "DELETE" payload including the SA(s) deleted in Step 1 (# of SPIs attribute set to the number of SAs deleted and SPI(s) attribute identifying the SA(s) deleted) if it wants to delete SA for a single protocol, e.g., ESP. If the PDIF wants to delete SAs for more than one protocol (e.g., IKE and ESP), the PDIF includes multiple delete payloads each containing SPI's corresponding to each of the protocols. The PDIF updates the related service information and/or status of the subscriber.
- Step 4 – When the IKE_SA with the MS is deleted, the PDIF informs the H-AAA that the session is being terminated by sending an Accounting-Request message with Acct-Status-Type set to Stop when RADIUS protocol [R3] is used, or a Session-Termination-Request (STR) Command and an Accounting-Request (ACR) Command when Diameter protocol [R2] is used.
- Step 5 – The H-AAA checks whether the user is known and whether it has an active session, if the check is successful it sends an Accounting-Response message with Acct-Status-Type set to Stop when RADIUS protocol [R3] is used, or a MS command and an ACA command when Diameter protocol [R2] is used.

NOTE: Steps 3 and 4 can occur in parallel.
The identified SA, IKE or Child, with the PDIF is released at the end of this procedure.

17.3.8.2 PDIF-initiated tunnel disconnection

This section describes the procedures of tunnel disconnection-initiated by the PDIF. PDIF decides to releases the SA (e.g., PDIF timeouts on not receiving any packet for a configurable amount of time, PDIF-initiated re-authentication procedures fail).

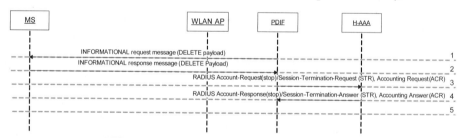

Figure 17.18: PDIF-initiated tunnel disconnect.

A description of the steps shown in Figure 17.18 is provided below:

- Step 1 – The PDIF sends an IKEv2 INFORMATIONAL request message to the MS containing a "DELETE" payload including the SA(s) to be deleted (# of SPIs attribute set to the number of SAs to be deleted and SPI(s) attribute identifying the SA(s) to be deleted) if it wants to delete SA for a single protocol, e.g., ESP. If the PDIF wants to delete SAs for more than one protocol (e.g., IKE and ESP), the PDIF includes multiple delete payloads each containing SPI's corresponding to each of the protocols.

- Step 2 – The MS sends an IKEv2 INFORMATIONAL response message to the PDIF containing a "DELETE" payload including the SA(s) deleted in Step 3 (# of SPIs attribute set to the number of SAs deleted and SPI(s) attribute identifying the SA(s) deleted) if it received a request to delete SA for a single protocol, e.g., ESP. If it received a request to delete SAs for more than one protocol (e.g., IKE and ESP), the MS includes multiple delete payloads each containing SPI's corresponding to each of the protocols. The PDIF updates the related service information and/or status of the subscriber.

- Step 3 – When the IKE_SA with the MS is deleted, the PDIF initiates a session termination exchange with the H-AAA. The PDIF informs the H-AAA that the session is being terminated by sending an Accounting-Request message with Acct-Status-Type set to Stop when RADIUS protocol [R3] is used, or a Session-Termination-Request (STR) Command and an Accounting-Request (ACR) Command when Diameter protocol [R2] is used.

- Step 4 – The H-AAA checks whether the user is known and whether it has an active session, if the check is successful it sends an Accounting-Response message with Acct- Status-Type set to Stop when RADIUS protocol [R3] is used, or a MS command and an ACA command when Diameter protocol [R2] is used.

- Step 5 – The interface is brought down and the application is notified. The application is subsequently not allowed to use this interface.

NOTE: Steps 1 and 3 can occur in parallel.

At the end of the procedure the identified SA is released and the application notified of the closure of the IPSec tunnel. The application subsequently needs to reinitiate establishment of the SA prior to packet transmission.

17.3.8.3 H-AAA-initiated tunnel disconnection

This section describes the procedures of tunnel disconnection-initiated by the H-AAA. H-AAA revokes the authentication credentials and triggers the PDIF to release the SA.

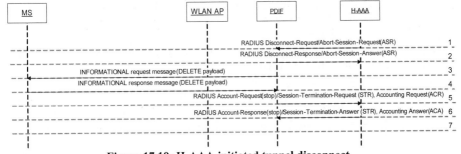

Figure 17.19: H-AAA-initiated tunnel disconnect.

A description of the steps shown in Figure 17.19 is provided below.

- Step 1 – The H-AAA initiates a session termination exchange with the PDIF to disconnect all IPsec tunnels (SA(s)) [R4]. The H-AAA informs the PDIF that the session is terminated by sending a Disconnect-Request message [R4] when RADIUS protocol is used, or an Abort-Session-Request (ASR) Command when Diameter protocol [R2] is used.

- Step 2 – The PDIF checks whether the user is known and whether it has an active session, if the check is successful it sends to the H-AAA a Disconnect-ACK message [R4] when RADIUS protocol is used, or an Abort-Session-Answer (ASA) Command when Diameter protocol [R2] is used.

- Step 3 – The PDIF sends an IKEv2 INFORMATIONAL request message to the MS containing a "DELETE" payload including all the SA(s) to be deleted (# of SPIs attribute set to the number of SAs to be deleted and SPI(s) attribute identifying the SA(s) to be deleted) if it needs to delete SA for a single protocol, e.g., ESP. If the PDIF needs to delete SAs for more than one protocol (e.g., IKE and ESP), the PDIF includes multiple delete payloads each containing SPI's corresponding to each of the protocols.

- Step 4 – The MS sends an IKEv2 INFORMATIONAL response message to the PDIF containing a "DELETE" payload including all the SA(s) deleted in Step 3 (# of SPIs attribute set to the number of SAs deleted and SPI(s) attribute identifying the SA(s) deleted) if it received a request to delete SA for a single protocol, e.g., ESP. If it received a request to delete SAs for more than one protocol (e.g., IKE and ESP), the MS includes multiple delete payloads each containing SPI's corresponding to each of the protocols. The PDIF updates the related service information and/or status of the subscriber.

- Step 5 – Upon reception of an IKEv2 INFORMATIONAL response message containing a "DELETE" payload, the PDIF informs the H-AAA that the IPsec tunnel has been terminated by sending an Accounting-Request message with Acct-Status-Type set to Stop when RADIUS protocol is used, or a Session-Termination-Request (STR) Command and an Accounting-Request (ACR) Command when Diameter protocol is used.

- Step 6 – The H-AAA sends an Accounting-Response message with Acct-Status-Type set to Stopwhen RADIUS protocol is used, or a MS command and an ACA command when Diameter protocol is used.

- Step 7 – The interface is brought down and the application is notified. The application is subsequently not allowed to use this interface.

Note that Steps 2, 3, and 5 in the call flow may happen in parallel. At the end of the procedure the identified SA is released and the application notified of the closure of the IPSec tunnel. The application subsequently needs to reinitiate establishment of the SA prior to packet transmission.

17.3.9 Application specific Child SA support

In order to support flows over WLAN it is not enough to such configure the QoS traffic streams with the WLAN AP. We also need a mechanism to filter the packets into the right streams. With the secure tunnel established with the PDIF, we need a mechanism to prioritize packets sent within the same tunnel. This will be very complex to achieve. Instead it is simpler to establish independent Child SAs to the PDIF with the QoS flows mapped to the appropriate secure tunnels.

There are scenarios where independent applications require different encryption mechanisms and hence will need to establish an independent Child SA with an independent configuration used to carry that applications traffic.

One more usage example is not to encrypt IP packets, between MS and PDIF, that carry SIP signaling, because these packets are already encrypted between MS and P-CSCF (SIP Server).

17.3.10 NAT Keep Alive and Dead-Peer Detection procedures

17.3.10.1 NAT Keep Alive

The peer behind NAT MUST send KeepAlive payloads every so often to avoid expiring the NAT mapping table entry. It is required to retain the mapping so that the mobile-terminated services can be delivered appropriately. The MS is typically provisioned with a timer that is used by the MS when it idle, i.e., not actively exchanging traffic. It does not need to send these keep alive packets when engaged in active traffic as the traffic keeps the NAT binding refreshed.

The NAT keep alive packet is a packet sent from the MS with the local IP address and port number to the PDIF. The payload contains the source and the destination port numbers, a length field, a checksum that is set to '0' disabling checksum verification and a octet set to '0xFF'. The source and destination port numbers must be 4500, which the same port used for UDP ESP encapsulation. The receiver, PDIF in this case should silently discard the packet.

17.3.10.2 Dead Peer Detection (DPD)

Both the MS and the PDIF nodes retain a timer expecting to hear from the other end to ensure that the node is still available. This is required by the MS so that it knows that the MS terminated services are still available to it. This is required by the PDIF so that it can periodically clean up the unneeded security associations. This procedure is typically run only for the IKE_SA as just this is enough to check on the availability of the peer.

The messages exchanged to execute DPD procedures are the INFORMATIONAL Request/Response with an empty payload. The empty payload is sent with the payload encrypted so that the peer can verify that the packet received is from the correct peer and is not some spurious packet. When a peer is detected to dead, each end initiates clearing of the IKE session.

17.3.11 Voice call establishment procedures

Before we talk about moving the call from one domain to another we briefly describe the call establishment procedures over the circuit-switched and the packet-switched domains.

The procedures on the circuit-switched side remains the same as before except for one behavior, Since a call can be routed either the circuit-switched or the packet-switched domain, the decision of where to page the MS cannot be done from the MSC, The call is sent to the VCC AS box (See section on Voice Call Continuity) to identify the domain to route the call to.

17.3.11.1 Voice call establishment procedures for the packet-switched domain

Adapting to the wireless environment, the number steps/exchanges involved in the call setup is optimized to allow for quicker call establishment times when compared with the extended procedures. Both the mobile-originated and the mobile-terminated call flows are shown below, with MS-1 as the originating the call and the MS-2 receiving the call.

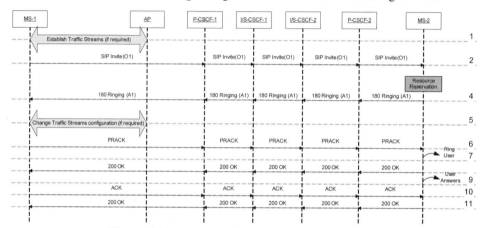

Figure 17.20: Optimized IMS call setup over WLAN.

A description of the steps shown in
Figure **17.20** is provided below.

- Step 1: When the AP supports QoS and admission control is mandated, then the MS establishes the SIP and RTP traffic stream with the AP. When QoS is enabled, but admission control is not enabled, the MS contends for the channel resources based on the parameters for voice traffic stream. If QoS is not enabled, then the MS used best-traffic to support signaling and voice traffic.
- Step 2: The MS-1 sends the SIP Invite message to MS-2. The message is sent the MS-1's P-CSCF, which forwards the message to MS-1's S-CSCF. Now the S-CSCF forwards the invite to the S-CSCF of MS-2, which in turn forwards the message to the P-CSCF of MS-2, assuming that the MS-2 is also located in the IMS domain. The MS-1 includes the vocoder and other flow related parameters in the message that it sends to MS-2. MS-1 may include one or more profiles based on what it supports, indicating that its pre-conditions are met to support all the profiles indicated in the message.
- Step 3: The MS-2 allocates the required resource.
- Step 4: MS-2 sends the 180 Ringing message to MS-1. The MS-2 chooses one amongst the proposed profiles by MS-1 based on its preference and resources allocated to MS-2.

- Step 5: MS-1 upon receiving the response may need to update its local resource allocation based on the profile choice made by the MS-2.
- Step 6: The MS-1 sends the provisional acknowledgement to MS-2.
- Step 7: MS-2, when it receives the provisional acknowledgement, it realizes the other end is ready to receive the call. The MS-2 rings the phone, altering the user of the incoming call.
- Step 8: Once MS-2 is ringing, it sends a 200 OK message to MS-1 indicating that.
- Step 9: The user answers the call.
- Step 10: MS-1 acknowledges the receipt of the 200 OK.

Once the user answers the call, the MS-2 sends a 200 OK acknowledging the initial Invite message sent by MS-1. This call is considered to be established after this step is complete and the two MSs exchange voice traffic.

17.3.12 Supporting mobility without the VCC feature

Before discussing the facilities offered by the VCC feature we consider the operational mechanisms of how the MS moves between the different domains prior to the availability of this feature. This section deals with the handling of mobility between the WLAN and 2G/3G systems when the voice call continuity feature is not available. This section addresses the mechanism of supporting mobility between Wireless Local Area Networks (WLAN) and Wireless Wide Area Networks (WWAN) (circuit-switched networks) prior to the availability of the Voice Call Continuity (VCC) feature. It assumes the existence of a network routing box that maintains the domain registration by the MS and routes the incoming call to the appropriate network.

Requirements that are to be handled in supporting mobility when the VCC feature is not available are identified below.

- The MS implementation shall adapt to the possible implementation choices made at the network.
- When dual registered, the MS shall monitor for incoming pages from both systems.
- The MS is operator provisioned with the preferred system to establish the connection when the page is received over both networks.
- The MS is operator provisioned with the preferred system to originate the voice call.
- When in voice traffic over a WLAN system, when the MS is moving out of the currently associated AP's coverage, it tries to continue the call over another neighboring AP.
  o Note that the call will drop if the move to the target AP is disruptive.
- Allow for supporting different modes of MS operation
  o Dual registered
  o Registered in only one domain at a time
- The MS shall receive incoming voice calls without any transitions gaps as the MS move in and out of WLAN coverage.
- The MS shall allow for both single and dual identity configurations.
- The MS stay dual registered only:

    o   During transition time periods as coverage of one radio is growing week and it needs to hand over the MS to the other network

    o   When partial set of services are available over WLAN.

Note that dual registration does not imply that the MS will monitor both systems. The network is expected to find the MS in one of the two domains. Support of dual registration in the network implies that the registration in one domain does not implicitly remove the registration from the other domain. Please refer to Appendix B for a discussion on single and dual identities.

17.3.12.1 Solutions supporting mobility

This section identifies the possible solutions and details each approach identifying the network implementation assumptions and MS requirements in supporting that solution. The solutions that apply to 3GPP and 3GPP2 are explicitly called out.

Solution 1: Page both the networks (WLAN and CS) simultaneously

When the MS is dual registered both the WLAN/IMS and CS networks, the network pages the MS on both domains simultaneously. The MS monitors and receives the page from both systems. The MS responds to the page based on the preferred system provisioned in the phone. The network processes the page response received from either domain.

This does waste network paging capacity. From the MS perspective, if a preferred system is identified by the operator and a Page is received over the other system, then the MS needs to wait for a certain time period to receive the Page on the preferred system. If no page is received over the preferred system, the MS responds to the page in the other domain. This solution applies to both 3GPP and 3GPP2.

Solution 2: Sequentially page the WLAN/IMS network and then the CS

This solution applies to both 3GPP and 3GPP2. When the MS is dual registered, it monitors only the preferred system. The network performs a sequential paging starting with the preferred system and on failing to get a response tries the other system.

A potential optimization that is suggested is when a page over the preferred system fails and is acquired over the other system, the network can optimize by not paging the MS over the preferred system until the MS performs another IMS registration over the preferred system. This is different from deregistering the MS from the preferred system. Subsequent pages will be tried over the non-preferred system and when it fails, retries the call over the preferred system as the MS has not yet performed an explicit deregistration and the registration timer has not expired as yet.

Assumptions on the network implementation is that the deployments will face the limitation that, until the IMS registration expires, the call will be tried over WLAN and when it fails it retries it over the CS. Sequential paging, based on the timeout value chosen for receiving the page response may result in the MS receiving the page over both domains. When this occurs, the network should account for receiving the page response on both domains. The network shall establish the call on the domain where the page response is received.

MS implementation issue seen is that the MS is monitoring only one domain at a time. The MS shall send the page response on the domain it received the page.

Supporting mobility between IMS and CS domains with single public/private identity without domain registration with VCC

This scenario describes the behavior of the MS moving between IMS and CS domains and where the MS is paged for MS terminated calls. The MS has a single public/private identity, is registered over the CS and the IMS domains and the MS prefers the IMS domains and monitors only the IMS domain. An incoming page triggers this sequence of events.

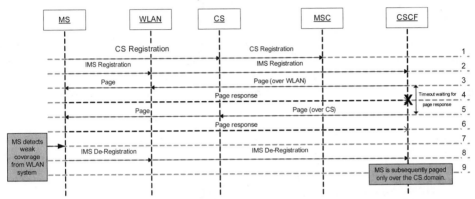

Figure 17.21: Supporting mobility between IMS and CS domains with single public/private identity without domain registration with VCC.

A description of the steps shown in Figure 17.21 is described below.

- Step 1: The MS performs a registration over the CS domain.
- Step 2: The MS enters WLAN coverage and performs an IMS registration.
- Step 3: The CSCF receives the incoming page, which is forwarded to the MS over the IMS/WLAN domain.
- Step 4: If the MS receives the page over IMS/WLAN it responds to the page. The call proceeds over the IMS domain.
- Step 5: If the CSCF did not receive the page response within a timeout period, it forwards the page through the media gateway to the CS domain.
- Step 6: The MS having gone out WLAN coverage is monitoring the CS domain. It receives the page and responds by sending the page response over the CS domain. The call is established over the CS domain.
- Step 7: A potential optimization that the MS can perform to avoid the delays incurred due to sequential paging under transition scenarios is to detect that the MS is going out of WLAN coverage and perform IMS deregistration to stop receiving it services over WLAN.
- Step 8: The MS sends a IMS de-registration while still under weakened WLAN coverage.
- Step 9: The MS is subsequently paged only over CS domain.
  At the end of the procedure the MS is paged over the appropriate domain based on it

current registration status. When failure occurs over the preferred domain the call is retried over the other domain. When a page over the preferred system fails and is acquired over the other system, the network can optimize by not paging the MS over the preferred system until the MS performs another IMS registration over the preferred system. This is different from deregistering the MS from the preferred system. Subsequent pages will be tried over the non-preferred system and when it fails, retries the call over the preferred system as the MS has not yet performed an explicit deregistration and the registration timer has not expired as yet.

Supporting mobility between IMS and CS domains with single public/private identity with domain registration with VCC

This scenario describes the behavior of the MS moving between IMS and CS domains and where the MS is paged for MS terminated calls. The MS has a single public/private identity, is registered over the CS and the IMS domains and the MS prefers the IMS domains and monitors only the IMS domain. An incoming page triggers the sequence of events.

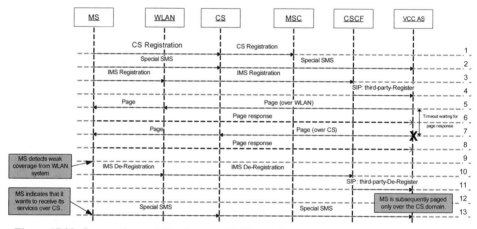

Figure 17.22: Supporting mobility between IMS and CS domains with single public/private identity with domain registration with VCC.

A description of the steps shown in Figure 17.22 is described below.

- Step 1: The MS performs a registration over the CS domain.
- Step 2: The MS sends a special SMS to the VCC AS to indicate the domain where the MS is willing to receive it incoming pages. SMS messages sometime take a very long time to reach the destination. This special SMS is indicated with a high priority so that it is delivered as soon as possible. The message is also tagged with the timestamp so that the VCC AS can compare this with the timestamp of the IMS registration. This is in case the MS moves back to the IMS domain from the CS domain and the special SMS reaches the VCC AS after the IMS registration.
- Step 3: The MS enters WLAN coverage and performs an IMS registration.

- Step 4: The CSCF forwards the IMS registration to the VCC AS to indicate that the MS is willing to receive it incoming pages over the IMS domain.
- Step 5: The CSCF receives the incoming page, which it forwards to the VCC AS. The VCC AS pages the MS over the IMS/WLAN domain.
- Step 6: If the MS receives the page over IMS/WLAN it responds to the page. The call proceeds over the IMS domain.
- Step 7: If the VCC AS did not receive the page response within a timeout period, it pages the MS over the CS domain. The VCC AS still expects the page response from the MS over the IMS/WLAN domain as it may be delayed more than the timeout period.
- Step 8: The MS having gone out WLAN coverage is monitoring the CS domain. It receives the page and responds by sending the page response over the CS domain. The call is established over the CS domain.
- Step 9: A potential optimization that the MS can perform to avoid the delays incurred due to sequential paging under transition scenarios is to detect that the MS is going out of WLAN coverage and perform IMS deregistration to stop receiving it services over WLAN.
- Step 10: The MS sends an IMS de-registration while still under weakened WLAN coverage.
- Step 11: The CSCF forwards the IMS de-registration to the VCC AS.
- Step 12: The MS is subsequently paged only over CS domain.
- Step 13: If the MS leaves the IMS/WLAN domain without having the chance to de-register, the MS uses the special SMS to indicate to the VCC AS that it wants to receives it incoming pages over the CS domain. If this special SMS is not supported and the MS did not perform an IMS de-registration, then the MS is paged over the IMS/WLAN domains fails and retries the call over the CS domain until the IMS registration times out.

At the end of the procedure the MS is paged over the appropriate domain based on it current registration status. When failure occurs over the preferred domain the call is retried over the other domain. When a page over the preferred system fails and is acquired over the other system, the network can optimize by not paging the MS over the preferred system until the MS performs another IMS registration over the preferred system. This is different from deregistering the MS from the preferred system. Subsequent pages will be tried over the non-preferred system and when it fails, retries the call over the preferred system as the MS has not yet performed an explicit deregistration and the registration timer has not expired as yet.

Supporting mobility between IMS/WLAN and IMS/WWAN domains with two private identities

This scenario describes the behavior of the MS moving between IMS/WLAN and IMS/WWAN domains and where the MS is paged for MS terminated calls. The MS has a single public identity with dual private identities, is registered over the IMS/WLAN and the IMS/WWAN domains with a single public identity and dual private identities and the MS prefers the IMS/WLAN domains and monitors only the IMS/WLAN domain.

Note that the details are still being worked out in the standards and have not been finalized. An incoming page triggers the following sequence of activities.

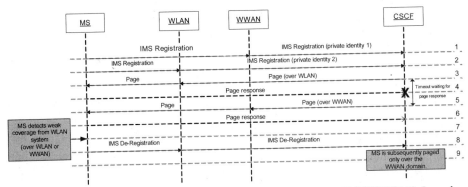

Figure 17.23: Supporting mobility between IMS/WLAN and IMS/WWAN domains with two private identities.

A description of the steps shown in Figure 17.23 is described below.

- Step1: The MS performs a registration over the IMS/WWAN domain with private identity 1.
- Step2: The MS enters WLAN coverage and performs an IMS/WLAN domain registration with private identity 2.
- Step 3: The CSCF receives the incoming page, which it forwards to the VCC AS. The VCC AS treats IMS/WLAN as the preferred domains pages the MS over the IMS/WLAN domain.
- Step 4: If the MS receives the page over IMS/WLAN it responds to the page. The call proceeds over the IMS domain.
- Step 5: If the CSCF did not receive the page response within a timeout period, it pages the MS over the IMS/WWAN domain.
- Step 6: It receives the page and responds by sending the page response over the IMS/WWAN domain. The call is established over the IMS/WWAN domain.
- Step 7: The MS having gone out of WLAN coverage is monitoring the IMS/WWAN domain. A potential optimization that the MS can perform to avoid the delays incurred due to sequential paging under transition scenarios is to detect that the MS is going out of WLAN coverage and perform IMS deregistration to stop receiving it services over WLAN.
- Step 8: The MS sends a IMS de-registration while still under weakened WLAN coverage. The MS when it loses the WLAN coverage prior to de-registration, can use the IMS/WWAN domain to de-register its IMS/WLAN domain if both WLAN and WWAN IMS belong to the same IMS core network.
- Step 9: The MS is subsequently paged only over IMS/WWAN domain.

At the end of the sequence the MS is paged over the appropriate domain based on it current registration status. When failure occurs over the preferred domain the call is retried over the other domain. When a page over the preferred system fails and is acquired over the other system, the network can optimize by not paging the MS over the preferred system until the MS performs another IMS registration over the preferred system. This is different from deregistering the MS from the preferred system. Subsequent pages will be tried over the non-preferred system and when it fails, retries the call over the preferred system as the MS has not yet performed an explicit deregistration and the registration timer has not expired as yet.

Supporting mobility between CS, IMS/WLAN and IMS/WWAN domains with two private identities without domain registration with VCC

This scenario describes the behavior of the MS moving between IMS/WLAN and IMS/WWAN domains and where the MS is paged for MS terminated calls. The MS has a single public identity with dual private identities, is registered over the CS, IMS/WLAN and the IMS/WWAN domains with a single public identity and dual private identities. The MS prefers the IMS/WLAN domains and monitors only the IMS/WLAN domain and the preference order of technologies for voice is IMS/WLAN followed by IMS/WWAN, which followed by CS.

An incoming page triggers the following sequence of events.

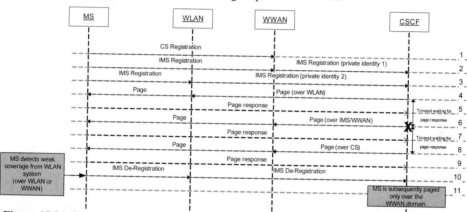

Figure 17.24: Supporting mobility between CS, IMS/WLAN and IMS/WWAN domains with two private identities without domain registration with VCC.

A description of the steps shown in Figure 17.24 is described below.

- Step 1: The MS performs a registration over the CS domain.
- Step 2: The MS performs a registration over the IMS/WWAN domain with private identity 1.
- Step 3: The MS enters WLAN coverage and performs an IMS/WLAN domain registration with private identity 2.
- Step 4: The CSCF receives the incoming page and it treats IMS/WLAN as the preferred domain and pages the MS over the IMS/WLAN domain.
- Step 5: If the MS receives the page over IMS/WLAN it responds to the page. The call proceeds over the IMS/WLAN domain.

- Step 6: If the CSCF did not receive the page response within a timeout period, it pages the MS over the IMS/WWAN domain.
- Step 7: The MS having gone out WLAN coverage is monitoring the IMS/WWAN domain. It receives the page and responds by sending the page response over the IMS/WWAN domain. The call is established over the IMS/WWAN domain.
- Step 8: If the CSCF did not receive the page response within a timeout period for the page sent over the IMS/WWAN domain, it pages the MS over the CS domain.
- Step 9: A potential optimization that the MS can perform to avoid the delays incurred due to sequential paging under transition scenarios is to detect that the MS is going out of WLAN coverage and perform IMS deregistration to stop receiving it services over WLAN. The MS sends a IMS de-registration message while still under weakened WLAN coverage. The MS when it loses the WLAN coverage prior to de-registration, can use the IMS/WWAN domain to de-register its IMS/WLAN domain if both WLAN and WWAN IMS belong to the same IMS core network.
- Step 10: The MS is subsequently paged only over IMS/WWAN or CS domains.

At the end of the procedures the MS is paged over the appropriate domain based on it current registration status. When failure occurs over the preferred domain the call is retried over the other domain. When a page over the preferred system fails and is acquired over the other system, the network can optimize by not paging the MS over the preferred system until the MS performs another IMS registration over the preferred system. This is different from deregistering the MS from the preferred system. Subsequent pages will be tried over the non-preferred system and when it fails, retries the call over the preferred system as the MS has not yet performed an explicit deregistration and the registration timer has not expired as yet.

Supporting mobility between CS, IMS/WLAN and IMS/WWAN domains with two private identities with domain registration with VCC

There are two parts to the VCC feature: 1) the domain registration and 2) active call transfer between the two domains. It is possible for only the domain registration to be supported initially without supporting the active call transfers.

This scenario describes the behavior of the MS moving between IMS/WLAN and IMS/WWAN domains and the domain the MS is paged for MS terminated calls. The MS has a single public identity with dual private identities, is registered over the CS, IMS/WLAN and the IMS/WWAN domains with a single public identity and dual private identities. The MS prefers the IMS/WLAN domains and monitors only the IMS/WLAN domain. The preferred order of technologies for voice is IMS/WLAN followed by IMS/WWAN, which followed by CS.

This is not a very likely scenario. The likely scenario is to support VS over IMS/WWAN and voice calls over IMS/WLAN and when that fails the call is retried over CS. An incoming page triggers the following sequence of events.

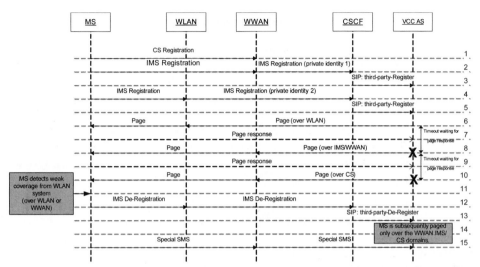

Figure 17.25: Supporting mobility between CS, IMS/WLAN and IMS/WWAN domains with two private identities with domain registration with VCC.

A description of the steps shown in Figure 17.25 is provided below.

- Step 1: The MS performs a registration over the CS domain.
- Step 2: The MS performs a registration over the IMS/WWAN domain with private identity 1.
- Step 3: The CSCF forwards the IMS/WWAN registration to the VCC AS to indicate that the MS is willing to receive it incoming pages over the IMS/WWAN domain.
- Step 4: The MS enters WLAN coverage and performs an IMS/WLAN domain registration with private identity 2.
- Step 5: The CSCF forwards the IMS/WLAN registration to the VCC AS to indicate that the MS is willing to receive it incoming pages over the IMS/WLAN domain.
- Step 6: The CSCF receives the incoming page, which it forwards to the VCC AS. The VCC AS treats IMS/WLAN as the preferred domains pages the MS over the IMS/WLAN domain.
- Step 7: If the MS receives the page over IMS/WLAN it responds to the page. The call proceeds over the IMS/WLAN domain.
- Step 8: If the VCC AS did not receive the page response within a timeout period, it pages the MS over the IMS/WWAN domain.
- Step 9: The MS having gone out WLAN coverage is monitoring the IMS/WWAN domain. It receives the page and responds by sending the page response over the IMS/WWAN domain. The call is established over the IMS/WWAN domain.
- Step 10: If the CSCF did not receive the page response within a timeout period for the page sent over the IMS/WWAN domain, it pages the MS over the CS domain.
- Step 11: A potential optimization that the MS can perform to avoid the delays incurred due to sequential paging under transition scenarios is to detect that the MS is going out

of WLAN coverage and perform IMS deregistration to stop receiving it services over WLAN.

- Step 12: The MS sends a IMS de-registration while still under weakened WLAN coverage. The MS when it loses the WLAN coverage prior to de-registration, can use the IMS/WWAN domain to de-register its IMS/WLAN domain if both WLAN and WWAN IMS belong to the same IMS core network.
- Step 13: The CSCF forwards the de-registration message to the VCC AS.
- Step 14: The MS is subsequently paged only over IMS/WWAN or the CS domains.
- Step 15: The MS can use the Special SMS to indicate to the network that it wishes to receive all its services over CS domain only. This should be rare since when the MS has CS coverage it should also have IMS/WWAN coverage.

At the end of the procedure the MS is paged over the appropriate domain based on it current registration status. When failure occurs over the preferred domain the call is retried over the other domain. When a page over the preferred system fails and is acquired over the other system, the network can optimize by not paging the MS over the preferred system until the MS performs another IMS registration over the preferred system. This is different from deregistering the MS from the preferred system. Subsequent pages will be tried over the non-preferred system and when it fails, retries the call over the preferred system as the MS has not yet performed an explicit deregistration and the registration timer has not expired as yet.

Solution 3: Registration in one domain removes the registration from the other

In this mechanism, when the MS registers over one system the MS's registration is removed from the other system. MS sends an SMS when it registers over CS. This mechanism is currently not supported by both 3GPP and 3GPP2. This solution is listed here for completeness.

Here the MS is paged only over one system. The MS monitors only one of the systems, the system it is currently registered with the network.

Assumption on the network implementation is that the network pages the MS on the currently registered domain. If it does not get a response it gives up assuming the MS is out of coverage. The MS needs to ensure that it performs/completes the registration each time it transitions systems.

A potential optimization is that the MS can measure the RSSI and when it crosses a threshold, deregisters with the IMS/WLAN network and powers down the WLAN module. This will allow for optimal paging behavior. This verification of the stability of the WLAN system needs to be done anyway to ensure reliability of the future calls over WLAN. All three suggested solutions will gain from this approach.

Solutions 1 and 2 are supported by both 3GPP and 3GPP2. In order to optimize the paging capacity, solution 2 is typically chosen for implementation.

Table 17.1: Tradeoff between the different solutions and recommendations.

Approach	Advantages	Disadvantages
Solution 1	Requires the minimum time to establish a call.	Requires the MS to monitor both systems. Must include additional logic to delay responding to a page on a system that is not preferred to ensure that the page is not received over the preferred system. Does consume system capacity as the page is duplicated over both systems.
Solution 2	Does not waste system capacity.	Incurs additional delay due to sequential paging.
Solution 3	The MS is not required to monitor both systems simultaneously for incoming calls.	Does not have a fall back mechanism when paging over one system fails. Requires repeated registration when switching between two domains.

17.3.13 Voice Call Continuity

With the small coverage area supported by WLANs, continuity of QoS applications becomes fundamental to deployments. Voice as a service will be one of the first services supported over WLAN as the operators will desire to offload user to the WLAN system, allowing for 2G/3G systems to support an increased customer base with existing deployments. This section describes the Voice Call Continuity (VCC) feature. The VCC feature consists of two parts: 1) domain registration itself and 2) transfer of the active call from WLAN to 2G/3G and 2G/3G to WLAN domains. As part of the active call transfers, the current standard only supports HO between the IMS/WLAN packet-switched domain and the 2G/3G circuit-switched domains.

The call transfer between two IMS packet-switched domains is still being studied in the standards bodies. The likely approach will be to enforce the support of Mobile-IP and using the Mobile-IP registrations from new system with the HA in transferring the call to the target packet domain. When Simple-IP is used by the MS, using proxy-MIP in supporting the mobility when possible is handled by the network. The exact mechanism used for the proxy-MIP is still being studied by the IETF standards bodies.

The rest of this section covers what is already supported in the standard of mobility between the IMS/WLAN domain and the 2G/3G circuit-switched systems. It provides details on how the MS moves between networks of different types, including scenarios when the MS has coverage from WLAN packet domain to 2G/3G packet domain.

Figure 17.26 provides the components and the interfaces involved in supporting the voice calls in the individual domain and the components interfacing between the two in transferring the call between the two domains. This figure also provides the interfaces for signaling and the media path used for carrying voice traffic.

Listed below is a typical set of assumptions made so that there is a good agreement over and beyond the standards in defining the implementation choices made in establishing the interaction between the MS and the network entities. These mainly reflect on the network/deployment configurations that enable early deployment of VCC without waiting for the full feature set to be available.

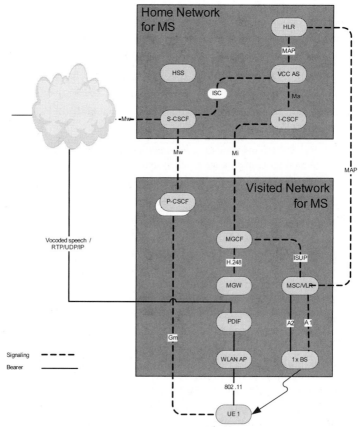

Figure 17.26: Nodes and interfaces involved in the VCC operation.

1. The MS will be provisioned with a valid VCC AS PSI (E.164 number) and the VCC AS node will be deployed along with the VoIP feature. The VCC AS is associated with the S-CSCF of the MS and is a single address that needs to be provisioned in the MS.

2. The MS does not remember its registration status across power cycles. This provides a mechanism to start with a clean slate periodically. The network employs the registration updates to refresh the MS information in its databases. This may cause some inconsistency of the registration status between the MS and the network

3. Support for dual registration is available. By this we mean that when the MS registers in one domain, the network does not actively proceed to remove the registration from the other domain.

4. If the current domain(s) the MS is operating cannot be clearly identified, it is left up to the network to the page the MS in the right network domain(s). For example, when the MS moves away from the PS domain and powered down, when it is powered up again, it appears over the CS domain. The MS does not remember the registration status across power cycles and the IMS registration and CS registrations are still valid, while the MS is monitoring only the CS domain in this case. The network can choose

between simultaneously paging the MS, or sequentially paging the MS (i.e., the network fails on not getting a page response from the MS. The network may make several attempts before it tries the other domain for sequential paging. Waiting time for the page response must be optimally chosen in the network to allow time for the MS to be paged in multiple domains. It is possible for the network to have prematurely paged the MS over the second domain and the page response comes from the first domain. The network must be capable of establishing the call over the domain it received the page response from. The MS under such scenarios must not respond to the page over both the domains. The network can also make an optimization of changing the MS's preferred domain when it receives a page response from the second domain the MS is currently expected to be paged over. The network does not page over the first domain until it hears back from the MS over that domain. It is also possible that the pages sent over the two domains with the above description to be completely independent and the MS has no way of correlating the page received over the CS domain and the PS/IMS domain. The network must take care to not page the MS over two domains with two independent calls simultaneously.

5. The IMS registration is retained in the network even when the MS moves to the circuit-switched domain and performs a registration and, for 3GPP2, it additionally sends a CS-ONLY registration in the SMS carrying the Domain Registration Status message to the VCC-AS. This indicates when the MS moves back to the IMS domain, the MS always assumes that the IMS registration is still intact as long as the registration timer has not expired.

6. The MS, upon reentering the IMS domain, retains the same IP address assignment and performs a re-registration. When that fails, the MS performs the IMS registration. When the IP address assigned to the MS has changed, then the MS goes through IMS registration.

7. As identified before, the SMS carrying the Domain Registration Status message function defined in this chapter applies only to 3GPP2 scenarios.

8. While in-traffic for a real-time application (e.g., VoIP) and when handing off from one AP to another, the MS shall assume that as long as the System ID remains the same, the HO will be non-disruptive to the higher layers. For example, the IP address and the authentication credentials will remain the same and the context of the IMS services can be retained across the HO to a new AP. Even when the local IP address changes when the MS moves APs, the IPSec assigned address that the MS is provided by the PDIF can be retained through MOBIKE procedures. Currently, when the MS moves APs, it will request for the same local IP address assignment as part of the DHCP procedures. The same address assignment can further be ensured by the network by supporting Proxy-MIP. With the varying mechanisms, each with its individual behavioral requirements, it is important to understand the deployment scenarios and push for the required solutions listed above solutions to make the inter-AP HO seamless. Note that in initial versions, the plan will be to HO the call to cellular as soon the currently associated AP goes bad even when there are other APs with the same SSID available to HO. This will change when the 802.11r feature of pre-authentication/fast HO support becomes available. The converse must also be stated. While in-traffic for a real-time application (e.g., VoIP) and when the System ID is different, the MS assumes that the HO to the new AP will be disruptive and will consider it to be the

edge of coverage for the WLAN RF. MS will further decides to retain the services over the new System ID if the system priority order indicates that the new WLAN system ID is preferred over the cellular system.

9. WLAN activity monitoring uses regular packet arrival to detect RF channel conditions. Discontinuous codecs are codecs that can shut off transmissions for extended periods of time without sending anything, including background traffic. When this occurs without the receiver being told about it, the receiver cannot distinguish between the transmitter not sending packet and the receiver in bad RF not receiving packets. Ideally, it is preferred that such discontinuous codecs not be supported. Note that this does not include EVRC type codecs that support DTX-ing of 1/8 rate packets and send packets with a certain periodicity during the DTX period which can be supported, detecting for those periodic packets received. EVRC in the DTX mode transmits one out of 12 1/8 packets.

10. Active call over the 2G/3G system is retained over that system even when the MS enters a valid WLAN coverage area. An active call refers to a call associated with any service over the 2G/3G system (e.g., voice and data calls). This will be true for initial deployments, but will be relaxed once the HO from the 2G/3G to the WLAN system is also supported.

11. When the MS transfers an active call from WLAN to 2G/3G-CS domain, the network shall stop sending packets over the source WLAN system prior to sending packets over the target system. Bi-casting of the packets over both the source and target system simultaneously shall not be supported. Once the network starts sending packets over the target system the network shall not send any packets over the source system. Several mechanisms were studied to make the transition as seamless as possible. This included bringing a conference bridge into the call and simulcasting the packets both in the source and the target domains. This, however, causes time synchronization issues of the received voice packets over both domains without a clearly defined point of when to move from the source to the target domain. The MS supports VoIP traffic over the WLAN system until one of the following conditions occur:

    i.   MS loses WLAN coverage before the HO is completed.
    ii. MS receives a SIP BYE from the other end during the HO of the voice call from the IMS/WLAN domain to the CS domain. See the VCC call flows detailed later in the chapter.
    iii. MS starts receiving non-NULL frames data (voice) frames over the 2G/3G network.
    iv. The MS abandons the HO to the 2G/3G procedure when the IMS/WLAN radio conditions improve and the current attempt to establish the connection over the 2G/3G system fails.

12. For 3GPP2, the network should delay sending the ServiceConnect message until it has heard back from the VCC AS or has timed out on not hearing back from VCC AS. The vocoder is switched to handling 3GPP traffic as soon as the connection is established and this procedure will minimize the voice traffic interruption experienced by the user.

13. For 3GPP, the network should delay sending the ConnectionComplete message until it has heard back from the VCC AS or has timed out on not hearing back from VCC AS. The vocoder is switched to handling 3GPP traffic as soon as the connection is

established and this procedure will minimize the voice traffic interruption experienced by the user.

14. The MS will always use service-based registration and will include the services that will be supported over the IMS domain. The service-based registration over the IMS domain offers the facility to indicate which services will be supported over the IMS domain. With this mechanism when the MS is CS registered and IMS registration is done using service-based registration, then some services are moved to the IMS domain with others retained over the CS domain. The MS will be considered dual registered when the MS sends Special SMS with CS-ONLY indication to the VCC AS. The details of the registration mechanisms are described later in the chapter. This allows for a mechanism to support services over both domains simultaneously with a subset of the supported services available in each domain.

15. Voice calls originated by a roaming VCC subscriber, using a called party number not in the international format and for which the home IMS network has no means to translate the called party number (e.g., local number) into proper routable format to the intended local destination, are not anchored in the IMS. If a call from a VCC subscriber is not anchored in the IMS, domain transfer is not supported for that call.

16. Since this a break-before-make HO (i.e., the traffic on the source side is terminated before the traffic on the target side is resumed), the vocoders used in the source and the target side are independent of each other. There is no bridging that needs to be performed between the source and the target vocoders as the call is transferred from one domain to another.

Having looked at the various assumptions and the discussion points on why these choices were made, let us look at the actual details involved in the registration process. We go on further to describe the voice call HO from WLAN to 2G/3G and from 2G/3G to WLAN.

17.3.14 Domain Registration

This section covers the procedures involved in the MS performing registrations once the MS has made a determination to associate with a particular domain.

17.3.14.1 IMS Registration in the IP-CAN domain

This section describes the scenario when the MS performs IMS registration over the IP-CAN domain. The MS at this point has acquired the address of the P-CSCF and has made a determination to support IMS services over this packet domain. This sequence of events is triggered when the MS enters the IP-CAN domain.

A description of the steps in Figure 17.27 is provided below.

- Step 1: MS sends a SIP service-based registration to the P-CSCF carrying the service tags for the services supported over the IMS domain.
- Step 2: P-CSCF contacts the I-CSCF, which queries the HSS to get the information on the user's S-CSCF.

- Step 3: Response is sent from the HSS providing details on how to reach the S-CSCF.
- Step 4: The P/I-CSCF forwards the SIP registration to the S-CSCF.
- Step 5: The S-CSCF requests the customer's profile information from the HSS.
- Step 6: The HSS responds with the user information.
- Step 7: S-CSCF sends the SIP 200 OK acknowledging the SIP registration.
- Step 8: The P/I-CSCF forwards the response to the MS.
- Step 9: The S-CSCF performs a third party registration with the VCC AS indicating that the call associated with the services associated with the IMS domain should be routed accordingly.
- Step 10: VCC AS confirms the registration to the S-CSCF.

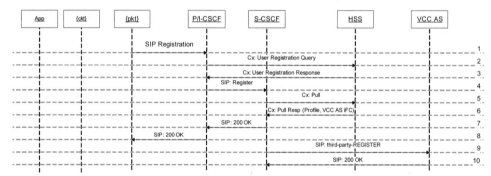

Figure 17.27: IMS Registration in the IP-CAN domain.

At the end of the procedure the MS considered to be associated the IP-CAN domain. Subsequent incoming pages are sent to the IP-CAN domain. When a page fails the network may retry the page over the CS domain.

17.3.14.2 Circuit-Switched Registration in the CS domain

This section describes the scenario when the MS performs CS registration over the 1X domain. This happens when the MS enters the CS domain and is currently not associated with the IP-CAN domain or is losing coverage of the IP-CAN domain.

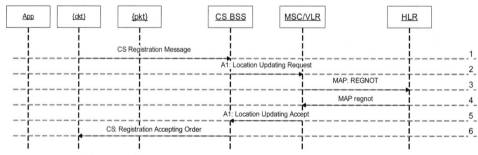

Figure 17.28: Circuit-Switched Registration in the CS domain.

A description of the steps involved in Figure 17.28 is provided below.

- Step 1: MS sends a CS Registration message to the circuit-switched Base Station.
- Step 2: The BS forwards the registration request to the MSC.
- Step 3: The MSC checks with the HLR on the user's profile.
- Step 4: The HLR provides the user information.
- Step 5: The MSC updates the ANSI-41 registration and sends an accept to the BS.
- Step 6: The BS forwards the Registration acceptance order to the MS.

At the end of the procedure, the MS is associated the CS domain and subsequent incoming pages are sent to the CS domain.

Domain Attachment Status Message

This section describes the scenario when the MS sends a special SMS to the VCC AS. The MS based on the previous two sections can register itself in the packet domain and/or the CS domain. With a lifetime associated with each domain, when the MS fades from a domain it is registered, it does not have a way to explicitly de-register itself from that domain.

One possible approach is to implicitly de-register the MS from all other domains when the MS registers itself in a particular domain. This suffers from the following handicaps:

- The MS cannot be associated with multiple domains, with each domain supporting a subset of the services.
- Registration in a domain takes time and it is costly to repeat such registrations particularly when ping-ponging between two domains.
- Given this, the registration of one domain does not remove the registrations on the other domains and the onus is on the network side to determine the best system to page the MS in.

The other option is to have the MS de-register from the domain it is leaving and register in the domain it is entering. The de-registering from the domain the MS is leaving may not be always be possible since it may fade away from the domain and may not have detected the condition early enough to initiate/complete the de-registration process.

The use of SMS carrying the Domain Attachment Status is an optimization introduced for scenarios when the MS moves away from the packet domain and enters the circuit-switched domain without de-registering itself from the packet domain (e.g., when the MS is currently associated with the WLAN domain and fades away from it and the system selection procedures results in the selection of the 1xRTT system). Note that the IMS domain supports service-based registration, while such a feature is not available over the circuit-switched systems. Registration in the ANSI-41 domain currently implies that all the services are supported over the circuit-switched domain. With the IMS registration still valid when the MS moves away from the packet-switched domain and into the circuit-switched domain, the registration with circuit-switched system leaves it ambiguous at the network as to which services are supported over which domain.

This SMS is sent from the circuit-switched domain 1xRTT when the MS moves in from the packet-switched domain without de-registration. The Domain-Attachment-Status message is sent by the MS to indicate to the VCC AS its attachment to a particular domain. This message indicates to the VCC AS whether the MS is attached to *CS-only*, *IMS-only*, or *both domains*. Currently, only the CS-only option is used with the other two options although in the standard does not have any practical application to it. This message allows the MS to tell the network to send the pages over the circuit-switched domain only even when the registration over the packet domain is still valid. This procedure is always initiated by the MS and the Special SMS is sent to the VCC-AS, the node that performs the routing of the calls.

There is a time ordering of Domain-Attachment-Status message and IMS registration. When MS moves into the circuit-switched domain and sends this SMS message, it indicates to the network that all the services are supported over the circuit-switched domain. When the MS subsequently sends the service-based IMS registration, these identified services are moved to the IMS domain with the remaining services still supported over the circuit-switched domain. Note that this procedure is independent of the domain registration itself. Note also that such a procedure is not needed when moving out of the circuit-switched domain into the packet-switched domain since service-based IMS registration is used to identify that services that are available over the packet-switched domain. This concept is introduced only in 3GPP2.

The following sequence of events is triggered when the MS powers up or decides to transition to the CS domain and the MS previously encountered a RF fade when in the IMS/WLAN domain.

When the MS encounters a fade in the RF channel and loses the IMS/WLAN coverage, it retains a flag indicating the occurrence of this event. This flag is cleared if the MS subsequently reenters the IMS/WLAN domain and successfully performs an IMS registration. When MS powers up or decided to transition to the CS domain and this flag is set, then the MS sends the Special SMS to the VCC-AS with the "CS-Only" indication. The VCC MS shall send a SMS on the CS network to notify the VCC AS. The "Destination Address" in the SMS message shall be set based on VDN.

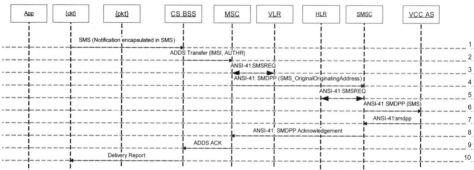

Figure 17.29: MS sending of the SMS with domain association over the CS domain.

A description of the steps shown in Figure 17.29 is provided below.

- Step 1: MS sends an SMS to the VCC AS indicating that the MS wishes to receive its incoming pages over the currently associated CS domain. Note that if a subsequent service-based IMS registration is performed, then the services other than the one supported over the IMS domain are routed to the CS domain.
- Step 2: The BS forwards the SMS message via the ADDS Transfer message to the MSC.
- Step 3: The MSC checks with the VRL on the transmission of the SMS for the user and update the billing information.
- Step 4: The MSC subsequently forwards the information to the SMSC.
- Step 5: The SMSC in turn checks with the HLR in locating the user's VCC AS.
- Step 6: The SMS message is forwarded to the VCC AS.
- Step 7: The acknowledges the receipt of the SMS to the SMSC.
- Step 8: The SMSC forwards the acknowledgement to the MSC.
- Step 9: MSC sends a ADDS acknowledgment to the BS confirming that the SMS has been delivered to the VCC AS.
- Step 10: The delivery report is forwarded to the MS.

This message is typically repeated multiple times if a response is not received from the VCC AS with a time gap between two attempts. At the end of the procedure the MS is registered in the CS domain and may or may not still be registered in the IMS domain. Subsequent incoming pages are sent only to the CS domain.

17.3.15 Active Call Handoff

Moving active calls between the WLAN and 2G/3G systems is one of the primary requirements in moving services to be provided over the WLAN system. This section describes the active call HO procedures between the WLAN and 2G/3G-CS domains, supporting mobility in both directions.

Active voice call HO from MMD/WLAN domain to the 3G CS network

This section describes the procedures involved in switching an active call from the IMS/WLAN domain to the CS/3G domain. This addresses the walk out of the house/office that has WLAN coverage scenario. HO in this direction is what is typically implemented first as it is assumed that the 2G/3G coverage is ubiquitous with hot spot coverage available with WLAN. For the walk into the house/office with WLAN coverage scenario, until the HO from the 2G/3G to the WLAN is supported, the call is retained in the cellular network and once the call ends the MS moves its services over to the WLAN domain when idle.

MS is active in one or more IMS voice originating or terminating session(s) at the time of initiation of Domain Transfer to CS. The current version of the standard does not support the transfer of multiple call contexts to the target side. Only the currently active call is transferred. There are times when actively switching between the two calls it is hard to predict which call is retained. The MS typically waits for packets to be received over one of the two calls and assumes that is the call to be retained and releases the other call. See details on when an active call HO from WLAN to 2G/3G system is triggered, detailed later in the chapter.

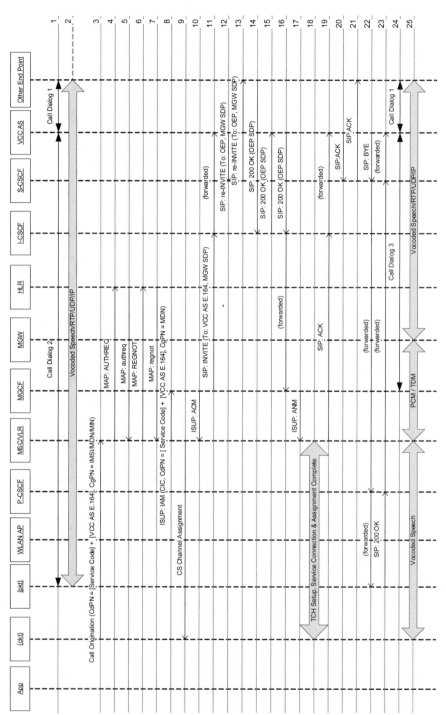

Figure 17.30: WLAN VoIP-to-1x CS voice HO.

1. The MS is on a voice call over the packet domain.
2. The MS exchanges voice packet over RTP/UDP/IP.
3. MS 1 sends a CS *Call Origination message* to the Visited MSC. MS 1 and includes the VCC AS E.164 number. The MS is provisioned with the VCC AS E.164 number.

   Currently the home network scenarios have been addressed in the standard. There needs to be mechanism introduce so that the MSC can route the VCC request to a local MGCF/MGW. This way, the bearer path does not have to traverse all the way back to the home network. Without such a mechanism, the MSC will treat the VCC call as regular CS call and route the bearer to the MGCF/MGW in the home network, the behavior of which will be suboptimal for routing traffic. Roaming scenarios are yet to be addressed by the standard.

   The MS also includes the IMSI/MDN/MIN, whichever is available in the calling party number. The MSC maps this to the MDN before the request is forwarded to the MGCF and VCC AS.
4. The Visited MSC initiates a 1x registration procedure on behalf of MS 1. The Visited MSC sends an MAP *AUTHREQ* message to MS 1's HLR to authenticate MS 1 prior to allowing registration and prior to allocating a 1x traffic channel to MS 1.
5. MS 1's HLR responds by sending an MAP *authreq* message to the Visited MSC.
6. The Visited MSC sends an MAP *REGNOT* message to MS 1's HLR.
7. MS 1's HLR responds by sending an MAP *regnot* message to the Visited MSC. Note, steps 4-7 are optional, depending on whether the MS has previously been 1x CS registered and authenticated.
8. Upon receiving the 1x Call Origination message from MS 1 in Step 3, and in parallel with Steps 4-7, the Visited MSC performs a translation on the Called Party ASCII Number field in the A1 CM Service Request message. The ISUP IAM message is routed to the MGCF in the home network.

   Note: For now we are assuming a 1-to-1 relationship between the Visited MGCF (which could actually be in any network) and the Visited MSC. This relationship can be based on the geographical location of the Visited MSC and Visited MGCF.

   The Visited MSC forms the ISUP IAM message by using the VCC AS E.164 number for the Called Party Number field and the E.164 number of MS 1 for the Calling Party Number field.

   Note that this step can happen in parallel with steps 4-7.
9. The Visited MSC sends a CS *Channel Assignment* message to MS 1.
10. Upon receipt of the IAM message in Step 8, the MGCF returns an *ACM* message to the MSC. The MGCF requests the MGW to create two terminations. The first termination is a TDM connection between the Visited MGW and the Visited MSC. The second termination is an RTP/UDP/IP ephemeral termination.
11. The Visited MGCF determines the VCC AS SIP URI via, for example, ENUM query and sends a SIP INVITE message via the I-CSCF to the VCC AS containing an SDP offer with the Visited MGW SDP information.
12. Anytime after Step 11, the VCC AS sends a SIP *re-INVITE* message to the S-CSCF containing an SDP offer with the Visited MGW SDP information. The VCC AS examines the From header of the SIP INVITE message that it received from the Visited MGCF in Step 11 to determine which subscriber is performing the WLAN VoIP-to-1x circuit-switched voice HO. Note that the VCC AS has already been put

into the call flow signaling path during session setup.

13. The S-CSCF forwards the re-INVITE to the far end network.

14. The OEP (IMS user or PSTN MGCF/MGW) modifies its RTP bearer termination with the Visited MGW SDP and sends a SIP *200 OK* message to the S-CSCF containing an SDP answer with the OEP SDP information in response to the SIP *re-INVITE* message that it received from the S-CSCF in Step 13.

15. The S-CSCF forwards the 200 OK to the VCC AS.

16. The VCC AS sends a SIP *200 OK* message via the I-CSCF to the Visited MGCF containing an SDP answer with the OEP SDP information in response to the SIP *INVITE* message that it received from the Visited MGCF in Step 11.

17. The Visited MGCF requests modification of the Visited MGW ephemeral termination with the OEP SDP information and instructs the Visited MGW to reserve/commit Remote resources. The Visited MGCF sends an ISUP ANM message to the Visited MSC.

18. Anytime after Step 9, the MS 1's reverse traffic channel is acquired, and Service Connection and Assignment are completed.

19. Anytime after Step 16, the Visited MGCF sends a SIP ACK message via the I-CSCF to the VCC AS in response to the SIP *200 OK* message that it received from the VCC AS in Step 12. This completes the establishment of SIP call dialog 3 between the MGCF and the VCC AS. The VCC AS records the location of MS 1 as present in the 1x CS domain. Note that this step is mandated to occur after the ISUP:ANM is received at the MSC from the VCC-AS.

20. The VCC AS sends a SIP ACK message to the S-CSCF in response to the SIP *200 OK* message that it received from the S-CSCF in Step 15.

21. The S-CSCF sends the ACK to the OEP.

22. The VCC AS sends a SIP *BYE* message to MS 1 via the S-CSCF to release SIP call dialog 2 between MS 1 and the VCC AS.
Note that this step can occur anytime after step 15.

23. MS 1 responds to the VCC AS with a SIP *200 OK*.

24. The existing Call Dialog 1 context with the Other End Point is used. The newly established Call Dialog 3 as part of the HO to the target side is used.

25. The Other End Point is still exchanging voice packets over RTP/UDP/IP with the MGW. The MGW exchanges PCM/TDM packets with the MSC. The MSC exchanges vocoded packets with the MS over the CS domain.

At the end of the procedure the call is handed over from the packet domain to the circuit-switched domain.

Having looked at the successful call transfer scenario we also look at what happens when the call transfer to the CS domain fails. A failure when the VCC AS node is not able to correlate the request received from the CS domain to an existing call in the IMS/WLAN domain is considered here.

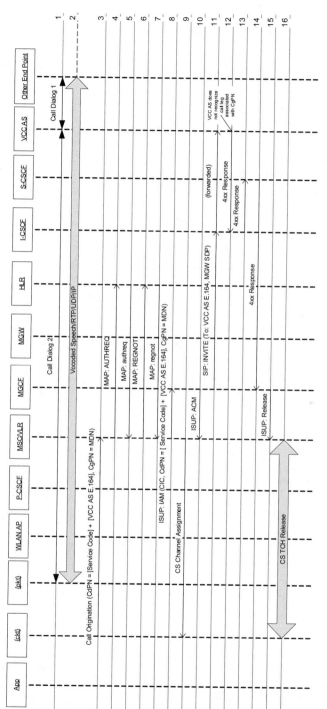

Figure 17.31: Failure scenario for handoff from WLAN VoIP to 1X-CS.

- Step 1 – 11: Follow the description provided in the main scenario.
- Step 12-14: The VCC AS does not recognize the call leg associated with the calling party number, since the call was not originally anchored at the VCC AS. The VCC AS sends a 4xx error response to the Visited MGCF in response to the INVITE, routed through the I/S CSCF.
- Step 15: The Visited MGCF converts the 4xx response to ISUP Release message and forwards it to the Visited MSC.
- Step 16: The Visited MSC sends back a failure notification (Release Order message) to the MS1 via the 1X BSC. The MS 1 responds back with a Release message to the Visited MSC.
  At this point, MS continues the call with the other end party in the IMS domain. At the end of the procedure the CS call establishment procedures fail and the MS continues the call with the other end party in the IMS domain.

Active voice call HO from MMD/WLAN domain to the 3G CS network

This section describes the procedures involved in switching an active call from the 2G/3G-CS domain to the IMS/WLAN domain. This addresses the walk into the house/office that has WLAN coverage scenario. HO in this direction is typically not implemented in the initial phases of the deployments.

MS is active in one or more CS voice originating or terminating session(s) at the time of initiation of Domain Transfer to IMS. Similar issues of which call is retained under the call hold/call waiting scenario as described in the HO in the other direction applies to this situation also.

See details on when an active call HO from 2G/3G to WLAN system is triggered detailed later in the chapter.

- Step 1: A 1x CS Voice call is established. When the call was initially setup, the VCC AS was anchored into the call (see CS Call Delivery and/or CS Call Origination scenarios).
- Step 2: The MS is exchanging voice packets with the other end.
- Step 3: Prior to the MS initiating a WLAN HO, the MS establishes the WLAN environment. We saw the details of this step shown in the section discussing the MS interaction with the PDIF node.
- Step 4: Conditionally, if MS is not already IMS registered, the MS performs IMS registration over the WLAN domain.
- Step 5: MS initiates a WLAN HO by initiating an IMS call to the VCC AS PSI (i.e., E164 number). The INVITE is sent from the MS to the P-CSCF and then to the S-CSCF.
- Step 6: The S-CSCF forwards the INVITE to the VCC AS based on filter criteria. The combination of the known E.164 number along with the P-Asserted-Identity value identify the INVITE are the key indicators that HO treatment has been initiated.
- Step 7: The VCC AS sends a re-INVITE message to the Other End Point (OEP) with the updated SDP for the MS call leg via the S-CSCF. Note: this OEP may be an MGCF if the other end is an ISUP termination or an I-CSCF for a SIP termination.

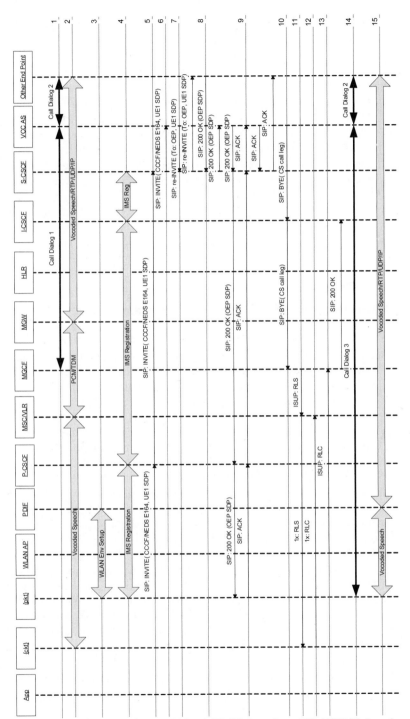

Figure 17.32: Voice call handoff from 1X CS domain to WLAN IMS domain.

- Step 8: The OEP acknowledges the re-INVITE with a 200 OK that is forwarded thru the IMS entities to MS.
- Step 9: The ACK from MS is forwarded thru the IMS entities to the OEP. This completes the establishment of call dialogue 3.
- Step 10: On reception of the ACK, the VCC AS clears the original 1x CS Voice call leg (call dialogue 1) by sending a BYE to the MGCF via the I-CSCF. Note: The MS is not expected to clear the 1x CS call leg; the MS waits for the network to clear the leg to ensure the network is established prior to clearing the 1x CS call leg.
- Step 11: The MSC initiates clearing of the 1X resources.
- Step 12: The MS responds to the request by clearing the call over the CS domain.
- Step 13: A 200 OK is sent back from the MGCF to the VCC AS.
- Step 14: Call dialog 3 is established between the MS PS domain and the VCC AS and the existing call domain 2 from before between the VCC AS and the other end is retained.
- Step 15: The MS is exchanging voice packets over the PS domain with the newly established path over the WLAN domain.

17.4 Mobility between the 2G/3G and WLAN domains

We conclude this chapter by providing details on how the MS switches between the different systems. The typical assumption will be that the 2G/3G systems have ubiquitous coverage with hotspot coverage available with WLAN. There are scenarios where the WLAN is used for coverage extension, but it does cause issues of reliably handing off the call to 2G/3G systems when the WLAN system starts to underperform.

When moving between these domains, the MS accounts for performing the proper registrations in the associated domains. For example, as the MS moves over the IMS domain and while still under coverage, performs an IMS de-registration. When it enters the circuit-switched domain, the MS performs an ANSI-41 registration. Additional for 3GPP2, when the MS enters the CS without performing a de-registration from the IMS domain, the MS sends an SMS to the VCC AS, providing the domain registration status and requesting all calls to be routed to the CS domain. When the MS requires retaining some services over the IMS domain and other services to be supported over the CS domain, the MS performs an ANSI-41 registration and service-based IMS registration to retain the specified services over the IMS and the remaining services to be supported over the CS domain.

Figure 17.33 shows the state transition that the MS goes through in moving between the different systems when idle. These state-transitions apply regardless of whether the VCC AS box is supported or not. A description of the states is provided in Table17.2.

Figure 17.34 shows the state transition that the MS goes through in moving between the different systems when idle. These state transitions apply only when the VCC AS box is supported. When the VCC AS box is not supported, then the active call is retained in the same domain until the call ends or the MS runs out of coverage. Once idle, the service is transferred to the appropriate domain and the idle state transition is used for this purpose. A description of the states is provided in Table 17.3.

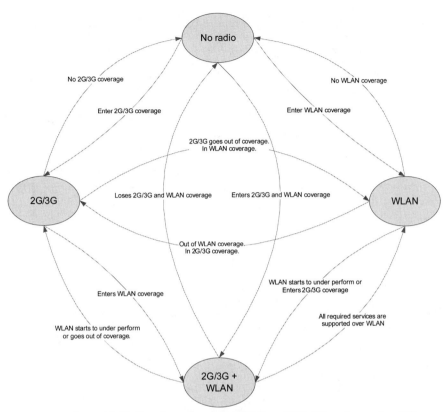

Figure 17.33: Handover between 2G/3G and WLAN systems when Idle.

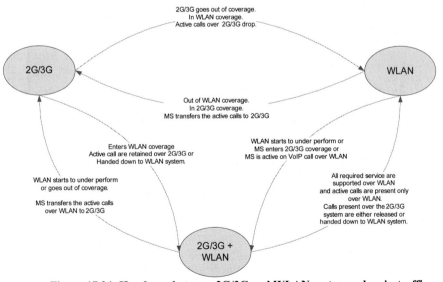

Figure 17.34: Handover between 2G/3G and WLAN systems when in-traffic.

Table 17.2: MS Association states when idle.

Association State	Registration Status	Coverage	Description
No Radio	Registrations may exist in one or both domains that will eventually timeout.	MS does not have any radio coverage.	The MS is considered to be in this state when it does not have any radio coverage or has been powered down. The MS performs a scan to find the 2G/3G/WLAN systems. When it reacquires a radio, the MS will perform registration.
2G/3G	Registered only on 2G/3G. Either registered in the CS/2G/3G domain or the PS/IMS/3G domain. If the MS is registered in the CS/2G/3G domain, the MS may also have an IMS/WLAN registration.	The MS has 2G/3G coverage but no WLAN coverage. The MS may also have WLAN coverage, but is shut down per operator policy.	The MS performs a scan to find the WLAN system.
2G/3G+WLAN	Registered on 2G/3G and WLAN.	MS has both 2G/3G and WLAN coverage.	The MS monitors both the 2G/3G and WLAN systems. Note that this may just be a transient state based on the mode of operation.
WLAN	Registered only on WLAN. The MS may also be registered in the CS/2G/3G domain. The registration may still exist in the CS domain when the MS moves away from the CS domain.	MS has WLAN coverage. The MS may also have 2G/3G coverage, but is shut down per operator policy.	MS when it detects bad WLAN RF or loses WLAN coverage, it tries to acquire and register in the 2G/3G network. The WLAN registration is retained for the transition period. During reactivation of the 2G/3G system, the MS shall use the previously associated 2G/3G system.

17.4.1 MS Operational Modes

WLAN can be deployed under different network configuration, particularly given the different types of 2G/3G available in the wireless markets. Any implementation needs to be flexible to adapt to the different deployments. With this is in mind the MS needs to have methods for acquiring the different technologies it can associated itself with – we term that as activation modes. When coverage of one or more technologies is available, the MS needs to have mechanisms to identify which technologies it will be associated with and which ones will be monitored, including HO and domain transfer procedures – we term that as operating modes. The operating modes typically depend on the type of service that is active. Described below are typical activation and operating modes used for voice service to enable the seamless continuity of voice calls across domains.

Table 17.3: MS Association states when in-traffic.

Association State	Registration Status	Coverage	Description
2G/3G	Registered only on 2G/3G. Either registered in the CS/2G/3G domain or the PS/IMS/3G domain. If the MS is registered in the CS/2G/3G domain, the MS may also have an IMS/WLAN registration.	MS has 2G/3G coverage but no WLAN coverage. The MS may also have WLAN coverage, but is shut down per operator policy.	The MS is active over the 2G/3G system.
2G/3G+WLAN	Registered on 2G/3G and WLAN.	MS has both 2G/3G and WLAN coverage.	The MS monitors both 2G/3G and WLAN system. Note that this may just be a transient state based on the mode of operation. The MS is active over 2G/3G and/or WLAN.
WLAN	Registered only on WLAN. The MS may also be registered in the CS/2G/3G domain. The registration may still exist in the CS domain when the MS moves away from the CS domain.	MS has WLAN coverage. The MS may also have 2G/3G coverage, but is shut down per operator policy.	MS when it detects bad WLAN RF or loses WLAN coverage, it tries to acquire and register in the 2G/3G network. The WLAN registration is retained for the transition period. During reactivation of the 2G/3G system, the MS shall use the previously associated 2G/3G system. The MS is active over WLAN system.

17.4.1.1 Activation modes

The MS is configured with one of two activation methods.

Activation Mode 1: When the MS is operating in this mode:

- MS manually selects the 2G/3G only RF or the WLAN only RF.
- When the MS leaves the coverage of the currently selected 2G/3G or WLAN RF, it considered to be out of coverage, even when the other technology coverage is available.
- Service supported will be restricted to the services that can be handled over the currently selected system.

Activation Mode 2: When the MS is operating in this mode:

- MS automatically selects the best available 2G/3G and WLAN system(s) based on policies and technology priorities.
- Policies in the MS may be defined based on inter technology preference order or based on service preference order, and sometimes potentially involving a combination of the two. There are also resource implications in the MS like battery life that govern the operation of the MS. Consider scenarios like all required services being supported by a single technology, making the MS's association with another technology although in coverage unnecessary. It is possible for the MS based on such policies to be associated with one technology or both the 2G/3G and WLAN systems simultaneously.

17.4.1.2 WLAN Operating Modes

The MS is configured with one of two operating modes:

Operating Mode 1: When associated with WLAN and this mode is enabled in the MS then:

- The MS monitors only the WLAN system.
- This applies only to Activation Mode 1.

Operating Mode 2: When associated with WLAN and this mode is enabled in the MS then:

- When idle and under good WLAN RF conditions, the MS monitors only the WLAN system.
- MS acquires the 2G/3G-CS when the active VoIP call is experiencing poor performance while retaining the call over the WLAN system. Once the 2G/3G-CS is acquired, the call is handed over from the WLAN to the 2G/3G-CS system. The dual coverage is retained only for the transition period.
- When idle and under bad WLAN RF conditions, the MS acquires the 2G/3G system. Upon acquisition of the 2G/3G system, the MS moves the services over to the 2G/3G system.
- This applies only to Activation Mode 2.

Operating Mode 3: When associated with WLAN and this mode is enabled in the MS then:

- When Idle and under good WLAN RF conditions, the MS monitors only the WLAN system.
- When associated with WLAN, the MS shall acquire the 2G/3G-CS system when the MS enters an active WLAN voice call. The MS does not register at this point with the 2G/3G-CS system. The dual coverage is retained only for the length of the VoIP call over WLAN.
- When the MS enters an active WLAN voice call and there is no 2G/3G-CS coverage available, then the MS shall retain only the WLAN coverage. In other words, when the 2G/3G coverage is not available or when the 2G/3G system is a packet domain only network, then the MS shall retain only the WLAN coverage.
- When the active VoIP call is experiencing poor performance over WLAN, HO of the call to the 2G/3G system is triggered.

- When the voice call ends and the WLAN system is still under good RF conditions, the MS stops monitoring the 2G/3G system.
- The 2G/3G system in not monitored when the MS is idle and the MS is deregistered from the 2G/3G system. This is a special case of Mode 1, when the trigger points to acquire the 2G/3G systems are set very aggressively.
- This applies only to Activation Mode 2.

17.4.2 Entry and Exit criteria for 2G/3G and WLAN systems

This section discusses the entry and exit criteria for the 2G/3G and WLAN systems. Note that both the exit criteria for source system and the entry criteria for the target system need to be satisfied for the HO to be triggered.

17.4.2.1 GSM/GPRS/EDGE

Entry and exit criteria are as specified in the 3GPP standard [18] and subject to the performance requirements in [17]. The entry and exit criteria depend upon the different scenarios and a description of one such a scenario is provided below. Please refer to [17], [18], and [19] for a detailed description.

Entry Criteria

The measured RSSI is better than ~ -102 dBm. The path loss criterion parameter C1 used for cell selection and reselection is defined by:

$$C1 = (A - Max(B,0))$$

where

- $A \doteq RLA\_C - RXLEV\_ACCESS\_MIN$
- $B = (MS\_TXPWR\_MAX\_CCH - P)$; except for the class 3 DCS 1 800 MS
  $= (MS\_TXPWR\_MAX\_CCH + POWER\ OFFSET - P)$; class 3 DCS 1 800 MS
- RXLEV_ACCESS_MIN = Minimum received signal level at the MS required for access to the system.
- MS_TXPWR_MAX_CCH = Maximum transmit power level an MS may use when accessing the system until otherwise commanded.
- POWER OFFSET = The power offset to be used in conjunction with the MS TXPWR MAX CCH parameter by the class 3 DCS 1 800 MS.
- P = Maximum RF output power of the MS.
- All values are expressed in dBm.

The path loss criterion (3GPP TS 03.22) is satisfied if C1 > 0.

The reselection criterion C2 is used for cell reselection only and is defined by:

$$C2 = C1 + CELL\_RESELECT\_OFFSET - TEMPORARY\ OFFSET *$$
$$H(PENALTY\_TIME - T)\ for\ PENALTY\_TIME \diamond 11111$$

$$C2 = C1 - CELL\_RESELECT\_OFFSET\ for\ PENALTY\_TIME = 11111$$

17.4.2.2 Exit Criteria

The downlink signaling failure criterion is based on the downlink signaling failure counter DSC. When the MS camps on a cell, DSC shall be initialized to a value equal to the nearest integer to 90/N where N is the BS_PA_MFRMS parameter for that cell (see 3GPP TS 05.02). Thereafter, whenever the MS attempts to decode a message in its paging subchannel; if a message is successfully decoded (BFI = 0) DSC is increased by 1, however never beyond the initial counter value as determined above, otherwise DSC is decreased by 4. When DSC ≤ 0, a downlink signalling failure shall be declared.

For GPRS, a MS in packet idle mode shall follow the same procedure. The counter DSC shall be initialized each time the MS leaves the packet transfer mode. When DRX period split is supported, the DSC shall be initialized to a value equal to the nearest integer to max(10, 90* NDRX), where NDRX is the average number of monitored blocks per multiframe according to its paging group (see 3GPP TS 05.02). In non-DRX mode, the MS shall only increment/decrement DSC for one block per DRX period according to its paging group. The exact position of these blocks is not essential, only the average rate.

NOTE: The network sends the paging subchannel for a given MS every BS_PA_MFRMS multiframes or, when DRX period split is supported, every 1/NDRX multiframes. The requirement for network transmission on the paging subchannel is specified in 3GPP TS 04.18 or 3GPP TS 04.60. The MS is required to attempt to decode a message every time its paging subchannel is sent. A downlink signalling failure shall result in cell reselection.

17.4.3 UMTS

Entry and exit criteria are as specified in 3GPP standard [19].

17.4.3.1 Entry Criteria

E_c/I_o should be above −18 dB <u>and</u> RSCP (pilot received power) above −111 dBm. Typical values are broadcast by the UMTS network.

17.4.3.2 Exit Criteria

When E_c/I_o is below −18 dB <u>or</u> RSCP below −111 dBm, this triggers a search for other available systems. The MS looks for other UMTS pilots and GSM/GPRS/EDGE systems. If there is no better cell to reselect to and the serving cell quality goes below above listed values, the MS would declare a DRX failure (unable to decode paging channel). If this condition persists for 12 seconds, the MS will declare out of service and look for service elsewhere.

17.4.4 1xRTT

17.4.4.1 Entry Criteria

Pilot received power = E_c/I_o + RX_AGC is above -89.9 dBm (cellular) or -92.9 dBm (PCS)

17.4.4.2 Exit Criteria

When Idle, E_c/I_o < -16dB, and the MS does not receive a message over the paging channel with successful CRC for 3 paging cycles, a system loss is declared.

When in traffic and HO to WLAN system in-traffic is supported, if E_c/I_o < -16 dB and the MS does not receive 12 consecutive forward link frames, then HO to WLAN is triggered.

17.4.5 1xEV-DO

17.4.5.1 Entry Criteria

SINR of the pilot > -9 dB.

17.4.5.2 Exit Criteria

When there are no Pilots above PilotDrop for PilotDropTimer time, the AT declares system lost. The MS is also monitoring the paging channel for periodic broadcast messages and consecutive errors in not receiving these messages will also trigger a system loss.

When in traffic and when HO to WLAN system in-traffic is supported, and MS is operating in DO-only mode, and HO to WLAN is preferred over 1X system, then if HRPD Coverage metric is less than a threshold (-7 dB) for a predefined period (4 sec) and 1xEV-DO Filtered DRC metric is less than 50%, then HO to WLAN is triggered.

17.4.6 WLAN

17.4.6.1 Entry Criteria

WLAN AP is acquired amongst the detected APs when the AP's RSSI is the strongest and is above -75 dBm.

17.4.6.2 Exit Criteria (with default values)

When idle and the filtered value of the RSSI < -95 dBm, the access point is considered to be not suitable to associate with.

When in traffic, the MS can measure additional metrics at the MAC level of the transmit frame loss based on the acknowledgments received from the AP and at the application level by measuring the packet error rate for applications like VoIP that have uniform packet arrivals. This typically triggers at an error rate of 8% to 10% to allow enough time for the VoIP call to be handed off to a better system.

17.5 MS is paged over the non-preferred domain

As detailed throughout this section, there is a lot of precaution taken to ensure that the registration status remains consistent between the MS and the network. It is still possible for inconsistencies to arise and the MS is paged over the lesser preferred domain.

This scenario addresses the case when such an inconsistency arises and the MS is paged over the non-preferred system. This typically will occur with sequential paging or the network has disabled the preferred system due to multiple paging failures over that system[4].

The MS is associated with dual technologies: System A and System B. The MS is registered to receive service X (e.g., Voice/VoIP) over System A and System B. The MS prefers to receive a service X over System A. The MS is listening for pages over both System A and System B.

The MS receives the page for Service X over System B, which triggers the following sequence of events.

1. The MS accepts the call received over the non-preferred system.
2. If a MS initiated call transfer of the in-traffic call to the preferred system is supported and the preferred system is still under coverage then, the call is moved the preferred system.
3. Otherwise, (i.e., transfer of the call to the preferred system is not supported)
   - The call is retained in the non-preferred system.
   - Once the call ends, the MS refreshes its registrations with the preferred system so that the network subsequently uses the preferred system to page the MS.
4. When supporting the call over the non-preferred (preferred) system, the MS, if capable, shall also monitor the preferred (non-preferred) system for incoming pages.

At the end of the procedure, the call is established over the non-preferred system. If the transfer of the call to the preferred system is supported, then the call is moved the preferred system if coverage from this system is still available. The MS refreshes its registrations with the preferred system, if required.

17.6 Conclusion

This chapter covered the details of what it takes for an independent RF system to tunnel into the 2G/3G system to provide the MS with the same service that the MS received when associated with the cellular system. A more specific instance of WLAN, which is the typical independent RF is used to describe the procedures. A description of the establishment/maintenance/releasing of the secure tunnel between the WLAN and the 2G/3G network is provided. Voice call establishment procedures over the WLAN domain with the optimized SIP negotiation procedure is illustrated to show the typical MS-originated and MS-terminated call flows. Mobility support between the WLAN and 2G/3G systems without the support of the Voice Call Continuity (VCC) feature is described to illustrate the phased support in the network. With the VCC support available in the network, a description of voice call continuity between 2G/3G and WLAN systems supporting the transfer of active calls in both directions is provided. The reader is given a feel for complexities involved in mobility support between the 2G/3G and WLAN systems

---

[4] This algorithm is network specific. The network may re-enable paging over the preferred system subsequently after a successful access is made over that technology.

when idle and with traffic – addressing which systems the MS is associated with, whether it is CS/IMS registered and domain over which each service is supported over at a given time.

Acknowledgements

To my colleague at Qualcomm, Amit Gil, for providing valuable insights in the structuring of this chapter.
To my wife Shubha for her support in all my endeavors and to my young son Suraj from whom I stole valuable play time to write this chapter.

17.7 References

[1]   WLAN 3G Interworking; Version 1.0; Dated: TBD; 3GPP2 X.P0028-100.
[2]   Access to Operator Service and Mobility for WLAN Interworking, Version 0.3, Dated: February 16, 2007; 3GPP2 X.P0028-200.
[3]   Voice Call Continuity between IMS and Circuit Switched Systems; Version 1.0; Dated February 16, 2007; 3GPP2 X.P0042-002-0.
[4]   3GPP TS   23.234: 3GPP system to Wireles Local Area Network (WLAN) interworking; System description. Version is 7.4.0, published as of 2006-12-11.
[5]   3GPP TS   24.234: 3GPP system to Wireless Local Area Network (WLAN) interworking; WLAN User Equipment (WLAN MS) to network protocols; Stage 3. Version is 7.4.0, published as of 2006-12-15.
[6]   3GPP TS 23.206: Voice Call Continuity (VCC) between Circuit Switched (CS) and IP Multimedia Subsystem (IMS); Stage 2. Version is 7.1.0, published as of 2006-12-11.
[7]   3GPP TS   24.206: Voice call continuity between Circuit Switched (CS) and IP Multimedia Subsystem (IMS); Stage 3. Version is 7.0.0, published as of 2006-12-08.
[8]   Internet Key Exchange (IKEv2) Protocol. December 2005. IETF: RFC 4306.
[9]   Diameter Base Protocol, September 2003. IETF: RFC 3588.
[10]   RADIUS Accounting, June 2000. IETF: RFC 2866.
[11]   Dynamic Authorization Extensions to Remote Authentication Dial In User Service (RADIUS), July 2003. IETF: RFC 3576.
[12]   Extensible Authentication Protocol Method for 3rd Generation Authentication and Key Agreeement (EAP-AKA), January 2006. IETF: RFC 4187.
[13]   Microsoft Vendor-specific RADIUS Attributes, March 1999. IETF: RFC 2548.
[14]   Negotiation of NAT-Traversal in the IKE. IETF: RFC 3947.
[15]   IKEv2 Clarifications and Implementation Guidelines. IETF: RFC 4718.
[16]   "The NetLMM Protocol", Gerardo Giaretta, 5-Oct-06, <draft-giaretta-netlmm-dt-protocol-02.txt>.
[17]   Technical Specification Group GSM/EDGE Radio Access Network; Radio transmission and reception, 3GPP TS 45.005 V6.14.0.
[18]   Technical Specification Group GSM/EDGE Radio Access Network; Radio subsystem link control, 3GPP TS 45.008 V6.18.0.

[19] Technical Specification Group Radio Access Network; Requirements for support of radio resource management (FDD), 3GPP TS 25.133 V6.16.0.
[20] IKEv2 Mobility and Multihoming Protocol (MOBIKE), June 2006. IETF: RFC 4555.
[21] Security Architecture for Internet Protocol IETF: RFC 2401.
[22] IP Authentication Header. IETF: RFC 2402.
[23] The Use of HMAC-MD5-96 within ESP and AH. IETF: RFC 2403.
[24] The use of HMAC-SHA1-96 within ESP and AH. IETF: RFC 2404.
[25] The ESP DES-CBC Cipher Algorithm with Explicit IV. IETF: RFC 2405.
[26] IP Encapsulating Security Protocol (ESP). IETF: RFC 2406.
[27] Internet IPSec Domain of Interpretation for ISAKMP. IETF: RFC 2407.
[28] Internet Security Association and Key Management Protocol. IETF: RFC 2408.
[29] The Internet Key Exchange. IETF: RFC 2409.
[30] The NULL encryption Algorithm and its use with IPSec. IETF: RFC 2410.
[31] IPSec Document roadmap. IETF: RFC 2411.
[32] The OAKLEY Key Determination Protocol. IETF: RFC 2412.
[33] The ESP CBC-Mode Cipher Algorithms. IETF: RFC 2451.
[34] AES-CBC Cipher Algorithm and its use with IPSec. IETF: RFC 3602.
[35] Negotiation of NAT-Traversal in the IKE. IETF: RFC 3947.
[36] UDP Encapsulation of IPsec ESP Packets. IETF: RFC 3948.
[37] Security Architecture for Internet Protocol. IETF: RFC 4301.
[38] IP Authentication Header. IETF: RFC 4302.
[39] Encapsulating Security Protocol. IETF: RFC 4303.
[40] IKEv2 Clarifications and Implementation Guidelines. IETF: RFC 4718.

17.8 Appendix A: NAT Types

Two kinds of network address translation exist. The type often popularly called simply "NAT" (also sometimes named "Network Address Port Translation" or "NAPT") refers to network address translation involving the mapping of port numbers, allowing multiple machines to share a single IP address. The other, a technically simpler form, - also called NAT or "basic NAT" or "static NAT" - involves only address translation, not port mapping. This requires an external IP address for each simultaneous connection.

NAT with port-translation comes in two sub-types: source address translation (source NAT), which re-writes the IP address of the computer which initiated the connection; and its counterpart, destination address translation (destination NAT). In practice, both are usually used together in coordination for two-way communication.

Full cone NAT is a NAT where all requests from the same internal IP address and port are mapped to the same external IP address and port. Furthermore, any external host can send a packet to the internal host, by sending a packet to the mapped external address. It is also known as "one-to-one NAT".

A *restricted cone NAT* is one where all requests from the same internal IP address and port are mapped to the same external IP address and port. Unlike a full cone NAT, an external host (with IP address X) can send a packet to the internal host only if the internal host had previously sent a packet to IP address X.

A *port restricted cone NAT* is like a restricted cone NAT, but the restriction includes port numbers. Specifically, an external host can send a packet, with source IP address X and source port P, to the internal host only if the internal host had previously sent a packet to IP address X and port P.

A *symmetric NAT* is a NAT where all requests from the same internal IP address and port to a specific destination IP address and port are mapped to the same external source IP address and port. If the same internal host sends a packet with the same source address and port to a different destination, a different mapping is used. Furthermore, only the external host that receives a packet can send a UDP packet back to the internal host.

NAT-TERM: IPv4 NAT; translates one IPv4 address into another IPv4 address to provide routing between private V4 and external V4 address realms.

NAT-PT: IPv4-v6 NAT; translates an IPv4 address into an IPv6 address, and vice versa, to provide routing between a V6 address realm and an external V4 address realm.

NATs exist primarily because of the shortage of IPv4 addresses, though there are other rationales. IP nodes that are "behind" a NAT have IP addresses that are not globally unique, but rather are assigned from some space that is unique within the network behind the NAT but that are likely to be reused by nodes behind other NATs. Generally, nodes behind NATs can communicate with other nodes behind the same NAT and with nodes with globally unique addresses, but not with nodes behind other NATs. There are exceptions to that rule. When those nodes make connections to nodes on the real Internet, the NAT gateway "translates" the IP source address to an address that will be routed back to the gateway. Messages to the gateway from the Internet have their destination addresses "translated" to the internal address that will route the packet to the correct end node.

17.8.1 What are the issues in handling VoIP without NAT Traversals?

- Firewalls allows for incoming traffic from the public domain only when the session is initiated from behind the firewall. This prevents the reception of incoming calls from the public domain.
- The VoIP signaling (e.g., SIP) messaging includes information of the private IP address and port numbers which cannot be recognized in the public domain and the connection setup fails because the packets are not routable.

17.9 Appendix B: Single and Dual Subscription

This section discusses the possible solutions of supporting mobility when the mobile uses a single public identity with a single and dual private identities that is used to perform IMS registrations over the WLAN and the WWAN domains.

Note that the mobile's registration in the CS domain is treated independent of its registration in the IMS domain(s). The mobile can be registered in the CS domain while it has registered using the public identity(ies) in the 3G and/or WLAN domains. The coordination between the CS domain and the IMS domain(s) is handled at the network. We consider below the dual and single subscription models.

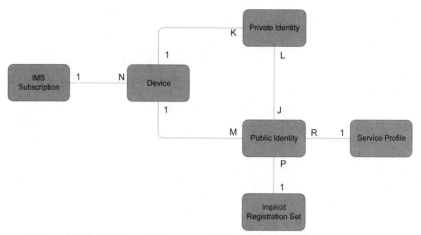

Figure 17.35: Relationship between different IMS identities and the device.

17.9.1 Dual subscription

By definition, each subscription will have independent public identities. The mobile has two different phone numbers, IP addresses, and two independent IMS registrations, potentially in independent IMS networks. The nature and deployment of dual independent public identities is not fully understood as yet.

Non-exhaustive possible variations of the dual identity are identified below:

- IMS registrations for WLAN and 3G domains in a single IMS network.
- IMS registration for WLAN and 3G domains in two independent IMS networks.

For both IMS registrations in a single or independent networks, the MS (based on the domain where it currently prefers to receive the call over) provides the call forwarding information to the IMS network from one number to the currently preferred number/domain where the call is to be rerouted, when supported.

When call forwarding between the two networks cannot be supported, the network has to reject the call forwarding request from the MS so that the MS is aware that it needs to monitor both domains.

When dual identity is used, there will be a single identity associated with each domain. Registration/Deregistration from one IMS domain does not impact the behavior on the other domain. It is expected that the MS will originate the call with the appropriate identity in the associated domain. It is also expected that the MS is paged with the appropriate identity in the domain where the mobile is paged.

Given that each public identity operates independently in their restricted domains, dual identities are not considered to be impacted by the availability of the VCC service and are not discussed further in this chapter.

17.9.2 Single Subscription

The use of single subscription restricts the mobile to a single IMS network. There are two possible variations here:

- with a single private identity and
- with two private identities.

When two private identities are used, at least one of the public identities used by the mobile needs to be same for the network to correlate the two registrations at the S-CSCF.

17.9.2.1 Single private identity

The mobile can be registered in only one IMS domain at a time. The mobile can be registered in the circuit-switched domain and the IMS domain simultaneously. The possible variations are:

- Registered in the CS domain.
- Registered in the IMS/3G domain.
- Registered in the IMS/WLAN domain.
- Registered in the CS domain and registered in the IMS/3G domain.
- Registered in the CS domain and registered in the IMS/WLAN domain.

17.9.2.2 Two private identities

The mobile can be registered in two IMS networks. The mobile can be registered in the circuit-switched domain and the IMS domain(s) simultaneously. The possible variations are:

- Registered in the CS domain.
- Using Private Identity 1 – Registered in the IMS/3G domain.
- Using Private Identity 2 – Registered in the IMS/WLAN domain.
- Registered in the CS domain + Using Private Identity 1 – Registered in the IMS/3G domain.
- Registered in the CS domain + Using Private Identity 2 – Registered in the IMS/WLAN domain.
- Using Private Identity 1 – Registered in the IMS/3G domain + Using Private Identity 2 – Registered in the IMS/WLAN domain.
- Registered in the CS domain + Using Private Identity 1 – Registered in the IMS/3G domain + Using Private Identity 2 – Registered in the IMS/WLAN domain.

The mobile uses the different private identities to perform two IMS registrations from two different domains. Each private identity has an independent IP address and authentication credentials. The mobile registers with each private identity to register with the IMS network using the single public identity. The network decides on the domain to page the mobile in based on the current IMS registrations.

17.10 Glossary

3G	3rd Generation systems; includes both 3GPP and 3GPP2 systems
3GPP	3rd Generation Partnership Project
3GPP2	3rd Generation Partnership Project – 2
AAA	Authentication, Authorization, Accounting
ACM	Address Complete Message
AN	Access Network
ANM	Answer Message
AP	Access Point (WLAN)
B-AAA	Broker AAA
CS	Circuit Switched; This refers to 1xRTT in 3GPP2 and UMTS CS domain in 3GPP.
CSCF	Call Session Control Function
DEA	Diameter EAP Answer
DER	Diameter EAP Request
Diameter	DIAMETER is an AAA protocol (Authentication, Authorization and Accounting) succeeding its predecessor RADIUS; Defined by RFC 3588
DNS	Domain Name System
EAP	Extensible Authentication Protocol
ENUM	The E.164 to Uniform Resource Identifiers (URI) Dynamic Delegation Discovery System (DDDS) Application – RFC 3761
FQDN	Fully Qualified Domain Name
H-AAA	Home AAA
HA	Home Agent
HO	Handoff or Handover
HSS	Home Subscriber Server
IAM	Initial Address Message
IKE	Internet Key Exchange
IGMP	Internet Group Membership Protocol
IMSI	International Mobile Station Identity
IP-CAN	IP Connectivity Access Network
ISUP	ISDN User Part
MAC	Message Authentication Codes
MAP	Mobility Application Part
MDN	Mobile Directory Number
MIN	Mobile Identification Number
MIP	Mobile IP
MLD	Multicast Listener Discovery
MMD	Multimedia Domain
MOBIKE	IKEv2 Mobility and Multihoming Protocol
MS	Mobile Station
MSK	Master Key Session
NAI	Network Access Identifier
NAS	Non Access Stratum
NAT	Network Address Translation
OEP	Other End Party
P-CSCF	Proxy-CSCF
PDIF	Packet Data Interworking Function
PLMN	Public Land Mobile Network
PRF	Pseudorandom Function
PS	Packet Switched
PSI	Public Service Identify (E.164 number)
Radius	Remote Authentication Dial-In User Service
REGNOT	Registration Notification
RSCP	Receive Signal Code Power ($E_c/I_o + RX\_AGC$)

RX_AGC	Total received power
SA	Security Association
TIA	Tunnel Inner Address
V-AAA	Visited AAA
VCC	Voice Call Continuity
VDN	VCC Domain Transfer Directory Number
WLAN	Wireless Local Area Network

18

Towards Service Continuity in Emerging Heterogeneous Mobile Networks

Sandro Grech, Henry Haverinen, Vijay Devarapalli, Jouni Mikkonen[a]

One noticeable trend for next generation mobile communication systems is the shift from the traditional vertically integrated approach in system design towards architectures that enable access to a set of common IP services through a variety of heterogeneous access technologies. With the introduction of multimode mobile devices, the next step in the integration of heterogeneous mobile networks, i.e. inter-access service continuity, gains relevance. This chapter reviews emerging heterogeneous access networks and their integration, and studies service continuity, including the relevance of Mobile IPv4 and Mobile IPv6 and other mobility solutions at, above, and below IP layer, in this context. We propose a new system architecture, which combines Mobile IP and IPsec to introduce multi-access mobility in a 3GPP operator environment. As part of the proposed framework we present new techniques for mobility security association bootstrapping and for enhancing the mobile device's NAPT traversal implementation. The selected architecture is evaluated using an experimental laboratory setup. The results are used to build up an analysis of the solution and draw up some conclusions on the viability of the selected architecture and associated performance.

18.1 Introduction

Mobile devices are exposed to an increasing selection of wireless access technologies ranging from wide range macro cellular technologies to short range local access technologies. While this leads to an increase in cost and complexity in mobile devices (when compared to single radio devices), it also enables several benefits such as the capability to:

- select the best suited technology (e.g., in terms of performance) for a specific context (e.g. availability of access technologies in a specific location),
- perform load balancing between access systems. This capability is particularly relevant in indoor locations where the propagation path to and from the base station involves in-building penetration losses. In this case, radio transmitters

---

[a] All authors currently affiliated with *Nokia Corporation*

typically need to switch to more robust modulation and coding schemes to compensate for the reduced SNR, leading to a degradation in system capacity and a reduction of the overall effective spectral efficiency of macro-cellular technologies.

- improve the economics of providing ubiquitous access to IP services to consumers, by allowing the operators to benefit from the availability of a range of access technologies that provide complementary coverage.

When faced with multiple access alternatives, a single subscriber credentials should be used to enable smooth roaming and seamless service availability. For operators providing access through the GSM family of standards and complementary access technologies such as WLAN, (U)SIM is a natural choice for subscriber management because it is widely deployed and enables roaming in existing GSM/GPRS networks. This is also reflected in the 3rd Generation partnership Project (3GPP) specifications, which introduced the 3GPP-WLAN interworking work item in release 6. By the time that work on 3GPP release 6 was starting, WLAN was gaining significant importance as a wireless access technology and 3GPP could not ignore its significance as a local-area complement to wide-area cellular packet access. 3GPP thus introduced WLAN access into the 3GPP architecture. 3GPP TR 22.934 introduces six interworking scenarios, representing various levels of integration between WLAN and 3GPP networks. 3GPP release 6 specifications cover the operation of scenarios 2 and 3. An overview of this architecture is given in [18]. In a nutshell, scenario 2 uses EAP-SIM [10] or EAP-AKA [16] to allow a WLAN access network to use AAA infrastructure that interfaces with 3GPP Home Subscriber Servers (HSS) to authenticate and authorize a SIM- or USIM- enabled device to use access network's resources. Scenario 3 introduces a Packet Data Gateway (PDG) adjacent to the Gateway GPRS Support Node (GGSN) to allow secure access to private services in the operator's home network. The PDG is essentially an IPsec gateway that is used to establish a secure tunnel through the WLAN access network between the mobile device and the service domain. The IPsec tunnel is established, managed and torn down using IKEv2 [4] with EAP-SIM or EAP-AKA methods. While scenario 2 is limited to public WLAN deployments, scenario 3 is generic enough so that it can be applied through any WLAN and non-WLAN access network. Work on further interworking scenarios will be taken up in 3GPP release 7.

In this chapter we start by studying the motivation for service continuity across heterogeneous mobile networks. We focus on 3GPP-WLAN interworking in particular. We present alternative approaches addressing mobility across cellular and WLAN. We then proceed to introduce the role of Mobile IPv4, Mobile IPv6 and other mobility solutions at or above IP and we propose an architecture based on the selected building blocks. Finally, using a laboratory testbed, we present an evaluation of the functionality and performance of the selected architecture.

18.2 Related Work

Besides the 3GPP-WLAN interworking work item in 3GPP, recently, there has been interest in other industry fora in utilizing non-cellular access networks to access a common set of communication services that would otherwise be accessible only through cellular access networks. The European Telecommunications Standards Institute (ETSI) and the Alliance for Telecommunications Industry Solutions (ATIS) are in the progress of specifying the access to communication services based on the IP Multimedia Subsystem (IMS) specified in 3GPP, through residential networks. The work in ETSI and ATIS, known as Next Generation Networks (NGN) focuses primarily on fixed residential access based on ADSL or cable. However, as the penetration of residential WLAN access continues to increase, an interesting variant of the technology will be the access of IMS services through residential WLAN. Even further, it should be possible to avoid disruption of the multimedia services (including VoIP) when a multimode mobile device switches across cellular and WLAN accesses. This is exactly the topic that is investigated in this chapter.

Providing communication services via the IMS platform is however not the only possible approach. Some operators have seen an opportunity gap for providing a solution based on the more traditional circuit-switched mobile communication systems, at least until the other solutions are mature enough and provide enough added benefit to justify the transition. An overview and analysis of the resulting solution, known as Unlicensed Mobile Access (UMA), is available in [22].

18.3 Proposed Architecture

18.3.1 Solution Space

The UMA solution adopts a model where cellular protocols are integrated on top of local area access technologies operating in the unlicensed frequency bands. UMA achieves mobility across unlicensed and cellular access, since in the UMA architecture WLAN is abstracted as a carrier for cellular protocols while mobility across BSC and UNC is treated like a traditional inter-BSC handover by the cellular mobility management protocols. On the other hand, when considering loose interworking between WLAN and cellular, as is the case in 3GPP-WLAN interworking, mobility across cellular and WLAN turns out to be a problem of IP address portability across access networks. The problem associated with the dual role of an IP address acting as both an identifier and a locator can be solved in various ways:

- utilizing a transport protocol that does not use IP address as an identifier (e.g. TLS [24] or DTLS [8]), or allows recovery from a change in underlying IP address (e.g. SCTP [21]).
- solving the problem where it occurs, i.e. at the IP layer by introducing a virtual link with an associated (semi-)static IP address that hides IP address changes associated with the physical links from the layers above IP. This is the approach taken by Mobile IPv4 [5] and Mobile IPv6 [7], [17].
- introducing an intermediate layer between network and transport layers that isolates transport protocols from the changes happening at the IP layer by introducing a static cryptographic host identifier. The Host Identity Payload (HIP

[20]) protocol provides such a solution through which the IP address retains only one of its functions, i.e. that of a locator.

TLS and DTLS are security protocols operating at the transport layer, but neither of them is yet widely adopted end-to-end. Thus, a solution based uniquely on the assumption that (D)TLS will be used would require the introduction of (D)TLS proxies or gateways with the primary interest of enabling mobility – which is not a good approach. In addition, (D)TLS protection would need to be applied over all access technologies including those that are considered already secure, for example through robust layer 2 encryption, as is the case in current cellular systems.

HIP provides an interesting approach for host mobility. Its main advantage is that security and mobility have been tightly integrated, and whereas Mobile IP can be seen as a fix to the current host mobility problem, HIP introduces a more elegant long-term solution by fixing the problem at the root cause by introducing an identifier-locator split. However, HIP assumes that both communicating endpoints support the protocol layer between network and transport layer. Consequently, at least in the shorter term, the applicability of HIP is limited to closed deployment environments, such as a mobile sensor network, where requirements can be set on both communication endpoints. As HIP matures and gains wider deployment we will probably start to see devices that support both HIP and traditional TCP/IP stacks, such that two communicating parties can first try to negotiate the operation of HIP, and subsequently use Mobile IP, if the operation fails. However, there will be a long time-span until HIP will be implemented in a sufficient number of devices such that it can be considered a stand-alone solution in a wide deployment scenario such as the Internet.

This leads us to the conclusion that we need to look at a mobility solution at layer 3. Building on top of 3GPP-WLAN scenario 3 interworking represents the full stretch of the problem space. In addition to basic inter-access service continuity, building on top of scenario 3 requires operation across and within security domains where IPsec needs to be dynamically enabled, disabled or maintained, according the to the requirements set by the source and target domain. This problem has been studied in IETF in [9] and a solution is proposed in [23]. The latter essentially proposes two layers of Mobile IP, one above, and one below IPsec. We argue that this approach is unnecessary complex and bears too much overhead, at least for the deployment scenario considered in this chapter. Instead we propose to use Mobile IP only when crossing access boundaries, whereas IP address changes that may occur while the mobile device is using IPsec can be handled using the IKE mobility extensions that are being specified in the IETF MOBIKE working group [15]. Such changes in IP address would occur for example if the mobile device were roaming across a large WLAN domain that spans multiple IP subnets.

The proposed architecture is illustrated in Figure 18.1. In order to support service continuity across bearers terminated at the GTP anchor point (GGSN) and bearers terminated at the IPsec anchor point (PDG) a new logical element (the Mobile IP Home Agent) is introduced in the architecture.

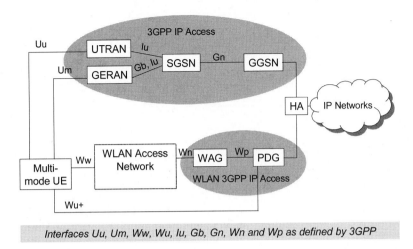

Interfaces Uu, Um, Ww, Wu, Iu, Gb, Gn, Wn and Wp as defined by 3GPP

Figure 18.1: Architecture overview.

When the multi-mode terminal is using GPRS, it registers the IP address of the Packet Data Protocol (PDP) Context as a co-located care-of address with the Home Agent. Applications use a separate home IP address, as usual in Mobile IP. No changes are required in the GPRS reference points or GPRS network elements. While in GPRS, the GPRS mobility management takes care of mobility and the care-of address does not change. The usual Mobile IP tunneling mechanisms are applied for user data packets.

When the terminal is using WLAN and a scenario 3 IPsec tunnel, it registers the Tunnel Inner Address (TIA), received from the PDG, as a co-located care-of address. The same home IP address is used by the applications. In this case, there are two nested tunnels for user datagrams; an innermost Mobile IP encapsulation and an outermost IPsec tunnel mode encapsulation. Finally, the scenario 3 IPsec tunnel on the Wu reference point is updated with IP mobility support through the MOBIKE protocol.

18.3.2 Mobility Security Association Bootstrapping

Both Mobile IPv4 and Mobile IPv6 require a mobility security association between the Mobile Node and the Home Agent. This security association is based on a shared key (in case of Mobile IPv6 with Internet Key Exchange a subscriber certificate is also possible as an alternative). In order to facilitate deployment and usability, bootstrapping these security associations needs to be automated. 3GPP has specified a general authentication and key distribution solution called Generic Authentication Architecture (GAA), in release 6 [1]. Using GAA, a shared symmetric key, identified with a bootstrapping transaction identifier (B-TID), can be provisioned to USIM-enabled device using USIM authentication. The bootstrapping operation is carried out between the mobile device and a Bootstrapping Function (BSF) using HTTP Digest AKA. A Network Application Function (NAF) verifies the credentials supplied by the client by communicating with the BSF using the Diameter protocol.

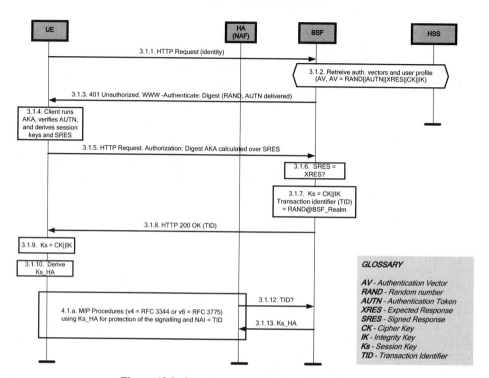

Figure 18.2: Security association bootstrapping.

As illustrated in the signaling flow of Figure 18.2, we propose a new way of applying the GAA framework for bootstrapping the mobility security associations in the context studied in this chapter. The GAA framework fits very well to Mobile IP environment, assuming that all the relevant mobile devices are USIM-enabled. In order to keep the HA agnostic of the provisioning solution, the NAF role can be delegated to a back-end AAA server, and use RFC 3957 [5] to derive a MN-HA key from the MN-AAA key generated through the GAA bootstrapping sequence.

18.3.3 Performance Optimization

So far we have discussed only the co-located mode of operation for Mobile IPv4. This is due to the fact that we cannot always rely on the existence of a Foreign Agent in a WLAN access network – indeed, there is a bigger likelihood that it will not be available. In order to reduce tunneling overhead inherent with the co-located mode of Mobile IPv4 operation, however, a Mobile IPv4 Foreign Agent can be co-located with the GGSN and PDG for (E)GPRS/WCDMA and WLAN scenario 3 access respectively. This kind of optimization is however not possible in Mobile IPv6. We thus argue that tunneling overhead is best handled using IP header compression techniques at overhead sensitive links such as cellular air interfaces. In addition, it is also possible to implement the Home Agent and access gateway in a configuration that allows one of the access links to represent the Mobile IP home link for the mobile device. This allows tunneling overhead to be avoided over that particular link, and applies for both Mobile IPv4 and Mobile IPv6.

18.3.4 Traversing Network Address and Port Translators

Network Address and Port Translators (NAPT or NAT) are commonly used in the IPv4 Internet. Both Mobile IPv4 and IPsec have been extended to support NAPT traversal.

[11] specifies the NAPT traversal extensions for Mobile IPv4. There is UDP encapsulation, and also a keep-alive mechanism. NAPT devices maintain mappings between the IP addresses and ports used in both sides of the NAPT, and when there is no traffic for some period of time, the NAPT mapping will expire. Due to the keep-alive mechanism, the terminal can send "dummy" echo requests to the home agent in the absence of any other traffic, so as to keep the mappings alive.

NAPT traversal extensions for IPsec are specified in [3]. Just as in Mobile IPv4 NAPT traversal, IPsec packets are UDP-encapsulated, and the client can send dummy packets to the IPsec gateway when there is no regular traffic to send.

In principle, the NAPT traversal solutions for Mobile IPv4 and IPsec work fine even though there are nested Mobile IPv4 and IPsec tunnels and several simultaneous keep-alive procedures. However, sending frequent keep-alive messages can have a very negative effect on the battery performance of a mobile device.

For cases when there are several nested tunnels, such as in the architecture of Figure 18.1, we propose a new way to avoid unnecessary keep-alive messages. The mobile device implementation should make sure that the uppermost tunneling layer in the terminal's stack uses the shortest keep-alive interval. This arrangement will effectively disable the keep-alive mechanisms of lower tunneling layers, as each lower layer will reset its keep-alive timer when the uppermost layer sends a keep-alive message. If the intervals can be set dynamically, for example as in [11] for the Mobile IP keepalive timer, then the mobile device implementation needs to explicitly provide the shortest interval to the uppermost tunneling layer.

Many NAPT devices apply a 30 second interval for UDP mappings, which implies a keep-alive interval of 20…25 seconds. In practice, the battery drain of such short keep-alive intervals will prevent the terminal from having an "always-online" mobile IP registration. How to improve the battery performance of IPv4 NAPT traversal mechanisms is an open research question. Besides migrating to IPv6, a possible solution might be the use of TCP encapsulation, as the NAPT devices use a longer lifetime for TCP mappings. However, TCP encapsulation would be cumbersome for UDP-based real-time applications.

18.4 Evaluation

After presenting the proposed architecture for (E)GPRS/WCDMA-WLAN service continuity, we now proceed to evaluate the availability, complexity, stability, QoS impact and performance of the solution. We perform this evaluation by setting up an experimental testbed as illustrated in Figure 18.3. We limit the testbed to the most basic operation, i.e. we do not yet support access using 3GPP-WLAN scenario 3 interworking (and consequently we do not evaluate IKEv2 mobility extensions). Also, the mobile device is manually configured with the Mobile IP home address, Home Agent IP address, and mobility security association parameters. The testbed supports IPv4 and IPv6 operation in isolation.

For Mobile IPv4 we use a laptop with WLAN/(E)GPRS/WCDMA PCMCIA cards running a Mobile IPv4 client on a Windows 2000 platform. The Mobile IPv6 counterpart is a Symbian OS plugin running on top of the Nokia 9500 (E)GPRS/WLAN terminal. The Home Agent for both Mobile IPv4 and Mobile IPv6 are the implementations from Helsinki University of Technology [13], [14].

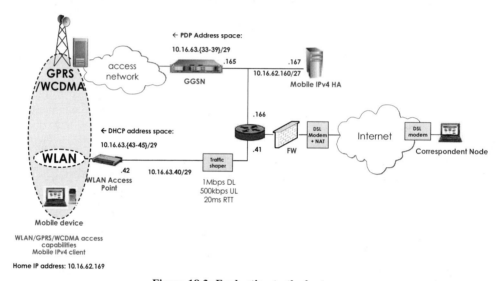

Figure 18.3: Evaluation testbed setup.

18.4.1 Results

We are particularly interested in the service continuity performance characteristic of the selected solution. In order to represent the results graphically we used a custom made tool that generates a continuous stream of packets and monitors the reception at the receiving end. The results are illustrated for GPRS/WLAN and WCDMA/WLAN mobility in Figures 18.4 and 18.5, respectively.

These results were generated using physical mobility of the mobile device, and represent the typical performance observed after repeating the test several times by moving in different directions and around different obstacles. The plots illustrate results for 30 kbps streams, characteristic of VoIP applications.

The results indicate that smooth transitions (i.e. no packet loss and no latency) across (E)GPRS and WLAN are achievable with one important condition. In these measurements we have maintained an active PDP context on the cellular side even while the mobile device is using WLAN. This PDP context does not consume radio resources, but facilitates a faster transition back to cellular access, when needed. It also avoids disrupting any ongoing communication while IPsec tunnels are set up over the Wu interface. From the cellular perspective the mobile device is inactive while it is using WLAN. The Traffic Block Flow establishment procedure is the only signaling that is required before the mobile device can start sending and receiving user-plane data over the (E)GPRS link. The latency incurred by this procedure is small enough not to cause any noticeable degradation to the performance of the transition.

Similar results are obtained for transitions across WCDMA and WLAN. These are illustrated in Figure 18.5. In this case the transitions from WCDMA to WLAN are also smooth, but transitions from WLAN to WCDMA incur a small outage period of about 2.5 seconds. This result is studied in more detail in the next section.

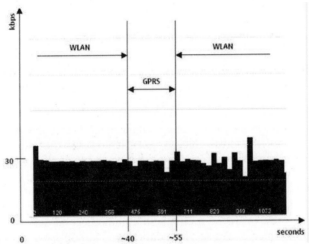

Figure 18.4: GPRS/WLAN Service continuity performance.

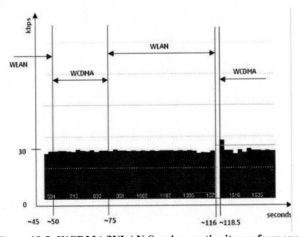

Figure 18.5: WCDMA/WLAN Service continuity performance.

18.4.2 Analysis and Discussion

One of the issues identified in the previous section is the transition outage during transitions from WLAN towards WCDMA. This can be explained with reference to the RRC state machine for a WCDMA UE [2]. CELL_DCH is the state in which a continuous stream of packets can be sent and received in the steady state. The limited resources associated with the CEL_DCH state are released after a configurable inactivity timer elapses. While the mobile device is using WLAN, the inactivity timer in WCDMA will start running, consequently the RRC state machine will cause the mobile device to release any dedicated resources. In turn, this will cause some delay in recovering these dedicated cellular resources when WLAN coverage is lost.

18.5 Conclusions

The first commercial multi-radio capable mobile devices have just recently been launched into the market. In order to maximize the utility of such devices there is a need to ensure that the mobile device is always connected through the best suited access technology. In order to avoid disrupting services during transitions across access technologies an inter-access device mobility solution is required.

While there are various ways on how to fulfill this requirement, Mobile IP is seen to be most suited for the deployment scenario considered in this chapter. As layers above IP that are resilient to IP address changes gain deployment the role of Mobile IP will start to diminish. In the meantime solutions for inter-access mobility below IP have also been specified. These solutions are based on tight interworking between access technologies, and arguably represent short-term solutions since they are based on the legacy voice-centric, circuit-switched communication infrastructure. Solutions based on IMS will enable a new wave of multimedia communication services, however some functionality such as inter-access service continuity is still missing. Providing a solution for this has been the subject of this chapter.

We argue that this gap is best filled with Mobile IP. We propose, evaluate and analyze a new system architecture based on this recommendation, incorporating the IPsec tunneling of the 3GPP WLAN interworking scenario 3. We propose the use of Generic Authentication Architecture for the bootstrapping of Mobile IP security associations. We observe that despite the fact that a lot of effort has been invested to optimize the Mobile IP handoff performance in IETF, much of this effort has been focused on Layer 3 and consequently access-agnostic. We note that access layer characteristics have significant effect on the performance on inter access service continuity.

18.6 References

[1] 3rd Generation Partnership Project (3GPP), "Generic Authentication Architecture; Generic Bootrsapping Architecture", 3GPP TS 33.230.

[2] 3rd Generation Partnership Project (3GPP), "Radio Resource Control (RRC) protocol specification", TS 25.331.

[3] A. Huttunen, B. Swander, V. Volpe, L. DiBurro, M. Stenberg, "UDP Encapsulation of IPsec ESP Packets", IETF RFC 3948, January 2005.

[4] C. Kaufman, "Internet Key Exchange (IKEv2) Protocol", work in progress, IETF draft-ietf-ipsec-ikev2-17, Oct. 2004.

[5] C. Perkins (ed.), "IP Mobility Support for IPv4", IETF RFC 3344, August 2002.

[6] C. Perkins, P. Calhoun, "Authentication, Authorization, and Accounting (AAA) Registration Keys for Mobile IPv4", IETF RFC 3957, March 2005.

[7] D. Johnson, C. Perkins, J. Arkko, "Mobility Support in IPv6", IETF RFC 3775, June 2004.

[8] E. Rescorla, M. Modadugu, "Datagram Transport Layer Security", work in progress, IETF draft-rescorla-dtls-04, April 2004.

[9] F. Adrangi et al., "Problem Statement and Solution Guidelines for Mobile IPv4 Traversal Across IPsec-based VPN Gateways", work in progress, IETF draft-ietf-mobileip-vpn-problem-statement-guide-03, Oct. 2004.

[10] H. Haverinen and J. Salowey, "Extensible Authentication Protocol method for GSM Subscriber Identity Modules (EAP-SIM)", work in progress, IETF draft-haverinen-pppext-eap-sim-16, Dec. 2004.

[11] H. Levkowetz, S. Vaarala, "Mobile IP Traversal of Network Address Translation (NAT) Devices", IETF RFC 3519, April 2003

[12] H. Soliman, "Dual Stack Mobile IP", work in progress, IETF draft-soliman-v4v6-mipv4-01, Oct. 2004.

[13] Helsinki University of Technology: Dynamics Mobile IPv4 implementation from Helsinki University of Technology. Available at: http://dynamics.sourceforge.net/ (Refereed April 2005).

[14] Helsinki University of Technology: MIPL – Mobile IPv6 for Linux. Available at: http://www.mobile-ipv6.org/ (Referred April 2005)

[15] IETF IKEv2 Mobility and Multihoming (MOBIKE) Working Group, available at: <http://www.ietf.org/html.charters/MOBIKE-charter.html>, Referred May 2005.

[16] J. Arkko and H. Haverinen, "EAP AKA Authentication", work in progress, IETF draft-arkko-pppext-eap-aka-13, Dec. 2004.

[17] J. Arkko, V. Devarapalli, F. Dupont, "Using IPsec to Protect Mobile IPv6 Signaling Between Mobile Nodes and Home Agents", IETF RFC 3776, June 2004

[18] K. Ahmavaara, H. Haverinen, and R. Pichna, "Interworking architecture between 3GPP and WLAN systems," IEEE Communications Magazine, vol. 41, no. 11, pp. 74–81, Nov. 2003.

[19] P. Eronen, "Mobility Protocol Options for IKEv2 (MOPO-IKE)", work in progress, IETF draft-eronen-MOBIKE-mopo-02, Feb. 2005.

[20] R. Moskowitz, P. Nikander, P. Jokela, T. Henderson, "Host Identity Protocol", work in progress, IETF draft-ietf-hip-base-02, Feb. 2005.

[21] R. Stewart et al., "Stream Control Transmission Protocol", IETF RFC 2960, October 2000.

[22] S. Grech, P. Eronen, "Implications of Unlicensed Mobile Access (UMA) for GSM Security". In proceedings of the first IEEE/CreateNet International Conference on

Security and Privacy for Emerging Areas in Communication Networks, Athens, Greece. September 5-9 2005.

[23] S. Vaarala and E. Klovning, "Mobile IPv4 Traversal Across IPsec-based VPN Gateways", work in progress, IETF draft-ietf-mip4-vpn-problem-solution-01

[24] T. Dierks, C. Allen, "The TLS protocol version 1.0", IETF RFC 2246, Jan. 1999.

[25] V. Devarapalli, "Mobile IPv6 Operation with IKEv2 and the revised IPsec Architecture", work in progress, IETF draft-ietf-mip6-ikev2-ipsec-01, Feb. 2005.

A Survey of Analytical Modeling for Cellular/WLAN Interworking

Enrique Stevens-Navarro, Chi Sun and Vincent W.S. Wong[a]

19.1 Introduction

A number of wireless technologies have evolved rapidly during the past decade. Mobile devices and gadgets (e.g., cellular phones, personal digital assistants (PDAs), laptops) supported by some of these technologies are becoming more and more important in people's everyday life. Wireless local area networks (WLANs) and cellular networks are two paradigms of such technologies in the present wireless realm.

WLAN, which is based on the IEEE 802.11 standards, is able to provide services with high data rate up to 11 Mbps (802.11b) or 54 Mbps (802.11a/g) at a relatively low access and deployment cost. Moreover, 802.11n, which is still under development, promises to offer a maximum data rate of up to 700 Mbps. However, the coverage area of WLAN is typically less than 100 meters, making it only suitable for hotspot regions such as hotels, libraries, airports, and coffee shops.

Compared to the WLAN, cellular networks cover a much larger area that provides ubiquitous access over several kilometers. Nevertheless, the supported service data rate of cellular networks such as GSM (Global System for Mobile Communications), GPRS (General Packet Radio Service), UMTS (Universal Mobile Telecommunication System), or CDMA2000 (Code Division Multiple Access 2000) only ranges from a few kbps to 2.4 Mbps. Furthermore, the cost of accessing and deploying cellular networks is much higher than that of the WLANs.

Driven by the complementary characteristics of these two wireless technologies (high-rate, low-cost, small coverage area of WLAN versus low-rate, high-cost, large coverage area of cellular network), a strong trend of combining them into one integrated system has emerged during the past years [1]-[6]. The inspiration behind it, which is also the motivation of the fourth-generation (4G) wireless networks, is to take advantage of both networks to provide mobile users with ubiquitous wireless access to very high data rate services. As a result, interworking mechanisms between WLANs and the cellular networks are crucial to achieve this vision.

In cellular/WLAN systems, users carry mobile devices which are equipped with multiple interfaces to establish connections with different available access networks. As the

---

[a] All authors currently affiliated with *The University of British Columbia*

users move within the coverage areas, they are able to switch connections among networks according to roaming agreements. Several manufacturers and vendors of mobile devices have begun to offer products in this area. The process of switching connections among networks is called handoff or handover. A handoff is defined as horizontal if it is between networks using the same access technology (e.g., between two adjacent WLANs, or between two neighboring cells in a cellular network). On the other hand, it is defined as vertical if it is between networks using different access technologies (e.g., from cellular network to a WLAN).

In order to investigate the performance of telecommunication systems, such as the envisioned cellular/WLAN systems with horizontal and vertical handoffs, analytical modeling can be used. Analytical models have been used for many years to design, plan, and evaluate the performance of cellular networks. Sets of equations can be derived to show the quantitative performance of a communication system. Mathematical models such as queuing theory, stochastic processes, linear programming, dynamic programming, convex optimization, game theory, and network calculus can be used in analyzing a communications system. Based on these models, network designers can be able to:

- Gain an understanding of the interaction between the processes within a system;
- Analyze the behavior of the protocols or algorithms under different loads;
- Predict the behavior of the system under changes of any of its components;
- Use the results of one model as an input of other models.

The use of analytical models is not the only approach to analyze the telecommunication systems. Two other approaches can be used: experimental and simulation. In the context of telecommunication systems, the first approach requires the set up of a test bed and may sometimes be considered to be costly. The second approach is becoming popular due to its flexibility to model complex systems.

The objective of this chapter is to first introduce and compare the architectures proposed for cellular/WLAN interworking. Then, we provide a survey of analytical models used to analyze and evaluate cellular/WLAN systems. We begin with some simple models and then present other models that can increase the accuracy of the results. For the sake of completeness, simulation models for cellular/WLAN interworking will also be and described. We conclude the chapter with a summary of open issues on modeling cellular/WLAN interworking.

19.2 Cellular/WLAN Interworking Architectures

The following subsections present the current interworking architectures proposed for the integration of cellular networks and WLANs.

19.2.1 Loose Coupling Architecture

The *loose coupling architecture* is shown in Fig. 19.1. This architecture places WLAN in a position that is parallel to other cellular core networks (e.g., GPRS, CDMA2000) for Internet access [1], [2]. This means that both WLANs and the cellular networks connect to the Internet directly, and they are connected to each other indirectly. The loose coupling of WLAN and the cellular network is carried out at the reference point between the GGSN/PDSN (for GPRS/CDMA2000 respectively) and the Internet [1]. There is no direct link between the WLAN and the cellular network. Therefore, the WLAN data will be transmitted to the Internet through the WLAN gateway, rather than passing through the cellular core network as in other architectures.

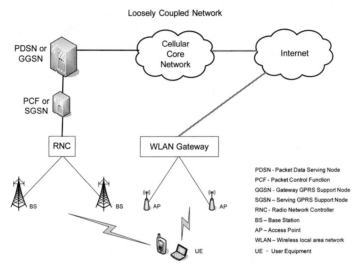

Figure 19.1: Loose Coupling Architecture.

Since the loose coupling scheme totally separates the data paths in WLANs and cellular networks, it is not necessary to include cellular technologies into WLAN. However, due to the fact that WLAN mostly uses Internet Engineering Task Force (IETF) protocols for authentication, billing and mobility management and certain cellular networks do not support these IETF protocols [2], some adaptations need to be made in order to let the two networks interoperate. First, the cellular network needs to employ specific Authentication, Authorization, and Accounting (AAA) servers for interworking with WLAN, while the WLAN gateway is required to support AAA services to interwork with the AAA servers of cellular network as well. Furthermore, the WLAN and the cellular network also require implementing mobile-IP for supporting mobility between the two networks.

Several benefits can be gained from the use of the loose coupling mechanism. Since WLAN and the cellular network are two separate elements in this architecture, they can be owned by different operators. This essentially allows the independent deployment of the two access networks [2]. The cellular operators can benefit from third-party owned WLAN deployments without significant capital investments, and the WLAN owners are able to choose their partners from a variety of cellular network carriers. Moreover, with the roaming agreements, customers are able to roam among different networks regardless of the ownership of them, and just obtain one integrated bill from their service providers instead of paying separately to each service provider when using different access technologies.

On the other hand, there is a drawback of the loose coupling architecture. The signaling between the two separate networks may take a longer time as it needs to traverse a relatively long pathway. As a result, the corresponding handoff latency and the packet loss rate may be high. Thus, the loose coupling architecture may not be able to provide seamless vertical handoff between WLANs and cellular networks.

19.2.2 Tight Coupling Architecture

The *tight coupling architecture* positions the WLAN in series with the cellular core network which is directly connected to the Internet [1], [2]. This logic aims to let all data originated from the WLAN to pass through the cellular core network before reaching the Internet. The tight coupling of the WLAN and cellular network is carried out at the reference point between the radio access network (RAN) and the SGSN/PCF [1] (for GPRS/CDMA2000 respectively) or the GGSN/PDSN [2] (for GPRS/CDMA2000 respectively), making WLAN appear as other cellular RAN in the cellular core network. By incorporating certain interworking functions at the WLAN gateway, the cellular core network will not differentiate a routing area from WLAN radio technology or a routing area from GPRS or CDMA2000 radio technology. The tight coupling architecture is shown in Fig. 19.2.

The benefits that can be gained from the use of tight coupling architecture are as follows: First, a large number of elements in the cellular network can be reached and reused by the WLAN. For example, the WLAN and the cellular core network can share the same user database, a single charging system, and the same AAA server. Moreover, many protocols from the cellular network can be used by the WLAN as well. Since the cellular network and WLAN are firmly coupled, seamless vertical handoff across two networks can be supported. Other important features such as enhanced security and guaranteed Quality-of-Service (QoS) are also available by the tight coupling scheme.

However, this approach also presents several disadvantages which may hamper its deployment. First, in order to support the vertical handoff between different RANs, the mobile terminals have to load the corresponding protocol stacks on top of their WLAN networking interfaces. Additionally, the WLAN gateway which interworks with the cellular core network needs to implement all the protocols (e.g., mobility management, session management, authentication, etc.) required in the cellular network. Currently, most WLAN terminals do not support the vertical handoff to cellular networks because they do not have the corresponding protocols on top of their WLAN networking cards. As a result, these WLAN terminals are not feasible for tight coupling scheme. Another drawback of this architecture is the cellular core network may become a network bottleneck, due to the high rate WLAN data traffic having to go through the cellular core network. Moreover, as the cellular core network directly interfaces with the WLAN, both networks must be owned by the same operator, resulting in a situation in which there is no support from third-party WLAN operators. Thus, in order to deploy the tight coupling architecture, the complexity and the high cost associated with the reconfiguration of the WLAN terminals, gateways and some key elements in the core cellular network will make it difficult to compete with WLAN-only service providers or the networks that are configured with loose coupling architecture.

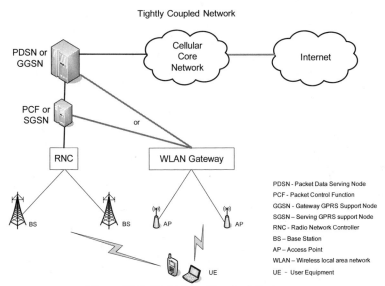

Figure 19.2: Tight Coupling Architecture.

19.2.3 Hybrid Coupling Architecture

The *hybrid coupling architecture*, by its name, uses both the loose and tight coupling architectures to better integrate the WLAN and the cellular network [7]. The motivation behind this mechanism is that although the loose coupling architecture is preferable compared to the tight coupling scheme, it still cannot support seamless service continuity during the vertical handoff between different RANs, thus resulting in long handoff latency and a high packet loss rate. On the other hand, the tight coupling architecture is able to provide users with guaranteed QoS and seamless mobility, but the cellular core network may become the bottleneck of the system since its capacity may not be enough for accommodating the high data rate traffic from the WLAN. As a result, the new hybrid coupling scheme emerges to handle this problem by differentiating the data paths from the WLAN to the Internet according to the type of data traffic, thus achieving guaranteed QoS and seamless mobility of the services [7]. The hybrid coupling architecture is shown in Fig. 19.3.

The main difference between the loose and tight coupling architectures is the path that the data traffic traverses before reaching the Internet. The hybrid coupling scheme differentiates the data traffic in terms of whether it is real-time or non real-time. The real-time traffic (e.g., voice over IP) typically demands lower bandwidth and seamless mobility support, so it is routed to the Internet using the tight coupling architecture. On the other hand, non real-time traffic (e.g., a large file on a FTP server) usually requires higher data rate but is able to bear reasonable delay. Therefore the traffic is routed in the network constructed by loose coupling architecture. It has been numerically shown in [7] that the hybrid coupling scheme is able to take advantage of both architectures. It achieves a lower packet loss probability than that of the tight coupling scheme and a lower handoff signaling cost than that of the loosely coupled scheme.

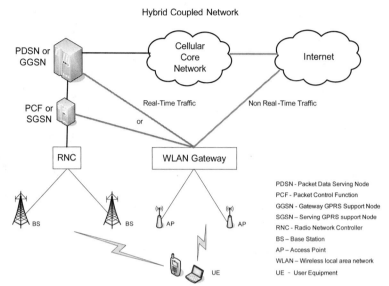

Figure 19.3: Hybrid Coupling Architecture.

19.2.4 IMS Architecture for 3GPP/3GPP2-WLAN Interworking

The IP Multimedia Subsystem (IMS) was originally designed by the 3rd Generation Partnership Project (3GPP) for delivering IP multimedia services to the GSM/GPRS network users [8]. It has been subsequently updated by telecommunication standard bodies such as 3GPP, 3GPP2, TISPAN (Telecoms & Internet converged Services & Protocols for Advanced Networks) and OMA (Open Mobile Alliance) to support networks other than GSM/GPRS. The spirit of the IMS is to act as an agent between different RANs (e.g., GPRS, CDMA2000, WLAN, PSTN) and the Internet to provide a common platform such that QoS and AAA charging are guaranteed for multimedia sessions established among them [9].

The IMS architecture has several layers. The *IMS 3GPP/3GPP2-WLAN architecture* is shown in Fig. 19.4. The device layer contains a large number of choices for mobile devices, most of which are able to access the IP network (e.g., Internet) by different networks. The IP network is connected to the IMS core network in the control layer, which is responsible for controlling all of the sessions and calls from all the network subscribers. The CSCFs (Call/Session Control Functions), which are Session Initiation Protocol (SIP) proxies and servers, are the core elements in the control layer. There are three types of CSCFs [8]. The P-CSCF (Proxy-CSCF) acts on behalf of the mobile devices in the IMS. The S-CSCF (Serving-CSCF) is responsible for the registration of the users and the call/session controls. The I-CSCF (Interrogating-CSCF) is used to interface peer networks and hide topologies between different operators. Another important entity in this layer is the Home Subscriber Server (HSS). HSS is a database that stores the information of each user within the network. This information consists of the user's profile (e.g., IP address). There are some other building blocks in the control layer such as Breakout Gateway Control Function (BGCF) and Media Resource Function (MRF). The BGCF is a SIP server that behaves as a routing node between an IMS terminal and a user in a circuit-switched network (i.e., PSTN), while the MRF provides the network with the ability to play and mix media streams, translate different codecs and do any kind of media-related analysis. The service layer of the IMS architecture contains application servers which are responsible for executing advanced services according to the user's profile and interfacing with the control layer.

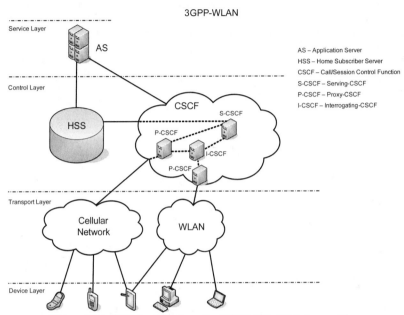

Figure 19.4: IMS Architecture for 3GPP-WLAN Interworking.

As described above, the I-CSCF in the IMS core network is responsible for interworking with other IP multimedia networks. Therefore, the interconnection of the WLAN and the 3GPP IMS network is carried out between the I-CSCF and the WLAN gateway. However, the problem is that the WLAN gateway needs to understand the protocols that the IMS is using and be able to translate them for the access points of its own network. So, a new component named SIP proxy is inserted between the I-CSCF and the WLAN gateway to solve this problem. The function of this proxy is to enable the communication between the I-CSCF and the WLAN gateway. By doing this, the users in the WLAN are able to interwork with the users in the 3GPP cellular networks.

Finally, 3GPP and 3GPP2, the 3G cellular standardization bodies, have been actively working in cellular/WLAN interworking [10]. The objective of their efforts aims to extend the packed data and IP multimedia services (e.g., IMS) to the WLAN environment. Several levels of integration have been proposed ranging from simple common billing and customer care to seamless mobility and session continuity. The combination of WLAN and 3G cellular network access technologies provides an excellent complementary coverage environment to offer data services to cellular subscribers.

19.3 Simple Models for Cellular/WLAN Interworking

In this section, we first describe how to model an integrated cellular/WLAN system using stochastic processes and queuing theory. These models are intended to act as initial approximations to a model for cellular/WLAN interworking. The first model considers the mobility of users, resource reservation, and the capacity of the networks, and is based in part on the model for an integrated cellular/WLAN system proposed in [11] using birth-death processes. After that, we describe a model that considers similar modeling assumptions on the duration of connections (i.e., exponentially distributed), but it is formulated using multi-dimensional Markov chains [12].

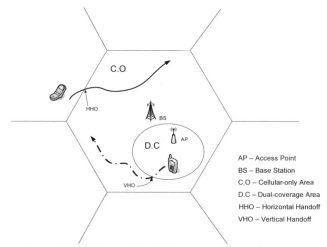

Figure 19.5: Cellular/WLAN Interworking coverage areas.

19.3.1 Cellular/WLAN Model using Birth-Death Processes

First, we need to define the two different coverage areas in cellular/WLAN interworking, namely: the cellular-only coverage (CO) area, and the dual cellular/WLAN coverage (DC) area. In this context, coverage implies service availability. Horizontal and vertical handoffs can occur between the different coverage areas. The coverage areas as well as the handoffs are shown in Fig. 19.5.

19.3.1.1 Model Assumptions

An integrated cellular/WLAN system where one or more WLANs may be deployed inside each cell of the cellular system is considered in [11]. The cellular system has M^c cells. Let A_i^c be the set of cells adjacent to cell i, W_i^c be the set of WLANs inside the coverage of cell i, A_k^w be the set of WLANs adjacent to WLAN k, and D_k^w be the set containing the overlaying cell of WLAN k (i.e., a DC area). The new connection arrival processes to cell i and WLAN k follow Poisson distribution with rates λ_i^c and λ_k^w, respectively, which are independent of other arrival processes. The channel holding time of a connection in cell i is an exponentially distributed random variable with mean $1/\mu_i^c$. The channel holding time in

WLAN k is exponentially distributed with mean $1/\mu_k^w$. Both holding times are independent of earlier arrival times and connection duration times. Each cell i of the cellular system has a capacity of C_i^c units of bandwidth, while each WLAN k has a capacity of C_k^w units of bandwidth. Finally, in order to keep the model to be simple, a connection may request only one unit of bandwidth (i.e., one channel).

19.3.1.2 Mobility Model

At the end of a channel holding time, a connection in cell i of the cellular system may terminate and leave the system with probability q_{iT}^c, or move within the system and continue in an adjacent cell or WLAN with probability $1 - q_{iT}^c$. The probability that a connection continues and moves to an adjacent cell of cell i or WLAN k inside cell i is given by:

$$1 - q_{iT}^c = \sum_{j \in A_i^c} q_{ij}^c + \sum_{k \in W_i^c} q_{ik}^c , \tag{1}$$

where q_{ij}^c is the probability for attempting a horizontal handoff to adjacent cell j, and q_{ik}^c is the probability for attempting a vertical handoff to WLAN k inside cell i.

On the other hand, at the end of a channel holding time of a connection in WLAN k, it may terminate and leave the system with probability q_{kT}^w. The probability that the connection continues and moves to the overlaying cell or to an adjacent WLAN of WLAN k is given by:

$$1 - q_{kT}^w = \sum_{l \in A_k^w} q_{kl}^w + \sum_{i \in D_k^w} q_{ki}^w , \tag{2}$$

where q_{kl}^w is the probability for attempting a horizontal handoff to adjacent WLAN l, and q_{ki}^w is the probability for attempting a vertical handoff to overlaying cell i.

19.3.1.3 Traffic Equations in the Cellular Network

Let n_i be the number of connections in cell i. The occupancy of a cell evolves according to a birth-death process independent of other cells. Note that the birth-death process is a single-dimension Markov chain. The process for cell i evolves with birth rate ρ_i^c. The death rate of cell i in state n_i is $n\mu_i^c$. The total traffic offered to cell i in state n_i is:

$$\rho_i^c = \lambda_i^c + \sum_{j \in A_i^c} \upsilon_{ji}^c + \sum_{k \in W_i^c} \upsilon_{ki}^w . \tag{3}$$

The term υ_{ji}^c is the horizontal handoff rate of cell j offered to adjacent cell i, and is given by:

$$\upsilon_{ji}^c = \lambda_j^c(1 - B_j^c)q_{ji}^c + \sum_{x \in A_j^c} \upsilon_{xj}^c(1 - B_{h_j}^c)q_{ji}^c + \sum_{y \in W_j^c} \upsilon_{yj}^w(1 - B_{h_j}^c)q_{ji}^c, \qquad (4)$$

where B_j^c and $B_{h_j}^c$ are the new connection blocking and handoff dropping probabilities in cell j, respectively.

The term υ_{ki}^w is the vertical handoff rate of WLAN k offered to overlay cell i, and is given by

$$\upsilon_{ki}^w = \lambda_k^w(1 - B_k^w)q_{ki}^w + \sum_{x \in A_k^w} \upsilon_{xk}^w(1 - B_{h_k}^w)q_{ki}^w + \sum_{y \in D_k^w} \upsilon_{yk}^w(1 - B_{h_k}^w)q_{ki}^w, \qquad (5)$$

where B_k^w and $B_{h_k}^w$ are the new connection blocking and handoff dropping probabilities in WLAN k, respectively.

From the birth-death process of cell i, the detailed balance equations are

$$P_i^c(n_i - 1)\rho_i^c = P_i^c(n_i)n_i\mu_i^c, \qquad (6)$$

where $P_i^c(n_i)$ is the stationary distribution that cell i is in state n_i, and is equal to

$$P_i^c(n_i) = \frac{(\rho_i^c)^{n_i}}{G_i^c n_i!(\mu_i^c)^{n_i}}, \quad 0 \le n_i \le C_i^c, \qquad (7)$$

where

$$G_i^c = \sum_{n_i=0}^{C_i^c} \frac{1}{n_i!}\left(\frac{\rho_i^c}{\mu_i^c}\right). \qquad (8)$$

19.3.1.4 Traffic Equations in the WLAN

Similarly, let m_k be the number of connections in WLAN k. The occupancy of WLAN k evolves with birth rate $\rho_k{}^w$ and death rate $m_k{}^w\mu_k{}^w$. The total traffic offered to WLAN k in state m_k is:

$$\rho_k^w = \lambda_k^w + \sum_{l \in A_k^w} \upsilon_{lk}^w + \sum_{j \in D_k^w} \upsilon_{jk}^w. \qquad (9)$$

The term υ_{lk}^w is the horizontal handoff rate of WLAN l offered to adjacent WLAN k, and is given by:

$$\upsilon_{lk}^w = \lambda_l^w (1 - B_l^w) q_{lk}^w + \sum_{x \in A_l^w} \upsilon_{xl}^w (1 - B_{h_l}^w) q_{lk}^w + \sum_{y \in D_l^w} \upsilon_{yl}^w (1 - B_{h_l}^w) q_{lk}^w . \quad (10)$$

The term υ_{jk}^w is the vertical handoff rate of cell j offered to WLAN k, for overlaying cells j and k, and is given by:

$$\upsilon_{jk}^w = \lambda_j^c (1 - B_j^c) R_{jk} q_{jk}^c + \sum_{x \in A_j^c} \upsilon_{xj}^c (1 - B_{h_j}^c) R_{jk} q_{jk}^c + \sum_{y \in W_j^c} \upsilon_{yj}^w (1 - B_{h_j}^c) R_{jk} q_{jk}^c , \quad (11)$$

where R_{jk} is the coverage factor between WLAN k and overlay cell j. It considers the coverage ratio between the radio coverage area of WLAN k and the radio coverage area of cell j with $0 < R_{jk} \le 1$.

The stationary distribution that WLAN k is in state m_k, denoted as $P_k^w(m_k)$, is similar to $P_i^c(n_i)$ for cell i, but with the corresponding WLAN parameters m_k, ρ_k^w, μ_k^w, C_k^w.

19.3.1.5 Performance Measures

The performance of the model can be evaluated in terms of the blocking probabilities of a new connection and the dropping probabilities of a handoff connection. Note that for the sake of simplicity, we do not use any admission control policy to differentiate such connections. Thus, the new connection blocking and handoff dropping probabilities are the same. Admission control can be included in the model easily by modifying the total traffic equations and stationary distributions. Interested readers may refer to [11] for additional details about it.

In the cellular system, using the stationary distribution $P_i^c(n_i)$, the blocking probabilities in cell i can be computed as:

$$B_i^c = B_{h_i}^c = P_i^c(C_i^c) . \quad (12)$$

In the WLANs, using the stationary distribution $P_k^w(m_k)$, the blocking probabilities in WLAN k can be computed as:

$$B_k^w = B_{h_k}^w = P_k^w(C_k^w) . \quad (13)$$

The traffic equations of the horizontal and vertical handoff rates define a set of fixed point equations that can be solved by using an iterative algorithm based on repeated substitutions until it converges to the fixed point. Similar algorithms have been used to estimate the blocking probabilities in cellular systems, and recently also for cellular/WLAN systems as in [15].

19.3.2 Cellular/WLAN Model using Multidimensional Markov Chains

In previous sections, we describe a model of a cellular/WLAN system where each cell and each WLAN are represented by a birth-death process (i.e., a single-dimensional Markov chain). In this section, we show a cellular/WLAN system using multi-dimensional Markov chains that was proposed to evaluate load sharing schemes in [12]. The model considers the traffic intensities, the number of available channels, the buffer sizes, and the coverage of WLANs. Note that in this model, the horizontal and vertical handoffs are not considered.

19.3.2.1 Model Assumptions

An integrated UMTS/WLAN system is considered. In the UMTS network, each connection can request one dedicated data channel (DCH). In the WLAN, the use of the Point Coordination Function (PCF) as MAC in infrastructure IEEE 802.11 WLAN is assumed. Each connection can request one logical channel (LCH) to the point coordinator. The connections that cannot access any DCH and LCH have to wait in the buffer of the corresponding system.

New connections arrive at the UMTS and the WLAN according to Poisson processes with rates λ_u and λ_w, respectively. Service times are exponentially distributed with mean $1/\mu$ in both networks. The behavior of the integrated UMTS/WLAN system is modeled as a four-dimensional (4-D) Markov chain. Let M and N represent the number of DCH in the UMTS and the number of LCH in the WLAN, respectively. Also, let B_u and B_w represent the buffer sizes in the UMTS and WLAN, respectively. The 4-D state representing the status of the system (m,n,x,y) is defined in $0 \leq m \leq M$, $0 \leq n \leq N$, $0 \leq x \leq B_u$, and $0 \leq y \leq B_w$. Based on the treatment of the incoming connections and the coverage overlap among the networks some of the states (m,n,x,y) may not exist [12].

Let us define the set E as the set of states existing in a state diagram for a particular load scheme and define the indicator function $\varepsilon(m,n,x,y)$ to indicate if a state (m,n,x,y) belongs to E as follows:

$$\varepsilon(m,n,x,y) = \begin{cases} 1, & \text{if } (m,n,x,y) \in E, \\ 0, & \text{otherwise.} \end{cases} \tag{14}$$

The probability that the system is in state (m,n,x,y) defined as $\Pr(m,n,x,y)$, the equilibrium equation for state (m,n,x,y) can be formed by solving the flow equations and the conservation among all state probabilities is given by:

$$\sum_{m=0}^{M} \sum_{n=0}^{N} \sum_{x=0}^{B_u} \sum_{y=0}^{B_w} \Pr(m,n,x,y)\varepsilon(m,n,x,y) = 1. \tag{15}$$

Recall that the model in [12] was proposed to evaluate the load sharing schemes named: block balancing (BB), fully sharing (FS) and reserved sharing (RS). Each varies in the treatment and assignment of channels once a new connection arrives to the networks. Interested readers may refer to [12] for additional details about the schemes. For the scope of this chapter, we just need to define N_C as the number of sharable LCHs in the WLAN, and is equal to:

$$N_C = \begin{cases} N, & \text{for } BB \text{ and } FS, \\ N_S, & \text{for } RS. \end{cases} \tag{16}$$

19.3.2.2 Performance Measures

The performance of the model can be evaluated in terms of the new connection blocking probabilities, and the connection buffering probabilities.

In the UMTS system, using the stationary distribution $\Pr(m,n,x,y)$, the new connection blocking probability can be computed as:

$$P_{Bu} = (1-c) \sum_{n=0}^{N_S-1} \Pr(M,n,B_u,0) + \sum_{n=N_S}^{N} \Pr(M,n,B_u,0) + \sum_{y=1}^{B_w} \Pr(M,N,B_u,y) \tag{17}$$

where c is the normalized WLAN coverage with respect to that of the UMTS.

In the WLAN, using the stationary distribution $\Pr(m,n,x,y)$, the new connection blocking probability can be computed as

$$P_{Bw} = \sum_{m=0}^{M} \Pr(m,N,0,B_w) + \sum_{x=1}^{B_u} (M,N,x,B_w) . \tag{18}$$

In the UMTS system, the connection buffering probability can be computed as

$$P_{Fu} = \sum_{x=0}^{B_u-1} \left[(1-c) \sum_{n=0}^{N_S-1} \Pr(M,n,x,0) + \sum_{n=N_S}^{N} \Pr(M,n,x,0) + \sum_{y=1}^{B_w} \Pr(M,N,x,y) \right]. \tag{19}$$

In the WLAN, the connection buffering probability can be computed as:

$$P_{Fw} = \sum_{y=0}^{B_w-1} \left[\sum_{m=0}^{M} \Pr(m,N,0,y) + \sum_{x=1}^{B_u} \Pr(M,N,x,y) \right]. \tag{20}$$

In [12], the assumption of exponential distribution for the channel holding and service time is considered. Models that allow general distributions will be discussed in Section 19.4. Also, other approaches to model the WLAN capacity and the mobility of users in the networks will be discussed.

19.4 Further Analytical Models for Cellular/WLAN Interworking

Section 19.3 introduced analytical models of cellular/WLAN interworking with assumptions made to simplify the analysis procedure. As a result, these models generally cannot achieve a very high accuracy level; however, they are able to calculate operation bounds. This section will describe more sophisticated analytical models proposed in the literature of cellular/WLAN interworking. Each of them aims to improve the accuracy of its modeling.

19.4.1 WLAN Capacity

Before investigating the cellular/WLAN system, the capacity of each network has to be analyzed in terms of its supporting voice and data services. This process can be performed relatively easily in the cellular network since it employs the centralized control and bandwidth reservation scheme. However in the WLAN, this analysis is more complicated due to the fact that it uses a contention-based access algorithm.

The model from [13] analyzes the WLAN capacity with the "per-flow queue with per-flow backoff" principle rather than the "per-node queue with per-node backoff" scheme. The reason for it is that the per-node based rule can be unfair. It is not effective supporting the various QoS requirements for the multimedia sessions. In order to analyze the WLAN capacity, a queuing model is first built with the following entities. Let n_v^w and n_d^w be the number of voice and data calls admitted into a WLAN. The packet arrival rate of the voice flow λ_v^p is assumed to be a constant. Note that it is assumed there are always traffic flows during a data call since the data files to be transmitted are usually pre-stored in a server. Furthermore, we denote the service rate for packets from one voice or data flow by μ_v^w or μ_d^w, respectively. Both μ_v^w and μ_d^w are functions of n_v^w and n_d^w. The transmission probability of a voice and data user in a slot is denoted by τ_v and τ_d, respectively.

After the queuing model is well described, the capacity analysis can be carried out in the following way. First, assume that the data transmission follows the RTS (request to send) – CTS (clear to send) – DATA – ACK scheme, while the voice transmissions do not use RTS/CTS since voice data packets are usually small. Then, the following parameters can be used for analyzing the capacity of the WLAN.

Let p_v be the probability that a voice user suffers a collision in a slot. Since a successful voice transmission happens only if all the other voice and data users are silent, this collision probability is given by:

$$p_v = 1 - P\{\text{a successful voice transmission}\}$$

$$= 1 - \left[1 - \frac{\lambda_v^p}{\mu_v^w}\tau_v\right]^{2n_v^w - 1}\left(1 - \tau_d\right)^{n_d^w}. \qquad (21)$$

Similarly, the collision probability of a data user, which is defined as p_d, is given by:

$$p_d = 1 - P\{\text{a successful data transmission}\}$$

$$= 1 - \left[1 - \frac{\lambda_v^p}{\mu_v^w}\tau_v\right]^{2n_v^w}(1-\tau_d)^{n_d^w-1} \tag{22}$$

Furthermore, let T_{sd}/T_{cd} be the duration of a successful/collided transmission from a data user, and T_{sv}/T_{cv} be the duration of a successful/collided transmission from a voice user. These four parameters can be interpreted by the durations of the DIFS, SIFS, RTS, CTS, CTS TIMEOUT, ACK and ACK TIMEOUT form the IEEE 802.11 MAC Distributed Coordination Function (DCF), and the transmission time of a data or voice frame from the user. Interested readers may refer to [13] for detailed expressions and additional considerations.

Moreover, the service rates μ_v^w and μ_d^w can be expressed in terms of the following parameters: the average number of total flows involved in a collision (k); the average backoff time of a voice or data flow ($\overline{W}_v$ or $\overline{W}_d$, respectively); and the average collision time of a frame in a voice or data flow ($\overline{T}_{cv}$ or $\overline{T}_{cd}$, respectively). Note that $\overline{W}_v$ is a function of p_v, and $\overline{W}_d$ is a function of p_d. $\overline{T}_{cv}$ can be obtained from T_{cv} and p_v, while $\overline{T}_{cd}$ can be obtained from T_{cd} and p_d.

Thus, we have

$$\frac{1}{\mu_v^w} = \left\{\left[\frac{(2n_v^w-1)\lambda_v^p}{\mu_v^w}+1\right]T_{sv} + n_d^w T_{sd} + \overline{W}_v + \frac{1}{k}\left[\left(\frac{(2n_v^w-1)\lambda_v^p}{\mu_v^w}+1\right)\overline{T}_{cv} + n_d^w\overline{T}_{cd}\right]\right\}^{-1}, \tag{23}$$

and

$$\frac{1}{\mu_d^w} = \left\{2n_v^w\frac{\lambda_v^p}{\mu_v^w}T_{sv} + n_d^w T_{sd} + \overline{W}_d + \frac{1}{k}\left[2n_v^w\frac{\lambda_v^p}{\mu_v^w}\overline{T}_{cv} + n_d^w\overline{T}_{cd}\right]\right\}^{-1}. \tag{24}$$

In addition, the service rate of a voice flow has to be greater than the voice arrival rate in order to support the real-time traffic. Therefore the following constraint should be met:

$$\mu_v^w > (1+\delta)\lambda_v^p. \tag{25}$$

where δ is a design parameter which can be determined from simulations

Finally, based on equations (21) to (25), the feasible set of (n_v^w, n_d^w) and each of its corresponding data service rates can be obtained. These results describe more accurately the capacity region of a WLAN using DCF as MAC.

19.4.2 Other Mobility Models

19.4.2.1 Non-uniform Mobility within a Single Cell

A uniform mobility model is not applicable in practice since the users behave differently in the CO area and the DC area (refer to Fig. 19.5). For example, as the WLANs are typically deployed in indoor environments, the mobility level of the users in the DC area is much lower than that of the CO area within the same cell. Thus, a non-uniform model which can differentiate the mobility characteristics of the users in the CO area from the ones in the DC area is necessary to achieve a higher accuracy for the system analysis.

For the sake of simplicity, a few assumptions are made in this model, proposed in [13]. First, there is only one overlaying WLAN network in each cell, and the cell is referred to as a cell cluster. Second, statistical equilibrium is assumed for the whole network, which means each cell cluster is statistically the same as others. Therefore, the analysis can be focused on a single cell. Moreover, three types of calls are considered in this model: new calls, horizontal handoff calls from neighboring cells, and vertical handoff calls between a WLAN and its overlaying cell.

Let T_r^{co} be the residence time of a user stays in the CO area before moving to its neighbor cells (i.e., a horizontal handoff) or to the DC area (i.e., a vertical handoff), and be exponentially distributed with parameter η^{co}. Similarly, let T_r^{dc} be the residence time of a user stays in the DC area before moving to the CO area (i.e., a vertical handoff), and be exponentially distributed with parameter η^{dc}. The probability that a user will move to its neighbor cells from the CO area after T_r^{co} is denoted by p^{cc}, and the probability that a user will move to the DC area from the CO area after T_r^{co} is denoted by p^{cw}. Note the probability of a user moving from the DC area to the CO area is 1.

Let T_{r1}^C denote the cell residence time of a user initiating a new call in the CO area or carrying a horizontal handoff call to the CO area, and T_{r2}^C denote the cell residence time of a user initiating a new call in the DC area. Referring to Fig. 19.6, T_{r1}^C and T_{r2}^C both follow a phase-type (PH) distribution, where each of the transient state (i.e., the dashed rectangle) follows a generalized hyper-exponential distribution with the probability density function (PDF) defined as

$$f_{T_r^{co}+T_r^{dc}}(t) = \frac{\eta^{dc}}{\eta^{dc}-\eta^{co}}\eta^{co}e^{-\eta^{co}t} + \frac{\eta^{co}}{\eta^{co}-\eta^{dc}}\eta^{dc}e^{-\eta^{dc}t}$$

$$= \frac{\eta^{dc}\cdot\eta^{co}}{\eta^{dc}-\eta^{co}}(e^{-\eta^{co}t}-e^{-\eta^{dc}t}) . \tag{26}$$

19.4.2.2 A Cell Residence Time Model for Two-Tier Integrated Wireless Networks

In the model proposed in [14], the assumptions are more realistic in the sense that there can be more than one WLAN access points inside each cell. Moreover, it is possible that a WLAN is crossing the boundary of two neighbor cells (refer to Fig. 19.7). The model also assumes the statistical equilibrium of the cells and considers three types of traffic as in Section 19.4.2.1.

This model is described by the following parameters. Referring to Fig. 19.8, A and B may either stand for WLAN and cellular network (i.e., DC area and CO area) respectively; A_i and B_i denote the access technology in use in state i. We define a_i to be the probability that a user moves to state $i+1$ (i.e., a vertical handoff), and b_i to be the probability that a user moves to its neighbor cell (i.e. a horizontal handoff). Moreover, k stands for the number of alternative visits of networks A and B, and T_i denotes the duration spent in each state. Generally, T_i follows a PH distribution. Note another assumption of this model is that k is a fixed number, while in practice it should be indefinite.

A new call initiated in C.O or handoff to C.O

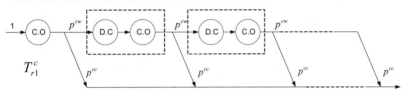

A new call initiated in D.C

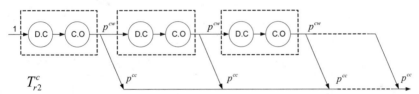

Figure 19.6: Modeling of the cell residence time of a user.

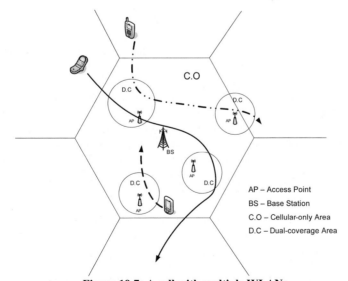

Figure 19.7: A cell with multiple WLANs.

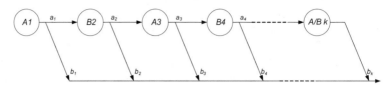

Figure 19.8: The mobility model.

From Figs. 19.7 and 19.8, the cell residence time of the user is composed of all the sojourn times spent in the WLANs (i.e., DC area) and the cellular networks in between (i.e., CO area) before it leaves the cell. Thus, the resultant cell residence time will also follow a PH distribution due to its closure property under summation.

The parameters in the model can be estimated based on the information collected from simulation. First, the initial technology (i.e., A or B), the WLAN duration (T_i in DC area), the inter-WLAN duration (T_i in CO area) and the number of switches between access networks (k) are measured for each cell visited by the user. Then, a_i and b_i can be calculated based on equation (27), where $N_c(i)$ denotes the number of cells in which exactly i vertical handoffs are carried out.

$$b_i = \frac{N_c(i-1)}{\sum_{j=i-1}^{\infty} N_c(j)} \quad \text{and} \quad a_i = 1 - b_i. \tag{27}$$

Furthermore, the PH distribution parameters can be estimated using some distribution fitting tools. For instance, according to the coefficient of variation (CoV) of each state, a state measurement can be fitted to a hyper-exponential distribution (CoV > 1), an exponential distribution (CoV = 1) or a hype-exponential distribution (CoV < 1).

19.4.3 Models with General Distributions

In most proposed analytical models, the time variables are usually assumed to have exponential distributions. However, this hypothesis is not valid for some recent mobile networks. As a result, a new modeling scheme is proposed by [15], in which the cell residence times follow general distributions. This model is based on the framework using Laplace transformation to deal with general distributions of the call and the cell residence times in hierarchical cellular networks developed in [16]. The corresponding handoff arrival rates and the channel holding times are derived and analyzed in the following subsections. As in other analytical models, one assumption for this model is that all the cells of the cellular network (i.e., 3G cells) and their overlaying WLANs are stochastically identical, which means the analysis can be focused on one 3G cell and it's overlaying WLANs.

Let H_{3G} (H_{WL}) be the call holding time of a 3G (WLAN) user with mean h_{3G} (h_{WL}), and R_{3G} (R_{WL}) be the cell residence time of a 3G user in a 3G cell (WLAN) with mean r_{3G} (r_{WL}). Moreover, let λ_{3G}^n, λ_{3G}^{hh} and λ_{3G}^{vh} be the arrival rates of new calls, horizontal handoff calls and vertical handoff (WLAN to 3G) calls in a 3G cell respectively, while λ_{WL}^n

and λ_{WL}^{vh} be the arrival rates of new calls and vertical handoff (3G to WLAN) calls in a WLAN respectively. Let X^r be the residual time of an arbitrary time variable X, of which the Laplace transformation is denoted by $f_X^*(s)$.

We define r as the ratio between the total areas of the overlaying WLANs and the area of the 3G cell. The weight factor w is necessary due to the fact that the user density under the WLANs is typically larger than that of the 3G cell. As a result, in order to calculate the probability that a user is under the WLAN, w is introduced with $w \geq 1$ and $0 \leq wr \leq 1$. Since wr denotes the population coverage rate of a 3G cell, p is used to substitute wr for convenience. Furthermore, we assume there are M channels in each 3G cell, and the maximum number of channels allowed for new calls is m.

19.4.3.1 Traffic Equations of Handoff Rates

There are three kinds of calls in a 3G cell: new calls, horizontal handoff calls and vertical handoff (i.e., from WLAN to 3G) calls. While in a WLAN, there are only two types of calls: new calls and vertical handoff (i.e., from 3G to WLAN) calls. The argument in [15] is that WLANs are usually deployed in hotspots fashion, of which the coverage areas in general do not overlap with each other.

Fig. 19.9 shows that there are three kinds of horizontal handoff calls in a 3G cell: new calls initiated in neighbor 3G cells carried on their sessions to the current cell (case a); horizontal handoff calls of neighbor 3G cells continue their sessions to the current cell (case b); vertical handoff (WLAN to 3G) calls keep their sessions on to the current cell (case c). Therefore, the horizontal handoff arrival rates in a 3G cell can be obtained from the following steady state equation, where $P_A^{~B}$ is the probability that a call of type B is currently in a network of type A:

$$\lambda_{3G}^{hh} = \lambda_{3G}^n \left(1 - P_{3G}^n\right) \mathbb{P}\left(R_{3G}^r < H_{3G}\right) + \lambda_{3G}^{hh}\left(1 - P_{3G}^{hh}\right)\mathbb{P}\left(R_{3G} < H_{3G}^r\right) + \lambda_{3G}^{vh}\left(1 - P_{3G}^{vh}\right)\mathbb{P}\left(R_{3G}^r < H_{3G}^r\right) \quad (28)$$

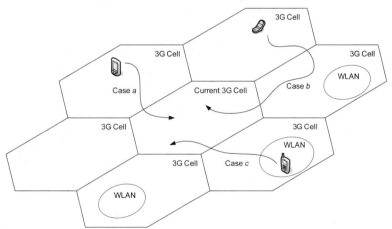

Figure 19.9: Three types of horizontal handoff calls in a 3G cell.

By taking the Laplace transformation of the probabilities in equation (28) in the following way

$$\mathrm{P}\!\left(R_{3G}^r < H_{3G}\right) = \int_0^\infty \int_t^\infty f_{H_{3G}}(\tau) f_{R_{3G}^r}(t)\,d\tau dt = \frac{r_{3G}\left[1 - f_{R_{3G}}^*(h_{3G})\right]}{h_{3G}}, \qquad (29)$$

$$\mathrm{P}\!\left(R_{3G} < H_{3G}^r\right) = \int_0^\infty \int_t^\infty f_{H_{3G}^r}(\tau) f_{R_{3G}}(t)\,d\tau dt = f_{R_{3G}}^*(h_{3G}), \qquad (30)$$

$$\mathrm{P}\!\left(R_{3G}^r < H_{3G}^r\right) = \int_0^\infty \int_t^\infty f_{H_{3G}^r}(\tau) f_{R_{3G}^r}(t)\,d\tau dt = \frac{r_{3G}\left[1 - f_{R_{3G}}^*(h_{3G})\right]}{h_{3G}}, \qquad (31)$$

We are able to obtain

$$\lambda_{3G}^{hh} = \left[\lambda_{3G}^n\left(1 - P_{3G}^n\right) + \lambda_{3G}^{vh}\left(1 - P_{3G}^{vh}\right)\right]\frac{r_{3G}\left[1 - f_{R_{3G}}^*(h_{3G})\right]}{h_{3G}\left[1 - \left(1 - P_{3G}^{hh}\right)f_{R_{3G}}^*(h_{3G})\right]}. \qquad (32)$$

Similarly, Fig. 19.10 shows that there are three types of vertical handoff calls in a WLAN: a new arrival call blocked by a 3G cell will attempt a direct vertical handoff to its overlaying WLAN if it is under its coverage (case *a*); otherwise, it will attempt an indirect vertical handoff (case *b*); a horizontal handoff call blocked by a 3G cell will also attempt an indirectly vertical handoff (case *c*). Here, direct handoff means that if a new call from a 3G user is blocked by the 3G cell, it can directly attempt a vertical handoff to the overlaying WLAN if it happens to be under the coverage area of that WLAN. Otherwise, for an indirect vertical handoff, the system will ask one of the ongoing calls which are under the WLAN coverage to perform a direct handoff to release a channel for the new call.

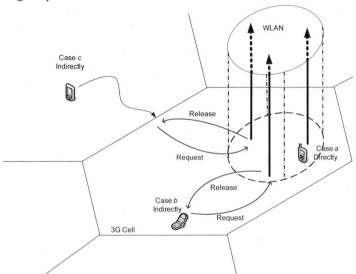

Figure 19.10: Three types of vertical handoff calls in a WLAN.

Hence, the vertical handoff arrival rate in a WLAN can be determined by the following equation, where $\overline{m}$ is the number of ongoing calls in the current 3G cell:

$$\lambda_{WL}^{vh} = \lambda_{3G}^n P_{3G}^n \left\{ p + (1-p)\left[1 - (1-p)^{\overline{m}}\right] \right\} + \lambda_{3G}^{hh} P_{3G}^{hh} \left[1 - (1-p)^M\right]. \tag{33}$$

In the same way, the vertical handoff arrival rate in a 3G cell can be calculated by the following equation:

$$\lambda_{3G}^{vh} = \lambda_{3G}^n P_{3G}^n p \left(1 - P_{WL}^{vh}\right) \mathbb{P}\left(R_{WL}^r < H_{3G}\right)$$
$$+ \left\{ \lambda_{3G}^n P_{3G}^n (1-p)\left[1 - (1-p)^{\overline{m}}\right] + \lambda_{3G}^{hh} P_{3G}^{hh} \left[1 - (1-p)^M\right] \right\}\left(1 - P_{WL}^{vh}\right) \mathbb{P}\left(R_{WL}^r < H_{3G}^r\right) \tag{34}$$

While using the Laplace transformation as follows:

$$\mathbb{P}\left(R_{WL}^r < H_{3G}\right) = \mathbb{P}\left(R_{WL}^r < H_{3G}^r\right) = \frac{r_{WL}\left[1 - f_{R_{WL}}^*(h_{3G})\right]}{h_{3G}}. \tag{35}$$

We are able to obtain

$$\lambda_{3G}^{vh} = \left\{ \lambda_{3G}^n P_{3G}^n \left\{ p + (1+p)\left[1 - (1-p)^{\overline{m}}\right] \right\} + \lambda_{3G}^{hh} P_{3G}^{hh} \left[1 - (1-p)^M\right] \right\} \cdot \left(1 - P_{WL}^{vh}\right) \frac{r_{WL}\left[1 - f_{R_{WL}}^*(h_{3G})\right]}{h_{3G}}. \tag{36}$$

As a result, the explicit solution of different handoff arrival rates can be obtained from equations (32), (33) and (36), or an iterative algorithm can be used to solve the fixed-point equations as in [11].

19.4.3.2 Channel Holding Times

Let T_{3G}^n, T_{3G}^{hh}, and T_{3G}^{vh} be the channel holding time for a new call, a horizontal handoff call, and a vertical handoff (WLAN to 3G) call of a 3G user respectively, while let T_{WL}^n be the channel holding time for a new call of a WLAN user and T_{WL}^{vh} be the channel holding time for a vertical handoff (3G to WLAN) call of a 3G user. From Fig. 19.7, the channel holding times for new calls, horizontal handoff calls and vertical handoff calls in a 3G cell can be described by:

$$T_{3G}^n = \min\left\{H_{3G}, R_{3G}^r\right\}, \tag{37}$$

$$T_{3G}^{hh} = \min\left\{H_{3G}^r, R_{3G}\right\}, \tag{38}$$

$$T_{3G}^{vh} = \min\left\{H_{3G}^r, R_{3G}^r\right\}. \tag{39}$$

By using Laplace transformation, the corresponding mean channel holding times can be obtained as follows:

$$E\left[T_{3G}^{n}\right]= E\left[T_{3G}^{vh}\right]=\frac{1}{h_{3G}}-\frac{r_{3G}}{h_{3G}^{2}}\left[1-f_{R_{3G}}^{*}\left(h_{3G}\right)\right], \tag{40}$$

$$E\left[T_{3G}^{hh}\right]=\frac{1}{h_{3G}}\left[1-f_{R_{3G}}^{*}\left(h_{3G}\right)\right]. \tag{41}$$

In the WLAN, since the users have very limited mobility, the channel holding times are assumed to be the same as call holding times. Thus we have

$$T_{WL}^{n}=H_{WL}^{n}, \tag{42}$$

$$E\left[T_{WL}^{n}\right]=\frac{1}{h_{WL}}. \tag{43}$$

From the channel holding time of a vertical handoff call of a 3G user in the WLAN, the corresponding mean channel holding time is given by:

$$E\left[T_{WL}^{vh}\right]=\frac{1}{h_{3G}}-\frac{r_{WL}}{h_{3G}^{2}}\left[1-f_{R_{WL}}^{*}\left(h_{3G}\right)\right]. \tag{44}$$

19.5 Simulation Models

The previous sections illustrate a number of analytical models of cellular/WLAN interworking. This section will present several models whose performances are evaluated mainly based on computer simulations rather than analytical approaches.

In [17], a system consists of a set of base stations (BSs) and access points (APs). Each mobile station (MS) is assumed to support only one wireless interface and can switch between WLAN and the cellular network. It is further assumed that there are three classes of users in this system, each of which has a unique user profile. The system is modeled and simulated using *GloMoSim* [18]. The performance of the system is evaluated in terms of the throughput of the network, the number of handoffs carried out per user, the average cost per byte and the total revenue generated in the system.

In [19], the system under consideration mainly focuses on one single cell cellular network and one overlapping WLAN. An MS is considered to move through the DC area a number of times with different speeds. The system is simulated using an IEEE 802.11 WLAN protocol simulator, which models the IEEE 802.11a specification. The BS of the cell and the AP of the WLAN are configured in the simulator with different coverage ranges and transmission rates such that the AP only supports the highest PHY mode with a maximal data rate of 54 Mbps, while the BS was fixed to the QPSK mode with a data rate of 12 Mbps. The simulation results provide the optimal handoff positions for the MS and evaluate the system in terms of its average throughput.

In [20], a Transmission Control Protocol (TCP) scheme is proposed for seamless vertical handoff between WLAN and the cellular network. The system architecture is similar to that of [19], and the performance of the proposed scheme is evaluated using the *ns-2* network simulator [21] in terms of the congestion window size and the TCP sequence number. In this simulation, the AP is configured with a data rate of 2 Mbps and an end-to-end RTT (Round Trip Time) equals to 100 ms, whereas the BS is emulated by another AP which is configured with a lower data rate (144 kbps) and a larger end-to-end RTT (300 ms). However, in [22], by using the same simulator (i.e., *ns-2*) and a similar system architecture, the BS is implemented based on the preamble-based TDMA protocol rather than emulated by another AP. This enhances the accuracy of the model and provides more reliable results from the simulation. The performance of this system is evaluated in terms of the bandwidth-delay product and the sequence number regarding to its vertical handoff mechanism for TCP congestion control.

The above models all use one simulator to evaluate their performances. In some cases (e.g., in the industry), multiple simulator modules are employed in the simulation process to achieve a higher level of accuracy in terms of the performance of the model. For example, in [23], different simulation tools for propagation modeling, site optimization, frequency planning, WLAN physical layer and cellular/WLAN capacity analysis are combined together in order to evaluate the throughput, power, data rate, coverage and the capacity of the integrated system.

Sometimes researchers will also build experimental prototypes of their models to test their performances. For instance, in [2], [5], the IOTA gateway and the corresponding client software are implemented to evaluate their proposed model. Similarly, in [24], an OmniCon prototype is built to demonstrate the ability of their model in achieving low handoff latency and close to zero packet loss. Furthermore, in [25], a testbed network which consists of real GSM/GPRS and IEEE 802.11b operators, network exchange nodes (i.e., routers) and IPv6 infrastructures is constructed in order to validate the concept of coordination and resource management of different RANs through their proposed hierarchical system.

19.6 Open Issues

Although this chapter presents a survey of analytical models for cellular/WLAN interworking that have been proposed in the literature, there are several open issues that need to be addressed and require further investigations. We briefly summarize some of them:

- **The modeling of traffic.** In the case of Internet data traffic, its self-similar and fractal nature need to be considered. In the case of IP multimedia traffic, its variable bandwidth requirements need to be addressed for admission control and resource management. Such considerations further complicate the analytical models.
- **The modeling of QoS in cellular/WLAN.** The different approaches to deal with guaranteed QoS in each network need to be considered. In the cellular network due to the resource reservation it can be done, but for the WLAN using contention-based

MAC there is no guaranteed QoS. To solve this, modeling of the IEEE 802.11e with QoS support is required.

- **The modeling of joint network operations.** Although for the case in that each network is operated by a different operator (i.e., loose coupling) it may not be totally possible. For the case of both networks being managed by the same operator (i.e., tight coupling) it is relevant and necessary to investigate about models that consider joint admission control, and resource sharing.

- **The model of pricing strategies.** The current charging models of cellular networks and WLANs are different. Cellular networks usually use a volume-based pricing, while WLANs use a flat-based pricing. Models that include both pricing strategies and that allow the evaluation of new ones are required.

- **Model validation from real-traces.** In order to validate the real accuracy and relevance of a particular model, in the ideal case, real traffic traces are required. However, due to the fact that cellular/WLAN interworking is starting to be deployed, the validation from real traffic traces is not possible at this time.

- **Accuracy versus complexity tradeoff.** Although new analytical models are required to include and address the open issues presented in this section. It is also important to investigate the tradeoff between accuracy and complexity of the models. A model can be very accurate, but very difficult or complicated to use, while on the other hand, other model can be less accurate and only provide bounds, but widely used for its simplicity.

19.7 Conclusions

This chapter presented a survey of analytical models for cellular/WLAN interworking. Different interworking architectures for the integration of cellular networks with WLANs were summarized. We started with several simple analytical models and then we introduced models that increase the modeling accuracy (e.g., mobility of users, WLAN capacity). Table 19.1 summarizes and organizes the information regarding each of the models. For each model, its main advantages and limitations are listed. For the sake of completeness, some simulation-based models and prototypes for cellular/WLAN interworking were also briefly described. Although an extensive list of the simulators that can be used for cellular/WLAN interworking is beyond the scope of this chapter, it is worth to mention two other simulators widely used for wireless networks: *OPNET* [26], and *OMNet ++* [27].

Finally, this chapter considered only the case of integration of cellular networks with WLANs. This case is only a particular interworking case of the wireless networks being deployed everywhere. There are also personal area networks (PANs) formed by Bluetooth-enabled mobile devices, and also other wide area networks (WANs) beside cellular networks, such as the IEEE 802.16 WiMax standards. The mixture of these wireless technologies with cellular networks and WLANs is referred to as *heterogeneous wireless networks*. Analytical models for heterogeneous wireless networks will be an active research area for the coming years.

Table 19.1: Advantages and drawbacks of various analytical models.

Model	Advantages	Limitations
[11]	• Considers both horizontal and vertical handoff. • Can be extended to include admission control and resource sharing. • Multiple WLANs in each cell.	• Uses exponential distributions for channel holding times. • Only considers single service.
[12]	• Buffers for connections to reduce blocking probability. • Considers three load schemes (i.e., BS, FS and RS).	• Uses exponential distributions for channel holding times. • Does not consider horizontal and/or vertical handoff. • WLAN only with PCF as MAC.
[13]	• Non-uniform mobility model (i.e., Phase-type distributed). • Accurate capacity for WLAN with DCF as MAC. • Considers admission control and resource sharing.	• Uses exponential distributions for data file size. • Assumes only one WLAN in each cell.
[14]	• Accurate cell residence time model (i.e., phase-type distributed). • Multiple WLANs in each cell.	• Does not consider admission control and/or resource sharing. • The number of vertical handoffs is truncated to a specific value. • Requires statistical parameter estimation (e.g., fitting tools).
[15]	• General distribution cell residence time (i.e., analyzed using Laplace Transform). • Considers admission control.	• Does not consider horizontal handoffs between WLANs. • Only considers single service.

19.8 References

[1] A. Salkintzis, C. Fors, and R. Pazhyannur, "WLAN-GPRS Integration for Next-Generation Mobile Data Networks," *IEEE Wireless Communications*, vol. 9, no. 5, pp. 112-124, October 2002.

[2] M. Buddhikot, G. Chandranmenon, S. Han, Y. Lee, S. Miller, and L. Salgarelli, "Integration of 802.11 and Third-Generation Wireless Data Networks," in *Proc. of IEEE Conference on Computer Communications (INFOCOM'03)*, San Francisco, CA, April 2003.

[3] M. Lott, M. Siebert, S. Bonjour, D. von Hugo, and M. Weckerle, "Interworking of WLAN and 3G systems," *IEE Proc. of Communications*, vol. 151, no. 5, pp. 507-513, October 2004.

[4] A. Salkintzis, "Interworking Techniques and Architectures for WLAN/3G Integration Toward 4G Mobile Data Networks," *IEEE Wireless Communications*, vol. 11, no. 3, pp. 50-61, June 2004.

[5] M. Buddhikot, G. Chandranmenon, S. Han, Y. Lee, S. Miller, and L. Salgarelli, "Design and Implementation of a WLAN/CDMA2000 Interworking Architecture," *IEEE Communications Magazine*, vol. 41, no. 11, pp 90-100, November 2003.

[6]   V. Varma, S. Ramesh, K. Wong, and J. Friedhoffer, "Mobility Management in Integrated UMTS/WLAN Networks," in *Proc. of IEEE International Conference on Communications (ICC'03)*, Anchorage, AK, May 2003.

[7]   J. Song, S. Lee, and D. Cho, "Hybrid Coupling Scheme for UMTS and Wireless LAN Interworking," in *Proc. of IEEE Vehicular Technology Conference (VTC'03-Fall)*, Orlando, FL, October 2003.

[8]   Gonzalo Camarillo and Miguel-Angel Garcia-Martin, *The 3G IP Multimedia Subsystem (IMS): Merging the Internet and the Cellular Worlds*. John Wiley & Sons, 2006.

[9]   F. Galan Marquez, M. Gomez Rodriguez, T. Robles Valladares, T. de Miguel, and L. Galindo, "Interworking of IP Multimedia Core Networks Between 3GPP and WLAN," *IEEE Wireless Communications*, vol. 12, no. 3, pp 58-65, June 2005.

[10]   K. Ahmavaara, H. Haverinen, and R. Pichna, "Interworking Architecture Between 3GPP and WLAN Systems," *IEEE Communications Magazine*, vol. 41, no. 11, pp 74-81, November 2003.

[11]   E. Stevens-Navarro and V. Wong, "Resource Sharing in an Integrated Wireless Cellular/WLAN System," in *Proc. of Canadian Conference on Electrical and Computer Engineering (CCECE'07)*, Vancouver, Canada, April 2007.

[12]   D. Chen, X. Wang, and A. Elhakeem, "Load Sharing With Buffering Over Heterogeneous Networks," in *Proc. of IEEE Vehicular Technology Conference (VTC'05-Fall)*, Dallas, TX, September 2005.

[13]   W. Song, H. Jiang, W. Zhuang, and A. Saleh, "Call Admission Control for Integrated Voice/Data Services in Cellular/WLAN Interworking," in *Proc. of IEEE International Conference on Communications (ICC'06)*, Istanbul, Turkey, June 2006.

[14]   A. Zahran, B. Liang, and A. Saleh, "Modeling and Performance Analysis of Beyond 3G Integrated Wireless Networks," in *Proc. of IEEE International Conference on Communications (ICC'06)*, Istanbul, Turkey, June 2006.

[15]   S. Tang and W. Li, "Performance Analysis of the 3G Network with Complementary WLANs," in *Proc. of IEEE Global Communications Conference (GLOBECOM'05)*, St. Louis, MO, November 2005.

[16]   K. Yeo and C. Jun, "Modeling and Analysis of Hierarchical Cellular Networks with General Distributions of Call and Cell Residence Times," *IEEE Transactions on Vehicular Technology*, vol. 51, no. 6, pp. 1361-1374, November 2002.

[17]   D. Joseph, B. Manoj, and C. Siva Ram Murthy, "Interoperability of Wi-Fi Hotspots and Cellular Networks," in *Proc. of ACM Workshop on Wireless Mobile Applications and Services on WLAN Hotspots (WMASH'04)*, Philadelphia, PA, October 2004.

[18]   GloMoSim, http://pcl.cs.ucla.edu/projects/glomosim/

[19]   S. Goebbels, M. Siebert, M. Schinnenburg, and M. Lott, "Simulative Evaluation of Location Aided Handover in Wireless Heterogeneous Systems," in *Proc. of IEEE International Symposium on Personal, Indoor and Mobile Radio Communications (PIMRC'04)*, Barcelona, Spain, September 2004.

[20]   S. Kim and J. Copeland, "TCP for Seamless Vertical Handoff in Hybrid Mobile Data Networks," in *Proc. of IEEE Global Communications Conference (GLOBECOM '03)*, San Francisco, CA, December 2003.

[21] The Network Simulator - ns-2, http://www.isi.edu/nsnam/ns

[22] Y. Gou, D. Pearce, and P. Mitchell, "A Receiver-based Vertical Handover Mechanism for TCP Congestion Control," *IEEE Transactions on Wireless Communications*, vol. 5, no. 10, pp. 2824-2833, October 2006.

[23] A. Doufexi, S. Armour, and A. Molina, "Hotspot Wireless LANs to Enhance the Performance of 3G and Beyond Cellular Networks," *IEEE Communications Magazine*, vol. 41, no. 7, pp. 58-65, July 2003.

[24] S. Sharma, I. Baek, Y. Dodia, and T. Chiueh, "OmniCon: A Mobile IP-based Vertical Handoff System for Wireless LAN and GPRS Links," in *Proc. of IEEE International Conference on Parallel Processing Workshops (ICPPW'04)*, Montreal, Canada, August 2004.

[25] G. Karetsos, S. Kyriazakos, E. Groustiotis, F. Giandomenico, and I. Mura, "A Hierarchical Radio Resource Management Framework for Integrating WLANs in Cellular Networking Environments," *IEEE Wireless Communications*, vol. 12, no. 6, pp. 11-17, December 2005.

[26] OPNET - http://www.opnet.com

[27] OMNet ++ - Discrete Event Simulation System, http://www.omnetpp.org

20

Coexistence of Unlicensed Wireless Networks

Stephen J. Shellhammer[a]

20.1 Introduction

There are a variety of unlicensed wireless networks that operate in shared frequency bands. Many of these networks are installed in portable and hand-held devices. Given that these devices are portable it is not uncommon for devices using one unlicensed network to be in close proximity to another device using another unlicensed wireless network. These unlicensed wireless networks may operate in the same frequency band and may even operate in the same channel within the band. This raises the issue of whether these two networks can coexist in the same location. This chapter describes this coexistence issue.

Section 20.2 gives an overview of the various unlicensed frequency bands in the United States. Many of these bands are also unlicensed bands in other countries. Section 20.3 is a short survey of some of the unlicensed wireless networks that share these frequency bands. Section 20.4 gives a short history of coexistence of unlicensed wireless networks. Section 20.5 illustrates how to evaluate coexistence of two unlicensed wireless networks. Section 20.6 describes several methods of improving coexistence. Sections 20.7 and 20.8 give example of evaluating coexistence of wireless networks, using two IEEE 802 standards as case studies. And finally Section 20.9 discusses the issue of dynamic spectrum access in which unlicensed wireless networks identify and use unused portions of licensed frequency bands.

20.2 Overview of Unlicensed Frequency Bands

Unlicensed wireless networks can operate in a variety of frequency bands. This section will briefly review these frequency bands and some of the key regulatory rules governing operation in these frequency bands. The US rules can be found in 0.

In the United States the most commonly used unlicensed frequency bands are the industrial, scientific and medical (ISM) bands. These bands are: 902-928 MHz, 2400-2483.5 MHz and 5725-5850 MHz. In addition to the ISM band there is the unlicensed National Information Infrastructure (U-NII) band, which consists of several sub-bands: 5150-5125, 5250-5350 and 5725-5850. Notice that the highest U-NII sub-band is also an ISM band. Hence it is possible to obtain FCC certification according to either set of rules.

---

[a] *Qualcomm, Inc and IEEE 802.19 Chair*

Recently there have been a few new bands that have been added to the list. In 2005 the 3650-3700 MHz band has been opened up for non-exclusive licensed use. Though wireless networks must be licensed it is on a non-exclusive basis. So it is possible for multiple networks to operating in the same geographical region. This non-exclusive nature of the frequency band makes it much like an unlicensed band.

The 60 GHz frequency band spans from 57 GHz to 64 GHz. Hence, this band offers significant bandwidth but at these high frequencies is poses challenges to the RF designer and the RF propagation characteristics tend to be mostly line-of-sight.

The FCC has indicated that it is intending to allow unlicensed operation in the unused television (TV) frequency bands beginning in February 2009, once the transition to digital television is complete. The preliminary FCC rules in this band require that these unlicensed devices use cognitive radio techniques to sense the spectrum and identify geographically unused TV channels for used by these wireless networks.

In the remainder of this section we will briefly review the FCC rules for the various unlicensed frequency bands.

20.2.1 ISM and U-NII Frequency Bands

Originally the FCC rules for the ISM bands required the use of spread spectrum transmission, with at least a 10 dB processing gain. However, a few years ago the FCC loosened the spreading requirements. The rules now require either the use of frequency hopping spread spectrum or the use of digital modulation. The phrase digital modulation is very broad and covers most modern wireless technologies.

In the 900 MHz frequency band the frequency hopping rules depend upon the bandwidth of the signal. If the bandwidth is less than 250 KHz then the system must hop over at least 50 hopping frequencies and cannot dwell on any frequency for more than 400 ms out of every 20 seconds. If the bandwidth is greater than 250 KHz then the system must hop on at least 25 hopping frequencies and cannot dwell on any hop frequency for more than 400 ms our of every 10 seconds. These rules are intended to ensure sufficient spreading and to keep the power spectral density uniform when averaged over time. Frequency hopping systems that hop on at least 50 hopping frequencies are permitted to transmit up to 1 watt of power while those systems that hop on between 25 and 49 channels can transmit up to 250 mW of power.

In the 2.4 GHz frequency band the frequency hopping rules differ from those in the 900 MHz band. Originally in the 2.4 GHz band a frequency hopping spread spectrum system was required to hop on at least 75 channels. However, a few years ago the required number of channels was dropped to a minimum of 15 if the system avoids certain channels to improve coexistence with another wireless network. This rules change was introduced to permit Bluetooth adaptive frequency hopping (AFH) which was introduced in Bluetooth version 1.1 to reduce interference between Bluetooth and frequency-static systems like IEEE 802.11. These systems that operate on at least 15 channels, but less than 75 channels, can transmit up to 125 mw while those systems that hop on at least 75 channels can transmit up to 1 watt.

The rules for frequency hopping systems in the 5 GHz band are much simpler. These systems must hop on at least 75 channels and must have a bandwidth of no more than 1

MHz. The system must not dwell more than 400 ms on any given channel within each 30 second interval.

The rules for systems that used digital modulation are not as elaborate as the rules for the frequency hopping systems. In any of the three ISM bands the bandwidth must be at least 500 KHz. The maximum conducted transmit power must be less than 1 watt. In addition, a directional antenna with an antenna gain of up to 6 dBi can be used. The combination of conducted transmit power and antenna gain can be represented by the effective isotropic radiated power. The combination of the 30 dB transmit power and the 6 dBi antenna gain gives an EIRP of 36 dBm. A higher gain antenna is used then the transmit power must be reduced so that the total EIRP is no larger than 36 dBm. There is an exception for point-to-point systems, but most networks operating in a point-to-multipoint mode of operation. Note, that power measured in dBm is the power in milliwatts, converted to dB. Hence if the power is P_{mw} milliwatts then the power in dBm is given by,

$$P_{dBm} = 10Log(P_{mw}) \tag{1}$$

The 5 GHz U-NII band is intended for wireless systems that use wideband digital modulation for high-speed mobile and fixed communication. In this context, high-rate means data rates of at least 1 Mb/s.

The maximum allowed transmit power is dependent on the bandwidth of the signal and also depends on which sub-band is used. In general, the power is the minimum of either a fixed value or a bandwidth dependent value. The maximum conducted transmit power for each of the sub-bands is summarized in Table 20.1. The bandwidth B is the bandwidth in MHz. Similarly, to the ISM bands a 6 dBi antenna gain is allowed, and any antenna gain above 6 dBi requires a corresponding decrease in transmit power. In fixed point-to-point systems devices a directional antenna gain of up to 23 dBi is permitted without any reduction in transmit power.

Table 20.1: Maximum conducted transmit power for U-NII sub-bands.

Frequency Band	Maximum Conducted Transmit Power
5.15-5.25 GHz	min(17, 4 + 10Log(B)) dBm
5.25-5.35 GHz and 5.47-5.4725 GHz	min(24, 11 + 10Log(B)) dBm
5.725-5.825 GHz	min(30, 17 + 10Log(B)) dBm

There are also additional requirements for operation in the middle two sub-bands to enable these U-NII bands to coexist with radar systems that share the same bands. The two required features are transmit power control (TPC) and dynamic frequency selection (DFS).

A device operating in those bands with an EIRP of 500 mw or more is required to support transmit power control. This is a feature that allows a device to reduce it's transmit power based on a message from the device receiving the signal. The device is required to be capable of lowering the EIRP to below 24 dBm.

In addition to TPC the devices must support dynamic frequency selection (DFS). A DFS-enabled network observes a given channel for the presence of a radar signal and if it detects a radar signal above a given detection threshold the U-NII device must evacuate that channel. This process of observing (listening to) the channel is also called *spectrum sensing* in the cognitive radio literature. A U-NII device with an EIRP of more than 200 mw must evacuate the channel if it detects a radar signal with a power of greater or equal to -62 dBm. A U-NII device with an EIRP less than 200 mw must evacuate the channel if it detects a radar signal with stronger than -64 dBm. The reason the higher power U-NII device is required to have a more sensitive detector is that it can cause more interference to the radar system. The radar signal power must be averaged over 1 microsecond reference to a 0 dBi sensing antenna.

There are timing requirements for DFS operation. Before operating in a channel a U-NII device must sense the channel for 60 seconds to determine that no radar systems are using the channel. If a U-NII device is using a channel it must also monitor the channel for radar signals, and if a radar signal is detected the U-NII device must cease transmission in that channel within 10 seconds. During those 10 seconds, only 200 ms of the time can be used for data transmission. However, control signals can be sent, that can be used to inform other devices in the network that the network is moving to a different channel. After evacuating a channel, the device cannot use that channel again for at least 30 minutes.

20.2.2 The 3650 MHz Frequency Band

In 2005 [2] the FCC issued a report and order (R&O) allowing wireless broadband services in the 3650-3700 MHz frequency band. Services in this band are based on a non-exclusive nationwide license. These licensing rules are a hybrid of unlicensed and licensed rules. In order to offer a service in this band one must obtain a nationwide license. Since it is a non-exclusive license others can also obtain a license and operate in the same region. Hence, the non-exclusive nature of the license shares characteristics with the unlicensed bands. One needs to be concerned about coexistence of wireless devices in the band that may be operated by another party and may be designed to another standard or specification. However, since it is licensed the government intends to keep a database of the location of registered stations. So if an interference issue arises it will be possible to access this database to identify possible sources of interference.

The rules for operation in this band are specifically designed to facilitate sharing of the spectrum by multiple wireless networks that may be designed to different specifications. The one key element that must be present in any device operating in this band is that the device must employ a contention-based protocol. The FCC definition of a contention-based protocol is given in [2],

> *Contention-based protocol: A protocol that allows multiple users to share the same spectrum by defining the events that must occur when two or more transmitters attempt to simultaneously access the same channel and establishing rules by which a transmitter provides reasonable opportunities for other transmitters to operate. Such a protocol may consist of procedures for initiating new transmissions, procedures for determining the state of the channel (available or unavailable), and procedures for managing retransmissions in the event of a busy channel.*

So we can see that in this band a contention based protocol consists of three procedures,

1. A procedure for initiating transmissions
2. A procedure to testing the state of the channel
3. A procedure for managing retransmissions in the event of a busy channel

All wireless networks utilize Procedure 1. The critical procedure is Procedure 2 for testing the state of the channel. The most common method for fulfilling Procedure 2 is to sense the channel by listening for certain features of a signal that may be present in the channel. These features can be as simple as the received signal power within the frequency band or more complex as the header of a packet being transmitted by a nearby station. A common phrase for such a protocol is a listen-before-talk (LBT) protocol, and a typical example of such a protocol is the carrier sense multiple access collision avoidance (CSMA/CA) protocol used in Wi-Fi networks. Finally, most wireless networks use Procedure 3 for retransmission of lost packets. These packets may be lost due to interference from other wireless systems or due to insufficient signal-to-noise ratio. This procedure is easily implemented by including a cyclic redundancy check (CRC) in the packet to validate the accuracy of the data and an index to identify the packet number, so as to not confuse a retransmission of a previous packet with a new packet.

20.2.3 VHF and UHF Television Frequency Bands

In 2004 [3] the FCC issued a notice of proposed rulemaking (NPRM) indicating its intention to allow unlicensed operation in unused television (TV) channels. These unlicensed systems are to use cognitive radio techniques to identify geographically unused TV channels. This opportunistic use of unused spectrum represents a new era in spectrum allocation. Given that spectrum is a precious resource that in some cases is not fully utilized the FCC is looking for more novel methods of spectrum access. In 2006 the FCC issued its first report and order (R&O) as well as a further NPRM on the use of unused TV channels [4]. The FCC intends to issue the final R&O in late 2007 and to permit use of unused TV channels after the completion of the DTV transmission in February 2009. At that point terrestrial TV transmission will be based on the advanced television standards committee (ATSC) digital TV standard. The devices operating in unused TV channels are referred to in FCC documents as *TV band devices*.

Currently the FCC has decided to allow fixed TV band devices. The further NPRM seeks comments are to whether portable TV band devices should be allowed.

The TV channels potentially available for use by a TV band device include both the very high frequency (VHF) and ultra high frequency (UHF) TV channels, with some exceptions. The VHF channels 2-12 are included in the set of usable channels. The lower UHF channels (14-20) are used in some metropolitan areas for public safely communications. The FCC has decided that those will not be allowed for use by portable TV band devices (assuming there final decision is to allow portable devices at all). The FCC is still considering whether to allow fixed TV band devices in channels 14-20, outside the areas that used these channels for public safety. UHF channels 21-36 and 38-51 will be available for use by TV band devices. Channel 37 is excluded since it is used for radio astronomy.

The initial rules for use of unused TV channels are similar to the dynamic frequency selection (DFS) rules that were introduced in the 5 GHz unlicensed band for avoiding interference to radars. The DFS parameters are of course different since in this case the objective is to avoid causing interference with TV broadcasts signals. The transmit power limits are similar to those used in the 2.4 GHz ISM band.

The maximum conducted transmit power is 1 watt. A directional gain antenna with a gain of up to 6 dBi is allowed without any reduction in transmit power. Any gain above 6 dBi requires a corresponding reduction in transmit power. Hence, the maximum EIRP is 36 dBm.

The TV band device must listen to a TV channel for at least 30 seconds to unsure that the TV channel is currently be used for TV broadcast. This interval is referred to as the *Channel Availability Check Time*. For a TV channel to be classified as available no TV signal stronger than -116 dBm may be present in the TV channel. This signal power level is referred to as the *minimum DFS detection threshold*. After operating in a TV channel the device must continue to monitor the channel for TV broadcast signals, at least every 10 seconds. This process is referred to as *in-service monitoring*. If a TV broadcast signal is detected above the minimum DFS detection threshold then the device must cease transmission on the channel within 10 seconds. And during those 10 seconds the maximum permitted time for data transmission is 200 ms. Other transmission for management and control signals are permitted during the 10 seconds. That allows a network of TV band devices to exchange information about moving the network off the current TV channel and over to another TV channel.

20.3 Survey of Unlicensed Wireless Networks

The purpose of this section is to give the reader a flavor of the various wireless networks operating in the unlicensed frequency bands. This section is not intended as a detailed description of each of the various wireless networks, just a brief summary. This section is organized by wireless network technology: WLAN, WPAN, WMAN, WRAN, etc.

20.3.1 Wireless Local Area Network (WLAN)

The most popular wireless local area network (WLAN) standard is IEEE 802.11, of which there have been a number of amendments. Some of these amendments have been to introduce new physical (PHY) layers, while other amendments have been to introduce new capabilities in the medium access control (MAC) layer. The original 802.11 standard issued approximately 10 years ago in 1997 [5]. The original standard included three different PHY layers. When the 802.11 standard was originally issued operation in the ISM bands required use of spread spectrum technology. There were two reasons for this FCC rule. First by spreading the signal the power spectral density of the unlicensed transmission would be spread out over a wider frequency range resulting in less interference to narrow band systems. Second, the unlicensed system is less susceptible to narrowband interference due to the processing gain benefit at the receiver. The two RF PHY layers in the original 802.11 standard were the frequency hopping spread spectrum (FHSS) PHY and the direct sequence spread spectrum (DSSS) PHY. Both of the RF PHY operate in the 2.4 GHz ISM frequency band. There was also an infrared (IR) optical PHY which was never implemented in the industry. The FHSS PHY transmits at 1 or 2 Mb/s, which hops over 79 1-MHz channels. This PHY is no longer popular; however, the Bluetooth SIG leveraged parts of this PHY in the Bluetooth specification. The DSSS PHY transmits at either 1 or 2 Mb/s using BPSK or QPSK respectively. The signal is spread with an 11-chip Barker sequence.

In 1999 the IEEE issued two high-rate PHY layer amendments: 802.11a and 802.11b. The 802.11b PHY [6], and extension of the original DSSS PHY, became quickly popular partial due to the establishment of the Wi-Fi alliance (originally called the wireless Ethernet Compatibility Alliance). The 802.11b PHY added 5.5 Mb/s and 11 Mb/s data rates to the original 1 and 2 Mb/s data rates. The bandwidth of the signal remained unchanged. By increasing the data rates while maintaining the signal bandwidth, the spread spectrum processing gain is reduced, resulting in signals that are more susceptible to interference. This is a common theme, that with limited bandwidth and a demand for higher data rates, the signals require higher SNR and hence are often much more susceptible to interference. On the other hand, higher data rates tend to lead to shorter packet durations, which with bursty interference shorter packets tend to be much less susceptible to interference.

Also in 1999 [7], the IEEE issued 802.11a which is an orthogonal frequency division multiplexing (OFDM) PHY operating in the 5 GHz frequency band. Due to the use of the 5 GHz frequency band this PHY initially did not become as popular due to the cost of RF circuits in that band at the time. However, 802.11a has now become quite popular. Also, the bandwidth available at 5 GHz is significantly greater than that at 2.4 GHz which makes this frequency band attractive. The 802.11a PHY supports data rates from 6 Mb/s up to 54 Mb/s. These data rates use modulations ranging from BPSK up to 64-QAM and code rates ranging from rate ½ to rate ¾.

Over the last several years the IEEE has been developing a higher-rate PHY layer using multiple input multiple output (MIMO) antenna technology and wider bandwidth signals (i.e., 40 MHz versus 20 MHz channels). This new PHY is currently only available as a draft since the standard has not been ratified by the IEEE. When ratified the amendment will be 802.11n. This new amendment to 802.11 not only includes a new PHY but also many enhancements to the MAC layer to improve the overall throughput available to the user. For many more details about the draft 802.11n see Chapter 8 of this book. IEEE 802.11n will be able to operate in either the 2.4 GHz or the 5 GHz frequency band. In terms of coexistence with other wireless devices in the band we can make several generalizations, prior to performing any simulations or detailed analysis. First, as an interferer, when the 20 MHz channels are used, 802.11n causes similar interference to the previous 802.11 PHYs. However, as a victim of interference this new PHY is more susceptible to interference, since the data rates have been increased within a given bandwidth. This increase in spectral efficiency comes at a price of an increase in the required SNR which results in higher sensitivity to interference from other wireless systems.

Finally, the IEEE is also currently developing a PHY for operation in the 3650 MHz frequency band. This band, from 3650 to 3700 MHz does not allow unlicensed wireless networks. However, it does allow non-exclusive licensed operation. Since the licenses are non-exclusive it is quite possible for multiple wireless networks, potentially designed according to different standards or specifications, to operate in the same region. The new amendment, 802.11y, is currently a draft since it has not yet been ratified by the IEEE. Operation in this band requires that the network use a contention-based protocol. A typical example of a contention-based protocol is carrier sense multiple access with collision avoidance (CSMA/CA) used in 802.11. To improve sharing of the spectrum with other wireless networks, the sensitivity of energy detection (ED) has been increased within 820.11y. The ED is used in the clear channel assessment (CCA) within the MAC in determining if the channel is currently busy.

20.3.2 Wireless Personal Area Network (WPAN)

There are a variety of wireless personal area networks. The nodes in these networks can be portable or fixed. Unlike in a WLAN where it is most common for one of the nodes to be in a fixed location and connected to an external network, it is quite common in a wireless personal area network (WPAN) for all the nodes to be portable. This portability makes it quite possible for a WPAN device to come into close proximity of a node of another wireless network (e.g., WLAN) since these portable WPAN devices are often carried around by the user. The typical range of a WPAN link it approximately 10 meters. In this subsection we will give a very brief overview of the various WPAN standards.

The first popular WPAN standard was Bluetooth, which was also standardized within the IEEE as 802.15.1 in 2002 and revised in 2005 [8]. Bluetooth uses a frequency hopping spread spectrum (FHSS) PHY that hops at up to 1600 hops per second. The hop set consists of 79 1-MHz channels across the 2.4 GHz band. This results in significant coexistence issues which resulted in the formation of an IEEE task group to address this coexistence issue [9]. A subsequent release of the Bluetooth specification included a new mode of operation referred to as adaptive frequency hopping. This allowed the Bluetooth network to select a subset of the 79 channels so to allow the network to "hop around" other frequency-static networks in the same vicinity.

Another WPAN becoming popular is ZigBee, whose PHY and MAC layers have been standardized within the IEEE as 802.15.4 [10]. There are two new amendments to this IEEE standard which are currently in draft form. The 802.15.4a draft [11] and the 802.15.4b draft [12] both introduce alternative PHY layers to the original standard. The original 802.15.4 standard includes three different PHYs, for three different frequency bands. The first PHY operates in the 868 MHz frequency band (in Europe). It has one data rate of 20 kb/s with BPSK modulation and uses DSSS with 15 chips per bit. There is only a single channel supported due to the limited bandwidth of this frequency band. The second PHY operates in the 915 MHz frequency band (in the US). It has a single data rate of 40 kb/s with BPSK modulation and used DSSS with 15 chips per bit. There are 11 channels supported in this band. The third PHY operates in the 2.4 GHz frequency band. It supports a single data rate of 250 kb/s with 16-ary offset QPSK (O-QPSK) modulation using DSSS 32 chips/symbol or equivalently 8 chips/bit. This PHY supports 16 channels.

The 802.15.4a amendment adds a chirp spread spectrum (CSS) PHY in the 2.4 GHz band and an ultrawideband (UWB) PHY in three frequency bands. The CSS PHY supports data rates of 250 kb/s and 1 Mb/s and 14 channels within the band. The UWB PHY operates in one of three frequency bands. The first frequency band is 150-650 MHz which is divided into two channels. The second band is 3.1-4.8 GHz which is divided into four sub-bands, each of which is divided into two channels, giving a total of 8 channels. The third band is 6.0-10.6 GHz which is divided into 11 sub-bands, each of which is divided into two channels, giving 22 channels.

The 802.15.4b revision introduces two new PHYs operating in the 868 and 915 MHz frequency bands. The amplitude shift keying (ASK) PHY operates in both bands. In the 868 MHz band it has a data rate of 250 kb/s with a chip rate of 400 kc/s. In the 915 MHz band the ASK PHY has a data rate of 250 kb/s with a chip rate of 1000 kc/s. The O-QPSK PHY, which in the original standard operated only in the 2.4 GHz frequency band, operates in the 868 and 915 MHz bands in the 802.15.4b revision. In the 868 MHz band it has a data rate of 100 kb/s with a chip rate of 400 kc/s and in the 915 MHz band it has a data rate of 250 kb/s with a chip rate of 1000 kc/s.

20.3.3 Wireless Metropolitan Area Network (WMAN)

Another wireless network that can be deployed in unlicensed bands is the IEEE 802.16 Wireless Metropolitan Area Network [13]. This wireless network, which is also know as WiMAX can be deployed in both licensed and unlicensed frequency bands (referred to as license-exempt bands in the standard). Initially the standard [13] was for a fixed wireless network, however, in 2005 an amendment to the standard was added which supports mobile operation [14]. There are several physical (PHY) layers included in the standard: single carrier (SC), orthogonal frequency division multiplexing (OFDM) and orthogonal frequency division multiple access (OFDMA).

The full standard supports both frequency division duplex (FDD) and time division duplex (TDD). The license-exempt section of the standard (referred to as WirelessHUMAN) only supports the TDD mode. This is a reasonable restriction since support for FDD requires the availability of a pair of channels with reasonable frequency separation to allow simultaneous transmission and reception. Since it is unlikely that one could guarantee the availability of such pairs of channels in an unlicensed band, it is reasonable to use TDD in unlicensed bands.

There are two unlicensed bands listed in the 2004 version of the standard [13]: 2.4 GHz and 5 GHz band. However, the new amendment [14] eliminated references to specific frequency bands of operation so as to allow for more flexibility. In practice, the 2.4 GHz and 5 GHz bands are likely the unlicensed band that will be used for unlicensed deployments. However, due to the transmit power limits in the unlicensed bands it is unclear if they will become as popular for deployment as the licensed bands.

The WirelessHUMAN version of the standard supports dynamic frequency selection (DFS) to avoid interference to radar systems in the 5 GHz frequency band, just as DFS is supported in 802.11 WLAN.

Unlike 802.11, the 802.16 WMAN does not perform carrier sensing before accessing the wireless channel. This makes coexistence of 802.16 with other unlicensed wireless systems a challenge. In order to improve the coexistence capability of 802.16 the IEEE has initiated a project to develop a new amendment to the standard. The scope of the 802.16h [15] project is,

This amendment specifies improved mechanisms, as policies and medium access control enhancements, to enable coexistence among license-exempt systems based on IEEE Standard 802.16 and to facilitate the coexistence of such systems with primary users.

The coexistence mechanisms being included in the 802.16h draft [16] can be divided into two categories: coordinated and uncoordinated mechanisms. The coordinated coexistence mechanism supports scheduling transmission for nearby cells. This approach was originally developed to improve coexistence between 802.16 systems, but could also be used to coexist with other systems. The draft also supports coexistence mechanisms that do not require coordination between networks. These types of coexistence mechanisms are aimed at improving coexistence of 802.16 with other networks, like 802.11.

The draft includes support for dynamic frequency selection (DFS) for sharing with primary licensed users (e.g., radar) similar to what is done in 802.11. The draft also supports dynamic channel selection (DCS) for automatic selection of the operating channel. DCS is similar to DFS but is designed for sharing with other wireless network with similar regulatory status. Unlike in DFS where the system must move to another channel if the primary system is detected, DCS can select a channel based on other criteria, like the performance of the 802.16h network.

In addition to DFS and DCS the draft also supports an uncoordinated coexistence protocol (UCP) which consists of three features: extended quiet periods (EQP), adaptive extended quiet periods (aEQP) and listen before talk (LBT).

EQP schedules quiet periods that are intended to allow other wireless networks access to the medium. If the other wireless network utilizes a listen before talk protocol, like 802.11, then the other wireless network will detect the extended quiet period and utilize that time for its own transmissions. Of course, if 802.16h scheduled a lot of quiet time which is not used by another wireless network then the performance of the 802.16h network is degraded for no reason. So a better approach is to be able to vary the amount of quiet time based on spectrum usage by other wireless networks. The adaptive extended quiet period (aEQP) does just that. The 802.16h network can increase or decrease its maximum duty cycle based on observations of the channel.

EQP is fixed and does not adapt to the channel utilization. aEQP is a slowly adapting technique that does adapt to the channel utilization. A faster adaptation method is listen-before-talk (LBT) which is also supported in the 802.16h draft. The LBT parameters in the 802.16h draft are modeled after the clear channel assessment times of 802.11. So for a 20 MHz bandwidth system the minimum listening time is 4 µs, for a 10 MHz system the minimum listening time is 8 µs and for a 5 MHz system the minimum listening time is 16 µs.

Members of IEEE 802 have recently begun evaluation of the coexistence of 802.11 and 802.16h with these new features. Given that these studies have just begun there are no definitive simulation results of yet.

For more information about the IEEE 802.16 standard see [17].

20.3.4 Wireless Regional Area Network (WRAN)

Another new unlicensed system that is being standardized within the IEEE is the 802.22 Wireless Regional Area Network (WRAN) [18]. The ongoing project is to develop a standard for unlicensed operation in geographically unused television (TV) channels. The wireless network is to use cognitive radio techniques to identify which TV are unoccupied by licensed systems (TV broadcasts and professional wireless microphones) so that the WRAN can safely utilize one of these channels. The unlicensed use of unoccupied channels is often referred to as dynamic spectrum access, which is discussed more in Section 20.9.

IEEE 802.22 is still under development so one cannot make definitive statements about the form of the standard once it is completed. However, it is possible to make some general observations about the ongoing standards development project. A WRAN is fixed wireless network intended to deliver wireless access to rural areas. The current development is based on an OFDMA PHY (similar in some ways to 802.16e) with support for time division duplex (TDD) operation. One of the new capabilities to be included in the standard is referred to as *spectrum sensing*. Spectrum sensing is the process of making observations of the RF spectrum to identify which TV channels are occupied by licensed systems and which are unoccupied.

Because an 802.22 WRAN will operate in TV band it has to coexist with both licensed and unlicensed wireless systems. However, the requirements for these two types of coexistence are quite different. The 802.22 WRAN is not permitted to cause harmful interference to licensed systems, which is a very strong requirement. The FCC is basically requiring the WRAN to not cause any harmful interference to a licensed system. On the other hand, the 802.22 WRAN has no regulatory requirement to coexist, to any specific level, with other unlicensed wireless network in the band. Hence, this coexistence with other unlicensed systems then becomes a practical matter of what is an acceptable level of interference between the WRAN and any other unlicensed wireless network in the band.

20.3.5 Cordless Telephones

Cordless phones are also popular wireless system in the unlicensed frequency bands. Originally, these cordless phones operated in the 900 MHz frequency band. However, as RF circuit technology developed these cordless phones have moved into the 2.4 GHz and 5 GHz bands. Cordless phones are typically designed to meet a proprietary specification. These cordless phones typically use either direct sequence spread spectrum (DSSS) or frequency hopping spread spectrum (FHSS) technology. These phones are often higher power than other unlicensed wireless systems, with power levels of several hundred milliwatts. The reason for using higher power is to increase the range, which is an important performance characteristic. Some of the parameters that effect coexistence of cordless phones and other unlicensed wireless systems can be found in [19].

20.4 History of Unlicensed Wireless Coexistence

As long as there have been wireless networks there have always been issues of coexistence. However, with the introduction of wireless networks in unlicensed bands the issue has become more acute. The issue began to receive significant press with the introduction of the Bluetooth technology around the same time that IEEE 802.11b networks were becoming quite popular. At the highest data rate the 802.11b PHY no longer has any spread spectrum processing gain, and as a result is susceptible to interference. Bluetooth hops quickly across the 2.4 GHz frequency band at up to 1600 hops/sec. Since Bluetooth hops at such a high rate there is a good probability that during an 802.11b packet transmission that the Bluetooth network will hop into the 802.11b channel. So if the Bluetooth device is near an 802.11b station a packet collision occurs. With Bluetooth being standardized within the IEEE during this time period this issue became quite significant. As a result the IEEE formed a task group to quantify the severity of the problem and offer solutions. This ultimately resulted in a published IEEE recommended practice [9]. The recommended practice included a coexistence assessment of Bluetooth and IEEE 802.11b. In addition to including a coexistence assessment the recommended practice includes recommendations on several methods for improving coexistence. Recently a book has been published describing methods of evaluating coexistence and suggesting several coexistence strategies [20].

It became apparent that the best solution to this problem was for the Bluetooth specification to be modified to allow the Bluetooth network to modify its hop set so that it could "hop around" the 802.11b network. Work on adaptive frequency hopping was done within the IEEE, but ultimately the Bluetooth specification needed to be updated by the Bluetooth SIG. The version of the Bluetooth specification was later standardized by the IEEE [8].

With the IEEE generating more wireless standards for operation in unlicensed frequency bands, in 2002 the IEEE established the IEEE 802.19 technical advisory group (TAG) on wireless coexistence. Now as new unlicensed wireless standards are developed the working group developing a standard generates a coexistence assurance document which is reviewed by the 820.19 TAG. The coexistence assurance document is an evaluation of how well the standard under development coexists with currently published standards. The 802.19 TAG has also begun development of a recommended practice on methods of evaluating the coexistence of unlicensed wireless networks [21].

20.5 How to Evaluate the Coexistence of Wireless Networks

How does one determine how well two wireless networks coexist? First it is necessary to describe the situation. What type of wireless networks are these? Now many stations are in each network? How close are the stations in one network to the stations in the other network? This description of the setup is often referred to as the *coexistence scenario*. So step one is to specify the coexistence scenario. The next step is to decide on now to measure the coexistence between these wireless networks. The values that are used to measure the coexistence of the wireless networks are often called *coexistence metrics*. There are a variety of metrics that can be used: packet error rate (PER), throughput, latency, etc. A useful illustration of how well two unlicensed wireless networks is a plot of the coexistence metric (e.g., PER) versus the separation the closest stations in the two networks. There are of course many possible coexistence metrics and also many other independent variables (e.g., physical separation) that can be varied when evaluating the coexistence metrics. Finally, after specifying the coexistence scenario and selecting the coexistence metric or metrics, we need a method of evaluating this metric. This involves modeling the two networks. The model can be based on a simulation, a set of analytic formulas, or some combination of analysis and simulation. In this section we will give an overview of a method that uses a combination of analytic and simulation methods.

In the remainder of this section we will give an illustration of a typical coexistence scenario for unlicensed wireless networks, select a typical coexistence metric and show one method of evaluating that metric.

Several of the factors that lead to concerns about coexistence between unlicensed wireless networks are the portability of the devices and the fact that they often share the same frequency band. These two factors lead to situation where one or more of the stations in one network are near one of the stations in the other network, while operating on the same channel. One example of this is a person who has one or two WPAN devices with him, possibly in his pockets, walks over near a WLAN station, possibly in his colleague's laptop computer. One of the first observations about this typical scenario we can make is that the interference between the two networks is dominated by only a few stations. That is not to say that there are not coexistence scenarios where there is interaction between many stations, but it turns out that this case where the interference is dominated by a few stations is a very common scenario. So in this section we will consider this typical coexistence scenario with few stations in each network.

The coexistence scenario we will consider here consists of a WLAN network and a WPAN network. Each network has only two stations. Though this seems like an artificially simple scenario it is not, since as was just described the interference between networks is often dominated by the stations in each network which are closest to one another. Figure 20.1 illustrates this coexistence scenario. The WLAN consists of an access point (AP) and a station (STA). The station is placed and the origin for convenience. The separation between the AP and the station is L meters. The WPAN consists of two stations, with the station closes to the WLAN station d meters away. In this scenario we will consider the impact of the interference of the WPAN on the WLAN. In this case we will assume that the second WPAN station is far enough away from the WLAN that its impact is much smaller than the station closes to the WLAN.

In this coexistence scenario we will consider that case in which both the WLAN and the WPAN are frequency static systems (i.e., not frequency hopping) and they are both operating on the same channel. This can occur for several reasons. In many cases the wireless networks are set of fixed frequencies and since the WPAN is highly mobile it can easily come close to a WLAN station on the same channel. And even if one of the wireless networks is sophisticated enough to modify its operating channel, there may be only a small set of channels, which are all used by the WLAN for example, and the WPAN must use one of these channels.

What should we choose as our coexistence metric? A very typical metric to use is the packet error rate (PER) of the WLAN link between the AP and the station. This is a reasonable choice for a coexistence metric for two reasons. Fundamentally, the interference from one network results in packet errors in the other network. This is the method in which the interference impacts the network. Secondly, many of the other possible coexistence metrics (e.g., throughput and latency) depend directly on the packet error rate. So for these reasons it makes for a reasonable choice as a coexistence metric.

Finally we must select a method of calculating the value of the coexistence metric under various conditions. One approach that can be taken is to build a simulation which includes the PHY and MAC layer of both networks. This approach is often used but development of this combined simulator can be time consuming. And often it is desirable to evaluate the coexistence of two networks during the standards development stage. So it is very useful to have methods that simplifies this process so that results can be obtained in a shorter timeframe. This section will give a brief summary of a method of estimating the packet error rate for this coexistence scenario. More details of this approach can be found in [22].

First it is necessary to set the transmit power levels of the WLAN and WPAN transmission. The distance between the WLAN AP and station should be set to a typical value. Of course, that value can be set to represent a station near the edge of coverage. The distance d between the WLAN and WPAN stations should be varied to see the effect of the separation. Then a path loss model is applied to determine the actual receive signal power and noise power levels. In this scenario the WLAN station at the origin is the WLAN station being impacted by the interference, so all the signal and noise values are measured for that station.

There are many possible path loss models. One that is popular for indoor applications uses a path loss exponent of 2 out to 8 meters and then a path loss exponent of 3.3 beyond that distance. In the 2.4 GHz frequency band the path loss formula is,

$$pl(d) = \begin{cases} 40.2 + 20Log_{10}(d) & 0.5m < d \le 8m \\ 58.5 + 33Log_{10}\left(\dfrac{d}{8}\right) & d > 8m \end{cases} \tag{2}$$

This path loss model is illustrated in Figure 20.2.

Given the transmit power for the WLAN signal and the WPAN interferer, the distances between nodes and the path loss formula one can easily calculate the signal to interference ratio γ, at the WLAN station located at the origin.

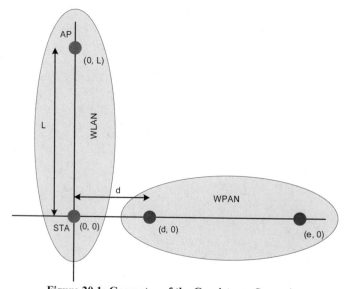

Figure 20.1: Geometry of the Coexistence Scenario.

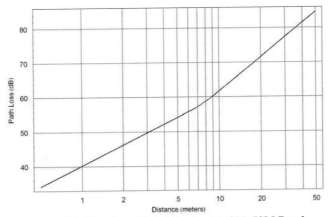

Figure 20.2: Path Loss Model for the 2.4 GHz ISM Band.

If the WPAN station generated a continuous transmission then the PER would be straightforward to determine. However, the WPAN transmissions occur in bursts. Before we deal with temporal variation in the WPAN transmissions let us consider the effect of a continuous transmission.

The interference from the WPAN causes errors at the WLAN receiver located at the origin. In this coexistence scenario the reception at the other WLAN receive is typically unaffected by the WPAN interference due to the power levels and the geometry of the scenario. The WLAN packet consists of a series of symbols. These modulation symbols encode information bits. In this section the analysis will be performed in terms of the WLAN symbol error rate (SER), however, the same approach can be applied in terms of the WLAN bit error rate. It is easier to visualize the analysis using SER so that is the approach that will be taken here.

It is necessary to determine the WLAN symbol error rate (SER), as a function of the signal to interference ratio (SIR). Since the WLAN typically supports multiple data rates, we need the SER versus γ for each data rate. This is best accomplished using a simulation of the WLAN receiver. Typically, in the standard development process it is common to have such a simulator available. It is important to make sure that if the WPAN signal bandwidth is larger than the WLAN bandwidth that the interference power level is scaled accordingly. For example, if the WPAN bandwidth was twice that of the WLAN then the SNR value we would use would be twice (3 dB higher) the SIR value since only half the interference passes through the WLAN receiver. So based on a simulation, or possibly a set of previously developed curves, we have the SER versus SIR which we will indicate as $p(\gamma)$ or in some case we drop the SIR argument and write the SER as just p.

Now we can address the bursty nature of the WPAN interference. We could of course build a simulator with bursty interference and run it for all the various cases, but that turns out to be unnecessary.

In general the WLAN packet structure consists of a header followed by data, as shown in Figure 20.3. Typically for most data rates the header is much more robust than the data portion of the packet and to first order can be ignored in this analysis. If however, the WLAN is at its lowest data rate and the robustness of the data is approximately the same as the header than we can consider the entire packet in total.

Figure 20.3: General WLAN Packet Structure.

The data portion is segmented into a series of symbols, as shown in Figure 20.4, with symbol duration T.

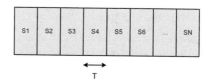

Figure 20.4: Data Portion of WLAN Packet.

As the WLAN is transmitting packets the WPAN is also sending packets. These WPAN packets represent interference to the WLAN station. These interference packets may be of various durations and may be separated by a fixed time or different times. There are many deterministic and stochastic models that can be used in this analysis.

Figure 20.5 illustrates the WLAN packet and several WPAN interference packets. As we can see there is a possibility that the WPAN interference packet can overlap in time with a portion or even the entire WLAN packet. This time overlap can cause a packet error. The longer the temporal overlap the higher the chance of a packet error.

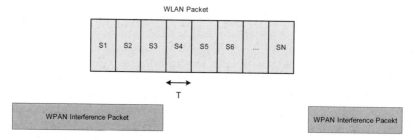

Figure 20.5: WLAN Packet and WPAN Interference Packets.

We can introduce some formula for calculation of the packet error rate. We label the event of a packet error as PE. The packet error rate (PER) is the probability of this event occurring,

$$PER = P(PE) \tag{3}$$

The probability of a packet error is dependent on the number of symbols that are interfered with by the WPAN interference packets. Depending on the time alignment of the WLAN and WPAN packets, the duration of the WPAN packet and their frequency of occurrence the number of WLAN symbol collisions will vary. Let us label the number of symbol collisions M. This M is a discrete random variable, whose probability mass function is $f_M(m)$. There are several methods of determining this probability mass function. If the WPAN packets occur with a fixed duration and a fixed repletion time then one can find a simple formula for the probability mass function [22]. If the WPAN packet duration and repletion times vary and one has difficulty finding a closed formed solution to the probability mass function one can run a simple simulation to find the probability mass function. Actually, the periodic WPAN packet transmission mode is a good one.

It has been shown that if the WPAN packet transmission are random with varying packet durations and repletion times that the ultimate PER is approximately the same as that resulting from periodic interference with a packet duration and duty cycle equal to average duration and duty cycle of the random transmission [22].

Once we have the probability mass function for the number of symbol collisions we can calculate the PER from the total probability function [23],

$$PER = P(PE) = \sum_{m=0}^{N} P(PE \mid m) f_M(m) \qquad (4)$$

Where N is the number of symbols in the WLAN packet. Now the probability of a packet error conditioned on m symbol collisions is simply,

$$P(PE \mid m) = 1 - (1 - p)^m \qquad (5)$$

Hence the PER formula can be written as,

$$PER = \sum_{m=0}^{N} [1 - (1 - p)^m] f_M(m) \qquad (6)$$

So the work all comes down to finding the SER curves as a function of SIR and the probability mass function for the number of symbol collisions.

We can illustrate this process with a very simple example. We consider a WLAN with two different data rates, the low data rate using BPSK and the high data rate using 64-QAM modulation.

In this illustration the WLAN packet contains a fixed message size independent of which of the two data rates is used. Hence when the low data rate is use the duration of the WLAN packet is longer in time than when the high data rate is used. This is not uncommon since often a WLAN message (MAC Frame) is set by other factors, like the size of the IP packet that the MAC frame is carrying.

The timing in this example is as follows. The WPAN is transmitting periodically with a 25% duty cycle. The duration of the WLAN packet, when transmitting at the low data rate, is the same as the duration of the WPAN packet. However, when the WLAN transmits at the high data rate the WLAN packet duration decreases. Since the WPAN packet transmissions are periodic it is possible to find a closed form solution for the probability mass function for the number of bit collisions. Using this probability mass function we can calculate the PER for the for different modulation schemes used in the WLAN.

Figure 20.6 shows the PER versus network separation distance for the two different WLAN data rates.

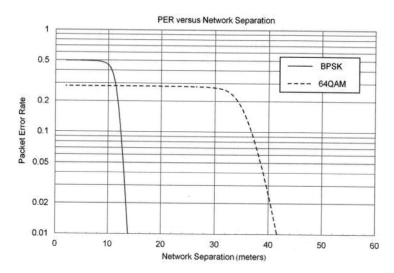

Figure 20.6: PER Curves.

We can make several observations from these curves. When the network separation is small the PER is worse for the low data rate case. This is because the longer packet duration results in a higher probability of a collision. And up close the SIR is so low almost any symbol collision causes a packet error. When the network separation is large the low data rate case performs better. This is because the SER for the low rate case drops quickly beyond a certain SIR and even though there is a significant temporal overlap it results it rarely causes a packet error.

The approach described in this section will be applied to a real-life example in Section 20.7.

20.6 Methods of Improving Coexistence

There are a variety of methods that can be used to improve coexistence between unlicensed wireless networks. In this section we will go over several methods.

One way to think about the coexistence of different wireless networks is as *uncoordinated multiple access*. In a multiple access technique the system is designed for sharing of the medium (spectrum) between stations in the network. However, in the unlicensed bands we have a variety of wireless networks designed to different specifications or standards. So you do not have the luxury of redesigning the other wireless network. However, many of the concepts used in multiple access apply in some sense to coexistence in unlicensed bands. In multiple access you have a variety of techniques designed to make multiple transmissions orthogonal, or nearly orthogonal. These techniques include time division multiple access, frequency division multiple access and code division multiple access. These same concepts apply in the unlicensed bands.

The original FCC rules for operation in the unlicensed frequency bands required the use of spread spectrum technology with a processing gain of at least 10 dB. This approach makes the system much less susceptible to interference. Frequency hopping spread spectrum technology is used effectively in Bluetooth. With the network hopping over 79 channels there is a processing gain of 19 dB, which is quite effective in making Bluetooth quite robust to interference. However, there is a fundamental limit to the processing gain you can use as these networks go to higher and higher data rates, in a fixed bandwidth. So we see that the original 802.11 included two PHYs each using spread spectrum with processing gain of over 10 dB. However, the demand for higher data rates resulted in the newer 802.11 PHYs instead of having a bandwidth higher than the data rate (hence having processing gain) they have in fact gone to higher modulation schemes resulting in much higher spectral efficiency (higher bits/Hz) which makes them more susceptible to interference. So spread spectrum is useful at lower data rates but it is difficult to use at higher data rates given the limited bandwidth of the channel.

Frequency division multiple access is very natural method of coexisting in an unlicensed band. Many systems are designed to operating in a portion of the band and hence if the two wireless networks operate on different channels, which do not overlap in frequency, you often get excellent coexistence. Of course the obvious problem arises when there are not enough channels. The other issue is the selection of the channel of operation. Many unlicensed wireless networks are operated by consumers who are not familiar with the technology and will operate the network on the default channel. On method of addressing this issue is to design the network to automatically select the channel of operation. This method is typically referred to as dynamic frequency selection (DFS). In some bands DFS is required so that the unlicensed wireless network can avoid operating in a channel which would cause harmful interference to a licensed wireless network nearby. This technology is more recently referred to as dynamic spectrum access (DSA) and is discussed in Section 20.9.

Similar to dynamic frequency selection in a frequency-static network we have adaptive frequency hopping (AFH) in a frequency hopping spread spectrum system. The updated Bluetooth specification (standardized by the IEEE in [8]) supports AFH. The primary motivation for introducing AFH in the Bluetooth specification was to improve coexistence with 802.11 networks. When AFH is enabled in a Bluetooth network all the nodes in the network can observe the spectral occupancy of the 79 hopping channels. The slave nodes can report to the master node there observations of which channels are occupied and which are vacant. The Bluetooth master then selects a subset of the 79 channels for use. The AFH protocol is designed to synchronize when the network switches to a new set of hopping channels. AFH enables Bluetooth to observe the spectral occupancy of the channels and then modify its hop set to "hop around" a nearby frequency-static network.

Having discussed code division and frequency division multiple access techniques, what can be done in the time domain? For two unsynchronized wireless networks it is difficult to use time division multiple access. However, there is one simple approach that can be used: fragment packets. As was shown earlier the longer the packets the higher the probability of a packet collision. So one method of improving robustness is to fragment a MAC frame and transmit a series of smaller packets. There is obviously an increase in overhead associated with fragmenting packets, so the approach requires a balance between the increased robustness versus the increased overhead associated with fragmentation.

If the stations of the different wireless networks are collocated in the same physical piece of hardware it is possible to coordinate some time domain techniques. Two approaches are recommended 802.15.2 [9]. In one approach each 802.11 station is collocated with a Bluetooth network master. The 802.11 network divides the time between the 802.11 beacon transmissions into two intervals, one for 802.11 transmissions and one for Bluetooth transmissions. The 802.11 stations collocated with the Bluetooth masters can indicate to the Bluetooth master when the Bluetooth network is permitted to transmit. Since Bluetooth is a polled system with the Bluetooth slaves only transmitting immediately after being polled from the Bluetooth master, no changes are required to the Bluetooth slaves. In addition the Bluetooth slaves do not have to be collocated in the same physical device with an 802.11 station. This approach eliminates simultaneous transmission of 802.11 and Bluetooth. The second technique recommended for the case when an 802.11 station and a Bluetooth node are collocated in the same physical device, is to time multiplex on a packet-by-packet basis. In this way, the 802.11 station would not schedule a transmission during a Bluetooth reception and visa versa. This approach enables scheduled transmissions based on priority between 802.11 and Bluetooth. For example, priority can be given to a Bluetooth voice packet over an 802.11 packet, while giving priority to 802.11 over Bluetooth data packets.

Some of these approaches require modification to the specification (e.g., AFH) while others can be implemented without modifications to the specification.

20.7 Coexistence Assessment – IEEE 802.15.4b

The 802.15.4b revision [12] to the 802.15.4 low-rate WPAN standard introduced several new PHYs. These new PHYs operate in 868 and 916 MHz frequency bands, as briefly described in Section 20.3.2. During development of the revision an evaluation was performed to see how well these new PHYs coexist with the previous PHYs in the same frequency bands [25]. This assessment makes for a rather straightforward example given the low complexity PHY.

For our purposes we will consider the introduction of the offset QPSK (O-QPSK) PHY operating in the 868 MHz frequency band (a European unlicensed band). The data rate is 100 kb/s using direct sequence spread spectrum with 400 kc/s spreading code. We consider the effect of a legacy 802.15.4 wireless network upon this new revised wireless network. Both are low-power networks with a typical transmit power of 1 mw (0 dBm). Another factor that significantly improves the coexistence of these networks is that low duty cycle used by these networks. Actually, in the 868 MHz frequency band there is a maximum transmit duty cycle of 1%. So the actual throughput of these networks is quite low. In this example the transmit duty cycle of 1% is used, with transmit power of 0 dBm. The path loss model is similar to that described in Section 20.5 with the parameters changes due to the operation in the 868 MHz band versus the 2.4 GHz band. The packet durations are comparable for the interfering legacy system and the victim new wireless network.

Applying that method described in Section 20.5 we obtain a PER curve for the O-QPSK system with the legacy interferer, as shown in Figure 20.7.

We observe that the worst case PER is about 1%. In this simple case with the interfering packet size approximately the same duration at the victim packet sizes, and for very low duty cycle the worst case PER turns out to be approximately equal to the duty cycle of the interfering network. This very low duty cycle leads to excellent coexistence. Of course the reason is that neither of the two networks is heavily utilizing the frequency band. We will see in the next section that when both networks attempt to share the same channel while attempting to maintain high throughput the story is different.

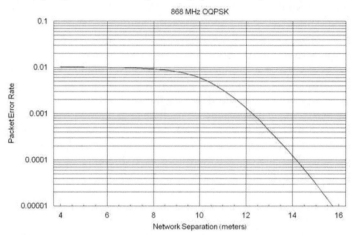

Figure 20.7: IEEE 802.15.4b O-QPSK PER curve with Legacy Interferer.

20.8 Coexistence Assessment – Draft IEEE 802.11n

On of the most significant amendments to the IEEE 802.11 standard has been underway for the last few years. The amendment will add a new PHY layer and make significant enhancements to the MAC layer. This amendment is currently a draft and has not yet been ratified by the IEEE [26]. For a detailed description of this draft amendment see Chapter 8. This section will highlight some key points about coexistence of 802.11n with other wireless networks. A detailed study of the coexistence of 802.11n with other unlicensed wireless networks is provided in [27]. A detailed study of the coexistence of 802.11n with Bluetooth is provided in Chapter 21.

During the development of this draft an assessment was made of the coexistence of this new draft with current IEEE standards [27]. Since 802.11n will be deployed in both the 2.4 GHz and 5 GHz unlicensed frequency bands there are a number of other unlicensed wireless networks that need to be considered when evaluation coexistence with 802.11n. For example, in the 2.4 GHz frequency band there is IEEE 802.15.1 (Bluetooth), IEEE 802.15.4 (ZigBee) and a variety of cordless telephones. In the 5 GHz frequency band there is IEEE 802.16 (WiMAX).

The 802.11n draft uses multiple input multiple output (MIMO) antenna technology. There are a wide range of modulation and coding schemes (MCS) that result in different data rates. Each MCS specifies the number of spatial streams, the modulation and code rate. In [27] one of the first steps in the evaluation is plotting the PER versus SNR for continuous co-channel interference. This of course represents an extreme worst case condition but it does give us insight into possible coexistence issue.

Three representative choices of MCS are MCS 0, MCS 7 and MCS 15. MCS 0 indicates a single spatial stream using BPSK modulation and a rate ½ convolutional code resulting in a data rate of 6.5 Mb/s. MCS 7 indicates a single spatial stream using 64-QAM modulation and a rate 5/6 convolutional code resulting in a data rate of 65 Mb/s. And finally MCS 15 indicates two spatial streams of using 64-QAM modulation and a rate 5/6 convolutional code resulting in a data rate of 130 Mb/s. All of these three cases utilize a 20 MHz bandwidth. Simulation results [27] with continuous interference show that the required SNR to attain a PER of 1% is approximately 12 dB for MCS 0, 31 dB for MCS 7 and 36 dB for MCS 15.

The observation that we can make is that the higher data rates require significantly higher SNR to maintain a reasonable PER. However, to offset this effect we recall that for a fixed message size the higher the data rate the shorter the duration of the packet. So higher data rates have many advantages at the cost of higher required SNR.

A new feature introduced in 802.11n is an aggregate packet that combines multiple MAC Frames into a single PHY packet. This results in longer packet durations than for non aggregated packets. The motivation for this is to reduce the effect of MAC and PHY overhead. Since if the data rate increases and the MAC and PHY overhead is the same duration in time, then the overhead increases percentage wise. There are several aggregation techniques, one of which can recover from a partial loss of data while the other aggregations cannot.

As an example from [27] we see the effect of interference from an 802.15.1 network (Bluetooth) that does not have adaptive frequency hopping (AFH) enabled. As one might expect the interference from a non-AFH Bluetooth network is significant.

Figure 21.11 in Chapter 21 shows the throughput of a simple two-node 802.11n network with interference from a nearby Bluetooth network without AFH enabled. There are curves for different Bluetooth duty cycles. When Bluetooth is off (occupancy of 0%) the 802.11n throughput continues to increase with larger and larger packet aggregation size. This makes sense since a longer aggregated packet leads to less overhead and hence higher throughput. However, when Bluetooth is on with a 100% duty cycle we see the throughput drop off at some point. This is due to the increased probability of a packet collision. This is the same concept as the packet fragmentation method described in Section 20.6. When we discussed packet fragmentation we pointed out how a shorter packet has a lower probability of collision, at the expense of higher throughput. In the case of packet fragmentation we are fragmenting a MAC Frame into pieces and transmitting the pieces separately. In the case of packet aggregation we are combining multiple MAC Frames into a single PHY packet. But in either case, with interference one can adjust the PHY packet size to make the best tradeoff between overhead and probability of packet collision.

The other obvious observation to make is that AFH should always be used with Bluetooth to avoid this kind of interference.

For more details on the coexistence of 802.11n and Bluetooth see Chapter 21. For more details on the coexistence of 802.11n and other wireless networks (IEEE 802.16, 802.15.4 and cordless telephones) see [27].

20.9 Dynamic Spectrum Access

As described in Section 20.2 the FCC is currently working on rules that will likely allow unlicensed use of the unused television channels. The process of an unlicensed wireless network observing the spectral occupancy of a set of channels and utilizing the channels unused by the licensed network(s) is referred to as dynamic spectrum access. The dynamic frequency selection in IEEE 802.11h is another example of dynamic spectrum access used to avoid interference with radar systems in the 5 GHz frequency band.

In response to the FCC actions regarding the TV bands the IEEE formed a working group on wireless regional area networks as mentioned in Section 20.3.4. The activities of this working group illustrate some of the challenges in building a dynamic spectrum access network. In this section we will briefly highlight one of those challenges.

One of the primary challenges is in meeting the spectrum sensing requirements being developed by the FCC. As described in Section 20.2.3 a station (either base station or client) must be able to sense (detect) the presence of an ATSC signal within a 6 MHz TV channel at -116 dBm. It is worth converting this signal power into a typical SNR value to illustrate the challenge.

The bandwidth of a TV channel is 6 MHz. If we assume a nominal receiver noise figure of 8 dB, then the noise level at the receiver is,

$$N = -174\,\mathrm{dBm} + 10 Log(6 \times 10^{6}) + 8\,\mathrm{dB} = -98.2\,\mathrm{dBm} \tag{7}$$

Hence, the SNR at the -116 dBm signal level is -17.8 dB, for a system noise figure of 8 dB. In some examples in this section a more conservative system noise figure of 11 dB will be used, since that is value is commonly used in the IEEE 802.22 working group.

Of course the reason for needing to sense at such an SNR value is that the wireless network may be in a faded location and must still be able to decide which TV channels are occupied by an ATSC transmission.

How does one detect an ATSC signal at this SNR? One way to think about this problem is to identify what signal characteristic you can use to detect the signal. The most obvious choice is the signal power (or equivalently the signal energy). Other signal features include unique characteristics of ATSC like the ATSC Data Field Sync pattern, the ATSC pilot tone and the shape of the spectrum. Another feature that can be used is the cyclostationary characteristics of the VSB modulation. Within the IEEE 802.22 working group there have been a number of presentations on sensing of ATSC using these various signal features. One can find these presentations on the IEEE 802.22 web site [28].

In this section we will illustrate the sensing of ATSC at negative SNR using two signal features: signal power and the ATSC Data Field Sync.

The most commonly referenced sensing technique is to observe the channel for a period of time, estimate the signal power (or energy) from that observation and then compare that estimate to a threshold. Typically the threshold is selected to limit the probability of false alarm (false alarm rate) to some specified value. In [29] an evaluation of the power (energy) detector was performed. A brief summary of that evaluation is given here. A more detailed derivation of the theoretical performance of the power detector can be found in [32].

After converting the signal from a TV channel to baseband the signal is complex sampled (in-phase and quadrature) at 6 Msamples/sec. The test statistic is the sum of the absolute value of these samples, scaled by the signal bandwidth and number of samples.

$$T = \frac{B}{M} \sum_{n=1}^{M} y(n) y^*(n) \tag{8}$$

The test statistic is a random variable whose probability density function depends on whether there is only noise present or whether there is both signal plus noise. Let us first consider the case when there is only noise present. We refer to that situation as hypothesis zero.

$$H_0: \quad y(n) = w(n) \tag{9}$$

Since the signal is sampled at the Nyquist rate the samples are independent and identically distributed Gaussian random variables. Under this condition the test statistic is,

$$T = \frac{B}{M} \sum_{n=1}^{M} w_R(n)^2 + w_I(n)^2 \tag{10}$$

So T is the sum of squares of $2M$ independent identically distributed Gaussian random variables. Hence T is a Chi-squared random variable with $2M$ degrees of freedom. Based on properties of Chi-squared random variables we can write down the mean and variance of T.

$$E(T) = N B \tag{11}$$

$$\text{var}(T) = \frac{(N B)^2}{M} \tag{12}$$

Where N is the noise power spectral density, B is the bandwidth and M is the number of samples. The parameter N was used instead of N_0 so that N can include the noise figure of the receiver and not just the thermal noise floor.

From this we can see why T was scaled by the bandwidth and the number of samples. We observe that the average value of T is the noise power in the bandwidth B. We also observe that as the number of samples increase the variance decreases. For a large number of samples (which is quite common) the test statistic can be approximated by a Gaussian random variable, by application of the Central Limit Theorem [23]. So under Hypothesis zero the test statistic has the following probability density function,

$$f_T(t) = N\left(N B, \frac{(N B)^2}{M} \right) \tag{13}$$

Here the notation $N(m, \sigma^2)$ indicated a Normal (Gaussian) random variable with mean m and variance σ^2. Similarly, the probability density function for the test statistic under Hypothesis one (when both signal and noise are present) is,

$$f_T(t) = N\left(P + BN, \frac{(P+NB)^2}{M}\right) \tag{14}$$

where P is the noise power (not PSD) within the bandwidth B.

The threshold is selected so that under Hypothesis zero the probability that the test statistic equals the specified probability of false alarm,

$$\gamma = NB\left(1 + \frac{Q^{-1}(P_{FA})}{\sqrt{M}}\right) \tag{15}$$

where Q(x) is the tail probability [23].

The probability of misdetection (one minus the probability of detection) under Hypothesis One is given by,

$$P_{MD} = Q\left(\frac{\sqrt{M}}{(P+NB)}[(P+NB)-\gamma]\right) \tag{16}$$

Plots of the theoretical and simulated probability of misdetection versus signal power are provided in the following figures. The number of samples was varied between 60 and 1200, corresponding to observation times of 10 μs to 200 μs respectively. For a probability of false alarm of 0.1 (10%) curves of the probability of misdetection (one minus the probability of detection) are shown in Figure 20.8. The solid lines represent the theoretical values and the discrete points represent simulation results. The theoretical values use the central limit theorem [23] and hence match best when the number of samples is large. The x-axis in this figure is the signal power and not the SNR. A conservative system noise figure of 11 dB (enough to include cable losses, etc) was used. So the noise level is -95.2 dB in this figure.

We see that for longer sensing times the power detector can detect at negative SNR. However, there is an issue with this approach. This sensing technique is detecting the difference between noise power and signal plus noise power. When the SNR is negative the signal plus noise power is less than 3 dB larger than the noise power. At an SNR of 0 dB the signal plus noise is 3 dB higher than the noise power. For very negative SNR the difference between the noise power and the signal plus noise power can easily be a fraction of a dB. Implicit in this detector is the fact that the detector knows the noise power. If there is an error in that knowledge of the noise power the detector breaks down at negative SNR.

In order to evaluate this effect of not having perfect knowledge of the noise power is to model the noise with some uncertainty [30][31][32]. Following the procedure in [31] we assume there is a total system noise figure of 11 dB, so the average noise power spectral density is given by,

$$\overline{N} = N_0 + 11 = -174 + 11 = -163 \ dBm \ / \ Hz \qquad (17)$$

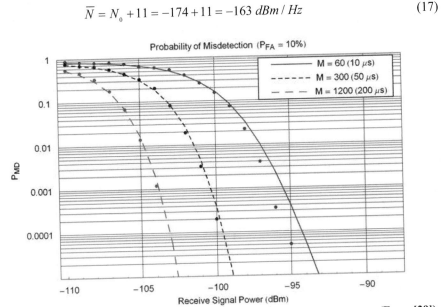

Figure 20.8: Probability of Misdetection versus Signal Power for Power Detector (From [29]).

However, the actual noise power spectral density is known within a tolerance of $\pm \Delta dB$,

$$N = \overline{N} \pm \Delta = -163 \pm \Delta \ dBm \ / \ Hz \qquad (18)$$

If we analyze and simulate the probability of misdetection with this noise uncertainty values ranging from 0 dB to 1 dB we see in Figure 20.9 that when there is noise uncertainty the power detector breaks down for negative SNRs.

Before considering other sensing techniques that utilize other signal features, let us look briefly at the effects of distributed sensing. Let us assume there are L sensors that effected by independent shadow fading from the TV transmitter. How can we use these independent measurements to improve our overall detection probability? We assume each local detector applies its detection technique and reports the results to a centralized processing location (i.e., the base station). These local decisions must be combined in some fashion into a global decision. Under our assumption that each of the local sensors sees independent shadow fading as well as independent noise, we can relate the probabilities of the local decisions and the global decision. We must first specify the method used at the base station for combining local decisions into a global decision. There are many possible methods that can be used. In this section we will consider the simplest and most conservative method. We will assume that if any of the sensors detects the TV signal then the global decision is that the signal is present. If we think each local decisions as a single bit, with a one representing that the signal is present and the zero representing that the signal is absent, then this technique can be thought of as a logical OR function. If any of the bits are a one (signal present) then the global decision is a one (signal present).

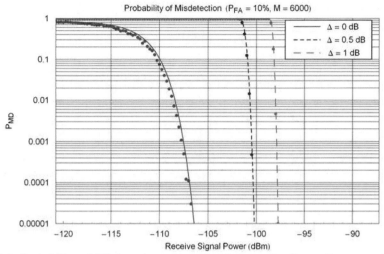

Figure 20.9: Probability of Misdetection versus Signal Power for Power Detector with Noise Uncertainty (From [31]).

Based on this logical OR function at the base station we can relate the local probability of false alarm (P_{LFA}) at each individual sensor, with the global probability of false alarm (P_{GFA}) at the base station after combining decisions.

$$(1 - P_{GFA}) = (1 - P_{LFA})^L \tag{19}$$

Similarly we can relate the local probability of misdetection to the global probability of misdetection. Since the only way a global misdetection can occur is if all the local decisions also were midsections, we can write,

$$P_{GMD} = P_{LMD}^L \tag{20}$$

Given this method of combining local decisions into global decisions we can evaluate the global misdetection probability. We can use the previous results for the power detector with noise uncertainty. We then consider L sensors with independent shadow fading. The fading is assumed to be lognormal with a standard deviation of 5.5 dB. The local detector threshold must be modified so that the local probability of false alarm is reduced so that the final global probability of false alarm matches the target value (10% in this case). By modifying the threshold the misdetection probability at any given sensor is worse than before, however, the global probability of misdetection is much improved. This is because all sensors must be in error for the final decision to be in error.

Figure 20.10 shows the global misdetection probability when the power detector is used. There are curves for various levels of noise uncertainty. We see that as the number of independent sensors increases the performance improves. So even though the individual sensor performance is poor the overall performance is much better than the individual sensor performance. The main limitation to this approach is the availability of sensors that see independent shadow fading. In a large WRAN deployment like in IEEE 802.22 there are likely to be multiple sensors separated by a large enough distance to result in sensors with independent shadow fading. However, in a smaller system, like in a WLAN deployment, many or all of the sensors are likely to be effected by the same shadow fading. For example, they may all be blocked by the same structure. So distributed sensing is a powerful tool but it relies on the availability of multiple independent sensors. The independence of noise and multipath fading is a valid assumption in virtually all cases. The independence of shadow fading requires a significant physical separation between sensors.

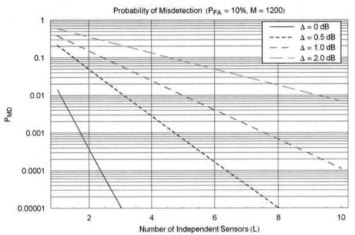

Figure 20.10: Probability of Misdetection with Noise Uncertainty and Shadow Fading using Multiple Independent Sensors (From [31]).

In cases where we cannot rely on having a significant number of independent sensors we need to spectrum sensing techniques that utilize other signal features other than just the signal power. In this section we will illustrate an example on one such technique. There are a wide variety of other techniques that can also be used.

One signal feature in the ATSC signal is the Data Field Sync pattern [33] that occurs once in every 313 ATSC segments. The Data Field Sync pattern is a PN sequence with good autocorrelation properties. One method of detecting the presence of an ATSC signal is to correlate the baseband signal with the expected PN sequence. The output of the correlator will generate a peak when there is a match and noise the rest of the time. The trick of course it to operate at negative SNR and so to be able to distinguish correct correlation matches with false matches due to noise. If the channel is observed for the length of an entire ATSC data field of 24.2 ms then the largest peak out of the correlator is an indication of whether an ATSC signal is present. So we can use the maximum of the absolute value of the correlator as a test statistic and compare it to threshold. The threshold is selected to so at to obtain a specified probability of false alarm.

If we observe the channel longer can we do better? Yes, but is requires a slightly more sophisticated approach. If there was perfect synchronization between the ATSC transmitter and the spectrum sensor then the true peak in the correlator output would occur in exactly the same part of the data field. However, to clock offset and jitter the time of the true peak in the correlator output can be offset by several samples. Also, the channel from the ATSC transmitter to the spectrum sensing node can change within the 24.2 ms, due to multipath even with small Doppler. So the polarity of the true peak in the correlator output can reverse polarity. We can address that issue by taking the absolute value of the correlator output.

The method proposed in [34] for processing multiple ATSC data fields is as follows. First correlate the baseband signal with the expected ATSC data field sync pattern. Take the absolute value of the correlator output. Select the N (e.g., 3) largest outputs during one ATSC data field. This is illustrated in Figure 20.11.

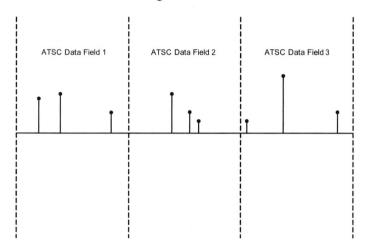

Figure 20.11: N Largest Peaks from the Correlator output (From [34]).

We then overlay the peaks from the different ATSC data fields and combine the peaks that are within a window of M samples, as show in Figure 20.12.

Then we select the largest peak remaining and compare it to a threshold. This method improves the sensing technique because true peaks, from the actual ATSC data field sync, tend to combine from each data field with high probability, while noise tends not to combine too frequently. The reason for this is that the noise peaks are random and occurs equally likely anywhere within the ATSC data field, and hence it is unlikely to have strong peaks in one data field occur at the same time (within a small window of time) in the next data field. So the peak combining process improves the probability of detection while very slightly increasing the probability of false alarm. The very small increase in the probability of false alarm can be offset by modifying the detector threshold.

This approach was simulated with 12 actual captured ATSC signal files, representing a broad range of multipath channels. The results from the signal files were averaged. Figure 20.13 shows the probability of misdetection curves for various sensing times, from one ATSC data field up to 16 ATSC data fields. We see that as the sensing duration is increase the peak combining technique results in lower probability of misdetection.

We see that it is possible to develop spectrum sensing techniques that can detect the presence of an ATSC signal, for example, by utilizing a specific signal feature. We also see that even with this approach and observing the channel for a duration equivalent to 16 ATSC data fields (16 x 24.2 = 387.2 ms) that the 90% probability of detection point (10% probability of misdetection) is at approximately -14 dB SNR.

Other sensing techniques are being considered within the IEEE and within various research organizations, in order to develop the best methods of spectrum sensing to enable effective dynamic spectrum access networks.

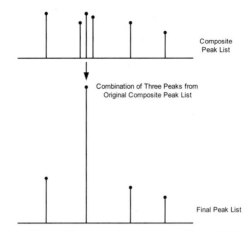

Figure 20.12: Peak Combining (From [34]).

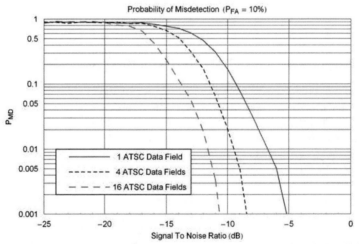

Figure 20.13: Probability of Misdetection using Peak Combining (From [34]).

20.10 Conclusions

An overview of how unlicensed wireless networks coexist is given in this chapter. We can see that there are many different unlicensed wireless networks all sharing a few frequency bands, which drives the need to understanding how well these networks coexist. A number of methods for improving coexistence were described. Finally, we described some aspects of dynamic spectrum access in which unlicensed wireless networks can utilize portions of the licensed spectrum unused by licensed networks.

20.11 References

[1] Federal Communication Commission, Title 47, *Code for federal regulations,* part 15.

[2] Federal Communication Commission, *Report and Order and Memorandum Opinion and Order: In the Matter of Wireless Operation in the 3650-3700 MHz Band,* Document FCC 05-56, March 2005.

[3] Federal Communication Commission, *Notice of Proposed Rulemaking: In the Matter of Unlicensed Operation in the TV Broadcast Bands,* FCC 04-113, May 2004.

[4] Federal Communication Commission, *First Report and Order and Further Notice of Proposed Rulemaking: In the Matter of Unlicensed Operation in the TV Broadcast Bands,* FCC 06-156, October 2006.

[5] IEEE Std 802.11, *Wireless LAN Medium Access Control (MAC) and Physical Layer (PHY) specifications,* 1997.

[6] IEEE Std 802.11b, *Wireless LAN Medium Access Control (MAC) and Physical Layer (PHY) specifications: Higher-Speed Physical Layer Extension in the 2.4 GHz Band,* 1999.

[7] IEEE Std 802.11a, *Wireless LAN Medium Access Control (MAC) and Physical Layer (PHY) specifications: High-speed Physical Layer in the 5 GHZ Band,* 1999.

[8] IEEE Std 802.15.1-2005, *Wireless medium access control (MAC) and physical layer (PHY) specifications for wireless personal area networks (WPANs),* 2005.

[9] IEEE Std 802.15.2-2003, *Coexistence of Wireless Personal Area Networks with Other Wireless Devices Operating in Unlicensed Frequency Bands,* 2003.

[10] IEEE Std 802.15.4-2003, *Wireless Medium Access Control (MAC) and Physical Layer (PHY) Specifications for Low-Rate Wireless Personal Area Networks (LR-WPANs),* 2003.

[11] IEEE Draft 802.15.4a/D3, *Wireless Medium Access Control (MAC) and Physical Layer (PHY) Specifications for Low-Rate Wireless Personal Area Networks (LR-WPANs),* 2006.

[12] IEEE Draft 802.15.4REVb/D6, *Wireless Medium Access Control (MAC) and Physical Layer (PHY) Specifications for Low-Rate Wireless Personal Area Networks (LR-WPANs),* 2006.

[13] IEEE Std 802.16-2004, *Air Interface for Fixed Broadband Wireless Access Systems,* 2004.

[14] IEEE Std 802.16e-2005, *Air Interface for Fixed Broadband Wireless Access Systems: Amendment 2: Physical and Medium Access Control Layers for Combined Fixed and Mobile Operation in Licensed Bands,* 2005.

[15] *IEEE Project Authorization Request: Amendment to IEEE Standard for Local and Metropolitan Area Networks - Part 16: Air Interface for Fixed Broadband Wireless Access Systems - Improved Coexistence Mechanisms for License-Exempt Operation,* 802.16h, 2004.

[16] IEEE Draft Std 802.16h, *Improved Coexistence Mechanisms for License-Exempt Operation,* D2, January 2007.

[17]   Carl Eklund, Roger B. Marks, Subbu Ponnuswamy, Kenneth L. Stanwood and Nico J.M. Van Waes, *WirelessMAN: Inside the IEEE 802.16 Standard for Wireless Metropolitan Area Networks*, IEEE Press, 2006.

[18]   *IEEE Project Authorization Request: Cognitive Wireless RAN Medium Access Control (MAC) and Physical Layer (PHY) specifications: Policies and procedures for operation in the TV Bands*, 802.22, 2004.

[19]   Stephen R. Whitesell, *Cordless Telephone Coexistence Considerations*, IEEE 802.19-05/26r0, July 2005.

[20]   Nada Golmie, *Coexistence in Wireless Networks: Challenges and System-Level Solutions in the Unlicensed Bands*, Cambridge University Press, 2006

[21]   *IEEE Project Authorization Request: Methods for assessing coexistence of wireless networks*, 802.19, 2006.

[22]   Steve Shellhammer, *Estimation of Packet Error Rate caused by Interference using Analytic Techniques – A Coexistence Assurance Methodology*, IEEE 802.19-05/28r2, October 2005.

[23]   A. Papoulis, *Probability, Random Variables, and Stochastic Processes*, Third Edition, McGraw Hill, 1991

[24]   J. Proakis and M. Salehi, *Communication Systems Engineering*, Second Edition, 2001.

[25]   Robert Poor, *Coexistence Assurance for 802.15.4b*, IEEE 802.15/05-632r0, November 2005.

[26]   IEEE Draft 802.11n, *Wireless LAN Medium Access Control (MAC) and Physical Layer (PHY) specifications: Enhancements for Higher Throughput*, D2.0, February 2007.

[27]   Eldad Perahia and Sheung Li, *p802.11n Coexistence Assurance Document*, IEEE 802.11-06/338r4, September 2006.

[28]   http://grouper.ieee.org/groups/802/22/.

[29]   Steve Shellhammer, *Performance of the Power Detector*, IEEE 802.22-06/75r0, May 2006.

[30]   Rahul Tandra, *Fundamental Limits of Detection in Low SNR*, Masters Thesis, University of California Berkeley, Spring 2005.

[31]   Steve Shellhammer and Rahul Tandra, *Performance of the Power Detector with Noise Uncertainty*, IEEE 802.22-06/134r0, July 2006.

[32]   Steve Shellhammer, Sai Nanadagopalan, Rahul Tandra and James Tomcik, *Performance of Power Detector Sensors of DTV Signals in IEEE 802.22 WRANs*, IEEE TAPAS Workshop, Boston MA, August 2006.

[33]   Advanced Television Standards Committee, *ATSC Digital Television Standard*, ATSC A/53E, Revision E with Amendment No. 1, December 2005.

[34]   Steve Shellhammer, *An ATSC Detector using Peak Combining*, IEEE 802.22-06/243r5, March 2007.

## 21

Coexistence of IEEE 802.11n and Bluetooth

Eldad Perahia[a]

The development of the IEEE 802.11n standard amendment enables MIMO-OFDM waveform transmission in the 2.4 GHz band. Additional PHY modifications relative to 802.11a/g include 40 MHz channels, additional data tones in 20MHz channels, and rate 5/6 coding. MAC enhancements include two types of frame aggregation. In this paper we model and simulate the sensitivity of an 802.11n device in the presence Bluetooth interference. Spatial and temporal properties of both systems are considered. Results are provided in terms of packet error rate, throughput, and required separation between devices.

21.1 Introduction

The most current draft of IEEE 802.11n (11n) enables MIMO-OFDM waveform transmission in the 2.4 GHz band [11]. Therefore, analysis of IEEE 802.11b/g coexistence with Bluetooth (BT) devices [1], [2], [3], and [5] must be extended to cover 11n. Not only will 11n extend the physical layer (PHY) for spatial division multiplexing with one to four spatial streams, but will also increase the data rate with additional data tones in 20MHz as compared to 802.11g, and rate 5/6 coding. In addition, 11n will create a new 40 MHz channel width, for more than double the data rate, relative to 20 MHz transmissions.

The scope of 11n is to increase the throughput of IEEE 802.11, not just the PHY data rate. In order to do so, the efficiency of the medium access control (MAC) layer must also be improved. Two types of frame aggregation have been developed: aggregate MAC protocol data unit (A-MPDU) and aggregate MAC service data unit (A-MSDU).

The principal focus of this analysis is geometric and temporal interferer modeling. A geometric analysis will demonstrate the necessary separation between 11n devices and the BT devices to avoid packet collisions. A PHY interference model with an 11n affected wireless network in the presence of BT devices will be presented. This is followed by a temporal packet collision analysis, in which we determine the probability of 11n and BT devices in close proximity of each other transmitting at coincidental times and frequencies. We will incorporate the new 11n packet structures based on frame aggregation. And last,

---

[a] Intel Corporation

the geometric and temporal packet collision analysis is combined to illustrate the overall throughput of the 11n devices as a function of location of the BT devices.

In Section 21.2, geometric analysis will be presented. Temporal analysis will be described in Section 21.3. This is followed by combined geometric and temporal in Section 21.4. And lastly, Section 21.5 contains concluding remarks.

21.2 Geometric Analysis

The basic PHY geometric model is given by an 11n client communicating with an 11n access point (AP) while simultaneously a nearby BT device is transmitting causing interference to the client. This is illustrated in Figure 21.1.

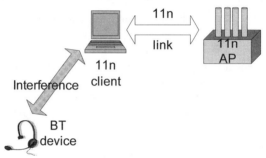

Figure 21.1: Basic PHY geometric model.

We initially assume with pure geometric analysis complete overlap of AP-client transmission and interference-client transmission in time and frequency. Our goal is to determine the separation necessary between the interfering BT device and 11n client to completely avoid interference. We define an "interference free" link as one that achieves a packet error rate (PER) of 1%.

To perform this analysis, we first specify the separation between the client and AP. The separation between client and AP sets the received signal level, and therefore the signal-to-noise ratio (SNR) of the link. An example of this analysis is given in the link budget in Figure 21.2.

In the example, the separation between the client and AP is 20 meters. This results in a total pathloss of 79 dB, based on the pathloss and shadow fading model in [7]. The pathloss combined with the effective isotropic radiated power (EIRP) and receiver antenna gain results in a received signal strength indication (RSSI) of -58dBm (with transmit power of 17 dBm and transmit and receive antenna gain of 2 dBi.)

$$RSSI = P_T + A_T - PL - SF + A_R$$

P_T : Transmit power

A_T, A_R : Transmit or receive antenna gain

PL : Path loss

SF : Shadow fading

11n Link

Tx Power	dBm	17
Tx antenna gain	dBi	2
pathloss		
frequency	GHz	2.4
distance	m	20
breakpoint	m	5
shadow fading before breakpoint		3
shadow fading after breakpoint		4
free space	dB	54.0
total loss	dB	79.1
Rx antenna gain	dBi	2
RSSI	dBm	-58.1
Noise Power		
NF	dB	6
BW	MHz	20
total	dBm	-95.0
Received SNR	dB	36.9
minimum allowable C/I		21.1
Allowable Receive Interference power	dBm	-79.2
Interferer		
Tx Power	dBm	0
Tx antenna gain	dBi	2
Pathloss	dB	83.2
separation from STA	m	26.2

Figure 21.2: AP – client link budget.

With a noise figure of 6 dB and a noise bandwidth of 20 MHz, the thermal noise power is 95 dBm.

We then specify the target modulation and coding scheme (MCS) for the client-AP link. From the target MCS, the required SNR at a PER equal to 1% can be derived. Figure 21.3 illustrates PHY simulation results for single stream, BPSK, rate ½ (MCS 0), single stream, 64-QAM, rate 5/6 (MCS 7), and two stream, 64-QAM, rate 5/6 (MCS 15) for 20MHz. MCS 0 provides the lowest data rate, longest range MCS in 20 MHz. MCS 7 provides the highest data rate for single stream transmission. MCS 15 provides the highest date rate for two stream transmission. The simulations to produce the PER curves include phase noise, frequency error, and power amplifier distortion, as described in [6] and [12]. The simulations incorporate a MIMO frequency-selective fading channel model, as described in [7], [9], and [10]. The simulations were performed in channel model B, which has a 15 nsec RMS delay spread. The PER results are given in Figure 21.3.

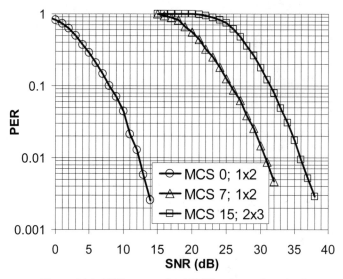

Figure 21.3: PER curves for 20MHz, channel model B.

For 40MHz channels, MCS 32 provides the lowest rate and longest range by duplicating the tones in the lower 20MHz channel into the upper 20MHz channel, providing frequency diversity. The modulation and coding scheme is BPSK, rate 1/2. MCS 0, 7, and 15 for 40MHz is same definition as for 20MHz. Figure 21.4 illustrates PHY simulation results for MCS 32, 0, 7, and 15 for 40 MHz.

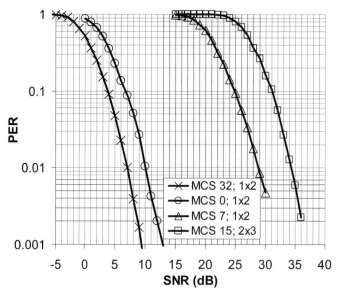

Figure 21.4: PER curves for 40MHz, channel model B.

Table 21.1 gives the required SNR at a PER of one percent. As can be seen, 40 MHz operation is slightly more robust than 20 MHz with comparable modulation and coding rate. The wider bandwidth provides more frequency diversity in a fading channel.

Table 21.1: Required SNR at one percent PER for channel model B.

MCS	# Tx Ant × # Rx Ant	BW (MHz)	Data Rate Mbps	Required SNR (dB)
32	1×2	40	6	7
0	1×2	20	6.5	12
0	1×2	40	13.5	10
7	1×2	20	65	31
7	1×2	40	135	29
15	2×3	20	130	35.5
15	2×3	40	270	34.5

For the example in Figure 21.2 we will use MCS 0 with a required SNR of 12dB. With a received SNR of 37dB, the link in the example exceeds this requirement. The next step is to determine the amount of interference that can be tolerated by the receiver. Figure 21.5 illustrates the simulation results of an 11n receiver in the presence of a BT interferer. The simulation assumes that the signal and interferer propagate though a channel based on the same model, but independent paths. The signal-to-interference ratio (SIR) is swept from 0 to 40 dB, with a fixed SNR of 40 dB. A high SNR operating point was selected to better isolate how the 11n receiver reacts to a BT interferer. A standard minimum mean square error (MMSE) receiver is modeled, with no additional interference mitigation techniques implemented. As can be seen in Figure 21.5, the higher order modulations are increasingly sensitive to interference, even narrow band as BT. In fact, a comparison between Figure 21.3 and Figure 21.5, show that the receiver performs much better in additive white Gaussian noise (AWGN) than in interference. Sensitivity of 40 MHz MCSs to BT interference are comparable to 20 MHz.

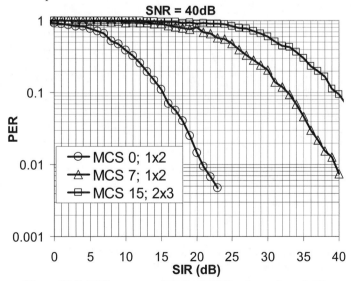

Figure 21.5: PER curve vs. SIR for 20MHz, channel model B.

For the example in Figure 21.2, a link with an SNR of 37 dB can tolerate an SIR of 21 dB with MCS 0 and 20MHz bandwidth. With the RSSI and minimum allowable carrier-to-interference ratio (C/I), the maximum allowable level of interference can be derived as follows:

$$Int = \frac{RSSI}{C/I}$$

The resulting maximum allowable interference level is -79 dBm. The minimum pathloss between the interferer and the 11n client is derived as follows:

$$PL_I = -(Int - A_R - SF_I - A_{T,I} - P_{T,I})$$

$P_{T,I}$: Interferer transmit power

A_R : STA receive antenna gain

$A_{T,I}$: Interferer transmit antenna gain

SF_I : Shadow fading on the interferer-client link

The resulting allowable pathloss equals 83dB. Since the pathloss equation is a function of range, we invert the pathloss equation to derive the necessary separation between the interferer and client. In this example the separation results in 26 meters.

This example derived the interferer–11n client separation based on a specific 11n client–AP separation and MCS. The following figure expands the analysis to span a range of separation between client–AP for MCS 0, 7, and 15 with 20MHz channel and channel model B.

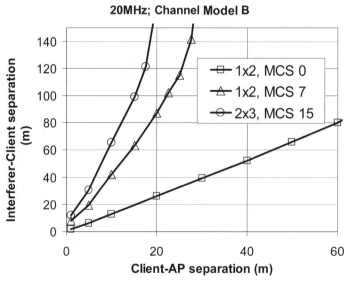

Figure 21.6: Required separation between the 11n client and interferer for interference free operation with 20MHz channel and channel model B.

As illustrated for each MCS, the required interferer–client separation for collision free performance is calculated based on the corresponding client–AP separation. As the client–AP separation increases, the required interferer–client separation increases. And with higher MSC, the sensitivity to interference increases resulting in larger required separation between interferer and client.

21.3 Temporal Analysis

In the previous section, geometric analysis assumed complete overlap of transmission of AP-client and interference-client in time and frequency. In this section we will investigate the probability of overlap based on analysis by [3] and [5].

We begin by highlighting the two new features in 11n that will most impact time and frequency properties of an 11n transmission. In order to increase efficiency, aggregation is used to increase packet lengths. A typical 11n packet exchange with aggregation and block acknowledgement (BA) during a transmission opportunity (TXOP) is illustrated in the figure below. With aggregation, longer packet lengths will lead to more time overlap with BT interferers.

A second feature in 11n is 40MHz channels, for more than double increase in PHY data rate. Occupying double the bandwidth, as illustrated in Figure 21.8, a 40 MHz 11n transmission will be more susceptible to BT frequency hops.

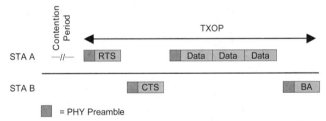

Figure 21.7: Typical 11n packet exchange with aggregation.

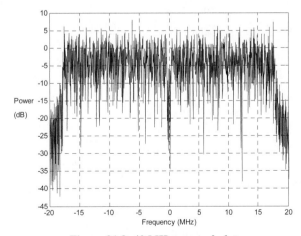

Figure 21.8: 40 MHz spectral plot.

A temporal collision occurs when neighboring interfering BT devices and 11n AP-client devices transmit packets which overlap in time. Figure 21.9 illustrates a BT packet stream overlapping with an 11n aggregated packet.

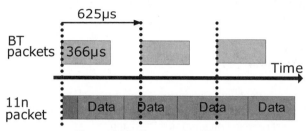

Figure 21.9: Temporal collision.

Since BT is a frequency hopped system, a packet collision only occurs if the frequency hop coincides with the 11n channel. This is illustrated in the Figure 21.10.

To calculate the probability of a collision, the joint probability of a temporal and frequency overlap must be computed. As a first step, we determine the number of BT packets would overlap with the 11n aggregated packet, and the probability of such an event occurring. Two types of aggregation are included in 11n, A-MSDU and A-MPDU. With A-MSDU aggregation, the entire aggregate is protected by a single frame check sequence (FCS). Therefore bit errors anywhere in the aggregate will cause all sub-frames to be lost and require retransmission. As such, A-MSDU aggregation follows the derivation in [5], as described below.

$$N = \mathrm{int}(L_{11n} / T)$$
$$\text{if } \mathrm{rem}(L_{11n}, L_{BT}) <= T - L_{BT}$$

N with probability of $P_{\mathrm{overlap}} = \dfrac{T - L_{BT} - \mathrm{rem}(L_{11n}, L_{BT})}{T}$

$N+1$ with probability of $1 - P_{\mathrm{overlap}}$

else

$N+2$ with probability of $P_{\mathrm{overlap}} = \dfrac{\mathrm{rem}(L_{11n}, L_{BT}) - (T - L_{BT})}{T}$

$N+1$ with probability of $1 - P_{\mathrm{overlap}}$

end

T : BT dwell period ($625\mu \sec$)

L_{11n} : 11n packet time on air

L_{BT} : BT packet time on air ($366\mu \sec$)

Next, we determine the probability of an overlap in frequency, P_f.

$$P_f \approx \begin{cases} 20/79 & \text{for 20 MHz} \\ 40/79 & \text{for 40 MHz} \end{cases}$$

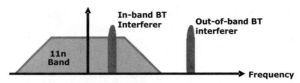

Figure 21.10: Frequency overlap.

The joint probability is given by the following [5]:

if $\mathrm{rem}(L_{11n}, L_{BT}) <= T - L_{BT}$

$$P_{\text{no collision}} = P_{\text{overlap}} \cdot \left(1 - P_f\right)^N + \left(1 - P_{\text{overlap}}\right) \cdot \left(1 - P_f\right)^{N+1}$$

else

$$P_{\text{no collision}} = P_{\text{overlap}} \cdot \left(1 - P_f\right)^{N+2} + \left(1 - P_{\text{overlap}}\right) \cdot \left(1 - P_f\right)^{N+1}$$

end

$$P_{\text{collision}} = 1 - P_{\text{no collision}}$$

Individual BT devices may not occupy all time slots. For all time slots to be occupied, a number of BT devices may be active in a BT picocell. The impact of partial utilization (or occupancy) within a BT picocell is accommodated by multiplying the above collision probability by the picocell's percentage utilization.

Given the probability of a collision, we need to determine the impact on system performance. Assume a collision causes a packet error. This will necessitate a retransmission, reducing throughput. The average time on air of an 11n aggregated packet exchange during a TXOP is:

DIFS + Avg backoff time + RTS + SIFS + CTS + SIFS + 11n Aggregated Packet + SIFS + BA
DIFS: distributed (coordination function) interframe space
SIFS: short interframe space

where the 11n aggregated packet time is:

Legacy Preamble + HT preamble + Data + Data + …
HT: high throughput

Throughput with no collisions with BT packets or other sources of packet errors is given by: (Information bits) / (Time on air); where information bits are defined as only those conveyed during the data portion of the aggregate.

Throughput with collision with BT packets and retransmission of entire aggregate is given as follows:

$$\text{Throughput} = \frac{B \cdot \left(1 - P_{\text{collision}}\right)}{T}$$

T : Time on air
B : Information bits

Figure 21.11 and Figure 21.12 illustrate the impact of aggregate packet length and BT utilization on 11n throughput for the 20 MHz data rate of 130 Mbps, and for the 40 MHz data rate of 270 Mbps. The curves with BT utilization equal to 0 (noted by the blue line with stars with the label "BT occup = 0%") illustrates throughput with no collisions. This demonstrates the dramatic improvement in efficiency derived from aggregation. However as the BT utilization increases to 100%, the higher the likelihood of a temporal overlap and a collision with longer packet lengths. With 40MHz, this effect is even more dramatic with double the probability of a BT device hopping in-band. As a note, with very low PHY data rates the effect is much less noticeable since individual data segments occupy so much time on air that there is little opportunity to aggregate many packets with typical TXOP lengths.

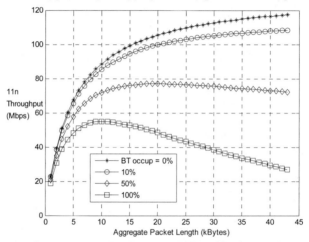

Figure 21.11: Impact of aggregate packet length and BT utilization on 11n throughput for 130 Mbps, 20 MHz mode with A-MSDU.

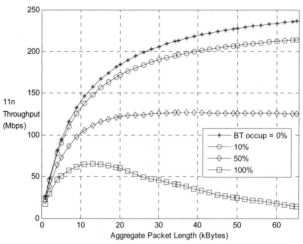

Figure 21.12: Impact of aggregate packet length and BT utilization on 11n throughput for 270 Mbps, 40 MHz mode with A-MSDU.

With A-MPDU aggregation, each MPDU contains its own FCS. In addition, each MPDU is preceded by a delimiter. Therefore portions of the transmission can be corrupted without losing all MPDUs. The delimiters make re-synchronization possible by scanning forward to the next valid delimiter. If a collision occurs with a BT device, only those effected MPDUs will require retransmission. The issue with high BT utilization and the drop in throughput as the aggregate packet length increases (as illustrated in Figure 21.11 and Figure 21.12) will be alleviated by only requiring retransmission of effected MPDUs. However, if the PHY preamble and header are corrupted by a collision with a BT packet, the entire aggregate will still be lost.

A simulation was constructed to model the frequency and time overlap between BT packets and individual MPDUs in an 11n A-MPDU aggregate. Only MPDUs which collided with BT packets were considered lost. If a BT packet collided with the PHY preamble, the entire aggregate was considered lost. The throughput results for 20 MHz with a data rate of 130 Mbps and for 40 MHz with a data rate of 270 Mbps are illustrated in Figure 21.13 and Figure 21.14. We demonstrate that even with 100% BT utilization, throughput increases with increasing aggregate length.

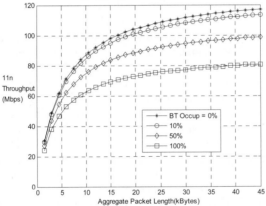

Figure 21.13: Impact of aggregate packet length and BT utilization on 11n throughput for 130 Mbps, 20 MHz mode, sub-packet length = 1500 B, with A-MPDU model.

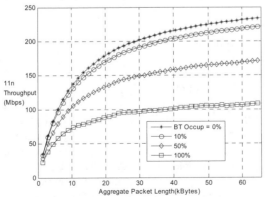

Figure 21.14: Impact of aggregate packet length and BT utilization on 11n throughput for 270 Mbps, 40 MHz mode, sub-packet length = 1500 B, with A-MPDU model.

As a final note on temporal and frequency overlap analysis, we address two features which will effect the frequency overlap between BT and 11n. The first feature is the inclusion of adaptive frequency hopping (AFH) in BT version 1.2 [13]. AFH changes the hopping pattern of the BT devices. The AFH enabled BT devices measure the interference on individual 1 MHz channels. The BT devices can then restrict its hopping pattern to skip channels with interference.

The use of AFH in the United States was made possible by the Federal Communications Commission (FCC) in 2002 when it began allowing frequency hopping between a minimum of 15 and maximum of 75 channels [13]. In other regulatory domains such as the European Telecommunications Standards Institute (ETSI), frequency hopping is also allowed with a minimum of 20 channels [13]. The hopping pattern in BT with AFH can be reduced to a minimum of 20 hops.

The benefit of AFH is subject to the ability to accurately measure interference in the channel. In addition, the efficacy of AFH depends on the overlap between WLAN cells and the location of the BT and 11n devices. Consider a three cell/AP 11n system with each occupying one of channels 1, 6, and 11, encompassing the entire 2.4 GHz band. At the triple point between the three cells, fairly equal power levels will be received by BT devices on each channel. It will be difficult for a BT device to select channels to skip. This is further exacerbated by an 11n system utilizing 40MHz channels. Obviously AFH will improve performance in many situations. The analysis presented here, without considering AFH, can be considered a conservative estimate of coexistence between these systems.

The second feature that will effect the frequency overlap between BT and 11n is an artifact of the switching mechanism between 40 MHz and 20 MHz in 2.4 GHz in 11n. Due to partial overlap between channels in 802.11 in 2.4GHz, scanning, detection, and reporting of overlapping 802.11 systems when operating in 40 MHz was added to the 11n draft amendment [11]. Clients are required to scan adjacent channels, and upon detection of an overlapping 802.11 system, report back a detection flag to the AP. This forces the AP to switch the basic service area back to 20 MHz operation. The intent is that in congested environments, 11n devices will revert back to 20 MHz operation. A by product of this feature is that an 11n device can send a detection flag to the AP for any reason. An 11n device which is "BT aware", may use this flag to restrict use of 40 MHz operation in the basic service area which will reduce the frequency overlap between the two systems.

21.4 Combined Geometric and Temporal Analysis

Geometric and temporal analysis can be combined to provide insight into the impact of client–interferer separation on throughput in an interference limited environment.

The following steps were taken to derive the curves for throughput as a function of client–interferer separation in Figure 21.15. As a first step, in the geometric analysis of Section 21.2, the 11n client–AP separation is set small enough to achieve a carrier-to-noise ratio (C/N) of 40dB to create an interference limited environment. With C/N fixed at 40dB, we sweep over a range of the C/I levels. With RSSI and C/I, the range of interference levels is calculated. Using the pathloss equation, the range of 11n client–interferer separation is derived from the interference level giving the x-axis values in Figure 21.15.

As a second step, we begin again with the C/N and C/I values from before. These are converted to a range of PER values from the waterfall curves for each MSC illustrated in Figure 21.5. To compute the joint probability of a packet error, these PER values are multiplied by the temporal probability of collision and the BT percent occupancy. The probability of collision is computed based on 11n packets with A-MSDU type aggregation, as described in Section 21.3. With a TXOP of 1.5ms, the aggregate packet length is approximately 19kB. Finally, the throughput is calculated with the joint probability of a packet error, as follows.

$$\text{Throughput} = \frac{B \cdot \left(1 - P_{\text{packet error}}\right)}{T}$$

T : Time on air

B : Information bits

Figure 21.15 illustrates the throughput with A-MSDU, 20MHz channel, and MCS 0, 7, and 15 as the separation between the 11n client and interferer increases. The results are based on an aggregate filling up a 1.5ms TXOP. The BT occupancy is assumed to be 50%. The C/N level is 40 dB, creating an interference limited environment.

Substantial throughputs are achieved even with small separation given a reasonable length aggregate and BT occupancy. As expected, as the separation increases between the 11n client and interferer, the throughput increases.

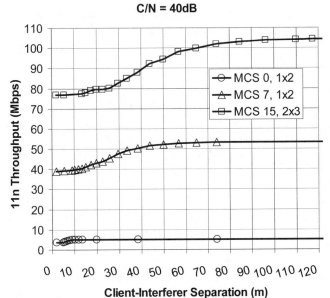

C/N = 40dB

Figure 21.15: 11n throughput with A-MSDU aggregation for 20 MHz.

The A-MPDU simulation constructed for temporal analysis was modified to include the joint probability of a packet error for combined geometric and temporal analysis. Figure 21.16 illustrates throughput as a function of 11n client – interferer separation with A-MPDU type aggregation, 20MHz channel, and MCS 0, 7, and 15.

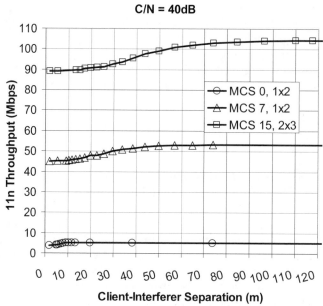

Figure 21.16: 11n throughput with A-MPDU aggregation for 20 MHz.

At small separation, A-MPDU provides 15% throughput improvement over A-MSDU. However, this improvement diminishes as the separation increases.

21.6 Conclusion

The IEEE 802.11n standard amendment [11] adds MIMO-OFDM transmission to the 2.4 GHz band. In addition, PHY modifications such as rate 5/6 coding, additional data tones in 20MHz, and 40 MHz channels increase the sensitivity to Bluetooth interference. To improve MAC efficiency, both A-MPDU and A-MSDU are used to increase the packet lengths. However, longer time-on-air also increase the probability of collision with a BT interferer. We demonstrated throughput ranging from 75 Mbps to 105 Mbps, with small to large separation between 11n device and BT interferer, with typical conditions of 1.5 ms TXOP length, two spatial streams, 64-QAM, rate 5/6, 20MHz channel, and 50% BT occupancy. At small separation between 11n device and BT interferer, A-MPDU provides for a 15% improvement over A-MSDU.

Similar to IEEE 802.11b/g devices, coexistence between 11n and BT can be further improved by incorporating interference mitigation techniques in an 11n device, such as interference cancellation and rate adaptation. If the 11n radio and the BT radio are co-located in the same device further hardware and software techniques are available to mitigate collisions, as an example time division multiplexing between the two radios. In addition, BT devices with AFH will also coexist better with 11n devices by avoiding channels utilized by an 11n network, if possible. 11n devices which are BT aware, may cause a basic service area to revert from 40 MHz to 20 MHz operation to reduce possible frequency overlap with BT devices.

Acknowledgement

The author thanks Tom Kenney for developing the simulation platform necessary to generate the 11n PHY simulations.

21.7 References

[1] Aguado, L.E., Wong, O'Farrell, "Coexistence Issues for 2.4 GHZ OFDM WLANs," 3G Mobile Communications Technologies, May 8-10, 2002.

[2] Doufexi, Angela, et. al., "An Investigation of the Impact of Bluetooth Interference on the Performance of 802.11g Wireless Local Area Networks," IEEE 0-7803-7757-5/03, 2003.

[3] Ennis, Greg, "Impact of Bluetooth on 802.11 Direct Sequence," IEEE 802.11-98/319, 1998.

[4] Shellhammer, Stephen, "An Analytic CA Model," IEEE 802.19-04/38r1, 2005.

[5] Zyren, Jim, "Extension of Bluetooth and 802.11 Direct Sequence Interference Model," IEEE 802.11-98/378, 1998.

[6] Perahia, Eldad, et. al., "Joint Proposal Team PHY Simulation Results," IEEE 802.11-06/67r2, 2006.

[7] Erceg, Vinko, et. al., "TGn Channel Models," IEEE 802.11-03/940r4, 2004.

[8] Perahia, Eldad, et. al. "p802.11n Coexistence Assurance Document", IEEE 802.11-06/0338r4, 2006.

[9] Kermoal, Jean Philippe, et. al., "A Stochastic MIMO Radio Channel Model With Experimental Validation," *IEEE Journal on Selected Areas in Communications*, Vol. 20, No. 6, AUGUST 2002.

[10] MATLAB© packages developed by AAU-CSys and customized by FUNDP-INFO, http://www.info.fundp.ac.be/~lsc/Research/IEEE_80211_HTSG_CMSC/distribution_t erms.html

[11] "IEEE P802.11n™/D2.0 Draft Amendment to STANDARD [FOR] Information Technology-Telecommunications and information exchange between systems-Local and Metropolitan networks-Specific requirements-Part 11: Wireless LAN Medium Access Control (MAC) and Physical Layer (PHY) specifications: Enhancements for Higher Throughput," February, 2007.

[12] Stephens, Adrian P. "IEEE 802.11 TGn Comparison Criteria," IEEE 802.11-03/0814r31, 2004.

[13] J. Wojtiuk, "Bluetooth and WiFi integration: Solving co-existence challenges," RF Design, October, 2004.

22

Measured WLANs: The First Step to Managed WLANs

Richard H. Paine[a]

The IEEE 802.11 effort (11k) to provide measurements has resulted in a request/response mechanism so end user devices and Access Points can obtain information from each other. In addition, the Management Information Base (MIB) serves as the repository of the information for use by upper layers. The mechanism for accessing the information in the MIB is by Object Identification (OID) addressing. This chapter provides an overview of the mechanisms and the use of the MIB to deliver more accurate and useful information for a more precise wireless environment. At publishing time, 11k had passed from Working Group Letter Ballot to Sponsor Ballot and therefore was still be subject to change until the specification is approved as a standard.

22.1 Introduction

One of the major difficulties with radio and wireless environments is the propensity for interference and radio physics to cause issues for the applications and users of these wireless systems. This propensity is what makes national regulatory control necessary, but there is much more to the issues than just regulatory control. In order to manage and control wireless, standards are needed and information is required to assess what to do about frequency allocations, radio physics problems, interference, and protocols needed to manage the exchange of data wirelessly. Measured wireless systems are the first step to managing the interference and radio physics issues in all wireless systems. In the case of the IEEE 802.11 Wireless Local Area Networks (WLAN), the measured WLAN is specified in the "k" amendment to 802.11.

The recognition of these issues in the Wireless LAN (WLAN) environments became a work item in IEEE 802.11 in 2002. The effort was to address the problems that were becoming evident in WLAN use in high density commercial, industrial, and multiple family dwelling environments. The characteristics of the WLAN radio are a range of approximately 100m, a power level of less than 100mW, and the PHY and MAC protocols to manage the exchange of packets over the unlicensed frequency bands used by the WLAN for a particular regulatory environment. In the case of commercial applications, many commercial stores with other commercial enterprises around them were addressing

---

[a] *Boeing and IEEE 802.11k Chair*

the issues of multiple companies having wireless LANs in the same radio coverage area (e.g., Starbucks in a commercial office building with companies using their own WLANs). In the case of industrial uses, there were instances of many tablets and laptops being used in a mobile manner in the same area with heavy use of the communications infrastructure for 3D CAD, video, and Voice over IP (VoIP). The work item in IEEE 802.11 became the 802.11k Task Group (TGk) and work began in 2002 on a standards submission.

The requirements for measurement included many different wireless environments including factories, coffee shops, airports, Wireless Information Service Providers (WISP) such as T-Mobile and Earthlink, airplanes, airplane WISPs, homes, emergency services, municipalities, apartment buildings, office buildings, VoIP, and Independent Basic Service Set (IBSS) or ad hoc WLAN networks. The requirements led to a set of principles. The requirements review, the principles, and a subsequent architecture addressed these issues by developing a series of requests and corresponding reports between wireless end points that enable the WLAN components to discern their radio environment and the radio environments of those components around them.

22.2 802.11k Measurements

IEEE 802.11k measurements enable end devices in IEEE 802.11 (STAs) to observe and gather data on radio link performance and on the radio environment, including non-802 interference emitters. A STA may choose to make measurements locally, request a measurement from another STA, or may be requested by another STA to make one or more measurements and return the results. Radio Measurement data is made available to STA management and upper protocol layers where it may be used by a range of applications. The measurements enable adjustment of STA operation to better suit the radio environment. The Radio Resource Measurement specification includes measurements that extend the capability, reliability, and maintainability of WLANs by providing standard measurements across different vendors, and provides the resulting measurement data to upper layers in the communications stack.

In addition to featuring standard measurements and delivering measurement information to upper layers, there are applications that require quantifiable radio environment measurements in order to attain the necessary performance levels. These applications include Voice over Internet Protocol (VoIP), Video over IP, location based applications, as well as applications requiring mitigation of harsh radio environments (multi-family dwellings, airplanes, factories, municipalities, etc.). The radio measurements of 802.11k address most of the existing issues in using unlicensed radio spectrum to meet the requirements of these emerging technologies. To address the mobility requirements of technologies, such as VoIP handoff and video streaming handoff, radio measurements such as Channel Load request/report, and the Neighbor Report request/report may be used to collect pre-handoff information which can drastically speed up handoffs between WLAN cells within the same BSS or ESS. By accessing and using this information, the STAs (either in the APs or in the individual devices) can make intelligent decisions about the most effective way to utilize the available spectrum, power, and bandwidth for its desired communications.

The request/response measurements are:

- Beacon
- Measurement Pilot
- Frame
- Channel Load
- Noise Histogram
- STA Statistics
- Location Configuration Information
- Neighbor Report
- Link Measurement
- Transmit Stream Measurement

The request-only mechanism is:

- Measurement Pause

These measurement mechanisms provide the capability for a STA to manage and query its radio environment, and to make appropriate assessments about its health and efficiency. It is the first step in making WLANs smart and capable of making appropriate decisions for fast transition, for mesh connectivity, and for managing the radio environment for all wireless devices.

22.2.1 Beacon

The Beacon request/report pair enables a STA to request from another STA a list of APs it can receive on a specified channel or channels. This measurement may be done by active mode (active scan), passive mode (passive scan), or beacon table modes. If the measurement request is accepted and is in passive mode, a duration timer is set and the measuring STA monitors the requested channel, measures beacon, probe response and measurement pilot power levels (RCPI), and logs all beacons, probe responses and measurement pilots received within the measurement duration. If the measurement request is in active mode, the measuring STA sends a probe request on the requested channel at the beginning of the measurement duration, then monitors the requested channel, measures beacon, probe response and measurement pilot power levels (RCPI), and logs all beacons, probe responses and measurement pilots received within the measurement duration. If the request is beacon table mode, then the measuring STA returns a Beacon Report containing the current contents of any stored beacon information for any supported channel with the requested SSID and BSSID without performing additional measurements.

22.2.2 Measurement Pilot

The Measurement Pilot frame is a compact management frame periodically transmitted by an AP with a relatively small interval as compared to a Beacon Interval. The Measurement Pilot frame is designed to provide a minimal set of information as compared to a Beacon frame to allow for the required small interval. The purpose of the Measurement Pilot frame is to provide timely information to a STA and satisfying the needs of dual-mode cellular/WiFi devices.

22.2.3 Frame

The frame request/report pair returns a picture of all the channel traffic and a count of all the frames received at the measuring STA. For each unique Transmitter Address, the STA reports the Transmitter Address, number of frames received from this transmitter, average power level (RCPI) for these frames, and BSSID of the transmitter.

22.2.4 Channel Load

The channel load request/report pair returns the channel utilization measurement as observed by the measuring STA.

22.2.5 Noise Histogram

The noise histogram request/report pair returns a power histogram measurement of non-802.11 noise power by sampling the channel when Clear Channel Assessment (CCA) indicates idle.

22.2.6 STA Statistics

The STA statistics request/report pair returns groups of values for STA counters and for BSS Average Access Delay. The STA counter group values include: transmitted fragment counts, multicast transmitted frame counts, failed counts, retry counts, multiple retry counts, frame duplicate counts, Request to Send (RTS) success counts, RTS failure counts, Acknowledgement (ACK) failure counts, received fragment counts, multicast received frame counts, FCS error counts, and transmitted frame counts. BSS Average Access Delay group values include: AP average access delay, average access delay for each access category, associated STA count, and channel utilization.

22.2.7 Location

The Location request/report pair returns a requested location in terms of latitude, longitude, and altitude. It includes types of altitude such as floors and permits various reporting resolutions. The requested location may be the location of the requestor (e.g. Where am I?) or the location of the reporting STA (e.g., Where are you?). The location formatting is based on the Internet Engineering Task Force's (IETF) RFC 3825.

22.2.8 Measurement Pause

The measurement pause request is defined, but no report comes back from this request. The measurement pause permits the inclusion of a quantified delay between the execution of individual measurements which are provided in a series within a measurement request frame. The measurement pause used as the last measurement in a frame provides control of the measurement period when measurement request frames are to be repeated.

22.2.9 Neighbor Report

The neighbor request/report is sent to an AP which returns a neighbor report containing information about known neighbor APs that are candidates for a 820.11 Basic Service Set (BSS) transition. Neighbor reports contain information from the table dot11RRMNeighborReportTable in the MIB concerning neighbor APs. This request/report pair enables a STA to gain information about the neighbors of the associated AP that could be used as roaming candidates.

22.2.10 Link Measurement

The link measurement request/report pair is like an RF ping in which a STA requests another STA to measure and report the link path loss and estimation of the link margin. This enables understanding the instantaneous capabilities of a link for streaming and QoS-type requirements.

22.2.11 Transmit Stream Measurement

The Transmit stream measurement is a request/report pair that enables a QoS STA to inquire of a peer QoS STA the condition of an ongoing traffic stream link between them. The Transmit Stream Measurement Report provides the transmit-side performance metrics for the measured traffic stream. Trigger conditions included in the Transmit Stream Measurement Request may initiate triggered Transmit Stream Measurement Reports upon detection of the trigger condition.

22.3 The 11k Interface to Upper Layers

The distinction between the 11k approach and previous approaches to measuring network devices is the use of the MIB as a library of Object IDs. This use of the MIB is described in the specification in the MIB.specification. The use of Simple Network Management Protocol (SNMP) and SNMP clients has held back WLAN network management progress for many years. The choices 802.11k had were to create yet another API and an application to go with it, or to use an object ID concept to gain information from the MIB without using SNMP. The 11k approach is to follow the object ID path for the STA to have instantaneous access to MIB information.

The dot11RRMRequest and dot11RRMReport portions of the SMT MIB provide access to the Radio Measurement service. By performing SET operations on the various dot11RRMRequest MIB objects, radio measurements may be initiated directly on the local STA or on any peer station within the same BSS. Subsequently, by performing GET operations on the various dot11RRMReport MIB objects the results of the requested measurements may be retrieved.

The use of the MIB access method enables a STA, which can be an individual 802.11 device or it can be an AP, to request and have returned measurements from other STAs. These measurements are made available to the STA for determining its most optimal configuration for its radio and network environment. The other major change that makes the 11k a very important change to the 802.11 specification is in the development and use of the Neighbor Report. The Neighbor Report enables an AP to determine which APs in its neighborhood are candidates for potential transition. Early research from the University of Maryland in the potential approaches for Radio Resource Measurement showed that this was very efficient and effective way to determine and enable transitions across a Basic Service Set (BSS) or an Extended Service Set (ESS).

22.4 Impact of the 11k Standard

So, what is the impact of the 11k standard? The measurements in the specification enable a vendor to develop very smart ways of dealing with interference and issues of efficiency in the network and protocol of the BSS and ESS. It will enable more efficient use of many devices and vendor implementations in the same area and enables secure handoff across subnet boundaries. The 11k measurement specification was designed to be the toolbox that could extend the capability of the end device to make intelligent decisions and, in addition, lead to an 11v (Wireless Network Management) control capability. The RRM specification does enable the toolbox so 11v can efficiently control wireless network capabilities for the enterprise or the WISP/ISP. The 802.11 will not only be the wireless that "just works", but it will also be the wireless that "just works very effectively".

23

Cognitive WLAN: A Better Architecture

Nestor Fesas[a]

23.1 Introduction

If you are reading this book, I probably do not have to convince you that wireless LANs (WLANs) are experiencing dramatic growth and are rapidly becoming an entrenched technology in every day life. Today, we see that WLANs have become a competitive differentiator for hotels, multi-tenant dwellings, coffee shops and other establishments where customers are expected and encouraged to stay for more than a few quick moments. Conversely, it is quickly becoming a negative distinction to not have WLAN services in those circumstances.

Like the personal computer revolution before it, as WLAN products permeate applications from the least significant to the mission critical, product lines are stratifying to serve those new and different market segments. However, with this increased popularity and diversity, comes increased deployment at the hands of those that do not understand, and should not need to, the intricacies of RF propagation and network management, often resulting in poor performance and disillusionment with the technology.

To date, two architectures have emerged as the leading approaches to WLAN implementation: *independent* and *dependent*. The *independent architecture* was first to arrive and is a logical extension of standard bridging practices followed in the wired LAN (Ethernet) world. With this approach, access points (APs) are treated as individually managed edge devices. Each AP is configured and managed independently from other peer APs irrespective of whether or not the APs belong to a specific administrative domain. This approach works well for relatively small deployments, but quickly becomes unmanageable, in proportion with the number of APs participating in the WLAN.

As WLANs have grown in size, efficient management techniques have become a requirement thus leading to the development of the WLAN controller and the *dependent architecture* that supports it. A WLAN controller[1] is a device or group of devices designed to centralize control of WLAN features and thus make WLAN management, deployment and security scaleable. Features that typically benefit from centralized control include authentication, authorization, management, intrusion detection, intrusion prevention, performance optimization and mobility among others. However, some aspects of WLANs

---

[a] Bandspeed, Inc
[1] No distinction is assumed here between a WLAN switch and WLAN controller. Although a distinction can be made based on functional partitioning, the terms are considered synonymous herein.

do not lend themselves well to centralization. One example is packet forwarding. Consider that wireless traffic transiting through a secure WLAN air interface must be processed somewhat extensively due to the unreliable nature of said interface as well as to the requirements of mobility and security features. Concentrating all the traffic for a group of APs across a single wireless controller introduces significant complexity, cost and a single point of failure.

Architecturally, a WLAN controller implements a portion of the 802.11 MAC protocol as well as other higher level features – the key idea behind a WLAN controller is that the features of the access point are partitioned in such a way that certain ones are processed at the controller rather than the access point. This partitioning provides the WLAN controller with its ability to consolidate and simplify the management and security of large WLANs. The exact functional split between controller and the access point is vendor specific.

The *dependent architecture* was a natural first step in addressing the scalability and management problems of WLANs. A controller can "see" all the traffic for the APs under its direct control. Some features that are relatively hard to implement with *independent AP architectures* become somewhat natural with the *dependent architecture*. For example, the secure hand off of a client VoIP session from one AP to another can be trivially managed by a WLAN controller that has access to the security keys and state information for the client and the two APs in question[2]. *Dependent architectures* have been well received in corporate environments and have formed the benchmark for performance and feature content to date.

However, once again emerging trends are forcing new requirements:

- Ubiquitous deployment of WLANs is exposing interference between neighboring WLANs as well as between WLANs and consumer products (microwave ovens, cordless phones, wireless cameras, etc);
- Convergence of voice, video and data networks is driving latency and jitter requirements along with the need for quality of service (QoS) to support VoIP and other multimedia applications;
- Price erosion continues unabated as WLANs penetrate ever more cost sensitive applications;
- The emergence of the Small/Medium Enterprise (SME) and the carrier managed home as distinct market segments is driving the requirement for increasingly sophisticated

---

[2] There are several well known complexities involved in the handoff of voice connections between 802.11 APs. These can be resolved with proper protocol design on both the handset and the AP, e.g., 802.11r. Alternatively, handsets can be eliminated from the handoff process altogether with techniques that emulate a single AP using multiple APs. This latter technique creates the illusion of complete RF coverage of a given geographic area by a single AP by duplicating certain WLAN parameters such as SSID and BSSID in beacon frames. Thus, the handoff decision is reduced to a an exchange of call specific parameters between APs (or data structures representing APs) and an agreed transition point where the target AP will commence servicing the connection and the source AP will cease doing so.

[3] The popularity of WLANs in the home is driving considerable interest with cable and telecom providers. One significant obstacle has been the carrier's inability to effectively manage and specifically troubleshoot WLAN connections remotely without incurring a service call (truck roll).

features and technologies that can be easily and quickly deployed by unskilled technicians and users[4].

These new requirements necessitate a new way of thinking about WLANs, their deployment, management and target audience.

23.2 Evolution of the Cognitive WLAN

WLANs were first deployed to solve the specific needs of niche markets, e.g., specific vertical markets like healthcare, warehousing and retail where mobility combined with a LAN connection is a key requirement.

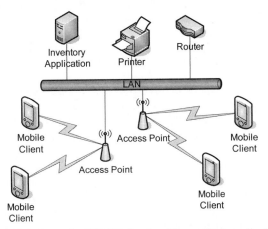

Figure 23.1: Legacy deployment of a WLAN for specific vertical application such as retail or warehousing. This approach is simple to implement but lacks centralized security and management.

The network architecture of these early deployments was relatively simple, with the access points serving primarily as layer 2 bridges between Ethernet (802.3 and its close relatives) and 802.11

As the benefits of WLAN became understood, a mass market ensued forcing a bifurcation into high function, high price solutions for enterprise markets and low function, low cost solutions for the high volume consumer markets. The large enterprise solutions are feature rich and provide extensive configurability at the expense of ease of use. The consumer solutions are easy (in relative terms) to setup and configure at the expense of being inflexible and feature poor. As WLANs proliferated, the weaknesses of each approach became apparent – a new architecture was needed.

---

[4] The popularity of WLANs in the home is driving considerable interest with cable and telecom providers. One significant obstacle has been the carrier's inability to effectively manage and specifically troubleshoot WLAN connections remotely without incurring a service call (truck roll).

23.2.1 WLAN Architecture – Independent APs

Fundamentally, WLAN APs are translational bridges (transparent bridges that interconnect dissimilar LANs[5])

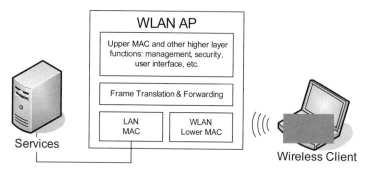

Figure 23.2: Typical architecture of an independent WLAN AP.

The architecture depicted in Figure 23.2 is a straightforward adoption of layer 2 bridging concepts. For networks consisting of a small number of APs, this approach is simple and straightforward. Each AP is physically installed and configured in turn. All the relevant parameters for the AP are stored internally to the device, e.g., passwords, security keys, RF parameters, etc.

This architecture also lends itself to very low cost implementation. Mature components are available at commodity pricing thus making inexpensive APs relatively abundant. The simplicity of this approach is both its strength and its weakness.. The problem is in its lack of scalability. When configuring a network based on this approach, each device must be individually manipulated. The workload and opportunity for error grows significantly with the number of APs in the network (not accounting for any cumulative effect of human fallibility). Imagine a network where the APs are spread across a campus, a town, a city or country. How does one manage such a network efficiently?

Regardless, this approach has some significant benefits:

- Low cost;
- Can be expanded one AP at a time;
- Simple management of the APs, typically through a command line interface (CLI) or a web GUI.

Some notable drawbacks are:

- Does not scale well to large deployments;
- Does not easily support advanced security, e.g., rogue detection;
- Typically does not support mobility[6] features.

---

[5] Seifert, Rich., "Bridging between Technologies.", *The Switch Book: The Complete Guide to LAN Switching Technology,* New York:: John Wiley & Sons, Inc., 2000. 108-109.

It is worth noting that the mobility limitation can be partially addressed, at least in principle, by adopting IEEE Std 802.11i™-2004, namely PMKSA caching. Moreover, work is underway in the IEEE to address mobility on a more comprehensive scale than is attempted in 802.11i, specifically IEEE P802.11r™/D7.0, July 2007.

23.2.2 WLAN Architecture – Dependent APs

To overcome the deficiencies of the *independent AP architecture* when managing large networks, one can concentrate the network's operational parameters in a centralized store and then disburse the information as needed, when needed.

With the *independent AP architecture* depicted previously, all decisions relating to WLAN activity happen at the access point. However, with the introduction of an encapsulation layer and a separation of the components depicted in Figure 23.2 between the Frame Translation and MAC layers, all functions of the AP from the Frame Translation layer upwards can be performed elsewhere in the network. This separation allows one device to consolidate all management, security, user interface and configuration functions and it also relegates the AP to the role of physical interface to the medium. By concentrating all the higher layer functions into the WLAN controller, management is simplified.

Many APs can be managed from a single WLAN controller thus reducing the administrator's workload associated with common tasks such as setting global security parameters or updating firmware revisions on many APs at once. An interesting side effect of the *Independent AP Architecture* is that sharing data such as security keys can become trivial. Consider the diagram in Figure 23.3. The controller manages all of the data structures for higher level functions such as authentication, association and security. Having control over these data structures allows the controller to transfer association information easily and efficiently between APs. The challenge to any transfer (handoff) of clients between APs in 802.11 based architectures as currently defined is that the clients must participate in that transition when certain key parameters are different between the APs in question such as channel or SSID. In these cases, the process of successfully and efficiently transferring a client from one AP to another depends greatly on the client. This is indeed a challenge in most environments where clients are of heterogeneous make, model and revision. The work currently underway within the IEEE on Draft Standard P802.11r should go a long way towards improving this situation. Just the same, it will take considerable time for this standard to be ratified and the millions of units already in the field to be upgraded.

The partition depicted in Figure 23.3 represents one way to achieve the benefits of the dependent AP architecture. The benefits of this approach are:

- Simplified management;
- Rich feature set;
- Scalability.

---

[6] In this context, mobility is defined as the low latency (less than fifty millisecond) connection handoff between APs, usually for the purpose of maintaining a call connection over a broad geographic area covered by multiple APs.

The principle drawbacks are:

- High cost due to the controller
- Performance bottleneck at the controller
- Adds an unnecessary layer to the time proven and well understood wiring closet architecture

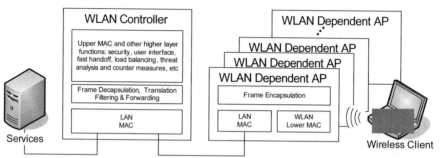

Figure 23.3: One variation of the dependent AP architecture. In this depiction, the lower MAC function is conducted by the lower MAC in the dependent AP. A different architectural partition could introduce a unified lower MAC function in the WLAN Controller to facilitate MAC features such as acknowledgement suppression for VoIP client handoff in blanket roaming deployments.

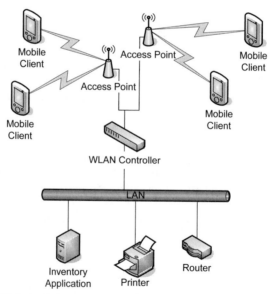

Figure 23.4: Simplified depiction of an Enterprise class WLAN based on the *dependent architecture*. A WLAN controller appliance is introduced between the LAN and the WLAN. This device improves manageability and security by consolidating those functions. When compared to the legacy *independent architecture*, the *dependent architecture* facilitates new features such as VoIP client handoff between APs.

23.3 Cognitive WLAN Architecture

Up to this point, we have focused our attention on aspects of wireless LANs pertaining to how easy, or not, they are to deploy and maintain. However, it would be much more satisfying if there was no need to contemplate such ideas. It would be infinitely better if WLAN access points could be sprinkled about a desired area much like lamps can be placed intuitively to provide light where needed. If a room seems unnecessarily dark, then add a reading lamp or a spot lamp or background lighting as the situation requires. This type of flexibility is the ultimate goal of a cognitive WLAN. However, as WLANs by necessity provide services beyond basic illumination, the situation demands added sophistication in the WLAN equipment.

23.3.1 Cognitive WLAN Goals

The central idea behind a Cognitive WLAN is to make WLANs truly easy and affordable to deploy and use. One way to facilitate this ideal is to blend the most desirable attributes of the *independent* and *dependant* architectures and to avoid the deficiencies of each. However, this alone is not enough. One must also pay close attention to and introduce concepts that make the deployment and management of complex services relatively easy. For example, RF interference although invisible has very real manifestations in terms of reduced performance and connection quality. Troubleshooting an interference problem is impossible for the average user. The system must have the inbuilt sophistication to recognize an RF interference problem when it occurs, save as much specific detail about the problem as possible in a log file and ultimately overcome the problem in a sensible way, say by alerting of the problem and switching to a clear channel if one is available.

In brief, a cognitive WLAN has the following high level design goals:

- Easy to deploy and configure by inexperienced users
- Easy to manage day to day without special diagnostic tools
- Provide advanced features
- Be self healing

To achieve a low cost, organically scaleable, feature rich and easy to manage cognitive WLAN, the system must have the following attributes:

- easy to deploy
- self configuring
- automatically adjust to changing RF conditions
- visibility into the RF environment
- RF event logging
- unified remote management
- arbitrary and organic scaling
- distributed control
- detect and thwart security threats

Earlier, we touched on the idea that WLANs are increasingly deployed by people without detailed knowledge of RF systems. This actuality drives the need for increased sophistication at the equipment. For example, when initially deployed, APs must discover as much as possible about their environment, adopt reasonable settings and provide a secure, basic functional setup. Consider the case where an AP is deployed by an accountant in a small office. At first, one AP is sufficient to meet the wireless needs of the office. The AP is unpacked, placed in a convenient location, connected to the user's Ethernet network and powered up. The AP attempts to acquire an IP configuration from the local DHCP server. Not finding a server, the AP adopts a default IP address after probing to establish that the address is available. The AP then scans the RF environment to select the best channel based on a combination of variables such as observed interferers, channel utilization, preferred frequency and regulatory limits. Once the RF parameters are configured, the AP commences operation using the default SSID and security settings.[7]

During its operational phase, the AP continuously monitors the RF environment, looking for interference sources and security threats, adjusting along the way by selecting operating frequencies that avoid the interferers and employing countermeasures respectively. These events are recorded in a log file for later analysis. For example, in a crowded band where the only available channel coincides with the frequencies used by a cordless phone or microwave oven, intermittent performance problems may be observed. Interference events recorded in the event log can provide the necessary insight to resolve the problem, say by replacing the cordless phone with one using compatible frequencies.

As the office WLAN capacity requirements grow, a second AP is added. The accountant unpacks, places, connects and powers the new unit as before. This time however the new AP is recognized by the existing one. Using the WLAN management interface, the accountant logically groups the new AP with the existing one. In so doing, the new AP establishes a management link with the first, adopting its configuration and operational policies and thus forming an AP cluster. The process is repeated as needed, one or more APs at a time. The resulting cluster collectively provides advanced services such as fast AP to AP handoff for voice connections over WLAN enabled telephones and source localization to establish the relative whereabouts of selected assets (given enough APs to adequately determine the location of the source signal).

Eventually, the small office system evolves into a medium or even large system encompassing many APs across different departments in geographically dispersed locations. The APs are over deployed by a small margin to provide redundancy. If one fails, its neighbors can pick up the load. The APs are arranged into super clusters to hierarchically structure the WLAN thus more efficiently handle the flow of management information between APs. Advanced unified remote management capabilities permit the clusters and super clusters to be managed at a distance. Invoking the management interface for a cluster or for a super cluster allows the administrator to alter policies and

---

[7] This scenario is open to some potential complexities such as the default SSID is in use by a neighboring AP or the selected default IP address belongs to a dormant device or how to determine the IP address of the first AP installed if the address is selected using DHCP. A little clever programming will cover much of the distance towards addressing these issues. For those cases not covered by these techniques, the user can revert to directly connecting the AP to a computer and addressing the configuration conflicts manually.

configuration details that apply to a single AP, a group of APs, the cluster or the whole WLAN.

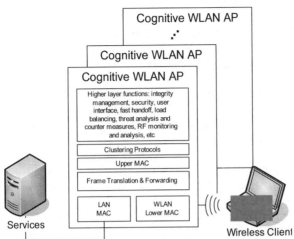

Figure 23.5: One variation on the cognitive WLAN architecture. The diagram shows that a cognitive WLAN AP builds on the basic blocks of the *independent* and *dependent* architectures. Key additions include clustering protocols, integrity management and RF analysis.

22.3.2 Components of a Cognitive WLAN

Three key components differentiate a cognitive WLAN from conventional *independent* and *dependent* WLANs: clustering, RF analysis and integrity management. These three components are the foundation upon which logic to fulfill the stated goals is built.

22.3.2.1 Clustering

The clustering protocols are responsible for establishing a cluster, discovering an existing cluster, registering APs with the cluster, monitoring, distribution of configuration information, the exchange of coordination messages and failover. In general, the clustering system provides the mechanisms through which collaboration can take place. For example, to provide a "roaming" service for voice calls within a building or to provide a rogue neutralizing feature for secure networks, it is useful to "know" where the handset or rogue is in rough terms and then to instruct cluster nodes (APs) within range of the handset or of the rogue to behave in a specific fashion. The clustering protocols enable the flow of control and status messages between the cluster participants.

22.3.2.2 RF Analysis

The RF analysis component monitors the RF environment, building and maintaining a database of observed RF phenomena in the process. This component is responsible for channel selection, interference avoidance and interference classification. Different approaches are possible to achieve the desired behavior for RF analysis. However, the most effective method involves the use of dedicated hardware to monitor the bands of interest in a continual fashion. The hardware can be as simple as a dedicated WLAN interface used primarily in the analysis function or it can be more sophisticated as is the case with devices on the market today that can process raw spectral data to provide detailed signal analysis and classification. This approach enables the AP to make better target channel selections when vacating channels either to avoid interference or to avoid interfering with a primary user, e.g., radar service.

One useful attribute of the RF analysis component is that it eliminates the need for site surveys. Previously, an RF site survey was required to ensure coverage in desired areas of a target site. A surveyor estimates AP placement using modeling tools and/or experience. To validate the deployment, the surveyor measures signal strength throughout the site, adds or moves APs where coverage is weak and modifies AP RF parameters to tune the signal footprint. This trial and error process is tedious, error prone, expensive and time consuming. A cognitive WLAN avoids the site survey step by using the RF component to measure signals from peer APs and adjust transmit power and operating channel accordingly. By placing the APs a little closer together than strictly necessary, the RF analysis component is able to adjust the transmit power and channel settings to accommodate the deployment environment

The RF analysis module is interesting in some other ways as well. With dedicated RF analysis hardware, one also gains the ability to conduct more precise measurements of RF signals. These measurements can enable future applications such as source localization which can be used to determine with a predetermined resolution, the location of an emitter. Location information of this type can be used to predict the location of a rogue device or of an emitter mounting a denial of service attack as examples.

22.3.2.3 Integrity Management

The integrity management component is responsible for maintaining configuration and policy integrity within the cluster. Specifically, the challenge to overcome is rooted in the distributed nature of the system. Distributed systems cannot inherently guarantee the availability of any cluster node (AP). The issue is that a node that is out of service when new configuration or policy settings take effect must acquire the new settings prior to returning to service. Consequently, a mechanism for ensuring that important information such as configuration and policy settings is unequivocally transferred successfully to each cluster member is required. Moreover, the integrity management function must also resolve conflicts stemming from incompatible user selections or from out of service nodes rejoining the cluster with dated configuration and policy information. For example, an administrator might choose to initiate a data service on a given set of wireless interfaces (WIFs) within a given set of APs in a cluster. Let us assume the administrator configures a data service on WIF-3 of each of the APs in a cluster, a voice roaming service on WIF-1 of each AP and an RF monitor (security and RF scanning) service on WIF-2. Furthermore, AP-3 goes out of service due to a malfunction that is corrected at a later time. If the system administrator updates the configuration settings for the data service, each in service AP will be adjusted to the new settings. However, AP-3 is out of service and will not receive the updated configuration. Left unchecked, AP-3 would initiate a data service on its WIF-3 with erroneous parameters as demonstrated in Figure 23.6 resulting in undefined system behavior. The integrity manager would prevent this situation from occurring.

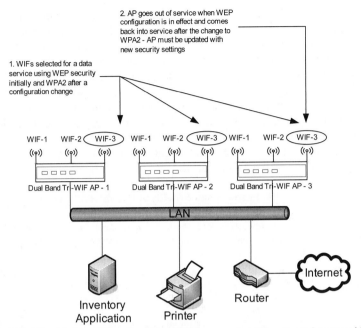

Figure 23.6: WLAN with 3 multi-radio APs. WIF 3 on each AP is selected to provide a WLAN access for data connections and is configured with WEP encryption. Later AP-3 suffers a fault and is taken out of service, after which APs AP-1 and AP-2 are reconfigured for WPA2 encryption. When AP-3 is returned to service, it will have an outdated configuration. The integrity management function is responsible for ensuring that AP-3 does not commence servicing clients until such time that its configuration has been updated with the latest information.

23.4 Features and Benefits of the Architecture

We have touched on some of the features and benefits of a cognitive WLAN already. However, there are a few key distinctions that merit further consideration. Table 23.1 summarizes these features and their corresponding benefits.

It is important to highlight that cognitive WLANs like all systems comprise a number of trade-offs. In exchange for the flexibility and intelligence embedded in the system, the APs require greater CPU and memory requirements than the conventional APs used with *independent* and *dependent* architectures. Conversely, the cost of computing continues to drop dramatically and AP designs with sufficient computational ability for cognitive WLAN applications are readily available today at highly competitive price points.

23.5 The Vision for the Future

Observing the downward price trend for WLAN equipment and the upward capability trend for computing hardware, it is straightforward to envision a future where WLANs become increasingly sophisticated to the point where they offer enough performance and ease of use to be considered a viable alternative to a conventional wired LAN. Some might argue that day is here. Envision if you will a cognitive WLAN where nodes are distributed conveniently about without too much regard for placement. The nodes are inexpensive enough that they are treated as consumables. They need little more than electrical power to initiate service. As each node comes to life, it discovers its neighbors, self organizes into a mesh, identifies its target cluster, adopts the specified operating configuration and policies, begins offering data and voice services and commences to guard against intruders and attackers, pinpointing their location and alerting the proper authorities when required. The individual pieces to realize this vision are largely in place today with the missing ones following quickly. Maybe the day isn't far off.

Table 23.1: Features and Benefits of Cognitive WLANs.

Feature	Benefit
Self Configuring	Automatically selects basic RF parameters such as transmit power and channel frequency; monitors the cognitive WLAN's operating bands for adverse conditions (e.g., interference) and adjusts the configuration accordingly.
Scalability	Network can grow from small single AP deployments to thousands of APs, either one AP at a time or in groups of APs, while using the existing wired LAN infrastructure.
Redundancy	Low cost AP hardware makes it possible to cost justify over deployment of APs to provide coverage redundancy
Unified Management	Management of the cluster is accomplished by effecting configuration changes at one point in the cluster – configuration changes are automatically distributed to all cluster participants in a guaranteed fashion.
RF Analysis and Classification	Analyses, classifies and records RF events in system event log; provides real time visualization (e.g., a spectrogram) of the AP's operating bands.
Advanced Features	All the same advanced features possible with controller based architectures but without the cost and complexity of the controller, e.g., fast handoff of VoIP or other QoS sensitive applications, intrusion detection and neutralization, etc.

24

Wi-Fi Range

Perry Correll and John DiGiovanni[a]

24.1 Introduction

In the wired Ethernet environment, distance limitations and data rates are fully defined. This is a result of specific transmitter and receiver standards and a controlled media, i.e. the wire. A controlled media (such as wired Ethernet) is the key point here because a defined data rate can be maintained over a specified distance.

Things change significantly with wireless communications and once again the key is a controlled media, or lack there of. Physical media will always return fixed results; distances and data rates can vary greatly when using Radio Frequency (RF) as the transmission medium. It is because of this "fluid" nature of RF that deploying a Wi-Fi network can be fraught with issues, miss met expectations and a generally unhappy group of users.

It is also important to note the range, or the coverage area of a Wi-Fi Access Point is impacted by several items including data rate, capacity, interference and other variables so there are many things to contend with when going wireless.

However, with an understanding of a few basic principles such as antenna design and gain along with some information on items that impact a Wi-Fi network, you will be in a position to better create a higher performing, longer range wireless network.

24.2 Defining Range and Coverage

Before the RF signal leaves the antenna, a digital signal processor will convert the data stream into complex symbols that carry it over the air as it is transmitted. From there is goes into a radio transceiver that translates those symbols to a specific carrier frequency. In the case of Wi-Fi, it is either the 2.4 GHz (for 802.11b/g) or 5 GHz (802.11a) frequency ranges. Those signals will then pass through a power amplifier on the way out of the transmitter's antenna. The antenna on the other end of the signal will pass that received signal down to a low noise amplifier of the receiver. This completes the radio connection between the two ends of a radio link.

---

[a] All authors currently affiliated with *Xirrus, Inc*

Between the two antennas, information will travel on the radio waves at the speed of light as it moves from transmitter to receiver. Radio frequency signal is modulated and encoded with data and a wavelength is represented by λ.

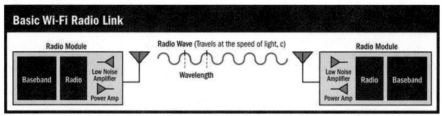

Figure 24.1: Basic Wi-Fi Radio Link.

When defining range, we define it as the maximum distance at which two radios can operate and maintain a connection. Therefore we can use simple geometry to determine the coverage area of an Access Point using the formula to determine the area of a circle $(\pi)r^2$ where the radius (r) is the range of the Wi-Fi signal. The coverage area of an Access Point is often referred to as a cell and these terms will be used interchangeable throughout this chapter.

As an example, the expected coverage area of a Wi-Fi device with a 300' operating range (r), would yield a 280,000 sqft. coverage area (Figure 24.2).

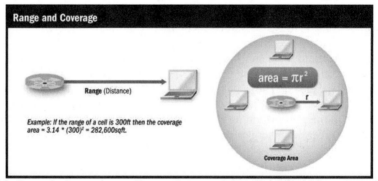

Figure 24.2: Range is the maximum distance between two radios for which a connection can be maintained. Coverage is the total area in which all radios can maintain a connection to an Access Point.

24.3 Range Basics

Intuitively, everyone understands that range is a function of data rate or simply put, the higher the data rate, the shorter the range. In order to understand what goes into determining the range of an Access Point, a few terms need to be defined and a basic understanding of the mathematics that goes into determining the distance by which a radio signal will travel needs to be provided.

In an open environment, or what is referred to as Free Space, Power varies inversely with the square of the distance between two points (the receiver and the transmitter). The stronger the Transmit Power, the higher the signal strength or Amplitude. Antenna Gain also increases Amplitude and will be further discussed in a subsequent section of this chapter.

While Gain and Power increase the distance a wireless signal can travel, the expected signal loss (Path Loss) between the transmitter and a receiver reduces it. Path Loss is the reduction in signal strength that a signal experiences as it travels through the air or through objects between the transmitter and receiver. The relative strength of that signal at the receiver is measured as the Received Signal Strength Indicator (RSSI). RSSI is normally expressed in dBm or as a numerical percentage. For clarification purposes, a dB (Decibel) is a measure of the ratio between two quantities and dBm is a Decibel with respect to milliwatts of power. By taking into account all of the gains and losses of a signal as it moves from a transmitter to a receiver, an overall Link Budget can be defined (Figure 24.3).

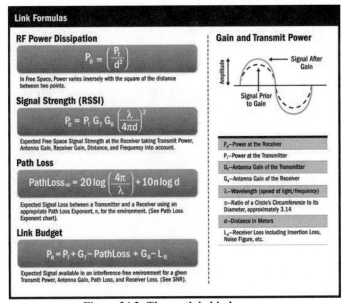

Figure 24.3: The math behind range.

24.4 Antenna Design

Antennas play a key role in determining the amount of range and coverage area of a Wi-Fi network so let's start with a short discussion on the physical layer of the Wi-Fi connection with an overview on antennas.

An Isotropic antenna has a radiation pattern of a perfect sphere. Imagine a device that has a power density equal in all directions (Figure 24.4). The Isotropic pattern forms the basis from which all other antennas are measured. One of the simplest antenna designs is the dipole, like the simple whip antenna on most cars. Where an Isotropic antenna pattern is spherical, the dipole has a radiation pattern of a torrid (like a donut). The largest amount of energy is being radiated perpendicular to the antenna, in most cases this is along the horizontal plane. You can visualize this by thinking of a pebble dropped into a calm pond, the wave patters extend from the center in all directions along a two dimensional plane. Note: Before we get in too much trouble with antenna designers, let us agree that physics say all antennas will radiate some energy in all directions; however the goal of antenna design is to focus the greatest amount of energy in an intended pattern and since this chapter is not intended to teach antenna design – the analogy works.

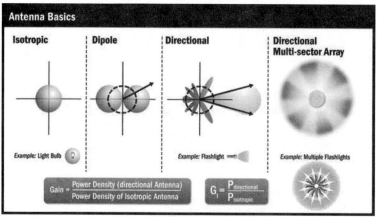

Figure 24.4: Antenna Basics.

Gain (also known as Amplification) is critical to improving the range of an antenna and therefore plays a critical part in determining (or extending) the range of a Wi-Fi network. Gain refers to an increase of the Amplitude or Signal Strength and comes in two forms; active and passive. Active Gain refers to an increase in power that is applied to the antenna where passive Gain is achieved by focusing the energy of the antenna in a particular direction. Gain is usually expressed as a ratio in dB's. It is the log of the ratio of the powers. A typical dipole antenna will have about 2 dBi, (i = isotropic) of Gain.

One of the advantages of a dipole, or any type of directional antenna is greater antenna Gain; this is a result of the RF energy pattern being focused vs. an isotropic design. Other types of antennas are more directional in design taking their radiated energy and squeezing it into a very narrow pattern (Figure 24.4). A good analogy here is to think of the isotropic antenna as a light bulb radiating energy equally in all directions; the directional antenna would be seen as more of a flash light with the light focused in one direction. And just like the flashlight, some energy (light) would also go in all directions so instead of having a spherical pattern, the light is focused into a beam which enables it to travel further.

Antenna Gain is bi-directional so it will amplify the signal as it is being transmitted and as it is received. So if a directional antenna is providing 6 dB Gain on transmit, it will also increase received sensitivity an equal amount so the antenna design of the Wi-Fi Access Point plays a critical role in the amount of range (coverage) delivered.

24.5 Range and Coverage

The vast majority of Access Point deployments today consist of products that use omni-directional antennas. For the most part, this type of deployment has served the market well for home use and light use in the enterprise and in public-access types of locations like airports and coffee shops. But with the increase in Wi-Fi users and the associated number of Access Points to support them, the omni-directional antenna becomes its own worst enemy in the battle to address improved range performance.

No matter what type of Access Point is used, their respective use of an omni-direction antenna that blasts RF energy in all directions becomes a barrier to the performance needed for today's Wi-Fi networks. This problem consists of a number of issues that all limit high-performance deployments: cell size, channel reuse, hidden nodes and multi-path.

As we've discussed, omni-directional antennas transmit and receive RF energy in all directions. Directional or sectored antennas focus RF energy into a single direction, thereby intensifying the strength of the signal (Gain) that is transmitted and increase the receiver sensitivity for traffic coming from the clients. Since directional antennas offer more Gain, they have the ability to transmit further and "listen better" to the signals of wireless stations (clients) therefore increasing the range and coverage of the Access Point in a given direction.

The drawback of a directional high gain antenna is that it does not cover the same area as a standard dipole antenna. The solution here is to arrange the directional antennas in a circular pattern and create an array of antennas. This provides the 360 degrees of coverage of a traditional dipole antenna, yet offers the range advantages of a directional antenna. Some sectored antenna systems have multiple radios and multiple antennas that allow for 360 degrees of increased coverage.

24.6 Range Limiting Factors

In section 24.3 we discussed how Gain and antenna design can be used to increase the range of a Wi-Fi Access Point. We also mentioned that physical and environmental factors will impact that signal, this section will examine several of those factors.

24.6.1 Interference

802.11b/g uses the 2.4GHz ISM band. Many other devices also operate in the 2.4GHz ISM band and interference causes data to be garbled forcing packets to be re-transmitted. This causes reduced end-user throughput and increased latency of data traversing the Wi-Fi network.

Conversely, the 5 GHz band for 802.11a is relatively clean from interfering devices. 802.11a is also deemed as the primary user of the spectrum. This disallows other types of wireless data devices in this band. Since the 2.4 GHz band is more susceptible to interference, it is highly recommended to migrate towards an 802.11a environment.

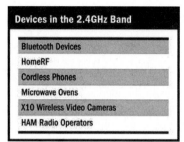

Figure 24.5: Devices in the 2.4 GHz Band.

24.6.2 Multipath

Omni-directional antennas, and to a lesser extent directional antennas generate vast amounts of RF energy that is transmitted into the Wi-Fi environment. Multipath occurs when signals bounce off multiple objects in the environment and are reflected back to the receiver.

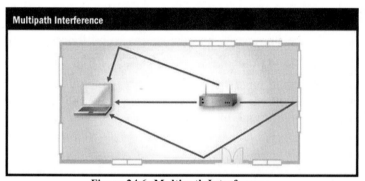

Figure 24.6: Multipath Interference.

The effect is that weaker "copies" of the original signal arrive slightly later than the primary signal. This causes inter-symbol interference and gets worse as the "delay spread", or time between the reception of the primary signal and secondary signal increases (Figure 24.7). The end result is corrupt packets that must be re-transmitted, lowering network performance.

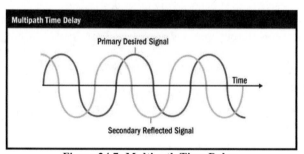

Figure 24.7: Multipath Time Delay.

Any type of reflected signal that can be additive or destructive to the original signal is identified as multipath interference. Figure 24.8 shows an example on how multipath is caused. As the signal strikes an objective, it can react in several ways creating reflection, scattering, refraction, diffraction or all of the above.

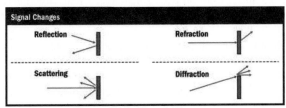

Figure 24.8: Signal Changes.

Reflection is simply when the signal is reflected back towards the transmitter. Scattering occurs when the signal is scattered back towards the transmitter into multiple new signals. Refraction occurs when the signal is bent as it passes through an object and Diffraction happens when the signal changes direction as it passes around an object.

In some cases, a strong enough signal received out of phase can essentially create a null, a spot where no signal is available. Yet only a few feet away you may have a strong signal. This is called a multipath null.

One important note here is the benefit of directional antennas in limiting multipath interference. As previously discussed, omni-directional Access Points inherently create large amounts of performance-robbing multipath. With a directional antenna, this problem is greatly reduced because RF energy is not blindly transmitted in all directions. RF signals are transmitted in the direction of the wireless client within a given sector and not in the direction opposite the wireless client that would otherwise come back as distorting multipath.

24.6.3 Attenuation

RF signal strength is reduced as it passes through various materials. This effect is referred to as Attenuation. As more Attenuation is applied to a signal, its effective range will be reduced. The amount of Attenuation will vary greatly based on the composition of the material the RF signal is passing through. Figure 24.9 shows typical attenuation through common office environments.

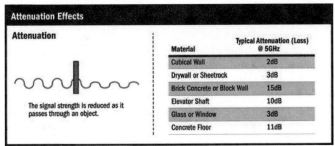

Figure 24.9: Attenuation Effects.

Cubical walls offer relatively low attenuation, in the 2 dB range, while concrete and brick walls will cause higher attenuation levels reducing the range of an Access Point. It is extremely important to consider not just the type of obstruction, but how many obstructions the RF signal must pass through when designing a Wi-Fi network.

A directional antenna offers an advantage over omni-directional antennas when it comes to attenuation as they are better able to penetrate different materials than traditional dipole antennas.

The bottom line to remember is that antenna selection and the physical environment of the facility have the biggest impact on range and coverage performance of an Access Point. However, it is important to note these are not the only factors involved; the 802.11 specification in and of itself creates issues that impact the overall performance of the Wi-Fi network.

24.6.4 Hidden Node

The 802.11 specification operates under a "collision avoidance" schema whereas clients must wait for the medium to be free before making a transmission. This basic premise creates a situation where two clients within a Wi-Fi cell (coverage area of the Access Point) are within range of the Access Point, but out of range of each other. A wireless station on one edge of a cell may not hear a station on the other side of the cell. Because of this, wireless stations will not be able to hear when the other is transmitting; incorrectly assuming the air is idle and begin to transmit its own packets. This will cause the two transmissions to collide requiring both stations to re-transmit greatly reducing the effective bandwidth within the cell.

A protection mechanism exists with in the 802.11 standard called CTS-RTS that can help address this issue requiring each client to ask for permission from the Access Point before transmitting. But the use of this protocol creates overhead on the network and will reduce overall performance by 30%. Another method commonly used to eliminate a hidden node issue is to reduce the range of the Access Point by decreasing the transmit power. By reducing an Access Point's range, it increases the probability that all clients within the cell will hear each other; but greatly increases the number of Access Points needed within the deployment.

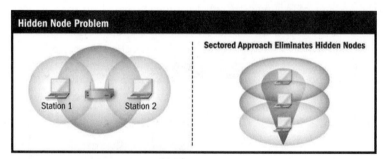

Figure 24.10: Station 1 and Station 2 cannot hear each other's transmissions in the omni-directional antenna example (above left) whereas stations can hear each other in the directional (sectored) antenna approach.

The use of a directional antenna over an omni-directional antenna will also eliminate the issue of Hidden Nodes because all wireless stations (clients) in a given RF sector are associated to the same Access Point; so they are geometrically within the same sector. Since the clients operate in the same sector, the hidden node problem is eliminated as all stations are able to hear each other and correctly determine when the air is busy or idle. This eliminates the performance-robbing issues found with legacy omni-directional Access Points and the use of the CTS-RTS protocol. It also has the added benefit of not increasing the number of Access Points or reducing their respective coverage areas.

24.7 Signal to Noise Ratio (SNR)

The range of an Access Point is a function of data rate. The notion that higher data rates do not appear to "travel" as far as the lower data rates is a function of the Signal to Noise Ratio (SNR) and not because the Access Point and the client can't necessarily "hear" each other.

SNR is the ratio of the desired signal to that of all other noise and interference as seen by a receiver (Figure 24.11). SNR is important as it determines which data rates can be correctly decoded in a wireless link. It is expressed in dB as a ratio. Figure 24.11 also contains a chart that shows the required SNR for 802.11a (5 GHz) data rates.

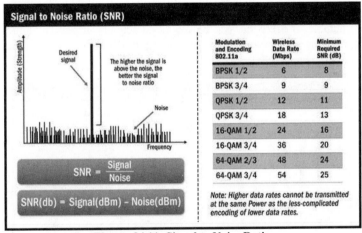

Modulation and Encoding 802.11a	Wireless Data Rate (Mbps)	Minimum Required SNR (dB)
BPSK 1/2	6	8
BPSK 3/4	9	9
QPSK 1/2	12	11
QPSK 3/4	18	13
16-QAM 1/2	24	16
16-QAM 3/4	36	20
64-QAM 2/3	48	24
64-QAM 3/4	54	25

Note: Higher data rates cannot be transmitted at the same Power as the less-complicated encoding of lower data rates.

$$SNR = \frac{Signal}{Noise}$$

$$SNR(db) = Signal(dBm) - Noise(dBm)$$

Figure: 24.11: Signal to Noise Ratio.

Figure 24.11 shows the received signal, the noise level, (or noise floor) and with these two values, the SNR can be determined. As data rates increase from 6 Mbps to 54 Mbps, more complex modulation and encoding methods are used that require a higher SNR to properly decode the signal.

Using the table in Figure 24.11, a 54 Mbps signal requires 25 dB of SNR so it will not be properly decoded at greater distances because as the signal moves further from the source, a greater amount of path loss (signal attenuated) occurs. Lower data rate transmissions, can be more easily decoded and as a result appear to "travel" farther.

As an example in an outdoor environment with just free space loss, a 6 Mbps signal can actually be decoded 7 times further away than a 54 Mbps signal. An indoor environment, due to obstructions, would only offer a 3X advantage in distance.

24.8 Range versus Capacity

Many of the initial Wi-Fi deployments focused on providing the maximum amount of coverage so the range of an Access Point was a critical factor in the purchase process. While still important, range becomes less of an issue as more clients connect to the network given the inherit bandwidth limitations of today's Access Points. It is important to realize that greater range is not always a positive thing. Sometimes it works against you depending on your application and the number of users on the network. As you increase the range of an Access Point, the coverage area increases and you will now be covering more users with a single device. As more users join the network, they will all be vying for the finite amount of bandwidth available within that cell.

As the size of the cell increases, clients at the edge of the cell will be using lower data rates and therefore consume more time on the wireless medium lowering the performance of all of the clients. Viewing Figure 24.12 shows the tradeoff between range and capacity. Example: Deploying an Access Point within 75 feet will allow operation at higher data rates and can support 10-12 clients with 2 Mbps of TCP throughput guaranteed for all users. As the cell size increases, clients that are further away will operate at lower data rates and the number of users that can be supported actually drops. The bottom line is to match cell size to the capacity and number of users in the area as well as the application requirements (date, voice, video).

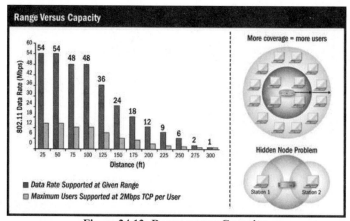

Figure 24.12: Range versus Capacity.

Here are a few simple rules to remember when deploying a Wi-Fi network using omni-directional antennas:

• As omni-directional coverage increases, the number of potential covered users increases as well as the amount of capacity required

- As the number of users increases, the available capacity or effective bandwidth per user decreases
- The larger the omni-directional cell size, the greater the chance that two stations will not hear each other's transmissions creating a Hidden Node problem
- A multi-sector, directional antenna approach can help provide both range and capacity and can also help mitigate hidden nodes

Directional antennas also help to alleviate the problems of cell capacity by focusing the RF energy in specific sectors. In this way, greater distance can be attained along the defined sector without also associating with stations outside the sector (Figure 24.13).

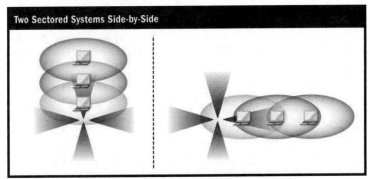

Figure 24.13: Two Sectored Systems Side-by-Side.

On the opposite end of the scale is the idea of shrinking cell sizes to increase capacity. First, shrinking the cell size of an Access Point by lowering the transmit power does not lower the transmit power of the wireless stations (clients) that are associated to it. Second, the client's transmit power and receiver settings are not under the control of the Access Point and do not change. By decreasing the transmit power of the Access Point, the overall cell size shrinks only slightly so the real size of the Wi-Fi cell is not just the transmission range of the Access Point. In fact, the real size of a Wi-Fi cell is the transmission range of the Access Point and the transmission range of all the wireless clients in that cell.

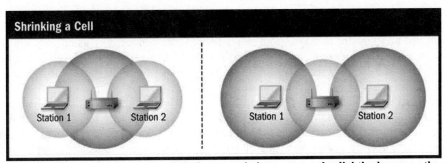

Figure 24.14: Decreasing the Access Point's transmission power only slightly decreases the size of the cell.

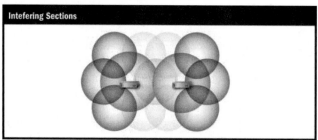

Figure 24.15: Decreasing the Access Point's transmission power causing client interference (between the two Access Points).

Shrinking cells also creates additional interference issues and limits range. When two Access Points are placed near each other on the same channel, they may not hear each other but the stations (clients) that associate to those Access Points may well interfere with stations on the adjacent cell and vice versa. Interfering stations are shown in between the two Access Points (Figure 24.15).

24.9 Site Surveys and Dead Spots

One of the more common, practical decisions that need to be made when deploying a Wi-Fi network is to decide whether or not to do a site survey. This decision hinges on how big and complex the Wi-Fi network will be; the more complex the network, the greater the need for a site survey.

The most obvious conclusion that can be drawn from this chapter is that Wi-Fi range and the resulting coverage area is "fluid". Wireless signals do not propagate the same way in all directions and are constantly impacted by factors in the environment. Unlike a wired network, it is far from static and that's simply a result of the medium which is why we always recommend a site survey. Obstructions including walls, cubicles and people offer varying degrees of Attenuation, which causes the wireless signal to be irregular. As a result, it's often necessary to perform a site survey to understand how the wireless signal will move within a facility before installing Access Points.

The goal of a site survey is to obtain enough information to determine the number and placement of Access Points to ensure proper coverage of the facility. Proper coverage can be defined in a number of ways, from making sure basic connectivity using very low data rates is available in certain areas to ensuring high data rate connectivity is available everywhere. At the very least, doing the site survey will force you to determine exactly what you want from your Wi-Fi network.

Not all site surveys are created equal. Two principle methods exist. The first is predictive meaning no real data collection takes place at the location. The predictive site survey is derived from mathematical calculations of how wireless signals *should* propagate throughout a facility given the types of building materials and objects that exist at the location. There are excellent tools available that perform this function but the accuracy of the site survey is largely determined by the quality of the data entered, i.e. cement floors vs. wood, etc. The old adage garbage in, garbage out rings true here.

The second type of site survey is completed by collecting actual readings of signal strength between an Access Point and a client at various points throughout the facility. This information is then plotted against a layout of the facility. While this method is considered more accurate, it is also more time consuming and therefore more expensive. Regardless of site survey methodology, both will provide excellent guidance in determining the number of Access Points and their respective placements. The site survey will also offer the added benefit of forcing a set of requirements to be agreed upon prior to the actual installation.

Another benefit to completing a site survey will help in identifying "dead spots" or areas within the facility that are difficult to cover. As we've discussed, using an Access Point with omni-directional antennas can yield more dead spots or multipath nulls. Directional antennas will greatly reduce the occurrence of dead spots as more antennas (and radios) will produce a richer environment of wireless signals increasing the likelihood of the client hearing the Access Point.

24.10 Future Technologies

There are many different wireless technologies; all of which tend to serve a particular purpose or where designed for specific application. For wireless local area networking, the 802.11 standard will continue to rein king for the foreseeable future. Remember that it took nearly 7 years for current iteration of Wi-Fi to become mainstream once it was ratified

As of the writing of this book, the next major transition that 802.11 will undergo will be what's being dubbed 802.11n. The IEEE expects to formally ratify 802.11n sometime in 2008 although products based on draft versions of the standard starting appearing in the market in late 2006.

Work on the 802.11n standard started in January of 2004 when the IEEE announced that it had formed a new Task Group (TGn) to develop the next amendment to the 802.11 standard for wireless local area networks. The goal of the IEEE when work began on 802.11n was to obtain TCP throughputs in excess of 100 Mbps.

802.11 Comparison Table					
	802.11b	802.11g	802.11a	802.11n	
IEEE Ratified	1999	2001	1999	Expected 2008	
Frequency	2.4 GHz	2.4 GHz	5 GHz	2.4 GHz	5 GHz
Non-overlapping Channels	3	3	12	3	12
Baseline Bandwidth Per Channel	11Mbps	54Mbps	54Mbps	65Mbps	65Mbps
Number of Spatial Streams	1	1	1	2, 3 or 4	2, 3 or 4
Channel Bonding	No	No	No	No	Yes
Max Bandwidth Per Channel	11Mbps	54Mbps	54Mbps	130Mbps	270Mbps

Figure 24.16: 802.11 Comparison Table.

In order achieve this goal, changes to the physical and MAC layers of 802.11 standard would need to be made. While 802.11n builds upon previous 802.11 standards, updates to the MAC layer were substantial as well as introducing MIMO (Multiple Input Multiple Output) signal processing.

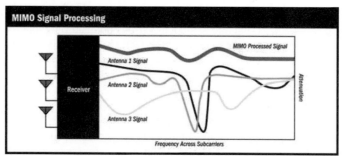

Figure 24.17: MIMO Signal Processing.

MIMO uses multiple transmitter and receiver antennas and takes advantage of multipath reflections to improve signal coherence that greatly increases receiver sensitivity. This extra sensitivity can be used for greater range or higher data rates. The newly enhanced signal (Figure 24.17) is the processed sum of individual antennas. Signal processing eliminates nulls and fading that any one antenna would see. MIMO Signal Processing is sophisticated enough to discern multiple spatial streams (Figure 24.18).

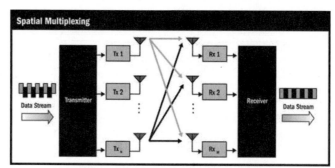

Figure 24.18: Spatial Multiplexing.

Spatial multiplexing transmits completely separate data streams on different antennas (in the same channel) that are recombined to produce new 802.11n data rates. Higher data rates are achieved by splitting the original data stream into separate data streams. Each separate stream is transmitted on a different antenna (in the same channel). MIMO signal processing at the receiver can detect and recover each stream. Streams are then recombined which yields higher data rates.

With changes to the 802.11 MAC and the use of MIMO signal processing, data rates can reach over 200Mbps and thus achieve the original IEEE objective for 802.11n. In addition to the increases in raw throughput, wireless networks using 802.11n will also see an increase in range and coverage but again, the range and vs. capacity arguments still hold true.

It is important to note the introduction of 802.11n will force organizations to review their entire wired infrastructure as the increases in bandwidth from 802.11n will create bottlenecks in other parts of the network. Equally important, MIMO signal processing will provide improvements in range even if it is only present on one half of the link, i.e. the client or the Access Point.

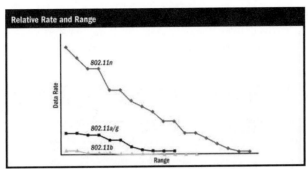

Figure 24.19: Relative Range and Rate.

24.11 Long Range Wi-Fi Case Study

Optimizing for both range and capacity can be achieved through the use of multiple radio Access Points that utilize high gain directional antennas. The following case study highlights how such a solution can be deployed quickly and in a very demanding environment: Interop New York.

Interop New York is a comprehensive IT event that helps IT and business decision-makers bring together the key components of a solid information strategy to drive efficiency and competitive advantage for their organizations. The event was held from September 18-22, 2006 at the Jacob K. Javitz Convention Center in New York and included *100 educational sessions and 150+ exhibitors* covering approximately *353,000 square feet*.

To supply the most reliable, high-speed networking services for exhibitors, conference rooms and attendees during the event, a group of innovative vendors (including Xirrus) and volunteers are selected, who come together to take on the ultimate network - creating the InteropNet.

By using Xirrus' LANPlanner software (RF prediction tool), it was determined that *four* Arrays would cover the entire conference (*lobby, classrooms, and Exhibit Hall totaling 350,000+ square feet*) and provide Wi-Fi connectivity for more than 3,400 users. However, due to the increasing popularity of the Interop conferences and Wi-Fi, the decision was made to overprovision the network by deploying eleven XS-3900 Arrays thereby delivering nearly *10 Gbps of Wi-Fi bandwidth for more than 11,000 potential clients*.

With the plan in place, the next step was to deploy the wireless network. Due to scheduling at the Javitz Convention Center, InteropNet was given two days to install the entire network. Other Wi-Fi providers would have collapsed at the task, but not Xirrus. The Xirrus XS-3900 Array packs 16 radios along with an onboard Gigabit Switch into a compact, ceiling mountable device that supports 802.11a, b, and g. By drastically reducing the amount of access devices required, InteropNet was able to deploy the Xirrus solution with a day allowing them to focus on other aspects of the network.

Today's traditional Wi-Fi networks cannot keep pace with the trajectory of needed wireless capacity and coverage for voice, video and data applications running over Wi-Fi. By combining a Wireless LAN Controller/Switch, up to 16 Integrated Access Points,

Gigabit Ethernet Switch, Wi-Fi Firewall and Multi-Sector Antenna System into one easy-to-manage device, Xirrus delivers the only long range, high performance Wi-Fi platform in the industry that can effectively extend wired network capabilities to wireless – without compromise.

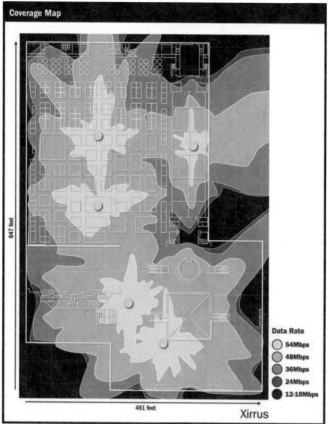

Figure 24.20: Coverage map (partial) of the Xirrus Wi-Fi network at the Javitz Convention Center in New York City.

24.12 Summary

Setting up a large scale Wi-Fi network is very different than its wired cousin Ethernet. Given the fluid nature of the wireless medium, coverage and performance can vary widely depending upon the environment. Most Access Points use an omni-directional antenna to propagate their signals which limit their ability to adequately deliver increased range.

With network administrators facing an ever-increasing demand for wireless connectivity, the traditional approach of deploying omni-directional Access Points needs to change in favor of utilizing devices that use directional, high gain antennas coupled with multiple radios in order to get the best range and capacity out of a Wi-Fi network.

25

An 802.11g WLAN SoC

Srenik Mehta, David Weber, Manolis Terrovitis, Keith Onodera, Michael Mack,
Brian Kaczynski, Hirad Samavati, Steve Jen, Weimin Si, MeeLan Lee, Kalwant
Singh, Suni Mendis, Paul Husted, Ning Zhang, Bill McFarland, David Su[a]
Teresa Meng, Bruce Wooley[b]

© 2005 IEEE. Reprinted with permission from *IEEE Journal of Solid-State Circuits*, Volume 40,
Issue 12, pp. 2483 - 2491, December 2005.

A single-chip IEEE 802.11g compliant wireless LAN system-on-a-chip (SoC) that
implements all RF, analog, digital PHY and MAC functions has been integrated in a 0.18-
μm CMOS technology. The IC transmits 0 dBm EVM-compliant output power for a 64
QAM OFDM signal. The overall receiver sensitivies are better than -92 dBm and -
73 dBm for data rates of 6 Mbps and 54 Mbps, respectively.

25.1 Introduction

The IEEE 802.11g specification [1], which was only ratified in June 2003, has become the
most widely deployed wireless local area network (WLAN) standard today. Its popularity
is due in large part to its support for higher data rates while maintaining backwards
compatibility to legacy IEEE 802.11b [2] WLANs. An IEEE 802.11g device achieves the
higher data rate when communicating with other 802.11g devices by using orthogonal
frequency division multiplexing (OFDM) modulation. When communicating with legacy
802.11b devices, it will revert back to either direct sequence spread spectrum (DSSS) or
complementary code keying (CCK) modulation. The standard uses 83.5-MHz of available
spectrum in the 2.4-GHz band and allows for three non-overlapping channels. The data
rates range from 1-2 Mbps using DSSS modulation, 5.5-11 Mbps using CCK modulation,
and 6-54 Mbps using OFDM modulation. As in the IEEE 802.11a specification [3], the
OFDM in 802.11g uses 52 sub-carriers, each of which can be modulated with BPSK,
QPSK, 16-QAM or 64-QAM.

The rapid adoption of IEEE 802.11g WLANs and their growing popularity in portable
applications such as PDAs and cellphones highlighted the need for a low-cost, small form
factor solution. This paper describes an integrated single-chip system-on-a-chip (SoC) that

---

[a] All authors currently affiliated with *Atheros Communications*
[b] All authors currently affiliated with *Stanford University*

can meet both the cost and form factor requirements by implementing all of the functions of an IEEE 802.11g WLAN system in a single 0.18-μm CMOS die. The integrated SoC combines the RF transceiver, analog baseband filters, data converters, digital baseband, physical layer (PHY), and medium access controller (MAC). The IC essentially converts the input RF signal to digital bits. In addition to reducing overall package cost, integration eliminates the area and power associated with driving package pins in a multi-chip implementation [4] [5]. Furthermore, the merging of the analog and digital blocks on the same chip enables a wide digital-analog interface that allows for the use of sophisticated digital signal processing and calibration techniques to mitigate analog and RF impairments [6]-[8]. These techniques include closed loop calibration of receiver DC offset, I/Q mismatch and transmitter carrier leak. However, SoC integration has significant challenges, such as noise isolation between noisy digital circuits and sensitive analog circuits. Special care must be taken to minimize coupling between high-swing digital switching I/Os and RF circuits processing signals many orders of magnitude smaller.

Section 25.2 of this paper addresses the overall SoC architecture and frequency plan. Section 25.3 focusses on the RF transceiver blocks: receiver, transmitter and synthesizer. Specific circuits in the receiver and transmitter are examined. Section 25.4 addresses advantages and challenges of SoC integration such as calibration. Finally, measurement results are shown in section 25.5.

25.2 Architecture

The overall SoC block diagram is shown in Fig. 25.1. The chip receives the RF signal from the antenna and down-converts it to baseband using inphase (I) and quadrature (Q) filters and analog-to-digital converters (ADC). The digital processing of the ADC outputs is done in the physical layer (PHY) which interfaces with the Medium Access Controller (MAC). The MAC interfaces to the PCI bus which connects directly to the host computer. In the transmitter, digital data from the host computer passes through the PCI bus, MAC, and PHY to drive a set of I and Q digital-to-analog converters (DACs). The DAC outputs are filtered and upconverted to RF. A single synthesizer generates the local oscillator (LO) signals for the receive and transmit blocks. Analog circuits in the transceiver are configured to operate in the various modes using a control block.

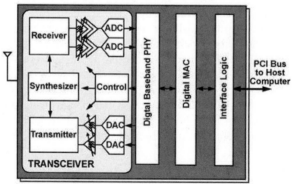

Figure 25.1: SoC Block Diagram.

The frequency plan of the transceiver, shown in Fig. 25.2, uses a sliding intermediate frequency (IF), similar to the IEEE 802.11a frequency plan in [4] and [9]. This IF is not fixed in frequency but changes depending on the desired RF channel in the 2.4-GHz band. This channel is first converted to an IF frequency at 1/3 the RF frequency using a local oscillator (LO) of 2/3 f_{RF}. The IF signal is then down-converted to baseband using an LO of 1/3 f_{RF}. Only one synthesizer is required because the 1/3 f_{RF} signal is generated using a simple divide-by-2 block from the 2/3 f_{RF} signal. In addition, voltage controlled oscillator (VCO) pulling is avoided because the LO and the RF frequencies are approximately 800-MHz apart.

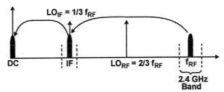

Figure 25.2: Sliding IF frequency plan.

Like a direct-conversion architecture, the dual conversion with sliding IF architecture produces baseband analog signals centered at DC and not at a low IF frequency. Even though DC offset from LO self mixing can be an issue in a dual conversion receiver, it is more easily controlled and calibrated. DC offset due to self mixing is introduced when the IF LO signal leaks to the input of the IF mixer and then self mixes to produce a DC output in the analog baseband. This on-chip IF leakage is more deterministic than its direct conversion counterpart in which the LO can leak off-chip to the antenna and be amplified by the low noise amplifier (LNA) before self mixing occurs. Unlike a direct conversion receiver, RF LO self mixing in a dual conversion architecture creates a DC offset at IF and does not introduce any impairment in the analog baseband.

25.3 Implementation

25.3.1 Receiver

Fig. 25.3 shows a detailed receiver block diagram. The RF signal from the antenna is amplified by two stages of RF gain. The LNA and the RF variable gain amplifier (RFVGA) both have programmable gain steps that can also provide attenuation for large signals [9]. An explicit image rejection filter is not needed because the 2.4-GHz input matching network and the on-chip, bandpass, inductively loaded front end stages provide approximately 40 dB of image attenuation at 800-MHz, which is adequate for WLAN applications. The programmable gain is controlled by digital logic that implements an automatic gain control (AGC) algorithm. The RFVGA output is down-converted to an IF of 800-MHz by the 2/3 f_{RF} LO. A transmitter feedback signal, which is *off* during normal operation, is injected at the output of the RFVGA for transmitter calibration. The output of the RF mixer is then down-converted to baseband with the 1/3 f_{RF} LO. The I/Q phases of the IF LO are generated from the 2/3f_{RF} LO by a simple divide-by-2 block consisting of master-slave differential D-flip-flops. The baseband path includes two sets of offset DACs to null the receiver DC offset and two gm-C biquads forming a fourth-order Butterworth filter response with programmable gain of 41dB in 1dB steps. The filter output is digitized by two 9-bit ADCs. The baseband gain and the DC offset DACs are controlled by the digital baseband processor, and the offset is calibrated and stored for different baseband gain settings.

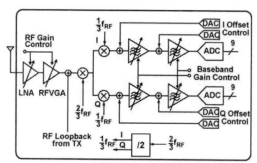

Figure 25.3: Receiver block diagram.

Fig. 25.4 shows the circuit diagram of the receive dual down-conversion mixer, which comprises two differential stages. The RF signal is down-converted to 800-MHz in the first stage using an NMOS Gilbert-cell type mixer topology. NMOS diode loads, which improve the IF gain stability over process and temperature, are used instead of bulky inductors. The NMOS load transistors are placed inside deep N-wells to allow a source-to-bulk connection that removes transistor body effects. The gain of this stage is gm_{in}/gm_{load} and is relatively constant over process and temperature. The reduction in gain variation afforded by this circuit reduces the amount of programmable gain required in the receiver. A common-mode feedback circuit biases the load transistors, setting the output common-mode voltage to allow direct coupling to the PMOS differential pairs of the IQ mixer and leaving enough headroom for the current source. Direct coupling avoids the large area and attenuation associated with AC-coupling capacitors. A PMOS Gilbert cell down-converts the IF signal to produce a baseband current that is then low-pass filtered by a programmable-gain fourth-order Butterworth gm-C filter with on-chip frequency tuning [10].

The measured overall receive chain noise figure, from the LNA input to the analog baseband output, is 5.5 dB. When operating as an integrated WLAN system, the overall receiver sensitivity is better than -92 dBm and -73 dBm for data rates of 6 Mbps and 54 Mbps, respectively.

25.3.2 Transmitter

The transmitter architecture is shown in Fig. 25.5. The digital baseband processor drives two 11-bit I and Q current-steering DACs. This resolution enables fine transmit power control in the digital domain. The DAC current sources are implemented with cascoded PMOS I/O devices with a 3.3-V supply. The current switches are driven by full-swing CMOS logic from a 1.8-V supply. The DAC output current drives a third-order current-mode Butterworth gm-C low-pass filter designed to attenuate the DAC spectral images. The baseband signal is first up-converted to an IF of 800-MHz with a 1/3 f_{RF} LO. The output is then amplified with an IFVGA stage that drives the RF up-conversion mixer. Image-reject mixers are not used in the RF up-conversion because the 800-MHz image, which is not in the FCC restricted band, is sufficiently filtered by the bandpass characteristics of the RF stages. The RFVGA and the power amplifier (PA) driver, which is capable of delivering a saturated output power of 12 dBm, provide additional gain control. The gains in the transmitter are adjusted through the use of a power leveling loop to keep the output power at a desired level. The maximum linear output power of 0 dBm is sufficient to drive an external power amplifier. The output of the RF mixer is buffered and fed to the receive chain for calibration purposes.

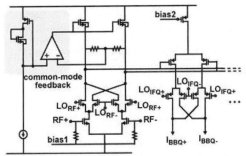

Figure 25.4: Receive dual down-conversion mixer.

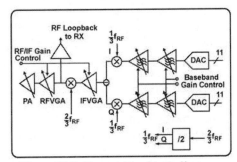

Figure 25.5: Transmitter block diagram.

Fig. 25.6 shows the schematic of the transmit dual up-conversion mixer. The current-mode baseband signal from the output of the transmit low-pass filters is fed to the IQ mixer. The up-converted current at IF is mirrored and amplified in the IFVGA. The IFVGA consists of multiple parallel stages to provide AC amplification while keeping the DC current constant in the RF mixer. This is accomplished by routing the signal from each stage to either the positive or negative output terminal of the RF output depending on the desired gain setting. The load of the RF mixer is of particular interest. In differential mode, the load is tuned to 2.4-GHz and consists of the differential inductor L_{dm} and any parasitic capacitance at output drains. Node "X" in Fig. 25.6 is an AC ground for differential-mode signals. For common-mode signals, L_{dc} acts like an RF choke and C_X is chosen to form a series resonant circuit with L_{cm} at 3.2-GHz. This arrangement creates a common-mode notch at 3.2GHz to attenuate the second harmonic of the LO, which can fall in a restricted band. Measurements show that an additional 5 to 6 dB of rejection is achieved with the common-mode circuit compared to a simple differential inductive load.

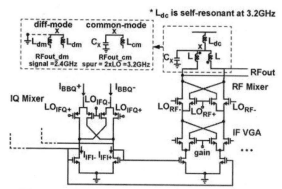

Figure 25.6: Transmit dual up-conversion mixer.

25.3.3 Synthesizer

The synthesizer architecture, shown in Fig. 25.7, uses a 40-MHz off-chip crystal in conjunction with an on-chip amplifier to create the reference clock for the synthesizer. This clock is divided down to 1.33-MHz with a divider to create the reference for the phase-frequency detector (PFD). The up and down output signals of the PFD drive a charge pump whose supply is regulated with a dedicated linear regulator. The charge pump output it filtered and applied to the VCO. A 16/17 dual modulus prescaler, P&S counters and re-timing logic are used in the feedback path to complete the phase locked loop. The VCO is a varactor-tuned inductively loaded LC tank. The VCO has a dedicated linear regulator to reduce noise coupling from other circuits through the supply. The VCO frequency is chosen to be $4/3\ f_{RF}$ instead of $2/3\ f_{RF}$ to allow for a smaller inductor value and hence smaller die area. The higher VCO frequency also allows for a higher reference frequency and thus a wider loop bandwidth with better phase noise performance. If the VCO frequency was $2/3\ f_{RF}$, then the reference frequency needed would be $1.33\text{MHz}/2$. Inductive tuning was not necessary in the high-speed dividers because of the high f_t available in the 0.18-μm technology. The loop filter of the synthesizer is integrated on-chip using poly resistors and the gate capacitance of NMOS transistors. The synthesizer phase noise measured at the transmitter output is -104 dBc/Hz at 100-kHz offset.

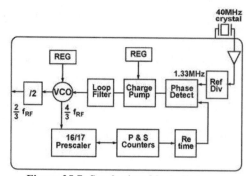

Figure 25.7: Synthesizer block diagram.

25.4 SoC Integration

25.4.1 Calibration

In a single-chip implementation, interconnection between the analog and digital blocks is no longer constrained by the limited number of package pins. This design leverages the high level of integration in a CMOS technology to implement calibrations of analog and RF circuits using digital signal processing. This calibration eases the matching requirements in the transceiver, thereby reducing the analog and RF power and area. Shown in Fig. 25.8 is the transceiver in DC offset calibration mode. During receiver calibration, the loopback path is initially shut off and the offset caused by the self mixing IF LO at 1/3 f_{RF} is digitized and sent to the digital baseband. The digital baseband processor does a binary search to find the correct offset DAC code to null the offset. Once the receiver DC offset is cancelled, the digital baseband logic can null out any IQ mismatch without any feedback into the analog or RF blocks. Any DC offset in the baseband circuits of the transmitter appears as carrier leak in the RF output. Transmit carrier leak calibration is performed after the receiver calibrations are completed. An RF loop back path is provided from the output of the transmit RF mixer to the input of the receive RF mixer. During calibration, a predetermined digital sequence is transmitted and looped back to the receiver. The digital baseband processor compares the received signal with the expected value to calculate the transmit DC offset. The computed offset is then subtracted from the digital input to the transmit DACs to null out the carrier leak.

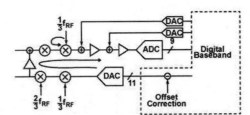

Figure 25.8: Digital calibration.

25.4.2 Noise isolation

A major challenge in the implementation of a single-chip radio is to prevent the corruption of sensitive analog and RF signals by digital switching noise. Although the RF and digital baseband circuits are widely separated in frequency, crosstalk can occur when the supply and bias voltages are modulated [11]. The generation of digital switching noise in the SoC is reduced through the extensive clock gating of unused digital logic. In addition, a careful parasitic capacitance extraction was done on critical clock traces to avoid over-sizing the digital clock buffers. Shields are used to provide ground return paths on switching nets. Sharp clock edges are avoided wherever possible because the amplitude of the associated high order harmonics of the clock decreases relatively slowly with frequency. Even the 61st harmonic of the 40-MHz reference clock, which falls at 2440-MHz, can affect in-band RF performance. Fully differential analog and RF circuits are employed to provide first-order rejection of common-mode switching noise coupled from the digital logic. Separate, or star-connected power supplies (supplies that only connect to other supplies at the pads) are used to reduce supply crosstalk. For the more sensitive circuits, such as those in the frequency synthesizer, further reduction of supply crosstalk is accomplished with on-chip voltage regulators. Similar to [12], a wide deep N-well isolation area of 150-μm, tied to a quiet analog supply pin, is used between the digital circuits and the RF transceiver in order to reduce substrate coupling. In addition, substrate ground rings that are tied to local grounds surround the individual RF blocks.

The PCI interface requires a large pin count so many RF friendly packages with a good backside ground could not be used. A 224-pin BGA with separate analog and digital ground substrate planes was chosen. Many of the noise isolation techniques that are applicable on-chip can be extended to the package design. Where possible, sensitive signal nodes are not routed to package pins. For example, the VCO control voltage node can be kept completely on-chip with an integrated loop filter. Sensitive nodes that must be connected to the package, such as the sensitive LNA inputs, are routed orthogonal to noisy digital output pins to reduce magnetic coupling.

25.5 Experimental Results

The SoC, integrated in a 0.18-μm five-layer metal CMOS process, occupies a total silicon area of 41 mm^2. A die micrograph is shown in Fig. 25.9. The LNA is placed in the top right corner for maximum separation from the switching digital logic as possible. The wide deep N-well isolation area is shown between the analog and digital sections. The measured performance of the SoC is summarized in Fig. 25.10. The single-chip radio meets or exceeds the IEEE 802.11g WLAN specification.

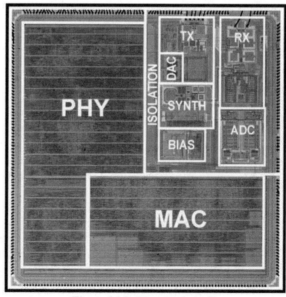

Figure 25.9: Die micrograph.

Transmit Mode Power Dissipation		
SoC	180mA @ 1.8V	50mA @ 3.3V
RF Transceiver alone (excludes DAC)	128mA @ 1.8V	10mA @ 3.3V
Receive Mode Power Dissipation		
SoC	175mA @ 1.8V	60mA @ 3.3V
RF Transceiver alone (excludes ADC)	65mA @ 1.8V	20mA @ 3.3V
Transmit EVM	-28dB @ -2dBm & 54 Mbps	
Receive Sensitivity	-94dBm @ 1Mbps	
	-92dBm @ 6Mbps	
	-73dBm @ 54Mbps	
Receive Chain Noise Figure	5.5dB	
Phase Noise (2.412GHz)	-104dBc/Hz @ 100kHz offset	
	-125dBc/Hz @ 1MHz offset	
Technology	0.18µm CMOS 1P/5M	
Die Size	41 sq mm	
RF Transceiver + ADC, DAC, PLL	11.5 sq mm	
Package	224-pin BGA	

Figure 25.10: Measured performance.

Fig. 25.11 shows the receiver sensitivity with no external LNA, RF switches, or filters for various data rates and frequency channels. The SoC achieves -75 dBm sensitivity for 54 Mbps, -94 dBm for 6 Mbps and -95 dBm for 1 Mbps. The sensitivity typically exceeds the IEEE spec by 10 dB. The overall sensitivity degrades by 2 dB close to 2440-MHz, the 61st harmonic of the system clock. All measurements are taken with the chip operating in an active cardbus card plugged into a notebook computer.

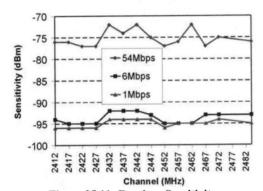

Figure 25.11: Receiver Sensitivity.

Fig. 25.12 shows the measured input third-order intercept point (IIP3) of the receiver as a function of its gain. The availability of reduced gain modes in the LNA and RFVGA significantly extends the receiver IIP3 at minimum gain, while providing for large RF gain to achieve good receiver sensitivity for small RF inputs.

Fig. 25.13 shows the transmit output spectrum in the CCK and OFDM modes. The output spectra conform to the required spectral mask with significant margin. The transmitter error vector magnitude (EVM) performance as a function of output power and channel is shown in Fig. 25.14. When using a typical external power amplifier with 25 dB or more gain, the chip transmit output is set to -4 dBm, where the EVM is -30 dB exceeding the 64QAM specification by 5 dB. As expected, the linearity of the transmitter does degrade at larger output powers but it meets the -25 dB specification even at an output power of -1 dBm.

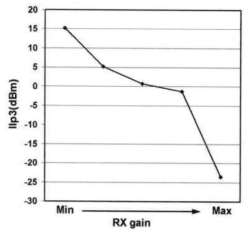

Figure 25.12: Receiver IIP3 versus RX gain.

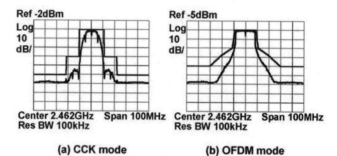

(a) CCK mode   (b) OFDM mode

Figure 25.13: Transmit spectrum.

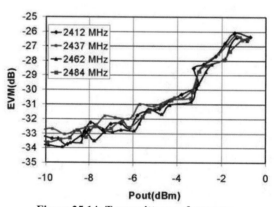

Figure 25.14: Transmitter performance.

Fig. 25.15 shows a synthesizer phase noise as measured at the 2.4-GHz RF output. The phase noise is -104 dBc at 100-kHz offset and -125 dBc at 1-MHz offset. This phase noise is not significantly degraded when measured with or without the digital logic toggling.

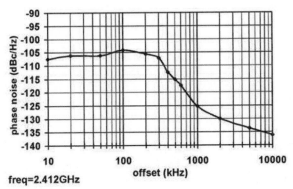

Figure 25.15: Synthesizer phase noise.

25.6 Conclusion

A single-chip IEEE 802.11g SoC has been implemented in standard digital CMOS. The WLAN system achieves receiver sensitivities better than -92 dBm and -73 dBm for 6 Mbps and 54 Mbps respectively as well as transmit EVM of -30 dB at -4 dBm output power. This SoC leverages CMOS integration of digital signal processing to mitigate analog and RF impairments. The measurements demonstrate that a WLAN SoC with RF and digital circuits integrated on the same silicon can achieve state-of-the-art performance.

Acknowledgments

The implementation of an integrated WLAN SoC requires a large engineering effort consisting of teams for RF, analog, digital, architecture, algorithms, software, and system design. The authors wish to acknowledge the effort of the entire wireless LAN team at Atheros.

25.7 References

[1] IEEE Standard 802.11g/D8.2, Wireless LAN MAC and PHY specifications: Further higher data rate extension in the 2.4 GHz band, New York: IEEE, 2003.
[2] IEEE Standard 802.11b/D8.0, Wireless LAN MAC and PHY specifications: Higher speed PHY extension in the 2.4 GHz band, New York: IEEE, 2001.
[3] IEEE Standard 802.11a-1999, Wireless LAN MAC and PHY specifications: High-speed physical layer in the 5 GHz band, New York: IEEE, 2000.
[4] D. Su, M. Zargari, P. Yue, S. Rabii, D. Weber, B. Kaczynski, S. Mehta, K. Singh, S. Mendis, B. Wooley, "A 5-GHz CMOS Transceiver for IEEE 802.11a Wireless LAN," *ISSCC Digest of Technical Papers*, pp. 92-93, Feb. 2002.

[5] J. Thomson, B. Baas, E. Cooper, J. Gilbert, G. Hsieh, P. Husted, A. Lokanathan, J. Kushin, D. McCracken, B. McFarland, T. Meng, D. Nakahira, S. Ng, M. Rattenhalli, J. Smith, R. Subramanian, L. Thon, Y. Wang, R. Yu, X. Zhang, "An integrated 802.11a baseband and MAC processor," *ISSCC Digest of Technical Papers*, pp. 126-127, Feb. 2002.

[6] J. Bouras, S. Bouras, T. Georgantas, N. Haralabidis, G. Kamoulakos, C. Kapnistis, S. Kavadias, Y. Kokolakis, P. Merakos, J. Rudell, S. Plevridis, I. Vassiliou, K. Vavelidis, A. Yamanaka, "A digitally calibrated 5.15-5.825GHz transceiver for 802.11a wireless LANs in 0.18μm CMOS," *ISSCC Digest of Technical Papers*, pp. 352-353, Feb. 2003.

[7] P. Zhang, T. Nguyen, C. Lam, D. Gambetta, C. Soorapanth, B. Cheng, S. Hart, I. Sever, T. Boiurdi, A. Tham, B. Razavi, "A direct conversion CMOS transceiver for IEEE 802.11a WLANs," *ISSCC Digest of Technical Papers*, pp. 354-355, Feb. 2003.

[8] A. Behzad, Z.M. Shi, S. Anand, L. Lin, K. Carter, M. Kappes, T. Lin, T. Nguyen, D. Yuan, S. Wu, Y.C. Wang, V. Fong, A. Rofougaran, "Direct-conversion CMOS transceiver with automatic frequency control for 802.11a wireless LANs," *ISSCC Digest of Technical Papers*, pp. 356-357, Feb. 2003.

[9] M. Zargari, M. Terrovitis, S. Jen, B. Kaczynski, M. Lee, M. Mack, S. Mehta, S. Mendis, K. Onodera, H. Samavati, W. Si, K. Singh, A. Tabatabaei, D. Weber, D. Su, B. Wooley, "A single-chip dual-band tri-mode CMOS transceiver for IEEE 802.11 a/b/g WLAN," *ISSCC Digest of Technical Papers*, pp. 95-96, Feb. 2004.

[10] S. Mehta, M. Zargari, S. Jen, B. Kaczynski, M. Lee, M. Mack, S. Mendis, K. Onodera, H. Samavati, W. Si, K. Singh, M. Terrovitis, D. Weber, D. Su, "A CMOS Dual-band Tri-Mode Chipset for IEEE 802.11a/b/g," *IEEE RFIC Symposium*, pp. 427-430, June 2003.

[11] M. Xu, D. Su, D. Shaeffer, T. Lee, B. Wooley., "Measuring and modeling the effects of substrate noise on the LNA for a CMOS GPS receiver," *IEEE J. Solid-State Circuits*, pp. 473-485, March 2001.

[12] P. Van Zeijl, J. Eikenbroek, P. Vervoort, S. Setty, J. Tangenberg, G. Shipton, E. Kooistra, I. Keekstra, and D. Belot, "A Bluetooth radio in 0.18mm CMOS," *ISSCC Digest of Technical Papers*, pp. 86-87, Feb. 2002.

26

Antenna Design for Portable Computers

Frank Caimi[a], David Wittwer[b], Anatoliy Ioffe[b] and Marin Stoytchev[c]

26.1 Introduction

The basic function of an antenna is to act as a transducer - transforming electrical signals into broadcast energy that can be recovered at a distance using amplification. As with any transducer, the ability to convert signals efficiently without undue physical restriction is paramount. To understand the physical, as well as electrical aspects of the conversion process, it is necessary to describe how an applied signal gives rise to a radiated electromagnetic field. Fortunately, Maxwell's equations govern all aspects of this problem, and radiation occurs when electrical charge is accelerated.

The field produced by an accelerating charge is produced in a direction perpendicular to the charge motion according to the equation:

$$E = \frac{q}{4\pi\varepsilon_0}\left[\frac{c^{-2}\kappa^{-3}}{r}\left(\hat{r}\times(\hat{r}-v/c)\times a\right)\right]$$

$$I = n\frac{dq}{dt} = \frac{dQ}{dt} \tag{1}$$

$$E \propto \frac{dI}{dt} \Rightarrow |E| \propto \sum_i \frac{1}{r_i}\frac{dI_i(t-r_i/c)}{dt}$$

where a is the acceleration, v is the velocity, c is the speed of light, and r is a vector to the point of measurement. For charges on a conductor, the velocity is much smaller than the speed of light (in the order of 10^{-5} m/s) so v/c is generally negligible in the calculation. This situation is as shown in Figure 26.1 for a time varying current I in a conductor.

The radiated power (E^2) is proportional to the square of the charge and the square of the acceleration. The field produced at a point is determined by the spatial integration over the length of the current carrying conductor, weighted inversely by the distance, r_i, from the conductor to each current element. There is no power flow in the direction of the

[a] *SkyCross, Inc*
[b] *Intel Corporation*
[c] *Rayspan Corporation*

acceleration. However, the electric field is in the same direction as the acceleration. Therefore a current flow in a single direction, say z, results in an electric field in the same direction, and is thus considered "polarized" in that direction.

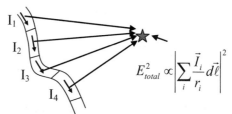

Figure 26.1: Electric field produced by accelerating charge (sinusoidal varying current I).

26.1.1 Source of Radiation

For a sinusoidal current variation of angular frequency ω and a short wire of length l, at a constant distance r from the measuring point, the electric field is perpendicular to the current flow and is given by:

$$E_\perp = \frac{j\omega \sin(\theta)}{rc^2} \int_{-l/2}^{l/2} I\, dz \tag{2}$$

where the angle θ is measured with respect to the z-axis. It is reasonable to suppose that the current is maximum I_0 in the center of the wire and decreases to zero at the wire ends. Assuming a linear dependence:

$$I = I_0 \exp(-j\omega t)\left[1 - \frac{|z|}{l}2\right] \tag{3}$$

Substituting (3) into (2) yields the familiar dipole equation:

$$E_\perp = \frac{j\pi \sin(\theta)}{c} I_0 \frac{l}{2\lambda r} \exp(-j\omega t) \tag{4}$$

where the power (E^2) radiated in a direction θ from the z-axis is clearly proportional to $\sin^2(\theta)$ and the square of the peak current $\tilde{I_0}$ The time averaged power is computed directly from this analysis and is given by:

$$\widehat{P} = \frac{\pi^2}{3c}\left[\frac{l}{\lambda}\right]^2 I_0^2 \tag{5}$$

Time varying currents therefore give rise to the radiation. The magnitude of the current and the length of the conductor in relation to the wavelength establish the total radiated power as well as the electric field strength at a given distance from the source.

26.1.2 Factors Affecting Small Antenna Design

Our understanding of radiation resulting from currents can be applied to the design of commercial antenna products. In addition, properties of the currents on the antenna can be used to examine the fundamental limits exhibited by all antennas. These limits are highlighted in the following sections.

26.1.2.1 Conductor Area

An antenna can be considered a lossy transmission line. A ½-wave dipole generally has an open circuit at one end and a connection to a low impedance line at the other end. The same is true in the case of a ¼-wave monopole over a ground plane. The open end is a high impedance point where the boundary conditions require that the current is zero. A forward wave propagates down the radiating element, reaches the end and is reflected back to the source producing a reflected wave. The magnitude of the current at different portions along the line is determined by the sum of the reflected and forward propagating waves and the boundary conditions. The voltage along the line is also affected by the boundary conditions. The net current and voltage along the line therefore varies as a function of position from the end. If the line is non-radiating and non-absorptive, we say the line is lossless. In this ideal case, the current must be zero at the line end, but the voltage in maximum. Further along the line, there is a point that is reached where the current is maximum and the voltage is minimum. The distance from the end where the current goes from zero to a maximum is ¼ wavelength in a lossless transmission line. Antenna current and voltage distributions depart from the ideal lossless transmission line since loss to radiation occurs along the length of the radiator.

Nevertheless, with this behavior the loss resistance of the conductors becomes important in the antenna design. Conductor resistance is an important factor in determining the antenna performance.

A model containing antenna resistance may be considered as composed of two dissipative portions - the EM wave radiation into space and the loss due to heating in the conductors making up the antenna; the so-called "I-squared R" loss. To achieve efficient radiation of energy into space with minimum heating of the antenna elements, the resistive loss in the antenna conductors must be minimal with respect to the radiation resistance. The relative value of the conductor resistive loss in relation to radiative loss can be controlled through engineering design. Poorly designed antennas will dissipate an excessive amount of power in the conductors relative to the radiated power, thereby reducing the power conversion efficiency of the antenna. A measure of the design quality is the "radiation efficiency". Small antennas can have efficiencies that range from small fractions of 1% for submarine antennas to nearly 100% for others.

Since high frequency currents flow at the outside of conductors and are governed by the skin effect, the effective resistance is related to excitation frequency and the conductor surface area. The basis for a good design is that conductors be of sufficiently large surface area to maintain low I^2R loss at points where the current is a maximum. The points where the current is zero, or small, matter less and can have lesser surface area.

26.1.2.2 Radiation Resistance

The radiation resistance of an antenna depends on its geometry, current distribution along the radiating elements, and wavelength. The radiation resistance is defined as the ratio of twice the average power radiated divided by the square of the peak current:

$$R_{rad} = 2 \frac{\hat{P}}{I_0{}^2} \tag{6}$$

For a short dipole, the resistance is given approximately by:

$$R_{rad,dipole} = \frac{2\pi^2}{3c} \left(\frac{l}{\lambda}\right)^2 \text{ s/cm} = 200 \ (l/\lambda)^2 \ \Omega \tag{7}$$

where L is the dipole length. Similarly for a loop antenna, the radiation resistance is much larger:

$$R_{rad,loop} = 300,000 \ (b/\lambda^2)^2 \ \Omega \tag{8}$$

where b is the loop radius. Clearly, the loop has a much higher radiation resistance at higher frequency as compared to a short dipole. An example calculation gives $R_{rad,dipole} = 5.5$ ohms at 1900 MHz for a length of 1 inch, and $R_{rad,loop} = 57$ ohms for a diameter of 1 inch.

As the antenna size is reduced, the radiation resistance becomes smaller in relation to the square of the length, or the loop area normalized to the wavelength according to equations (7) and (8).

The radiation resistance can be increased by mutual coupling; for instance, a) when multiple (N) turns are used in a loop, b) when the loop is wound on a magnetic material, and c) when a design employing multiple (N) inductively coupled radiating elements is used. The latter case pertains to some meander line antennas and to the Goubau [1] antenna, where $N = 2$ and 4, respectively.

It can be shown that the radiation resistance is raised by a factor of N^2, allowing the antenna to accommodate a higher resistance so that a smaller conductor area can be used for a given radiation efficiency.

26.1.2.3 Radiation Efficiency

The antenna radiation efficiency, η, is determined by the conductor loss R in relation to the radiation resistance R_{rad}.

$$\eta = R_{rad}/(R_{rad} + R) \tag{9}$$

The reactive portion of the antenna impedance seen at the terminals is capacitive for a dipole and inductive for a loop. The dipole requires a series inductor to cancel the reactive portion of the antenna impedance, i.e., for resonance/impedance matching. The loop typically requires a series capacitance to cancel the inductive reactance at the frequency of operation; i.e., to create a resonance condition where the radiation resistance plus loss resistance is seen at the input terminals. For the dipole, the inductor or so-called loading coil is best placed above the base of the antenna, as the current is high at the base where "I squared R" losses in the coil are more substantial.

For high radiation efficiency, the radiation resistance should be much larger than the loss resistance. For broad bandwidth, the radiation resistance should be large in relation to the reactive part of the unmatched antenna impedance (the stored energy) as required by the definition of Q.

26.1.2.4 Antenna Q

As an antenna becomes smaller in relation to the wavelength, the stored energy increases, representing an increase in the reactive portion of the antenna impedance as seen at the feedpoint. The antenna quality factor, Q, can be loosely considered as the ratio of the reactive term of the impedance to the resistive or dissipative portion.

$$Q \sim (\text{reactance/resistance}) \sim f/\Delta f \qquad (10)$$

A higher Q implies a narrower bandwidth. So, as the antenna becomes smaller the reactive portion increases in relation to the resistive portion, increasing the Q and reducing the bandwidth. This behavior is readily observed in the Chu-Harrington curve in Figure 2. In the Chu-Harrington space, the best antennas achieve radiation efficiency approaching 100% and fall just above the 100% efficiency curve indicating the reactance to resistance ratio (i.e., Q) is minimal. That means that the antenna exhibits the broadest instantaneous bandwidth without using an additional tuning mechanism.

26.1.3 Fundamental Limits of Electrically Small Antennas

It is well known among antenna engineers that the design of a small antenna will be a tradeoff between its dimensions and its electrical performance, and furthermore that physical laws determine the ultimate limitations inherent in any design. The physical limits imposed on a design may be simply stated: "Bandwidth and Antenna Size are inversely related", and "Antenna Gain and Size are directly related". This means that the maximal gain of an antenna can be enhanced somewhat by varying the geometry and that the bandwidth can similarly be increased. A third degree of dimensionality is also afforded the antenna designer: The efficiency of the antenna can be purposely degraded to additionally increase the bandwidth, but it will reduce the gain.

26.1.3.1 Chu-Harrington Limit on Q

The relationship that describes the ultimate size versus bandwidth capability of an antenna was developed in several seminal papers by Harold Wheeler [2] and L. J. Chu [3], and later by Roger Harrington [4].

In 1946, Wheeler introduced the concept of the *"radiansphere"* and the volume relation to the maximum power factor achievable by the antenna. He related the energy within the *radiansphere* and outside the antenna volume to the fundamental limitation of the antenna power factor. The power factor, it is reasoned, is proportional to the antenna volume and also a shape factor. The nominal bandwidth is given as the power factor p multiplied by the resonance frequency f_0. We have the fundamental relationship:

$$V \propto p \propto \frac{BW}{f_0} \tag{11}$$

In 1948, Chu extended Wheeler's analysis and expressed the fields for an omnidirectional antenna in terms of spherical wave functions and found limits for the antenna quality factor, Q, the maximum gain, G_m, and their ratio, G/Q. In this approach, he was able to use a partial fraction expansion of the wave impedance of spherical modes that exist outside the smallest circumscribing sphere surrounding the antenna to obtain an equivalent ladder network from which the Q could be found by conventional circuit analysis. In 1964, Collin and Rothschild presented a method for evaluating the minimum Q without using the equivalent network. Later in 1959, Harrington related the effects of antenna size, quality factor, and gain for the near and far field diffraction zones for linearly and circularly polarized waves, and also treated the case where the antenna efficiency is less than 100%.

In 1969, Fante [5] extended these results to multimode antennas. Additional work carried out by others [6 - 7] from 1969 through 2001 was directed to obtain exact expressions for the antenna Q over an expanded size range. These efforts led to the equation:

$$Q = \frac{1}{(ka)^3} + \frac{1}{ka} \tag{12}$$

Here k is $2\pi/\lambda$ and a is the radius of a sphere containing the antenna and its associated current distribution. It is important to realize in applying this equation that currents can exist on a counterpoise for a radiating antenna element, and that the physical size of the antenna can actually be much larger. For instance, a small size of a resonator can be designed to excite currents on a larger structure, as observed with mobile phone antennas in the 800-900 MHz band, which use the entire PCB as a radiating structure. In addition, the radiation efficiency is often included as a factor on the right side of (12), indicating that lower Q, and therefore greater bandwidth, is achievable by reducing the efficiency η.

26.1.3.2 Fundamental Gain Limitation

The gain of a small dipole antenna in free space is subject to fundamental limits. This can be seen by considering the antenna in the limit as the size is reduced in comparison to the wavelength. Ultimately, the antenna would appear as a point source, having uniform radiation in all directions. The maximum normal gain limit is given by Harrington [4]:

$$G = (ka)^2 + 2(ka) \tag{13}$$

This equation applies to an inherently omni-directional antenna of maximum dimension $2a$. It can be seen that increasing the aperture size a to multiples of the wavelength can provide increased gain. However, this is seldom practical for embedded antennas in portable wireless devices where ka values of 0.5 or less are typical.

Small antennas where $ka \ll 1$, are subject to high Q-values, but can sustain larger gain values than specified by equation (13): the so-called super-gain regime. For practical small antennas of reasonable Q (~10), however, the gain is very near that of a dipole, or approximately 1.5. In decibels, this is approximately 2 dBi (referred to an isotropic radiator). Many small antennas exhibit multimode behavior and are more often omni-directional, having a decibel gain close to 0 dBi.

26.1.3.3 Qualification Metrics

The evaluation criterion for small, embedded antennas is usually specified as a minimum radiation efficiency over the desired operational bandwidth. This is often a practical criterion, since the gain achievable is nominally less than 2 dBi due to the antenna size, and because antenna patterns are generally either non-descript (approach that of a spherical radiator) or are determined primarily by the platform on which they are used.

If a wireless device is intended to be used on a single floor or in the "enterprise", and if it is operable in a more-or-less fixed orientation, then additional criteria may apply. Typically, a minimum gain value is specified for specific percentage coverage in a reference plane. A preferred reference for single floor operation is the horizontal plane, requiring antenna range measurements versus azimuth. Typically, the minimum gain is specified for 80% coverage and with a specified polarization.

26.1.3.4 Q-Volume Space

It is the role of the antenna designer to select an antenna design that achieves high volumetric efficiency i.e., small size for a given bandwidth. A comparative measure of an antenna's volumetric efficiency is often expressed on a volume versus bandwidth graph in relation to the theoretically determined limit. The so-called Chu-Harrington relationship[2-6] establishes the limiting values of bandwidth versus volume for a 100% efficient antenna - one that radiates all of the power applied to its feed terminals. When comparing antennas, it is desirable to select a design that will provide the most effective use of physical volume for the bandwidth required. This is easily seen from Figure 26.2 that illustrates the relationship of various antenna technologies to the theoretical limitation defining bandwidth relative to frequency and physical volume relative to wavelength.

Generally, the best antennas achieve performance gains through an effective and complete utilization of the antenna volume, and are "space-filling" while minimizing loss. Fractal antennas [8 - 9] make this as a claim for bandwidth, although other designs [13], which are space-filling accomplish large bandwidth and high efficiency. Several recent papers verify the Chu-Harrington limit for space filling designs [10 - 12] and for electrically small, open-ended antennas [12].

Recent hypotheses [7] over the specific form of the Chu-Harrington relationship indicate that multimode excitation of an antenna may allow greater bandwidth than predicted by the conventional theory, and need to be verified experimentally. Future experiments may be able to support these hypotheses and such verification would be a significant result from a scientific standpoint.

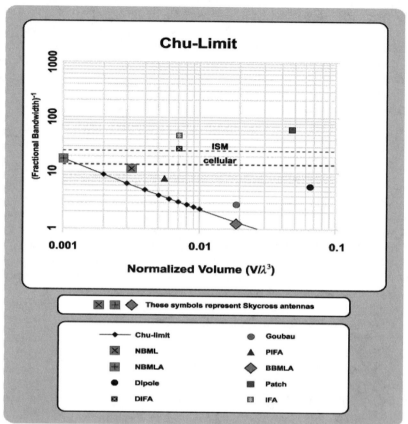

Figure 26.2: Antenna comparison - size versus bandwidth. The antennas illustrated at the leftmost portion of the figure indicate small physical volume, while maintaining sufficient bandwidth to cover the required frequency bands associated with the various wireless services (also indicated). The Broadband MLA (BBMLA) provides 800 - 2500 MHz bandwidth that is unmatched in size (volume) by other antennas.

26.1.3.5 Q-Volume Space: Example Antennas

A variety of examples can be given for antennas having different size and bandwidth. For the purposes of the Chu-Harrington comparison, antenna size can be described as the antenna volume or in terms of the radius of a sphere enclosing the antenna and its dominant current distribution. The latter approach is easier to visualize and is illustrated in Figure 26.3. The curve shown (equation 12) represents the size limit for a 100% efficient antenna. Different common antenna designs are illustrated as points in the space, as well as several proprietary designs [13]. The thick dipole is one example of a space inefficient design, while others such as the Goubau [1], FourSquare [14], and Cubic Meanderline [13] antennas (MLA), exhibit better bandwidth for their respective sizes.

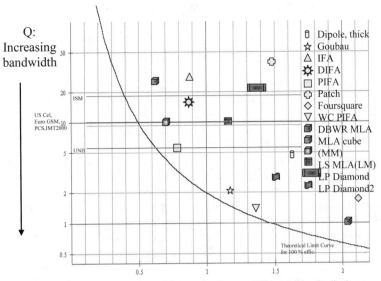

Figure 26.3: Antenna *Q* versus Spherical Radius normalized to wavelength for common antenna designs.

An illustration of the dependency of antenna bandwidth on physical antenna size is shown in Table 26.1 for a common design developed for the Wi-Fi band at 2400 MHz. In order to cover the required bandwidth of 100 MHz, the antenna size is nominally 14 mm or greater.

Table 26.1: Computation of Antenna Bandwidth and Q at 2450 MHz versus Antenna Size.

F (MHz)	λ(mm)	λ(in)	BW(MHz)	Q	a (mm)	a(in)	ka	Qest	size(mm)	size (in)
2450	122.4	4.8	22	111.4	4.1	0.16	0.210275	112.3118	8.2	0.32
2450	122.4	4.8	25	98.0	4.3	0.17	0.220533	97.76996	8.6	0.34
2450	122.4	4.8	30	81.7	4.6	0.18	0.235919	80.39632	9.2	0.36
2450	122.4	4.8	50	49.0	5.5	0.22	0.282077	48.10035	11	0.43
2450	122.4	4.8	100	24.5	7.0	0.28	0.359007	24.39734	14	0.55
2450	122.4	4.8	200	12.3	9.0	0.35	0.46158	12.33502	18	0.71

26.1.4 WLAN/WWAN Antenna Requirements

The proliferation of Wi-Fi connectivity has brought with it the need to embed antenna solutions for a variety of platforms. Recently, the most common integration effort has been associated with the laptop and notebook or smaller portable computers. Historically, this includes the PCMCIA card, the USB dongle, and the totally integrated solution with the antenna mounted in notebook base or screen. Each of these efforts has represented a learning process for the industry - specifically with respect to production performance variability, radiation efficiency, pattern coverage, and cost. Cost reduction has been the trend in all aspects of WLAN/WWAN integration with specific minimum requirements for performance. More recent popular platforms include the Express 34 card and embedded antennas for Personal Entertainment Devices.

Several issues exist relative to the integration of Wi-Fi antennas into notebook or smaller platform computers - specifically choice of location and resulting pattern coverage, radiation efficiency, and antenna design factors relating to cost. A description of the pitfalls, successes, and performance expectations achievable with certain designs and potential solutions for improvement is desirable.

WLAN/WWAN antennas used in remote devices generally must meet minimum requirements for VSWR. VSWR < 2.0 or better over the entire operating frequency range is desirable from an efficiency and harmonic conformance standpoint. In some cases, VSWR < 1.5 is stipulated. In addition, there has been a trend on behalf of manufacturers to use one antenna that can operate in any available band, so that one antenna could fit all devices that would nominally operate from 2.3-2.4, 2.5 and 4.9-5.9 GHz. More recently, with the adoption 1xEVDO, HSPA, and WiMAX standards, the one antenna solution has included the 824-960 MHz dual bands, the 1710-2170 MHz bands, as well as GPS, and expanded coverage from 2.3-2.7 GHz. Specifications for radiation efficiency although useful, are usually supplanted with a specification for percent angular coverage with gain above a specified threshold in a desired plane. Antenna placement becomes an issue for uniform coverage, as well as for schemes utilizing pattern or polarization diversity.

26.1.4.1 Secondary Design Considerations

Selection of the antenna element design is predicated upon several factors, but the cost/performance trade-off is perhaps the most significant. Form factor and fit are also issues that drive the design and size of the antenna solution to the limit. Related to these issues, a variety of locations have been tried ranging from behind the display screen to the sides of the base to the sides and top of the display. The selection of the antenna location determines the pattern, gain, transmission line loss and dominant polarization. In addition, with the drive toward thinner and larger area displays, as well as the inclusion of other feature components, the antenna size and therefore bandwidth has generally been stressed. This has resulted in production issues where tuning at the low band is adversely affected by antenna fine scale position variability with respect to the dielectric and metal surrounding structures. As a result, antenna testing and/or production binning is sometimes required to meet performance requirements. Reducing the antenna sensitivity to these effects by employing specific broadband designs or dielectric loading can be curative but may increase cost or overall performance.

Production assembly costs can also be an issue with associated antenna mounting hardware adding to labor intensity. Every effort is made on behalf of the manufacturers to minimize these costs. Recent designs utilize an adhesive backed multi-band, or broadband monopole antenna pairs that are suited for direct integration with a radio chipset in the display unit. This approach eliminates the 0.5 meter transmission line and associated loss (~3 dB at 5 GHz) with LCD mounted radio implementations.

26.1.4.2 Antenna Location Selection

Choice of antenna location is usually limited in any embedded design due to industrial design space constraints and interference considerations. Since diversity is standard in the 802.11, WiMAX, and 1xEVDO protocols, antenna location is important in establishing performance of the system in specific operating environments. Some typical examples of antenna placement options in a notebook computer are shown in Figure 26.4.

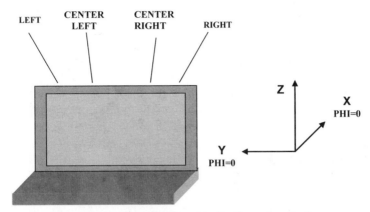

Figure 26.4: Antenna Mounting Configuration Options.

Since the height in the z-dimension is limited, antennas having current distributions along the y-direction are typical. Examples are transmission-line antennas, such as the inverted "F" antenna (IFA) and multi-resonant inverted L variants. Slot antennas having primary radiative currents along the z-direction also have utility in this application, since the display screen and laptop lid are electrically conductive.

26.1.4.3 Example Gain and Radiation Patterns

The authors have conducted extensive simulation and testing for the F- and L-designs in standard 15" laptop lid configurations and present some typical data for the positions specified in Figure 26.4. Simulated far-field radiation plots at 2400 MHz using antennas fitting an 18 x 3 x 5 mm form factor are shown in Figure 26.5.

Note that the azimuth cuts ($\theta = 90$) are nearly symmetric with greatest field strength at the left- and right-most positions. These positions also allow for radiation in the z-direction for multi-floor coverage. Simulations and measurements have also been made for antennas rotated about the y-axis. These studies also indicate that more complex pattern diversity schemes may be suited to this application.

The plots in Figure 26.5 are for total field components rather than for field components in Horizontal (x, y) or Vertical (z) directions. Analyses and tests indicate dominant horizontal with lesser vertical contribution. The total field coverage is shown in Figure 26.6 for a multi-band dielectrically loaded L-antenna where 4-branch pattern diversity combining has been used to indicate the total coverage contribution from each antenna. Note that 90% coverage is available with a provided gain above -4 dBi. If the vertical field component is considered, only 40% coverage is achieved at the same gain threshold and a maximum of 80% coverage is available at -10 dBi.

In Figure 26.7, another antenna arrangement is used, providing better pattern diversity coverage. In this case a 2 dB advantage is observed on the average when compared to Figure 26.6.

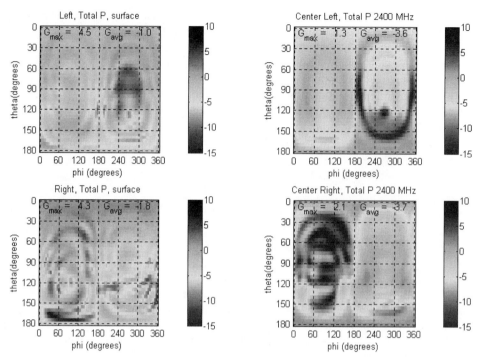

Figure 26.5: Far Field Simulated Patterns for Antenna Locations (2400 MHz, total field). (Clockwise from upper left: Left, Center Left, Center Right, Right).

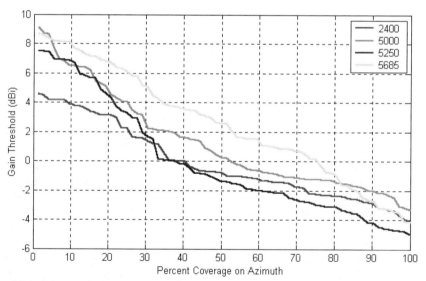

Figure 26.6: Azimuth Gain Distribution (Total P): Percent AZ-coverage versus gain threshold (total field) for standard antenna configuration.

The benefits of selecting an optimized antenna location and orientation are obvious when comparing the contributions from properly placed antennas, as shown in Figure 26.8. Here each separate antenna pattern was measured in the AZ plane. The composite pattern resulting from diversity combining is also shown. The combined pattern is nearly circular with and average gain near zero dBi.

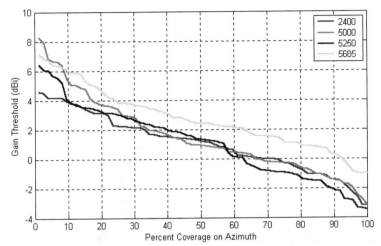

Figure 26.7: Azimuth Gain Distribution (Total P): Percent AZ-coverage versus gain threshold (total field) for antenna configuration providing optimum pattern diversity.

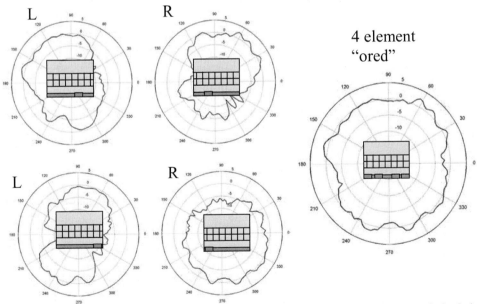

Figure 26.8: Separate azimuthal patterns (total field) for each antenna in an optimized 4-branch configuration for 802.11a. Rightmost plot shows the net combined pattern when used with a 4-branch diversity scheme. ("ored" in the text in top right corner).

Range measurements for a similar configuration are shown in Figure 26.9 illustrating the utility of a two-branch pattern-diversity combining scheme. When both patterns are combined, a nearly complete spherical coverage is obtained.

Multi-band slot antennas or broadband monopoles have been devised to provide a greater vertical field component [15 - 16] and some industry leaders have driven more complex pattern diversity configurations.

Manufacturing considerations suggest designs that are suited for rapid integration into a variety of platforms with minimal impact due to detuning effects from nearby objects either in assembly or use. In order to achieve this degree of reliability, some manufacturers have opted for broadband or reconfigurable antennas with integrated radio chipsets and these are expected to gain favor in future applications.

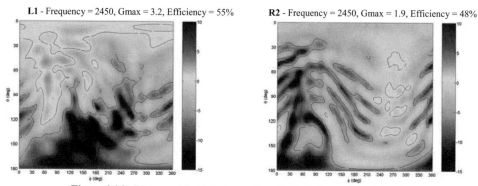

L1 - Frequency = 2450, Gmax = 3.2, Efficiency = 55%   **R2** - Frequency = 2450, Gmax = 1.9, Efficiency = 48%

Figure 26.9: Right and Left Antenna Far Field Patterns (total field).

26.2 Power Statistics of Small Scale Fading in Rayleigh Radio Channels

Perhaps one of the most surprising observations of radio wave communications can be made at the terminals of the receiving antenna. Gone are the carefully crafted waveforms, which have been sent by the transmitter. Instead, at the point of detection rapidly varying waveforms are observed, often with large signal variations; scrambling the original message to a greater or smaller degree. In general, the received signal is composed of multiple delayed and scaled copies of the original transmitted signal. The resulting power observed at the receiver demonstrates rapid variations about a central value. The magnitude variation can be in excess of 30 dB.

Figure 26.10 illustrates the variations in power received by two antennas. We note the deep fading phenomena of the processes in time. It is convenient to introduce the normalized power, $\hat{P} = P / \langle P \rangle$. This normalization has the advantage of making the distribution unit mean which facilitates comparison between distributions having different average powers. The PDF and CDF can then be expressed in closed form as:

$$f_{\hat{P}}(\hat{p}) = e^{-\hat{p}} \tag{14}$$

$$F_{\hat{P}}(\hat{p}) = 1 - e^{-\hat{p}} \tag{15}$$

These functions are displayed in Figure 26.11 below.

It is important to note in Figure 26.11a that the most probable value of a Chi-square power distribution is zero! The CDF in Figure 26.11b plots the likelihood of measuring a power value less than or equal to a measured power value (ordinate). For example, the power values of -10 dB or below will occur with the probability of 10%. We term this point the 10% threshold power. Similarly, 1% of the measured power values will occur -20 dB below the average power value. One can easily understand that the CDF is important in choosing receiver thresholds. As a result, the term "Outage Probability" often used to refer to the abscissa of the CDF function with our example demonstrating the 10% and 1% outage probabilities.

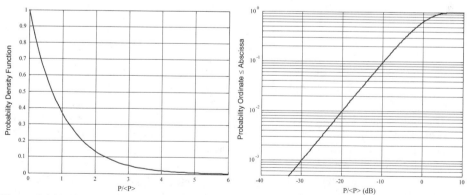

Figure 26.10: Received Power Versus Time in a Rayleigh Channel.

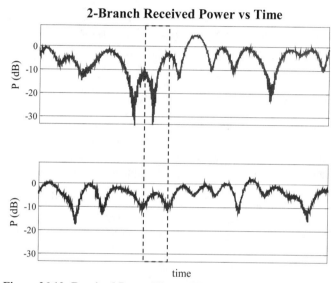

Figure 26.11: (a) Probability density function (PDF) (Chi-square), (b) Cumulative distribution function (CDF) of the received power.

This model allows us to write the power received by an ideal antenna as a random process, $\hat{P}$, such that a sequence of this process in time (Figure 26.10) can be written as

$$\left\{ \hat{P_1} , \hat{P_2} , \hat{P_3} , ..., \hat{P_n} \right\} \qquad (16)$$

Further, each $\hat{P_i}$ is distributed according to (14) and (15).

We postulate that this model remains valid only so long as the associated statistics remain representative of the scattering environment. Cases in which the model may not be valid include changes in the regime of scattering caused by moving from environments characterized by Rayleigh statistics (distant, obstructed locations) to those that are not (close, line of sight locations). Given the large dynamic range of the received power statistics, it is not hard to understand why communications engineers have worked hard to devise radio architectures which mitigate to this environment.

26.3 Diversity Architectures

Digital communication coding schemes require a minimum received power to provide tolerable error rates. In general, radio architectures seek to maximize the received power through manipulation of one or more copies of the received signal. Antenna diversity has been widely and successfully used to combat channel fading by providing an alternate received signal from an additional antenna.

In its simplest form (switched diversity) a single receiver utilizes two antennas by means of an RF switch. The additional antenna provides not only another copy of the signal, but its position, polarization and directivity can be exploited as well.

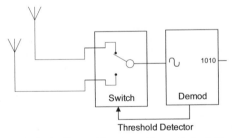

Figure 26.12: Radio architectures for switched diversity.

The system uses the signal received by the primary antenna until its power falls below a fixed threshold established *a priori*. In that case, the rest of the antennas are sampled until finding a signal which satisfies the performance requirements. The selection diversity approach simultaneously monitors the signals received by all antennas and selects the one that provides best system performance. The disadvantage in implementing such technique is that the system requires multiple radios so that all signals are received simultaneously. Here, we consider a hybrid between these two approaches. In the case of WLAN communications the channel is slowly varying. As a result, there is sufficient time to sample the signals of all receive antennas in a short time interval before packet reception.

For each frame, the receiver is able to select the branch containing the most power. This scheme allows the performance of selection diversity as specified above, while having a simple system requiring only one radio. We initiate our understanding of diversity architectures by considering a simple two-antenna configuration.

The power sequence available from each antenna can be written in the form

$$\left\{\hat{P}_1^{(1)}, \hat{P}_2^{(1)}, \hat{P}_3^{(1)}, \ldots, \hat{P}_n^{(1)}\right\}$$
$$\left\{\hat{P}_1^{(2)}, \hat{P}_2^{(2)}, \hat{P}_3^{(2)}, \ldots, \hat{P}_n^{(2)}\right\} \tag{17}$$

where the superscript is used to denote the antenna number. We now assume that the sequence $\hat{P}^{(1)}$ is independent from the sequence $\hat{P}^{(2)}$. The statistics of each antenna path (branch) have the same PDF and CDF as shown in Figure 26.13. However, the process of selecting the maximum power value from all the branches as a function of time results in a different distribution. This distribution is written as

$$\left\{S(2)_1, S(2)_2, \ldots, S(2)_n\right\} \tag{18}$$

where each element in the new sequence is defined as the maximum value of the two branch elements,

$$S(2)_i = max\left(\hat{P}_i^{(1)}, \hat{P}_i^{(2)}\right) \tag{19}$$

The PDF and CDF for this new sequence is now plotted and compared with those of the individual branches.

Choosing the branch with the largest power value at each instance of time reduces the probability of receiving a zero power value from the most likely (single branch) event to an unlikely event!

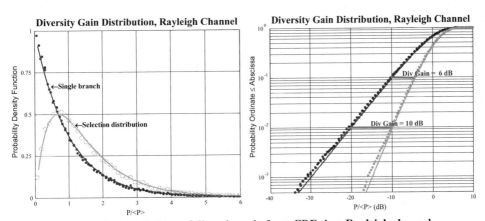

Figure 26.13: Definition of diversity gain from CDFs in a Rayleigh channel.

Further, one observes the normalized power at the 10% outage probability has increased from -10 dB to -4 dB; a 6 dB improvement. This improvement in power is attributed to the diversity radio architecture and is thus termed "Diveristy Gain", G_{div}. The value of diversity gain is dependent on outage probably used to compute the increased average power. In Figure 26.13, we see that the diversity gain is 6 dB at the 10% outage probability threshold but is 10 dB at the 1% outage probability threshold. The choice of comparison at 10% outage probability is mostly used in experimental studies where an insufficient number of points can be collected to provide consistent results at the 1% outage probability. However, numerical studies can readily generate sufficient samples to yield accurate values at the 1% outage probability [17]. More compactly, we denote the choice of outage probability, v (10% in our demonstration), and the outage power of a single branch, $P_0^{(1)}$, such that $P\left[\hat{P} \le P_0^{(1)}\right] = v$. Similarly, for two branch diversity, the threshold power, $P_0^{(2)}$, is determined so that $P\left[\hat{P} \le P_0^{(2)}\right] = v$. The diversity gain may now be rigorously defined as the ratio of those values,

$$G_{div} = \frac{P_0^{(2)}}{P_0^{(1)}} \tag{20}$$

or as their difference in dB.

A box chart below displays the diversity gain as the height of each box, the power of each individual branch aligned with the bottom edge of the box and the threshold power of the diversity branch aligned with the top of the box.

This representation highlights not only the diversity gain of the antenna and receiver configuration but also the resulting threshold power at the desired outage probability. As we will soon see, it is possible to encounter cases with large diversity gain numbers that still yield low threshold power values.

Figure 26.15 illustrates the PDF and CDF functions for $N = 2, 3, 4, \ldots$, etc. obtained using the distributions derived above.

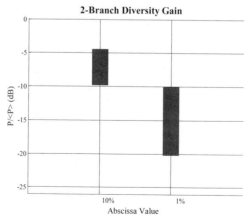

Figure 26.14: Diversity Gain Box Chart.

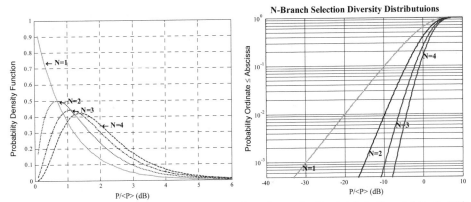

Figure 26.15: Power statistics of multi-antenna diversity under Rayleigh fading, (a) Probability distribution function (PDF), (b) Cumulative distribution function (CDF).

These figures clearly show the difference in power statistics obtained from two, three and four selective diversity branches. Increasing the number of antennas increases the mean power and the threshold power for any outage value. Specifically, the threshold power at 10% outage increases from -10 dB for a single branch to -4 dB for two, -2 dB for three, and -1 dB for four branches. From these relationships in the outage power, we can predict the expected diversity gain for a system operating in uncorrelated Rayleigh fading conditions.

For more than two-branch antenna diversity ($N_b > 2$), we extend the definition in (20) to N_b branches. Assuming $P_0^{(1)}$ is the smallest received power, the diversity gain is then defined as the maximum ratio of the values,

$$G_{div} = max\left\{\frac{P_0^{(2)}}{P_0^{(1)}}, \frac{P_0^{(3)}}{P_0^{(1)}}, ..., \frac{P_0^{(N_b)}}{P_0^{(1)}}\right\} \tag{21}$$

or as their difference on a dB scale.

Again restricting our analysis to $v = 10\%$ outage probability, the diversity gains for up to four branches are shown in Figure 26.16 below.

The box chart clearly shows a saturation trend as the number of antennas is increased. The largest percentage improvement occurs with the addition of just a single antenna to form a two-branch diversity system. Exponentially decreasing improvements are achieved as the number of antennas increases.

26.4 Rician Channel Power Statistics

The discussion thus far has been limited to channels in which the received power is comprised of independent random partial plane wave components. This regime of propagation is typically found far from the transmitter in rich multipath environments. However, in some cases the transmitter is near or the environment does not provide rich multipath and the channel cannot be described by Rayleigh statistics. These environments can be characterized by Rician statistics. The Rician distribution describes the variability of received channel power about its mean and is parameterized by the factor K. We write the Rician random variable as a sum of two field components

$$E_t = E_c + E_r \qquad (22)$$

where E_c and E_r are the constant (deterministic) and random (diffuse) components, respectively. The Rician K-factor is commonly defined as the ratio of constant to diffuse power. We confirm this observation below:

$$K = \frac{|E_c|^2}{\left\langle |E_r|^2 \right\rangle} \qquad (23)$$

The probability and cumulative distribution functions are expressed in terms of K as:

$$f_P(p) = (K+1)e^{-(K+Kp+p)} I_0 \left(2\sqrt{pK(K+1)} \right) \qquad (24)$$

$$F_P(p) = 1 - Q_1 \left(\sqrt{2K}, \sqrt{2p(K+1)} \right) \qquad (25)$$

The probability and cumulative distribution functions are shown in Figure 26.17 illustrating distributions for $K \in \{0,1,4\}$. One should note that the mostly likely received power value tends to the mean as the K-factor is increased. This behavior is intuitively gratifying since we expect the received power to be very near the average when no scattering is present (i.e., high K-factor, line of sight channels). Finally, we identify the K-factor as a quantitative definition of the channel condition; independent of user perception. Environments perceived by a bystander to be "line-of-sight" may, in fact, contain a significant degree of multipath, exhibiting a power distribution with a low K-factor. This case most often occurs indoors when the transmitter and receiver are close together, yet the power at the receiver is composed of many reflections from surrounding objects.

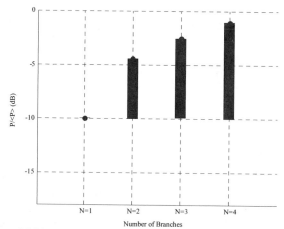

Figure 26.16: Diversity gain as a function of number of antennas.

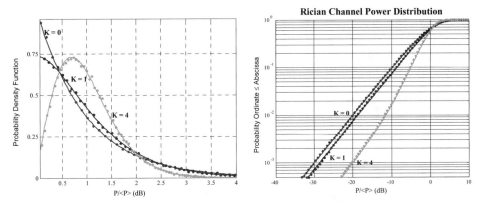

Figure 26.17: PDF and CDF of Rician distributions with varying *K*-factor.

We note that the probability and cumulative distribution functions now are not written in terms of s^2 and $2\sigma^2$ specifically, making it unnecessary to re-derive them for a normalized power distribution. Given normalized power values $\left(|E_c|^2 = 1, \left\langle |E_r|^2 \right\rangle = 1 \right)$, the total received electric field can be expressly written as a function of the parameter K as:

$$E_t = \sqrt{\frac{K}{K+1}} E_c + \sqrt{\frac{1}{K+1}} E_r \qquad (26)$$

This notation compactly illustrates that the Rayleigh fading channel is a special case of the Rician with the K-factor set to 0.

Rician power statistics provide the capability of modeling a wide range of scattering regimes. As an example, experiments have shown [18 - 20] that for indoor home environments at 2.4 GHz and 5 GHz, the K-factor falls in the range of 0-4.

Figure 26.18 illustrates the impact of the K-factor on the distribution. When $K = 0$, the magnitude of the constant component is zero, and we recover the Rayleigh envelope distribution with zero mean. As K increases, the constant component begins to dominate the distribution leading to less variation about the variable's mean value.

Characterization of indoor channels as a function of K-factor allows us to evaluate the performance of diversity architectures in these environments.

26.4.1 Diversity Gain of Omni-Antennas in Rician Channels

Substituting the Rician distribution for power ensembles in (17), we follow the same selection rule as in (19) to arrive at the two-branch diversity power ensemble. We plot the probability and cumulative distribution functions for this new sequence for the values of $K \in \{1,4\}$ below.

Considering the single-branch distributions, we note that in comparison to the Rayleigh PDF, the Rician $K = 1$ PDF has reduced the probability of zero power. On the CDF we note an increase in single-branch threshold power from -10 dB (Rayleigh) to -9 dB (Rician $K = 1$). Similarly to the Rayleigh results, we observe the threshold power at the

10% outage power has increased from -9 dB to -4 dB; a 5 dB improvement. We now set the K-factor to 4 and observe the diversity results.

Here the most likely power value of the single-branch distribution has shifted much farther from 0, driving the 10% outage power higher to -5 dB. The diversity branch offers a 10% outage power of -2 dB, representing a 3 dB improvement.

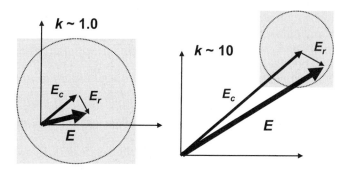

Figure 26.18: Vector representation of the Rician envelope distribution with K~1.0 and K~10.

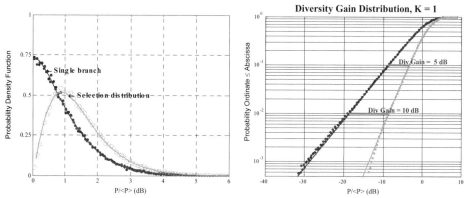

Figure 26.19: PDF and CDF of two omni-antenna diversity in Rician (K=1) channels.

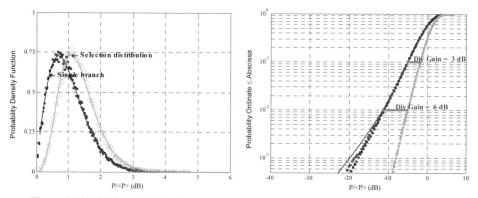

Figure 26.20: PDF and CDF of two omni-antenna diversity in Rician (K=4) channels.

We summarize the results for Rician channels using a box chart, which displays the diversity gain as the height of each box, the power of each individual branch aligned with the bottom edge of the box and the power of the diversity branch aligned with the top of the box. Along the x-axis we vary the Rician K-factor.

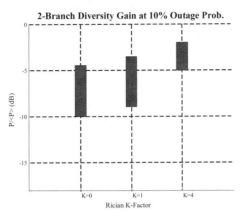

Figure 26.21: Two omni-antenna diversity gain in channels characterized by varying K-factor.

We note that $K = 0$ corresponds to a Rayleigh channel. When comparing the $K = 0$ and $K = 4$ cases (Figure 26.21), the increased K-factor results in a higher threshold power (upper boundary of the box). In fact, one could interpret the diversity gain in the $K = 0$ case as recovering the lost threshold power due to multipath; which is equivalent to a single branch (lower boundary of the box) in the $K = 4$ environment. From these results, we conclude that with increasing K the outage power for a single branch increases (top of box). This result is intuitively gratifying: as the power in the signal's constant component increases, the receiver is more likely to observe a stronger signal. However, the diversity gain (box height) is inversely proportional to K: with a stronger constant component in the signal, the variation about its mean decreases, and the likelihood of observing greater signal strength with a second antenna element decreases. This simple example quantifies the engineering trade-off in diversity architecture design. Systems operating in a high K-factor environment should expect modest performance improvement (when compared to Rayleigh). In these cases, the use of a second antenna may not be economically feasible.

26.4.2 Diversity Gains of Multiple Antennas under Rician Fading

Similar diversity results have been obtained for multiple antenna configurations in Rician scattering environments. Figure 26.22 below illustrates the N-branch diversity gain results, where diversity gain is defined in (21). In the figure, we consider the 12 possible combinations of $N_b \in \{1,2,3,4\}$ and $K \in \{0,1,4\}$, with N_b varying along the x-axis. For each value of N_b, we plot the diversity gain results in sets, corresponding to the Rician K-factor (increasing from left to right within a set). For each box, the bottom is aligned with the power threshold of a single branch, the top with the power threshold of the diversity branch, and the height of the box corresponds to diversity gain (in dB). The Rayleigh channel results, which correspond to the $K = 0$ subset of the box plots, are provided as reference.

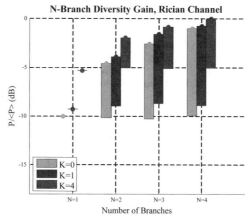

Figure 26.22: Diversity gain as a function of number of antennas (N_b = 1, 2, 3, 4) in Rician fading conditions (K = 0, 1, 4).

We observe the impact of Rician channel statistics on the results within each set corresponding to a given value of N_b. The single-branch threshold power and the diversity-branch threshold power increase with K. However, diversity gain decreases with K. The increased K factor reduces the impact of additional antennas on system performance.

We now consider the impact of increasing the diversity order. We observe that for a given K-factor, the threshold power of the diversity branch increases with increasing diversity order. However, the incremental increase due to each new additional antenna decreases dramatically. Given a constant cost of implementation of an additional receiver chain, the cost to increase threshold power by a single dB rises with increasing diversity order.

26.5 Conclusion

The design of WLAN communication devices for best performance at lowest possible cost must involve a systems design approach that considers the radiation characteristics of the antenna elements, their interaction, as well as the channel characterization. The problem is not a simple one. A comprehensive design methodology, which considers the effects of channel statistics on signals derived from alternative antenna configurations, enables the system architect to optimize receiver for operation in multipath environments encountered in various operational scenarios. This chapter has served as a brief introduction to the problem.

26.6 References

[1] Goubau, G., "Multi-element monopole antenna", *Proc. Workshop on Electrically Small Antennas, ECOM*, 63-67, 1976.
[2] Wheeler, H., "Small Antennas," *IEEE Antennas and Propagation*, Vol. AP-23, 4, pp. 462 - 469, 1975.

[3]   Chu, L. J., "Physical limitations of omni-directional antennas," *Journal Applied Physics 19*, pp. 1163-1175, Dec. 1948.

[4]   Harrington, R. F., "Effects of Antenna Size on Gain, Bandwidth, and Efficiency," *J. Nat. Bur. Stand.*, v 64-D, pp. 1-12, 1960

[5]   Fante, R. L., "Quality Factor of General Ideal Antennas," *IEEE Antennas and Propagation*, Vol. AP-17, pp. 151 - 155, Mar. 1969.

[6]   McClean, J. S., "A Re-examination of the Fundamental Limits on the Radiation Q of Electrically Small Antennas," *IEEE Antennas and Propagation*, Vol. 44, pp. 672-675, May 1996.

[7]   Grimes, D. and Grimes, C. A., "Minimum Q of Electrically Small Antennas: A critical Review," *Microwave and Optical Technology Letters*, 28, No. 3, Feb. 5, 2001.

[8]   Cohen, N., "Fractal Antenna Applications in Wireless Telecommunications," Proc. Electronics Industries Forum of New England, 1997, pp. 43-49.

[9]   Puenta-Baliarda, J. Romeau, R. Pous, and A. Cardama, "On the Behavior of the Sierpinski Multiband Fractal Antenna," *IEEE Trans. Antennas and Propagation*, Vol. AP-46, pp. 517 - 524, 1998.

[10] Stuart, H.R., Tran, C., "Near Chu-Limit Performance in Electrically Small Antennas Based upon Metamaterial Spherical Resonators, *IEEE APS Conference*, Albuquerque, NM, session 404, 13 July, 2006.

[11] Puenta-Baliarda, Romeau, J., Pous, R., and Cardama, A., "The Koch Monopole: A small Fractal Antenna," *IEEE Trans. Antennas and Propagation*, Vol. AP-48, No. 11, pp. 1773 - 1781, 2000.

[12] Best, S. R., "A Discussion of the Quality Factor of Electrically Matched Small Wire Antennas," *IEEE Trans. Antennas and Propagation*, Vol. 53, No. 1, pp 502 - 508, Jan. 2005.

[13] US Patent #5,790,080, "Meander Line Loaded Antenna," August 4, 1998.

[14] Suh, S.Y, Stutzman, W. L., Davis, W. A, "Low Profile, Dual Polarized, Broadband Antennas," *IEEE Antennas and Propagation Society International Symposium*, 2003, Volume 2, Issue 22-27, pp. 256 - 259, June 2003.

[15] Behdad, N., and Sarabandi, K., "Dual Resonator Slot Antennas for Wireless Applications," *IEEE AP-S Int. Symp. Digest*, Monterey, California, June 20-25, 2004.

[16] www.skycross.com

[17] Turkmani, A.M.D., Arowojolu, A.A., Jefford, P.A., Kellet, C.J., "An experimental evaluation of the performance of two-branch space and polarization diversity schemes at 1800 MHz", *IEEE Trans. Vehicular Technology*, Vol. 44, pp. 318 - 326, 1995.

[18] Wittwer, D.C., Stoytchev, M.S., "Diversity Performance of Integrated Desktop Antennas in a Home Environment", *Proc. IEEE Antennas and Propagation*, Summer 2005.

[19] Dietrich, C.B., Jr., Dietze, K., Nealy, J.R, and Stutzman, W.L., "Spatial, Polarization, and Pattern Diversity for Wireless Handheld Terminals," *IEEE Trans. Antennas Propagation*, Vol. 49, pp. 1271 - 1281, Sept. 2001.

[20] Ioffe, A., Stoytchev, M.S., "Wideband channel measurements and the characterization of the home environment", *Conference Proceedings IEEE Antennas and Propagation Society*, July 2006.

Service Control and Service Management of Wi-Fi Hotspots

Jasbir Singh[a]

27.1 Wi-Fi Hotspots Introduction

Wi-Fi hotspots are wireless *Local Area Network* (*LAN*) locations that provide broadband Internet access and *Virtual Private Network* (*VPN*)[1] access from a location. One or more access points can cover a single hotspot location. It enables customers at a hotspot to use their wireless-enabled laptop, PDA (personal digital assistant) or cell phones to access the Internet with a secure connection. While the costs of portable devices continue to decline, the popularity of Wi-Fi technology and the acceptance of Wi-Fi in the marketplace continue to increase.

The size of hotspots can range from a single room to many square miles of overlapping hotspots. Hotspots are often located at restaurants, train stations, airports, libraries, coffee shops, bookstores, and other public places. Today, many universities and schools have wireless networks deployed on their campus.

27.2 Brief History of Hotspots

The concept of Wi-Fi hotspots was first proposed by Brett Stewart at the NetWorld/InterOp conference in the San Francisco Moscone Center in August 1993. Stewart, instead of using the term 'hotspot', referred to them as public accessible wireless LANs. The term "Hotspot" was first introduced by Nokia in 1998.

27.2.1 Overview of Commercial Hotspots

Commercial hotspots are now deployed in places such as Internet cafes, coffee houses (commonly called Wi-Fi-cafés), hotels, and airports around the world. These business establishments may charge the customers for the service, but some hotels provide the

---

[a] *Pronto Networks, Inc*
[1] A virtual private network (VPN) is a private communications network often used within a company, or by several companies or organizations, to communicate confidentially over a publicly accessible network.

service for free to guests as an added amenity. A commercial hotspot may have one or more of the following features:

- The end users are redirected to a Captive Portal[2] for authentication and payment.
- The payment options can be through a credit card, or other online payment services such as PayPal[3].
- Choices include using an offline "scratch card" or "voucher" which provide a authentication code and password to obtain access.
- The end users are allowed free access to certain websites – these free sites are called a "walled" garden site or a "white-listed" site.

Figure 27.1: A Wi-Fi enabled Cafetaria.

Many Wi-Fi networks support roaming, in which a mobile client station such as a laptop computer can move from one hotspot to another as the user moves around a building or area.

27.2.2 Overview of Free Hotspots

While many groups, communities and cities have set up free Wi-Fi hotspots in order to provide open access, such hotspots still require some amount of authentication. Thus, we differentiate open hotspots from free hotspots. Open hotspots are wireless LANs that allow users to connect to the wireless signal and get broadband access. Free hotspots, on the other hand, require the user to go through some level of authentication, i.e., enter a username password, or some other method of validating the credentials of the user.

Many municipalities have joined with local community groups to help expand free Wi-Fi hotspots. Independent community groups have set up Wi-Fi hotspots using volunteer efforts and donations.

---

[2] The *captive portal* technique redirects an HTTP client on a network to see a customized web page (usually for authentication purposes) before surfing the Internet.
[3] *PayPal* and *World Pay* perform payment processing for e-commerce business by allowing payments and money transfers to be made through the Internet.

OLSR[4] is one of the protocols used to set up free hotspots, while some of the other free hotspots use static routing.

27.2.3 Wi-Fi Hotspot Signal Range

The range of the wireless signal in a hotspot depends on several factors, including power output of the wireless card or router; strength of the receiving wireless card or cards, obstructions from buildings or trees which may be in the way of the transmitting path, walls, etc. Wi-Fi hotspots generally provide a range of about 75 to 150 feet in a typical home or office. In an open environment like a park or an empty warehouse, a Wi-Fi network may provide a range of up to 1,000 feet or more. It primarily depends upon the strength of the signal. The stronger the signals the larger the area it will cover. The range of the Wi-Fi access also depends on the kind of antenna used. With the right antennas and optimal placement, a range of up to a mile is possible.

Using Wi-Fi technology (802.11b or 802.11a), a "gradual degradation" in range occurs. This implies that the transmission rate becomes slower as you move further away from the access point. For example, with Wi-Fi 802.11b technology, within 100 feet of the access point, the Wi-Fi radio in your laptop computer will get about 11 Mbps data rate. As you move farther away, that rate will drop down to 5.5 Mbps, then to 2 Mbps and finally to 1 Mbps.

Hotspot with an extended Access Point

Figure 27.2: Hotspot with an extended access point.

27.2.4 Advantages of Wi-Fi

The advantages of Wi-Fi are numerous. Some of them include:

- Wi-Fi uses unlicensed radio spectrum and does not require regulatory approval for individual installations.
- Allows LANs to be deployed without cabling, thus reducing the costs of network deployment and expansion.
- Ideal in locations where running physical cables would be impractical, such as outdoor areas and historical buildings.
- Wi-Fi products are widely available in the market.

---

[4] Optimized Link State Routing (OLSR) is a protocol used to connect mobile ad-hoc networks, sometimes called wireless mesh networks.

- Different brands of access points and client network interfaces are interoperable at a basic level of service.
- Competition amongst vendors has lowered prices considerably since their inception.

Additionally, Many access points and network interfaces support various degrees of encryption to protect traffic from interception.

27.3 Service Management - Overview

Service Management is an important component of the Operations Support Systems (OSS) responsible for services rendered to the customer. These are the functions that are executed behind the scenes that enable the smooth operation and management of the network. These functions are required to ensure that the network is capable of supporting the demands of the users throughout the lifecycle of the service, ranging from the user ordering the service to the user getting a bill for the service or a post-use satisfaction survey. The services include order management, inventory management, provisioning and activation, network topology management and maintenance, and stability/performance diagnostics of communication service providers and their networks. However, almost all service providers provide a combination of product and services with emphasis on efficient service management techniques, which help generate more revenue.

A service management system automates manual operations of the network, rendering services and support to the customer, making these processes more efficient and error-free. Service providers cannot focus only on technology and their internal organization. They have to consider and constantly improve the quality of services provided to the customers to develop a continuing beneficial relationship with the customers.

27.3.1 Hotspot Service Management

Service providers are deploying metro-scale broadband wireless hotspot networks for both public safety and other internal productivity enhancement applications. These hotspot networks are increasingly viewed as a vehicle for bringing broadband access to a wide range of users and are a platform to deploy advanced multimedia applications and content-based services. The responsibility to provide secure and guaranteed connectivity at these hotspots, that meets regulatory and commercial interests lies with the Service Provider.

This chapter will describe how Wi-Fi hotspot networks can incorporate service control and service management platforms to provide these features. The technical aspects of the network that need to be considered are also described here.

Wi-Fi Hotspot Service Management requires automated systems, and personnel with the following areas of expertise:

- Support for unique hotspot environments combined with the efficiencies of a standardized solution.
- Efficiencies and cost-savings of a comprehensive, integrated service management solution.

- Assurance that customer data shared in support of service management processes will be protected.
- Support for a wide range of clients' evolving business strategies and complex IT infrastructures.
- Service providers require expanding the services, as and when the provider decides to change its services offered or the network expands beyond the scalability limits.
- Ongoing support, security, diagnostics, and crisis aversion/recovery must be accessible at all times.

27.3.2 Importance of Service Management

With mounting customer expectations, the quality of services offered by the service providers should be fast and accurate. Service management drives all aspects of management related to planning, providing, monitoring, coordinating, and reporting quality of service.

Starting with the service level agreement, service management also establishes the basis for Wi-Fi service delivery process such as availability, capacity, security, and production operations management, as well as support processes such as problem management, configuration management, and release and maintenance management.

Service Providers can ensure rendering high quality and cost-effective services to their customers by efficiently managing service demands and managing them throughout the service lifecycle.

27.3.3 Network and Data Management Services

There are many software solutions available in the market for network monitoring and management for LAN, WAN, VoIP, VPN, Wi-Fi, and network security technologies. These software solutions also provide comprehensive data and server management services that include systems and database administration, application monitoring, and related services.

Using powerful customizable tools and applications along with methodologies and processes service providers can deliver high-quality services to increase network/data availability and optimize performance.

Most of these remote network and data management services are usually enabled by Network Operations Centers (NOCs). The services offered by NOC include:

- Tracking problems and performance;
- Driving problem resolution;
- Performing software and firmware updates;
- Monitoring and managing servers;
- Enhancing network security.

The software provides monthly and online reports, as well as gives information on the portal that allows customers to view various activities in real time.

The user-friendly online customer interface facilitates a more comprehensive relationship with the customers. Clients get the information they need at any time, without the burden of building and maintaining their own operations centers. By receiving comprehensive service-level reporting, clients understand whether or not they are receiving targeted levels of service and what actions must be taken to address deficiencies.

27.3.4 Obtaining a Network Address

To obtain IP addresses for use within an organization's network, the network administrator has to contact the ISP, which will provide the IP addresses from the block of addresses allocated to that ISP. For example, an address block 339.34.15.0/10 may have been allocated to the ISP. The ISP, in turn, will divide this address block into equal-sized smaller address blocks and allocate one of these address blocks to each of the organizations that are supported by this ISP.

27.3.5 Obtaining a Host Address

The organization can assign individual IP addresses to the host and outer interfaces that are associated with the organization from the block of addresses obtained from the ISP.

There are two ways in which a host can be assigned an IP address, they are:

- Manual configuration. The IP address is manually configured into the host.
- Dynamic Host Configuration Protocol (DHCP). The host obtains an IP address automatically through DHCP.

DHCP is popularly referred to as plug-and-play protocol because of its ability to automate the network-related aspects of connecting a host to a network. Wireless LANs use DHCP very widely, where hosts are constantly joining and leaving the network. For example, an employee who carries a laptop from the cafeteria to a conference room to the lobby is most likely to be connecting into a new network at each of the different locations, and hence will need a new IP address at each location.

Figure 27.3 below illustrates three IP networks and allocation of IP address.

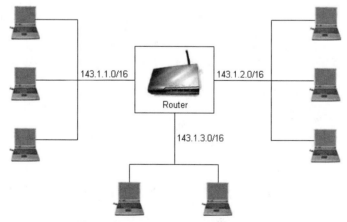

Figure 27.3: Allocation of IP Addresses.

The three end users in the left portion of the router interface, to which they are connected, will have an IP address of the form 143.1.1.xxx. The network address is 143.1.1.0 and "/16" data is the network mask. Any additional hosts attached to the 143.1.1.0/16 network would be required to have an address of the form 143.1.1.xxx.

27.3.6 Authentication, Authorization and Accounting for Hotspots

AAA (Authentication, Authorization and Accounting) are the functions in IP-based networking that control the computer resources the end users have access to and keep track of the sessions of end users over a network.

Authentication identifies which entity (user or device) can gain access to the network. *Authorization* identifies the boundaries or limits of the access provided to each user or entity. In some cases, the user or entity may have authorization to restricted resources and functionality in the network. *Accounting* tracks the duration of the access, and accounts for the time used and data transferred during the session. This information can subsequently be used for billing purposes.

27.3.6.1 Authentication

Every end user is required to authenticate to access the Internet in most controlled and "mixed-use" wireless networks. Authentication is the process of identifying an individual, usually based on a *Username* and *Password*. Authentication is based on the idea that each individual user will have unique information that sets him or her apart from other users. Some of the different ways for authentication are given below:

- **Open System Authentication:** This type of authentication allows any device to link to the network. This type of authentication is used in locations where the need for easy authentication is more important than the security of the communicating channel.
- **Shared Key Authentication:** This type of authentication is for end users devices, which access the Internet through Access Points. The device and the AP share the same secret key. The end user's device is authenticated using shared key authentication process. The figure below illustrates the authentication process that follows a standard request and response procedure.
- **802.1x Authentication:** It is an authentication method wherein the end user devices initiate a management protocol to authenticate themselves to access the network. 802.1x applies the Extensible Authentication Protocol (EAP), to provide this authentication mechanism. Using this type of authentication, the user has to install a software "supplicant" on his device. The user associates with the AP, which is the authenticator. The supplicant will request for the access, the authenticator acts as a relay and forwards the request to an authentication server, which is typically a RADIUS server. The server queries its database and checks the credentials and authorization of the end user, validates the information with the supplicant and the network access is granted. If the validation fails then the AP will not pass on any of the user's traffic and the network access is denied.

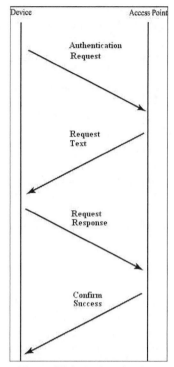

Figure 27.4: Authentication.

27.3.6.2 Authorization

This is the process of granting or denying a user access to network resources once the user has been authenticated through the *Username* and *Password*. The amount of information and the amount of services the user has access to depend on the user's authorization level.

27.3.6.3 Accounting

This is the process of keeping track of a user's activity while accessing the network resources, including the amount of time spent in the network, the services accessed and the amount of data transferred during the session. Accounting data is used for trend analysis, capacity planning, billing, auditing, and cost allocation.

27.3.7 User Login Page

A login page facilitates the end user to enter their username and password for authentication. If the username and password does not match the records in the database, then the end user will be denied access to the wireless Internet/network. A username is a unique sequence of characters the user chooses to represent him. A password is another unique sequence of characters that provides the user with the key to the system and is kept as a secret from others.

Typically, the following steps need to be followed to get a login page displayed on the end user's wireless enabled device:

- The first-time user has to power up the laptop or PDA and enter the SSID corresponding to the one configured on the Service Controller. The correct SSID can be selected manually to access the desired WLAN in the same area. In a typical deployment of WLAN, the SSID is used as an identifier of the WLAN. It is quite common these days for the user device's network interface card to automatically select the SSID and associate with that SSID. However, the SSID selected by the device may or may not be the one desired by the user. Hence, the user needs to verify that the desired SSID has been used to associate to the network.
- The returning user (subscriber who has already registered) has to open a Web browser on the laptop/PDA and try accessing any of the common sites like www.google.com or www.yahoo.com, i.e., any site that has not been designated as a "white-listed" or "walled garden" site, then the login splash page is displayed to the subscriber. This login page can vary from WISP to WISP and within a WISP it can vary from location to location. The subscriber's access credentials is verified here and on successful authentication, the requested page is displayed and simultaneously a window with a *Logout* button is also provided, which is used to terminate the session.

27.4 QoS Services

Data traffic access can be controlled by configuring each user's session via QoS provisioning to restrict or define the types of services (i.e., http, https, ftp, etc.) that are permitted to the user. This ensures efficient bandwidth management while delivering voice, text, or video traffic to the end user. The QoS parameters form the building blocks for network planning. QoS becomes very important in bandwidth-limited wireless networks. The contracts, which specify the QoS are called Service Level Agreement (SLA). The SLA specifies the end-to-end performance to which the client is entitled over a specified period of time.

Services like HTTP[5], FTP[6], VoIP[7] can be provided to the customer with varied network performance. The performance will depend on the different SLA's associated to different services for different customer segments.

27.4.1 Per User QoS

Based on the identity of the user, and the terms of the service, a specific set of traffic management SLAs are associated with the user. Once this information has been

---

[5] Hypertext Transfer Protocol (HTTP) is a method used to transfer or convey information on the World Wide Web.
[6] File Transfer Protocol (FTP) or is used to connect two computers over the Internet so that the user of one computer can transfer files and perform file commands on the other computer.
[7] Voice over Internet Protocol (VoIP), IP Telephony, Internet telephony, Broadband telephony, Broadband Phone and Voice over Broadband is the routing of voice conversations over the Internet or through any other IP-based network.

communicated from the authentication system (such as a RADIUS server or the Operations Support System), the access gateway can enforce these traffic management parameters on the user's traffic.

In a public WLAN, it is normal that different users with different SLAs connect simultaneously to the network. This means that the access gateway would be required to concurrently support multiple traffic management policies at the user (i.e., each IP address) level of granularity. It is not uncommon to expect the level of granularity to be at the service level (i.e., per TCP port, etc.) since service providers require the flexibility to design customized services to meet their users' needs. For example, a service could be defined as an IM-only (Instant Messenger) service. In this case, only those TCP ports would be opened up for this type of user.

To identify, which QoS to enforce for the specific IP address that is allocated to a user who is requesting access, the QoS values are associated with the user, not the IP address he gets from the mesh network. Thus, when a user logs in, the server communicates to the AP the QoS values that need to be enforced. Based on the Service Level Agreement of the user, the server will assign a service class to the user and it will get mapped to the QoS values.

27.4.2 Service Level Agreement (SLA)

Service Level Agreements are contracts that quantify the service given to the customer. An agreement is established based on bandwidth usage, which is also known as Bandwidth Throttling[8], service availability, latency, throughput, and outage notification requirements. An acceptable level of service provider performance with respect to the above listed metrics is guaranteed to the customer.

27.4.2.1 Packages

The price plans offered to the customers can be divided into granular units called Packages. A price plan can comprise of many such packages. A package will have at least one package fee or one service rate. A package fee will be a standard fee that has no measurement unit, while a service rate will be measured with an activity unit. Service rates can be qualified as Peak/Off Peak services based on the time of usage. Package fees can be qualified as recurring or one-time charges.

27.4.2.2 Price Plan

The price plan will be the actual product rate plan ultimately presented to the customer. A price plan can have multiple packages based on various dates, as long as none of the dates are overlapping.

---

[8] Bandwidth throttling is a method of ensuring a bandwidth intensive device, such as a server, will limit ("throttle") the quantity of data it transmits and/or accepts within a specified period of time. Bandwidth throttling helps provide quality of service (QoS) by limiting network congestion and server crashes.

The customer at the location can have the option to purchase either *Plans* or *Prepaid Cards*. Plans can be categorized into different plans, for example: *Subscription Plan, Pay-Go Plan, etc.*

In the *Subscription Plan*, the customer will be billed as per the pay period chosen (for example, Monthly, Quarterly, Half-Yearly, Yearly) whereas in the *Pay-Go Plan* the customer will be billed immediately after the user session ends or the charges will be added to the final room bill, in case of a stay at a hotel. Subscription price plans can be created only if the payment type is defined as "credit card".

27.4.2.3 Prepaid Cards

Prepaid services are very convenient and the preferred choice for hourly or daily passes. The OSS can provide a flexible prepaid management system with the ability to manage real time balance updates supporting *time based* prepaid cards as well as *data volume* based prepaid cards.

The customer at the location can be given the option to purchase either *Plans* or *Prepaid Cards*. In case of a hotel, the customer can have the option of being charged directly to the room or to pay by credit card.

27.4.3 Billing and Payment

The process of aggregating customer records for a billing cycle and creating an invoice is called Billing. In the payment process, the list of invoices are generated during the billing process is retrieved from the repository and an attempt is made to charge the corresponding customer's preferred credit card for the invoiced amount. At the end of payment processing, bulk email notifications are sent to the customers indicating successful or unsuccessful payment process.

27.4.4 Network Monitoring

Network monitoring is the first step towards efficient network management control. It is critical to monitor network hardware and software components like router, switches, AP's, firewalls, authentication servers, etc. The status of these devices helps in determining whether the AP is available in the network or not. This is done by polling the AP at regular intervals. The pinging utility is the tool used to poll the device, which is also known as heartbeat monitoring. In case of AP outages, the network administrator is notified via email. Faults in devices can be immediately detected, isolated and fixed. The performance of the AP on CPU and memory utilization can be remotely gauged. This information can be used for base lining the performance of the AP with respect to the current load and the maximum load it can handle, as per its specifications.

27.4.5 Customer Care

Customer Care describes an integrated suite of tools that provides basic customer management capabilities to meet the minimum system requirements that a Customer Care Representative of the service provider would require to manage the regular business operations like user administration, technical support, accounting, and usage history.

The customer care section can have the following basic capabilities implemented in the back-office support systems:

- Complete account management;
- Configuration of user accounts;
- Incident creation and tracking;
- Historical and online usage monitoring.

27.4.6 Reports

Reports helps the WISP and NOC engineers get the complete information that they require for monitoring and managing the whole process of accounts and billing. The option to convert these reports into a printable *pdf* or *csv* format can also be provided. The reports can be broadly divided into different sections as given below:

- NOC and WISP Reports.
- Audit Trail Report - This section will provide information regarding the date and time of creation and modification of audit trails and also the details of the administration personnel who executes these changes.
- Location Report - This section will provide information pertaining to the particular location.

27.4.7 Alerts

Alert management will send alert messages to the concerned support personnel at the NOC and WISP level, depending on the nature of the incident that has happened in the APs. At the WISP level, access to only those controllers under a specified WISP is provided. It involves setting up groups and recipients that handle the different types of alert messages received, based on the pre-defined rules. A list of alerts based on different search criteria can also be viewed.

27.4.8 Contracts and Tariffs

Contracts and Tariffs provide NOC level setups, the details about business contracts, roaming contracts, and revenue sharing.

27.5 Wi-Fi Network Designs

27.5.1 Centralized Control over Services for Wi-Fi Networks

Vendors/Service Providers face challenges from various fronts in deploying a wireless access network. These include regulatory, technology choices, network planning and network operational issues. Network operational issues now include designing and maintaining a network that is more than a "single-use" private enterprise network that happens to be wireless. This expands the scope of requirements related to customer management, accounting and billing, and network and service guarantees.

The software solution - OSS will allow the Vendors/Service Providers to have a centralized control on the solution by placing the servers required to provide service at the central NOC (Network Operations Center), where all the access and management functions can be controlled. The figure below shows the deployment scenario where the servers are installed at the central NOC that manage service controllers and gateways in the network. This management functionality can also be extended to the management and monitoring of the mesh APs that are connected to the controller using SNMP[9] and other API-based methods, thus simplifying the operation and management of the network.

Among other challenges faced by the communities, the IT staff now has to consider customer support issues, billing and trouble ticketing issues, and bandwidth and service level guarantees for the various users of the network.

Integrating with external applications, roaming partners, etc., provides an additional level of complexity to the Vendors/Service Providers. The need for providing flexible business models that can adapt to the changing network and commercial dynamics also provides fresh challenges to the Vendors/Service Providers, each with its own level of support requirements from the network.

This solution will be multi-tiered, enabling municipalities to sublet its network, if desired, to other network operators. A large carrier, for example, may want to offer its own Wi-Fi service brand within the municipality's hot zone. Wi-Fi providers support Virtual Network Operators (VNOs) as well as multiple VLANs[10] enable third party operators to offer their own branded service leveraging the existing wireless infrastructure of the city.

The Wi-Fi providers' broadband wireless platform supports VLANs and specific QoS assignments per user, which enable the network to be separated for both public and private use keeping in mind their specific needs. Multiple SSIDs[11] and thus multiple VLANs can be assigned to the same gateway in various situations. Some of these SSIDs could be hidden, e.g., for auto authentication for the public safety and fire department etc.

An OSS Solution can also support multiple service plans/offerings on the same VLAN (i.e., Captive Portal Page). The Service Gateway can support multiple VLANs with a customized Login Page per VLAN. Even within a single login page (i.e., the same VLAN), the system can simultaneously support different types of users. As an example, the first type of user can have subscription-based login. The second can be for prepaid users. A third option is to have an authentication from external repository, such as an external RADIUS, or part of a Roaming Network, etc. Each service option, i.e., service plan can have its own QoS/SLA.

---

[9] The simple network management protocol (SNMP) forms part of the internet protocol suite as defined by the Internet Engineering Task Force (IETF). SNMP is used by network management systems to monitor network-attached devices for conditions that warrant administrative attention. It consists of a set of standards for network management, including an Application Layer protocol, a database schema, and a set of data objects.

[10] A virtual LAN, commonly known as a vLAN or as a VLAN, is a method of creating independent logical networks within a physical network.

[11] In a Wi-Fi wireless LAN, a service set identifier (SSID) is a code attached to all packets on a wireless network to identify each packet as part of that network.

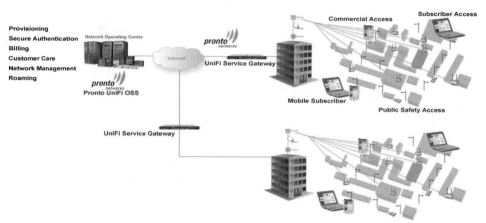

Figure 27.5: Centralized Hotspot Control.

Multi-tiered rating and service plans enable both retail and wholesale offerings with support for a variety of authentication and payment mechanisms. Credit cards, pre-paid cards, monthly subscription plans and bandwidth usage plans are also supported, as well as roaming settlements with aggregators such as Boingo Wireless®, iPass, and GRIC. The authentication options can vary from prepaid authentication to USB key based authentication.

This is very useful when there is a requirement to have different IP address ranges for different VLAN subscribers. Using VLANs, the network can be logically segmented to support different domains of users; secure access by government agencies, secure/open access for public users, residential, business users, etc.

All the target users can have a separate logical access to the network, and bandwidth rates that provide quality of service and allocation guarantees. In addition to supporting VLANs, which secure city networks by keeping them separate from general-purpose use, the OSS solution provides a SSL[12]-encrypted registration and authentication process and supports corporate VPN clients to allow secure, encrypted access to one's corporate LAN.

The controllers can also be configured to support multiple VLANs on its LAN interface. This allows the service provider to offer different captive portals, splash pages, and associated service authentication options on each VLAN.

The customer portal represents the brand image of the service provider to the end users. Keeping this in view, the application will provide sophisticated customization capabilities that customize the look and feel at the WISP, franchisee, and location levels for different VLANs.

27.5.3 Additional revenue generation opportunities

A well-designed Operations Support System can enable Wi-Fi service providers to introduce custom service offerings, and to meet their marketing strategies across targeted

---

[12] Secure Sockets Layer (SSL), are cryptographic protocols which provide secure communications on the Internet for such things as web browsing, e-mail, Internet faxing, and other data transfers.

segments, by supporting various authentication choices, billing models, payment mechanisms, charging principles, and custom branding options including co-branding with various different entities.

Custom branding options enable the Service Providers to own customers and have them associated with a known level of performance and assurance. The users are directed to a service provider portal where they can use their user ID and password for authentication purposes. This portal, called a Splash Page, has various links to sites that can be accessed even without authentication. These "Walled Garden"[13] sites serve as an opportunity for the operator to co-brand their services and enhance revenue with sponsor channels and information specific to the location. One can offer additional customized services based on types of users, including seasonal messages, promotions, etc. Multiple Splash pages are supported such that a different set of users on different VLANs can be configured to receive different pages and hence different service offerings, based on the SSID/VLAN they are associated in the WLAN network, as described earlier in this chapter.

The Customer Registration portal represents the brand image of the service provider to the end users. Keeping this in mind, the OSS can provide capabilities that allows for customization of the registration process at various levels. The customer portal web application can be customized on the following counts:

- Style sheet modification at various levels, e.g., support for customization of fonts, position of the images, etc.
- WISP/Municipality specific branding, and if the end-user is accessing the Customer Registration portal after authentication, location-specific branding is also possible, where content is customized at a static or a dynamic level.

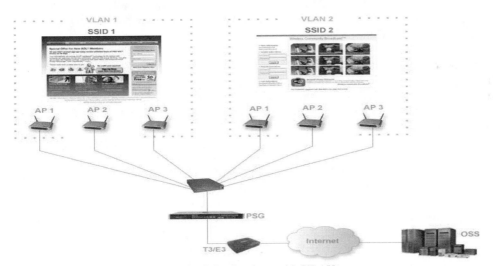

Figure 27.6: Deployment with VLANs.

---

[13] On the Internet, a walled garden refers to a browsing environment that controls the information and Web sites the user is able to access.

- International language support, e.g., the customer portal content can be displayed in any language. This feature enables presentation of labels and error messages in Web applications in multiple languages.
- Multiple authentication realms - RADIUS, LDAP, SMS messages, 802.1x, etc.

The solution can provide the WISP or municipalities the capability to have their own logos/branding on the login page (i.e., the Captive Portal/Splash Page). These WISPs can provide these options to their various franchisees that can further let their various locations add advertisements on the login page for additional revenue opportunities.

Additionally, the solution can provide the option of adding Walled Garden links on the login page; these can be the links for partners or for the service provider's websites for which no subscriber authentication is required. Any subscriber without valid login credentials can still browse through these pages.

When a subscriber powers up the laptop or PDA and enters the SSID corresponding to the one configured on the controller, as defined by the WISP, he/she will get connected. When the subscriber attempts to access the Internet prior to authentication, the controller redirects the request to its own internal Web server located on the controller, from where it will pick up the initialization templates for loading the Splash Page onto the subscriber's device. Now the subscriber can enter his *Username* and *Password* to get authenticated for wireless Internet access in that particular hotspot. The initialization template stored in the controller is a standard template, which gets loaded with the WISP specific files also stored in the OSS.

The Splash Page is an HTML page comprising of multiple "cgi-bin" programs for providing different types of authentication options to the subscribers. The initial authentication screen can vary from WISP to WISP and within a WISP it can vary from location to location. A window with a Logout button will be provided that can be used by the user to terminate the session.

27.5.4 Session Intercepts and "HotLining"

Wi-Fi service providers also desire the ability to broadcast public service announcements and emergency information to their constituents and the wireless infrastructure now provides an additional avenue of dissemination.

The OSS can be equipped with the ability to have "session intercept pages" being presented to the users logged in at the time, enabling for the dissemination of critical information for various emergency situations. For example, in case of lost/missing children, these intercepts can be used for sending out Amber Alerts. Homeland Security Alerts can be sent for information related to emergencies, and other public service announcements. These intercepts can also be used for service provider approved and initiated snap surveys requesting opinions and feedback from their constituents. This capability can also be used for WiMax "Hotlining" applications.

The Hotlining feature will provide a WiMax operator with the capability to efficiently address issues with users that would otherwise be unauthorized to access packet data services. When a problem occurs such that a user may no longer be authorized to use the packet data service, a wireless operator using this feature may hotline the user, and upon the successful resolution of the problem, returns the user's packet data services to normal. When a user is hotlined, their packet data service is redirected to a Hotline Application (HLA), which notifies the user of the reason(s) that they have been hotlined for and offers them a means to address the concerns meanwhile blocking access to normal packet data services. Reasons for hotlining a user can be to block the prepaid users whose account has been depleted; or users who have billing issues such as expiration of a credit card; or users who have been suspected of fraudulent use.

As a result, hotlining performs the following fundamental activities:

- Blocking normal packet data usage;
- Notifying subscriber that packet data usage is blocked;
- Directing subscriber to rectify blockage;
- Restoring normal operations when the user has rectified issues that triggered the hotlining of their service.

Or:

- Terminate service if the user failed to address the issues that triggered the hotlining of their service.

Hotlining would further provide consistency across all applications that utilize the packet data service, lowering the operating costs.

27.6 Lawful Intercept for Communications Assistance for Law Enforcement Act (CALEA)

The main purpose of CALEA is to define a telecommunications carrier's duty to assist law enforcement agencies with the lawful interception of communications and the collection of call-identifying information in a constantly changing telecommunications environment.

Operations Support Systems are required to support CALEA using its "session intercept" capabilities. An OSS equips law enforcement agencies with the ability to capture the Internet packets traversing the network managed by an OSS. These packets can then be routed to law enforcement agencies.

The OSS collects all URLs/IP addresses requested by the Legal Agencies and dumps them in a log with a timestamp. The administrator can optionally specify a copy IP address for the lawfully intercepted IP addresses (LIIP). The controller's WAN bandwidth can be configured to allow enough bandwidth partition. All traffic to and from the user is sent to this LIIP.

27.7 Security in Wi-Fi

Security concern in wireless networks is of great importance as the data is vulnerable to risks and threats from spam, viruses, etc., Strong security mechanisms adopted by service providers can combat this threat to a great extent. SSL protocol based communication protects various types of confidential information in transit. The information in transit may include username, password, credit card details, system administrator's credentials, private data, etc., on the public network. Wireless LAN switches generate alert notifications on various security events like Rogue AP detection, Rogue AP client detection, etc. Further security can be provided for authentication and authorization, audit and logging, password controls, web application session management, and data encryption. Secure services will be based on IPSec based authentication, VPN pass through, secure HTML login page, and support for public and private IP's, IP routing with NAT, etc. Client mails can be protected by the implementation of SMTP-AUTH proxy.

27.7.1 Security Benefits

27.7.1.1 Data privacy

A private connection is established over the public channel by using data encryption. The data is protected from interception and readable only by the intended recipient.

27.7.1.2 Authentication

Authentication is provided between communicating parties by the use of standard key encryption.

27.7.1.3 Reliability

Reliability and data integrity are provided. Data cannot be tampered during a session.

27.8 Future Technologies

Hotspot service providers believe VoIP will be the most popular add-on at their hotspots. Many Wi-Fi companies are diversifying into Mobile Hotspots, Managed Services Platforms, and Wi-Max. There is also a sudden surge in the tie-ups between the service providers and Virtual Network Operators for business and operations benefits.

27.8.1 Mobile Hotspots

Wi-Fi is being extended to give broadband Internet connectivity to passengers in transit beyond the reach of conventional Wi-Fi hotspots. Passengers traveling on trains, commuter rail, buses, and ferries should be able to use their Wi-Fi equipped laptops and PDAs to access Internet services, email, and enterprise networks, enjoying the convenience and improved productivity that comes with high-speed Internet connectivity while traveling.

27.8.2 Managed Services Platform

The vision for enterprises and municipalities is to offer simple and mobile Internet connectivity to their people. This vision can be realized by adopting the Managed Service Platforms to deliver Wi-Fi communication to the end users. This will make the people more productive, collaborative, and effective in their work and their personal lives. This concept helps to build and manage Wi-Fi networks more quickly and economically, with fewer problems.

Managed Service Providers offers a single contact point for 24/7 monitoring, management and customer support of large-scale Wi-Fi networks, for both enterprises and municipal configurations. These types of providers also simplify project management and minimize risk significantly and reduce capital and operational expenses. They also offer conventional pricing with determined periodic costs with no hidden expenses and specific service level agreements and performance metrics.

27.8.3 Wi-Max

The wireless communication industry is adopting Wi-Max technology, which is gradually becoming popular. Wi-Max is a term coined to describe a standard and interoperable implementation of IEEE 802.16 wireless networks. Wi-Max is stable under overload and over subscription. It is also more bandwidth efficient. It also allows the base station to control QoS parameters by balancing the time-slot assignments among the application needs of the subscriber stations.

27.8.4 Virtual Network Operators

Virtual Network Operators are operators who do not own their own spectrum and usually do not have their own network infrastructure. These types of operators have business arrangements with other Internet service providers for their customers to use the service provider's services.

The customers cannot distinguish any significant differences in service or network performance. The important benefit for service providers to get along with Virtual Network Operators is to expand their customer base at a minimal cost.

27.9 Summary

Generally, the most common usage of Wi-Fi technology is for laptop users to gain Internet access in locations such as airports, coffee shops, and so on, where Wi-Fi technology can be used to help consumers in their pursuit of work-based or recreational Internet usage. The main limitation of Wi-Fi technology is that it experiences interference and this technology is highly complex compared to their wired counterparts. This limitation is overridden by the fact that Wi-Fi does not restrict the users to be bound to a spot, but allows the user to browse while on the move.

Worldwide Wi-Fi hotspot revenue was $969 million in 2005, which is predicted to rise to $3.46 billion in 2009, as reported by In-Stat. This rapid growth will occur as the number of hotspot locations nearly doubles in size from 100,000 in 2005 to almost 200,000 by the end of 2009, the high-tech market research firm says. This explosive demand has service providers deploying wireless access points in every possible place. These large numbers of deployments pose multiple challenges to the service providers to overcome and accomplish complete, end to end Wi-Fi hotspot service management.

Service providers can implement successful service management strategies satisfying subscribers' requirements better. Deploying a services platform that leverages any Wi-Fi network can ease cities in the implementation of broadband applications and services such as centralized control over services and applications for "Mixed-Use" networks, virtual partitioning of wireless network for different types of users, session intercepts for public service announcements, community alerts, and other location-based messages. It also enables multimedia applications and content-based services for a wide range of users.

Using a properly designed Operations Support System, service providers can provide the back office operational functions, including network monitoring, 24/7-customer service, billing and revenue distribution, reports, and system maintenance.

28

Hot Spots: Public Access using 802.11

James Keeler[a]

28.1 Introduction

A Wi-Fi "Hot Spot" is the common term used to describe locations that provide public Internet access using IEEE 802.11a,b or g wireless Ethernet, otherwise known as "Wi-Fi" (Wireless Fidelity). Wi-Fi uses a wireless Ethernet protocol that comes in various flavors including the most popular 802.11b/g, both of which operate in the 2.4 GHz frequency band of unlicensed spectrum. 802.11b can transmit/receive up to 11 megabits/second and 802.11g can transmit/receive up to 54 megabits/second. The new proposed 802.11n standard uses multi-path capabilities to transmit/receive at up to 155 megabits/second. 802.11a is a close cousin to 802.11g, with transmit/receive speeds of up to 54 megabits/second, but 802.11a runs in the 5 GHz unlicensed band. Most hotspots provide 802.11b and 802.11b/g (dual-mode capable) access – that is, the access points have radios that are able to function with either 802.11b or 802.11g connections. Some access points have tri-mode radios that run in 802.11a/b/g modes.

Wi-Fi has become extremely popular. Whereas 8 years ago, 802.11b was on the fringe edge of techno-geek toys with only a few thousands of devices sold, now more than 90% of the mobile computers shipped by Dell, HP and IBM (Lenovo) come with 802.11b/g built in (many also with 802.11a forming tri-mode radios). Hence, there are hundreds of millions of laptop computers with Wi-Fi built in. Those that do not have it built in can add it on with a PCMCIA card or USB drive device for less than $30.00. ABI Research reports that 200 million Wi-Fi chipsets were shipped in 2006, bringing the total to 500 million and that shipments will reach over 1 billion by 2012 [1]. Many enterprises have installed 802.11 networks in office buildings to enable employees to work more efficiently throughout the buildings. Moreover, millions of people have purchased wireless access points for their homes so that they can use mobile devices in their homes.

Since Wi-Fi is so common in the home and office, it is only natural that people who travel with their Wi-Fi enabled mobile computing devices would like to use these devices on the road. Thus, the Wi-Fi "hotspot" was born. Hotspots are locations that enable users to connect to the Internet via Wi-Fi at a public venue. These locations are commonly

- Hotels;

---

[a] *Wayport, Inc*

- Airports;
- Restaurants;
- Coffee Shops;
- Bookstores and other retail establishments.

The number of hotspots has grown substantially over the past several years. There are several databases that list hotspot locations. For example, as of this writing, Jiwire's database [2] lists 142,751 Wi-Fi locations worldwide. Skyhook also has a database of access points that it has compiled in several cities with over 7,000,000 Wi-Fi access points in the database. Many of the access points in the Skyhook database use secure connections that do not allow public use, and hence are not "hotspots", whereas many others have no access control and allow for anyone that happens to be in range of the access point to connect – these are typically from a home or small business that has connected a wireless access point to their cable modem or DSL line without locking out other users (via WEP or WPA), and may not have intentionally set up the access point to allow others to use it. Hence, we do not refer to these "accidental" open access points as "hotspots".

Of the sites that are more commonly known as hotspots, they come in several different flavors. The most important distinction is whether there is a charge for the connection or not ("paid" versus "free" hotspots). Some Wi-Fi service providers such as T-Mobile and Wayport charge for Internet access at some locations. The access is usually for a 24-hour period with prices ranging from $2.95 at a McDonald's restaurant to $14.95 at premium hotels in the US, with international pricing varying greatly (some places charge as much as USD40/day!). Bundled "all-you-can-eat" plans can be obtained from providers such as T-Mobile, Boingo Wireless®, IPASS, etc. that aggregate connections from multiple providers onto a single billing plan. Of the approximately 50,000 hotspots in the US, a large fraction of those are managed by T-Mobile (>7,000) and Wayport (>13,000). Some of the sites managed by Wayport provide free access, but the vast majority are paid hotspots.

The Jiwire database lists 17,296 free hotspots internationally (out of 142,751, or 88% are paid), and 13,411 free in the US (out of 50,053, or 73% are paid). The breakdown for types of locations from Jiwire is:

- Hotel/Resort 36,830;
- Restaurant 28,590;
- Cafe 19,036;
- Store/Shopping Mall 16,178;
- Other 13,878;
- Pub 5,918;
- Office Building 2,601;
- Library 2,370;
- Gas Station 1,741;
- Airport 1,679.

This list shows that the highest concentration of hotspots is in hotels/resorts. This is not surprising given that many business travelers carry Wi-Fi enabled mobile computers and spend a disproportionate amount of time in hotels compared to the other destinations. It should be pointed out that the counting of hotspots can be misleading. A single hotel may have over 1,000 rooms and may have hundreds of access points. A single hotel is counted in most databases as a single hotspot whereas a restaurant or café typically has only one access point. Similarly, airports typically have many access points (several dozen at Dallas Fort Worth airport, for example). From a use standpoint, hotels and airports are king. Wayport, for example, provides Internet access in approximately 190,000 hotel rooms in about 600 hotels. Use in hotels is many times larger than in the 8,700 McDonald's restaurants that it serves, whereas the number of hotspots is disproportionate – 600 to 8,700.

The number of hotspots deployed over the past several years has increased dramatically. As one example, view the deployment of hotspots in hotels, airports and retail establishments by Wayport shown in Figure 28.1.

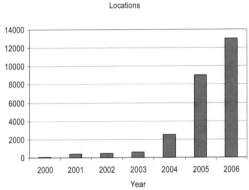

Figure 28.1: Growth of hotspots provided by Wayport as of June 30 of each year. Note that the hotspots in the early years were mainly hotels which may have hundreds of rooms whereas in later years they were mainly retail establishments with a single access point.

Figure 28.1 shows the rapid growth in hotspot deployment over the past several years. In the early years, very few customers used high speed Internet access (at a hotel, the use rate was less than 1% of customers in 2000). However, as more people demanded broadband connections and the number of locations increased, the use rate correspondingly increased. Now, the use rate is well above 20% in many hotels and as high as 70% at some locations. Figure 28.2 shows the (non-cumulative) quarterly growth of uses (each use is a single 24-hour paid for connection at a site).

28.2 Access Control

Another very important differentiating factor in hotspots is whether the hotspot provides access control or not. All paid hotspots have access control – that is, a user is not allowed to gain Internet access until the appropriate payment is made. Some free hotspots also provide access control. This can be via a simple password or simply by accepting some terms and conditions. Other free hotspots are "wide open" meaning that anyone can use the network without restriction. In this case, no authentication or control is enforced. These "wide open" sites can present several problems (as discussed in more detail below).

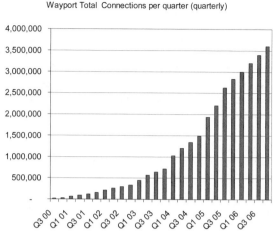

Figure 28.2: **Non-cumulative quarterly use at Wayport-provided hotspots on a per-quarter basis. Two factors are at work here, use rate (percent of people using the network) increased, and the number of locations also increased.**

Access control is usually done via a gateway that controls access to the Internet via a firewall rule set. In the case of access control, the process of getting connected at a hotspot goes like this:

1) The user's Wi-Fi network interface card associates with an access point by selecting an SSID (service set identifier) from a list of available SSIDs.
2) If the user is configured for DHCP, the computer obtains an IP address from the gateway (usually this is a private, RFC1918-space address)
3) If the device is configured for a static IP address, the gateway will masquerade the address by performing network address translation of that address to another address.
4) DNS servers are assigned.
5) The user opens a browser going to their homepage.
6) The gateway detects Web proxy settings and listens for an http request on proxied ports.
7) The firewall prevents the connection to the homepage and performs a 302-re-direct to a "splash page" (also called a forced first page).
8) The user interacts with this Web page to either purchase or otherwise accept terms and conditions.
9) After the appropriate access/payment credentials and/or terms and conditions are accepted, the firewall rule set is changed to allow access to the Internet at large for the time period purchased (typically 24 hours). Some sites charge per minute.
10) If the user wishes to use e-mail, many gateways provide transparent SMTP proxy to an SMTP e-mail server, or provide instructions on how to set up e-mail to use the server.
11) If the user wishes to use a VPN, they are provided an option (usually on the first page) to use a public IP address rather than a 1918-space private address because some VPNs do not function well with network address translation.
12) The access control is usually done via the MAC address of the network interface card. Thus, if the user changes locations (e.g., goes to a different part of the building), the

session continues without interruption. Some gateways also use cookies to determine that the user has a valid session in case they switch from wireless to wired connection (or vise-versa).

13) At the end of the session, the firewall rule set is set back to the default, preventing the user from accessing the Internet until another session is purchased (some gateways offer multiple contiguous sessions or an option of continuing the current session).

Note that this is a fairly complex set of requirements. There are many gateway providers that have figured out this "plug and play" connection process to make it relatively easy for customers to use their computers at a hotspot.

In the case of no access control, there is no firewall and no Web-browser interaction. The user gets DHCP (no masquerade for Static IP) and is able to use the Internet without any browser interaction. This is usually done with a consumer-grade access point with 1918-space (private) IP addresses. In the case of paid hotspots, the payment methods vary from provider to provider, but the most common are:

- Credit Card (to a secure, https server);
- Payment to hotel folio (via room number and name);
- Membership with the hotspot provider (via username and password);
- Coupon (promotional or handed out by the venue staff);
- Prepaid card (discounted pre-paid sessions);
- Amenity – included with some membership in a hotel frequent traveler or other member program;
- Roaming – using username and password from another provider to access at the local site.

Each of these methods usually requires a connection to different billing servers that communicate with the gateway on property. In the case of roaming, the username and password credentials are usually entered in a Web page tied to a RADIUS proxy server that communicates with the user's RADIUS realm server which transmits an authorization accept or deny message back to the provider's RADIUS proxy server which then transmits the session accept or deny message back to the gateway. One advantage of using roaming systems is to allow for aggregation of multiple providers under a single billing statement. For example, IPASS, Boingo Wireless®, and Fiberlink aggregate access at hotspots from many different hotspot providers and the user can use the same username/password at all sites with a single bill at the end of the month. Several telecommunication providers have tied this capability to the user's phone or DSL subscription (e.g., Sprint, AT&T, T-Mobile, etc., allow the user to use many hotspots with the bill showing up on their monthly phone bill). Many of these providers have an "all you can eat" plan allowing unlimited use at many different sites for a fixed monthly payment. In some cases, there may be an up-charge for roaming at certain sites.

In addition to aggregation of hotspots and simplified billing, most aggregators also provide a client for the user to connect to a hotspot. These clients provide a list of hotspots, and, in essence, function like a Web robot as if the user were entering the credentials into a Web page). Some of them provide a VPN connection or automatically launch a VPN to enhance security.

28.3 New Trends

There are many new trends in Wi-Fi that impact hotspot providers and users. Some of the major trends are:

1)   Proliferation of Wi-Fi devices;
2)   Multi-purpose access points;
3)   Municipal Wi-Fi;
4)   Free Wi-Fi/ Advertising;
5)   WiMax.

This section will discuss some of the salient features and challenges presented by these trends.

28.3.1 Trends in Wi-Fi Enabled Devices

Whereas the hotspot industry grew mainly serving notebook computers, the Wi-Fi industry is rapidly changing. Wi-Fi is being incorporated into hundreds of new devices including:

- PDAs (Palm™, Trio™, iPac™, etc.);
- Gaming devices (such as Nintendo DS™, Sony Playstation Portable™, etc.);
- Digital cameras;
- Media players (iPod™, Zune™, Gigabeat™, etc.);
- Dual mode phones (Wi-Fi and cellular combined phones);
- Wi-Fi enabled phones (e.g., using Skype);
- RFID tags.

One of the biggest problems that hotspot service providers and Wi-Fi-enabled equipment manufacturers face is how to get these devices connected at hotspots. If the device has a browser, then the browser-based connection process described above can be used (albeit very awkwardly unless the device has a keyboard). However, if the device does not have a browser, there is no way for the user to interact with the gateway and perform the Web-based acceptance function. Especially in a paid hotspot, if the provider allows a Wi-Fi device to access the network without using the Web-based authentication and payment method, then it would be relatively easy for other customers to get around the Web-based method and spoof system to avoid payment.

One way of getting these devices to work is for the hotspot providers and equipment manufacturers to work in coordination to enable the devices to work at hotspots. One excellent example of this collaboration is the work done by Wayport, Nintendo, and McDonald's to enable the Nintendo DS Wi-Fi enabled games to work at all of the McDonald's Wayport hotspots. This was actually a fairly complicated integration project, but took only 3 months to deliver to market (August 15 to November 15, 2005). As with any device that attempts to access the Internet at a McDonald's network, a firewall rule is in place that blocks the access. Wayport engineers worked with Nintendo engineers to develop a system that would recognize and authenticate the device on the network. A major constraint of the design was that it had to appear seamless to the user. All the user has to do is select play an online game from the game menu. The Nintendo DS game software was modified to recognize a McDonald's site by searching for the standard Wayport_Access SSID. Wayport set up another non-broadcast SSID for the Nintendo device to connect. After association with this SSID and on receiving a DHCP IP address, the DS makes a HTTP request and receives a response that contains a URL to post validation credentials to. The DS then makes a request with this information to a central server that validates the DS device and posts the validated credential response back to the gateway. The gateway accepts the response and modifies the firewall rules to allow the device out onto the Internet.

Note the similarity of the DS authentication system to the roaming authentication described above. In fact, it is nearly identical. This is one way to allow these devices onto the network, but requires a central authentication server. Clearly this could be turned into a standard method for other device manufacturers to provide access for their devices as well. Wayport currently has plans to develop a freely available software development kit and specification so that any manufacturer can add authentication to their device. Note that the business terms have been left out – if this is a paid hotspot the provider may work with the manufacturer to develop business terms for allowing the device onto the network (e.g., pay some amount per use or pay a flat fee).

Whereas the authentication described above works well for certain devices, other devices have no simple means of performing authentication with a central server. Access control for these devices is much more difficult (see for example [3]), but various methods are being proposed for recognizing the devices and granting access based on information available on the device.

28.3.2 Trends Multi-Purpose Access

Many higher-end commercial access points are capable of supporting multiple SSIDs. Each SSID can be tied to a different VLAN and can have different security settings for each system. For example, some access points with an 802.11b/g radio can allow for up to 16 SSIDs, each SSID can be independently set to broadcast or non-broadcast mode, and each SSID can be tied to a different virtual LAN. This allows venues to deploy a single wireless network and use it for many different purposes. For example one SSID may be configured for public access tied to a public VLAN with no encryption or security whereas a second SSID may be set to non-broadcast mode with encryption and security (either via WEP or WPA/WPA2). Examples of applications for this multi-purpose configuration abound. Hotels commonly are configured with public access on one SSID and a private network on the same set of access points to enable VoIP devices such as Vocera or Spectralink. Airports may run a private network alongside a public network. Some service providers enable an open public SSID along with a public secure connection via WPA on another SSID. Hospitals use these networks for several applications including VoIP, asset tracking, and secure communications for hospital equipment. McDonald's currently is configured with 4 different SSIDs each with different purposes (3 public plus one private plus several different wired applications running alongside the wireless applications on separate VLANs). Configuration and management of these multi-purpose networks requires some serious network engineering and firewall rule configurations to ensure that the applications perform properly, have the correct QoS, and the correct security. The benefits of such multi-purpose installations are obvious: Instead of creating multiple networks for the multiple applications, a single network can serve multiple purposes (this also minimizes RF pollution from having too many access points in a single venue).

28.4 Architecture

The architecture of Wi-Fi hotspots is different for different venues. Hotels typically will have both wired and wireless access provided to guests, whereas airports will be wireless only. Retail establishments such as McDonald's, Barnes & Noble, Borders, Starbucks, etc. provide wireless access only. Typical architectures for hotels and airports thus differ from the architecture in a retail establishment. Hotels and airports require a much more extensive 802.3 (Ethernet) switchplant to distribute the access to the rooms as well as to distribute the connections to the access points (in some cases, hundreds of access points are required to cover a large hotel). Typically, these access points use power over Ethernet to provide power to the device so that a local power connection is not required at the site of the access point. Note that the length of the cables between an access point and an Ethernet switch is typically shorter than 100 meters, so switches need to be distributed in equipment closets (IDFs) throughout the facility. Figure 28.3 shows a typical architecture in a hotel or airport (airports lacking the wired connections in the rooms). It also shows the off-property connection to the Internet.

In some hotels private applications are also supported (such as local VoIP, environmental controls, etc.), and some airports have multiple wireless connections running on different VLANs.

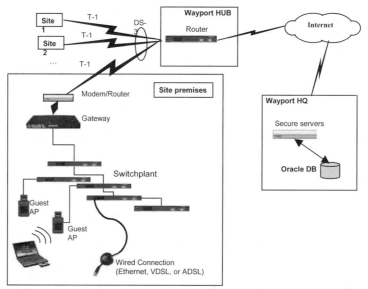

Figure 28.3: Typical network architecture at Wayport-provided hotels and airports (airports lacking wired connections). Access points are connected through a standard Ethernet switchplant to a head-end switch and to a gateway that provides access control.

Retail establishments range in complexity from a simple single access point to much more complex applications involving multiple wireless services and multiple back office services. As an example of a typical sophisticated multi-purpose retail deployment, Figure 28.4 shows an architecture that supports four wireless applications and several wired applications for the back office (private applications). In this case, the gateway provides quality of service and firewall policies to ensure that the private applications run correctly. Some of the back office applications that are common are transport for credit/debit card, video surveillance transport, manager computer connection, electronic learning system connection, VoIP, etc. This example also shows a private VPN (IPSec) and connection to the corporate datacenter. Also in this example, one of the wireless connections supports a Wi-Fi device (Nintendo DS™) that does not have a browser. Authentication for the device is provided by custom code on the device and the gateway. Note that this is a very complex network with different QoS and security requirements for each different application.

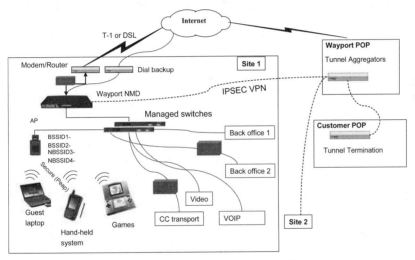

Figure 28.4: Architecture of a retail establishment showing multi-purpose application support. In this example, four different SSIDs are supported (three public applications, one private), and an IPSec VPN tunnel with aggregation and transport to the corporate datacenter shown.

28.5 Trends in Municipal Wi-Fi

One of the fastest growing trends in recent years has been the deployment of citywide Wi-Fi networks, otherwise known as municipal Wi-Fi. Typically these networks are set up with mesh-enabled access points that allow for wireless backhaul of the signals to several root nodes within the network. The access points of the mesh networks are usually mounted outside (e.g., on city light or power poles). Several major cities have announced plans for citywide deployment including Philadelphia, Anaheim, San Francisco, etc. Earthlink has been quite active in the muni-Wi-Fi business sector and offers Wi-Fi broadband connections to subscribers in the city on various pricing plans with discounts to the city for network access in exchange for access to light poles. Whereas the eventual plans for deployment will take more time to roll out, the initial 6 square mile pilot was deployed in Anaheim last summer with eventual plans for covering 50 square miles. In the initial deployment, some of the issues with propagation of Wi-Fi signal in the hills of Anaheim surfaced as difficulties that would have to be overcome. One of the problems that has surfaced has to do with the difficulty of getting decent signal-to-noise ratio of the Wi-Fi signal from outside to the inside of a building.

Several cities have jumped on board with muni-Wi-Fi plans. Tropos announced in January [4] that the 500th city to use their mesh solution had signed a contract. It should be noted, however, that whereas there are major new initiatives being signed every month, the initial results of these deployments are mixed with even some initial success stories reporting ongoing troubles [5]. Moreover, there are several legal issues regarding the muni-Wi-Fi systems being deployed. For example, Pennsylvania passed a state law requiring a waiver from local broadband providers (with Philadelphia grandfathered in). This has been challenged at the Federal level with the Communications, Consumer's Choice, and Broadband Deployment Act in the Senate and a similar bill with different wording proposed in the House of Representatives. It is unclear at this time what the future holds on the legal, business, and technical front for these networks.

28.6 Trends in Advertising on Wi-Fi Networks

One of the most interesting business models that has been proposed is to fund access to Wi-Fi networks via advertising. The basic idea is similar to how radio and television programs are funded through advertising – sponsored advertisements could be presented to a user (e.g., on the splash page of a web-browser connection), and this advertising revenue stream can be used to fund the Wi-Fi network. Moreover, seeing as a Wi-Fi signal can be localized, the advertising can be targeted towards local businesses. Imagine being at a hotel and seeing an advertisement for a restaurant 3 blocks away. Google and Earthlink have teamed up in San Francisco to pilot this kind of system. It should be noted, however, that there are several technical challenges to be overcome. One clear issue is how does this kind of a business model work for non-browser-based access as described above? Also, this is not really a new idea and there are several patents around local advertising and services in a Wi-Fi environment [6].

28.7 Trends in WiMax

In addition to Wi-Fi networks, WiMax (802.16d) is a new technology that can run over private or public RF channels. The new mobile WiMax standard (802.16e) is also potentially disruptive. Some have even gone as far to say that WiMax may be the death of Wi-Fi. Whereas WiMax (in the 802.16d version) is an excellent last-mile connectivity alternative to DSL, cable, T-1 or other broadband methods of distribution, it is not well suited for mobile computing access. It is an excellent alternative for muni-Wi-Fi deployments, and may present a strong competitive alternative to these networks. 802.16e is designed to address the mobile computing issues, but it has a long way to go to catch the economic curve of 802.11 with 500 million chipsets already in production and costs decreasing every year.

28.8 Hotspot dangers and issues

As with many things on the Internet, there are dangers associated with hotspots. Some of the dangers are the same as you would find, for example, with a cable modem at your house, and some are particular to hotspots. In this section, we discuss some of the dangers and discuss some steps that the user can take to protect himself/herself at an access point, and in other cases discuss the steps that the hotspot provider should take to protect the customer. The items discussed in this section are:

- Phishing;
- Fake hotspots;
- Worms, viruses;
- Snorting;
- Port isolation;
- Preventing network degradation;
- DMCA;
- CALEA.

28.8.1 Phishing

One of the most troubling aspects of hotspots is phishing. This is a scam set up by nefarious "hackers" to try to steal identities or credentials from users. The basic idea of phishing is that a hacker will set up an artificial website that pretends to be a legitimate website. A couple of years ago, one of the most famous phishing schemes created a fake website for major credit card or bank sites. To work the scheme, the hacker would then send out spam e-mail to millions of users informing them that there was some kind of problem with their account and to click on a link to go to the company's website to correct the problem. The hackers would spoof the "from" field of the e-mail so that it look as if it came from the bank or credit card agency. The unsuspecting user would click on the link and think that the site was actually their real site. I received dozens of these spam e-mails and I clicked on the links just to check them out. Indeed they went to sites that looked and behaved just like the real site. The sites even had URLs that looked like a legitimate site and https signed certificates that looked legitimate – this is one of the holes that certificates and DNS entries have – they can have substitute character sets that look just like the legitimate characters (DNS, in particular has many other holes, but we do not have the space to talk about that here). Unless the user is particularly careful, the user may enter their credit card or account information in hopes of correcting the problem with the account. If they enter this information, the hacker has stolen their identity (at least for that account). This whole process is known as "phishing". It is a criminal and fraudulent activity and it is a worldwide problem.

- In the case of hotspots, the phishing scheme is slightly different. In this case, a hacker can set up a fake SSID and pretend to be a legitimate provider. The hacker will configure a computer-to-computer ad-hoc network to appear as a legitimate hotspot (e.g., by broadcasting "tmobile" or "Wayport_Access"). The user may associate with this SSID, and the hacker then fakes the response to look like the legitimate site by mocking up Web pages similar to the legitimate provider's site. For a detailed analysis on how this is done, see [7]. Just like in the normal phishing scheme, the hacker can then gain identity information from the user.

How can a user protect himself/herself from these schemes? Just as in any phishing scheme, only enter sensitive information on an https page with a signed certificate from a legitimate signing authority and check that the certificate is issued to the company that is the hotspot service provider. Another way is to use a roaming client that will only allow you to connect at sites that are in the database of recognized sites (this is done by checking the MAC address of the AP that the user is associating with).

It should be noted that this kind of fake hotspot is a local attack – meaning that someone has to be within 300 feet of you to make it work (or with a high-gain antenna at a longer distance). From a threat standpoint, it is relatively minor in the world of phishing – by setting up a fake bank site a hacker can have potentially millions of unsuspecting customers go to that site, whereas with a fake hotspot, only a few people a day that wander into range can be trapped. The economics favor the mass phishing schemes, hence, they are much more prevalent.

28.8.2 Fake hotspots

A close cousin to the phishing scheme is for a hacker to set up an SSID such as "freeWiFi". This is easier to set up, and many people fall for it. Rather than faking out a provider's Web page, the hacker can connect via NetBios or other protocols to the customer's computer to take advantage of many vulnerabilities in the operating system, or, in some cases, scan the computer because the user's computer allows file sharing (yes, it happens!). The hacker can then launch man-in-the-middle attacks on the user, and/or install worms/viruses/Trojans on the user's computer. How to prevent this?

1) Do not connect to a non-legitimate hotspot;
2) Run a personal firewall that blocks NetBios and other known vulnerable ports/protocols;
3) Turn off file sharing (or at least only allow sharing for known, legitimate users);
4) Install all relevant operating system updates and security packs;
5) Install and run an anti-virus program and keep it up-to-date.

28.8.3 Worms and Viruses

Anytime you connect at a public hotspot, you may be getting more than you bargained for. There are many different worms and viruses that have been created over the years that can easily propagate in a non-protected environment. The worms will scan all ports and/or IP addresses looking for other computers on the local subnet. If it finds another computer it will try to install itself using one of the operating system vulnerabilities. Some famous past examples of this are the Sobig.f and SQLSlammer worms. The hotspot provider should configure firewalls and port protection to prevent this, but the user can protect himself from this as well by following the same steps 1-5 listed above. Note that the worms and viruses also present a tremendous problem for hotspot providers in that these critters can send out an enormous amount of traffic on the network. One computer can spoof thousands of IP addresses scanning 64k ports for vulnerabilities. If other computers become infected they can cause an unintended denial of service attack for any legitimate users on the network. Sophisticated hotspot providers have hardened their systems against these types of attacks but the casual hotspot may be a much more dangerous place.

28.8.4 Snorting

So you are a user at a legitimate provider's hotspot and have followed steps 1-5 above. You are safe, right? Wrong. You are not safe. Hotspots have the added vulnerability that the packets are flying through the air rather than down a cable. That means that anyone in the near vicinity may be listening. There are several public-domain programs such as AirSnort (hence the term snorting) that allow someone near another user to listen in on all of your traffic (it is like having a conversation on a party line, or similar to the un-encrypted old analog cell phones). These snorting programs allow someone to see everything that you are doing on the Wi-Fi connection. How to prevent this? You cannot. But you can make the information useless by encrypting it. Either use a VPN or only enter sensitive information over encrypted applications such as SSL (https). Some hotspot providers take care of this for you by offering WPA connections (an encrypted Wi-Fi channel) but this is not offered by all providers (and very few users actually use this as it is considered difficult to set up). Thus, we have a 6th step that the user can take to protect himself:

6) Run a VPN or use secure applications for sensitive information (encrypted e-mail, SSL, etc).

28.8.5 Port Isolation

One of the things that hotspot service providers can do to minimize some of the problems with worms/viruses/hackers is to configure access points with port isolation. This configuration builds a logical relationship between the bridging of a wireless port connected with a wireless client (PC) at the MAC layer and prevents any packets from one wireless client to be transmitted to another client. This prevents worms/viruses from propagating across a network and also prevents file sharing from device to device. That is, without port isolation, the wireless client connections to the AP act as a hub, whereas with port isolation they act as a switch and the packets only go to the intended device (even broadcast packets are limited to the intended device).

Sophisticated hotspot providers configure access points with port isolation, but many lower-end sites do not provide this feature (and older access points do not support this feature). It should also be pointed out that if the access point is configured for multiple SSIDs and for multiple applications, the configuration can become quite complex. For example, in the case of one SSID configured for public access (blocking broadcast packets between clients) and another SSID is configured for local VoIP (allowing broadcast packets), the network configuration of the switches becomes a non-trivial network engineering exercise in the case of multiple access points.

28.8.6 Peer-to-peer file sharing

Another class of applications that can negatively affect the performance of a network is peer-to-peer file sharing applications such as BitTorrent, Kazaa, LimeWire, etc. These "p2p" applications are typically used to download movies, MP3 files, and other content by using a peer-to-peer program installed on a customer's computer. The p2p programs will seek out other p2p computers on the network and serve to download as well as upload files. Depending on the demand for the files and the number of clients connecting, the p2p programs can increase the bandwidth use at a fantastic rate until all available bandwidth is utilized. Figure 28.5 shows an example of a single BitTorrent session using bandwidth at a hotel.

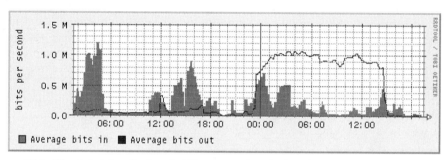

Figure 28.5: Chart showing the bandwidth use at a hotel wherein a BitTorrent session started around 0:00. This session rapidly transferred files upstream consuming over 1 Mbps. The session was ended about 12 hours later.

From Figure 28.5, one can see the affects on a property with low utilization due to few users. In this case, there was no negative impact on the site because the site still had available bandwidth. However, if many users happen to be using p2p at the same time, the bandwidth demands can be overwhelming, leaving little available bandwidth for other applications.

28.8.7 DMCA

Not only is a service provider responsible for making sure that the bandwidth available to customers is well managed, there are legal requirements to prevent unlawful sharing of files. Service providers are responsible for compliance with the Digital Millennium Copyright Act (DMCA) of 1998. This law requires service providers to take steps to protect copyright-protected digital files. Examples of these files are movies, songs, and other digital content that movie production or music production sites wish to protect from free file sharing. Upon receiving a DMCA violation notification, the service provider has a window of time to stop the file sharing and to obtain a "safe haven" to avoid DMCA fines. This can be done by blocking ports via firewalls, or blocking the user from using the network in extreme cases. The fines for DMCA violations can be substantial (tens of thousands of dollars or more). Wayport responds to dozens of these notifications per week.

28.8.8 "Whack a mole"

Early methods for suppressing the negative effects of p2p programs involved firewall rules and bandwidth management policies for blocking or limiting p2p program traffic. For example, Kazaa and BitTorrent run over well-known ports, and firewall rule policies can be used in conjunction with a bandwidth management policy to limit the traffic over those ports to a maximum (e.g., 256kbps), or to block the traffic altogether. Some of the standard ports used by common file sharing programs are listed below:

```
Gnutella    : 6346, 6347, 5634
Napster     : 8875, 4444,5555,6666,7777,8888,6600-6699
Kazaa       : 1214
eDonkey     : 4661, 4662, 4663, 4665
Aimster     : 5025
Soulseek    : 2234-2239
BitTorrent  : 6881
```

However, smart users can modify the programs to use different ports. Thus this can become a game much like the "whack-a-mole" game found in many arcades wherein a mole pops out of a hole and you whack it on the head, but then it pops out of another hole. After some time playing this game, smart users eventually realize that they can use port 80, the port that is commonly used for standard Web traffic and the game is over – service providers cannot block port 80 without negatively affecting all Web applications. Clearly service providers must find a better way of managing this traffic to protect legitimate applications. Fortunately, there is a new technology available that is capable of getting around this game. The next section describes this solution in detail.

A more sophisticated approach is to do Layer-7 packet inspection, categorization and bandwidth shaping. In this approach, the traffic is categorized and all peer-to-peer file sharing traffic is put into a shared bucket with limited bandwidth. This is a very good solution to the problem in the sense that it still allows users to run the peer-to-peer applications but prevents the applications from consuming the entire bandwidth. Note that at a casual hotspot that has no service provider, the network is wide open. There is typically no management of bandwidth or of the device configuration. However, Wayport has incorporated Layer-7 packet categorization and shaping in a hub architecture. There are several vendors that provide packet inspection and traffic categorization (without any endorsement of a particular product, see for example, Packeteer http://www.packeteer.com and Pocera http://www.proceranetworks.com). All of these devices function on similar principles in that they inspect the traffic at Layer 7 based on telltale signatures in the application protocol. From these signatures, they can classify the traffic into many different categories (e.g., e-mail, Web-browsing, streaming video, BitTorrent, etc.). Once the traffic is categorized, the traffic can be put into different quality of service "buckets", e.g., one can limit the total amount of peer-to-peer traffic to be say 1/8 of the total circuit bandwidth, or suppress it entirely. This approach can be applied in a hub-based method, or at the gateway at the edge.

28.8.9 CALEA

In addition to DMCA, the Communications Assistance for Law Enforcement Act (CALEA) requires service providers to cooperate with law enforcement agencies in the case of a warrant or subpoena for information. The new regulations that go into affect on May 14, 2007 require assistance by ISPs for VoIP traffic and other Internet data. These new regulations are quite complex to comply with and require sophisticated network infrastructure at the facilities based telecommunications level. Coordination between telecommunications providers and service providers will likely need to coordinate with each other to meet these requirements.

28.9 Summary

In summary, there is a wide variance in hotspots from size (single access point at a café to hundreds of access points at a large hotel) to configuration (with or without access control). The number of hotspots has increased tremendously over the past five years and with the number of Wi-Fi devices increasing to over a billion in the next 5 years, it is expected that Wi-Fi access via hotspots and hotzones will also increase. Nevertheless, the future holds many challenges for hotspot service providers as the types of devices change from laptops to devices without browsers. New connection methods and business models will emerge in the hotspot environment to enable these devices to access the Internet at different sites. New technologies such as 802.11n, WiMax and others will surface and change the landscape of how hotspots will be deployed. Location-based services and advertising are expected to play a major role in the future of hotspots. In addition, hotspot service providers will face new challenges to providing quality of service and protecting customers at the hotspots.

28.10 References

[1] Feb 13, 2007 http://www.abiresearch.com/abiprdisplay.jsp?pressid=809
[2] Feb 23, 2007 http://www.jiwire.com/search-hotspot-locations.htm
[3] http://www.muniwireless.com/article/articleview/5718/1/6/
[4] http://www.tropos.com/news/pressreleases/2007_01_31.php
[5] http://www.tmcnet.com/usubmit/2006/06/29/1700685.htm
[6] http://patft.uspto.gov/netacgi/nph-
     Parser?Sect1=PTO2&Sect2=HITOFF&u=%2Fnetahtml%2FPTO%2Fsearch-
     adv.htm&r=9&p=1&f=G&l=50&d=PTXT&S1=Wayport.ASNM.&OS=AN/Wayport
     &RS=AN/Wayport
[7] http://www.ethicalhacker.net/content/view/66/24

29

Strategies for Maximizing Access to Public Commercial Hot Spots[a]

With all the ongoing commotion about Wi-Fi access gravitating towards free since the inception of the hotspot industry, it might come as a surprise that most networks in high profile locations remain for-fee, commercial hotspots. Among the top airports, fewer than 10% have chosen to deploy free Wi-Fi. Even within the airline club rooms, most of the clubs charge for service – or outsource the service to an operator who charges for service.

Similarly, outside of the business economy hotels (think about the second and third tier offerings from the prominent hotel brands), Wi-Fi access in major hotels remains fee-based, with some frequent traveler programs subsidizing this service for their members. Some hotels are returning to fee-based access after enduring what they consider a failed experiment in providing free access.

While there will always be a place in the market for free Wi-Fi, the explosive growth and exponentially increasing usage of fee-based Wi-Fi demonstrates the viability of a market that was projected to be on its last legs in 2003.

In order to capitalize on this opportunity, it is best to assess the factors that will maximize the chances of success, which generally means generating meaningful revenue. Fee-based Wi-Fi depends heavily on business travelers – users with a real need for Internet access – who are willing to pay for service.

In this regard, the first rule of real estate applies to commercial hotspots as well, "Location, Location, Location." Airports are the dominant locations for usage, followed by hotels. Cafes, restaurants and other retail and public facilities are well behind the leading location categories for usage, perhaps with the exception of a massive chain of coffee shops that are sufficiently ubiquitous as to be convenient for business travelers (who also have a caffeine addiction to feed). Without the global presence that comes with massive corporate expansion, restaurants and coffee shops have a diminished chance of generating significant revenue, though they can generate enough to be worthwhile endeavor.

29.1 Retail Service Offerings

To be a commercial hotspot, there must be a fee. In order to maximize revenue, different products should be offered at different price points. Depending on the location, users may

---

[a] *Boingo Wireless, Inc®*

connect for 20 minutes or several hours. The average session in an airport is about an hour. For hotels, the average sessions are several hours in length, or more typically multiple 1-hour or 2-hour sessions over the course of a stay (after check-in, before bed, prior to check-out in the morning). Cafes run the gamut from 15-20 minutes to several hours; in large part it depends on whether it is a location that servers business travelers – with a quick in and out – or local residents using the service as DSL replacement.

29.1.1 Day passes

The industry standard for access in North America is the day pass. This is typically 24 hours of access in a hotspot. The "day pass" makes a tremendous amount of sense in a hotel, where the user will typically be present for 8 hours or more. In an airport, where the length of stay is considerably shorter, 24 hours seems unnecessarily long. However, network operators who control numerous hotspots with broad distribution frequently define the "day pass" as 24 hours on the entire network – Internet access in all associated hotspots within that time period. That way a day pass is good at your departing airport, arriving airport and your hotel at the destination – as long as all connects occur within the same 24-hour period.

29.1.2 Time-based passes

In Europe and Asia, it is more common to see time-based Internet access, such as 1-hour, 2-hour or 4-hour passes. These tend to be convenient lengths of time for users, but are typically priced more in line with North American day pass rates. It is common that a "day pass equivalent" (24 hours) in Europe will cost two or three times the going rate in North America. In many instances, there is no pre-set duration. Instead, fees are charged on a per-minute basis. This ultimately proves more expensive than the day pass, especially in locations with longer sessions, such as hotels. As Internet access has historically gravitated away from metered access (per-minute), time-based passes too may run their course for the majority of hotspots within the next couple of years.

29.1.3 Subscriptions

Increasingly, Wi-Fi users are turning to monthly subscriptions for their Internet access. As the number of hotspots continues to proliferate, users are looking for more economical means to access the Internet over Wi-Fi hotspots on an ongoing basis. This is especially true for business travelers who access multiple times per business trip – hotels, airports, cafes, convention centers, train stations and other public locations. In general, monthly subscriptions can be had for two to five times the cost of the average day pass. Keys to finding a subscription that saves the user access fees is to identify a network operator or aggregator, whose service provides you with access in locations where you frequently travel. Some operators tend to specialize in airports, some in hotels and others in coffee shops or other specialty category. Frequently, aggregators will sign roaming access agreements with these varied network operators and can provide the largest number of access options.

29.1.4 PIN-based or Pre-paid Access

Some operators find it useful to provide Pre-paid access via PIN-based mechanisms (e.g., purchase 5 connects in advance and use a card-based PIN to log in 5 times). This is more common in Europe and Asia than it is in the Americas. For broad-based pre-paid access, there are infrastructure issues to address in the context of generating tens or hundreds of thousands of cards with registration PINs that would then need to be distributed through retail channels to be available to users.

This functionality can also be used by the network operator to provide guest access to users or to provide promotional giveaways for "One Free Connect." If you use the PIN-based logins for promotional purposes only, you don't face the same production and distribution issues that are inherent to a pre-paid access business.

29.1.5 Private services

In many large-scale venues where multiple concessionaires provide category-specific services, it is frequently incumbent on the Wi-Fi service provider to also provide Internet access services to other concessionaires. These access services can be used by other concessionaires for point-of-sale support, corporate backend services or venue-specific private services (e.g., security video feeds).

29.2 Neutral Host Network Configurations (Roaming)

The capital expenditure and operating expenditure required to deliver enterprise class services outlined above is not trivial. In order to maximize revenue to support those expenditures, network operators need to think beyond the services provided directly to various end-user groups, and ensure that they can provide services indirectly to much larger groups of end-users through roaming agreements with other network operators and aggregators. This is most commonly referred to as a "neutral host network", and is nearly always mandated as an operating practice in public-run venues such as airports, train stations and city- and state-owned facilities.

29.2.1 Neutral Host Overview

The fundamental principle behind neutral-host networks is that the hotspot operator endeavors to provide access to the end users of numerous other hotspot operators or aggregators. This is achieved by implementing network access agreements and performing authentication integrations with the partner providers. These network access agreements are frequently bi-lateral, wherein the users of one hotspot operator can roam onto a partner network's hotspots and vice versa.

These agreements are most fruitful when both parties have network assets that are of interest to each other. Alternatively, operators unwilling to grant bilateral roaming or aggregators without network to trade are at a disadvantage and are frequently subject to minimum fees for the right to access this network, because they provide no direct benefit back to the network operator allowing access beyond generating connect fees from their users. Because of the real cost of this integration and ongoing maintenance, this kind of roaming access is not open to all comers as the ongoing fees for access would be cost-prohibitive to smaller companies.

29.2.2 Universal Access Method (UAM) Roaming

The most common point of integration for roaming access via a neutral host network is via the "walled garden", which is the controlled set of web pages that are served up in a hotspot by the access controller. Until a user successfully logs in, the access controller restricts the user's access to any Internet pages beyond the local pages in the walled garden. Using a walled garden to control logins is referred to as UAM authentication. Everyone with a Wi-Fi connection and an Internet browser is able to log in to the hotspot.

For roaming partner access, the walled garden typically has a login section for "Roaming Partners" or "Roaming Login." This can be a link that takes the user to a separate login page where they can choose their roaming provider (home Wi-Fi account provider), or a pull-down menu with a list of the roaming partners that allows the user to login with their home account credentials (username and password) on the walled garden itself.

29.2.3 Smart Client Roaming

Many network operators or aggregators also have smart clients – software applications that manage the complexity of a highly-fragmented public Wi-Fi space. They recognize networks with which the user's provider has roaming access agreements and completes the login for the user with minimal interaction. This takes the onus off the user in recognizing partner roaming networks and knowing how to navigate the "Roaming Partner" links within the myriad walled gardens they might encounter. Smart clients greatly simplify Wi-Fi roaming for the end-user.

The challenges of Smart Client Roaming from a neutral-host perspective is that it involves additional functional integration and testing, such that the smart client is able to recognize the partner network and can complete an authentication against the hotspot through inherent knowledge it has or has acquired through software updates. The most common methods of performing smart client are through de facto industry standards for smart client roaming. The two most common are WISPr and GIS. Outside of these two "standards", there are countless one-off approaches that require significant flexibility in smart clients to be able to adopt and implement custom login methods to complete a roaming authentication.

29.2.4 Smart Client Roaming "Standards" – WISPr and GIS

The "standards" for smart client aren't really standards, but recommendations for best practices that have been widely adopted throughout the industry. This has greatly simplified smart client integration efforts; however, it presents its own challenges. Since the specifications are recommended best practices and not actual requirements, there are almost as many varied implementations of the "standards" as there are hotspot operators who claim to be standard-compliant.

29.2.4.1 WISPr

Developed and maintained by the Wi-Fi Alliance, WISPr is the most comprehensive set of specifications for enabling smart client roaming through industry-supported best practices. It covers not only the authentication aspect, but the backend tracking perspective as it defines which attributes are to be shared through the RADIUS traffic that completes the login.

29.2.4.2 GIS

GIS is effectively a subset of the WISPr recommendations that was published for public consumption by iPass. Whereas WISPr includes nearly comprehensive recommendations for numerous aspects of the authentication, authorization and accounting exchanges, GIS is more like a lowest common denominator – the simplest way to implement smart client roaming.

29.3 Provisions for Enabling Devices

Beyond using neutral-host strategies to ensure the widest possible user base from which to draw and generate revenue to cover capital and operating expenditures, it behooves the network operator to also ensure that the maximum number of devices possible can access the hotspot.

29.3.1 Wi-Fi Enabled Device Overview

While the developing years of commercial Wi-Fi access have been driven by laptop computers, the future growth will be driven by Wi-Fi enabled specialty devices that benefit from a broadband network connection. These devices will be used by customers in hotels, airports, coffee shops and other public spaces, especially as municipal networks get built out across the major cities in the world. The near omni-presence of Wi-Fi, with its high-speed, low-cost price/performance ratios, will provide next generation devices with cheap, fast Internet access.

29.3.2 Laptops

This has been the workhorse for the commercial Wi-Fi industry since the nascent days of the marketplace. Business travelers with their laptops from IBM and Dell utilize the airport and hotel networks as they travel from city to city on their capitalistic adventures, while emos with their Apple notebooks and gamers with their Alienware laptops hang out in coffee shops ingesting various forms of caffeine and carbohydrates while hammering away on the Wi-Fi network. This category is well-covered. Any standard hotspot has been designed and optimized for these devices.

29.3.3 PDAs

PDAs have traditionally been the red-headed stepbrother or stepsister to laptops. Optimizing hotspots for PDAs typically involves a number of steps that many hotspot operators do not undertake because of the effort relative to the return. There are not very many Wi-Fi enabled PDA users that are consistent commercial Wi-Fi users. In part, this has historically been due to a relative dearth of Wi-Fi enabled PDAs in the market. That is changing as the latest generation of devices are as likely to have Wi-Fi as not. The dual-mode smartphones (see below) that combine Wi-Fi with 3G, and also incorporate traditional PDA functions will also accelerate the growth of usage in this marketplace.

Steps to optimize your hotspot for PDAs

1. Set redirects for the devices via user-agent string recognition from the device's browsers to present PDA-specific content.
2. Create a secondary set of signup/login pages that have been optimized for screens that are 240 pixels wide by 320 pixels high or smaller.

A cautionary note, many times PDA browsers support a subset of the technology that laptop browsers have. Some basic UAM functions in commercial hotspots that are commonly enabled via JavaScript or other scripted functions may not work for all PDAs.

29.3.4 Dual-Mode Mobile Phones and Wi-Fi VoIP Phones

This category should probably be broken into web-enabled phones – or devices that are designed to browse the World Wide Web – and single-functions phones – or devices that are designed to do one thing, and one thing only.

Among the former, there are Symbian-based, Linux-based and Windows Mobile-based smartphones, as well as devices such as the iPhone from Apple, that combine the mobility of a cellphone with the broadband connectivity of Wi-Fi. This enables the user to access the optimum network based on task and availability. Like PDAs, though these devices typically include Internet browsers, many lack comprehensive Internet libraries or technology support required to perform all the UAM functions necessary for login.

Similarly, the single-function devices – like the Wi-Fi phones from Belkin, NetGear, SMC and other networking device manufacturers – rarely have Web browsers. They are designed to do exactly what they do – in this instance make VoIP calls – and do not include the myriad applications you might find in a smartphone. In this case, accessing a commercial hotspot requires embedded functionality that can navigate the complex, fragmented public hotspot space.

In both instances, there are industry initiatives that provide lightweight code that can be included as an application or integrated into the phone's OS or middleware that performs the authentication functions usually managed via the hotspot's walled garden. One such example is the open source Wi-Fi toolkit that Boingo designed to be embedded into devices to automate the hotspot login function. Belkin has integrated this functionality into one of its Wi-Fi phones. Another example is Boingo Wireless' software for mobile devices, which is based on this embedded toolkit. This software is a fully functional, user-friendly software application that provides access to tens of thousands of hotspots, instead of a developer-ready toolkit for enabling the same access.

29.3.5 Other Mobile Devices (cameras, MP3 players, gaming consoles)

As more and more users get acclimated to using Wi-Fi with their laptops, PDAs and smartphones, they will come to realize that they own other devices that could benefit from a broadband network connection. Some of these mobile devices are already coming to market with early-market implementations of integrated Wi-Fi. Some of these implementations have begun to show the possibilities, but few really exploit the breadth of opportunity to create an optimal user experience.

Imagine sitting in your hotel room while on vacation and automatically uploading your photos to your online photo albums while you're still traveling. Imagine downloading just-purchased songs to your MP3 player while finishing your latte at your favorite coffee shop. Think about playing someone halfway around the world in a multi-player game on your mobile gaming console while you finish up dinner at your local fast food restaurant.

All of these scenarios exist today. The challenge is that the devices enabled as such are limited in product lines and haven't reached critical mass in market (they will), and most of these devices are saddled with the same challenges and shortcomings exhibited in PDAs and Dual-mode phones. Their Internet technologies are incomplete, and frequently they lack a browser. Fortunately, these shortcomings can be overcome with widely available toolkits and SDKs, including the open source Boingo Embedded Wi-Fi Toolkit, which allows developers to download, port and test the code on their prototypes without any licensing fees or intellectual property infringement issues.

30

A Discussion of 802.11 for Sensor Networks

William Merrill, Dustin McIntire, Josef Kriegl, and Aidan Doyle[a]

Systems coupling embedded computing and sensing have vast potential, particularly when wirelessly networked [1, 2, 3]. However, the focus of much of the literature for wireless sensor network is on idealized systems with potentially millions of members (see for example [4, 5]), with minimal power, weight, and size. This focus on idealized systems can lead to discounting the use of current WLAN technologies such as 802.11, or WiFi, in sensor network applications. In this chapter we discuss the applicability of WiFi, as the pre-eminent WLAN technology, to wireless sensor network applications. We begin this discussion with an introduction to wireless sensor networks. Then we describe how the adhoc capabilities and communication efficiency of 802.11 radios are suited to certain sensor network applications. Finally, to illustrate a type of sensor network for which 802.11 radio properties are appropriate we provide an overview of a prototype wireless network for sonobuoys developed by Sensoria Corporation and Exponent Corporation, which was demonstrated using 802.11b radios.

30.1 Introduction

One application of ubiquitous computing is to create autonomous or semi-autonomous systems that monitor and report changes in the physical environment. When communicating wirelessly, this large range of systems is called wireless sensor networks. These networks are envisioned as large numbers of individual sensing "nodes", each connected to its neighbors wirelessly, and networked together to enable communication, coordination, and collaboration. This interaction is envisioned between groups of nodes locally as well as across the network to entities potentially on external networks. An abstract illustration of a wireless sensor network is presented in Figure 30.1.

Wireless sensor networks represent a broad class of systems which often encompass very different requirements. However the wireless aspect of these systems leads to two common components:

---

[a] All authors currently affiliated with *Tranzeo Wireless Technologies USA, Inc* (*formerly Sensoria Corporation*) Wireless Fabric is a trademark of Tranzeo Wireless Technologies. All other trademarks are the property of their respective owners.

Figure 30.1: Concept of a small wireless sensor network, with wireless links shown as lines and embedded sensor nodes as cylinders.

1. Each node operates using an internal battery and so in most cases is not "wired to power".
2. Each node communicates via a "wireless" radio. Optical and acoustic communication has also been considered for wireless sensor networks however in this chapter we focus solely on radio communication.

Most discussion of wireless sensor applications centers on very low unit costs to support large volumes of tiny sensor nodes intended to last months if not years [3, 4]. This assumption leads to designs that minimize the power consumption of each node and in conjunction limit the complexity and data capacity of the radio technology used [6, 7]. In fact a vast body of literature exists presenting and discussing networking and radio designs targeted towards very low power and low capacity systems, see for example [1, 2, 3]. However, as this chapter illustrates there are many applications of wireless sensor network systems that do not require extreme low power, require more than 100kbps of communication capacity, and in fact are ideal for conventional WLAN technology. Prior to discussing an example wireless sensor network based on conventional WLAN technology, the following sections provide an overview of general radio considerations for a wireless sensor network, and discuss the general suitability of the 802.11 WLAN to different types of sensor networks.

30.2 Sensor Network Radio Considerations

Wireless sensor networks represent a wide class of applications, which encompass a wide range of radio requirements. A variety of radios and radio standards have been developed for wireless sensor network systems. This section provides a comparison of 802.11 to a few of the existing wireless sensor network radios and discusses general radio requirements for wireless sensor networks. Following the overview of sensor network radios in this section, the next two sections, consider why the adhoc mode of 802.11, and why the energy efficiency of the higher throughput operation of 802.11 are appropriate for certain sensor network applications.

30.2.1 Sensor Network Specific Radios and Standards

Many sensor network radios and networking technologies have been proposed and developed (for an overview see any of the following [1, 2, 3]). For example, the Zigbee standard [6] has been designed for low power operation, particularly to support duty cycling. In addition, the Bluetooth standard is also considered for some sensor networks as for example [8]. In fact an IEEE working group encompassing both Bluetooth and Zigbee is developing radio standards suitable for many wireless sensor network applications [7]. A number of proprietary radios, targeted for industrial and military also have been developed (for example [9]). In general these radios are designed for lower transmit power, and lower throughput than 802.11 radios, to minimize power consumption of the radios, and are often designed for a specific class of applications, such as unattended ground sensors [10].

Much sensor network radio and networking development focuses on the use of standardized radio and or networking technologies. This focus can be split in one of two areas: either developing a standard from the ground up, focusing on minimal energy use, for short range applications such as sensor network; or taking existing standards and discussing how those standards will support sensor networks. This chapter discusses the second area, specifically using WiFi radios. However, unlike most of the other wireless sensor network radios discussion (such as Bluetooth [8], or the development work of the IP6 over low power WPAN (6lopan) working group [11] our emphasis is not on a minimum power radio.

A primary motivation for the development of sensor network standards is the inefficiency of IP networking, for low data rate and small packet size communications. For applications with low data rate sensor updates corresponding to events that occur at a period of seconds or longer, the per packet overhead of Internet Protocols, particularly when aggregated with overhead from PHY and MAC layers can be overwhelming. For example in an Internet Protocol solution operating over WiFi, per packet overhead approaching 100 bytes is not uncommon. As a further complication, it is well demonstrated that the predominant IP protocol, TCP, operates poorly over lossy as opposed to congested links [12], increasing the difficulty of using IP protocols over a wireless, and particularly multi-hop environment. However, as this chapter intends to demonstrate, for specific applications, wireless protocols developed to support Internet protocols, such as WiFi, are still very appropriate for a class of wireless sensor networks. Particularly, when those networks are designed with the benefits and costs of a specific radio, networking, and transport protocol in mind.

30.2.2 Radio Range and Network Scalability

Sensor networks are often envisioned at a scale of thousands to millions of communicating sensor nodes. Often with nodes locally collaborating and reporting status information to a much smaller number of traffic sinks. While the potential for such systems in the future is vast, current adhoc wireless networking technologies frequently have difficulty at a scale of tens to hundreds of nodes, as opposed to millions. These difficulties are often exacerbated in wireless applications in which each node may be deployed in a variety of environments, and as a result may experience vastly different radio propagation considerations. For example, consider a 2.4GHz radio that provides 90dB of path loss margin, a reasonable assumption for many sensor network radios as illustrated in more detail later in Table 30.1. If the near field of this radio's antenna is relatively un-obstructed, communication out to 1m requires approximately 40dB of path loss. However, the remaining communication range can then be vastly different for the remaining 50dB of path loss. Propagation is often modeled with an empirical path loss exponent [13] ranging from 1.5 to 6. This translates to a maximum communication range variation from two kilometers to seven meters for 50dB of path loss.

Considering a semi-mobile sensor network, or one in which part of the sensor nodes are elevated (on building roofs, UAVs, or balloons). Nodes that are airborne will have clear communication with neighbors (up to 2km for a 90dB link margin at 2.4GHz), while similar nodes that may be lying on the ground, and will experience closer to fourth power path loss in communicating with other low lying nodes, may only effectively communicate with neighbors within 20m or so. As a result even for a single application, and static deployment with homogeneous radios, vast variations in network topology could be anticipated due to antenna placement. It is easy to imagine wireless sensor network deployments in which a few sensor nodes are in communication with most of the other sensor nodes, and a few only in communication over many hops to reach a common information destination. This means the medium access control and networking must be flexible for a wide variety of network topologies.

Network scalability is strongly dictated by the radio traffic sent in addition to the node density as a function of communication range. Million node sensor networks are often considered to only be communicating very rarely due to stringent duty cycling. Similarly, if the traffic is primarily only between node's within one radio hop, then the huge overhead of relaying traffic is avoided [14, 15]. To achieve scalability most sensor networks include capability to process incoming data to only pass on relevant information, and are designed to send locally processed information to one or a few data sinks. An example of a common scalable traffic pattern is to design communication capacity to support limited local communication (for collaboration over at most a few radio hops from each sensor node) in addition to rare information updates to one or a few information sinks, possibly representing gateways out of the sensor network. Network scalability is dictated by who may communicate, and if anticipated traffic patterns and communication ranges will overwhelm available network resources. Each time a node relays a packet it uses radio spectrum that may overlap with that available to the sender and the receiver of that packet, and possibly the sender and receiver when they relay. As a result variation in the communication range of each radio hop can strongly influence network scalability, particularly if large amounts of traffic flow over multiple hops. The influence on scalability of traffic patterns coupled with communication range variation can lead to wide variability in network scalability, even within a single wireless sensor network deployment type.

Wireless sensor network traffic is not simply a function of the number of sensors. Often sensors are designed to collaborate. As an example the sensor traffic on a network could be made invariant to the sensor node density if sensor nodes are aware of the location of neighboring nodes, and use that information to only report at an average sampling density over a region. Thus even as the node density changes it maybe possible to maintain a constant sensor reporting density (hence constant traffic load), using local sensor collaboration, and as a result maintain a scalable network even at high node densities. If the control overhead for a sensor network is low, then sensor nodes could collaborate to maintain efficient radio spectrum usage, independent of the density of sensor nodes in an area. However, as the radio propagation may strongly vary due to small variations in the environment, and similarly the desired sampling density of each sensor may also change due to environmental variation, often a wireless sensor network must include significant scalability margin in its design.

An example of the implications the variation in communication range and node density can have for scalability of the sensor network system, is examined later in the chapter when we present results from a buoy wireless sensor network, in which buoys may drift and clump to provide vastly different node densities in a dynamic system. However, first in the next section we discuss how the flexibility and low overhead of adhoc mode of 802.11 radios are well suited to enabling scalability in sensor networks.

30.3 802.11 Ad-hoc Mode: An Enabler for Sensor Networks

Wireless sensor networks are often envisioned with large numbers of sensor "nodes". Large deployments are considered both to enable extensive spatial sampling and to realize manufacturing cost savings when amortizing design or development costs. In addition, many deployment environments do not have a communication infrastructure that can be easily leveraged [16]. As a result, many wireless sensor networks are designed using multiple hops within the network, where each sensor node relaying information for other nodes, effectively forming a mesh network. Thus the medium access control (MAC) layer of the radio used must support multi-hop communication.

A multi-hop or mesh network imposes constraints on the MAC used in the sensor network radio [14, 15, 17]. TDMA or other centralized MACs require coordinated access of each of the neighboring nodes to the radio channel. As the scale of the network increases the complexity of the required MAC coordination increases. As a result various techniques have been developed to enable multi-hop scalability of the MAC to minimize collisions [18, 19]. In cases when the communication efficiency of the radio is paramount, use of these collision-mitigation techniques are important as they allow efficient use of the radio channel, up to 100% capacity. However, where radio channel usage will often be less than 100%, fully distributed MACs such as 802.11's channel sense multiple access with collision avoidance (CSMA/CA) are still effective. In addition, the CSMA MAC easily scale across large networks [20].

As an example of how MAC complexity may become difficult to scale, consider the idea of duty cycling radios, or alternatively periodic communication loss within the network. If a centralized MAC is used, then when a node shuts down its radio to save power to maintain efficient channel usage, this status must be coordinated, potentially across multiple communication hops as the potential MAC interference range will invariably be larger than the radio communication range (assuming the signaling mechanism requires an signal to noise ratio (SNR) of greater than zero). In many sensor network algorithms, as minimizing energy usage is paramount, duty cycling is central to the system operation, and as a result the MAC is designed specifically for the sensor network topology. However, being able to limit the complexity is a necessary part of the design in order to develop and test within a limited time frame using available components. Yet, as discussed in the next sub-section, many sensor network applications result in widely varying network topologies. Thus restricting the MAC control complexity is particularly important if inter-node collaboration may result in unanticipated emergent behavior at the system level due to interactions of the MAC and power control. The

CSMA/CA MAC limit interaction complexity at the cost of communication efficiency when the radio channel is fully utilized.

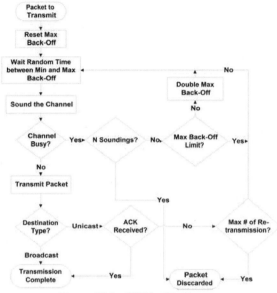

Figure 30.2: CSMA-CA diagram.

The 802.11 MAC is a CSMA/CA communication scheme in which each node that wishes to transmit first senses the channel for a short period of time, and then transmits if it determines the channel is available. Each time a radio with a packet to send detects that the channel is busy, it waits an additional period of time randomly distributed up to a maximum back-off window, before again sounding the channel to determine if it is busy. The maximum back-off window size is doubled every time the channel is detected as busy, up to a set maximum delay between channel soundings. The channel is sounded to determine if it is clear to send up to a maximum of N times. After N soundings without a transmission the packet is treated as sent. Since broadcast packets are not acknowledged N unsuccessful soundings result in a broadcast packet being dropped, while for a unicast packet this process is repeated until a successful acknowledgement is received, or until the maximum number of retransmission attempts are made. Each retransmission may consist of N soundings, or S<N soundings and a successful transmission. After the maximum number of retransmission attempts without a received acknowledgment the packet is dropped by the MAC layer. This CSMA/CA decision cycle for each packet is illustrated in Figure 30.2. The CSMA/CA MAC is designed to share a channel efficiently between nodes as long as the total traffic on the channel is well below the maximum that channel could theoretically support. Each radio determines based on its own sounding if there is channel contention. A large amount of traffic on the channel will increase perceived contention, and increase the likelihood of a collision based on inaccurate channel soundings. Inaccurate channel soundings could be due to the sounding timing granularity, or due to hidden terminal problems or gray zones as discussed in [21] and [22]. Thus contention and the cost of collisions results in an effective upper limit to the throughput generally well below the

quoted signaling rate. For example the effective throughput distributed over all senders, assuming a 1 Mbps signaling rate in a shared channel (not counting per packet overhead) is generally between 500kbps and 900kbps, depending on average packet sizes ([23], [24], [25]). This upper limit is a result of the perceived congestion at the MAC, additional limits on throughput are dependent on the SNR through packet loss (bit error rate) and the signaling rate that SNR will support.

To provide an example of the efficiency loss of the 802.11 CSMA/CA MAC we follow the analysis of [25] using settings consistent with an 802.11b radio. We simplify our analysis to assume that every packet is eventually sent (thus packets are never timed out in our simplified example) and that all packets sent are the same size. With these limiting conditions the 802.11b CSMA available throughput is shown in Figure 30.3. This figure is based on the additional assumption of a set number of nodes sharing the same channel, and sending packets at an average rate as shown. It is clear from this figure that the number of nodes sharing the channel scales with the traffic on the channel. This analysis does not include MAC maintenance traffic. However in 802.11 adhoc mode, the only maintenance traffic is the beacons, which scale easily as the number of nodes increases, and if a static 802.11 cell id is chosen can be turned off all together. The traffic on the channel dictates the usable communication rather than the number of nodes at least for 802.11b ad-hoc mode (IBSS operation) as opposed to infrastructure (BSS operation) mode. When operated in the infrastructure mode, additional control traffic is sent by each access point so that the total amount of contention will scale with not just the amount of data traffic, but also the number of nodes whose radios are active. Figure 30.3 demonstrates the cost of saturating the channel (the total available goodput drops off substantially due to collisions) and that in ad-hoc mode this saturation is driven by the traffic and not the number of nodes, at least with more than ten nodes sharing the channel. Figure 30.3 is based on the simplification presented in [25] that every packet will be eventually transmitted, it illustrates the qualitative effect on throughput of saturating the channel with packets, to account for the qualitative throughput or the packet latency during saturation requires considering the retransmission limit considered in [26] and is outside the scope of this chapter.

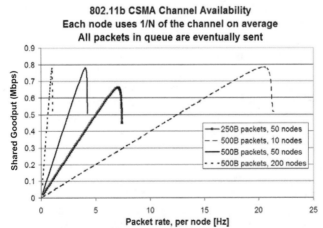

Figure 30.3: Simulation of how throughput scales with the number of users in the 802.11b 1Mbps CSMA MAC.

As a result of the MAC scalability with the anticipation of large numbers of sensor nodes used, and the limited interaction of the MAC with duty cycling the 802.11 adhoc mode is often very well suited to sensor network applications. The 802.11 adhoc mode is particularly suited to applications where the number of nodes sharing a wireless channel may vary dramatically depending on the propagation environment or varying node deployment density, such as the buoy sensor network discussed at the end of this chapter.

30.4 802.11 Power Usage Suitability for Sensor Networks

The most common reason 802.11 radios are not considered for wireless sensor networks, is that power used by the radios is considered to be too large for the sensor node's energy budget. However, this reason assumes that the other components of the sensor node require significantly less power when active, and that the duty cycling of the radio cannot achieve energy efficiency comparable to other radio options. Each of these assumptions is addressed in the paragraphs below.

The energy used by a radio is relative to the energy required for other components of each sensor node. At a fundamental level we separate a sensor node into the following subsystems: processing sensing, communications, and actuation. The power requirements of most 802.11 radios, as they approach Watts are often well outside the envisioned power usage (when active) of many postulated sensor nodes. As an example consider a sensor node: with an 8-bit microprocessor, including passive sensors reporting at a rate of a Hertz or less, that does not require energy for deployment, and that is stationary after deployment. However, in many applications higher energy budgets are required to support each of these component areas. Examples include:

- The need for a standard operating system able to support low cost, rapid development and to support complex processing tasks and autonomic behavior on each sensor node. This may lead to requirements for a 32bit microprocessor with significant processing capability, and a potential processor energy budget on the order of Watts.
- High bandwidth data storage requirements.
- Active sensor components such as the active sonobuoys discussed later within this chapter, or high performance seismic sensors.
- Mobility requirements that cannot be met passively. Often the energy budget required for autonomic ground or air mobile sensor platforms over the sensor node lifetime dwarfs the energy use of 802.11 radios over that entire lifetime.
- The need for dynamically tasked sensor platforms supporting multiple sensing applications simultaneously (multi-user multi-sensor support).

As a result, even if always active, a radio that requires a few Watts may still represent only a small fraction of the available energy budget. In the wireless sensor network buoy field discussed in the following section the energy budget for the active sensing, and for the local buoy processing were each comparable to the energy budget for the 802.11 radio, even with an RF amplifier to support the 802.11 buoy radio's 30dBm transmit power.

Another consideration in evaluating a radio choice for a wireless sensor network is that higher power does not necessarily mean lower energy efficiency ([27] [28]). For example in [29] the author shows that with appropriate duty cycling an 802.11 uses less energy to transmit the same information, than a lower power Zigbee radio. 802.11 radios as compared to Zigbee send at a higher data rate with more efficient spectral coding, so 802.11 radios send the same information in much shorter time. This is illustrated in Table 30.1 taken from [29]. To minimize the energy cost of a higher power radio in a low power sensor network application, duty cycling is needed that amortizes the energy cost of starting, synchronizing, and stopping each radio. However, for other applications where large amounts of information need to be transferred, the lower per bit energy costs of 802.11 may also make it a competitive choice for example when a wireless sensor network enters an active phase, where the entire network must collaborate or report quickly. As a result, depending on the amount of information to be transferred and the communication latency required, with judicious duty cycling 802.11 radios may provide a viable, or even best, option for many sensor network applications, even when limited energy budgets are required for an application. 802.11 radios may be appropriate either as one of multiple radio options within a wireless sensor node, as in [29], or as the only radio in an appropriate wireless sensor network application such as the sonobuoy network discussed next.

Table 30.1: Efficiency comparison of an 802.11g and Zigbee radio, from [29].

	802.11g	Zigbee
Chipset	Atheros 5006XS	CC2420
Output Power	16dbm	0dbm
Path Loss	90bB	
Rx Sensitivity (BER 10^{-5})	-74dbm@48Mbps	-90dbm@250Kbps
Tx Power (Max Output)	1320mW	57mW
Rx Power	924mW	65mW
Total Power	2.24W	122mW
Effective Throughput	24Mbps	125Kbps
Efficiency (nJ/bit)	93	976

30.5 A WLAN for a Field of Sonobuoys

To illustrate the application of 802.11 radios in sensor networks we consider a WLAN connecting a field of sonobuoys. Specifically, we focus on the prototypes developed in the Network Enabled Anti-Submarine Warfare (NEASW) program to demonstrate the feasibility of a networking a field of sonobuoys.

The United States Navy's anti-submarine warfare (ASW) task force is exploring new submarine detection techniques. For example collaborative and distributed sensing techniques within fields of networked sensors have the potential to revolutionize detection capabilities by improving detection probability, reducing false alarm rates, and minimizing the logistics burden of large buoy fields ([30]). To explore the capability of a sonobuoy network the US Navy funded development and testing of an eighty node field of sonobuoys, built leveraging commercial off the shelf (COTS) components. The U.S. Navy's Task Force ASW funded Exponent Corporations to develop the NEASW program. This program developed a prototype sonobuoy system that demonstrated potential benefits of networked sonobuoys for submarine detection. Sensoria Corporation was awarded a sub-contract within this program to develop the NEASW networking hardware and software components, to demonstrate robust, networked communication between sonobuoys. A simple example of the potential benefit for multi-sonobuoy collaboration in a multi-static sonar is illustrated in Figure 30.4.

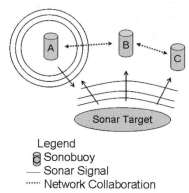

Legend
Sonobuoy
— Sonar Signal
····· Network Collaboration

Figure 30.4: Illustration of collaboration in a multi-static sonar.

To demonstrate the feasibility of a network of sonobuoys within NEASW, 802.11 radios were selected as the best COTS choice. A prototype NEASW system was developed and the results of the network testing are presented here. For the sonobuoy network tests discussed the buoy-to-buoy ranges were scaled to the specific properties of the WiFi radios used in the prototype the system. Due to the military's operation requirements an alternative radio would be developed to meet the specific bandwidth, range, and jam-resistance necessary for the tactical application of NEASW. To maintain the scalability, multi-hop communication, and resilience needed for a sonobuoy wireless sensor network this new radio would require many properties common with WiFi (e.g., the MAC layer). However, to support the specific bandwidth, communication range, and processing needs of the sonar sensors a deployed NEASW system would use a radio operating at a different frequency, waveform, data bandwidth, and operating range than the prototype system discussed here.

This section presents an overview of the NEASW prototype wireless sensor network. The NEASW network was built using 802.11 radios and mobile adhoc networking (MANET) technology. The NEASW sensor network was designed to support hundreds to thousands of sonobuoys. Each sonobuoy was designed using a sonobuoy provided by Sparton Corporation, retrofitted with additional local processing, 802.11b radios, a mobile adhoc networking (MANET) implementation, and a delay or disruption tolerant networking (DTN) transport layer. Table 30.2 provides a summary of the NEASW wireless sensor network application and prototype demonstration system.

This section provides an overview of the prototype NEASW WLAN. First we describe the requirements of the system, and summarize why we chose the 802.11b radio standard. Next we provide an overview of the: network processing platform within each buoy, the MANET implementation connecting the buoys, and the disruption tolerant transport layer developed with NEASW. Finally at the end of this section we present a short summary of the NEASW network prototype test results.

Table 30.2: An overview of the NEASW program.

WLAN Objectives
- Network buoys to support development of collaborative submarine detection algorithms
- A solution that scales to protect hundreds of square miles of ocean
- Operate in heavy sea state, with graceful degradation with higher seas
- Network buoys, ships, UAVs, and potentially land based command and control elements

Solution Description
- Autonomous multi-hop communication robust to buoy loss and movement
- Disruption tolerant transportation protocol to ensure information delivery on rough seas with intermittent links due to wave motion
- Internet protocol (IP) communication over 802.11b radio links
- Multicast support with efficient communication between multiple groups
- Embedded hardware platform integrated into a Sparton A size sonobuoy

Results of Network Prototype Testing
- Demonstrated the network in multiple tests with up to eighty buoys, multiple ships and aircraft off the coasts of Hawaii and California
- Demonstrated mobile operation as network components move at different speeds -- buoys (slow), ships (moderate), and small aircraft (fast)
- Demonstrated network operation and graceful degradation up to sea state five
- Reliably relayed data over tens of hops
- Integrated into hundreds of prototype buoys

30.5.1 The Requirements of the NEASW WLAN

The goal of the NEASW system was to demonstrate how a network might support collaborative signal processing between sonobuoys. The program was funded to provide an example of how state of the art commercial communications could increase the scale, autonomy, and usefulness of fields of sonobuoys. To show that the demonstrated system might be useful to support a range of collaborative signal processing algorithms, general requirements were put in place. These general requirements were to demonstrate communication: of the order of hundreds of kilobits per second, over an inter-buoy spacing range of a few kilometers, reliably within a timescale of tens of seconds, between a field of hundreds of sonobuoys and a human operator, and to a ship or land based command and control center. These requirements were created based on discussions of the potential requirements of a range of collaborative signal processing algorithms. They are a representative example on what a collaborative signal processing solution could require rather than the requirements of the final system. The NEASW demonstration system was designed and built within a few phases each lasting only a few months by leveraging COTS components available within that timeline. Existing radio technologies were considered for how they could:

- Provide inter-sonobuoy communication ranges of a few kilometers using buoy antenna heights of only a few meters.
- Provide communication throughput that supported average data from all the sonobuoy to the operator up to 250kbs on the WLAN. This goodput needed to be maintained at deployment scales of hundreds of sonobuoys.
- Maintain throughout and provide resilient communications as links are periodically lost and reformed on the time scale of wave motion [31].

- Provide a standard networking capability over which MANET and DTN technologies could be easily integrated.
- Support buoy lifetimes of multiple days at a power comparable to the existing DICASS sonobuoy [32].

Within the NEASW program 802.11 radios represented the best choice to demonstrate a system meeting these criteria. 802.11b radios operating at the lowest 1Mbps data rate provided the most mature technology with the flexibility, capacity and range needed, that was available within the time-frame. Extensive additional discussion of why 802.11b was chosen for this system is presented in [33].

At first glance 802.11 may seem a poor choice if only from considering the range of the radio. Often 802.11 radios are quoted with ranges on the order of 10s to 100s of meters, and the NEASW system representative requirements called for inter-buoy (inter-node) spacing of kilometers. In addition, to build a reliably connected network requires more than just a single connection at each buoy, so that the node spacing must be a fraction of the radio range. However, 802.11 radios were determined to be the best available radios in the summer of 2004 to support three goals: communication far enough to build a connected network, communication at a fast enough rate to pass information between buoys during the short periods when wave height enables connections between buoys on rough seas, and energy efficient enough to operate on battery power during the multiple days it takes for the buoys to drift out of position. As a representative set of requirements for various ASW signal processing techniques the NEASW system was designed for a reliable 2km range supporting an average goodput of 250kbps over the network to a system controller for a lifetime of three days. Further discussion on why 802.11 radios were selected to meet each of these goals is provided in the following sub-sections.

30.5.2 Radio Range on the Ocean Surface

A distributed field of buoys is in constant motion, both locally and globally. Locally, buoy orientation changes as the buoys bob and tilt due to wave motion. Globally, surface currents move the buoys. As a result, buoys will have intermittent radio contact due to orientation changes and temporary wave obstacles; in addition, buoys will drift in and out of range, without necessarily maintaining the same neighbors.

A-size sonobuoys are limited to radio antenna heights that can be supported by the buoy float; at most a few meters above the water line. Airborne relays are designed into the NEASW system, but as a gateway out of the system, and the buoys must still communicate when airborne assets are not present. Thus the NEASW demonstration radio was chosen to allow reliable communication using antennas about two meters above the water. Depending on the collaborative sonar application, a variety of inter-buoy ranges maybe desired. However, for the NEASW demonstrations, 2km was chosen based on the available radio power, radio frequency, and antenna height.

To explore radio range on the ocean surface we consider the final 802.11b link budget for the prototype NEASW buoy of Table 30.3. At 2.4GHz based on the Friis transmission equation this link budget allows free space communication to past 12km. This illustrates the long ranges 802.11 is capable of even without high gain antennas using the lowest card sensitivity (-95dBm) and near the FCC maximum radiated power. The 30dBM transmit power and 5.5dB antenna gain of the final NEASW prototype system is close to the FCC limit of 36dBm EIRP. However, as demonstrated in multiple tests within NEASW and in the literature (see for example [33, 34, 35]) propagation near the surface of the ocean is not comparable to free space propagation. A more effective, yet still simplified propagation model simulates the moving antenna height compared to the intervening average wave height and considers only a two-ray model of propagation, mixed with periodic obstructions blocking all communications. The two-ray model approximates the Fresnel zone reflections from intervening waves as those off a planar surface with the electrical properties of sea water (for a discussion of the model see [13, 36]). The two-ray model is illustrated in Figure 30.5. The results of this model in comparison with an early NEASW test, discussed in detail in [33], are shown in Figure 30.6. Within this figure the similar trend between the simple model of sinusoidal antenna height, and the measured data is clear to 1km (this test used 1.5m antennas and a 26dBm EIRP system).

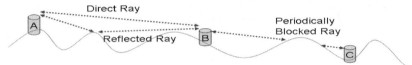

Figure 30.5: Illustration of two ray model on the ocean surface.

After initial NEASW tests demonstrated that 1km communication ranges were achievable on the ocean, the NEASW system was improved to use 36dBm EIRP by integrating an 802.11b amplifier, and increasing the antenna height to 2m. Figure 30.7 illustrates the range at which reliable inter-buoy links were obtained in a thirty eight buoy test. Each buoy broadcasts 68 byte HELLO packets periodically. Figure 30.7 shows the fraction of expected HELLO packets received by all buoys as a function of the buoy separation at when deployed. This figure shows the fraction of HELLOs received with high enough SNR to represent a reliable communication link, and the total fraction received. Figure 30.7 demonstrates the NEASW potential of communication past 2km, and that care must taken to ensure information delivery (this data is based on broadcast 802.11 packets which do not use an ACK), even at buoy spacing of less than 2km. The overhead cost of ensuring delivery, i.e., of achieving the 250kbps desired application layer throughput or goodput, is discussed in the following sub-section.

Table 30.3: NEASW 802.11 link budget example.

Transmitted Power	**30**	dBm
Transmitter Feed Losses	*-0.50*	dB
Transmit Antenna Gain (omnidirectional)	5.5	dBi
Polarization Loss	*-3.00*	dB
Receive Antenna Gain (omnidirectional)	5.5	dBi
Receiver Feed Loss	*-0.50*	dB
Receiver Sensitivity	*-95.00*	dBm
Fade Margin	*10.00*	dB
Allowable Propagation Path Loss	**122.0**	dB

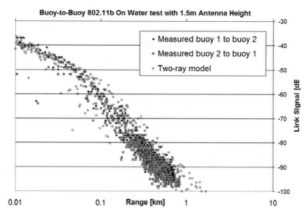

Figure 30.6: Comparison of measured and modeled signal on the ocean with 1.75m significant wave heights and 7s wave period between two model buoys.

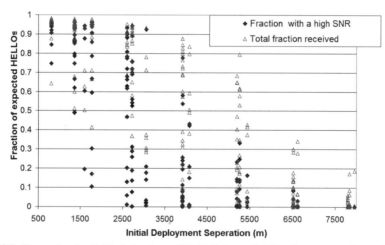

Figure 30.7: Example of reliable inter-buoy 802.11 link availability in sea state 3 as a function of link range in meters.

30.5.3 Inter-Buoy Communication Throughput

The sonobuoy system was designed to provide a perimeter detection system covering large areas of the ocean. Each sonobuoy monitors its vicinity of the ocean and communicates with a command and control center. Each communication link traverses the multi-hop network connecting sonobuoys on the perimeter, until reaching either a UAV gateway or the ship gateway back to the control center. This communication must be flexible and support hundreds of kilobits of data between the control center and the sonobuoy field, including various data types. A few hundred kbps data rate is needed to pull significant acoustic results out of the buoy field, as a result 250kbps was chosen as a general requirement for the NEASW demonstration. To show how traffic on a model NEASW network could scale Figure 30.8 illustrates traffic volume as a function of the nodes communicating over a link. This figure uses a NEASW traffic model discussed in detail in [33] and [34], and illustrates the increased traffic as the buoy fields extends in a line, such that one buoy is relaying for large numbers of other buoys. However this figure neglects overhead costs of the chosen protocol. The figure also neglects protocol overhead and the interruption overhead due to disruption of the communication flow by wave motion. The impact of each of these types of overhead is considered in the paragraphs below.

Using the Internet Protocol results in a per-packet overhead due to the IP packet header. In addition, choosing 802.11b results in additional per-packet overhead at the MAC and PHY layer. As discussed in [37] the total of these overheads is on the order of 100Bytes for each packet depending on the specific protocol used. This overhead is significant for small packets. In addition, since the buoy field uses a single shared channel and relays information back over the buoy multi-hop network, each time a data packet is relayed within interference range of another relayed transmission this duplicate use of the channel throughput reduces the available communications capacity. This "relay" overhead often dwarfs the IP, MAC, and PHY header overhead, in the shared communication channel. See [14], [15], [17], or [38] for further discussion relay overhead. A simplified interference range of twice the communication range is often used to half the goodput of a relay channel. However, as is illustrated in Figure 30.7 on the ocean occasional communication at much longer than twice the "reliable" 2km range is frequently observed. In fact in our testing, detectable 802.11 collision avoidance has been empirically demonstrated outdoors to five times the reliable communication link range of the higher data rate 802.11a signaling [38].

Another significant limit on the communications capacity of the NEASW system is the wave shadowing of links. Wave motion occurs at a periodicity within an order of magnitude of ten seconds. Our system is designed for antenna heights of 2m, but must operate in sea states in which the wave height often exceeds two meters. In fact our eighty buoy NEASW network test was cut short for safety reasons because the sea state in the test range exceeded sea state four, with waves observed approaching heights of 10m. Thus for any link between buoys, communication over the WLAN will be disrupted (shadowed) for seconds at a time by intervening waves. As wave disruption of links gets worse, communication must gracefully degrade. Many possibilities exist to adapt communication around wave motion, since buoys are often far enough apart so that the shadowing of any two links may be approximated as uncorrelated. Thus a quickly adapting system might be able to choose between different links at any instant in time. However, in the short development time frame of the NEASW program, our approach was to use a simple solution by over provisioning the network capacity to meet the goodput requirements when relaying along intermittent links. Since information on the network could all tolerate delays of more than ten seconds, the MANET routing was tuned to be insensitive to wave motion based intermittency at this time scale, and a transport protocol developed to enable reliable communication over periodically disrupted links. Each of these solutions is discussed further on sub-sections below.

As a result of wave motion and protocol overhead, the capacity of the radio specified must be well above the required application goodput. For NEASW this drove our choice of 802.11 as a high capacity radio, whose quoted 1Mbps minimum data rate (or about

600kbps effective one link goodput) is well in excess of the data requirements of a few hundred buoys. Example data requirements for the NEASW system are shown in Figure 30.8. From this figure we see that to support approximately 200 buoys, the scale envisioned for segments of the NEASW network, require around 100kbps of "goodput". However, to provide margin for overhead, for goodput limits through wave obstruction, and for scalable operations as buoys clump together and send more traffic in each local radio channel, a high capacity radio was required.

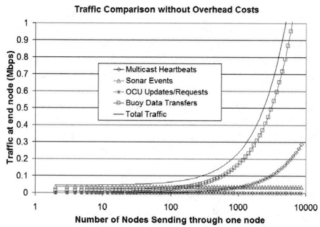

Figure 30.8: Illustration of how traffic could aggregate on a NEASW WLAN assuming all traffic aggregates over one last hop.

30.5.4 Buoy Energy Availability

The useful lifetime of a buoy in the NEASW network is limited by the speed at which buoys drift out of the area of interest, and the speed at which buoys drift apart. Based on simulations and empirical measurements of sonobuoy drift [33] a buoy lifetime of three days was determined as a reasonable buoy lifetime. When compared to many envisioned applications of battery operated sensor networks, this is a relatively short operation lifetime. In fact, while 802.11 radios are often considered to be high powered (Watts instead of milliwatts) as illustrated in the power budget of Table 30.4, the radio, RF amplifier, and network processor use only about 2/3 of the energy budgeted for just the active sonar system. In addition the size and weight of the A size sonobuoy and acoustic sensors meant that significant reductions in battery size and weight would result in only incremental system size and weight reductions. The power budget shown in Table 30.4 is based on the energy available with the current salt water battery technology in legacy sonobuoys.

As a result within the NEASW demonstration system, while the three day lifetime required reasonably low power electronics, it did not require the energy vigilance often associated with wireless sensor networks, and hence even without complex energy optimization, 802.11 radios fit well with the energy needs of the system.

Table 30.4: NEASW sonobuoy power budget.

Item	Power (W)	Energy (W hour)
Network Processor and Radio (50% duty cycle @ peak power)	1.9	57
General purpose processor	0.7	21
I/O board	1.5	45
RF Amplifier (@ 20% duty cycle)	1.5	45
Upper Electronics Subtotal	**5.6**	**168**
Lower Electronics and 100 seconds of active ping time	3.9	117
System Total	**9.5**	**285**

30.5.5 NEASW Hardware and Software System

The NEASW prototype sonobuoys were built using COTS electronics, to update the network capabilities of the DICASS SSQ-62 sonobuoy. Within the DICASS sonobuoy the UHF radio upper electronics package was replaced, and a digital interface board developed to integrate all legacy capabilities of the upper electronics package. The replaced DICASS UHF radios supported only direct low data rate communication with a P3 plane, thus to demonstrate the benefits of a mesh network, this was replaced with a WiFi radio. Within the updated upper electronics package each sonobuoy used an IEEE 802.11b radio and amplifier for RF communication. To maximize inter-buoy communication range during demonstrations each buoy was statically configured to use the 1Mbps 802.11b signaling rate, with the MANET routing software verifying the suitability of each link in a route based on testing at the 1Mbps rate. The radio was interfaced to an embedded processing platform, running a Linux OS, which hosted the Sensoria networking software. All communication between the sonobuoys used the IP protocol. Figure 30.9 is a picture of the updated sonobuoy upper electronics stack. This figure shows the I/O board, network processor, and onboard CF storage. Each buoy also included a PCMCIA 802.11b radio from the Demarc Technology Group, placed on the underside of the board stack of Figure 30.9.

Figure 30.9: NEASW upper electronics stack.

The NEASW buoy hardware was designed in collaboration with Sparton Corporation to integrate into one of their existing buoy lines. The communication hardware consisted of a Sensoria network processing platform (called the network processor) and a custom circuit card that interfaced with the buoy sensor system. NEASW hardware was integrated with several hundred prototype Sparton A size sonobuoys. To support NEASW testing over three hundred sonobuoys were built, not including multiple test and evaluation systems which also incorporated the buoy upper electronics package.

To minimize the disruptive effects of wave motion and drift the buoys were designed with elevated two-meter antennas and multiple layers of networking software supporting disruption tolerant relayed communication between buoys. Specifically, two areas of improved network operation were developed and demonstrated in NEASW:

1. A robust, self-assembling, self-healing network that ensured communication routes between each buoy and multiple users at different locations as buoys moved (a mobile adhoc network or MANET).
2. A delay-tolerant, reliable transport mechanism to assure multi-hop communications even in sea states where inter-buoy links are intermittent.

These technologies are summarized in the following sub-sections.

30.5.6 NEASW Mobile Adhoc Network (MANET)

At Sensoria Corporation we developed the NEASW network to adapt to slowly drifting links, yet maintain communication routes over links that were disrupted for seconds at a time, about every ten seconds. To implement this system quickly we started with an available AODV implementation that supported multicast [39-42]) then modified this solution to enhance its robustness to disruptive links. Early in NEASW program we compared the suitability of routing protocols including OSPF [43] and the MANET protocols AODV [42], OLSR [44], and ODMRP [45]. This comparison used the default implementations of each protocol available in the QualNet™ Simulation software. The AODV protocol was selected as a starting point for the NEASW MANET development based on simulations of representative traffic models with network sizes. In these simulations AODV provided the most robust packet delivery across links with outage periods on the order of seconds, with overhead scalable to network sizes of thousands of buoys, at the cost of increasing delivery latency, an acceptable trade-off for the NEASW application.

Key features of the MANET developed in NEASW include:

- Route creating and maintenance using route requests (RREQ), route replies (RREP), and route error (RERR) messages as specified in the AODV RFC [42].
- We explicitly required hello packets to verify the bi-directional link state and link quality at a 1Mbps signaling rate to each one hop neighbors prior to using that neighbor to form part of a route. This is analogous, but not equivalent, to the use of hellos in OLSR [44].
- An efficient multicast tree is built for each multicast group, by rewriting and modifying the multicast implementation described in [40].

- In NEASW the vast majority of traffic was between buoys and gateways to the command and control centers. As a result proactive route discovery and maintenance was added with gateway advertisements. Each advertisement embeds source route (as in DSR, [46]) information in each AODV hello packet. This proactive advertisement of each gateway, and buoys served by each gateway limited the amount of AODV overhead traffic, and minimized route update latency on the ocean. The gateway advertisement process consists of two parts:
  1. the route advertisement (RADV) is sent from a gateway node, and relayed out to a pre-set hop radius
  2. to enable the gateway to maintain up-to-date routes to each node, each node receiving a RADV generates a reverse advertisement (REVA), which is relayed back to the RADV generating gateway

  Further details of the MANET implementation for NEASW are provided in [47].

30.5.7 NEASW Disruption Tolerant Transport Mechanism

Buoy surface to surface communication over the ocean may be blocked intermittently by waves on a temporal scale consistent with the wave motion. As a result in order to enhance the robustness of our network, which enables buoy-to-buoy communication over the ocean, at Sensoria Corporation we added the capability to store and forward unicast data flows across multiple network hops as they become available with our Temporal Flooding for Intermittent Link Mitigation (TFILM) algorithm. TFILM was developed and refined during testing conducted within the NEASW program. TFILM is also built on prior work such as the algorithmic recommendations provided by the Internet Engineering Delay Tolerant Research Group [48, 49]. TFILM enables groups of up to MTU-sized packets to be sent over intermittent wireless links, by providing per packet acknowledgements and retransmissions, and packet custody transfer over routes determined by the MANET routing protocol operating on the network. The core of the TFILM protocol is a light-weight implementation of repeated transmission of packets along each link of a unicast route, until they are acknowledged. Additionally, TFILM is designed to work with a MANET, so that if a route changes or disappears the entire packet of data will still be transferred successfully once the route is rediscovered, with built in mechanisms to reduce duplicate transmissions due to route churn.

A high level representation of the TFILM operation is shown in Figure 30.10. This figure illustrates a number of features of the TFILM algorithm including:

- Each packet is acknowledged at each hop at the transport layer. TFILM aggregates acknowledgements to reduce its overhead.
- Packet custody is transferred to the next hop following reception of acknowledgement messages. A list of recent custody transfers is maintained to minimize forwarding of duplicate packets or looping.
- A node with custody of a packet will initiate route discovery to that packet's destination (through the MANET) if a route is lost or does not exist.
- While TFILM is designed for reliable transfer of information, data will time out and be dropped if an intermediate node with custody loses all connections for a predefined period of time, usually set to tens of minutes.

- TFILM is dependent on the MANET router to discover and maintain routes. A TFILM process will resend packets over a valid route until it receives an acknowledgement for a packet from the next hop on the valid route, or until the route over which it is sending is invalidated.
- Communication between the TFILM daemon and an application is enabled via a reliable TCP socket. This socket may be locally generated, or the connection may be established from another node.
- TFILM can accept multiple packets from the same application, from different applications, and from multiple packets addressed to different destinations. Each application uses a distinct application id for sending packets and another for receiving a flow of packets.
- A single pending queue is associated with all packets at each step along a route; however, two priories are supported to accelerate the transfer of high priority alerts.
- Multiple client applications can receive the same packet, assuming they have registered with the application id encoded in the packet's header.
- Once a packet is sent from the main queue, it is maintained in the packet processing queue until either: an ACK for that packet is received signifying a transfer of custody of that packet to the next hop, a route is lost to the packets final destination at which point it is moved back into the pending queue, or the route associated with the packet is changed and the packet is moved into tail of the main queue.

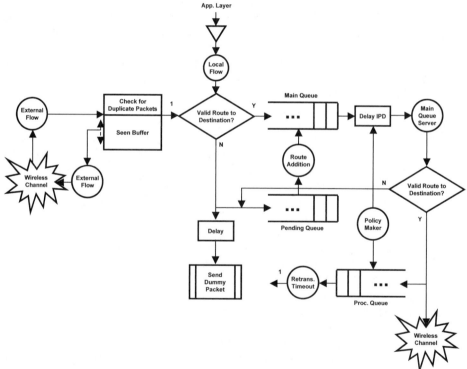

Figure 30.10: High level TFILM data and program flow chart.

30.5.8 NEASW WLAN Demonstration and Results

The NEASW WLAN was tested off the coast of Hawaii and off the coast of California in multiple tests of up to eighty sonobuoys over the summer and fall of 2005. These tests were used in a spiral development process to improve the buoy hardware and software components during each test, thus variability in test results was observed as components were changed. To illustrate the capabilities, and current limitations of the prototype sonobuoy WLAN, selected results are provided in this sub-section.

Figure 30.11: NEASW buoy.

To illustrate the testing conducted we will briefly summarize the usual test process. Prior to a test all sonobuoys were placed on one or more fishing vessels. Both emulated buoys (upper electronics packages with a simulated float and sensor design to ease buoy retrieval and reuse) and modified DICASS buoys were used in tests. The buoys were then dropped off the end of the fishing vessel at predetermined GPS coordinates. An example of a floating modified DICASS buoy after deployment is shown in Figure 30.11. Scenarios with the buoys turned on prior to leaving the dock, and turned on only when placed in the water were both tested. The vessels or multiple vessels would drop the buoys in the ocean, for example in the pattern shown in Figure 30.12, which is within an hour of the start of a 78 buoy test off the coast of Kauai, Hawaii. During each test, multiple operators would query the buoy field, from one or more of the fishing vessels and from a land station on an overlooking cliff. Buoy location and status, as well as sensor events would be reported from each buoy to all users. This communication consisted of tests with multicast data, unicast data (both UDP and TCP – although the TCP protocol used was only viable within one or two wireless hops), as well as with the TFILM transport protocol. In addition, users would query sensor data from selected buoys over the course of a test. Each NEASW network test was split into two parts:

1. Testing the network with an airborne relay. For this relay a small private plane was used on which a buoy radio and network processor, relayed from the buoy network to a point-to-point directional link provided by a Motorola Canopy radio.

2. Testing the network without an airborne relay. This tested the connectivity and throughput of the field to buoy radios on vessels and to buoy radios on land connected via high gain directional antennas.

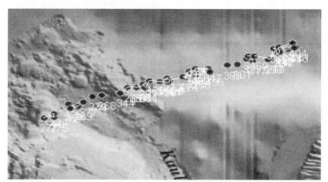

Figure 30.12: Operator view during a 78 buoy test spread over 40km.

Testing lasted from four to eight hours depending on sea conditions, and on available light to maximize the chances that all buoys could be re-collected even if a buoy's electronics failed after deployment. NEASW testing illustrated the following performance:

- Network scalability was demonstrated with 78 buoys, including maintaining communication when all buoys are within one hop, and when deployed along a perimeter that extended over more than ten network hops.
- Effective communication over routes in excess of six hops was observed with up to 2km spacing per hop. Information latency increased with the network hops, particularly past five hops but the system enabled reliable communication. Figure 30.13 illustrates this multi-hop communication by showing the cumulative distribution of the communication latency for 50kB example messages from buoys to an operator during the first ninety minutes of the 78 buoy test.
- Total throughput to a user over the buoy network exceeded 250kbps in many instances. Figure 30.14 illustrates this by showing the data into and out of a user terminal on one of the fishing vessels which was acting as a local control center. This data includes multicast, UDP unicast, TFILM DTN data, as well as a few short TCP data flows. The high variability in instantaneous throughput is mainly due to the traffic pattern selected. During the test traffic was user driven rather than scripted.
- Dynamic routing, adjusting for both the aerial relay and slow buoy drift, was demonstrated, as the buoys adapted to the presence or absence of the airplane. In addition this included the ability to support multiple gateways, out of the buoy network.

In general during NEASW network testing the sonobuoys maintained multi-hop communication in excess of 100kbps, with many instances above 250kbps observed, from all the buoys back to each active command and control center, with inter-buoy links being reliable out to 2km. Control centers were tested both on board the fishing vessels used for deployment, and at the land-based test location via the airborne relay and high gain directional antennas. Buoy autonomy in forming and maintaining communication routes, and in monitoring and reporting status and position were demonstrated. Testing occurred over a variety of sea conditions, with most of the system problems occurring in sea states of three or higher. Common inter-dependent system problems included:

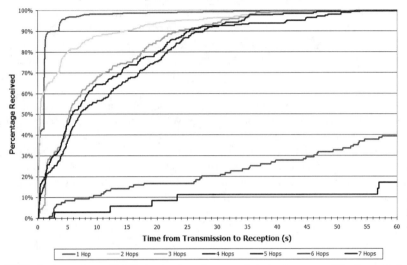

Figure 30.13: Latency for a 50 kByte message from each buoy to a user. The legend shows the network hops the information traveled over.

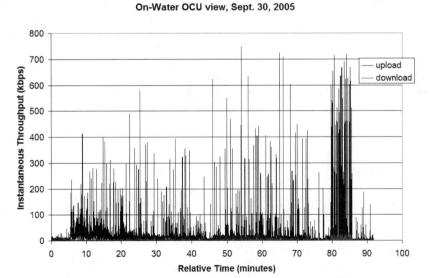

Figure 30.14: Data sent and received by a user operating on a fishing vessel over the first ninety minutes of the 78 buoy test. This shows per second average data at the user terminal. The increase at 80 minutes represents users taking advantage of increased availability due to the presence of an airborne relay.

- Gaps forming in the buoy string exceeding beyond a few kilometers and disrupting buoy-to-buoy communications. This resulted in link disruption and large latencies when high hop count paths were used.
- Bottlenecks in buoy-to-buoy communications causing TFILM queues to back up. This could result in dropping low priority messages and delivery delays.
- Route variability, both between buoys and through the airborne links, often interacted with TFILM to cause some duplicate packets and occasional delays in traffic delivery.
- Hardware reliability problems for a few buoys as they were manhandled on board vessels and due to wave action. A few buoys powered down as a result of this during the tests.

In general, the NEASW system was resilient to these problems, with none of the problems preventing data from flowing over the WLAN. Nearby buoys adapted to reform the WLAN and to ensure communication across the network.

The NEASW system provides an example of a wireless sensor network demonstration, which benefited from the use of 802.11 WLAN. Use of 802.11 suited this application due to the short development time frame of the application and due to the energy, capacity, and range capabilities of 802.11 WLANs.

30.6 Conclusion

A huge variety of applications of wireless sensor networks exist. In many of these applications, minimizing energy, cost and size is paramount. A wide variety of wireless sensor networks can also benefit from the flexibility, efficiency, and capability of current WLAN technologies such as 802.11 radios. 802.11 radios are appropriate for both demonstrating sensor network applications using low cost off the shelf hardware, and for a subset of applications in which the size, power, and capacity trade-offs intrinsic to 802.11 radios are well suited. This chapter provided a discussion of why WiFi radios are a good choice for many wireless sensor network systems. It presented a demonstration wireless sonobuoy network, for which 802.11 are radios are well suited. In this chapter, we explored the potential of the communication range, adhoc operation, and energy use of 802.11 radios for use in wireless sensor network applications.

30.7 References

[1] N. Bulusu and S. Jha editors, *Wireless Sensor Networks: A Systems Perspective*, Artech House Publishers, July 31, 2005.
[2] W. J. Kaiser and G. Pottie, *Principles of Embedded Networked Systems Design*, Cambridge University Press, 2005.
[3] I. F. Akyildiz et. al., "Wireless Sensor Networks: A Survey", *Computer Networks*. Vol. 38, No. 4. 2002. Pages 393-422.
[4] J. M. Kahn, R. H. Katz, and K. S. J. Pister, "Mobile Networking for Smart Dust", *ACM/IEEE Intl. Conf. on Mobile Computing and Networking* (MobiCom 99), Seattle, WA, August 17-19, 1999.

[5] R. Min, et. al., "Energy-Centric Enabling Technologies for Wireless Sensor Networks", *IEEE Wireless Communications*, August 2002, pp. 28-39.

[6] Anon., *Zigbee Specification*, Zigbee Alliance, ZigBee Document 053474r06, Version 1.0, December 14, 2004.

[7] Anon., *IEEE Standard 802.15.4-2003: Wireless Medium Access Control (MAC) and Physical Layer (PHY) specifications for Low Rate Wireless Personal Area Networks (LR-WPANs)*, IEEE, New York, Oct. 1, 2003.

[8] M. Leopold, M. B. Dydensborg, and P. Bonnet, "Bluetooth and sensor networks: a reality check", *Proceedings of the 1st international conference on Embedded networked sensor systems*, Los Angeles, 2003, pp. 103-113.

[9] Anon, "WIT2410 Roaming Frequency Hopping Transceiver OEM Module", Cirronet Corporation product brief, available at http://www.cirronet.com/pdf/brochure_wit2410.pdf.

[10] R. Tobin. "US Army's BLUE Radio", in *Proc. SPIE, Unattended Ground Sensor Technologies and Applications*, volume 5090, Orlando, Florida, April 2003.

[11] N. Kushalnagar, G. Montenegro, and C. Shumacher, "6LoWPAN: Overview, Assumptions, Problem Statement and Goals", IETF Network Working Group Internet-Draft work in progress, Nov. 8, 2006. available at: http://www.ietf.org/internet-drafts/draft-ietf-6lowpan-problem-06.txt

[12] M. Gerla, K. Tang, and R. Bagrodia, "TCP performance in wireless multi-hop networks", *Proceedings of IEEE WMCSA'99*, New Orleans, LA, February 1999.

[13] S. R. Saunders, *Antennas and Propagation for Wireless Communication Systems*, John Wiley and Sons, New York 1999, pp. 94-97.

[14] J. Li, et. al., "Capacity of Ad Hoc wireless Networks", *MobiCom 2001*, Rome, Italy, pp. 61-69.

[15] J. Jun and M.L. Sichitiu, "The nominal capacity of wireless mesh networks", *IEEE Wireless Communications*, 10 (5) (2003), pp. 8-14.

[16] D. Estrin, D. Culler, K. Pister, and G. Sukhatme, "Connecting the Physical World with Pervasive Networks", *IEEE Pervasive Computing*, Vol. 1, No. 1, pp. 59-69, January 2002.

[17] P. Gupta and P. R. Kumar. "The Capacity of Wireless Networks". *IEEE Transactions on Information Theory*, 46(2):388-404, March 2000.

[18] D. Maniezzo et. al., "T-MAH: A Token Passing MAC protocol for Ad Hoc Networks", *MedHocNet2002*, Chia, Sardegna, Italy, September 2002.

[19] L. Bao and J. J. Garcia-Luna-Aceves, "A new approach to channel access scheduling for Ad Hoc networks", *The Seventh Annual International Conference on Mobile Computing and Networking 2001* (2001) pp. 210–221.

[20] K. Sohrabi, et al., "Methods for Scalable Self-Assembly of Ad Hoc Wireless Sensor Networks", *IEEE Transactions on Mobile Computing*, Vol.. 3, No. 4, Oct.-Dec. 2004, pp. 317-331.

[21] Ware, Christopher, Tadeusz Wysocki and Joe Chicaro, "On the Hidden Terminal Jamming Problem in IEEE 802.11 Mobile Ad Hoc Networks", *Proceedings of IEEE ICC '01*. June 2001, pp. 261–265.

[22] Lundgren, H. et. al., "Coping with Communication Gray Zones in IEEE 802.11b based Ad hoc Networks", *WoWMoM'02*, Sept. 28, 2002, Atlanta, GA.

[23] Anon., *Part 11 Wireless LAN Medium Access Control (MAC) and Physical Layer (PHY) specifications. IEEE Standard 802.11-1999*, IEEE standard reaffirmation June 12, 2003.

[24] Anon., *Part 11 Wireless LAN Medium Access Control (MAC) and Physical Layer (PHY) specifications. High-speed Physical Layer in the 5 GHz Band IEEE Standard 802.11a-1999*, IEEE standard reaffirmation June 12, 2003.

[25] G. Bianchi, "Performance Analysis of the IEEE 802.11 Distributed Coordination Function", *IEEE Journal on Selected Areas in Communications*, Vol. 18, No. 3, March 2000.

[26] H. Wu, et. al., "Performance of Reliable Transport Protocol over IEEE 802.11 Wireless LAN: Analysis and Enhancement", *IEEE INFOCOM'2002*.

[27] D. McIntire, "Energy Benefits of 32-bit Microprocessor Wireless Sensing Systems" Sensoria whitepaper, Dec. 2004 available at http://www.sensoria.com

[28] Anon., "Low Power Advantages of 802.11a/g vs. 802.11b", Texas Instruments White Paper, December 2003. Available at: http://focus.ti.com/pdfs/vf/bband/80211_wp_lowpower.pdf.

[29] D. McIntire, et. al., "The low power energy aware processing (LEAP) embedded networked sensor system", *Proceedings of the fifth international conference on Information processing in sensor networks*, pp. 449-457. April 2006.

[30] R. Niu et. al., "Decision Fusion in a Wireless Sensor Network with a Large Number of Sensors", *7th International Conference on Information Fusion*, June 28 to July 1, 2004, Stockholm, Sweden.

[31] O. T. Magoon and J. M. Hemsley editors, *Ocean Wave Measurement and Analysis*, published by the American Society of Civil Engineers, New York, 1993.

[32] See for example the Federation of American Scientists overview of DICASS sonobuoys at: http://www.fas.org/man/dod-101/sys/ship/weaps/an-ssq-62.htm last accessed on December 20, 2006.

[33] M. Dunn, R. Kremer, and T. Hromadka, "Network-Enabled Anti-Submarine Warfare System Demonstration Plan Report Technical Volume", Exponent Corporation Technical Report, May 28, 2004.

[34] Anon., "Test Report for End-to-end Demonstration of NEASW Communications (3.4.1.3a)", Exponent Corporation Technical Report, November 16, 2005.

[35] C. I. Beard, "Coherent and Incoherent Scattering of Microwaves from the Ocean", *IRE Transactions on Antennas and Propagation*, Sept. 1961, pp. 470-483.

[36] W. M. Merrill, et. al., "Quantifying Short Range Surface-to-Surface Communication Links", *IEEE Antennas and Propagation Magazine*, Vol. 46, No. 3, June 2004, pp 36-46.

[37] W. M. Merrill, "Network Scalability Discussion Document for Network Enabled Anti-Submarine Warfare (NEASW)", Sensoria Corporation Technical Report, April 18, 2005.

[38] W. M. Merrill, "A Simple Model for Multi-Hop 802.11 Throughput", Sensoria Corporation Whitepaper, June 2006.

[39] E. M. Royer and C. E. Perkins: "Multicast Operation of the Ad-hoc On-Demand Distance Vector Routing Protocol", In *Proc. of the 5th annual ACM/IEEE International Conference on Mobile Computing and Networking (MobiCom)*, pages 207 - 218, Seattle, WA, August 1999.

[40] Anon, "Multicast Extensions of AODV (MAODV)" patch to the Uppsala University AODV implementation provided by the University of Maryland, available at http://www.isr.umd.edu/CSHCN/research/maodv/MAODV-UMD.html.

[41] Uppsala University open source implementation of AODV, available at: http://core.it.uu.se/core/index.php/AODV-UU

[42] C. Perkins, E. Belding-Royer, and S. Das, "Ad hoc On-Demand Distance Vector (AODV) Routing", RFC 3561.

[43] J. T. Moy, *OSPF Complete Implementation*, Addison-Wesley, Boston, 2001.

[44] Clausen, T. and P. Jacquet. "Optimized Link State Routing Protocol (OLSR). Network Working Group Request for Comments (RFC) 3626", 2003. Available at: http://www.faqs.org/rfcs/rfc3626.html.

[45] S. J. Lee, W. Su, and M. Gerla, "On-Demand Multicast Routing Protocol (ODMRP) for Ad Hoc Networks", Internet Draft of the IETF MANET Working Group, January 2000, available online at: http://www.cs.ucla.edu/NRL/wireless/PAPER/draft-ietf-manet-odmrp-02.txt.

[46] D. B. Johnson, "Routing in Ad Hoc Networks of Mobile Hosts", *Proceedings of the Workshop on Mobile Computing Systems and Applications*, pp. 158-163, IEEE Computer Society, Santa Cruz, CA, December 1994.

[47] W. M. Merrill, "Software Design Document for Network Enabled Anti-Submarine Warfare (NEASW)", Sensoria Corporation Technical Report, October 28, 2005.

[48] Anon., Internet Research Task Force Delay Tolerant Networking Research Group (DTNRG), see for example: http://www.dtnrg.org/wiki.

[49] S. Burleigh, et. al., "Delay-tolerant networking: an approach to interplanetary Internet", *IEEE Communications Magazine*, Volume 41, Issue 6, June 2003, pp. 128 – 136.

31

Wi-Fi based Tracking Systems[a]

The advent of accurate, low-cost tracking solutions coupled with management requirements that demand to know where high-value assets are at all times have created a market for location systems that use Wi-Fi as their infrastructure. This chapter examines some of the issues and challenges associated with Wi-Fi location technology.

31.1 "Where's my stuff?"

One of the most sophisticated and promising applications that can grow out of Wi-Fi LAN implementation is the ability to track the location of assets and personnel within campus and enterprise environments. Network-based tracking uses a combination of network-centric computers, radio tags or other wireless devices, base stations and application software to locate, track and monitor assets and personnel in real time. For the purpose of this book, we are only concerned with the ones which use existing Wi-Fi networks as their communications infrastructure.

This technology is a marvelous advantage over the manual process of searching for misplaced items and manual inventory. And, in critical healthcare and first-responder situations, the ability to immediately locate life-saving equipment or know the whereabouts of key personnel can be the difference between life and death.

31.2 Benefits of Tracking

Some important benefits of positioning and tracking systems include:

- Track computers or assets without being in 'line-of-sight' inside or outside;
- Monitor real time information via the corporate intranet (or remotely via an internet browser;
- Solve expensive logistics and, in the case of hospitals, safety and liability problems by instantly locating high-value assets or the technically proficient people that are required to operate them;
- Maintain a complete log of movements for auditing, security, maintenance and usage analysis;

---

[a] *Ekahau Corporation*

- Generate real time inventory reports of all tagged assets, thus saving time and money used in asset location;
- Improve workflow and therefore help improve process optimization, increase margins, secure assets, and assure that technical equipment is regularly maintained and ready for service.
- Tailor information to individual visitors in museum and education settings as the system "knows" their location and can provide appropriate localized information.

31.3 What exactly are RFID and RTLS?

RFID (radio frequency identification) incorporates the use of electromagnetic or electrostatic coupling in the radio frequency (RF) portion of the electromagnetic spectrum to uniquely identify an object, animal, or person. Its advantage over simpler identification technologies, such as bar coding, is that RFID does not require direct contact or line-of-sight scanning.

An RFID system consists of three components: an antenna and transceiver (often combined into one reader) and a transponder (the tag). The antenna uses radio frequency waves to transmit a signal that activates the transponder. When activated, the tag transmits data back to the antenna. The data is used to notify a programmable logic controller that an action should occur.

The action could be as simple as raising an access gate or as complicated as interfacing with a database to carry out a monetary transaction.

Low-frequency RFID systems (30 KHz to 500 KHz) have short transmission ranges (generally less than six feet). High-frequency RFID systems (850 MHz to 950 MHz and 2.4 GHz to 2.5 GHz) offer longer transmission ranges (more than 90 feet).

A real time location system (RTLS) goes a step further. It uses radio frequency technology, such as 802.11 Wi-Fi, to automatically transmit the physical location of tagged objects on a continual basis. A system of this type requires some type of RFID tag to be attached to each object that needs to be tracked and RF transmitters/receivers located throughout the facility to determine the location and send information to computerized tracking system.

Some RTLS solutions still require the use of proprietary readers but the most sophisticated systems use software-only solutions to report in.

31.3.1 A Review of RFID/RTLS Development

Before we get into the details of tracking over Wi-Fi, we need to review the history of this sector.

Various sorts of identification and tracking technologies have been around for decades, but tracking via Wi-Fi networks have only been in existence for about five years.

In the early days, the technology used various means of reading a signal as it passed some sort of portal. This has been used for years in grocery stores using bar code readers, for example.

31.3.2 Passive RFID

Passive RFID describes the next advance over bar code readers. This utilizes a small RFID tag attached to an item that moves past a portal or reader. The reader "excites" the otherwise dormant tag, which causes it to create a record in the reader that shows that X item has passed Y point. A typical application would record the movement of a package that has moved through a warehouse door and onto a truck.

31.3.3 Active RFID

The problem with passive RFID is that it must pass some sort of "chokepoint" to register its movement. How do you get items that never move to report themselves in? Enter the Active Tag. The Active RFID tag is smarter than the passive tag, and more expensive. Indeed many of these tags are quite smart, able to process a fair amount of information regarding its payload. Therefore an active RFID tag has an internal battery to power its processor.

31.3.4 RTLS System

The smartest of the solutions in this class is the Real-Time Location System (RTLS). These systems fit generally in the Active RFID category, though technically they are their own technology. When we speak of Wi-Fi enabled RFID systems, we are really talking about RTLS.

The Wi-Fi RTLS market, though less than $20 million in 2005, will grow to more than 1.6 billion by 2010, according to the Yankee Group. In this space there are three vigorous players: Aeroscout, Ekahau, and Pango. Each has its own proprietary way of dealing with the challenge of locating assets.

31.3.5 How RTLS works?

All positioning technologies make use of the fact that measurable transmission characteristics of wireless networks will vary with respect to the physical location where the measurement is done. This variability is caused by factors such as the distance and the angle between the location of the transmitter and the location of the receiver, and the properties of the surrounding physical environment, such as the location, shape and material of reflecting/absorbing surfaces like walls, furniture etc.

Unfortunately, measuring these location-dependent signals accurately is difficult mainly because of the so called multipath problem (Rayleigh fading): the signal measurements are inherently noisy as the radio signals travel between the transmitter and receiver along several alternative paths, and each path is affected by different environmental factors. What is worse, some of the environmental factors are constantly changing due to the presence of humans, variation in air humidity, and so on. The situation is especially problematic when there is no line-of-sight between the transmitter and the receiver which can mean the signal traveling distance is not necessarily the same as the direct physical distance.

When building a system for positioning, the two most important technological issues are the following:

- First, one needs to decide the location-dependent variables to be used for positioning, and how to obtain the required measurements. The most commonly used location-sensitive variables in positioning are timing-based variables like timed difference of arrival (TDOA), angle of arrival (AOA), and signal-strength-related variables like the received signal strength indicator (RSSI).
- Second, one needs to decide how to use the received measurements for location estimation. For solving these two sub-problems, Ekahau has developed methods that are both theoretically and empirically validated. In the following section, we will describe the Ekahau approach in more detail, and compare it to alternative solutions.

31.4 Summary of Timing-based Technologies

A common drawback in the timing-based approaches, such as TDOA, is that they all require expensive special-purpose hardware to be added in the network infrastructure.

31.4.1 The angle of arrival approach (AOA)

In the angle of arrival (AOA) approach to positioning the location-dependent variable is the signal direction, or more precisely, the angle at which the signal arrives to special multi-element directional antennas placed at the base stations. Naturally, a single directional antenna gives only the bearing, not the distance, of the transmitter. This means that several directional antennas located well apart are required for positioning.

For accurate positioning the angle measurements need to be relatively accurate, but as discussed earlier, achieving high accuracy measurements in wireless networks is difficult because of the multipath problem and other factors affecting the signals. Consequently, the main drawback of this approach is the price of the directional antennas to be added in the network, as the required high-precision angle-sensitive measuring equipment is relatively expensive.

Another problem with the AOA approach is that great care needs to be taken in network planning so that the base stations and the directional antennas are placed in such a manner that positioning is possible at any location (problems occur whenever two base stations are approximately in the same direction, but at a different distance). On the other hand, if the network with the directional antennas is too optimized for positioning, the resulting setup may be a poor solution for data transfer, which of course is usually the primary function of wireless networks.

Because of the complexity of the AOA approach, there are no relevant commercial implementations of the system.

31.5 Positioning Alternatives

Ekahau uses the received signal strength indicator (RSSI) as the basis for positioning and a probabilistic framework for estimating the location of the tracked item. Because RSSI data is a standard in all Wi-Fi networks, regardless of infrastructure vendor, the Ekahau positioning and tracking system can be used without any changes to an existing Wi-Fi network infrastructure. Another benefit is that signal strength values change relatively smoothly with respect to changes in location, which means that the RSSI method is not as sensitive to measuring errors as the timing-based or AOA approaches.

When using the RSSI signals, two alternatives are possible: one can either measure base station (access point) transmission strength from the perspective of the mobile device, or one can measure the transmission strength of the mobile device from the perspective of the base station. Ekahau has chosen to use the first approach because signal measurement from the client perspective provides the most accurate, comprehensive measurement of the WLAN, including Wi-Fi characteristics such as multi-path. In addition, signals transmitted from the WLAN infrastructure tend to be stronger and more consistent than signals transmitted from mobile devices.

31.5.1 Location estimation

Once the location-dependent variable has been chosen as the basis for positioning, the next question is to ask how to derive reliable location estimates from transmitted data. More precisely, the goal is to build a predictive mathematical model, which receives the chosen type of signal information (RSSI in this case) as an input, and outputs the estimated of location of the mobile device in question in meaningful coordinates. Location variable can be discrete or nominal (like "room B226" or "lobby"), or continuous (x, y, and z measured in pixels or distance units).

31.5.2 The Cell ID approach

The simplest solution to this problem is the so-called Cell-ID positioning. In this approach the current location is assumed to be somewhere within the coverage area of the base station to which the mobile terminal is currently associated (the base station from where the current feed of data is arriving over the wireless radio way). As the associated base station is often the same as the station with the strongest detected signal at the current location, and as the strongest signal comes often from the nearest base station, this location estimate makes intuitively some sense.

However, it is obvious that the accuracy can even in optimal conditions only be relative to the distances between the base stations. In Wi-Fi environments this can be dozens of meters and the associated base station is not necessarily the station with the strongest signal. Also the strongest signal may not come from the nearest base station. Consequently, the Cell-ID approach is generally quite inaccurate and unreliable and thus not feasible for practical applications, unless the requirements for the positioning accuracy is extremely low.

These problems with the Cell-ID approach can be overcome to some degree by decreasing the transmitting power of the signal sources so that the signal can be detected only at the immediate vicinity of the transmitter. This is the idea behind the radio frequency identification (RFID) tags and similar devices. However, in order to enable accurate positioning everywhere, a huge number of RFID readers should be placed around the environment so that their individual coverage areas together would cover the whole area. On the other hand, if positioning is required only at designated spots (like for example only at the doorways), then this type of an approach can be feasible, and has been adopted in many warehousing applications and similar environments (with the cost of purchasing the dedicated RFID hardware and software).

31.5.3 Triangulation-based approaches

Another commonly used approach for solving the positioning problem is based on the idea of triangulation. In this approach, one constructs a function that outputs the distance to each base station, given the measured signals. If the distance to at least three base stations can be determined, then the intersection of the three circles around the base stations at the estimated distances gives the current location of the mobile terminal.

However, when applied in practice, the triangulation-type approaches face severe problems. The main problem is that the signal measurements are inherently noisy due the multipath problem and other factors, as discussed in the beginning of the section. If one or more of the distance estimates are of poor quality, then the circles drawn around the base stations may not intersect at all, in which case the method does not produce a location estimate at all. The approach also breaks down completely if signals from less than three base stations are observable. This is a known problem with utilizing the GPS approach in places with high surrounding obstacles. Furthermore, the locations of the base stations needs to be known, otherwise positioning becomes totally impossible.

31.5.4 Ekahau Estimation Technology

In order to overcome the above problems, Ekahau has developed a probabilistic positioning framework, where the world is taken to be probabilistic, not deterministic, and one accepts the fact that the measured signals are inherently noisy.

A probabilistic model assigns a probability for each possible location (L) given observations (O) consisting of the RSSI of each channel:

$$P(L|O) = P(O|L)P(L)/P(O) \qquad (1)$$

where
$P(O|L)$ is the conditional probability of obtaining observations O at location L.
$P(L)$ is the prior probability of location L.
$P(O)$ is a normalizing constant.

The formula above is an example of an application of a mathematical theorem known as the Bayes rule. Based on probability theory, this Bayes theorem gives a formal way to quantify uncertainty, and it defines a rule for refining a hypothesis by factoring in additional evidence and background information, and leads to a number representing the degree of probability that the hypothesis (in our case, location estimate) is true.

In order to make the theoretically elegant probabilistic framework work in practice, a number of important problems need to be solved. For solving these problems, Ekahau has developed two innovative approaches, site calibration and rail tracking.

31.5.5 Ekahau Site Calibration™

As can be seen from the formula above, the key issue in the Ekahau approach to positioning is to estimate the probabilistic distribution of measured signals "O" in different locations "L"; in other words, we need to determine the conditional probabilities "P(O|L)."

For building the distributions, Ekahau technology uses a representative sample of site-specific calibration measurements as input. Using this type of calibration data and elaborate state-of-the art machine learning algorithms, it is possible to construct a site-specific model of the environment very quickly and easily. Note that with this type of an approach, the system does not need to be told anything about the wireless environment explicitly, not even the locations of the base stations.

In Ekahau Site Calibration™, a site-specific model of the radio network is created by collecting sample points from different site locations. The collection is done by measuring the signals using a laptop PC and calibration software. Each sample point contains received signal intensity (RSSI) and the related map coordinates, stored in an area-specific Positioning Model for accurate tracking.

31.5.6 Ekahau Rail Tracking™

The probabilistic positioning framework described above can be further improved by making the following observations.

First, it is intuitively clear that current location of the tracked person or asset is very probably near the place where the terminal was one or two seconds ago. So it is clear that historic location information can be used in a predictive way to further enhance positioning accuracy. This results in a process where not only the current observations are considered, but the full history of observations is taken into account when locating the mobile device or user.

Second, if one wishes to leverage device location history in determining the current device location, it is a sound practice to distinguish legal paths from illegal paths in the tracking model. Illegal paths meaning, for instance, paths going through walls, or paths outside of the building where tracking is not possible (on an upper floor for example). In this way, the introduction of tracking rails enhances the probabilistic framework by defining areas and pathways of higher location probability. The Ekahau Rail Tracking allows application managers to draw the navigational routes or rails to a layout of the navigation area. The rails drawn by the manager are used by the Ekahau Positioning Engine™ in estimating the locations of objects. The Ekahau Positioning Engine uses the rails for adjusting the location estimates to match the possible navigational routes of the site and thus improves better overall accuracy. All these problems are solved in the Ekahau RTLS through the patented concept of rail tracking.

31.5.7 Normalizing RSSI scales

The parameters used for estimating the location of tracked devices are the RSSI values read from the network interface cards (NIC). This means that also the accuracy of positioning depends on the accuracy of the RSSI values. As the quality and accuracy of the values depend on the Wi-Fi chip set used, the choice of NIC card generates error that directly affects to the accuracy of the position estimates.

To overcome the differences between NIC cards, Ekahau has invented and patented a technology for normalizing the RSSI values received. The normalization is done in the server and minimizes the error caused by the differences between the accuracy of the NIC cards. Additionally the same technology can be used for normalizing values received when the tracked devices is enclosed or attached to an object that may affect the signal such as hospital bed, forklift or any other large device.

31.5.8 Configuring Access Points

Although the probabilistic positioning framework can estimate a device's location when using just one access point, for best-possible results, it is recommended that at least three overlapping access points be audible at each location in the positioning area. In relatively open areas, less than three overlapping channels may lead to degraded location resolution (two or more sample points can have very similar access point/signal strength combinations). More overlapping access points provide better resolution, although overall accuracy will not increase significantly when exceeding six overlapping access points. Further documentation on WLAN design considerations is available directly through Ekahau, Inc.

Due to the nature of the high-frequency signal (2.4 GHz) and typically limited antenna receive strength in many 802.11 devices, positioning resolution and stability tends to be better near an access point.

In practice, all areas of a floor do not require the most precise level of accuracy, so a good rule is to place an access point in each area where the best resolution is required. If initial calibration and positioning results do not correspond with requirements, adding overlapping access points is not always the best solution. It may be possible to rearrange the existing access points in a way that increases resolution in significant areas.

If access points (or external antennas) are moved after calibration, the affects areas will have be recalibrated. All sample points, which are within the original and new coverage area of the moved access point(s) would have to be deleted and recorded again.

Sometimes removing access points from the positioning model may improve positioning results. After calibration, the software allows the user to see coverage area contours for each access point, from which the calibration device has received signals. If an access point's coverage looks flat or if it has more than one peak, this probably degrades accuracy (several distant sample points may have similar signal values). In addition, access points which are administered by outside parties should not be used since location, antenna placement and transmit power could be changed without notice. An analysis tool can be used to select the optimal access point combination.

Table 31.1: Location Tracking Technology Comparison.

	Ekahau	RFID	TDOA/EOTD	CELL-ID
Accuracy	1 - 3 m	50 cm - 2 m [a]	3 - 50 m	10 - 50 m [b]
Continuous tracking	Yes	No	Yes	Yes
Proprietary hardware	Tag	Reader and Tag	Base stations and Tags	No
Availability	Wi-Fi	Readers required	Dedicated network	Everywhere
Software-based	Yes	No	No	Yes
Cost	Low	Medium to High	High	Low
Roll out time	Fast	Slow	Slow	Fast
Frequencies	2.4 GHz	50 kHz-2.5 GHz	Any	800, 1900MHz
Line of sight	Not required	Not required	Not required	Not required

[a] Requires very short distances between the reader and the tag (passive tags function up to 50 centimeters, active tags up to 2 meters). Range may be reduced if reading at high speed is needed.
[b] The accuracy of the technology is highly dependent on access point density.

31.5.9 Adding "dummy" access points

One simple way of improving accuracy is to install beacon or "dummy" access points, which are useful to the positioning engine but do not connect to the backbone.

Almost any (even low-cost consumer-grade units) access point can be used for this purpose. When placing these dummies, make sure they do not sure the same ESSID as the access points that are used for data transfer. In addition, it is recommended that you use a different channel for the dummies. This is a security measure even though the dummies are only used to transfer the probe data. In addition, the "Wireless Zero Configuration" option found in some Windows versions is recommended to be disabled to avoid the operating system from roaming to unwanted networks.

31.5.10 Associating with the Host Network

Much has been written about the advantages and disadvantages of requiring a location system to associate with the Wi-Fi infrastructure that hosts the transmission of data.

The argument is made that NOT associating the radio tags with the network is preferable because it alleviates the traffic burden on the network. In actuality, just the opposite is true. If the traffic burden is the reason for avoiding association, then this point is moot because the amount of traffic being communicated by the outlying tags is so trivial that clients see very little impact (Ekahau claims that a typical burst from a tag is only about 60 bytes).

The other factor is that RTLS asset tracking are typically off more than they are on. This is by design to conserve battery life, and tags wake up to beacon their location based on configurable timing parameters. In some cases, tags may only wake up based on a periodic update, or based on movement. In many cases, hours, and indeed, days may go by before a unit associates to report its location. In this scenario, the unit wakes itself up, sends its payload, and returns to a sleep state until the next event.

But the most obvious reason for having the location system associate itself with its host network is that the network now "knows" its "members" and can therefore accommodate the more sophisticated features of a tracking system, including two-way communication session. Not to be associated with the host network means that tags are just beaconing somewhere in the environment with the hope of being listened to.

31.6 Conclusion

This chapter has attempted to address some of the challenges and possibilities of Wi-Fi enabled Real-Time Location. We have looked at the various methods that are being offered to achieve reasonable accountability of high-value assets. We have used the Ekahau example as a means of explaining how a system could bring value to a company that seeks to gain automated management control of its assets.

32

Building the Mobile Computing Environment through Context-Aware Service Management[a]

A whole new thinking is needed for mobile computing. We call this the Mobile Computing Environment. This encompasses a deep understanding of requirements for the mobile user and the specific mechanisms required to effectively supply services (a combination of applications and resources pertaining to those applications) to a broad variety of mobile devices. This chapter provides an overview of this emerging field, together with an outline of Appear's solution for this new environment.

32.1 Introduction

As high-performance computing in small form factor devices arrives, a shift from traditional browser-based interfaces to a combination with full self-contained services has become increasingly evident. These same mobile devices are standardized and inexpensive, creating whole new opportunities to cost-effectively computerize groups of mobile workers.

These services offer superior interactivity, work both online and offline and utilize the onboard processing power to execute locally, saving both bandwidth and precious battery life. To support these services, new powerful distribution mechanisms are needed to allow for automatic installation, filtering and execution over wireless networks - a service profiling and provisioning scheme that makes discovery, download and installation as natural as sending an email. These distribution mechanisms need to take into account several dimensions of the user's context (i.e., the information that describes the situation of a person or entity such as its location, time, profile, available bandwidth, language, and device type), when determining which services, data or content a user requires.

This new mobile environment has to enable the mobile user with capabilities to complete work actively and provide background updates when the mobile user is on the move, thus providing mobile users with the best tools needed for their current task.

To cater to these needs we present a context-aware service platform (call Appear IQ from Appear) for wireless networks. It features two powerful tools for the mobile user; the patented Click & Run™ technology, enabling enterprises to distribute a combination of full applications, content and services over-the-air to smartphones, PDAs, laptops, tablet PCs,

---

[a] *Appear*

and UMPC (Ultra Mobile PCs) in a single click; and the Appear Synchronization module, which always keeps the services up to date, whenever the mobile device is in network coverage.

32.2 Trends

As wireless users demand more and more data services in their handheld devices, the role of the device and the requirements on the hardware, undergo profound changes. The PDAs and smartphones launched today are not simply phones or electronic organizers, but powerful handheld computers with potent processors, a lot of memory and complete service environments.

Service developers will benefit immensely from the open interfaces, high level of modularity and extensibility of the new handheld computers. Regardless of which operating system they are targeting, services can be offered across multiple standards and devices globally. The major benefit and direct consequence of this for end-users, is that a flood of services will be made available for handheld devices, i.e. both legacy and new mobile applications.

As wireless devices become increasingly powerful more users will access services and the Internet using handhelds. Thus it is necessary to consider how this fundamental change - impacts the ideal usage model. By analyzing the way user behavior changes from one type of use to the other, the following patterns have been identified:

32.2.1 Maximizing employee process efficiency and time savings

The objective in a mobile work environment is for each employee to be served the necessary and relevant information to perform their work. This maximizes the efficiency of the employee and effectiveness of customer interactions. To be efficient this has to be based on a system that anticipates the particular employees work requirements. A context aware solution solves this problem.

32.2.2 Mobile users behave fundamentally different from Desktop users

- *Wireless users react to tasks, but are NOT proactive.* Wireless users are reactive and respond to a message or discover some issue that needs to be resolved. Automatic push mechanisms are necessary. Desktop users have the luxury of being able to select information to work on.
- *Wireless users focus on the work task or customer, NOT the computer.* Wireless users tend to use their devices more as a tool while focusing on some completely different task, e.g. fixing the escalator. Access to services must be as simple as possible. Desktop users are generally centered on the actual device (computer), e.g. preparing an electronic spreadsheet.
- *Wireless users need to be served the right information, they need the tool to anticipate their requirements, NOT surfing and searching.* Mobile workers are confronted with the situation of fixing a specific problem such as the wing of a plane or answering a

customer question about the late arrival of a particular train. This does not provide time for surfing or searching. Direct access to correctly anticipated services is critical.

- *Employees need the work tools, NOT a specific technology.* In mobile deployments it can be observed that employees need a combination of content, web services and rich applications. Rich applications examples include 1) applications that are transactional for example, requiring authorization, 2) streaming media such as security camera feeds, 3) interactive applications such as war games in a defense environment, 4) real time information such as traffic information, 5) Voice over IP. Mobile employees are not interested in the source or type of service, just the functionality. Multiple services must be manageable.

- *Dynamic, time and event-related work, NOT static.* Employees may need a specific drawing to solve a maintenance problem that has not occurred for a year. This data or applications need to be anticipated and delivered on the fly.

- *Serve enabling applications, NOT download.* Mobile users want the service to work. They can for example not be expected to download the Acrobat software when they are sent a PDF file. These types of related or enabling requirements must be automated.

- *Simply understood, NOT complex.* Blue-collar employees that are currently being computerized, because of the larger and more cost efficient access to standardized mobile devices and widespread connectivity, often have minimal experience with computers. This places large requirements on ease of training. A computer novice must be able to work productively quickly. Users must see a benefit to themselves and their ability to perform their tasks.

32.2.3 Mobile devices are not Laptops

- *Limited battery.* Traditional desktop users can always count on continuous power supply; wireless users have to cope with a limited battery life. Users cannot keep drawing on batteries by sending request to a central server; they need to perform operations locally, in offline mode.

- *Interface restrictions.* The desktop interface is much richer than its wireless counterpart, both in terms of presentation and input capabilities. Use of simple icons is necessary.

- *Typing on mobile devices is limited.* The desktop interface is larger and therefore suited for a real keyboard. The small form factor of the mobile device allows for only limited typing. The use of touch screen technology is preferable.

- *Heterogeneous fleet of devices.* Rapid product cycles, make it difficult for enterprises to buy large numbers of the same generation of device, due to progressive implementation of projects over time. Additionally, professional workers often want to have a say in the type of device they use. Many types of devices must be supported.

- *Mobile devices are sometimes shared and frequently replaced.* In an environment where employees work in shifts, considerable expense can be mitigated by sharing devices. This requires the distribution of applications and services based on user log-in. If a device is lost or needs to be replace the specific's user configuration must be able to be repopulated in a new device.

- *Physical security of the applications and device.* Because very sensitive information may be used on the mobile device at work, the employer may need to ensure that the data is no longer stored on the device outside of work. Additionally, if the device is lost it must be excluded from further access to services. Forced removal of applications as well as not distributing applications to rogue devices is important.

32.2.4 Mobile devices to adapt to various types of connectivity

- *Sporadic bandwidth.* While the desktop users enjoy a reliable broadband Internet connection, the wireless users experience somewhat different access methods; although they are likely to have high bandwidth wireless connections in hotspots, it is sporadic. Outside hotspot coverage they will rely on expensive cellular networks with limited bandwidth. Being able to predict the type of bandwidth and therefore what type of application will run effectively in that particular bandwidth is essential.
- *Sporadic connectivity.* By their intrinsic nature, as mobile users move between or outside of different types of networks, it will require use of services when the mobile devices are not connected to the network. Ubiquitous homogenous wireless coverage is not yet a reality. This requires the ability to store services locally for continued work outside of coverage or when connection is lost. Working in offline mode is necessary. This also implies a requirement for cross network functionality in the use of mobile services.
- *Cost of connectivity.* Today what may be possible over a GPRS or 3G network may not be cost effective for the enterprise. Thus solutions such as web browsing and thin clients that require constant connection are not practical and may often be cost prohibitive.

32.2.5 Mobile services to have a different life cycle then desktop services

- *Resources adaptation.* While a desktop service can require a user to move between different resources, e.g. files, search-engine, a mobile service has to be able to provide the mobile user a direct and effective way of solving the task handed to the user. The service needs to be pre-loaded with the right resources, when the service is being launched.
- *Configuration adaptation.* A desktop is normally the work tool of one single user, who has his or hers service environment setup as required, and normally has a supporting IT-department whom can support and configure the desktop environment for the desktop user. A mobile user on the other hand mostly is remotely located and is normally not educated to change the configuration for different tasks, and will require the system to be able to change the configuration as the work task is changing,
- *Run-time adaptation.* When a mobile service has been launched the contextual environment will change and it is in many cases important that the service has the capability to change with the environment. For example, the location might change, and the application will have to adapt itself for this new environment. In order for a rapid development of application these functions has to be available to the service, and cannot be expected to be developed each time a new service is created,

- *Resource synchronization.* While a desktop service always can rely on constant connectivity a mobile service has to be supported with the capability to upgrade its resources, or service offering, whenever a network becomes available. In most workforce scenario the only network a mobile service has connection to with is at its home base, where all relevant resources have to be synchronized to and from the device.

The high-performance computing made possible by new handheld devices and by these differences in ideal usage, imply that a combination of full applications and other services are needed and will become much more common. This strongly requires a model centered on adaptive user interfaces that provisions anticipated essential services for each employee. In additional there are problems of implementing mobile solutions in silos where information cannot be easily shared between applications and where much of the infrastructure needs to be recreated for each new solution. Appear's middleware directly addresses this issue by providing a standard set of services using context to react to changing business requirements. It eliminates the problem of silo implementations.

Web browsing is, and still will be, a vital part of the way wireless services are used, but it is evident that it needs a strong partner in full applications that consumes less battery and works both online and offline. Just as end-users prefer dedicated applications like word processors for specific tasks on their desktops, they will have similar needs on their handheld devices to a greater extent. In addition, the Web browsing approach requires the same use of context parameters to meet the unique behavioral needs of the mobile worker instead of the searching paradigm.

32.3 The Basic Components in the Appear Platform

Appear's platform consists of a number of building blocks, which together are centered on solving the mobility requirements described above. The main components in the platform are the foundation features, over-the-air service provisioning, over-the-air service synchronization, adaptive client and a contextual interface for context-aware services. All of these features are covered in this white paper, including a short description of two context-aware services, which shows full context-aware service integration.

32.3.1 Foundation features

The base layer of the Appear platform provides all the features (in picture denoted "Support") needed by any flexible software platform, including authentication, usage statistics, billing, network adaptation, resource caching, device session management, etc. All of these features are shared by the main modules of the platform.

The core foundation feature of the platform is the Context-awareness environment, which gathers all known context information about a device and produces a context profile for that device. The main components are Context Domain, Context Engine and Context Profile. The Context Domain is fed with Context Parameters, listeners that measure real-world attributes that are transformed into Context values. Context parameters include physical location, date/time, device type, user roles, network IP address range, user locale, language etc.

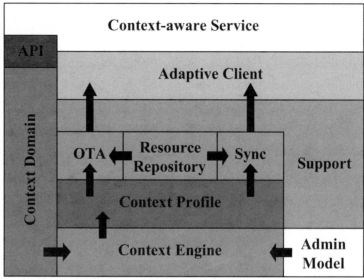

Figure 32.1: Diagram of the Appear context-awareness environment.

The context domain can also be extended with deployment of unique parameters, e.g. temperature, available battery, etc. The context engine maps the administrator model on to the context domain and produces a unique context profile for each device. This profile maps up a description of the services that should be available on the device and passes this on to the Over-the-air Service Provisioning and Over-the-air Service Synchronization modules. These modules then compare the current state of the device with the Context Profile set out for the device, and orders the adaptive client to update the services on the device. Information of context is used through the entire service life-cycle of the service: *selection* of service set based on the context profile, *filtering* of individual services, *enhancement* of services at boot or runtime and constant feeding of context information to services during execution to allow *service adaptation*.

The Administration model is a description of the relationship between different context parameters that can be found in the Context Domain, i.e. the business model the context engine shall map up profiles for. The resource repository holds all the resources a service consists of and the resources a service will need during run-time, such as configuration, data-files and common-runtime libraries.

On top of the foundation layer, the Appear platform provides modules enabling different areas of functionality that are all dramatically improved compared to traditional implementations due to the fact that they can leverage context information. The modules include Over-the-Air service provisioning (OTA), Over-the-Air service synchronization (Sync) and also Context-aware services such as Context-aware tracking and Context-aware communication.

32.4 Scientific theories vs. practical implementations

Context-aware services and computing is an established area of research within computer science in general and data communications and human-computer interaction in particular. Computer and behavioral scientists have explored the concepts of applications (or set of applications) dynamically being adapted to context from several different perspectives for over a decade. However, up until now, no one has succeeded in harnessing this powerful concept as a commercial product and brining it to market.

Technology has developed in the direction of context-aware services by several related strands of research, beginning with Weiser's (1991) vision of "ubiquitous computing" (now often referred to as "pervasive computing"). Additional research communities have been created around further developments such as "augmented reality" (Mackay et al., 1993) "tangible interfaces" (Ishii, 1997), "wearable computers" (Bass et al., 1997) "cooperative buildings" (Streitz, Konomi, and Burkhardt, 1998), etc. Today, annual conferences on ubiquitous computing are organized around the world.

Context awareness is fine in theory. The issue is figuring out how to get it to work in practice. There are a number of areas where theory and a practical solution come into conflict with each other. In this section, we will bring up the three most challenging areas.

The first challenge is that context-aware computing completely redefines the basic notions of interface and interaction. Research questions abound: What role does context play in real-world tasks? How can this be extended to a technological domain? What can the computation really do for end-users? How can end-users interact with a representation of context and yet maintain adequate control? How can software make end-users both more productive and less consumed by the technology itself? Appear's way of solving this issue is to have a user process that complies with all of the mobile requirements presented above providing an understandable and easy to use client interface.

The second challenge is to find a trade-off between which tasks that can be handled by the administrator and which is best handled by the system. Context-awareness quickly becomes an administration nightmare if the whole process of setting up a valid context set. As an example let us assume that we have one context parameter in the context domain, with ten different values. This is definitely not very hard for an administrator to map up and create relations for, however as soon as this number grows it quickly becomes very hard to oversee the administrator model. Just when the context parameter has grown to three (with each giving 10 values) this increases to a 1000 permutations for the administrator to handle. The other extreme view on the problem is to have a context aware system derive all possible relationships in the context domain and let the context engine perform matching against this set of relations compared to the criteria's set for each service. The problem with this approach is that this is increasingly difficult as the number of parameters grow. In the real world, one has to factor in hierarchical relationships of the objects of various parameters and cost of computing in a distributed network. Appear's way of solving this issue is to take out the domain specific knowledge from the computing problem and hand this to the administrator and then transform this administrator model into domain rules for the system to obey by. This gives the administrator control over the process and provides boundaries for the system to operate on. Even with these boundaries it still is a heavy computing load the context-aware system has to perform, which made the introduction of parallel computing necessary. The main factor determining the workload is the number of context parameters and the number of devices that exists in the system. Together with the third challenge Appear has chosen a distributed architecture to cope with these challenges.

The third challenge is the problems of creating a flexible enough solution to support incorporation of these features into existing telecom and Internet infrastructure. Appear realized these fundamental problems early on in the commercialization of its technology and concluded that solving these would be top priority and yield considerable competitive advantage.

Appear decided that the most promising approach to context-aware computing was a distributed software architecture where responsibility for gathering, processing and usage of context parameters would be divided between several different network components, now aptly named Appear Server, Appear Proxy, Appear Publisher and Appear Client. Around these components a model of context awareness and context-aware services was created, where real-world Internet and wireless networking issues could be handled through configuration of the basic components.

The Appear Server is the central point of administration, where all configurations, statistics and device sessions are kept. The Appear server is the organizer of the system. However, during run-time the Appear server has a minimum number of operations to perform, with the exception of organizing the other components.

The Appear Proxy is the run-time engine in the system. All processing and run-time operation is conducted by the proxy. In order for the system to work properly each proxy needs to be able to operate without the other proxies in the network.

The Appear Publisher is the smallest components in the system, and is normally placed together with the Appear Proxy. The Appear Publisher has two main functions; (1) to detect when new devices comes into the new network, and direct them to the closest Appear Proxy, and (2) to alert the Appear Client of real-time changes of the Context Profile and Context Domain. The Appear Client is the end-user interface that performs all Context-aware operations on the end-user device.

32.5 Context-aware Over-the-Air Service Provisioning (OTA)

Context-aware Over-the-Air Service Provisioning is a distribution mechanism for provisioning services based on a Context Profile. The distribution can be both pushing services to the devices, or to provide the end-user with a description of the tools that are available to the end-user. In either case the distribution mechanisms need to be fully automated, using client-side intelligence to install, configure and when necessary delete services for the users.

Due to the differences outlined above, Context-aware service provisioning solutions must be adapted to small screens and easy to use while on the move. The Appear Client is adapted to these criteria, however the services provisioned by the system also need to be adapted for these new form factors, in order for the services to fulfill the mobile user's requirements. Last but not least, they must also be able to take the new dimensions of physical location, time, and other tailored contextual parameters into account when distributing services; some services may be exclusively provided only in certain areas or to certain types of users and a certain time. Combinations of these parameters become particularly powerful in creating efficiencies in delivering the right service to the right employee at the right time.

Mechanisms to achieve this are called Context-aware Service Provisioning and can be defined as the method by which services are made available to the user. This includes download, context filtering, installation; context based configuration, execution, termination and deletion of the services and associated software from the wireless device.

Context-aware Service Provisioning with the Appear Solution can be divided into six steps:

- *Device detection.* In order for the system to keep a context domain of the device it has to have a relation with the device. Since the device moves from network to network, some public, others closed, it is the network that has to discover the client and provide the client with the configuration needed in a particular network. Appear provides this function by using the Appear Publisher to monitor the network for new devices and provide the client with network specific configurations when they appear in the network.
- *Service discovery.* Once a device has associated itself to the system, the system needs to make up a profile of the device. From the Context Profile of the device the OTA maps all available resources to the Context Profile. The Context Profile then compiles an offering to the Appear Client. The Appear Client presents this offering as icons in its user-interface. When a user needs a service they click on an icon, which starts the adaptation process of the service.
- *Download.* The first step in the adaptation process is to download all needed resources. This includes binaries, data-files and configurations.
- *Install.* When all resources are available on the device, the adaptation process performs installation of the resources.
- *Execute service.* When all resources are in place, and they have been properly installed, the last part of the adaptation process starts. The execution of the service starts with setting up the correct configuration for the service and the interaction with the operating system to launch the service. Once the service has been launched the application executes outside of the Appear Client in a system process by itself.
- *Discard service.* When a user leaves the network or if context condition has changed for a service, e.g. the device has moved, the Appear client will clean up resources not available anymore and shutdown services that are not allowed to continue outside of given context profile.

In addition to this, several other requirements have been taken into consideration when developing the Appear context-aware service provisioning system. For instance, the wireless arena of today is fragmented with several different access technologies and a plethora of handheld devices. These facts, combined with the differences between desktop and mobile use of services outlined above, lead to a number of key features provided by the Appear software platform.

- *Network agnostic.* The Appear software platform is designed to be truly network agnostic, in the sense that it only relies on IP-networking. All of the foundation features are implemented and it uses only IP-based techniques, which allows the platform to be used in a large variety of networks, and to support cross network operations. An Appear Server can for example at the same time serve both a WiFi and a GPRS network. The context listeners, the feeders of the context domain, will in some situations not be network agnostics, since their nature is that they measure network specific parameters. For example a WiFi and GPRS cross deployment would need to use different context listeners to determine device QoS, since this is a network specific parameter. The Appear Software Platform is designed to handle multiple listeners. It includes bandwidth optimizations for narrow bandwidth communications with capabilities such as compression and checkpoint restart (where a failed connection is restarted from the point of failure without having to go back to the beginning of the session or the start of a file).

- *Device agnostic.* There is a wide range of different handheld devices available on the market, which requires the service distribution to be cross device independent. Appear caters to these requirements by providing a device independent Appear Proxy-Client interface, which uses XML over HTTP, and two different Appear Client implementations. One java based client and one .Net client. These clients provide the same functionality, but for different device platforms. The java-client has a native interface to handle the interaction with the operating system, which makes it easy to port the java-client to any operating system. The .Net client is a Microsoft specific implementation of the Appear Client, which runs on any Microsoft platform, including Windows Mobile. It provides the same functionality as the java based client, however it does not require any java virtual machine, which is a constraint on some mobile devices.

- *Multiple Service distribution.* From an end-users point of view, the worker is interested in the use of services, as a tool for his work. He is not interested in how the tool (service) ends up on the device or if the chosen service fits the environment that exists on or around his device. Still the end-user will find himself in different environments that will affect the required services he has chosen. The Appear software platform handles all of his needs and provides him with the correct tool, which has been adapted to his unique environment. Each service can be setup to provide different levels of service depending on the environment, e.g. a security alert can be sent as a single picture when located on a low-bandwidth network, and will start providing streamed video as soon as the service comes into a high-bandwidth network.

- *Single-click execution.* The ease of use is one of the most important aspects of Appear software platform, which minimizes the end-user's required knowledge. When users enter a wireless network their devices will automatically be detected and available services displayed as new icons. To use any of the services, the users only need to click once to download, install, configure and run the complete service.

- *Independent applications.* Many of the services, enterprises want to deploy already exist today, so it is vital that the platform behaves transparent to these legacy-services. The provisioned service will not have to implement any Appear platform specific interface; they will be provisioned as is. However, as the use of context-aware services grows, the need for services that can react on its environment (i.e. its context domain) is becoming more and more relevant. Appear satisfies these needs by providing a generic interface to the context domain.

The user experience is very compelling and the interface is easy to use and learn. A typical workforce scenario in the Appear software platform might look like this (see Figure 2): A maintenance worker walks up to a train set that he is supposed to service. He uses his wireless device to look for available applications. The Appear service provisioning system automatically detects his presence and notifies him of available services by showing them as new icons on his PDA.

The maintenance worker sees that one of the icons represents the service manual for this particular train set and click on it to automatically download and install an interactive guide showing him what to do. As the maintenance worker leaves the area after completing his work, the service manual is automatically removed from the handheld device to save space and to avoid using stale information later. Optional storage should is available if the maintenance worker wants to use the service offline as well.

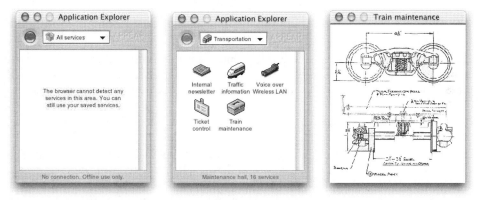

Figure 32.2: A typical end-user scenario where a maintenance worker automatically discovers available services on-site and can download and install interactive service manuals, etc. with a single click.

32.6 Context-aware Over-the-Air Service Synchronization

When synchronizing with the Appear platform it is possible to take advantage of the context awareness. It is for example possible to customize the synchronization behavior depending on who the user is, what responsibilities the user has and even the location of the user. One way of using the context is to only allow synchronization of sensitive sales data when the user is in the office, but allowing synchronization of task lists etc. in public hotspots.

The Appear synchronization module is built to automatically detect files that have changed and that need to be loaded onto the client devices. There is no need for the administrator to keep track of changed files; only the files that have actually changed will be synchronized. This makes it easy to allow external systems to run batch jobs to update the files that should be synchronized. The external batch jobs may very well update all files, but only the files where the content has actually been changed since the last batch job will be synchronized over to the client devices.

It is possible to synchronize files from the client devices back into the Appear platform. The manner in which files are stored in the Appear platform is very flexible and can easily be customized to fit the needs of external applications. There is a powerful way of using keywords to specify special paths for different users, locations and dates. The Appear synchronization solution also provides a simple way for external applications to get a complete history of what has been uploaded from the client devices.

The Appear synchronization solution also has a framework for client-side actions that can be executed locally on the device upon synchronization. These actions implement a wide range of features including file operations, profiling and device management (hardware inventory, remote upgrade, backup, forced uninstall of software, registry settings, etc.).

32.7 Context-aware Service

Most of the services that exist today are existing legacy services. However, as the field of context awareness is maturing the need for context aware services will grow. Appear can already today provide an interface for service developers that will bring context information to their services. Two interesting examples of such services are described here.

32.7.1 Context-aware tracking and dynamic change of application behavior

Context can also be fed directly into applications. For example, a maintenance worker can position himself in a station (in order to find the specific piece of equipment he needs to service) and he should be able to activate a map of the station. This map should display his current location as well as his target location. As he moves inside the station, the map should behave accordingly. In that case, the context parameters are not only used to present the right applications to the users, but also to modify dynamically the applications' behavior.

Another example: an employee is accessing his route-planning application in a train station to answer a customer question. When he starts the route-planning component, it takes into account the station where he is, since the system knows where he is located. In that case, the context parameters are not only used to present the right applications to the users, but also to modify dynamically the applications' behavior. The resulting improved ergonomics contributes to ease of use of the system.

32.7.2 Context aware Voice over IP

One of the most promising context-aware service is context aware Voice over IP. For example, with a context aware VoIP service calling the closest security officer is simply a click on an icon. The Appear solution has knowledge about all security officers (their role in the system) and the location and can then provide all users who could have an interest to call a security officer with that service.

When the calling user clicks on the icon, then system will make a match between the calling device context and match them towards all services that can provide a security call. This will match on the closest security resource and the Appear Client can launch the VoIP client with the configuration to call the security officer. All of this without any intervention from the user, who in a single click gets access to a security officer.

32.7.3 Implementing context-awareness

By making networking an integral part of the context-awareness model, Appear created the basis needed to solve the large number of real-world problems related to network structure and administration, network performance and reliability, security, cost, geographical distribution in data centers, etc. Building on the distributed architecture, Appear has also managed to solve these complex problems; The Appear platform scales from small private networks to large carrier networks, and can be upgraded from smaller deployments as new sites are added or improved performance is needed.

32.8 Networking and Appear Products

When deploying the context-aware services, three major degrees of freedom with regards to network are included in the Appear context-awareness model: (1) Component Distribution, (2) Device Detection, and (3) Change Publishing.

- *Component Distribution* is the issue of where in your network components best fit, and the impact they have on your network. The impact on your network can be measured in space, cost, bandwidth, geographical distribution, component administration, etc. The Appear platform supports both completely centralized configurations and fully distributed configurations. It also supports a wide range of hybrid semi-centralized or semi-distributed configurations in between.
- *Device Detection* is the first phase that takes place when a client device comes into coverage of an Appear enabled network. The Device Detection mechanism identifies the network and provides the Appear client with enough information of the network in order for the client to present itself and make a correct association to the Appear system and start gathering context parameters.
- *Change Publishing* is the mechanism that provides the Appear client with information on context changes and related information on changes in service or service offering availability.

It should be noted that the Device Detection and Change Publishing mechanisms are closely tied to each other and have to be considered in tandem when making a judgment of which configuration is best suited for each network.

32.8.1 Network Configuration

The Appear context-aware platform is designed to support the whole range of network configurations from a totally centralized deployment all the way to a fully distributed system. Decisions regarding where to put different components depend on the current network architecture, available resources and fault-tolerance requirements that are placed on the system.

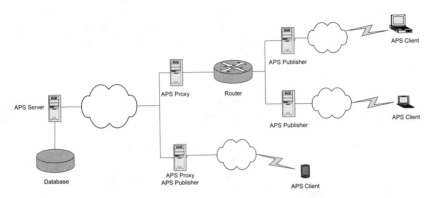

Figure 32.3: A distributed APS deployment.

The Appear Server and Appear Proxy are located in the provider network, and the Appear Client resides on the end-users' devices. In the network there will only be one Appear Server (or a cluster of Appear Servers acting as one Appear Server), but many Appear proxies, depending on the load and structure of the network.

When deploying an Appear system there are a number of factors that need to be taken into consideration; such as physical space, network bandwidth, geographical spread, component administration and fault tolerance of the system. However, it is important to notice that each component can be moved or reused if the requirements are changed and one configuration can easily be changed during operations.

The Appear Server shall always be placed in the back-end of the network, preferably in a Network Operation Centre (NOC) where the status and operation of the system can be supervised during operation. The differences come when we start looking at the placement of the Appear Proxy, since this component can reside wherever it is best suitable in the network, the only constraint on the network is that the clients are able to connect to the Appear Proxy. So the Appear Proxy can be placed as far out as possible or moved all the way back to the NOC; it can also be integrated together with the Appear Server and operate on the same physical machine as the Appear Server. So to get a better understanding of where to place the Appear Proxy we need to look at the different factors that controls where the Proxy is best placed.

First, the end-user experience is normally best if the Appear Proxy is placed as close to the end user as possible, or at least before any latency is introduced in the network (e.g. if the Appear system is deployed in a fragmented network where the network experience's different latency and/or bandwidth, the Appear Proxy should be placed between the low bandwidth/ high latency network and the Appear Server). This will reduce latency introduced by the network, both by the local network and by the back-end network.

If the network is built up of expensive leased lines the Appear Proxy can decrease the required bandwidth, and therefore decrease the cost of operating the Appear system. However, moving out the Appear Proxy into the network introduces new expenses to the operation of the network, such as hardware cost, location owner cost and remote administration. The Appear Proxy is fully administrated by the Appear Server, however if the operation fails on the hardware, maintenance personal has to send to the remote site, which increases the cost of operating the Appear system.

Deciding where to place the Appear proxy depends on the three general factors described above, and needs to be investigated from network to network. However the most common scenario is a combination of all three; the cost of infrastructure ownership has to be weighted against end-user experience. In any case, technically, it can be placed anywhere in the network.

32.8.2 Device Detection and Change Publishing

The Change Publishing mechanism can be configured to be either active or passive in its operation, which provides the administrator with two different ways of notifying the clients of changes, which to use depends on the network architecture used when deploying the Appear system.

The first configuration option is to use the optional Appear Publisher component to publish changes to all affected clients; this component is a very effective way to communicate to all clients without introducing a lot of traffic in the network, since this communication uses a one-to-many mechanism available in all IP based networks. The publishing protocol is today a proprietary Appear protocol based on IP multicast or broadcast. However, this will in the future be replaced by the IETF Zero Conf protocol. This configuration will require Zero Conf package or the APS multicast to be routed in the network or that the local publisher is placed in the local network.

The second configuration option for Device Detection and Change Publishing is based on standard DHCP (Dynamic Host Configuration Protocol) and Domain Name System (DNS), which provides the client with an anchor point to where the local Appear Proxy is located. When an Appear Client comes into the network it will make a DHCP request for the local network configuration. In the configuration there will be a reference to the local DNS server, which has to provide a reference to the local Appear Proxy/Proxies. This removes the need for introducing the Appear Publisher components.

Table 32.1: Configuration options used by different networks.

	Appear Publisher	DHCP/DNS
Device Detection	Active	Active
Change Publishing	Passive	Active

Typically the Appear Publisher scenario is to prefer in small to medium networks, where the dynamic behavior of services is much more frequent, which is mostly found in enterprise deployments. In larger carrier deployment, typical public networks, the DNS configuration is to prefer, since this removes the need for routing multicast, broadcast or Zero Conf packages, and the need to introduce a large number of Appear Publishers in the network.

32.9 Some Case Studies

32.9.1 Stockholm Metro

Using the Stockholm metro as an example, this section explains how intelligently connecting mobile workers in the public transport sector can increase and improve the flow of information, improve staff safety as well as lead to happier and more satisfied customers.

The very nature of public transport systems requires that the people on the ground - customer service representatives, ticket collectors and maintenance workers - must have fast and reliable access to real-time information in order to enable passengers to travel from A to B as smoothly as possible with the least amount of complications and disruptions to their planned journeys.

Internationally renowned for its efficiency, Stockholm's metro system plays an essential part in the functioning of the city's daily transport network. With over 100 metro stations, three railway lines with more than 50 stations and five tram lines with 98 stations the network is expanding to accommodate increased demand and support the city's development. Today the network currently supports more than 2 million passenger journeys per day and provides a daily service for almost 700,000 passengers.

Given these statistics, The Stockholm Local Transportation Authority (SL) which manages the network infrastructure and Veolia which is contracted to take responsibility of rolling stock and passenger operations established three goals to work towards: increasing the number of passengers travelling on the metro; increasing passenger satisfaction; and ensuring greater safety for ground staff in direct contact with the public.

With these goals in mind, SL and Veolia embarked on a journey to implement an innovative IT project based on a wireless broadband network and equip their station personnel with PDAs. Not happy with simply connecting personnel wirelessly to a central information system, SL and Veolia further decided to specially design a solution based on context-awareness. In other words an intelligent solution that could provide personalised and location based information and applications to their employees on the ground.

Considering that projects of this scope had not been tested, the next phase that had to be taken into account was planning. As such, a project team comprising of managers from SL and Veolia as well as technology experts including mobile middleware providers Appear was formed. The main task was to understand the needs of the organization and its employees and come up with a solution that could help SL and Veolia achieve their goals without having to substantially increase staffing levels. In productive and close cooperation the group drew up a system that would allow station staff to answer passenger questions as fast and as accurately as possible whilst giving staff the benefit of added security and safety on the ground using location-based technology.

32.9.1.1 The Solution

In November 2003 Gullmarsplan became the test ground for the first deployment of SL's and Veolia's new wireless solution drawing on the technology from Cisco, Intel, Microsoft, Fujitsu-Siemens and the innovative context-aware software from Appear.

Five wireless access points were fixed throughout the station giving coverage to the stations 3 floors, the platforms and ticket hall as well as the bus terminal outside the station. Using a Wi-Fi network from Cisco as a backbone, the technology from Appear makes it capable to deliver data to end-users according to location, time and who is making the request in a manner that minimises the need for browsing. Furthermore, the network allows the location of the user to be determined by pinpointing which station they are in and where in that station they are – such as the ticket hall, train platform or corridor.

Presently staff employees use PDAs to access information but the nature of the solution means that it has been designed to be both device and network agnostic. This has the extra benefit of compatibility and integration as the solution works with smartphones, PDAs or laptops and with either Wi-Fi or cellular networks such as GPRS and UMTS.

Information on the user's device is organised by tasks, such as a travel planner, local area map or nearest transport links such as buses. Each task appears on the device as a distinctive icon. Moreover, the tasks that appear on the screen are dependent upon the location of the device. This is due to the simple fact that the types of queries will vary according to where the member of staff is standing. Therefore the employee always has access to the right information and applications that are relevant to the situation at hand.

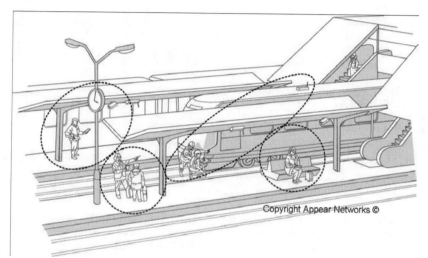

Figure 32.4: The metro station showing various access points. Depending on the access point different applications will be made available to the end user determined by preset parameters such as time, location, task, role of employee.

A key asset of the solution is its user-friendliness and the fact that everything is controlled centrally. Using the solution's Click and Run technology only relevant applications are automatically installed and non-relevant applications uninstalled. The only action the employee needs to take to access information is to click on the correct icon, for example journey planner, latest timetable disruptions or map.

Furthermore, using the positioning engine, a safety alarm has also been installed allowing employees to trigger an alarm call to the central control office. The control office can then immediately identify where in the station that employee is, and, if needed, send extra support or additional information.

32.9.1.2 The impact

Depending on the level of infrastructure already in place, the payback on the technology investment can at first seem high. However, by making a wireless solution work intelligently for the organization returns on investment can be seen within a year, particularly with operational costs and efficiencies.

Since deployment in December 2005 the benefits for SL and Veolia have been substantial.

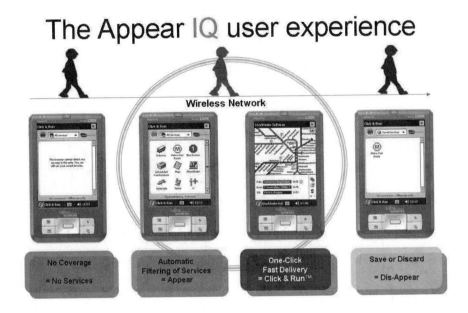

Figure 32.5: The end-user has access to a menu which is dependent upon his/her context. The services and applications available automatically change if the end-user moves from one access point to another or if the context of the end-user's situation changes.

- **Better information flow** – prior to the deployment employees lacked quick access to accurate traffic information for customers. This not only deterred customers from using the metro but it also made employees nervous about dealing with customer queries in times of disruptions or timetable changes. Now employees no longer have to refer customers to the station's information centre, or at its worst avoid customer queries, as wireless access to the central information systems enables them to answer customer queries quickly, correctly and confidently.
- **Better fault reporting and maintenance** – One of the areas marked for improvement prior to the solution was maintenance. After deployment staff can instantly notify the relevant department of faults or breakages in trains and stations via their handheld devices.
- **Improved security for employees** – employees were nervous about being confronted by abusive and violent passengers and other anti-social behaviour. On finding themselves in isolated and threatening situations staff can now trigger a silent alarm to the central operating room. Using the location based Wi-Fi technology security staff are automatically notified of the correct location.
- **Improved customer satisfaction** – since deployment customer numbers have been increasing steadily as better information has allowed travellers to re-plan their journeys in the event of a disruption. Staff can quickly direct customers to the nearest

alternative means of transport leading to customers that are happy to return to using public transport services.

Intelligently connecting the mobile worker using context-aware technology has changed the way SL and Veolia operate. The results have seen a streamlining of business processes worth 1.6 million euros to the organization, increased staff productivity, for example 15 staff can now handle 48,000 customer queries per year, plus a noticeable 7% increase in passenger numbers. The solution has kept IT costs under control and generated a return on investment in less than a year. More specifically, a successfully computerized mobile workforce was implemented after less than 6 hours of training, demonstrating how simple and efficient it is to adopt the technology.

32.9.2 Southwest Florida International Airport (SWFIA) Context-Aware Wireless Emergency Response

SWFIA serves more than 7 million passengers annually and is one of the busiest airports in the USA. SWFIA'S new modern terminal has deployed a context-based state-of-the-art wireless solution to ensure the targeted delivery of emergency response and security information to first on-the-scene mobile responders such as the fire department, security guards and maintenance workers.

32.9.2.1 Fast Emergency Response is Critical

Achieving a fast response to critical emergency and security situations such as fire alarms is paramount to the smooth running of every airport. Mobile airport personnel who are responsible for responding to activated fire alarms and security alerts must have access to the right information when and where they need it. Traditional processes of emergency response information sent through dispatch and operations result in slow response times due to the highly complex airport environment. In many cases emergency information is not targeted to the most relevant response team, critical details concerning the incident such as a fire alarm map are not communicated, plus related information is often not included or is inadequately described. SWFIA thus needed a context-aware mobile solution that could overcome these challenges. A mobile solution that could dynamically react to the demanding emergency and security situations that arise in airports everyday.

32.9.2.2 A Context-Aware Wireless Emergency Response Solution

SWFIA deployed a context-aware wireless emergency response solution based on the Appear IQ platform. Using the Appear Context Engine the solution rapidly and dynamically personalizes emergency and security alerts plus automatically delivers critical information to the most relevant personnel via their mobile devices. For example, if a fire alarm is set off, alarm information is sent to the emergency response centre. At the same time the system also notifies the closest and most relevant response team to the alert via their mobile devices providing them with the precise location of the fire alarm. Additional critical information such as related maps locating ventilation ducts are also pushed to the mobile devices.

Automatically selecting and pushing precise alert information to the most relevant first time responders such as the fire department, security personnel or maintenance workers gives SWFIA the ability to respond rapidly to emergency and security situations. A situation's status can be analyzed more effectively leading to quicker decisions and a response within the critical "three minute window".

32.9.2.3 The Value of Context

- Context ensures that the most relevant emergency response teams are automatically notified of an emergency or security. Central operations also receive notification of the alert.
- Context ensures that first on-the-scene response teams have access to rich information concerning the nature of the alert or emergency situation.
- Context ensures better analysis, decision-making and response times due to the automatic provisioning of critical information relevant to the alert or emergency alarm. For example, the location of a specific fire alarm and relevant maps such as ventilation ducts and which areas are affected.
- Context ensures greater passenger security as well as minimal operational disruptions to arrivals and departures due to targeted information flow to key personnel and emergency response teams.
- Context ensures that wireless communication technologies in a highly complex environment such as an airport are effective, simplified and user friendly.

Appear's Context-Aware technology is transforming the daily operations of Southwest Florida International Airport. Faster response times and more accurate analysis of daily emergency and security alerts enable the airport to continue normal operations with minimal disruptions. There is also a compelling reason for faster response to fire alerts. For SWFIA, their policy is to evacuate the area around an alert (if it cannot be proved to be a false alarm) after 3 minutes. If an area is evacuated, that will significantly impact airport operations and cost a significant amount of money to the airport in terms of lost business, customer confidence etc. Before the Appear solution, it sometime could take up to 30 or 40 minutes for the fire department to respond. With the Appear solution allow faster response by the closest able person, the speed of response has been quicker, the number of evacuations has fallen and the airport is saving a significant amount of money.

The Appear IQ platform is also laying the foundations for future pioneering IT and mobile projects. Using the software platform additional security services and applications can be integrated. For example, in the event of a security breach, video feeds from surveillance cameras can be automatically sent to the mobile devices of the nearest security personnel giving them the correct information they need to assess and react to the situation. Relevant information can also be sent to other relevant personnel. For instance, if the security breach has resulted in damage that requires immediate attention such as perimeter fencing, a message can be sent to the nearest maintenance team.

"The Appear IQ platform will close the time gap between an event and closure by providing each specialised incident responder with the specific information they need to perform their tasks." Robert C. Smallback, Director Information Technology, Lee County Port Authority

32.9.3 RATP

The Régie Autonome des Transports Parisiens (RATP), the Paris city transportation system, operates four different systems throughout Paris including the Metro, railway, tramway and bus systems. Employing over 44,000 people (700 in the IT department), RATP is considered one of the world's most innovative transportation networks.

32.9.3.1 Improving Operations and Customer Service

The RATP subway system was interested in the deployment of a user friendly wireless solution that would increase staff productivity, make employees feel more valued, and dramatically increase customer service levels. Ever-increasing numbers of mobile employees in the stations required fast and easy access to real-time customer information, such as service schedules, disruption information, points of interest in the area, maps, timetables, and security & maintenance information.

RATP wants to ensure that:

- Customers can access the information they require through customer service representatives on site
- Up to the minute information regarding the status of the system is obtainable immediately in the areas affected
- Area specific information is easily acquired by the relevant personnel
- Efficiency is increased with regards to the management of facilities and trains

32.9.3.2 A Context-Aware Solution

A Cisco end-to-end WLAN infrastructure provides high speed connectivity inside the RATP stations to fleets of WLAN-enabled devices such as Pocket PCs. The Appear Context Engine recognizes the context of each individual employee, such as location, profile, the time of day, etc. Appear IQ's context-aware software dynamically reacts to the changing requirements of RATP's mobile employees and ensures that they always have access to the right information when they need it and without them having to request it.

The solution provides possibilities for further customer care/service, security and operational applications such as staff management and equipment maintenance.

32.9.3.3 The Value of Context

- Context ensures that employees are able access relevant information in one single click so that customer requests can be answered accurately and quickly.
- Context ensures that mobile content is always up-to-date according to the time of day or year. Changes to fare prices and timetables are automatically updated, for example, weekend rates and night timetables.
- Context ensures that existing applications are enhanced. For example, the "from" field in the route planner application is automatically filled in according to the location of the employee.

- Context ensures that employees are automatically informed which station   specific information or services are available, such as the local employee phonebook   or   the station employee dashboard.
- Context ensures that employees can access the most up-to-date information available in real-time, for example the actual departure time of "bus no. 12". When offline the system automatically switches to the theoretical timetable.

32.9.3.4 Return on Investment

The sharing of a mobile infrastructure (hardware and middleware) greatly reduces the cost of ownership of the entire mobile solution. In addition, the context-aware software, the highly ergonomic user interface and the high performance network dramatically reduces the time required to access the required information. This has increased employee efficiency and greatly enhanced customer service levels.

RATP is currently rolling out the deployment within several of its departments, including frontline personnel in the tramway, subway, local trains and buses. Various categories of personnel are being equipped with devices, including customer care representatives, security personnel and maintenance staff.

"This installation of advanced Wi-Fi services in a subway station is unprecedented. Our initial goal was to help our employees provide better answers to the customers' requests in order to improve quality of service along their journey. Now, La Defense RATP employees can easily access proximity services to provide real-time traffic information and local information frequently asked by travelers." Philippe Vappereau, Manager, Traveler Information System Department

32.9.4 Dutch Rail (NS) Mobile Frontliners

Dutch Rail (NS) operates one of the busiest rail transport systems in the world. With more than 385 full-service rail stations throughout the Netherlands, the company's 25,000 employees operate approximately one million passenger journeys per day. NS continually strives to provide the most reliable and customer-friendly passenger rail transport in Europe.

32.9.4.1 Operational and Productivity Gains using Context

Improving operational efficiency and increasing employee productivity were key drivers when NS decided to implement an advanced wireless network in 50 "hub" stations with Appear's Context-Aware software. Using context, 10,000 mobile employees now have a user-friendly solution "pushing" exactly the information they need, when and where they need it. The traditional fixed-location employee sign-in/out process is now wireless with context ensuring that the process is restricted to specific "home" stations. The previously complicated fixed-location synchronization procedure is now a reliable and automated process with context ensuring that critical information is distributed using the best connectivity route. Context has ensured true mobility at NS as the wireless solution dynamically reacts to the changing requirements of the mobile employees and their environment.

32.9.4.2 A Context-Aware Solution

An advanced distributed architecture using Cisco end-to-end WLAN infrastructure provides high speed connectivity to a fleet of 10,000 Wi-Fi enabled Pocket PCs. The NS Desktop, based on the Appear Context Engine, enables the push and synchronization of personalized real-time information over the wireless network to the employees' handheld devices. Whenever the user enters or moves between authorized wireless zones, the Appear software dynamically reacts to the employees' situation and context, for example profile, job task and location. The addition of the Appear Synchronization Module automates a complex synchronization process based on context that supports file upload and download, version control and release management. A device lockdown feature uses context to remove unwanted applications and processes that do not fit into employees' profiles.

32.9.4.3 The Value of Context

- Context ensures that employees receive relevant prompts. Ticket collectors receive prompts to update information once a certain number of fines have been collected.
- Context ensures that train drivers receive critical information relevant to their specific journeys such as speed limits prior to boarding the train.
- Context ensures that the complex synchronization process occurs automatically when new information is available, for example changes in shift rosters, journey information or security reports.
- Context ensures the dynamic appearance of relevant icons informing employees which services are available in the area, for example printing facilities.
- Context ensures that employees can only sign in and out at specific "home" station by using the proxy name for identification. The WiFi network determines their location.
- Context is improving employee efficiency by 25 minutes per day per employee and cost savings estimated at EUR 8 million per year.

Appear's Context-Aware technology has transformed the day-to-day working environment of mobile employees at Dutch Rail. Operational processes have become more effective and employee productivity has increased dramatically. Possible future uses of Appear's Context-Aware technology at Dutch Rail can ensure that equipment and infrastructure can be used more effectively and intelligently. This will take the distribution of context-aware information to the next level with the possibility of intelligent trains, intelligent e-ticketing and intelligent information displays.

32.10 References

[1] Bass, L., Kasabach, C., Martin, R., Siewoirek, D., Smailagic, A., and Stivorik, J., "The Design of a Wearable Computer", *Proceedings of the CHI'97 Conference on Human Factors in Computing Systems*, pp. 139-146, 1997.

[2] Ishii, H. and Ullmer, B., "Tangible bits: Towards seamless interfaces between people, bits, and atoms", *Proceedings of the CHI'97 Conference on Human Factors in Computing Systems*, pp. 234-241, 1997.

[3] Mackay, W.E., Velay, G., Carter, K., Ma, C., and Pagani, D., "Augmenting Reality: Adding Computational Dimensions to Paper", *Communications of the ACM*, 36, 7, pp. 96-97, 1993.

[4] Streitz, N. A., Konomi, S., Burkhardt, H-J. (Eds.), *Cooperative Buildings: Integrating information, organization, and architecture.* Berlin: Springer, 1998.

[5] Weiser, M., "The computer for the 21st century", *Scientific American*, 265, pp. 94-104, September 1991.

33

Experiments Using Small Unmanned Aircraft to Augment a Mobile Ad Hoc Network

Timothy Brown, Brian Argrow, Eric Frew, Cory Dixon,
Daniel Henkel, Jack Elston and Harvey Gates[a]

Small unmanned aircraft (UA) are an ideal addition to mobile ad hoc networking. An ad hoc network allows any two nodes to communicate either directly or through an arbitrary number of other nodes that act as relays. Ad hoc networks that include UA improve traditional ground-based networking through the added connectivity provided by the more prominent UA. The networking also extends the operational scope of the overall Unmanned Aircraft System (UAS) beyond the limits of point-to-point and centralized communication architectures. As well, the ad hoc network increases the UAS operational range as communication can be extended across the set of networked UA and ground nodes. While these capabilities support a wide variety of applications, little prior work has fielded and tested the capabilities of such a system in practice. This chapter describes the implementation of a wireless mobile ad hoc network with radio nodes mounted at fixed sites, on ground vehicles, and in UA. The radio is an IEEE 802.11b/g (WiFi) wireless interface and is controlled by an embedded computer. The ad hoc routing protocol is an implementation of the Dynamic Source Routing (DSR) protocol. A network monitoring architecture is embedded into the nodes for detailed performance analysis and characterization. The following sections describe the network components in detail and provide performance data measured at a large-scale outdoor test bed.

---

[a] All authors currently affiliated with *University of Colorado at Boulder*

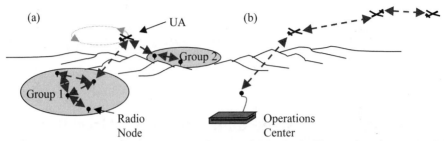

Figure 33.1: Two UAS deployment scenarios, (a) Scenario 1: UA used to increase ad hoc ground node connectivity and connects disconnected groups, (b) Scenario 2: ad hoc networking between UA increases operational range.

33.1 Introduction

Communication networks between and through aerial vehicles are a mainstay of current battlefield communications [1]. Present systems use specialized high-cost radios that operate in designated military radio bands and are mounted on high-cost manned and unmanned aircraft. Small low-cost Commercial Off-The-Shelf (COTS) radio equipment combined with powerful computer processing can be mounted on small (~10 kg) Unmanned Aircraft (UA) and has the potential to revolutionize battlefield communication and open up many scientific and commercial applications. (Following Ref. [1], the acronym UA (Unmanned Aircraft) is used instead of UAV, when referring to the flying component of an Unmanned Aircraft System (UAS)). One example COTS technology is the IEEE 802.11 wireless LANs (so called WiFi) which connect wireless mobile nodes to a fixed infrastructure and is being widely deployed, including in UAS applications [28]. More interesting applications are possible when the different mobile nodes connect to each other in peer-to-peer ad hoc meshed wireless networks [13],[30].

There is a current push towards the development of simple, small, low-cost UAS that work individually and in cooperative teams to accomplish complex and dangerous missions. UAS have been applied to numerous applications including diffuse gas and plume detection [5],[10], coordinated search [27],[29] and reconnaissance [7], in situ atmospheric sensing [2],[23], as agents in the battlefield [1],[15], and as components in command and control architectures [1]. With smaller, cheaper UAS that have limited sensor capabilities, cooperative control relies heavily on communication with neighbors [3],[18],[19],[26],[27]. By forming a multi-hop communication network, information can be efficiently shared among the UA, increasing the overall mission capabilities and operational range of the UAS.

Two motivating scenarios are shown in Figure 33.1. In Scenario 1, an ad hoc network of ground nodes is initially disconnected because of distance and/or terrain. When a UA is introduced, it provides connectivity among the ground nodes as an ad hoc network relay, since its altitude enables effective line-of-sight communication to each of the formerly disconnected entities. In Scenario 2, the power and payload constraints of the small UA limit its communication and travel ranges, which limit the operational range of the whole UAS. Relaying messages in an ad hoc network between multiple UA extends the communication range and thus the operational range of the system.

The development of communication technology, in conjunction with increased capabilities in embedded microprocessors, has enabled unmanned flight through the removal of the human pilot from the aircraft cockpit [1]. As the applications of UAS expand, the principal issues of communication technologies are flexibility, adaptability, and controllability of the information/data flows. Future systems will be net-centric and rely on mobile ad hoc networking to provide real-time connectivity among cooperating agents and to provide backhaul of sensor and telemetry data from mobile nodes to a data repository or command center.

In mobile ad hoc networking, wireless nodes cooperate to relay packets over multiple relay hops from source to destination [13]. Typical ad hoc approaches consider nodes with similar capabilities that move in random patterns relative to one another over time and the communication is separated from the other activity of the node. In contrast to that, a critical aspect of net-centric UAS operation is the tight coupling between communications, mobility, and the completion of tasks [25].

While there has been a significant amount of research in mobile ad hoc networks (MANETs), few of these have specifically exploited the capabilities of UAS [3],[4],[8],[19],[31],[32]. Reference [31] extends the Landmark Ad Hoc Routing (LANMAR) protocol to a hierarchical structure using high-quality backbone UAS links to improve communication performance. In Ref. [32], LANMAR is extended into a multi-cast framework presented as the Multicast-enabled Landmark Ad Hoc Routing (M-LANMAR) protocol. They show in simulation studies that M-LANMAR provides efficient and reliable multicast compared with the application of a "flat" multicast scheme that does not exploit team formation. In their simulations they use the Reference Point Group Mobility model [9] which does not accurately model the motion of UA, i.e., nodes are treated as point masses in the simulation that do not need to maintain forward speed for flight and ignore turning rate constraints as well.

This paper describes the AUGNet (Ad hoc UAS-Ground Network), a wireless architecture and test bed program developed at the University of Colorado, Boulder (UCB) for the application of IEEE 802.11 LAN principles and standards in a peer-to-peer ad hoc (meshed) network environment. Small UA, along with fixed and mobile ground platforms are the core elements of the mobile meshed network. Full-scale outdoor tests were conducted at the Table Mountain Field Site located approximately 15 km north of the UCB campus. The test site is within a Radio Quiet Zone that covers the approximately 3 km × 4 km area of the Table Mountain mesa. The UA flights are confined to the airspace above the Radio Frequency (RF) test site while mobile test vehicle excursions cover an area up to 6 km × 8 km (3.7 mile × 5.0 mile) within and surrounding this radio range. Figure 33.2 shows an aerial view of the Table Mountain Field Site along with the primary AUGNet components.

A unique feature of the AUGNet system, compared to other multiple-vehicle unmanned aircraft systems, is the implementation of a true mobile ad-hoc communication mesh network, construction of the UAS around the networked communication, and extensive monitoring of the network traffic. Cooperative control with small fixed-wing UA had previously been demonstrated using multicast protocols through a centralized ground station [20],[24]. Peer-to-peer wireless communication with UA was demonstrated in Ref. [3] and distributed task assignment was performed by a three-vehicle UAS using ad-hoc

Figure 33.2: Ad hoc UAS-Ground Network (AUGNet) combining COTS net-working technology with small UAS: (a) aerial view of the Table Mountain Field Site, (b) ladder-mounted ground node, (c) automobile mounted node, and (d) airborne nodes embedded inside three Ares UA.

networking [17]. While demonstrating successful deployment of a MANET on a UAS, neither reference provides an extensive discussion of the performance of the network.

The following sections present the AUGNet architecture, details of the test bed components, and results that quantify the functionality and performance of the AUGNet concept.

33.2 AUGNet Architecture

An AUGNet poses several challenges since the ad hoc radio nodes can be in a variety of configurations. These configurations include nodes mounted on high poles at fixed-sites, mobile nodes mounted on ground vehicles, personnel-carried nodes, and airborne UA nodes. Thus the design must be modular to apply to all of these configurations while the UA nodes require special consideration. For typical operating conditions the UA might communicate with most of the ground nodes. Because of fewer-hop paths, ad hoc routing algorithms based on the hop-count metric (e.g., DSR) will tend to route the traffic through the UA, creating a communications bottleneck limited to the UA-node bandwidth. The UA may also have additional radio links that can interfere with the ad hoc communication. As a comparison, note that the current application with small UA differs significantly for the ad hoc networking described in Ref. [19] where the UA is large, flies at high-altitude (60 kft, 18 km), and has a separate radio for UA-ground communication.

Figure 3 shows the components of the AUGNet Architecture. It also shows the data flow from the field site to remote users and the data repository (monitor server) on the

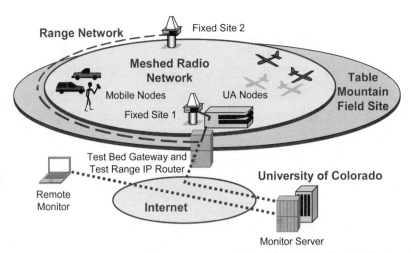

Figure 33.3: The AUGNet Architecture: A mobile ad hoc network with fixed and mobile ground nodes, and airborne nodes. Network monitoring and test data archiving are provided through Internet access from the Table Mountain Field Site infrastructure.

UCB campus. The primary components are: 1) communications hardware, 2) ad hoc network software, 3) test bed monitoring architecture including remote monitoring through the Internet, and 4) unmanned aircraft. These components are now described in detail.

33.2.1 The Mesh Network Node (MNN)

The ad hoc network software combined with the communication hardware is denoted as the mesh network node (MNN). The MNN hardware components, displayed in Figure 33.4(a), are a Soekris Engineering Model 4511 single board computer (100-MHz 486 processor, 64-MB RAM, 256-MB flash memory), an Orinoco 802.11b PCMCIA card, a Fidelity Comtech bidirectional amplifier with up to 1W output, and a Garmin Model 35-HVS GPS receiver. The core of each MNN is identical, only the packaging differs depending on

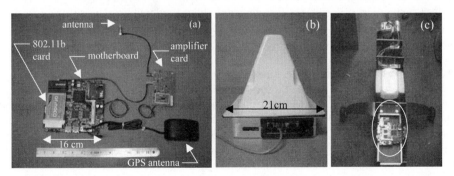

Figure 33.4: (a) Mesh Network Node (MNN) hardware components, (b) MNN mounted in environmental enclosure for fixed and ground-vehicle deployment, (c) MNN mounted in UA.

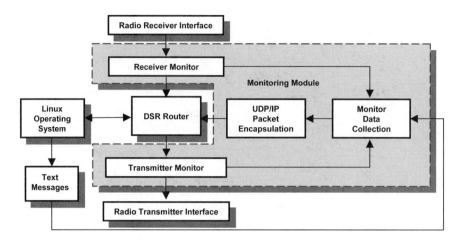

Figure 33.5: Test-Bed Monitor Agent Module.

whether the MNN is mounted at a fixed ground site, on a ground vehicle, or in a UA as shown in Figure 33.4b and Figure 33.4c. This commonality greatly simplifies the software and hardware development.

The MNN runs the dynamic source routing protocol (DSR) [21] with communication to other nodes via 802.11b. DSR is chosen because its routing is on-demand so a traffic source only seeks a route to a destination when it has data to send. Thus, nodes do not waste bandwidth trying to establish routes they do not need at the moment. When a node needs to send a packet, it initiates a *route request* process among nodes in the network to establish a route. DSR uses source routing whereby a packet's source node precisely specifies, in the header of each packet, which route the packet is to follow. Several parameters, such as route request timeout, route cache timeout, ACK timeout, send buffer lifetime, can be modified to adapt DSR to different network environments. DSR was implemented using the Click modular router [12],[22] and allows for the modification of the protocol as needed for testing. The software runs under the Linux operating system and has been ported to several different Linux distributions. The Linux WISP-Dist distribution (leaf.sourceforge.net) is used in the experiments reported here since it requires only 8 MB of flash memory and is suitable for our single-board computers. The software has been ported to a number of other devices including laptop, desktop, and iPaq handheld computers.

33.2.2 The Test-Bed Monitoring Agent

The test bed monitoring architecture consists of a monitoring agent on each MNN and a data collection protocol. Figure 33.5 shows the Monitor Module and its relationship to the other MNN components. The node Monitor Data Collection Subsystem (MDCS) in the Monitor Module is under the control of the Linux Operating System where it receives text messages concerning its operational mode. The MDCS is connected to the MNN receiver where it extracts and logs the type, size, route, sequence, and timestamp of each incoming packet that is destined for the DSR router. The MDCS also monitors the transmission from the router comparing the router input to its output for the purpose of tracking throughput delays and lost or damaged packets that are discarded by the router. This information allows each packet to be tracked as it passes through the network, including its physical (node) position versus time. A periodic report with packet and node information is sent to an ad hoc network gateway then to the data repository.

Figure 33.6 is a computer screenshot of the web-based interface used to visualize and explore network activity during an experiment or to analyze the data afterward. The GUI has four major fields. The top bar provides time and control information similar to a PC music or video application program. The screen displays a situational map of all the MNN locations and active links. A control panel provides the viewer with various test-data alternatives, map options, and status information selections. The data graphics portion of the GUI provides flexible data display. For example, one can display the metrics of a particular node such as data packet throughput as a function of time or it can focus on types of traffic such as User Datagram Protocol (UDP) versus Transmission Control Protocol (TCP), and can analyze source and destination data.

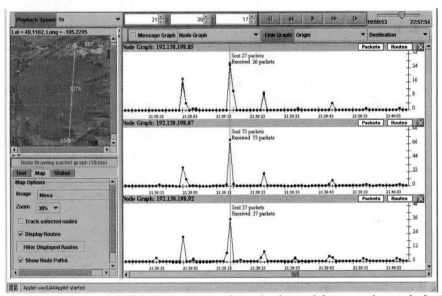

Figure 33.6: Web-based GUI for remote network monitoring and data experiment playback.

33.2.3 The Ares UA

The Ares UA was designed and constructed at UCB specifically for test bed operations. The primary criteria for the design were configuration flexibility, multi-hour (> 3 hr) endurance, and a payload capacity of at least 5 kg (10 lb) to carry the MNN and other payloads. The airframe is based on the Senior Telemaster, a design that is well-known to radio-control (RC) pilots for its benign flying characteristics and payload capacity. The fuselage was expanded to accommodate a 1-gal (3.8-L) fuel tank and a payload volume of 7 in × 15 in × 10 in (17.8 cm × 38.1 cm × 25.4 cm). Except for the vertical tail, the airframe skin is made entirely of carbon fiber composite with structural members, such as bulkheads and ribs, made from plywood laminated with carbon-fiber composite. The vertical tail is constructed of non-conducting fiberglass composite to house a 900-MHz antenna. The Ares is powered by a 5-hp two-stroke engine and has a cruise speed of 30 m/s, and a top speed of 54 m/s at about 1800-m AMSL (the Table Mountain test range airfield is at an altitude of about 1680-m AMSL). An added benefit of in-house fabrication is that copies are readily made from the original molds. This also makes for a quick turnaround time if repairs are needed. Four Ares airframes were built for the initial AUGNet communications experiments. A three-view and isometric view of the Ares UA are shown in Figure 33.7(a). Figure 33.7(b) shows the antenna placements on the airframe. Antenna 1 (72-MHz) is a conventional RC antenna used for the experiments described in this chapter. The 900-MHz Antenna 2 provides a backup link between the PiccoloPlus autopilot and its ground station (Cloud Cap Technology www.cloudcaptech.com). The primary autopilot-to-ground link is through the ad hoc 802.11b network using the 2.4-GHz MNN Antenna 5. The custom ground station runs a GUI for command, control, and communication (C3) through the meshed network. Although this C3 capability has been demonstrated [14], the 72-MHz RC link described earlier was used for all the experiments reported in this chapter before it was permanently removed (results for autonomous flights controlled through the 802.11b ad hoc network are presented in Ref. [16]).

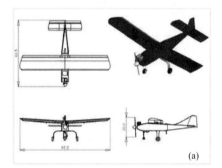 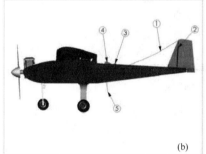

(a)   (b)

Figure 33.7: Ares UA: (a) Three-view dimensions are inches, (b) antennas, (1) 72 MHz, (2) 900 MHz, (3) GNC GPS 1227.60 MHz and 1575.42-MHz, (4) MNN GPS 1227.60 MHz and 1575.42-MHz, (5) 2.4 GHz.

Measurements with a spectrum analyzer and a Berkeley Variatronics Yellowjacket WLAN analyzer showed no discernable interference power between the 2.4-GHz 802.11b and the 900 MHz radios. Even with antennas placed within a few centimeters, no excessive packet loss was observed for either system. Antennas 3 and 4 shown in Figure 33.7 are the GPS antennas for the autopilot and MNN respectively. Although both were used simultaneously in the experiments, the MNN Antenna 4 has since been permanently removed and both the MNN and autopilot share the remaining Antenna 3.

33.3 Test Bed Experiment Plan

The experiments used 5 ground nodes (Figure 33.4b), 3 UA nodes (Figure 33.4c), and 2 laptop-based nodes. Figure 33.8 presents a simplified view that emphasizes the data routing, including an alternative back-haul link through Iridium satellite phones.

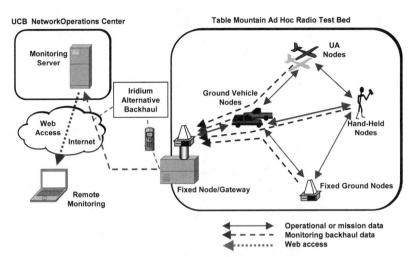

Figure 33.8: Data collection and monitoring architecture.

Ad hoc network performance depends on several factors including the path length (in number of hops), the quality of the links, whether nodes are fixed or mobile, and whether they involve a UA. An ad hoc network communicates between the source and destination either directly (a one-hop path) or via one or more intermediate relay nodes (a multi-hop path). A series of experiments were constructed to understand the significance of these factors. The experiments tested the performance with and without mobility, as well as with and without the UA for one to five hops. These experiments were organized as shown in Figure 33.9.

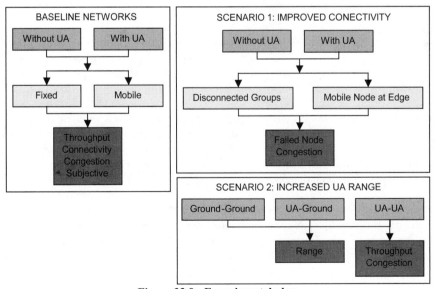

Figure 33.9: Experimental plan.

Each experiment consisted of a specific node deployment configuration combined with a specific set of tests. The node deployment dictates the physical distribution of nodes on the test range and their individual mobility patterns. The locations of the fixed MNN nodes and the paths of the mobile ground and airborne nodes are shown on the map of the field site in Figure 33.10. Base equipment and Internet backhaul connections are located in the permanent structures labeled FS1 and FS2 in Figure 33.10.

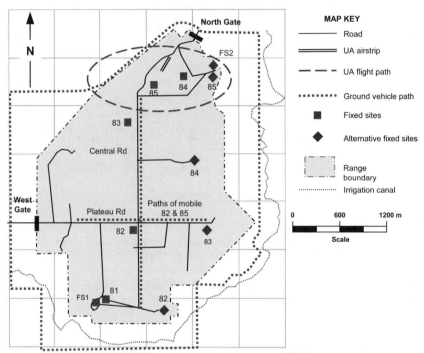

Figure 33.10: MNN deployment locations at the Table Mountain Field Site.

Six different tests can be run during any given deployment, although not all were run with each experiment. These tests are described below.

Throughput: Purpose – to test the throughput that can be achieved when no other traffic is present. This test uses the netperf[1] utility to measure throughput between node pairs over a 5 second period. A script is given a set of source destination pairs and the throughput is measured between each pair one at a time. The default is to measure between every source and destination pair in both directions.

Connectivity: Purpose – to measure the ability for node pairs to send packets to each other when the network is lightly loaded. Each node sends pings once per second to a random destination node. Every 20 seconds a new destination node is randomly chosen. Ping success and round trip delay statistics are collected for each pair.

---

[1] Originally developed by HP, now freely available at http://www.netperf.org.

Congestion: Purpose – to measure delays and throughputs when there are competing data streams in the network. Each node picks a random destination and either measures the throughput with netperf for 10 seconds or pings once per second. The schedule of pings vs. throughput is chosen so that 2 competing throughputs are always in the network at the same time. Delay and throughput statistics are collected for each pair.

Subjective: Purpose – to assess the performance of typical network applications as perceived by a user. A user on the test bed attempts typical web tasks such as downloading a webpage or a voice over IP connection. The user records the usability of these applications relative to T1 and dialup connections.

Node Failure: Purpose – to measure the ability of the network to route around node failures. The network is setup so that traffic has to pass through a specific node. The specified node alternates between shutting down its interface for 60 seconds and bringing it up for 60 seconds. Performance is measured during connectivity periods when the node is down, when the node is up, and during the transitions from down to up and up to down.

Range: Purpose – to measure the throughput as a function of separation between two nodes. This test uses the netperf utility to measure the throughput from a source to a destination every 5 seconds. In each interval the GPS coordinates of the two nodes is recorded so that throughput can be correlated with range.

33.4 Experiment Set 1, Baseline Network Measurements

The goal of these experiments is to understand the capabilities of the underlying ad hoc network with fixed and with mobile nodes, and the role that the UA plays in this performance. The result is the four experiments in Figure 33.9 where the throughput, connectivity, congestion, and subjective tests were run in each experiment.

33.4.1 Experiment 1.1, Fixed Ground Nodes

Six nodes numbered 1 to 6 are mounted 2 m above the ground and arranged to form a 5-hop network (1 to 2 to ... to 6). Node spacing and local terrain are used to enforce a 5-hop network without short cuts, for example, node 3 can only communicate reliably with nodes 2 and 4.[2] The specific placement is shown in Figure 33.11 where nodes 1 to 6 are labeled FS1, MNN1, ..., MNN5. Nodes are powered up with no scenario or location specific configuration in their communication software and then the four tests are run.

Purpose: Throughput is bounded by the 2-Mbps ad hoc network channel rate. Actual throughput, even between direct neighbors, will be less because of the 802.11, DSR, IP,

---

[2] Although the terrain is generally flat there are some variations. The top of the mesa has a slight bowing so that MN1 and MN3 are not line of sight. The vicinity of MNN4 and MNN5 is an excavated area below the main mesa level. MNN5 is placed close to and below the rim of this excavation, while MNN4 is placed at the far side of this excavation so that MNN3 has better signal to MNN4 than to the closer but shielded MNN5. Finally, FS1 and FS2 are to the side of the mesa slightly and below the level of the top.

and TCP overhead. Further, the throughput decreases with more hops because of interactions between packets sent on nearby hops. Therefore, the throughput test is designed to establish maximum throughputs as a function of the number of hops in the network. The connectivity test establishes the typical round trip delay and communication availability between nodes. In this case availability should be at or very near 100% since, by design, the nodes are placed to form a connected network. The delay test measures the increase in delay with the number of hops in the route. The congestion test establishes how competition for network resources impacts throughput. With two competing flows, throughput is expected to be reduced by more than half because the resources are split between the two and there is contention overhead. The subjective test should demonstrate whether the network is useful from an end user's perspective.

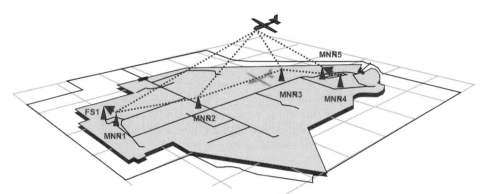

Figure 33.11: Baseline Network Measurements with and without the UA.

33.4.2 Experiment 1.2, Mobile Ground Nodes

In this test configuration, two of the nodes (MNN2 and MNN5) are placed on top of vehicles and driven during the tests at speeds around 25km/h. MNN5 cycles North and South and MNN2 cycles East and West as indicated in Figure 33.10. Otherwise the experiment is identical to the Fixed Ground experiment.

Purpose: Mobility increases routing dynamics as nodes move into and out of range. It increases jitter as the network pauses to find new routes when old routes are no longer valid. The hop count will change between a source and destinations over time. Throughput and availability will decrease while delay increases as more communication time is devoted to control packets and route error recovery. Some end-user applications may have perceptible degradations. This experiment establishes the degree of these effects.

33.4.3 Experiment 1.3, Fixed Ground Nodes with UA

In this test configuration, a UA flies over the test bed during the experiment. Otherwise the experiment is identical to the Fixed Ground experiment.

Purpose: The UA can be used occasionally when routes have too many hops or other routes are not available. The UA also introduces dynamics and will tend to interfere with all the nodes. This experiment establishes whether the UA is a net benefit to the network.

33.4.4 Experiment 1.4, Mobile Ground Nodes with UA

In this test configuration, a UA flies over the test bed during the experiment. Otherwise the experiment is identical to the Mobile Ground Nodes experiment.

Purpose: The longer range and more stable UA to ground links can be used to maintain connectivity. Whether the ad hoc routing can take advantage of these links is established in this experiment.

33.5 Experiment Set 2, Scenario 1: Improved Connectivity

These experiments are designed to show the role of the UA in improving connectivity among ground nodes.

33.5.1 Experiment 2.1, Mobile Node at Edge

Nodes are setup around the top of Table Mountain at the alternate sites shown in Figure 33.10. A mobile vehicle-mounted node is driven around the network on the roads surrounding Table Mountain. The connectivity test is activated while the node circuits the mountain. Although placement of the nodes helps provide more coverage, the mountain has rough and irregular sides. Five nodes, no matter how well placed, do not provide complete coverage around the entire base of the mountain.

Purpose: The mobile node can route traffic dynamically through nodes on Table Mountain as they come into and out of range. This experiment demonstrates the limits of ad hoc networks to connect to nodes moving at the fringe of the network's collective coverage.

33.5.2 Experiment 2.2, Mobile Node at Edge with UA

The deployment is the same as the Mobile Node at Edge except that a UA-mounted node flies above Table Mountain.

Purpose: This experiment shows whether the UA adds any significant coverage at the fringes in rough terrain

33.5.3 Experiment 2.3, Disconnected Groups

Nodes are set up on Table Mountain at the 6 primary sites in Figure 33.12. Once connectivity has been established, MNN3 shuts down its interface. When MNN3 is down, nodes MNN1 and MNN2 have poor connectivity to nodes MNN4 and MNN5.

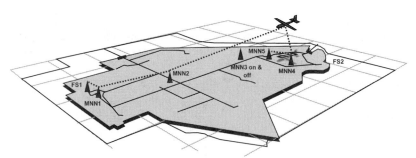

Figure 33.12: Connectivity tests with UA.

Purpose: This experiment provides a baseline for the next experiment. When MNN3 is down, the connectivity between the two separated groups will be zero or near zero.

33.5.4 Experiment 2.4, Disconnected Groups with UA

The deployment is the same as the Disconnected Groups except that a UA-mounted node flies above Table Mountain (see Figure 33.12).

Purpose: This experiment will show whether the UA can provide better connectivity over flat terrain.

33.6 Experiment Set 3: Scenario 2: Increased UA Range

These experiments are designed to specifically test the ability of the ad hoc routing to improve communications to and among UA.

33.6.1 Experiment 3.1, Ground-Ground Range

A single fixed MNN is set up at one end of the mesa. The range test is initiated between a ground-vehicle-mounted node and the fixed MNN. The vehicle is driven slowly around the test range collecting throughput data at different sender and receiver separations.

Purpose: Throughput is near maximum at short range, then falls off as distance increases. Throughput is also more variable as the range increases. So, this experiment does not measure an absolute range, rather it measures the effect of range on throughput for nodes near the ground.

33.6.2 Experiment 3.2, UA-Ground Range

One or two MNN are placed outside of FS2 to provide backhaul connectivity to the UA operations area. The range test is initiated between a ground-vehicle-mounted node and a UA-mounted node. A UA-mounted node flies around the UAS operations field. A ground-vehicle-mounted node drives around the top of the mesa and on the roads in the area of the mesa. During range measurements, the non-mobile nodes have their interface shut down to force all routes directly between the UA and mobile node.

Purpose: Generally range increases as the height above the ground increases. This measures the effect of range on throughput when one of the nodes is a UA.

33.6.3 Experiment 3.3, UA-UA Range

A second UA is added to the UA-Ground Range setup. The range test is initiated between the two UA, and different separations are maintained for a specified interval. To achieve longer ranges, one plane is flown from an airfield approximately 7-km south of the test range airfield.

Purpose: This experiment measures the effect of range on throughput when both of the nodes are UA.

33.6.4 Experiment 3.4, Three-UA Flight and Communications

Three planes are flown simultaneously and the communication behavior among the UA as well as with a ground node (MNN5) is observed as shown in Figure 33.13. UA fly in separate altitudes and orbits so that link distances will vary.

Purpose: This experiment measures throughput among a typical small group of UA and measures the effect of UA dynamics.

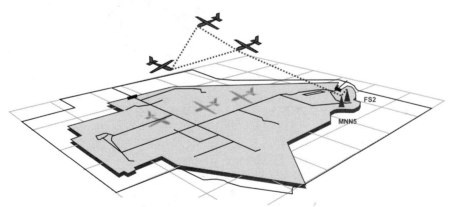

Figure 33.13: Three-UA flight and communication experiment.

33.7 Experimentation Results

This section describes the main experimental results. They are broken down into communication range, throughput, connectivity, mobility, and subjective tests. It should be emphasized here that the results are determined by the combination of 802.11 interfaces, DSR routing software, and MNN hardware used. Changing one or all of these aspects would have an effect on performance. Therefore these results indicate typical values within a system and are not absolute performance measures of the individual elements.

33.7.1 Communication Range

Since packets can be received occasionally even at long node separation distances, the communication range can be defined in a number of ways. The clearest indication is to look at the ability to communicate as a function of distance. Figure 33.14 shows throughput samples at different ranges for Ground node to Ground node and UA to Ground node communication. The Ground to Ground throughput falls off between 1 and 2 km. The UA to Ground throughput falls off between 2 and 4 km. This suggests that the UA doubles the communication range. However, variability increases when introducing the UA which can be attributed to UA dynamics.

During an additional field test, it was noted that placing a node on the edge of the mesa and driving to a distant hill that is in line-of-site provided reliable 1 Mbps throughputs at a distance of 10 km. Unlike the data in Figure 33.14, the sender and receiver were high above the terrain along the propagation path between them and stationary. As such this is the maximum communication range experienced throughout all our experimentation using ordinary omnidirectional antennas.[3]

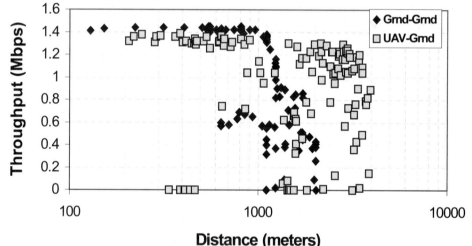

Distance (meters)

Figure 33.14: Throughput samples at different ranges.

The UA to UA range proved to be less than expected. When the UA were flown within a few kilometers of each other throughputs were around 1 Mbps. To measure larger distances, we used a second airfield located 7 km distant. No reliable throughputs were measured at this distance. At this range, the links are weaker and thus more susceptible to signal variations due to the UA maneuvering. Another cause that is hypothesized is that the second airfield is closer to residential areas and thus has more interference from other 802.11 devices. These effects are exacerbated by the DSR protocol which stops communication to search for a new route whenever too many packets are lost on existing routes.

---

[3] 200 km (125 miles) has been achieved mountain-top to mountain-top using large high-gain antennas [33].

33.7.2 Throughput

The nodes are placed in a linear topology on the ground such that MNN1 to MNN2 is one hop; MNN1 to MNN3 is two hops; and so on. These hop counts are nominal since the routing may discover shortcuts that skip nodes. However, in ground-based networks these shortcuts are generally weak with enough packet errors to force the network routing to select a stable path through the nominal number of hops. The throughput test measured the throughput between every pair of nodes and the throughput vs. the nominal number of hops in the stationary ground network was computed. The test was repeated with a UA-mounted node. In this case sustainable shortcuts through the UA could be formed since the UA-to-ground effective communication range is longer than the range between two ground nodes. The goal was to see if the network can discover these shortcuts and if they help throughput.

The throughput with and without the UA is shown in Figure 33.15. For the network without UA the black bars represent the average throughputs of all possible 1-hop, 2-hop, ..., 5-hop paths. The gray bars show the throughputs that can be achieved between the same source-destination pair with the help of the UA. The data points without the UA show the throughput falls off by a factor of two to three with each additional hop. This is a known phenomenon in ad hoc networks with a small span (five or less hops) [6]. As expected the UA makes no difference on one and two hop paths since paths between ground nodes that pass through the UA are at least two hops. For three and four hop paths the UA is able to maintain the end-to-end throughput close to the two hop throughput indicating that the network is able to find and sustain the two hop paths through the UA.

We note that the UA throughputs have a higher variance on average. We attribute this to the UA maneuvering. As the UA turns it can bank away from a target node so that the ground node is not in the main lobe of the dipole antenna. To observe this effect we measured the losses in low rate packet streams sent over 20-second intervals between every pair of ground nodes with and without the UA. The results are shown in Figure 33.16. Without the UA, given the fixed stable and connected network, no packets are ever lost. With the UA, shorter routes are formed through the UA. Loss samples with the UA experience occasional losses as high as 50% whereas the majority of samples have no

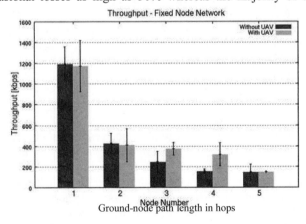

Figure 33.15: Throughput data with and without the UAV node vs. the nominal number of hops. Error bars are one error deviation.

losses at all. This reinforces the variability seen with the UA in the range experiments in Figure 33.15.

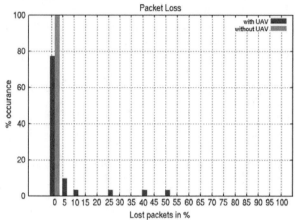

Figure 16: Loss rate over 20-second intervals with and without the UAV.

33.7.3 Delay

The delay depends on the number of hops, the hardware and software implementing the DSR, and the node mobility. Using ground based nodes on Soekris net4511 SBCs and a kernel space implementation of DSR, the delay is 13 ms per hop on average when the network is stable. This value is primarily a function of the processing speed of the node and whether the DSR is being implemented in kernel or user space. For instance, laboratory experiments using high-speed laptops and kernel modules show one-hop delays of 3 ms. With a user-space based DSR implementation, the delays are significantly increased.

With mobile nodes, the delays are larger, mainly because of the nature of DSR routing and our chosen parameters. Delays in the 100s of milliseconds were measured in the cases with mobile nodes and no UA. The mobile ground nodes would have periods where only longer range links were available with intermittent connectivity that would require multiple attempts to find routes. The UA can support longer range links that bridge the mobile nodes with the result that these outliers are eliminated and the delays are similar to the fixed ground node case.

33.7.4 Congestion

Competing traffic flows must share the network resources and as a consequence each flow has lower throughput compared to the throughput of each flow when sent one at a time. The throughput in the fixed node case was reduced on average by 29% when the competing flows overlapped, but had different sources and destinations. When the flows shared a source or destination, the throughput was reduced on average by 44%. In the three-UA experiment, the UA-UA throughput was reduced on average by 32% where the competing flows were either UA to UA, UA to Ground, or Ground to Ground flows. For the three-UA experiment, we treated all flows as one group since with three planes two competing UA-

UA flows necessarily have a common source or destination. Further, route overlap depends on the UA dynamics and could change over the course of a measurement.

It might be expected that two overlapping flows would divide the bandwidth resources so that each has a throughput that is reduced by 50%. However, a flow typically has multiple hops from source to destination. Different hops in the flow are already competing and so the sharing is not simply between the two flows but the hops within the flows. The total contention between hops does not necessarily double when the number of flows doubles.

The results above are averages. The variability across individual runs is high. The standard deviation is 27%, 27%, and 40% for each of the averages reported above. A close look at the data reveals that the reductions are not uniform between the two competing flows. One of the two competing flows was often much more affected than the other flow, even when the two flows had the same number of hops and relative interference. This suggests that the ad hoc routing combined with 802.11 is unfair and additional measures are required for fairness.

33.7.5 Connectivity

For the experiment with the mobile node at the network edge a node is driven around the mesa coming in and out of coverage of fixed nodes placed at the mesa edge. Packets are sent to the mobile vehicle and the losses recorded. This experiment tests the ability of the network to find routes to a mobile node with intermittent coverage. Table 33.1 shows that the UA has a dramatic positive effect on packet completion, only 4% of the packets are lost with the UA compared to 71% lost without the UA. Interestingly, the lower loss with the UA comes with only 19% of the packets actually passing through the UA. Without the UA the coverage to the vehicle has many gaps without a route. Each time a route can not be found, the router waits longer and longer before probing with a new route request; eventually reaching a state with waits up to 30 seconds between requests.[4] In this state the vehicle misses significant connectivity periods between route requests and many packets are lost. The UA coverage is also intermittent, but, provides enough additional connectivity to keep the interval between route requests short so that more connection opportunities (with or without the UA) are found by the router.

Table 33.1: Percentage of packets reaching destination with and without the UA.

	with UA	without UA
% packets lost	4	71
% packets received not routed through UA	77	29
% packets received routed through UA	19	N/A

The disconnected groups experiment sends packets between nodes when the nodes are divided into two groups separated so that packets sent between the groups suffered high loss rate. The packet loss rate between groups is 83%. With the addition of the UA the loss rate is reduced to 31%. A fixed ground node (i.e., MNN 3) could have also filled this gap.

---

[4] This so-called "back-off" mechanism is designed to manage congestion in overload situations. The router did not distinguish between packet losses due to poor links and packet losses due to congestion.

The UA did not fly directly over the gap but flew above the airfield as indicated in Figure 33.12. Similarly in the mobile node at edge experiment, the UA circled over the mesa without regard to any specific loss in connectivity. These results show that the presence of the UA was able to improve connectivity over the mesa without having to actively consider connectivity. The connectivity is improved but not optimal. Further work is exploring an active interaction between the connectivity and UA mobility [11].

33.7.6 Subjective Tests

While measuring throughput and delay gives good quantitative values for network performance, it does not say much about the end user experience while using the network for day-to-day activities. Browsing the web, downloading files, or using streamed media is a subjective experience. Two tests were designed to capture subjective impressions: a web-browsing test and a voice quality test.

For the web-browsing test the candidate browses a website consisting of several pages with a different size image on each of them, namely 10kB, 100kB, 300kB and 500kB. The web pages are served by the small-footprint, single-threaded web-server Boa (www.boa.org) installed on the gateway so as to minimize the impact on gateway performance. The voice quality test evaluates the subjective perception of a voice conversation carried out between two test candidates using laptops running DSR. Since the laptops used ordinary PCMCIA interface cards which operate at lower power and with a less efficient antenna than the MNN placed on ladders, they generally could only communicate to the network when in the vicinity to one of the MNN or the gateway. The open-source, Linux-based SIP-softphone Linphone (www.linphone.org) proved to be stable and user-friendly. It supports several voice codecs and enables adjustment of SIP and RTP parameters to compensate for changes in network performance.

Preliminary experiments proved quite successful in terms of satisfying end-user expectations in a reliable, well-performing network. With a well setup network of fixed nodes browsing web pages from as far as six hops away can be compared to surfing the Internet on a fast dial-up connection. However, the pictures are rendered in increasingly sporadic bursts with increasing hop count and when changing positions within the network the delay incurred by finding a new route with DSR can cause up to 30 second delays as the router waits for the old route to time out. Occasionally, downloads of pictures larger than 300kB stall noticeably halfway through the page, spoiling a user's browsing experience. With a hybrid network of stationary and mobile nodes browsing becomes choppier as nodes move out of reach and new routes to the web-server have to be found more frequently.

Voice quality as tested from the gateway to a laptop moving around the test bed was found to be exceptionally good up to three hops with no noticeable end-to-end voice delay. New routes formed automatically and voice contact was re-established without having to re-dial or restart the phone application, although there were gaps in the speech if the user moved out of range of the network. At a distance of four hops, voice streams became choppy and a meaningful conversation was not possible. As seen with web-browsing, the time to discover a new route can also considerably impair voice conversations. The impact of the UA node still has to be investigated, but throughput and delay results from other experiments suggest an improvement in user experience is expected.

33.8 Conclusion

The AUGNet concept is an extension of traditional ad hoc networks whereby a UA provides additional connectivity for ground nodes. The experiments in this chapter present a detailed performance evaluation of this role. Specifically, we show that the UA has longer range and better connectivity with the ground nodes. For a UA-supported network, this translates into shorter routes with better throughput. It also improves connectivity to nodes at the edge of network coverage. Subjective tests showed that the ad hoc network supported web browsing for up to six hops when the node positions remained static. Using the same setup, a real-time voice application worked well up to three hops.

The results presented here are for the particular combination of 802.11, DSR routing, and the MNN radio hardware. No optimizations for either the UA or the different traffic types were made. Some experiments suggest that better tuned parameters and faster hardware would improve performance. Other radio interfaces and routing software could also improve performance.

These experiments show that relatively high throughputs, on the order of 1Mbps, are possible between different UA and between UA and ground nodes. The availability of fully meshed networking is enabling us to realize robust communication architectures. We are currently using the AUGNet system for multi-vehicle cooperation; remote command and control of UA; as well as real-time vehicle telemetry.

Acknowledgments

This work was supported by L3-Communications ComCept Division. Thanks to the National Telecommunications and Information Administration for providing the Table Mountain site. Thanks to Jesse Himmelstein and Gerald Jones for support of the monitoring database. Thanks to Jake Nelson, Philip Nies, and Bill Pisano for their part in the design, construction, and operation of the UA. Thanks to Sushant Jadhav, Roshan Thekkekunnel, and Marc Kessler for testing support.

33.9 References

[1] Office of the Secretary of Defense, "Unmanned Aircraft Systems Roadmap, 2005-2030," http://www.acq.osd.mil/usd/Roadmap%20Final2.pdf.

[2] B. Argrow, D. Lawrence, E. Rasmussen, "UAV Systems for Sensor Dispersal, Telemetry, and Visualization in Hazardous Environments," in *43rd Aerospace Sciences Meeting and Exhibit*. Reno, NV, 2005.

[3] R.J. Bamberger, D.H. Scheidt, R.C. Hawthorne, O. Farrag, M.J. White, "Wireless Network Communications Architecture for Swarms of Small UAVs," in *AIAA 3rd "Unmanned Unlimited" Technical Conference, Workshop, and Exhibit*, Chicago, IL. 2004.

[4] P. Basu, J. Redi, V. Shurbanov, "Coordinated flocking of UAVs for improved connectivity of mobile ground nodes," in *MILCOM 2004*, 2004.

[5] A.L. Bertozzi, M. Kemp, D. Marthaler, "Determining environmental boundaries: asynchronous communication and physical scales," in *Cooperative Control*. A Post-

Workshop Volume 2003 Block Island Workshop on Cooperative Control, 9-11 June 2004, 2004, pp. 25-42.

[6] J. Bicket, D. Aguayo, S. Biswas, R. Morris, "Architecture and evaluation of an unplanned 802.11b mesh network," *Proc. of the 11th annual international conference on Mobile computing and networking* pp. 31-42, August 2005.

[7] P.K. Branch, "Unmanned vehicles - mainstays of future airborne reconnaissance," *Proc SPIE Int Soc Opt Eng,* vol. 3751, pp. 39-40, 1999.

[8] T.X Brown, S. Doshi, S. Jadhav, J. Himmelstein, "Test Bed for a Wireless Network on Small UAVs," *AIAA 3rd Unmanned Unlimited Technical Conference*, Chicago, IL, 20-23 Sep 2004.

[9] T. Camp, J. Boleng, V. Davies, "A Survey of Mobility Models for Ad Hoc Network Research," *Wireless Communications & Mobile Computing (WCMC): Special issue on Mobile Ad Hoc Networking: Research, Trends and Applications*, vol. 2, pp. 5, 2002.

[10] Y. Chen, K. Moore, Z. Song, "Diffusion Boundary Determination and Zone Control Via Mobile Actuator-Sensor Networks (MAS-net) – Challenges and Opportunities," presented at *SPIE Conference on Intelligent Computing: Theory and Applications II*, Orlando, FL, 2004.

[11] C. Dixon, E. W. Frew, B. Argrow. "Electronic Leashing of an Unmanned Aircraft to a Radio Source." *44th IEEE Conference on Decision and Control*, Seville, Spain, December 2005.

[12] S. Doshi, S. Bhandare, T.X Brown, "An On-demand minimum energy routing protocol for a wireless ad hoc network," *Mobile Computing and Communications Review*, vol. 6, no. 2, July 2002.

[13] C. Elliott, B. Heile, "Self-organizing, self-healing wireless networks," *Personal Wireless Communications*, 2000 IEEE International Conference on, 17-20 Dec. 2000, Page(s): 355-362.

[14] J. Elston, E.W. Frew, B. Argrow, "Networked UAV Communication, Command, and Control," AIAA-2006-6465, *AIAA Guidance, Navigation, and Control Conference*, Keystone, CO, August 2006

[15] Y. Eun, H. Bang, "Cooperative control of multiple UCAVs for suppression of enemy air defense," in Collection of Technical Papers - *AIAA 3rd "Unmanned-Unlimited" Technical Conference, Workshop, and Exhibit*, Sep 20-23 2004, 2004, pp. 617-630.

[16] E.W. Frew, C. Dixon, J. Elston, B. Argrow, T.X Brown. "Networked Communication, Command, and Control of an Unmanned Aircraft System." Under revision for *AIAA Journal of Aerospace Computing, Information, and Communication*, April 2007.

[17] E. Frew, S. Spry, T. McGee, "Flight Demonstrations of Self-Directed Collaborative Navigation of Small Unmanned Aircraft," *AIAA 3rd Unmanned Unlimited Technical Conference, Workshop, & Exhibit*, Chicago, IL, 20-23 Sep 2004.

[18] P. Gaudiano, B. Shargel, E. Bonabeau, B. Clough, "Swarm Intelligence: A New C2 Paradigm with an Application to Control of Swarms of UAVs," in *8th ICCRTS Command and Control Research and Technology Symposium*. Washington, D.C., 2003.

[19] D.L. Gu, G. Pei, H. Ly, M. Gerla, B. Zhang, X. Hong, "UAV Aided Intelligent Routing for Ad-Hoc Wireless Network in Single-Area Theater," *Wireless Communications and Networking Conference, 2000*. WCNC. 2000 IEEE, Volume: 3, 23-28 Sept. 2000 Page(s): 1220 -1225 vol.3.

[20] J. How, E. King, Y. Kuwata, "Flight demonstrations of cooperative control for UAV teams," *AIAA 3rd "Unmanned-Unlimited" Technical Conference, Workshop, and Exhibit, Sep 20-23 2004,* Chicago, IL, 20-23 Sep 2004, pp. 505-513.

[21] D. Johnson, D. Maltz, "Dynamic Source Routing in Ad HocWireless Networks," *Mobile Computing,* Chapter 5, pp. 153-181, Kluwer Academic Publishers, 1996. also IETF RFC 4728 http://www.ietf.org/rfc/rfc4728.txt

[22] E. Kohler, R. Morris, B. Chen, J. Jannotti, M. F. Kaashoek, "The click modular router," *ACM Transactions on Computer Systems,* vol. 18, no. 3, pp. 263–297, August 2000. http://www.pdos.lcs.mit.edu/click.

[23] J. Maslanik, J. Curry, S. Drobot, and G. Holland, "Observations of sea ice using a low-cost unpiloted aerial vehicle," presented at *16th IAHR International Symposium on Sea Ice, Int.,* 2002.

[24] D.R. Nelson, T.W. McLain, R.S. Christiansen, "Initial Experiments in Cooperative Control of Unmanned Air Vehicles," *AIAA 3rd "Unmanned Unlimited" Technical Conference, Workshop, and Exhibit,* Chicago, IL, 20-23 Sep 2004.

[25] H.G. Nguyen, N. Manouk, A. Verma, H.R. Everett, "Autonomous mobile communication relays," in *Unmanned Ground Vehicle Technology IV,* Apr 2-3 2002, 2002, pp. 50-57.

[26] S. Rathinam, M. Zennaro, T. Mak, R. Sengupta, "An Architecture for UAV Team Control," in *5th IFAC Symposium on Intelligent Autonomous Vehicles.* Lisbon, Portugal, 2004.

[27] J. Schlecht, K. Altenburg, B.M. Ahmed, K.E. Nygard, "Decentralized search by unmanned air vehicles using local communication," in *Proc. of the International Conference on Artificial Intelligence,* IC-AI 2003, Jun 23-26 2003, pp. 757-762.

[28] K. Schulze, J. Buescher, "A Scalable, Economic Autonomous Flight Control and Guidance Package for UAVs," *2nd AIAA "Unmanned Unlimited" Systems, Technologies, and Operations—Aerospace,* 15-18 September 2003, San Diego, California.

[29] P.B. Sujit, D. Ghose, "Search using multiple UAVs with flight time constraints," *IEEE Trans. Aerospace Electron. Syst.,* vol. 40, pp. 491-509, 04/. 2004.

[30] C.-K. Toh, *Ad Hoc Mobile Wireless Networks: Protocols and Systems,* Prentice Hall, 336p., 2001.

[31] K. Xu, X. Hong, M. Gerla, H. Ly, D.L. Gu, "Landmark routing in large wireless battlefield networks using UAVs," in *MILCOM 2001:* Communications for Network-Centric Operations: Creating the Information Force, Oct 28-31 2001, pp. 230-234.

[32] Y. Yi, X. Hong, M. Gerla, "Scalable Team Multicast in Wireless Ad hoc Networks Exploiting Coordinated Motion," presented at *4th International Workshop on Networked Group Communication,* Boston, MA, 2002.

[33] K. Zetter, "Hackers Annihilate Wi-Fi Record," *Wired News,* August 2, 2005. http://www.wired.com/news/wireless/ 0,1382,68395,00.html

## 34

Ultra-Wideband Wireless Technology

Kazimierz Siwiak and Yasaman Bahreini[a]

Ultra-Wideband (UWB) signaling technology is a modern wireless technique crafted to comply with recent regulations permitting UWB technology. Historically UWB, once called impulse radio, was defined by very short baseband signals that are transmitted and received without a radio frequency (RF) carrier in the usual sense. The technique reuses previously allocated RF bands by spreading the energy thinly in a wide spectrum, thus having a minimal impact on incumbent spectrum users. Regulations and Recommendations have been written in a way that restricted the permitted operating frequency ranges along with the emission levels, but remained silent on the modulation and signal characteristics. Hence in addition to pulse-based UWB technology, conventional technologies such as OFDM have been exploited under the rules. This chapter will expand on pulse-UWB, particularly at very high data rates, wherein the bandwidth of the signal is directly related to the inverse of the emitted pulse duration. Applications of UWB devices are presented, and potential use cases are described. It is shown that short-pulse low-power techniques have enabled practical through-wall radars, centimeter-precision 3-D positioning, and communications capabilities at the high data rates and with exceptional spatial capacities.

34.1 Introduction

Ultra-Wideband (UWB) wireless signaling is essentially the art of generating, modulating, emitting and detecting signals that inherently occupy large bandwidths. Wide band transmissions date back to the infancy of wireless technology. They include the wireless experiments of Heinrich Hertz in the 1880s, Alexander Popov in the 1890s, and later the 100 year old trans-Atlantic spark gap "impulse" transmissions of Guglielmo Marconi. Early radio circuits were comprised solely of passive electrical components, no tubes or transistors, and hence lacked the means to efficiently receive the short transient impulsive energy. At the time, the best method of separating radio users was by control over the emission bandwidth using tuned circuits as recognized by Popov and Nikolai Tesla. Spurred on by regulations mandating narrow bandwidths, radios were subsequently developed along narrow band frequency selective analog techniques. This led to voice broadcasting and telephony – and recently to digital telephony and wireless data. Through

---

[a] All authors are consultants

the years, a small cadre of scientists has worked to develop and refine impulse and pulse technologies. The origin of modern UWB technology stems from work in time-domain electromagnetics begun in the early 1960s to fully describe the transient behavior of a certain classes of microwave networks. By 1970 the primary focus in impulse radio research was on impulse radar techniques and government sponsored projects. Through the late 1980's, the technology was alternately referred to as baseband, carrier-free or impulse radio – the term "ultra-wideband" was not applied until approximately 1989 by the U.S. Department of Defense. By that time UWB theory, techniques, and many hardware approaches had experienced nearly 30 years of extensive development, see [1], culminating in today's UWB technology [2]. In late 1980s digital techniques began to mature to the point where modern low power impulse radio communications and precision radar could be demonstrated as a commercially practical technology. Digital impulse radio, the modern echo of Hertz's century-old transmissions, now emerges under the banner "Ultra-Wideband" radio.

34.2 Regulators Weigh in on UWB

Prior to 2002 UWB was understood to mean "impulse radio" and there was no lower frequency limit except what was dictated by suitable antennas. UWB was understood to mean systems having a fractional bandwidth of 25% or more, and the bandwidth was the consequence of the short pulses employed. Regulations, however, changed this landscape of UWB technology. On 14 February 2002 an FCC Report and Order legalized unlicensed UWB emission in the US. The full report, issued on 22 April 2002 [3] subsequently became part of US 47 CFR Part 15 [4], spelled out the rules for UWB in the US. That ruling placed a "hard" lower frequency corner at 3.1 GHz. Consequently "impulses" were no longer possible for communications devices; UWB pulses were band-limited by the rule. The FCC regulations further permitted bandwidths as small as 500 MHz in the 3.1 to 10.6 GHz range, and neither the signal shape nor the modulation was defined. It is possible as a result to operate under the UWB rules using conventional radio technologies so long as the basic bandwidth requirement of 500 MHz was met, and the spectral limits are not exceeded. Following six meetings of studying the impact of devices using ultra-wideband technology on systems operating within radio communication services, the ITU-R issued a Report and Recommendation [5] which recognized that applications using UWB technology may benefit sectors such as public protection, construction, engineering, science, medical, consumer applications, information technology, multimedia entertainment and transportation. The ITU-R Recommendation offers a summary of studies related to the impact of devices using UWB technology on radiocommunication services. The ITU-R recommended that administrations may consider the results of its studies in order to assess the impact of devices using UWB technology on allocated radiocommunication services when developing their national UWB regulations. The ITU-R Recommendation lists a summary of the US FCC rules and rules applicable to Japan. Later, the European Commission issued a Decision [6] applicable to the Community States. Figure 34.1 shows a comparison of the US, Japan and EC emission limits above 1 GHz.

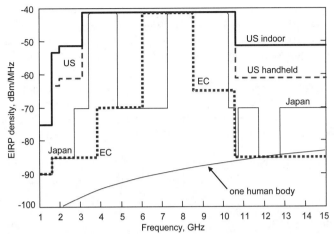

Figure 34.1: US, Japan and EC limits for UWB communications devices compared.

The US indoor rules are the most liberal. The only common spectrum among US, EC and Japan is in the range 7.25 to 8.5 GHz. The permitted emission levels are commensurate with the unintentional radio emissions of common electrical devices, computers, and electrical equipment. For comparison, one living human body emits about 130 watts of radiation following the black-body curve. In a room temperature environment this radiation can be approximated closely by

$$P_{body} = 20\log(f_{GHz}) - 106.7 \quad \text{dBm/MHz} \tag{1}$$

at frequencies below about 20,000 GHz [8]. In Figure 34.1 we can see that radiation from one human body exceeds the EC and the Japan emission limits at frequencies above about 12.2 GHz. This kind of comparison is meant to establish how weak the permitted UWB emissions are and is an indicator of the measurement challenges in demonstrating compliance at these low levels.

The following sections provide some more details, based on of the US and Japan approaches to UWB limits described in the ITUR-R Recommendation SM.1757 [5], as well as the recent EC Decision [6] on UWB technology.

34.2.1 Summary of US Rules on UWB

Five classes of UWB devices were identified by the FCC, each with its own operational limitations. However we will concern ourselves here mainly with the "communications" category. The specifics of the US rules are in [4], [3], however, the general technical requirements applicable to devices using UWB technology include:

–   Devices using UWB technology may not be employed for the operation of toys, or on board an aircraft, a ship or a satellite.
–   For devices using UWB technology where the frequency f_M, is above 960 MHz, there is a limit of 0 dBm EIRP on the peak level of the emissions contained within a 50 MHz bandwidth centered on f_M.

– Radiated emission levels at and below 960 MHz are based on measurements employing a CISPR quasi-peak detector. Radiated emission levels above 960 MHz are based on RMS average measurements using a spectrum analyzer with a resolution bandwidth of 1 MHz and an averaging time of 1 ms or less.

Table 34.1 summarizes the US FCC rules for UWB communications devices. The UWB bandwidth of a UWB communications system must be contained between 3,100 and 10,600 MHz. The Outdoor category is for UWB devices that are relatively small and primarily hand-held while being operated, and do not employ a fixed infrastructure.

Table 34.1: Summary of US rules for UWB communications devices.

	Indoor Systems		*Hand-held and outdoor systems*	
Radiated emission limits of resolution bandwidth of 1 MHz (dBm/MHz)	*Frequency*	*EIRP*	*Frequency*	*EIRP*
	960-1610	−75.3	960-1610	−75.3
	1610-1990	−53.3	1610-1990	−63.3
	1990-3100	−51.3	1990-3100	−61.3
	3100-10600	−41.3	3100-10600	−41.3
	Above 10600	−51.3	Above 10600	−61.3
Limits for resolution bandwidth of no less than 1 kHz	*Frequency*	*EIRP*	*Frequency*	*EIRP*
	1164-1240	−85.3	1164-1240	−85.3
	1559-1610	−85.3	1559-1610	−85.3

Intentionally radiated EIRP levels must be below −41.3 dBm/MHz to access the UWB spectrum. A UWB transmitter is defined as one that emits a UWB signal with bandwidth of the lesser of 500 MHz or 20% bandwidth, as measured at the −10 dB point.

34.2.2 Summary of UWB Rules Specific to Japan

The UWB emission rules specific to Japan are shown in Table 34.2 and Figure 34.1. The ITU-R Recommendations contain preliminary rules for use in Japan. Under those rules UWB devices are limited to only indoor use. A lower frequency band (3,400 – 4,800 MHz) provided that UWB devices emit at no more than −41.3 dBm/MHz and that UWB devices are equipped with interference avoidance techniques such as "detect and avoid" (DAA). Specific DAA techniques and rules are not defined, and the technology presents challenges.

A higher frequency band (7,250–10,250 MHz) is provided where UWB devices could emit at up to −41.3 dBm/MHz. In these permitted bands the peak power in a 50 MHz bandwidth is limited to 0 dBm.

34.2.3 Summary of the EC Decision on UWB

A recent decision [6] by the European Commission (EC) provides the basis for introducing UWB technology in the EC nations. The European Commission (EC) recognizes that it is important to establish regulatory conditions which will encourage the development of economically viable markets for applications of ultra-wideband technology as commercial opportunities arise. The purpose of EC Decision is to allow the use of the radio spectrum by equipment using UWB technology and to harmonize the conditions of its use in the

Community. Emissions are permitted in the bands shown in Table 34.3 subject to non-interference with incumbent radio communication services.

Table 34.2: Summary of UWB Limits applicable to Japan.

Frequency GHz)	Maximum mean EIRP density (dBm/MHz)
Below 1.6	−90.0
1.6 to 2.7	−85.0
2.7 to 3.4	−70.0
3.4 to 4.8	− 41.3 *(with detect and avoid technology)*
4.8 to 7.25	−70.0
7.25 to 10.25	−41.3
10.25 to 10.6	−70.0
10.6 to 11.7	-70.0
11.7 to 12.75	−85.0
above 12.75	−70.0

Table 34.3: Maximum EIRP according to the EC Decision.

Frequency range (GHz)	Maximum mean EIRP density (dBm/MHz)	Maximum peak EIRP density (dBm/50MHz)
Below 1.6	−90.0	−50.0
1.6 to 3.4	−85.0	−45.0
3.4 to 3.8	−85.0	−45.0
3.8 to 4.2	−70.0	−30.0
4.2 to 4.8	− 41.3 *(until Dec 31, 2010)* −70.0 *(beyond Dec 31, 2010)*	0.0 *(until Dec 31, 2010)* − 30.0 *(beyond Dec 31, 2010)*
4.8 to 6.0	−70.0	−30.0
6.0 to 8.5	−41.3	0.0
8.5 to 10.6	−65.0	−25.0
Above 10.6	−85.0	−45.0

For the purposes of the Decision, "equipment using ultra-wideband technology" means equipment incorporating short-range radio-communication technology involving the intentional generation and transmission of radio-frequency energy that spreads over a frequency range wider than 50 MHz, which may overlap several frequency bands allocated to radiocommunication services. The EIRP density is limited to –41.3 dBm/MHz in the 3.4 – 4.8 GHz bands provided that a low duty cycle restriction is applied in which the sum of all transmitted signals is less than 5% of the time each second and less than 0.5% of the time each hour, and provided that each transmitted signal does not exceed 5 milliseconds. Equipment using ultra-wideband technology may also be allowed to use the radio spectrum with EIRP limits other than those set out in Table 34.3 provided that appropriate mitigation techniques are applied with the result that the equipment achieves at least an equivalent level of protection to that provided by the limits in set out in Table 34.3.

34.2.4 UWB Definitions

UWB wavelets or pulses are bursts of electromagnetic energy wherein the wavelet duration in time is inversely related to its occupied bandwidth. For example, a 1 ns wavelet occupies 1 GHz of spectrum, a 2 ns pulse occupies 500 MHz; while a 0.13 nanosecond pulse instantaneously occupies 7.5 GHz of spectrum. Examples of these UWB pulse and their spectra are shown in Figure 34.2. The "zero crossing" rate of a wavelet determines where the center frequency of the spectrum wavelet bandwidth will appear. Finally the envelope shape of the wavelet (and to some extent the zero crossing structure) determine the detail of the distribution of the energy within the spectrum. Conventional technologies like OFDM (orthogonal frequency division multiplexing) can be configured to meet the UWB spectrum access rules by aggregating multiple narrow band modulated carriers.

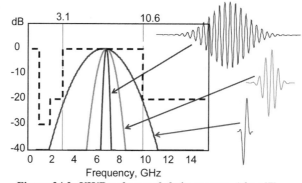

Figure 34.2: UWB pulses and their spectra. After [7].

The following terms and definitions are from the FCC rules [3] and ITU-R Report and Recommendations [5].

UWB: Ultra-wideband technology: technology for short-range radio communication, involving the intentional generation and transmission of radio-frequency energy that spreads over a very large frequency range, which may overlap several frequency bands allocated to radio communication services. Devices using UWB technology typically have intentional radiation from the antenna that at any point in time, has a fractional bandwidth equal to or greater than 0.20 or has a UWB bandwidth equal to or greater than 500 MHz, regardless of the fractional bandwidth.

UWB Bandwidth: the frequency band bounded by the points that are 10 dB below the highest radiated emission, as based on the complete transmission system including the antenna. The upper boundary is designated f_H and the lower boundary is designated f_L. The frequency at which the highest radiated emission occurs is designated f_M. The center frequency f_C equals $(f_H + f_L)/2$.

UWB Fractional Bandwidth equals $2(f_H - f_L)/(f_H + f_L)$.

UWB Impulse: a surge of unidirectional polarity that is often used to excite a UWB band-limiting filter whose output, when radiated, is a UWB pulse.

UWB Pulse: a radiated short transient UWB signal whose time duration is nominally the reciprocal of its bandwidth.

UWB Transmitter is an intentional radiator that at any point in time has a fractional bandwidth equal to or greater than 0.20 or has a UWB bandwidth equal to or greater than 500 MHz, regardless of the fractional bandwidth.

Equipment Using UWB Technology [6] means equipment incorporating technology involving the intentional generation and transmission of radio-frequency energy that spreads over a frequency range wider than 50 MHz, which may overlap several frequency bands allocated to radiocommunication services.

Wideband Transmitter is an intentional radiator which has a −10 dB bandwidth of at least 50 MHz.

Wide Band transmitters are permitted anywhere in the EC defined spectrum as adopted by EC administration, and in the spectrum between 5925 and 7250 MHz the US.

34.3 UWB Radio Technologies

UWB systems have been implemented in a wide range of technologies such as Continuous Pulse UWB (C-UWB), Multiband Pulses, Multiband OFDM, TM-UWB, TDR-UWB as shown in Table 34.4. C-UWB has been implemented by Pulse~Link, Inc. under the trade name CWave^{TM} as a very high data rate technology which can top 2,700 Mbps in a bandwidth of about 1.4 GHz. The IEEE P802.15.4a Standard [9] is also defined for various data rates between 0.1 and 27.24 Mbps in bandwidth between 0.5 and 1.36 GHz. Freescale has developed a C-UWB technology called Direct Sequence UWB (DS-UWB) which delivers a 112 Mbps in a 1.4 GHz bandwidth. Another method of accessing the UWB spectrum is based on Orthogonal Frequency Division Multiplexing (OFDM) which is being developed by WiMedia Alliance, and is described in another Chapter. Still another UWB technique involves a multi-band pulse technology using Time Division/Frequency Division Multiple Access (TD/FDMA) pulse approach which was implemented as Spectral Keying^{TM} by General Atomics. One of the first commercial implementations of UWB is Time-Modulated (TM-UWB) under the name PulseON^{TM} from Time Domain Corp. Finally a very simple approach called Transmitted Delayed Reference (TDR-UWB) was developed at General Electric [10].

Modulation efficiency is the needed energy per bit to noise density ratio for achieving a bit error rate (BER) no greater than a specified value. A BER less than 10^{-3} at the input to the forward error corrector is used for comparison in the Table. Table 34.4 lists various modulations that have been applied to UWB systems. System designers wishing to estimate performance in realistic environments should consider the additional propagation losses usually encountered in practical environments. Performance in multipath and shadowing conditions should also be considered.

Table 34.4: Many UWB communications systems are possible under the rules.

	Continuous Pulse UWB	Multiband Pulses	Multiband OFDM	TM-UWB	TDR-UWB
Implemented as	CWave[1] P802.15.4a[2] DS-UWB	Spectral Keying[3]	WiMedia[4]	PulsON[5]	Transmitted Delayed Reference
Bands	1–16	3–10	3–13	1	1
Bandwidth, GHz	0.5–1.5	1.2–5.2	1.5–6.9	2	2
Frequency Bands, GHz	3.1–5.1 5.8–10.6	3.1–5.0 4.9–10.6	3.1–4.8 3.1–10.6	3.1–5.2	3.1–5.2
Modulation	Pulse shape and polarity	Pulse band, shape and polarity	OFDM-QAM, DCM	TM-PPM	Differential polarity
Modulation efficiency at 10^{-3} BER	4–8 dB	5 dB	4–8 dB	9–15 dB	9–15 dB
Error correction codes	LDPC, Convolutional, Reed-Solomon,	Convolutional and Reed-Solomon	Reed-Solomon	various	various
Capabilities:	20 to 2700 Mbps in 1.4 GHz BW *Also: Cwave over home CATV networks*	120 Mbps in 1.4 GHz BW to 1 Gbps in 5.2 GHz BW	53 Mbps to 480 Mbps in 1.5 GHz BW	10 to 40 Mbps	10 Mbps

Notes: All trademarks are owned by their respective companies.
[1]CWave is a trademark of Pulse~Link, Inc., Carlsbad CA.
[2]802 is a trademark of IEEE, Piscataway NJ.
[3]Spectral Keying is a trademark of General Atomics Corp., San Diego CA.
[4]WiMedia is a trademark of WiMedia Alliance, San Ramon CA.
[5]PulsON is a trademark of Time Domain Corp., Huntsville AL.

34.3.1 Continuous-pulse UWB Technology

Pulse UWB wherein the bandwidth is the reciprocal of the pulse duration can be implemented as a continuous sequence of pulses sent at a rate equal to the pulse bandwidth. This results in Continuous-pulse UWB (C-UWB) sent at a very high "chip" rate. One implementation, capable of data rates as high as 2.7 Gbps in a 1.4 GHz bandwidth, is CWave™ technology from Pulse~Link. Another, low data rate implementation, is described in the IEEE 802.15.4a specification [9]. In C-UWB the baseband reference pulse has nominal chip duration of T_c which normally ranges from 0.5 to 2 ns producing bandwidths between 2 GHz and 500 MHz respectively. The baseband reference pulse spectrum for the complete transmission system including a matched filter receiver is defined by

$$X(f) = \begin{cases} T_c & \text{for} & 0 \le |f| \le \dfrac{1-\beta}{2T_c} \\[2ex] \dfrac{T_c}{2}\left\{1 + \cos\left[\dfrac{\pi T_c}{\beta}\left(|f| - \dfrac{1-\beta}{2T_c}\right)\right]\right\} & \text{for} & \dfrac{1-\beta}{2T_c} \le |f| \le \dfrac{1+\beta}{2T_c} \\[2ex] 0 & \text{for} & |f| \ge \dfrac{1+\beta}{2T_c} \end{cases} \tag{2}$$

where f is the frequency, T_c is the pulse duration, and β is the roll-off factor which is typically between 0.3 and 0.7. The transmitter reference pulse may be defined in the time domain as the impulse response of the root raised cosine filter, the square root of the filter spectrum described above, and is

$$r(t) = \frac{4\beta}{\pi\sqrt{T_c}} \frac{\cos\left(\dfrac{\pi t(1+\beta)}{T_c}\right) + \dfrac{T_c}{4\beta t}\sin\left(\dfrac{\pi t(1-\beta)}{T_c}\right)}{\left(1 - \left(\dfrac{4t\beta}{T_c}\right)^2\right)} \tag{3}$$

The transmitted pulse shape $p_{TX}(t)$ is constrained by the shape of its cross correlation function with a standard reference pulse $r(t)$. For the purposes of testing a transmitter pulse for compliance we define the cross correlation $X(\tau)$ of the transmitter pulse $p_{TX}(t)$ with $r(t)$ as

$$X(\tau) = \frac{1}{\sqrt{P_{TX}R}} \int_{-\infty}^{\infty} r(t) p_{TX}(t+\tau) dt \tag{4}$$

where P_{TX} is the energy in the transmitter pulse found from the time integral of the square of $p_{TX}(t)$, and R is the energy in the reference pulse found from the time integral of the square of $r(t)$. The cross correlation $X(\tau)$ for a compliant transmitter is greater than 0.7071 for a continuous range of τ surrounding the peak cross correlation value, and the range of τ should be equal to at least 26% of the reference pulse width. In addition, the remaining side-lobes of the correlation function should be less than or equal to 0.3. While the measurement described here occurs on the pulse envelope as if shaping is done at baseband, it is not intended or implied that pulse shaping occurs only at baseband.

Conceptually, the reference pulse is translated to the operating frequency by multiplication with $\cos(2\pi ft)$ and by $\sin(2\pi ft)$ to obtain two essentially orthogonal pulses $r_I(t)$ and $r_Q(t)$. In practice, filtering techniques and direct pulse synthesis may be used. Either of the pulses can be polarity modulated in a fashion analogous to BPSK in conventional radio. Both pulses $r_I(t)$ and $r_Q(t)$ can each be independently polarity modulated with the signals added together to form a 4-level encoding scheme analogous to QPSK in conventional radio. Signal generation and modulation, as well as other UWB pulse designs are further described in [2], [11], [9].

34.3.2 CWave™ UWB Technology

CWave™ is a continuous pulse or C-UWB technology using 0.741 ns long pulses sent at a 1.35 GHz rate to implement a high data rate communication system. The pulses are defined by Equations (2) – (4) with β=0.3 and are polarity modulated. Golay spreading codes in combination with low density parity code (LDPC) forward error correction are employed to achieve a range of data rates from 20 to 1350 Mbps. The use of two orthogonal pulses enables a doubling of the data rates in the same RF bandwidth to 2700 Mbps. The technology has been successfully demonstrated by transmission between UWB antennas as well as over home CATV networks where it is suitable for transporting IEEE 1394 signaling at application layer throughput rates of more than 400 Mbps with simple polarity modulation and up to 800 Mbps with orthogonal signaling. The top-level description is shown in Figure 34.3.

In the transmitter, the PHY protocol data unit (PPDU) elements are scrambled, encoded, spread and formatted, then mapped into symbols and modulated onto the final waveform for transmission over the cable system. Upon reception, the waveform is de-modulated, and the chip stream is de-spread, decoded and de-scrambled for presentation to the MAC. Dataflow in the transmitter and receiver are illustrated by Figure 34.3.

A chip (or pulse) is the fundamental PHY signaling unit and is transmitted at a fixed rate of 1.35 Gcps for either one pulse shape or a parallel pair of orthogonal pulses. Parallel sequences of chips using orthogonal UWB pulses double the combined chip rate to 2.7 Gcps. The information content of a sequence of chip varies according to the FEC rate, the spread factor, and pulse orthogonality.

Referring to Figure 34.3, the Data Input bit stream is first scrambled, then forward error correction (FEC) using Low Density Parity Check (LDPC) algorithm is applied to the scrambled bits. The number of output bits (coded bits) equals the inverse of the FEC rate, thus a rate 0.5 FEC doubles the number of transmitted bits. This encoded bit stream is then "spread" by the spreading code sequence factor of 1, 2, 4, 8, or 64. Spreading is the dot-multiplication of each coded bit by a contiguous set of chips chosen based on a given spreading code sequence. The ratio of the symbol duration in time to the chip duration in time is the processing gain.

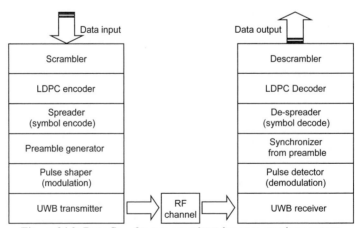

Figure 34.3: Data flow from transmitter input to receiver output.

Finally, symbols are polarity encoded before transmission over the RF channel media. The RF channel may be either wireless or wired. Polarity encoding is equivalent to Binary Phase-Shift keying (BPSK) in conventional radio technology. Optionally, sequences of two orthogonal UWB pulses (or wave shapes) can be used in a manner analogous to Quadrature Phase-Shift Keying (QPSK) of conventional radio technology. The data rate on a channel of given bandwidth may be doubled by using the combination of both orthogonal pulses and polarity modulation.

The receive process mirrors the steps of the transmission process and converts the received chips back to the information bits at the Data Output port.

Table 34.5: CWave modes and corresponding data rates using polarity modulation.

Transmit Mode	Data Rate (Mbps)	FEC Factor	Spread Factor
1	1350	1	1
2	900	2/3	1
3	675	1/2	1
4	450	2/3	2
5	338	1/2	2
6	225	2/3	4
7	169	1/2	4
8	113	2/3	8
9	84	1/2	8
10	21	1	64

The data rates in Table 34.5 are doubled to a maximum of 2700 Mbps in the same RF bandwidth when a set of two orthogonal pulses are used in a manner equivalent to QPSK in conventional signaling.

Some of the CWave timing parameters are shown in Table 34.6. The chip rate is 1.35 Gcps, thus each chip or UWB wavelet described by Equations (2) – (4) is 741 ps in duration. Chips are grouped into a block size of 24,576 chips of time duration of 18.2 ms. Data symbols can comprise 1 to 64 chips per symbol depending on data rate and processing gain.

Table 34.6: Some CWave timing parameters.

Parameter	Value	Units	Description
R_{chip}	1350	MHz	Chip rate
T_{chip}	741	ps	Chip duration
N_{block}	24576	chips	Block size
Tb_{Block}	18.2 μs	μs	Block duration
N_{dsym}	1 2 4 8 64	chips	Data symbol size

The Medium Access Control (MAC) layer in CWave technology is based on the IEEE 802.15.3 and IEEE 802.15.3b specifications, see [12], [13]. 802.15.3 is the IEEE standard for high data rate wireless personal area networks designed to provide isochronous distribution of multimedia content, such as video and music. The base 802.15.3 standard, see [14], uses a wireless physical layer (PHY) in the 2.4 GHz ISM band. The standard intends to provide a level of Quality of Service (QoS) suitable for a home multimedia wireless network.

34.3.3 IEEE P802.15.4a UWB Technology

The UWB physical layer (PHY) in the IEEE 802.15.4a specification [9] describes a low data rate technology which uses the C-UWB approach to achieve a wide range of data rate options. This specification is an extension of the low rate PHY in IEEE 802.15.4 Standard [15] which is described by [16]. The UWB PHY waveform is based the band-limited data pulses of Equations (2) – (4) with the parameter β =0.6. The chip duration T_c is very nearly, but not necessarily equal to the inverse of the UWB pulse bandwidth. This is because further band filtering may be applied at the RF frequency.

Table 34.7 shows the 802.15.4a UWB PHY frequency band plan.

Table 34.7: P802.15.4a UWB PHY bands.

Band	Frequency range	Channels
below 1 GHz	250 MHz to 750 MHz	channel: 0
low-band	3.1 GHz to 4.8 GHz	channels: 1-4
high-band	6.0 GHz to 10.6 GHz	channels: 5-12

A combination of burst position modulation (BPM) and polarity modulation (analogous to BPSK) is used to support both coherent and non-coherent receivers using a common signaling scheme. Each channel can support at least two complex channels that have unique length 31 preamble codes. The combination of a channel and a preamble code is termed a complex channel.

This UWB PHY supports a rich set of options in the sixteen channels defined for operation as shown in Table 34.8. Various data rates are supported through the use of variable-length bursts. Data rates from 0.11 to 27.24 Mbps. The various data rates are supported through the use of variable-length bursts. The combination of a channel and a preamble code is termed a complex channel and compliant device implement support for at least one of the channels (0, 3 or 9) in Table 34.7 along with two unique length 31 preamble codes.

The encoding process is similar to the one shown in Figure 34.3, and comprises the following steps:

1) Perform Reed-Solomon encoding
2) Assemble the PHY header
3) Add check bits
4) Perform further convolutional coding (in some instances at the 27 Mbps data rate this step bypassed)
5) Modulate and spread the data
6) Produce preamble field from the SYNC field (used for AGC convergence, diversity selection, timing acquisition, coarse frequency acquisition)

7) Transmit the data over the RF channel.

Table 34.8: P802.15.4a UWB channel allocations.

Channel Number	Center Frequency, MHz	Bandwidth, MHz	Comment
0	499.2	499.2	Mandatory "below 1 GHz"
1	3494.4	499.2	Optional
2	3993.6	499.2	Optional
3	4492.8	499.2	Mandatory in "low-band"
4	3993.6	1331.2	Optional
5	6489.6	499.2	Optional
6	6988.8	499.2	Optional
7	6489.6	1081.6	Optional
8	7488.0	499.2	Optional
9	7987.2	499.2	Mandatory in "high-band"
10	8486.4	499.2	Optional
11	7987.2	1331.2	Optional
12	8985.6	499.2	Optional
13	9484.8	499.2	Optional
14	9984.0	499.2	Optional
15	9484.8	1354.97	Optional

The reception process mirrors and reverses the transmitter operations. In the BPM-BPSK modulation, a UWB PHY symbol is capable of carrying two bits of information. One bit is used to determine the position of a burst of pulses while the second bit is used to modulate the polarity of this same burst.

34.3.4 Direct Sequence UWB

Direct Sequence UWB (DS-UWB) is a continuous pulse UWB technology from Freescale Semiconductor, which was acquired through Motorola from XtremeSpectrum, Inc. The implementation is largely analog-based and has roots in secure and military communications systems. The DS-UWB approach implements a wavelet tailored to occupy the spectrum between 3.1 to 5.1 GHz, and proposes another wavelet, which occupies the 5.8 to 10.6 GHz band. The two bands can be used independently or together to provide a range of options for system deployment. The DS-UWB system uses a Multi-level Bi-Orthogonal Keying (M-BOK) modulation or M-BOK in combination with pulse polarity and pulse shape depending on the required data rate. The UWB bandwidth spreading is accomplished by sending the wavelets at a 1.36–1.38 Gcps chip rate. The chip rate is varied among multiple users in the same network, enabling a rapid user signal acquisition. Modulation of the sequence of wavelets "whitens" the spectrum, and disrupts regularities that would otherwise result in spectral lines. M-BOK modulation comprises 24 and 32 length ternary orthogonal sequences (1, 0, +1) of wavelets. A UWB system comprising polarity modulated set of ternary codes is described by Lakkis and Bahreini [17]. Either 1, 2, 3 or 6 bits are sent with each code symbol. For example, 64-BOK modulation takes 6 chips at a time ($2^6 = 64$ BOK combinations) to form a symbol. The 24 length ternary codes are used with 2-BOK, 4-BOK, and 8-BOK, while 32 length codes are used with 64-BOK. DS-UWB symbols are sent at 42.75 MSym/s resulting in a channel rate of 256.5 Mbps. The data are encoded with a rate 0.44 error correction code resulting in the 112 Mbps information data rate. A second set of wavelets that are orthogonal to the original wavelets may be sent on top of the original wavelets in a manner analogous to QPSK in conventional radio systems to double the data rate in the same bandwidth. A 112 Mbps DS-UWB system has been offered commercially.

34.3.5 TM-UWB Technology

Time-modulated UWB was developed by Larry Fullerton [18] at Time Domain. It is a form of pulse-position modulation, see Figure 34.4, and is implemented as PulsON^{TM} technology.

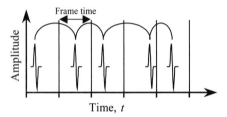

Figure 34.4: PN coded UWB waveform sequence in time. After [7].

Originally developed prior to the FCC Report and Order, early TM-UWB systems operated with impulses applied directly to a UWB antenna resulting in a pulse having a broad emitted spectrum centered somewhat below 2 GHz. As a pioneering UWB technology, TM-UWB is studied extensively in the literature, see for example [19]. Following the FCC ruling, TM-UWB was re-spun for use above 3.1 GHz. TM-UWB transmitters emit ultra-short pulses with tightly controlled pulse-to-pulse intervals. The waveform pulses are typically less than 1 nanosecond in duration, and are centered near 4 GHz, with pulse-to-pulse intervals of between 25 and 1000 nanoseconds. The pulse-to-pulse interval is varied on a pulse-by-pulse basis in accordance with two components: an information signal, and a channel code. The TM-UWB receiver directly converts the received RF signal into a baseband analog output signal with multiple pulses integrated to form a digital signal. A single bit of information is spread over multiple wavelets, providing a way of scaling the energy content of a data bit with the data rate. The receiver coherently sums the proper number of pulses to recover the transmitted information.

TM-UWB systems use a fine pulse shift modulation by positioning the pulse one quarter cycle early or late relative to the nominal PN coded location, or by pulse polarity. Multilevel pulse position modulation may be used to provide higher data rates in a given RF bandwidth. The error probability of fine-shift modulation in additive white Gaussian noise (AWGN) follows the same behavior as conventional orthogonal or on-off keying (OOK). When coherently detected the error probability of pulse polarity modulation in AWGN follows the same behavior as that of conventional BPSK or antipodal signaling. Because TM-UWB modulation is based on accurate timing, the method is especially suitable for accurate distance determination. FCC certified TM-UWB devices demonstrating communications, distancing and 2-d radar are commercially available form Time Domain Corp.

34.3.6 A Multiband OFDM Approach to Utilizing UWB Spectrum

Multiband OFDM in the UWB spectrum does not employ short pulses to meet the UWB regulatory bandwidth requirement, but rather comprises an ensemble of QAM modulated sinusoidal carriers spanning at least 500 MHz of bandwidth. One system, from WiMedia Alliance, comprises long bursts equal to the OFDM symbol duration, where the symbol duration defines the OFDM carrier spacing, while the total OFDM bandwidth results from the aggregation of those carriers. A total of 122 carriers spaced every 4.125 MHz span about 510 MHz; the OFDM symbol has a duration of 242.42 ns which equals the reciprocal of the carrier spacing. Symbols are hopped one per frequency band every 312.5 ns over at least 3 bands each separated by 528 MHz. WiMedia OFDM uses QPSK modulation for data rates up to 200 Mbps and dual carrier modulation (DCM) for 320 to 480 Mbps. DCM involves sending the same data, but with a rotated constellation, simultaneously on two of the OFDM carriers that are separated by at least 200 MHz. The MAC architecture, see [20], is based on two specifications – Certified Wireless USB, and WiMedia – that were developed to share the same communications channel over PHY layer. The WiMedia Alliance approach is described in Chapter 35.

34.3.7 A Multi-band Approach to UWB

The Time Division / Frequency Division Multiple Access approach uses multi-band pulses that are centered at frequencies spaced by, for example, 550 MHz. In this example the pulses are 3 ns long and occupy a 700 MHz bandwidth when measured at points 10 dB below the peak value. The pulse wavelets for the lower four bands are shown in Figure 34.5. The corresponding spectral occupancy is portrayed in Figure 34.6.

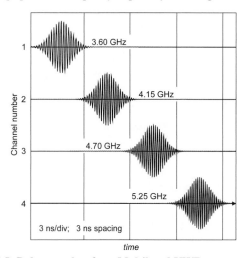

Figure 34.5: Pulse wavelets for a Multiband UWB system. After [7].

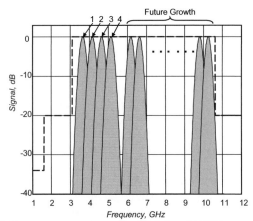

Figure 34.6: Spectrum occupancy of a Multiband UWB system. After [7].

Multipath is mitigated by the time interval between pulses that appear on the same channel. The Multiband system used pulses that are separated in frequency as well as time. Spectral Keying™, (a trademark of General Atomics) is a particular implementation of multi-band UWB from General Atomics and is described in [21] and [22]. Spectral Keying utilizes between 2 and 10 overlapping bands in the 3.1 to 9 GHz range.

34.3.8 TRD-UWB Technology

A method of transmitting pulses that include a self-reference pulse for easy detection is exemplified by Transmitted Delay Reference (TRD-UWB) and described in [10]. The method employs differentially encoded pulse pairs sent at a precise, but possibly multiple spacings D. The system is shown in the simplified block diagram of Figure 34.7.

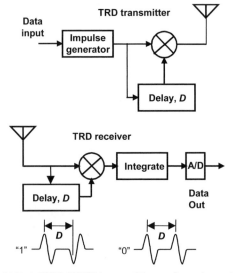

Figure 34.7: A TRD-UWB transmitter and receiver. After [7].

The transmitter sends a pair of pulses separated by one of several possible delays D, and differentially encoded by pulse polarity. The pulses, including propagation induced multipath replicas, are received and detected using a simple self-correlator with one input fed directly and another input delayed by D. The receiver resembles a conventional DPSK receiver which in AWGN exhibits an error probability P_D, see [23], of

$$P_D = \frac{1}{2}\exp\left(-\gamma_b \frac{N-1}{N}\right) \tag{5}$$

where $N>1$ is the number of differentially encoded pulses in a sequence and γ_b is the SNR per bit. The integration interval is sufficiently long to rake in some multipath energy. TRD-UWB modulation performance is compared with other modulations in Table 34.3.

One channelization method employing TRD-UWB, see [10], has $N=2$ and employs a family of delays D_i. Pulse pair sequences of these delay combinations comprise the channelization.

34.4 UWB Short Pulse Radiation and Reception

Radiated fields of pulses are found by solving Maxwell's equations in space and time. We define the geometry of an that supports surface currents $J(r',\tau)$ with a retarded time variable $\tau = t - R/c$. The solution for an arbitrarily shaped antenna is well beyond the scope of this work, and approximations will be given based on [2] and on [24]. Numerical solutions for specific antenna shapes can be found using the FDTD numerical method, as described in [25]. The solutions given below are general, and also apply to narrow band signals. Narrow band solutions are always approximately sinusoidal, hence narrow band solutions are always sine waves.

34.4.1 UWB Wavelet Radiation

A general expression for the magnetic far-zone field is based on the geometry in Figure 34.8.

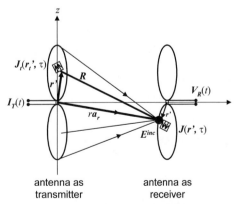

Figure 34.8: Wide band dipoles.

Using the space-time integral equation formulation is

$$H(r, t) = \frac{1}{4\pi rc} \int_V \frac{\partial}{\partial \tau} \, J(r', \tau) \times a_r \, dV' \qquad (6)$$

This integral reduces to a surface integral over current density J, in the *retarded time variable* $\tau = t - R/c$, where c is the speed of propagation, and a_r is a unit vector pointing in the direction of radiation. Note that the partial derivative in τ includes both time *and* delay. The radiated electromagnetic fields H and E are related to a time-delay derivative of the antenna current density. Because of this, the radiated field signal shape will be different from the signal shape supplied to the antenna feed point. Furthermore, when antenna has a realistic physical extent, the signal shape will vary as a function of look angle from the antenna. *This is a key difference from narrow band solutions where the signal shape everywhere is sinusoidal.*

34.4.2 Radiation of a UWB Elementary Dipole and Loop

We initially look at the infinitesimal radiating current and infinitesimal loop current. The fields surrounding an infinitesimal current element of length Δh can be derived from the Boit-Savart Law generalized by Jefimenko to time varying solutions. The magnetic field written in terms a current $I(t)$ oriented along the z axis is

$$\mathbf{H_D}(r, t) = \frac{\Delta h \sin(\theta)}{4\pi r} \left[\frac{I(\tau)}{r} + \frac{1}{c} \frac{\partial}{\partial t} I(\tau) \right] \hat{\varphi} \qquad (7)$$

and the corresponding electric field is

$$\mathbf{E_D}(r, t) = \frac{\Delta h}{4\pi\varepsilon_0 r} \left[\left(\frac{\int I(\tau) dt}{r^2} + \frac{I(\tau)}{cr} \right) \left(2\cos(\theta)\hat{r} + \sin(\theta)\hat{\theta} \right) + \frac{\sin(\theta)}{c^2} \frac{\partial}{\partial t} I(\tau)\hat{\theta} \right] \qquad (8)$$

Radiation from the infinitesimal loop current is found from the duality principal by first making a substitution selected so that the dipole elementary current and the current loop are represented by the same function,

$$\frac{1}{c} \frac{\partial}{\partial t} I(\tau)\Delta S \Rightarrow \Delta h I(\tau) \qquad (9)$$

where ΔS is the infinitesimal loop area in the x-y plane surrounded by the loop current. Thus the infinitesimal loop fields *for the same source current I(t) in the loop as in the dipole* are

$$\mathbf{E_L}(r, t) = \frac{\Delta S \sin(\theta)}{4\pi\varepsilon_0 rc} \left[\frac{\partial}{\partial t} \frac{I(\tau)}{rc} + \frac{\partial^2}{\partial t^2} \frac{I(\tau)}{c^2} \right] \hat{\varphi} \qquad (10)$$

and

$$\mathbf{H_L}(r,t) = \frac{\Delta S}{4\pi r}\left[\left(\frac{I(\tau)}{r^2} + \frac{\partial}{\partial t}\frac{I(\tau)}{cr}\right)\left(2\cos(\theta)\hat{\mathbf{r}} + \sin(\theta)\hat{\mathbf{\theta}}\right) + \sin(\theta)\frac{\partial^2}{\partial t^2}\frac{I(\tau)}{c^2}\hat{\mathbf{\theta}}\right] \quad (11)$$

Only the $1/r$ terms survive into the far field radiation region.

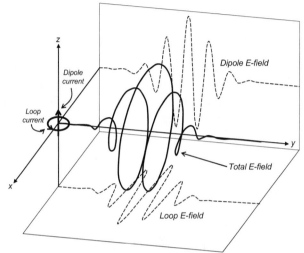

Figure 34.9: A traveling circularly polarized UWB pulse.

The dipole and loop fields are in *space quadrature*, but their representations in time are also orthogonal so they are also in *time quadrature*. This has interesting consequences in that the superposition of an infinitesimal loop and dipole with the same current excitation produces a rotating electric field vector, see Figure 34.9, everywhere in space for the duration of time that the field vector exists, hence it defines circular polarization for UWB pulse transmission.

Solving for the radiated fields of an arbitrarily shaped antenna is beyond the scope of the present work. However, for a dipole of length $2h_a$ we can write an approximate solution for the magnetic fields radiated from the dipole aligned with the z axis

$$H_\phi(r,t) = \frac{\sin(\theta)}{4\pi rc}\frac{h_a}{2}\frac{\partial}{\partial t}\left\{I_z(t) + I_z(t-[1-\cos(\theta)]h_a/c) + I_z(t-[1+\cos(\theta)]h_a/c)\right\} \quad (12)$$

where $I_z(t)$ is the dipole feed point current. The usual spherical (r,θ,ϕ) coordinates are employed, and the $\sin(\theta)$ term is the projection of the z-directed current density on the θ direction at the observation point. The electric far-field is

$$E_\phi(r,t) = -\eta_0 H_\phi(r,t) \quad (13)$$

The fields in Equation (12) appear to emanate from three sources, see Figure 34.8: one at the feed point, and one at each dipole end. The total field comprises time derivative components that are delayed by t as well as by $\cos(\theta)h_a/c$. Thus two processes contribute to the field signal shape, a time and distance delay. *Both processes tend to lengthen the signal in time, hence both processes will contribute to a narrowing of the radiated bandwidth compared to the bandwidth of the signal supplied to the antenna.* Furthermore, the angle dependent delay term $\cos(\theta)h_a/c$ in Equation (12) means that the radiated signal shape as a function of time will vary depending on where the observation point is relative to the antenna axis.

34.4.3 Receiving UWB Wavelets

A good understanding of the electromagnetic field receiving process by a dipole antenna follows from the Lorentz Force Law. That Law relates the force F exerted on a charge q by the incident electric E and magnetic H fields on an antenna having length Δh. For the infinitesimal antenna

$$F = Eq + q\mu_0 v \times H \tag{14}$$

Since the antenna velocity $v=0$, the magnetic field term drops out. The force per charge acting over the incremental antenna length has the units of voltage so

$$V_R(t) = -\int_{-\Delta h/2}^{\Delta h/2} \frac{1}{q} F \cdot dl = -\int_{-\Delta h/2}^{\Delta h/2} E^{inc} \cdot dl = -E^{inc}(t)\Delta h \tag{15}$$

giving the result that the open circuit received voltage on a point antenna is proportional to the electric field incident on it.

Again the complete electromagnetic solution for an arbitrary receiving antenna is beyond the current scope, but a reasonable approximation for the voltage received by a wide band dipole, validated by FDTD analysis [25], is

$$V_R(t) = h_a(E_z(t) + E_z(t - [1 - \cos(\theta)]h_a/c) + E_z(t - [1 + \cos(\theta)]h_a/c) \tag{16}$$

The received signal V_R is a weighted sum of and time-delayed electric fields incident on various parts of the antenna, which in turn are time-delay derivatives of the transmitted current I_T. Again it is evident the that the angle of arrival of the signal will have an impact on the shape of the received voltage versus time.

34.5 Propagation of UWB Signals

Electromagnetic fields propagate in free space as expanding spherical waves. Hence an inverse square law applies to the far field power density. In cluttered environments the wavelets additionally shed energy to multipath reflections leading to a multipath dispersion term. That dispersion term gives increases the propagation law to approximately inverse 3rd power inside buildings. The propagation channel dispersion term is also related to the maximum energy available for rake receiver gain. A theoretical analysis leads to the SBY propagation law that further predicts the possible rake (equalization) gain. A ray tracing approach is shown which gives a deterministic in-room model. Measurements, on the other hand, lead to several stochastic propagation models for UWB signaling.

34.5.1 The SBY Propagation Model

A propagation model in multipath, the SBY model [26], for the strongest impulse is based on theory and measurements in [27]. The path gain P_G between 0 dBi antennas is weighted by receiver antenna aperture $(c^2/4\pi f_m^2)$, where f_m is the geometric mean of the low and high frequency band edges of the UWB pulse, and c is the velocity of propagation. The SBY path attenuation comprise a free space term modified by a multipath dependent term and is

$$P_{SBY} = 10\log\left[\left(\frac{c}{4\pi d f_m}\right)^2\left(1 - e^{-(d_t/d)^{n-2}}\right)\right] \tag{17}$$

where n is the propagation law beyond a breakpoint distance d_t.

Nominally, and based on measurements, $n=3$. The term in the second parentheses represents wavelet attenuation due to the time dispersion of multipath occurring beyond the breakpoint distance d_t. Using a small argument approximation to the exponential term, the resulting $(d_t/d)^{n-2}$ term, with $n=3$, gives rise to an overall inverse 3rd power propagation law inside buildings. Equation (17) is especially useful for modeling propagation in short range indoor personal area networks as exemplified by the IEEE802.15.4a standard. The model generalizes to narrow band cases and to sinusoidal signals.

34.5.2 Relation to Maximum Rake Gain

The maximum average rake gain is defined by the ratio of 'total energy density' to 'single pulse energy density' which is the second parentheses in Equation (8), and on average is

$$G_{max} = 10\log\left(1 - e^{-(d_t/d)^{n-2}}\right) \tag{18}$$

Based on some indoor measurements, $d_t=1$ m and $n=3$, so $G_{max} = 10\log(d/d_t)$, and G_{max} increases with d as $10\log(d)$. This means that on the average, the maximum rake gain (perfect channel equalization) results from the total collection of all the multipath dispersed energy increases with distance, and is given by Equation (18).

34.5.3 An In-Room Ray Tracing Model

An in-room deterministic model [28] and [24] was designed to study UWB propagation on the personal area network scale. The model is based on ray-tracing the principal propagation paths within a single room as pictured in Figure 34.10. The model captures a line-of-sight (LOS) in-room direct component seen in Figure 34.11, and multipath is derived from the 13 primary reflections.

The 13 reflections include a floor reflection, 4 primary wall surface reflection, 4 reflections from the wall inner layer, and 4 double bounce corner reflection. Some of these reflections are pictured in Figure 34.10. Figure 34.11 shows the channel impulse response for one specific realization of locations of the transmitter (*Tx*) and receiver (*Rx*) antennas.

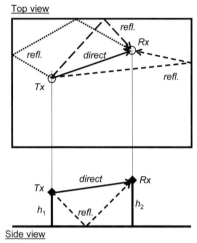

Figure 34.10: An in-room propagation scenario.

It was found empirically for a large number of realization that the impulse response decays approximately exponentially and that the rms delay spread is related linearly to the maximum corner to corner room dimension. Specifically, rms delay spread is approximately

$$\tau_{rms} = 0.2D/c \qquad (19)$$

where c is the speed of light, and D is the maximum corner to corner dimension in the room.

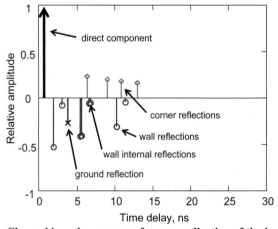

Figure 34.11: Channel impulse response for one realization of the in-room model.

Multiple realizations of the channel model are constructed by randomly selecting the location of the transmitting and receiving antennas within the room. The near-exponential decay in the channel impulse responses is consistent with measurements.

34.5.4 Statistical Propagation Models

Several stochastic propagation models are available for UWB signal paths. One, developed for the IEEE P802.15 standards group, is an extensive collaborative effort which produced a comprehensive statistical model for UWB propagation channels that is valid for a frequency range from 3-10 GHz. The model, see [29], is based on measurements and simulations in the following environments: residential indoor office indoor built-up outdoor, industrial indoor, farm environments, and body area networks. It includes frequency dependencies of the path attenuation as well as several generalizations of the Saleh-Valenzuela model, see [30], including mixed Poisson times of arrival and delay-dependent cluster decay constants. A separate model, see [28], is provided for frequencies below 1 GHz. The model can thus be used for realistic performance assessment of UWB systems. The model was accepted by the IEEE 802.15.4a Task Group as standard model for evaluation of UWB system proposals.

Another model, see [31], proposes a spatially averaged diffusion mechanism. The model propagates signal energy distribution in time by

$$S(t) = K_1 e^{m_1 t} + K_2 e^{m_2 t} \tag{20}$$

where coefficients m_1 and m_2 are negative. For same-room propagation $K_1 + K_2 = 1$, while for propagation into other rooms $K_1 - K_2 = 0$. The model approximates exponential distribution in same-room propagation and is consistent with the deterministic in-room model described in the previous section.

34.6 Recovering UWB Impulse Energy

UWB signals are received and recovered by implementing both a receiver filter function with its impulse response $h(t)$ as well as a sampling function $p(t)$ which might further contain the channel equalization. The resulting receiver implementation efficiency e_c is

$$e_c = 10 \log \left[\frac{\left| \int \int s(\tau) h(\tau\text{-}t)\, d\tau\; p(t)\, dt \right|^2}{\int s(t)^2 dt \int \left| \int p(\tau) h(\tau\text{-}t)\, d\tau \right|^2 dt} \right] \tag{21}$$

and is maximized when

$$C \int p(\tau)h(\tau\text{-}t)\, d\tau = s(t) \tag{22}$$

provided $h(t)$ is causal and where C is the rms value of $s(t)$. The efficiency e_c can typically range from –9 dB for a simple single-sample receiver operating in multipath, to less than one decibel for receivers implementing suitable channel equalization.

34.7 A UWB Link Performance

In a UWB link where a transmitter and antenna provide an EIRP of P_{TX}, and a receiver has sensitivity S_{RX}, the total path gain is the difference between P_{TX} and S_{RX}. This sets the maximum propagation attenuation P_L which determines the maximum communications distance. A transmitter complying with a –41.3 dBm/MHz EIRP limit, and operating over an equivalent bandwidth of about 1.3 GHz emits about $P_{TX} = -10$ dBm. A companion UWB receiver operating in AWGN, referenced to a data bandwidth W Hz, with noise figure, implementation loss and margin totaling L dB, and operating at an SNR dB signal-to-noise ratio per bit, has a sensitivity

$$S_{RX} = -174 + 10 \log(W) + L + SNR \quad \text{dBm} \tag{23}$$

The propagation term P_L is evaluated at a distance d=1 m using equation (17) and the resulting system gain SG, including a receiver antenna gain of G_{RX} dBi, is

$$SG = P_{TX} - S_{RX} + P_L + G_{RX} \quad \text{dB} \tag{24}$$

When $f_c = 4$ GHz then $P_L = -45$ dB, and using a data bandwidth of $W = 100$ MHz, losses $L = 6$ dB, $SNR = 7$ dB, and $G_{RX} = 5$ dBi, the receiver sensitivity is $S_{RX} = -81$ dBm, and the system gain at one meter is $SG = 31$ dB.

Using Equation (17) with d_f=1, a 31 dB system gain at 1 m permits a median range of approximately 11 m without rake gain. A "perfect rake" receiver, one with perfect channel equalization, would permit a median range of more than 31 meters. Practical implementations would result in a median range performance between 10 and 30 meters at a 100 Mbps throughput data rate.

34.8 UWB Applications and Target Markets

Wireless UWB is restricted to be a low transmit power technology. Naturally, the wireless applications are restricted to short ranges. UWB technology has also been applied to band-limited coax and power-line media for in-home multimedia and data distribution applications, see [32]. Ultra-wideband, by its mere property of occupying a wide frequency spectrum, uniquely enables a) high-rate communication, and b) high-precision ranging applications. In the next sub-sections we discuss a summary of applications and potential target markets for UWB.

34.8.1 Radar and Imaging

Pulsed UWB technology has roots in military open-air stealth radar applications. The stealth property of various flavors of the UWB signals are due to the property that the wideband low power level signal can appear as the background noise relative to an unintended narrow band receiver. Through-wall radar, perimeter and fence security, as well as imaging applications and active RFID tags have been suitably enabled using pulse systems, see [33]. RFID tags, precision radar and location technology products are also available, see [34]. "Security Bubble" applications using UWB systems can be configured to create a security "bubble" in the fashion of a bi-static radar to detect penetration of the bubble walls. UWB has an available vehicular radar band in the frequency range 22 GHz to 29 GHz for use in collision avoidance and parking aid applications.

34.8.2 High-Rate Wireless Communication

Since UWB can occupy a wide frequency range, high-bit-rate communication is a natural application space for this technology. High rates can be achieved without the need to resort to complex modulation (high number of bit per Hertz) schemes.

In the era of high-speed processors, multimedia and High Definition Television (HDTV), portable movie and MP3 players, a number of factors point to an increasing demand of "wireless" transfer speeds. These factors and market trends are:

a) Moore's law for processor speed
b) the explosive growth of digital media that gets created and shared in home/office environments
c) the advances in storage technology and availability of high storage capacities used in portable consumer electronic devices.

UWB is suited to respond to the ever-increasing demand for bandwidth as a personal-space wireless technology that untangles the consumer from interconnecting wires. Personal computer (PC) centric cable replacement applications have evolved through Wireless USB Forum (W-USB), see [35], that aims to enable high-speed (up to 480 Mbps) wireless connectivity to external hard-drives, printers, cameras/camcorders, personal video recorders, and portable media players, see [36]. Media centric cable replacement applications have evolved to enable wireless streaming of the multimedia content from/to the set top box and media centers to wide-screen displays, HD-DVD/Blu-ray players, projectors, and portable multimedia players.

34.8.3 Low-Rate Communications and High-Precision Location and Tracking

As the mobility of people and objects increases, up-to-date and precise information about their location becomes a relevant market need. While GPS and some E911 technologies promise to deliver some level of accuracy outdoors, current indoor tracking technologies remain relatively scarce and have accuracies on the order of 3 to 10 meters. UWB possesses a wide bandwidth, and hence a fine timing granularity with which to estimate the arrival of triangulated transponder signals, and can enable 3D location positioning applications to within a few centimeters. In addition to asset tracking, office/home automation and security applications, low-rate UWB has found home in RFID applications for commercial and medical applications. Another target market segment is low-cost long-battery-life sensor networks, which is a focus of IEEE 802.15.4a standard group [9] and Aether Wire & Cable [37] and [38].

34.8.4 High-Rate Communication through Wired Media

UWB technology can be applied to power lines and coaxial cables. This effectively enables reliable connectivity at high speeds in excess of 400 Mbps at the application layer within the home and small office using the existing home infrastructure. The UWB signals are out of band to existing CATV and satellite services and provide add on networking capability.

This integrated wire line and wireless technology provides tremendous wireless networking bandwidth but that also extends content security all the way from the cable provider's head-end offices out to a variety of wirelessly networked devices. The concept turns the home entertainment center into a wireless hub and networking gateway, see [39].

34.9 Summary and Conclusions

Ultra-Wideband (UWB) as a modern wireless technology has evolved to take advantage of recent regulations permitting UWB technology. Once thought of as an impulse radio, modern UWB redefines itself as a modern license-free spectrum access technique wherein spectrum is shared by spreading the energy thinly enough so as to minimal impact the existing spectrum users. Regulations and Recommendations have restricted the spectrum and emission levels, but otherwise have remained silent on transmission characteristics. Hence conventional radio technologies are being exploited under the rules along with pulse technologies. In this Chapter we saw many varieties of pulse-UWB wherein the bandwidth of the signal is directly related to the inverse of the emitted pulse duration. We focused on those pulse techniques which enabled communication systems operating at very high data rates – both radiated and on cables. Applications of UWB devices were presented which exploit the unique properties of wide-band pulses. The short-pulse low-power techniques have enabled products such as through-wall radars, centimeter-precision 3-D positioning, and communications capabilities at exceptionally high data rates. UWB technology has been adopted by several Administrations around the world, including the USA, Japan and in the European Community. A Report and Recommendations have been produced by the ITU-R. Products based on the technology are in the market place.

About the Authors

Kazimierz "Kai" Siwiak (k.siwiak@ieee.org) is founder of TimeDerivative, Inc., a wireless consultancy helping business and technical clients with UWB, MIMO techniques, antennas, propagation, and EMC. While working at Motorola, he was named a Dan Noble Fellow and received the Silver Quill Award. He holds more than 30 US patents and has published extensively. A Senior Member of the IEEE, he served on ETSI and IEEE 802 committees tasked with developing radio telecommunications standards. He received his

BS.E.E and M.S.E.E degrees from the Polytechnic University of Brooklyn and his Ph.D. from Florida Atlantic University.

Yasaman Bahreini (ybahreini@ieee.org) is a consultant focusing on technology development and business executive leadership in the communications industry. Specializing in physical layer (PHY) architectures, she has served on various industry standards committees developing specifications for 3G, and wireless local and personal area systems. She holds 1 US patent and her most recent contributions have been on UWB activities within various standard groups. She was a U.S. delegate to the ITU-R on UWB technology. Yasaman is a Senior Member of the IEEE and holds B.S. and M.S. degrees in electrical engineering from George Washington University.

34.10 References

[1]   R. J. Fontana, A history of UWB, Online: <www.multispectral.com/history.html>, 30 Jan. 2004.
[2]   K. Siwiak and D. McKeown, *Ultra-wideband Radio Technology*, UK: Wiley, 2004.
[3]   US 47 CFR Part15 UWB Operations FCC Report and Order, 22 April 2002.
[4]   US 47 CFR Part15, Radio Frequency Devices, 14 August 2006.
[5]   ITU-R RECOMMENDATION SM.1757, "Impact of devices using ultra-wideband technology on systems operating within radiocommunication services," 2006, Geneva Switzerland.
[6]   Commission of European Communities, "Commission Decision on allowing the use of the radio spectrum for equipment using ultra-wideband technology in a harmonised manner in the Community," Brussels, Belgium, 21 February 2007.
[7]   K. Siwiak and D. McKeown, "UWB Radio Technology in Wireless PANs," tutorial, Birmingham, AL, 7 May 2002. Online: <www.ieee.org/organizations/eab/icet/presentations.htm>.
[8]   K. Siwiak, "UWB Radiation Compared with Human Black Body Radiation," Online: <timederivative.com/2005-04-032rX-UWB&black-body-radiation.pdf>, March 20, 2005.
[9]   *IEEE Standard 802.15.4a*, IEEE Press: NJ, 2006.
[10] R. Hoctor, "Transmitted-reference, Delay-hopped Ultra-Wideband Communications," *Forum on Ultra-Wide Band*, Hillsboro, OR, Online: <www.ieee.or.com/IEEEProgramCommittee/uwb/uwb.html>, Oct 11, 2001.
[11] K. Siwiak; "Ultra-wideband high data-rate communications," *United States Patent*, No. 7,190,729, March 13, 2007.
[12] *IEEE Standard 802.15.3-2003*, Wireless Medium Access Control (MAC) and Physical Layer (PHY) Specifications for Wireless High Rate Personal Networks, IEEE Press: NJ, 2003.
[13] *IEEE Standard 802.15.3b-2005*, Wireless Medium Access Control (MAC) and Physical Layer (PHY) Specifications for Wireless High Rate Personal Networks— Amendment 1: MAC Sublayer, IEEE Press: NJ, 2005.
[14] J. K. Gilb, *Wireless Multimedia: A Guide to the IEEE 802.15.3 Standard*, IEEE Press: NJ, 2003.

[15] *IEEE Standard 802.15.4* Standard, "Wireless Medium Access Control (MAC) and Physical Layer (PHY) Specifications for Low-Rate Wireless Personal Area Networks LR-WPANs", IEEE Press: NJ, 1 October 2003.

[16] E. H. Callaway, *Wireless Sensor Networks: Architectures and Protocols,* Auerbach Publications (CRC Press): New York, N.Y., 2003.

[17] I. Lakkis and Y. Bahreini, "Ultra-wideband pulse modulation system and method," *United States Patent*, No. 7,190,722, March 13, 2007.

[18] L. Fullerton, "Time Domain Transmission System," *United States Patent,* No. 4,813,057, 14 Mar. 1989.

[19] M. Z. Win and R. A. Scholtz, "Impulse Radio: How it Works," *IEEE Comm. Letters*, Vol. 2, No. 1, Jan. 1998.

[20] Larry Taylor, "MAC Designs for UWB Systems," a chapter in *Ultra-Wideband Systems*, Oxford, UK: Elsevier, 2006.

[21] N. K. Askar, S. C. Lin, H .D. Pfister, G. E. Rogerson, and D. S. Furuno, "Spectral Keying™: A Novel Modulation Scheme for UWB Systems", *Second Ultra Wideband Systems and Technologies (UWBST) Conference*, Reston, VA, November 16-19, 2003.

[22] N. K. Askar, S. C. Lin, and D. S. Furuno, "Spectral Keying™: A Novel Modulation Scheme for UWB Systems", a chapter in *Ultra-Wideband Systems*, Oxford, UK: Elsevier, 2006.

[23] J. G. Proakis, *Digital Communications*, New York, NY: McGraw-Hill, 1983.

[24] K. Siwiak and Y. Bahreini, *Radiowave Propagation and Antennas for Personal Communications, Third Edition*, Norwood MA: Artech House, 2007.

[25] K. Siwiak, T. M. Babij, and Z. Yang, "FDTD simulations of ultra-wideband impulse transmissions," *IEEE Radio and Wireless Conference: RAWCON2001,* Online: <rawcon.org>, Boston, MA, August 19-22, 2001.

[26] K. Siwiak, H. Bertoni, and S. Yano, "On the relation between multipath and wave propagation attenuation," *Electronic Letters*, 9th January 2003, Volume 39 Number 1, pp. 142-143.

[27] Yano S. M.: "Investigating the ultra-wideband Indoor wireless channel," *Proc. IEEE VTC2002 Spring Conf.*, May 7-9, 2002, Birmingham, AL, Vol. 3, pp. 1200-1204.

[28] "UWB Channel Modeling for under 1 GHz" *IEEE P802.15 Working Group for WPANs*, IEEE document P802.15-04/505-TG3a, Dec, 2002.

[29] A. F. Molisch, D. Cassioli, C. -C Chong, S. Emami, A. Fort, B. Kannan, J. Karedal, J. Kunisch, H. G. Schantz, K. Siwiak, M. Z. Win, "A Comprehensive Standardized Model for Ultra-wideband Propagation Channels," *IEEE Transactions on Antennas and Propagation*, Volume: 54, Issue: 11, Part 1, pp. 3151-3166, Nov. 2006.

[30] A. A. M. Saleh and R. A. Valenzuela, "A statistical model for indoor multipath propagation," *IEEE J. Select. Areas Commun.*, vol. 5, no. 2, pp. 128–137, Feb. 1987.

[31] M. A. Nemati, R. A. Scholtz, "A Diffusion Model for UWB Indoor Propagation," *MILCOM 2004*, October 31 - November 3, 2004, Monterey CA.

[32] Coax technology, Online: <www.pulselink.net/technology/coax.htm>, Mar. 26, 2007.

[33] Time Domain Products, Online: <www.timedomain.com/products/>, Mar. 26 2007.

[34] Multispectral Solutions, Inc., RFID tags and radar applications, Online: <www.multispectral.com/products.html>, March 25 2007.

[35] Wireless USB Forum, Online: <www.usb.org/developers/wusb/ >, Mar. 25, 2007

[36] R. Aiello, "Commercial Applications," a chapter in *Ultra-Wideband Systems*, Oxford, UK: Elsevier, 2006.

[37] Aether Wire & Cable, Online: <www.aetherwire.com/> March 25, 2007.

[38] R. A. Fleming and C. E. Kushner, "Spread Spectrum Localizers," *United States Patent,* No. 5,748,891, 5 May 1998.

[39] Hybrid Coax technology, Online: <www.pulselink.net/technology/hybrid.htm>, March 26, 2007.

35

High-rate WPAN

Dagnachew Birru and Vasanth Gaddam[a]

35.1 Introduction

With the adoption of the 3.1-10.6 GHz band by the Federal Communications Commission (FCC) for commercial wireless communication, Ultra Wide-Band (UWB) has attracted both scientific and commercial interest and is now emerging as a promising technology for high-speed (several hundred Mbps and beyond)) short-range wireless communications for home and office networking applications. Sometimes also referred to as impulse radio, UWB communication systems operate across a wide range of spectrum relative to the center frequency. The radiated energy, occupying a large bandwidth (typically measured in GHz), is often made sufficiently small that it can facilitate co-existence with other devices without causing significant harmful interference to them. Advantages of current UWB implementations include low-cost, low-power, and resilience to multi-path interference.

Driven largely by UWB technology, high-rate short-range Wireless Personal Area Networks (WPANs) are expected to find wide use in the coming several years. Companies are already announcing chipset solutions complying with some of the industrial de-facto standards. A band hopping OFDM based Physical (PHY) layer, has gained wide industry support as a standard for high-rate WPANs. This proposal was initially developed by Multi-Band OFDM Alliance (MBOA) and presented as a candidate for 802.15.3a WPAN PHY layer [1]. It was later adopted and was further developed by WiMedia[1] Alliance. The version 1.0 of the WiMedia PHY specification together with a newly developed ad-hoc distributed MAC specification has been standardized by Ecma as a high-rate WPAN standard [2]. One of the earliest applications being envisioned for this technology is cable replacement for Universal Serial Bus (USB) devices. Therefore, one of the initial goals of the WPAN system was to provide data rates similar to the current version of the USB

---

[a] All authors currently affiliated with *Philips Research North America*

[1] In this chapter, we use the terms MBOA, WiMedia, and Ecma specifications interchangeably to refer to the multi-band OFDM based high rate WPAN specification. WiMedia refers to the UWB common radio platform for WPANs defined and supported by WiMedia™ Alliance, http://www.wimedia.org.

standard, USB 2.0. USB 2.0 can support data rates up to 480 Mb/s. Other applications such as high data rate video streaming, communication between peripherals, mobile phones and medical systems will also be made possible by this specification.

This chapter focuses on the WiMedia PHY layer system. In Section 35.2, we provide some background information on the need for high data rates and then discuss some applications areas and their requirements. We also discuss the current world-wide regulatory scenario with regard to UWB spectrum allocation. We then briefly describe the benefits of UWB communication in Section 35.3. A brief overview of the WiMedia/Ecma MAC is presented in Section 35.4 followed with an overview of the PHY specification in Section 35.5 with a particular emphasis on unique/novel aspects of the standard. In Section 35.6, we describe some advanced receiver algorithms that will enable the reception of a packet in adverse channel conditions with superior performance/QoS. Multiple antenna solutions are briefly reviewed in this section. In Sections 35.7 and 35.8, we review some next generation UWB systems and some new enabling technologies for future multi-Gb/s short range wireless communications.

35.2 Trends and Application Scenarios

35.2.1 The Increasing Need for Higher Date Rates

The past few decades have witnessed a tremendous increase in the use of connected devices. End users' requirement for freedom of movement (mobility), convenience and more productivity is fueling the proliferation of wireless systems. A clear evidence of this can be observed from the trend of the data rate in the past several years. Figure 35.1 shows the trend seen in the past few decades for some of the connectivity technologies. This trend is enabled by developments in integrated circuit (IC) implementations such as in CMOS process. Increase in speed of circuits, memory depth and disk space has enabled and fueled the need for more connectivity. Given the trends seen in the past, there is no reason to expect that the trends will stop any time soon. The following section describes some of the applications that are poised to fuel the growth of high-rate WPAN systems.

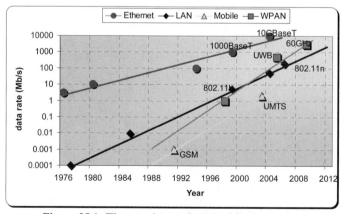

Figure 35.1: The ever increasing trend in data-rate.

35.2.2 Applications areas

The application areas of WPAN can be broadly classified into 5 application areas as shown in the following figure.

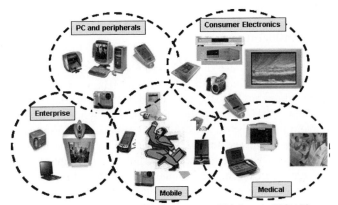

Figure 35.2: Broad application areas of high-rate WPANs.

From data throughput/networking perspective, the application areas for high-rate WPANs can also be classified into three domains:

Application Domain 1: Burst Throughput – This is like the current networked applications where the average throughput is low but there are several devices in the network that utilize high-speed transfer per device for a small duration of time. Some of the typical applications are: laptop docking station, general network to corporate network, and compressed video transfer at 20 Mbps. Here the overall network capacity is key and thus requires little MAC and PHY overhead.

Application Domain 2: Sustained Throughput Streaming – Typical applications here are point-to-point links such as uncompressed video streaming (e.g. HDMI) that require the use of the full bandwidth for the duration of streaming. This is generally used as a cable replacement for existing applications in multimedia devices.

Application Domain 3: Fast Bulk Data Transfer – Example applications here include: swapping the entire contents of a portable multimedia device such as an iPod® with 60 GB hard disk, with that of new content from a PC, uploading of images and video from a digital camera or a camcorder, and synchronizing a portable device with a desktop. These applications require that transfers are done as quickly as possible. The hard disk access time will be the main bottleneck for this type of applications. For example, current SATA drives support speeds up to 3 Gbps only.

Wireless technology is beginning to make its presence felt in almost all of the above segments, a trend that will likely intensify over the next several years. Looking further we anticipate that the demand for even higher data rates (multi-gigabit per second) will come from some of the following short-range applications:

- Higher resolution display port interfaces
- 3-D video streaming
- Virtual reality entertainment
- Ultra-fast content transfer (e.g., high-resolution movies)
- High-speed (image) sensors

35.2.3 Regulatory Considerations

FCC in the US has allowed the use of the 3.1-10.6 GHz spectrum for commercial use of UWB with transmitted power limited to -41.25 dBm/MHz. Other regions such as Europe and Japan are also allowing UWB, but with much tighter requirements. Especially, frequency bands below 5 GHz are expected to operate at either reduced power and/or by using detect and avoid mechanisms. The 5 GHz UNII band is also virtually un-usable due to interference concerns from and to 802.11 devices. Overall, the lower frequencies are more attractive both from RF implementation and link-budget point of view. However, for worldwide use, operating in the higher frequencies will also be necessary to satisfy the emerging regulatory requirements. For a more detailed discussion, see Chapter 34.

35.3 Benefits of Wide-Band Transmission

35.3.1 From Multipath Performance Perspective

Narrowband receivers see large fluctuation in received power when operating in multi-path channel environments. This is mainly due to the flat fading nature of the signal caused by large delay spreads. Wideband systems not only are less prone to flat fading, but also provide greater multi-path diversity. Figure 35.3 shows an analysis of an experimental data performed to investigate the differences in multi-path fading for a narrowband system and a wideband system. This figure plots the difference in received power between two receive antennas separated by 1-2cm when the transmitted signal occupies 500 MHz (wideband) and about 4 MHz (narrowband). The transmitting antenna is at a line-of-sight (LOS) link to the receive antennas. This figure shows that the received power of wideband systems do not exhibit significant dependence on location of antennas compared to that of narrowband systems. In fact, one can conclude that wideband systems can provide more than 4 dB fading margin 40% of the time. This will translate into a more reliable and stable link that is less dependent on multi-path fading or small movements in antenna location.

35.3.2 From Interference Perspective

Wide bandwidth systems also have potential good benefits for narrow-band interference. If the interfering signal does not saturate the front end, then the impact of such interference is low and can be further minimized by using advanced signal processing techniques as those described in Section 35.6.8. If the interfering signal causes saturation of the front end, then a narrowband notching filter can be used to minimize the effects of non-linear distortions caused by the interferer.

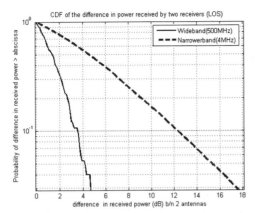

Figure 35.3: CDF of the difference between the received energy of two antenna equi-distance from the transmitter (LOS).

When the bandwidth is large, the transmitter power is spread across several frequencies, resulting in smaller interference to other narrowband devices. This property makes UWB particularly very attractive for some application such as medical systems where interference has to be avoided for mission critical operations.

35.4 Overview of the WiMedia 1.0 MAC

A brief overview of the WiMedia MAC is provided in this section. For more in-depth information, the reader is encouraged to refer to the Ecma standard [2].

The WiMedia MAC architecture is based on a fully distributed structure. This means that no device acts as a central coordinator. All devices provide all the necessary MAC functions as determined by the application. This eliminates the need for a network infrastructure by distributing functions across all nodes.

A superframe structure [3] is used as a basic timing structure to exchange frames. Figure 35.4 shows the super-frame structure of the MAC. A super frame starts with a beacon period. There are three basic ways of accessing the medium: beacon period based, reservation based, and contention based. During the beacon period, devices send beacon frames only. During reservations, devices participating in the reservation send frames according to the reservation rules defined in the standard. Outside these two schemes, devices may send frames using prioritized contention based access method. The beacon frames are intended to be received and interpreted by all devices in the network. Thus, this frame is transmitted using the lowest data rate mode. In general, the MAC provides features such as

- data communication between devices within a channel (physical/logical channel)
- a distributed reservation based channel access
- a prioritized contention based channel access
- mechanisms for rate adaptation, power management, secure communication, and measuring distance between devices

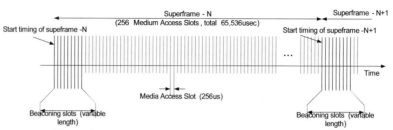

Figure 35.4: Ecma/WiMedia MAC superframe structure.

To reduce the frame error rate, the standard supports fragmentation of frames that can be reassembled at the receiver side. It supports 3 types of acknowledgements to verify the correct delivery of a frame based on the needs of the application: No-ACK, Imm-ACK and B-ACK policies. The No-ACK policy is suitable for frames that do not require guaranteed delivery. The Imm-ACK policy is based on acknowledging each frame. The B-ACK acknowledges multiple frames at a time. If a source device does not receive the requested acknowledgement, then it may re-transmit the frame.

35.5 Overview of the WiMedia 1.0 PHY

35.5.1 Introduction

The WiMedia PHY is based on a 128-carrier OFDM system with band-hoping option. The data rates range from 53.3 Mbps up to 480 Mbps. Different data rates are achieved through frequency-domain spreading (FDS), time-domain spreading (TDS), and forward error correction coding (FEC). Figure 35.5 shows the simplified block diagram of a WiMedia OFDM modulator. The IFFT module receives a 128-point vector for frequency-domain (FD) preamble, header and payload symbols in frequency domain and transforms them into time domain. The header and payload data are spread and coded according to the desired mode (see Section 35.5.4). The time-domain (TD) preamble symbols are already defined in time, so these symbols are not processed by the IFFT module. The 128-point time domain vector at the output of MUX2 is then zero padded with 37 samples to form a 165-sample vector and represents one OFDM symbol. The complex data vector is then serialized and transformed into analog domain using a digital-to-analog converter (DAC) operating at, say, 528 MHz. The resultant analog signal is low-pass filtered and then mixed up in frequency using the center frequency of the band for that particular symbol. An up-sampling digital filtering with a 2-times over sampling DAC may be a better implementation to relax analog filtering requirements. The following subsections describe an overview of some of key features of the WiMedia PHY. The readers are encouraged to consult the Ecma document [2] for a complete description of the specification.

35.5.2 Band Hopping and Time Frequency Codes (TFCs)

Due to the unique combination of power limitations (imposed by regulatory bodies) and the availability of wide spectrum (greater than 1 GHz worldwide), the UWB band (3.1-10.6GHz) is ideally suited for short range high data-rate applications. However, using a large bandwidth would correspondingly increase system complexity and cost due to the need for high speed circuits (such as A/D and D/A converters, filters, base-band processing modules, etc). The WiMedia PHY solves the issue of wide band requirement to a certain extent by using band hopping technique. This process is referred to as time-frequency hopping and involves changing the center frequency of the signal according to a pre-defined pattern. The bandwidth of the signal in each hop is 528 MHz and the signal can hop in a maximum of 3 bands resulting in the total occupied bandwidth of about 1.5 GHz. As a result of this scheme, the bandwidth requirements on the analog front-end are relaxed. Band-hopping also enables simultaneous operation of multiple links in this band within the same location providing increased spatial capacity.

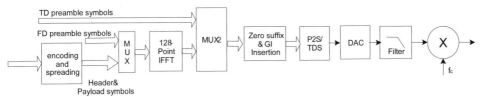

Figure 35.5: Simplified block diagram of WiMedia transmitter (represents the functionality of OFDM modulator).

In order to enable band-hopping mechanism, the allocated band for UWB operation is divided into 14 sub-bands as shown in Figure 35.6. The bands are further clustered into 5 band-groups, with 4 of them containing three bands each and one (Band Group #5) containing 2 bands. Additional band-groups are also defined to meet other regulatory domain requirements.

Once a device selects a band-group then it can hop only within its allocated bands, e.g. devices operating in band-group #1 can only hop among bands 1, 2 and 3. The particular hopping pattern used is determined by the Time Frequency Code (TFCs) assigned to the device. Figure 35.7 shows the table of the hopping patterns for different TFCs in band-groups 1 through 5. For example, band-group 1 and TFC 1 in this table indicates that the first OFDM symbol is transmitted in band #1, the second symbol in band #2 and so on and so forth. This pattern repeats every six OFDM symbols. Similarly, TFC 3 in band-group 1 indicates that the first two symbols are transmitted in band #1, the next two symbols are transmitted in band #2 and the following two symbols are transmitted in band #3 and then the pattern repeats every six symbols. Band hopping is disabled for TFCs 5, 6 and 7 and therefore these TFCs are referred to as Fixed Frequency Interleaved (FFI) modes. On the other hand, TFCs 1 to 4 are referred to as Time Frequency Interleaved (TFI) modes.

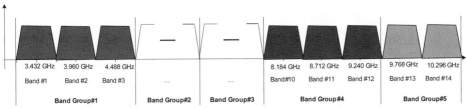

Figure 35.6: Grouping of bands into band groups.

TFC Number	Band Group #1						Band Group #2						Band Group #3						Band Group #4						Band Group #5					
1	1	2	3	4	5	6	4	5	6	4	5	6	7	8	9	7	8	9	10	11	12	10	11	12						
2	1	3	2	1	3	2	4	6	5	4	6	5	7	9	8	7	9	8	10	12	11	10	12	11						
3	1	1	2	2	3	3	4	4	5	5	6	6	7	7	8	8	9	9	10	10	11	11	12	12						
4	1	1	3	3	2	2	4	4	6	6	5	5	7	7	9	9	8	8	10	10	12	12	11	11						
5	1	1	1	1	1	1	4	4	4	4	4	4	7	7	7	7	7	7	10	10	10	10	10	10	13	13	13	13	13	13
6	2	2	2	2	2	2	5	5	5	5	5	5	8	8	8	8	8	8	11	11	11	11	11	11	14	14	14	14	14	14
7	3	3	3	3	3	3	6	6	6	6	6	6	9	9	9	9	9	9	12	12	12	12	12	12						

Figure 35.7: TFCs for different Band Groups. The values in the cells represent the band numbers.

35.5.3 WiMedia OFDM Parameters and Sub-carrier Allocation Scheme

The choice of OFDM parameters depend to a certain extent on the channel environment encountered in the application domain. For WPANs, the channel modeling sub-committee [6] of the IEEE 802.15.3a task group has recommended a modified Saleh-Valenzuela (SV) model [7] based on literature review and channel measurement efforts of different groups. It has also provided parameters for four different channel characteristics labeled as CM1, CM2, CM3 and CM4 to help in evaluating and comparing different modulation schemes. The RMS delay spreads of these channel models vary from 5.28 ns for the case of CM1 to 25 ns for the case of CM4. Figure 35.8 shows one realization from each of the four channel characteristics defined by the group.

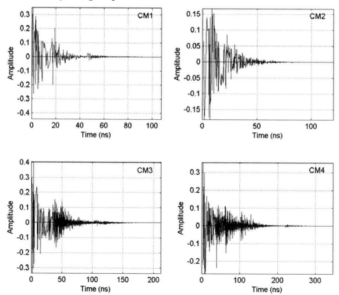

Figure 35.8: Channel impulse response for one realization each of the four channel characteristics.

Taking into account the indoor channel characteristics and the application requirements, the WiMedia OFDM proposal specifies a 128-point FFT with a 37-point guard interval (GI). Sufficient guard interval is required in OFDM systems to avoid inter-symbol interference (ISI). This combination of FFT and GI parameters minimizes transceiver complexity and at the same time does not sacrifice too much bit-rate efficiency due to the guard interval [8]. The sub-carrier spacing is specified as 4.125 MHz which results in a signal bandwidth of 528 MHz per each hop. For the case of TFI modes, where hopping is enabled, the total bandwidth would be 1.584 GHz. Table 35.1 summarizes some of the timing related parameters of the WiMedia PHY.

Table 35.1: Timing related parameters for the multi-band OFDM system.

Parameter	Value
Number of total sub-carriers, N_{FFT}	128
Number of data sub-carriers, N_{SD}	100
Number of pilot sub-carriers, N_{SP}	12
Number of Guard & NULL sub-carriers, N_{SG}	16
Sampling frequency, f_s	528 MHz
Sub-carrier frequency spacing (= f_s/N_{FFT}), Δ_F	4.125 MHz
IFFT/FFT period (= $1/\Delta_F$), T_{FFT}	242.42 ns
Zero suffix duration (= 32/528 MHz), T_{ZP}	60. 61 ns
Guard interval duration (= 5/528 MHz), T_{GI}	9.47 ns
Symbol period (= $T_{FFT} + T_{ZP} + T_{GI}$), T_{SYM}	312.5 ns
OFDM symbol rate (= $1/T_{SYM}$), f_{SYM}	3.2 MHz

Figure 35.9 shows the sub-carrier allocation scheme for the WiMedia PHY. The guard/NULL sub-carriers (except for the one at the DC) are specified at the band-edges to relax filtering requirements. The 12 pilot sub-carriers are spread across the band uniformly and enable fine frequency/phase tracking. The remaining 100 sub-carriers are designated as data sub-carriers and are used to carry coded and modulated payload data.

Figure 35.9: Sub-carrier allocation scheme for the WiMedia 1.0 PHY. Horizontal axis represents logical frequency bins.

35.5.4 PLCP Layer

The Physical Layer Convergence Protocol (PLCP) layer converts a PHY Service Data Unit (PSDU) to PLCP Protocol Data Unit (PPDU) by adding a preamble and a header. Similar to other packet based communication protocols, the WiMedia PPDU, shown in Figure 35.10, consists of three fields namely: the PLCP preamble, the PLCP header and the PSDU. The PLCP preamble portion is used for burst detection, synchronization and channel estimation purposes while the PLCP header provides information such as the frame length, rate, etc., that is used in the decoding of the payload portion of the frame.

Table 35.2 lists the different fields of the PHY header and their location. The PSDU consists of the payload bits, Frame Check Sequence (FCS), tail bits and some pad bits.

35.5.4.1 PLCP Preamble

WiMedia specification defines two forms of PLCP preambles: 1) a standard PLCP preamble consisting of 30 OFDM symbols, and 2) a burst PLCP preamble consisting of 18 OFDM symbols. The burst preamble is only used for high data rate streaming modes, while the standard preamble is used in all other cases. Burst preamble helps in reducing the overheads associated with preambles.

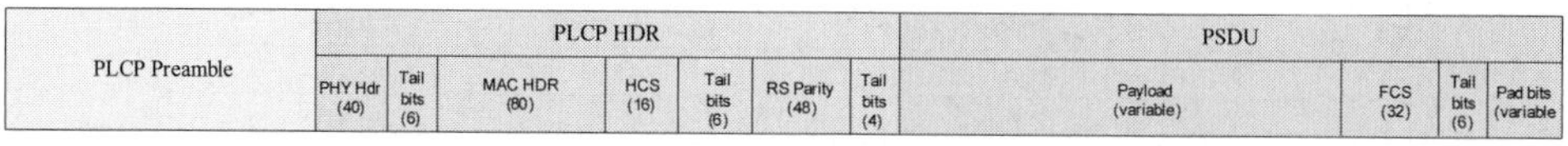

PLCP Preamble	PLCP HDR							PSDU			
	PHY Hdr (40)	Tail bits (6)	MAC HDR (80)	HCS (16)	Tail bits (6)	RS Parity (48)	Tail bits (4)	Payload (variable)	FCS (32)	Tail bits (6)	Pad bits (variable

Figure 35.10: PPDU frame format. The numbers in the parenthesis represent the number of bits for that particular field.

Table 35.2: PHY header contents.

Field Name	Bit Positions	Number of Bits	Description
Reserved	0-2, 20-21, 24-25, 32-39	15	
Rate	3-7	5	Determines data rate
Length	8-19	12	Length of PSDU in octets – 0 - 4095
Scrambler Init	22-23	2	Seed to initialize the scrambler
Burst Mode	26	1	Burst mode indicator
Preamble Type	27	1	Preamble type for next packet
TX TFC	28-30	3	TFC number, 1-7
Band Group LSB	31	1	Least significant bit of band-group

The PLCP preamble consists of a time domain (TD) portion and a frequency domain (FD) portion. The TD preamble is 24 symbols long in standard mode and 12 symbols long in burst mode and is used in burst detection and synchronization (timing and frequency offset estimation). The FD preamble is 6 symbols in duration and is mainly used for channel estimation. Figure 35.11 shows the standard mode preamble format for different TFCs.

A unique TD preamble is defined for each of the TFCs. The transmitter selects a sequence based on its TFC and replicates it multiple times to form the TD preamble. The polarity of some of these symbols is inverted to enable the (end of) frame detection. A cover sequence is used to specify the polarity of all the TD preamble symbols. The same base sequence associated with a particular TFC is also used in the burst mode. However, the cover sequence that is used to define the polarity of the symbols is different from that of the standard preamble.

The FD preamble is formed by taking a base sequence defined in the frequency domain and repeating it for 6 times (represented as symbols CE1, CE2, CE4, CE5 and CE6 in Figure 35.11). The same FD base sequence is used for all the TFCs. The order of the bands is cycled through the FD preamble. For example, in the case of TFC #1, the symbols CE1 and CE4 are transmitted in band 1, symbols CE2 and CE5 are transmitted in band 2 and symbols CE3 and CE6 are transmitted in band 3. The FD preamble is used by the receiver for channel estimation. For the case of TFI modes, the receiver has two symbols to average the channel estimation. For the case of FFI modes, since all the symbols are transmitted in only one band, the receiver can average the channel estimate across all the 6 FD symbols.

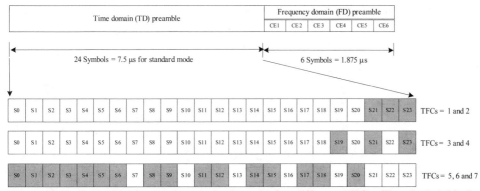

Figure 35.11: Preamble format in standard mode for different TFCs. The shaded blocks represent symbols with inverted polarity.

35.5.4.2 PLCP Header

The PLCP header consists of the PHY header, MAC header, Frame Check Sequence (FCS), parity bytes derived from a $t = 3$ (23, 17) Reed-Solomon coder and tail bits. The encoding steps for the header bits are shown in Figure 35.12. First the MAC header bits and the FCS bits are scrambled, which are subsequently appended to the PHY header bits and are provided as input to the RS encoder. The 6 parity bytes derived from the RS encoder are inserted in the PLCP header as shown in Figure 35.10. The tail bits are inserted so as to enable the convolutional encoder to reach a deterministic state. The PLCP header thus formed is encoded using a rate-1/3 constraint-length 7 convolutional encoder. The generator polynomials of the encoder are given as 133_8, 165_8 and 171_8. The encoded bits are then interleaved by a 3-stage interleaver. The interleaved bits are then mapped using QPSK constellation. The QPSK symbols are grouped into blocks of 100 symbols to form 6 such blocks. Pilot and guard tones are inserted in each of these blocks to form the 128-point input to the IFFT module.

Figure 35.12: Encoding of the PLCP header bits.

35.5.4.3 Encoding of PSDU

The encoding process for the PSDU is shown in Figure 35.13. The PSDU bits, FCS bits and the pad bits are first randomized using a data scrambler. The scrambler initialization is determined by the PLCP-Scrambler field in the PHY header. The pad bits are inserted in order to make the total number of bits a multiple of interleaver size.

The scrambled bits are encoded using a rate-1/3 convolutional coder. The encoded bits are then punctured to generate higher rate codes with R = 1/2, 5/8 or 3/4. The puncturing patterns for these rates are defined in Table 35.3. In this table, the variables A, B and C represent the 3-bit output of the convolutional coder for each input bit, and the sub-script represents the time index of the bits. Decoding by Viterbi algorithm is recommended. At the receiver, zeros are inserted in the locations of the punctured bits before the Viterbi decoder.

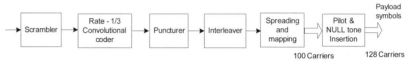

Figure 35.13: Encoding modules for the PSDU.

Table 35.3: Puncturing patterns for R = 1/2, 5/8 and 3/4.

Code rate	R = 1/2	R = 5/8	R = 3/4
Convolutional coder output	$A_1\cancel{B_1}C_1$	$A_1B_1\cancel{C_1}$ $A_2B_2C_2$ $A_3B_3\cancel{C_3}$ $A_4B_4C_4$ $A_5B_5\cancel{C_5}$	$A_1B_1\cancel{C_1}$ $A_2B_2C_2$ $\cancel{A_3}B_3C_3$
Puncturer output and Bit-inserter input	A_1C_1	$A_1B_1C_2\ A_3B_3C_4\ A_5B_5$	$A_1B_1C_2C_3$
Decoder input	$A_1 0 C_1$	$A_1B_1 0$ $00C_2$ $A_3B_3 0$ $00C_4$ $A_5B_5 0$	$A_1B_1 0$ $00C_2$ $00C_3$

Similar to the PLCP header bits, the coded and punctured PSDU bits are also interleaved by a 3-stage interleaver. The interleaved bits are mapped into symbols using either a Gray coded QPSK constellation or a Dual Carrier Modulation (DCM) scheme based on the data rate.

35.5.4.4 Dual Carrier Modulation (DCM)

For higher rate modes (i.e., for 320 Mbps and above), the coding gain is lower and as such the performance of these modes is prone to suffer in the presence of interference and frequency selective fading. The frequency diversity gains are not as good as the one with the lower rates. In order to improve the performance of the higher rate modes, the standard specifies spreading of two symbols over two widely separated carriers. Spreading in this instance does not involve bandwidth expansion or data rate reduction. However, it provides additional frequency diversity gains. The DCM process is defined below: Divide each set of N_{CBPS} (number of coded bits per OFDM symbol) bits of the DCM module input into m (= $N_{CBPS}/4$) groups, each consisting of 4 bits $\{u_{1,n}, u_{2,n}, u_{3,n}, u_{4,n}\}$. For high data rate modes, m is equal to 50.

$u_{i,n} \in \{0,1\}, i = 1,2,3,4$ and $n = 0,1,2,...,m-1$

The bits $u_{i,n}$ are converted to bipolar symbols $x_{i,n}$ by the following equation

$$x_{i,n} = 2*u_{i,n} - 1 \qquad (1)$$

$x_{i,n} \in \{-1,+1\}, i = 1,2,3,4$ and $n = 0,1,2,...,m-1$

Dual carrier spreading operation is then performed on $x_{i,n}$ according to the following equation .

$$\begin{bmatrix} y_n \\ y_{n+m} \end{bmatrix} = \frac{1}{\sqrt{10}} \begin{bmatrix} 2 & 1 \\ 1 & -2 \end{bmatrix} \begin{bmatrix} x_{1,n} + jx_{3,n} \\ x_{2,n} + jx_{4,n} \end{bmatrix}, \quad n = 0,1,2,...,m-1 \qquad (2)$$

The above described spreading operation results in a 16-QAM constellation. The complex symbol block $\{y_n; n = 0$ to $99\}$ is then combined with the pilot tones and the guard tones and then sent to the IFFT block.

35.5.4.5 Frequency Domain and Time Domain Spreading

In addition to puncturing, frequency domain spreading (FDS) and time domain spreading (TDS) are also used to obtain different data rates. FDS involves sending the same symbol on two different sub-carriers that are widely separated. FDS is only used with rates 53.3 Mbps and 80 Mbps.

TD spreading is performed by sending the information contained in an OFDM symbol in the subsequent symbol as well. In order to ensure a flat power spectral density, a cover sequence is used for the second symbol. The cover sequence is derived from a 127-length PN sequence and on each symbol one element from this sequence is selected. Time-domain spreading is usually done in the time-domain at the transmitter in order to save power. At the receiver, time-domain de-spreading is performed either in time-domain or in frequency-domain depending on the TFC and the rate parameters. TDS is used only for rates 200 Mbps and below and is defined as follows:

If $S_{2n}(k)$ represents an OFDM symbol prior to IFFT of a symbol to be transmitted at time $2n$, then the symbol $S_{2n+1}(k)$ is given as

(i) For PLCP header and rates 53.3 Mbps and 80 Mbps

$$S_{2n+1}(k) = P_s(n')S_{2n}(k) \qquad (3)$$

(ii) For rates 106.7 Mbps, 160 Mbps and 200 Mbps

$$S_{2n+1}(k) = P_s(n')\{|S_{2n}(N-k)| \exp(j(\frac{\pi}{2} - \phi_{N-k}))\} \qquad (4)$$

where n' is a function of symbol number n and P_s is a pseudorandom cover sequence. $k \in [0, N-1]$ is the sub-carrier index, N is FFT/IFFT size and is equal to 128. φ_k represents the phase of the symbol on sub-carrier k.

35.5.5 OFDM Modulation

The kth OFDM base-band symbol after IFFT and zero-suffix insertion is represented by the following equation

$$
r_k(t) = \begin{cases} \sum_{n=-N_{ST}/2}^{N_{ST}/2} C_n \exp(j2\pi n\Delta_f(t-T_Z)) & t \in [T_Z, T_{FFT} + T_Z] \\ 0 & t \in otherwise \end{cases}
\tag{5}
$$

where $T_z = T_{zp} + T_{GI}$. The coefficients C_n are derived from the data, pilots or the training sequences as described in earlier sub-sections. Table 35.1 defines the parameters and the specific values to be used for the MBOA PHY. The transmitted RF signal can be represented as

$$
r(t) = \text{Re} \left\{ \sum_{k=0}^{N_{pckt}} r_k(t - kT_{sym}) \exp(j2\pi f_c(q(k))t) \right\}
\tag{6}
$$

where N_{pckt} is the total number of symbols in a packet, $f_c(K)$ is the center frequency of the Kth band and $q(k)$ represents the hopping function.

35.6 Receiver Algorithms

35.6.1 Top-level Structure

A simplified diagram of the receiver is shown in Figure 35.14. The RF unit contains LNA, mixer, gain controls, band-limiting filter, frequency synthesizer and ADC. For detailed design consideration of the RF part, the reader is encouraged to refer to sources such as [9]. This section focuses on algorithms for the baseband part.

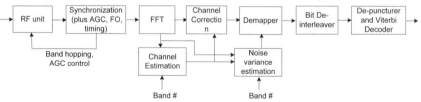

Figure 35.14: Simplified Block diagram of a TFI-OFDM receiver including the channel estimation and noise variance estimation units.

The baseband part contains synchronization block, gain control, frequency offset estimation (FO) and correction, sampling clock mismatch estimation and correction (not shown in figure), FFT, channel estimation and correction, common-phase error estimation and tracking (not shown in the figure), demapper, de-interleaver, and Viterbi decoder. In addition to the standard receiver blocks, one could also include advanced algorithms to improve the performance of the system. These include in-band interference mitigation by estimating the noise variance for each sub carrier and implementing the full ML decoder. The following sections provide details of some of these algorithms.

35.6.2 Synchronization

35.6.2.1 Introduction

Packet based communications systems rely on robust detection of the beginning of a packet. This is accomplished by transmitting a preamble sequence at the transmitter and employing a robust detection scheme at the receiver. The detection of the preamble needs to be more robust than that of the data portion of the packet, i.e., it should be at least 3 dB better compared to the lowest data rate mode (53.3 Mbps mode). In addition, accurate estimation of the timing of the OFDM symbol (window of the OFDM symbol) and frequency offset play an important role on the performance of the receiver. In this sub-section, preamble detection, frequency-offset estimation and timing estimation techniques are described for efficient implementation of hierarchical preamble adopted in WiMedia PHY.

Synchronization is performed on the time-domain preamble. This preamble is derived from a hierarchical binary sequence of the form

$$[a_0 B, a_1 B, \cdots, a_{L-1} B] \tag{7}$$

where $B = \{b_0, \ldots, b_{M-1}\}$ and $A = \{A_0, \ldots, A_{L-1}\}$ are binary sequences with $M = 8$ and $L = 16$. Looking at the sequence described in the Ecma standard, it may appear that the sequence is quite different from the above hierarchical sequence. A close look at the sequence reveals that the sign of the sequence follows a hierarchical sequence for the most part. This approximation (taking the sign of the samples) will lead to simpler receiver design with little performance impact. Thus, the synchronization algorithm will be based on detecting the hidden binary hierarchical sequence by easily reconstructing the related hierarchical sequences from the sign of the preamble. Several correlation techniques are described below in this regard.

35.6.2.2 Auto-correlation

The most frequent correlation technique used in WLANs/WPANs is a delayed auto-correlation on the received signal. The conventional delayed auto-correlation at sample index m can be expressed as

$$f(m) = \sum_{k=0}^{J-1} r(m-k) r^*(m-D-k) \tag{8}$$

where $r(m)$ is the received sample, D is the delay between successive sequences, J is the correlation window, and '*' denotes complex conjugate. The received signal is modeled as below

$$r(m) = x(m) e^{-j(2\pi\varepsilon T m + \alpha)} + n(m) \tag{9}$$

where $x(m)$ is the convolution result of the channel impulse response and the transmitted sequence, ε is the frequency error, T is the sampling rate, α denotes the phase error between the transmitter and receiver oscillators, and $n(m)$ denotes noise (or unwanted interference).

In the above model, we have intentionally neglected sampling clock error since the impact of sampling clock error on the performance of the correlation is negligible. Substituting (9) in (8), we obtain

$$f(m) = e^{-j2\pi\varepsilon TD} \sum_{k=0}^{J-1} x(m-k)x^*(m-D-k) + N(m) \qquad (10)$$

where $N(m)$ is the undesired signal term. Assuming that the channel is static, the ideal auto correlation peak occurs when $x(k) = x(k+D)$. Using this, when the ideal peak occurs, we calculate

$$f(m) = e^{-j2\pi\varepsilon TD} \sum_{k=0}^{J-1} |x(m-k)|^2 + N(m) \qquad (11)$$

Notice that the magnitude of the autocorrelation is independent of frequency error. This well-known behavior makes this technique extremely robust against frequency error. In addition, if the inter-symbol phase rotation εTD is small, then the real portion of $f(m)$ contains useful information that is adequate for peak detection. The imaginary component of $f(m)$ will be dominated by the undesired signal. A straightforward application of delayed correlation to the multi-band system is not efficient. First, it does not exploit the sequence property. Thus, it is '*blind*' to the type of the sequence that is being transmitted and thus is expected to perform poorly under Simultaneously Operating Pico nets (SOP). Additional processing will be needed to identify the sequence. Second, it does not perform well under low SNR, narrow-band interference, and/or DC offset conditions. Some form of cross-correlation is thus needed for fast acquisition under such conditions. The alternative solution is to exploit the hierarchical nature of the preamble. In the following sub sections, two methods are presented that exploit the hierarchical nature of the sequences. The main idea behind these methods is to first perform correlation over sequence B (i.e., de-spread sequence B) and then do correlation over sequence A.

35.6.2.3 Hierarchical Delayed-correlation

The hierarchical delayed-correlation can be described with

$$f(m) = \sum_{l=0}^{L-2} c_l \sum_{k=0}^{M-1} r(m-Ml-k)r^*(m-Ml-M-k) \qquad (12)$$

where the sequence $C = \{c_0, ..., c_{L-2}\}$ is obtained from A as follows

$$c_l = a_l a_{l+1} \qquad (13)$$

In a similar way as (11), the ideal correlation peak occurs when $x(k) = x(k+M)$. Thus, at the desired correlation peak (12) yields

$$f(m) = e^{-j2\pi\varepsilon TM} \sum_{l=0}^{L-2} c_l \sum_{k=0}^{M-1} |x(m - Ml - k)|^2 + N(m) \qquad (14)$$

Notice here that the magnitude of this correlation result is also independent of frequency and phase error. Thus, the hierarchical delayed-correlation inherits the performance benefits of the conventional auto correlation algorithm in terms of its robustness to frequency/phase errors. The imaginary portion of $f(m)$ does not contain useful information for peak detection since phase rotation εTM is small. This leads to considerable reduction in implementation complexity since only the real component of (12) needs to be computed. The implementation of this method is fairly simple. Figure 35.15 shows a high-level diagram of one way of implementing this algorithm.

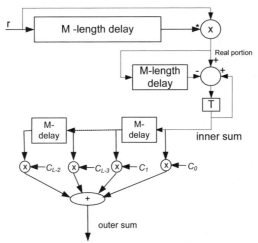

Figure 35.15: Top-level diagram of hierarchical delayed correlator.

35.6.2.4 Hierarchical Cross-Correlation (H-Xcorr)

Despite the attractiveness of implementation simplicity, the hierarchical delayed-correlator proposed above has a drawback: it does not make use of the spreading sequence B. As a result, it is blind to the contents of this sequence and thus it shares some of the weakness of conventional auto-correlators. An alternative technique is to use Hierarchical Cross-Correlation concept as follows:

$$f(m) = \sum_{l=0}^{L-1} a_l \sum_{k=0}^{M-1} r(m - Ml - k) b_k \qquad (15)$$

Substituting for $r(m)$ from Equation (9) yields

$$f(m) = e^{-j(2\pi\varepsilon Tm + \alpha)} \sum_{l=0}^{L-1} a_l \sum_{k=0}^{M-1} x(m - Ml - k) e^{j2\pi\varepsilon T(Ml+k)} b_k$$
$$+ \tilde{N}(m) \qquad (16)$$

The inner product de-spreads the B sequence while the outer sum de-spreads the $A-$ sequence. We notice that this method inherits the properties of cross-correlation techniques in that the result depends on frequency offset due to the terms $e^{j2\pi\varepsilon T(Ml+k)}$ and $e^{-j(2\pi\varepsilon Tm+\alpha)}$. Assuming a 40 parts per million (ppm) frequency offset error (i.e., 20ppm on TX and RX clocks), a sampling rate of 528 MHz and a center RF frequency of 5 GHz, we find

$$\varepsilon = \frac{40\times10^{-6}\times5\times10^{9}}{500\times10^{6}} = \frac{200\times10^{3}}{500\times10^{6}} = 400\times10^{-6}$$

Using this result, the expected maximum phase rotation that impacts the inner cross correlator can be computed as

$$360 \times \{M\cdot(L-1) + M - 1\} \times 400 \times 10^{-6} = 360 \times 127 \times 400 \times 10^{-6}$$

This is about 18 degrees. Thus, the maximum phase change in the inner cross-correlation part is relatively small and therefore its impact is negligible. The impact of the second term, $e^{-j(2\pi\varepsilon Tm+\alpha)}$, can be avoided by using a second-stage correlator described below. On the other hand, evaluating the magnitude of $f(m)$ will also make this method insensitive to this term.

While this method shows some sensitivity to frequency error, it provides other benefits such as improved performance in the presence of interference (noise, SOP, narrowband, DC offset). Nevertheless, the complexity of the implementation increases. The following figure shows a block diagram of the hierarchical cross-correlator.

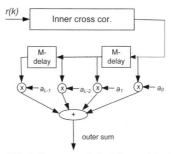

Figure 35.16: Simplified diagram of the hierarchical cross-correlator.

35.6.2.5 Second-stage Correlator

The performance of the synchronizer can be improved greatly by employing a second stage correlation using the output of the H-Xcorr as an input to it. This second step is in fact part of the MMSE detector described in Section 35.6.2.6. Notice that for the H-Xcorr, peak occurs when

$$f(m) = f(m-D)e^{j2\pi\varepsilon TD} + \delta(m) \qquad (17)$$

where D is the number of samples between subsequent sequences of the repetitive preambles, and $\delta(m)$ is the difference between the noise terms of $f(m)$ and $f(m-D)$. In the absence of undesired signal (interference) and zero frequency error, $f(m)$ will be equal to the channel impulse response. Thus, in principle, assuming static channels, one can employ auto-correlation across the impulse response of the channel as follows

$$y(m) = \sum_{l=0}^{Z-1} f(m-l) f^*(m-D-l) \tag{18}$$

where Z is the number of samples not greater than the delay spread of the channel. At the desired peak, the use of (17) in the above equation yields

$$y(m) = e^{-j2\pi\varepsilon TD} \sum_{l=0}^{Z-1} \left| f(m-l) \right|^2 + v(m) \tag{19}$$

where $v(m)$ is the noise term. Notice that the above processing accomplishes a number of objectives in one step. First, it provides the frequency error directly,

$$\varepsilon \approx -\frac{angle(y(m_{peak}))}{2\pi TD} \tag{20}$$

where m_{peak} is the value of m at which $|y(m)|^2$ is maximum. Secondly, the peak of $|y(m)|^2$, i.e. $|y(m_{peak})|^2$, coincides with the peak of the sum of the energy of the impulse response of the channel over the window Z. This information is very useful to set the start of the FFT window for an OFDM system. Thirdly, the real part of the peak is used for frame sync detection. And finally, of course, it provides the peak for burst detection purposes.

35.6.2.6 Burst Detection

Burst detection is accomplished by evaluating the correlation output. The most frequent technique used for burst detection is based on comparing the magnitude of the correlation output to a certain threshold. The threshold value is a function of the noise level, AGC setting and expected signal strength. Usually, the AGC is set to its maximum value at the beginning to catch the weak signals. Since this method is based on threshold, it is naturally sensitive to the threshold value and hence its performance can be impacted by noise. An MMSE-based peak detector on the output of the H-Xcorr can be used to provide robust performance that is not so sensitive to input signal levels. As indicated above and will be shown below, the second-stage correlator performs part of the computations needed for this detector. The MMSE detector using the output of the H-Xcorr can be described by

$$\arg\min_m \left\{ \sum_{l=0}^{Z-1} \left| f(m-l)e^{j\beta} - f(m-D-l) \right|^2 \right\} \tag{21}$$

where $\beta = 2\pi\varepsilon TD$. Further simplification of the above equation yields

$$\arg\min_m \left\{ \sum_{l=0}^{Z-1} \begin{aligned} &\left(\left| f(m-l) \right|^2 + \left| f(m-D-l) \right|^2 - \right. \\ &\left. 2\,\mathrm{Re}[f(m-l)f^*(m-D-l)e^{j\beta}] \right) \end{aligned} \right\} \tag{22}$$

Using (17), (18), and, (19) we estimate

$$e^{j\beta} \approx \frac{\sum_{l=0}^{Z-1} f(m-l) f^*(m-D-l)}{\sum_{l=0}^{Z-1} |f(m-D-l)|^2} \tag{23}$$

Substituting (23) in (22) and on further simplification, we get

$$\arg\min_{m} \left\{ h_1(m) + h_2(m) - 2\frac{|\widehat{f}(m)|^2}{h_2(m)} \right\} \tag{24}$$

where

$$h_1(m) = \sum_{l=0}^{Z-1} |f(m-l)|^2, \text{ and } h_2(m) = \sum_{l=0}^{Z-1} |f(m-l-D)|^2$$

From the above equations, we notice that in the absence of additive interference (AWGN, etc.)

$$|y(m)|^2 \leq \frac{1}{2}\left[h_1(m) + h_2(m)\right]h_2(m) \tag{25}$$

where the equality holds at the peak. $y(m)$ is computed by the second stage correlator described above. Ideally, it would be sufficient to check the equality condition to determine if the input signal contains the required preamble. Nevertheless, in practical systems, the equality condition is not generally true due to additive noise term. Thus, we can check the following condition to determine if the input signal is composed of a valid preamble.

$$|y(m)|^2 \geq k_1 \frac{1}{2}\left[h_1(m) + h_2(m)\right]h_2(m) \tag{26}$$

where k_1 is a constant, $k_1 < 1.0$. In general, k_1 is related to the input SNR, the lower the SNR, the lower the value of k_1. However, since information about the SNR is not available, k_1 is set to the lowest value that makes it sensitive to trigger on weak signals.

In order to evaluate the performance of this system, several simulations were carried out using the following conditions. The transmitter is implemented according to the Ecma specification. The channel models specified by IEEE 802.15.3a task group [4] are used to model multi-path channel environments. In particular, the non-line-of-sight channel models of CM2 and CM4 are chosen to stress the algorithms. The simulated packet structure contains a few random symbols at the beginning of the transmission to mimic random receiver startup. The transmitted signal is convolved with the impulse response of the channel from one of the channel models. White Gaussian noise is then added to the resulting signal. Frequency offset is also added to the transmitted signal. The simulation results shown below are for complete acquisition: i.e., AGC, initial burst detection followed with frame sync detection. The receiver model is a complete chain (see Figure 35.14) including models for RF, filtering, ADC/DAC, etc. The results are for the MMSE detector with the H-Xcorr. Each point in the graphs is derived by averaging over 50,000 noise realizations (using 100 different channel realizations and 500 packets for each channel).

Figure 35.17 and Figure 35.18 show the simulation results for the probability of misdetection for TFC1 and TFC3 respectively. Notice that the performance for CM2 is better than that for CM4. This is largely due to the impact of 3-fold increase of RMS delay spread in the case of CM4 compared to that of CM2. Also noticeable is the impact of k_1. Reducing k_1 improves performance for these TFC modes. Nevertheless, higher values of k_1 still provide sufficient performance, providing $< 10^{-5}$ probability of misdetection rate at 10 m for the 106.7 Mbps mode. This compares favorably well with the 8% packet error rate used as a metric of performance for this data rate at this distance. Figure 35.19 illustrates the SOP performance for TFC2 and TFC1. Notice the absence of false detection for the simulated conditions. The TFC code itself also contributes to the rather good performance.

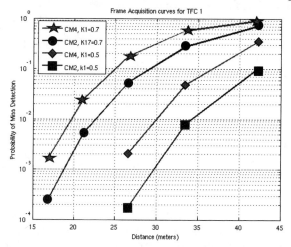

Figure 35.17: Simulated Prob. of misdetection curves for TFC 1.

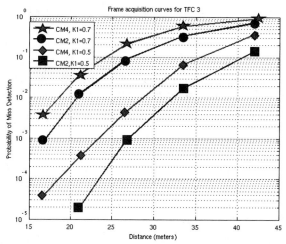

Figure 35.18: Simulated Prob. of misdetection curves for TFC 3.

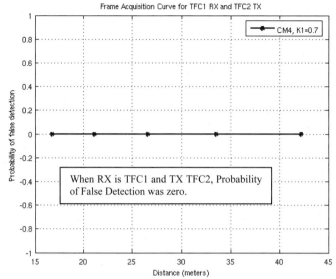

Figure 35.19: Simulated Prob. of false detection curve for SOP: TX is on TFC2 and RX is on TFC1.

35.6.2.7 Frequency Offset Estimation

As mentioned in Section 35.6.2.5, the frequency offset can be estimated using,

$$\varepsilon \approx -\frac{angle(y(m_{peak}))}{2\pi TD} \tag{27}$$

The frequency error estimation can further be improved by averaging the estimation over the number of preamble symbols that are available after burst detection as

$$\varepsilon \approx -\frac{angle(\sum_{k=0}^{W-1} y(m_{peak} - Dk) \times sign[Re\{y(m_{peak} - Dk)\}])}{2\pi TD} \tag{28}$$

where $sign[Re\{y(m_{peak} - Dk)\}]$ computes the binary cover sequence of the preamble. Further performance improvement may be possible by averaging the estimator across all the bands, but weighted according to the center frequency. However, overall performance may degrade if there is interference in one or more of the bands.

35.6.2.8 Timing (Optimum Start of the FFT Window)

It is well known that the optimal FFT window is the one that includes the maximum channel energy in the window. As mentioned earlier, the peak of $|y(m)|^2$ corresponds to the peak of the sum of the impulse response of the channel within a window of Z. Thus, the peak of this correlation window forms the reference for the start of the FFT window.

35.6.2.9 DC Offset, Narrow-Band Interference (NBI) and SOP

Typically, some residual DC signal comes from the RF/ADC front end. It can easily be seen that the output of the cross-correlator for DC input is zero if the reference sequence (preamble) is zero-mean.

For slowly varying narrowband interference, the behavior of H-Xcorr is similar to that for DC input, i.e. relatively insensitive to slowly varying NBI. However, for rapidly varying NBI, its performance depends on the cross-correlation property of the local reference sequence and the interferer. Nevertheless, considering the pseudo-randomness nature of the sequence and deterministic nature of the interference, the output of the H-Xcorr for this input will be significantly less than the power of the interferer and thus very little performance impact is expected.

The use of H-Corr combined with the MMSE detector reduces the effect of unwanted interfering signals, such as SOPs. This was illustrated in the simulation results provided in the preceding section.

35.6.3 AGC

AGC control for the FFI modes (non-hopping modes) is not significantly different from conventional techniques. However, the TFI modes require consideration of per-band AGC gain setting or a single AGC gain setting. Especially, if there is strong in-band interference from narrow-band devices or other UWB devices, then per-band AGC gain setting can provide better performance. A dynamically switchable AGC design is also not that difficult with today's CMOS/SiGe designs.

35.6.4 Timing Error Tracking

Differences in sampling clock rate between the transmitter and the receiver results in phase rotation of individual carriers that is proportional to their index and the error. The outer carriers will be the ones that will be affected the most. There will also be impact on the amplitude of sub-carriers. But, the impact on amplitude is usually negligible. For very short packet lengths, mismatch in sampling clocks can be accommodated with just phase rotation only. Nevertheless, for larger packet lengths, the timing error needs to be tracked. Algorithms such as those described in [10] can be used for tracking of sampling clock error. The sampling rate error is derived from the frequency offset error estimation. This assumes that the drift in the clock within the length of a packet is minimal, which is a reasonable assumption.

35.6.5 Common-Phase Error Estimation and Tracking

The common-phase error calculation and tracking is applied after FFT. The main purpose of this tracking loop is to compensate for any residual frequency error not compensated by the frequency error compensation part. The common-phase error can be estimated on a band by band basis using the pilot carriers differentially as follows.

$$\theta_k \approx \alpha \ angle(\sum_p S_{p,k} S^*_{p,k+1}) + \theta_{k-1} \qquad (29)$$

where $S_{p,k}$ is the pth pilot of OFDM symbol k in a band, α is proportionality constant, $\alpha < 1$. This estimation is based on the approximation that

$$S_{p,k} = R_{p,k}(\hat{H}_p P_p)^* \approx |(\hat{H}_p P_p)|^2 \ e^{j2\pi k f_\Delta - j\theta_{k-1}} \qquad (30)$$

where P_p is the transmitted pilot sequence, and H_p is the channel frequency response, and f_Δ is the residual frequency error.

35.6.6 Zero-padded OFDM Guard Processing

In the WiMedia specification, the OFDM symbol is zero padded (ZP). Thus, the standard OFDM process of removing the cyclic portion at the receiver is not valid for zero-padded OFDM. For zero-padded systems, more advanced equalization techniques can be employed to get the best performance [5]. Nevertheless, those equalization techniques are expensive to implement. An alternative, but, suboptimal, technique exists that allow the use of conventional 1-tap equalization techniques. For example, an overlap-and-add of a portion of the samples can be used. With this kind of technique, the noise power increases and results in performance penalty. However, for UWB, this loss is recovered from the reduced overhead of transmitted power (zero overhead power due to ZP). The figure below illustrates pre-symbol overlap-and-add. One could also employ post–symbol overlap-and-add where the first 32 samples after switching are added to the last 32 samples, and then take the last 128 samples for FFT. This removes the 1-symbol latency in implementation. For the non-hopping modes, the switching time is not needed and thus a total of 37 samples can be used for this step.

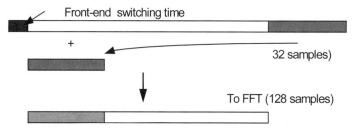

Figure 35.20: Illustration of overlap and add (pre).

35.6.7 Channel Estimation and Equalization

A WiMedia receiver uses the FD preamble (CE symbols) to estimate the channel frequency response and then uses these estimates to equalize the header and payload symbols.

35.6.7.1 Channel Estimation for Different TFCs

The received symbols (after the FFT block) can be represented as

$$R_n(k) = H_m(k)S_n(k) + N_n(k), \tag{31}$$

where $k \in [0,127]$ is the sub-carrier index, n is the OFDM symbol number, and $m \in \{1,2,3\}$ is the sub-band index. m is a function of the TFC number (shown in Figure 35.7) and the symbol number n.

$$m = \begin{cases} \mod(n,3)+1 & TFC = \{1,2\} \\ \mod\left(\left\lfloor \dfrac{n}{2} \right\rfloor, 3\right)+1 & TFC = \{3,4\} \\ TFC-4 & TFC = \{5,6,7\} \end{cases} \tag{32}$$

where the function $mod(a,b)$ represents the remainder of a/b and $\lfloor x \rfloor$ represents the integer part of x. $H_m(k)$ represents the channel frequency response for sub-carrier k on band m, $S_n(k)$ and $R_n(k)$ represent the transmitted and received symbols respectively in the frequency domain. The term $N_n(k)$ represent the white noise component on sub-carrier k.

For the FD preamble (CE symbols), $S_n(k) = A(k)$, where $A(k)$ is a known training sequence. A simple estimate of the channel can be derived by dividing the received symbol by the training sequence and averaging over the number of symbols transmitted in that band.

$$\hat{H}_{CE,m}(k) = \frac{1}{P} \sum_{p=1}^{P} \frac{R_{n(m,p)}(k)}{S_{n(m,p)}(k)} \tag{33}$$

where

$$P = \begin{cases} 2 & TFC = \{1,2,3,4\} \\ 6 & TFC = \{5,6,7\} \end{cases} \tag{34}$$

and $n(m,p)$ is given as

$$n(m,p) = \begin{cases} m+3(p-1) & TFC = \{1,2\} \\ 2m+p-2 & TFC = \{3,4\} \\ p & TFC = \{5,6,7\} \end{cases} \tag{35}$$

$\hat{H}_{CE,m}(k)$ represents the channel estimate on sub-carrier k in sub-band m derived from CE symbols. Substituting for $R_n(k)$ and $S_n(k)$ in Equation (33) and assuming that the channel is static for the duration of the packet, we get

$$\hat{H}_{CE,m}(k) = \frac{1}{P} \sum_{p=1}^{P} \left(H_m(k) + \frac{N_{n(m,p)}(k)}{A(k)} \right) = H_m(k) + \frac{1}{P \times A(k)} \sum_{p=1}^{P} N_{n(m,p)}(k) \tag{36}$$

As can be observed from the above equations, the quality of the channel estimates can be improved by increasing the number of terms (i.e. P) in the summation.

The channel estimates as derived above can be further fine-tuned by using some prior information about the channel environment. For OFDM systems, one such method is implemented by limiting the length of the channel impulse response equal to the length of the zero-suffix.

With the assumption that the channel impulse response is less than the length of the guard internal, we can truncate the channel impulse response to the length of the suffix and derive the least squares solution as described in [11] for each of the bands and is given by the following equation

$$\hat{\underline{H}}_{CE,m} = F(\mathrm{E}^H \mathrm{E})^{-1} \mathrm{E}^H \underline{R}_{n(m,p)},$$ (37)

where $\hat{\underline{H}}_{CE,m} = [\hat{H}_{CE,m}(0), \hat{H}_{CE,m}(1), \hat{H}_{CE,m}(2), ..., \hat{H}_{CE,m}(N-1)]^T$ and

$\underline{R}_{n(m,p)} = [R_{n(m,p)}(0), R_{n(m,p)}(1), R_{n(m,p)}(2), ..., R_{n(m,p)}(N-1)]^T$, N is FFT size and m is the frequency band. F is the truncated Fourier matrix of size $N \times N_L$ and $\mathrm{E} = AF$, where A is $N \times N$ diagonal matrix with the known transmitted symbols a_k in the diagonal. N_L is channel length and is equal to 32 for this system.

35.6.7.2 Channel Equalization

Channel equalization is applied to all the symbols of the packet starting from the first header symbol as shown in the following equation.

$$X_n(k) = G_m(k)R_n(k)$$ (38)

where $X_n(k)$ is the output of the equalizer, and $G_m(k)$ is derived from $\hat{H}_{CE,m}(k)$ based on the equalization scheme adopted for the system.

Figure 35.21 and Figure 35.22 compare the performance of four different data-rate modes in CM1 and CM4 channel environments respectively. The simulation were run on 100 different channel realizations for each of the channel models and was simulated for 200 packets with a packet length of 1024 bytes for each of these channel realizations. The ADC precision was 5 bits and the simulations also included a frequency offset of 200 KHz. It can be observed that the performance of the lower data rate modes are almost similar in both channel conditions while that of the 480 Mbps mode shows some performance loss in CM4 channel as compared to its performance in CM1 channel.

35.6.8 In-band Interference Mitigation

The bandwidth of WiMedia UWB system is about 1.5 GHz and since UWB devices do not have exclusive use of this band, there is a high probability that there will be interference from other narrowband devices operating in this band. Under narrowband interference scenario, some of the sub-carriers will be affected severely and this will degrade the overall performance of the system.

In the WiMedia system, multiple access is attained by using the frequency hopping sequence. As a result of this scheme, one pico-net might cause interference to another pico-net if the pico-nets are in close range. In this scenario, one or more subbands of the desired signal are affected by the interferer. Both the pico-net interference and the narrowband interference will manifest as white noise to the desired signal at the receiver.

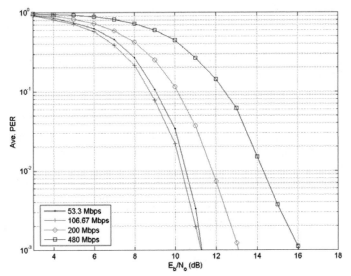

Figure 35.21: Ave. PER v/s. E_b/N_o for different data rates in CM1 channel environment. Mean PER over 90 best channel realizations. Shadowing is included.

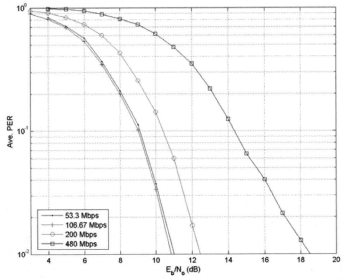

Figure 35.22: Ave. PER v/s. E_b/N_o for different data rates in CM4 channel environment. Mean PER over 90 best channel realizations. Shadowing is included.

Usually, in a coded system, it is assumed that all the symbols have same SNR (or white noise), and therefore this term is removed from the soft metrics derived in metric calculation unit of the error correction unit (Viterbi decoder). However, this is not the case in reality due to the above mentioned in-band interference scenarios. Under these conditions, scaling the metrics for the Viterbi decoder by a term proportional to the noise power will help in improving the receiver performance. This requires us to estimate the noise variance in each of the sub-carriers for all the CE symbols. Note that we do not have prior information about the interferer and therefore we cannot average the noise variance estimates across two symbols within the six CE symbols.

After deriving the channel estimates, the noise variance in each sub-carrier can be calculated using the following equation

$$\sigma_{CE,n}^2(k) = \left| R_n(k) - A(k)\hat{H}_{CE,m}(k) \right|^2 \tag{39}$$

where $n \in \{1,2,3,4,5,6\}$. $R_n(k)$ and $A(k)$ represent the received and transmitted symbols respectively. $\hat{H}_{CE,m}(k)$ represents the channel estimate on sub-carrier k in sub-band m derived from CE symbols.

In order to improve the reliability of the noise variance estimates, we can derive noise variance in each symbol by averaging $\sigma_{CE,n}^2(k)$ over all the sub-carriers as shown in the following equation

$$\sigma_{CE,n}^2 = \frac{1}{N} \sum_{k=0}^{N-1} \sigma_{CE,n}^2(k) \tag{40}$$

For the case of QPSK mapping, the noise variance estimates can then be used to scale the output of the equalizer for header and payload symbols as shown below:

$$Y_n(k) = \frac{X_n(k)}{\sigma_{CE,\mathrm{mod}(n-1,6)+1}^2(k)} \quad \text{or} \tag{41}$$

$$Y_n(k) = \frac{X_n(k)}{\sigma_{CE,\mathrm{mod}(n-1,6)+1}^2} \tag{42}$$

where $X_n(k)$ is the output of the equalizer and is defined by Equation (38).

The scaled metrics $Y_n(k)$ are then de-interleaved and sent to the Viterbi decoder, where they are used in decoding the bits. When a channel introduces narrowband or pico-net interference then estimating the noise variance and using those values in the metric calculation helps in improving the performance of the system. As mentioned earlier, a narrowband interferer might affect only a few carriers in a band. Therefore, these sub-carriers might experience more noise than the rest of the sub-carriers. In this scenario, it would be useful to estimate noise variance in each sub-carrier and then scale the metrics with this parameter. In the case of simultaneously operating pico-nets (SOPs), all the sub-carriers in a band will be affected by the interference and as a result will have the same noise variance. However, different sub-bands may have different noise variance. In this scenario, it is recommended to average noise variance estimates over the entire band as shown in Equation (40) and then using it to scale the metrics corresponding to that band.

Figure 35.23 shows the estimated noise power in the three bands for TFC = 1 with and without piconet interference in CM1 channel. The simulation parameters are same as the ones described in Section 35.6.7. In addition, for SOP multi-path simulations five different channel realizations were used for the reference and the interferer links thus providing 25 different combinations and the PER was averaged across these 25 simulations. The top plot shows the estimated noise power vs. SNR without piconet interference. It can be observed that all the three bands have similar noise power numbers and that the estimated noise power is inversely proportional to SNR. The bottom sub-plots shows the estimated noise power vs. SIR for SNR = 10 dB in the presence of one pico-net interferer in CM1 multi-path channel environment. The interfering pico-net has TFC = 2 and it is perfectly synchronized with the desired signal. When piconet interference is present, then some of the bands will have more noise power compared to the other bands. In this case, due to the particular band sequence used by the desired {1,2,3,1,2,3} and the interferer {1,3,2,1,3,2} signals, band 1 experiences more noise than the other two bands, as can be seen in the plot. It can also be observed that the estimated noise power is inversely proportional to the Signal to Interference power ratio (SIR).

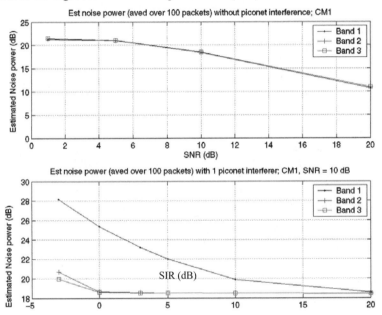

Figure 35.23: Estimated noise power vs. SNR and SIR for channel CM1 with and without pico-net interference.

Figure 35.24 compares the performance of a WiMedia receiver in the presence of pico-net interference using different receiver algorithms. In the figures, the term d_{int}/d_{ref} refers to the ratio of the distances of the interferer and the reference links. The legend NE0 represents a standard receiver which does not scale the metrics by noise variance estimates, while legend NE1 represents the above described receiver which estimates the noise variance in each of the sub-bands and then uses them to scale the soft-metrics. The desired signal's TFC number is 1 and the interferer signal's TFC numbers are 2, 3 and 4 in that order. It can be observed from the plot that scaling the metrics by noise variance estimates provides some gains when compared with the performance of a standard receiver. In the case of one pico-net interferer and at 10% Packet Error Rate (PER), we realize a gain of about 3.0 dB.

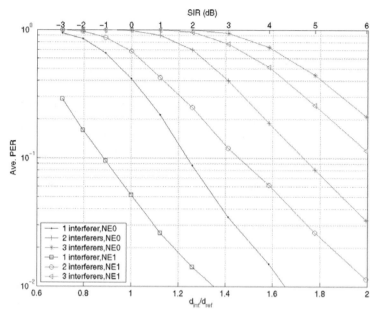

Figure 35.24: Average PER vs. d_{int}/d_{ref} (SIR) in the presence of piconet interference; CM3 channel, E_b/N_0 = 8.75 dB, 106.7 Mbps mode. NE0 – Without scaling by the noise variance estimates, NE1 – With scaling by the noise variance estimates.

35.6.9 Improved Channel Estimation and Noise Variance Estimation

The channel estimates and noise variance estimates derived in Equations (37) and (39) can be further improved during the header and payload symbols. In order to update the estimates, we need to know the information sequence. An estimate of the transmitted data can be generated either by slicing the output of the FFT or by using the output of the Viterbi decoder. Since the channel is assumed static for the duration of packet, updating the estimates over payload symbols does not provide significant performance gains (assuming that the in-band interference is present for the complete duration of the packet). In addition, the complexity and the latency associated with the update over the payload symbols is considerably more than the update over header symbols. The header is usually transmitted at the lowest data rate and therefore it is more resilient to the channel errors than the payload symbols. In addition, the receiver can process the payload symbols only after the header is decoded. This implies that the reference symbols for the header symbols can be generated and the estimates updated before we start processing the payload symbols.

In the WiMedia system, 12 OFDM symbols, with at least four symbols in each band, are used to transmit the header information. This is twice as many symbols as the CE symbols. Therefore, the estimates derived from the header symbols would be more reliable than the ones derived from the CE symbols, assuming that the header bits have been decoded correctly.

The channel estimates $\hat{H}_{HDR,m}(k)$ can be derived from the header symbols as shown in the following equation:

$$\hat{H}_{HDR,m}(k) = \frac{1}{P}\sum_{p=1}^{P}\frac{R_{n(m,p)}(k)}{\hat{S}_{n(m,p)}(k)} \tag{43}$$

$\hat{S}_{n(m,p)}(k)$ represents the estimate of the transmitted symbol $S_{n(m,p)}(k)$ and is derived from either the slicer output or the Viterbi decoder output. $R_{n(m,p)}(k)$ represents the received (after FFT block) header symbols. Substituting Equation (31) in Equation (43), we get

$$\hat{H}_{HDR,m}(k) = \frac{1}{P}\sum_{p=1}^{P}\left(H_m(k)+\frac{N_{n(m,p)}(k)}{\hat{S}_{n(m,p)}(k)}\right) = H_m(k)+\frac{1}{P}\sum_{p=1}^{P}\frac{N_{n(m,p)}(k)}{\hat{S}_{n(m,p)}(k)} \tag{44}$$

where

$$P = \begin{matrix} 4 & TFC = \{1,2,3,4\} \\ 12 & TFC = \{5,6,7\} \end{matrix} \tag{45}$$

and $n(m,p)$ is given as

$$n(m,p) = \begin{matrix} m+3(p-1) & TFC = \{1,2\} \\ 2(m-1)+p+4\left\lfloor\dfrac{p-1}{2}\right\rfloor & TFC = \{3,4\} \\ p & TFC = \{5,6,7\} \end{matrix} \tag{46}$$

The noise variance in each sub-carrier can then be calculated using the following equation:

$$\sigma_{HDR,n}^2(k) = \frac{1}{2}\left[\left|R_n(k)-\hat{S}_n(k)\hat{H}_{HDR,m}(k)\right|^2 + \left|R_{n+6}(k)-\hat{S}_{n+6}(k)\hat{H}_{HDR,m}(k)\right|^2\right] \tag{47}$$

where $n \in \{1,2,3,4,5,6\}$

We can then derive noise variance in each band by averaging $\sigma_{HDR,m}^2(k)$ over the entire band as shown in the following equation

$$\sigma_{HDR,n}^2 = \frac{1}{N}\sum_{k=0}^{N-1}\sigma_{HDR,n}^2(k) \tag{48}$$

The channel estimates and the noise variance estimates derived from the CE symbols and the header symbols can be averaged as shown below

$$\hat{H}_m^2 = w_m\hat{H}_{HDR,m}^2 + (1-w_m)\hat{H}_{CE,m}^2 \tag{49}$$

$$\sigma_n^2 = w_n\sigma_{HDR,n}^2 + (1-w_n)\sigma_{CE,n}^2 \tag{50}$$

where w_m and $w_n \in [0,1]$ represent weighting factors.

These new estimates can then be used to equalize and scale the payload symbols according to Equations (37) and (42) respectively.

35.6.9.1 Generating the reference header symbols from the slicer output

In the WiMedia proposal, the OFDM sub-carriers are modulated using QPSK mapping for the PLCP header and the lower data rate modes. This enables the receiver to make reliable decisions on the symbols by using a simple slicer. The output of the slicer is represented by the following equation:

$$\hat{b} = sign(X_n(k)) = sign(G_m(k)R_n(k)) \tag{51}$$

The estimated bits are then mapped into symbols to form $\hat{S}_n(k)$, which can then be used in Equations (44) and (47) to update the estimates.

35.6.9.2 Generating the reference header symbols from the Viterbi decoder output (iterative decoding)

The performance of the receiver depends on the reliability of the estimates which in turn depends on the accuracy of the estimates of the transmitted symbols $S_n(k)$. Using slicer outputs to derive the transmitted symbols will not maximize the gains achievable by using received symbols to update the estimates. On the other hand, using iterative decoding techniques maximizes the gains at the cost of increased complexity. In this method, channel response and noise variance are estimated on the CE symbols in the first pass. These estimates are used to equalize and scale the received data. After the Viterbi decoder, the bits are encoded back to generate the reference symbols ($\hat{S}_n$). The channel estimates and the noise variance estimates are updated according to Equations (49) and (50) respectively. In the second pass, the updated channel estimates $\hat{H}_m$ and noise variance estimates σ_n^2 are used to equalize and scale the received data. The computational complexity of this receiver is very high compared to a standard receiver. One way to reduce the computational complexity is to consider Viterbi decoder decisions of header symbols only. The header symbols are decoded prior to the decoding of the rest of the packet. This enables us to compute reference symbols $\hat{S}_n$ corresponding to the header bits and then update the channel estimate and the noise variance estimate accordingly and use them for the payload symbols.

Figure 35.25 shows the block diagram of the proposed receiver that includes the update module. It includes a switch to select either the estimates derived from CE symbols (position 1) or the estimates derived from the HDR symbols (position 2). The header symbols always use the estimates derived from the CE symbols, so in this case (i.e. for the 12 header symbols) the switch is in position 1. The equalized and scaled header symbols are processed by the 'despread & demapper', 'bit de-interleaver', 'Viterbi decoder' and 'RS decoder' blocks to decode the transmitted header bits. The decoded bits are then RS encoded, convolutional encoded, bit interleaved and mapped to generate the reference symbols for the update module. This block then uses the generated reference symbols along with the received symbols and the estimates derived from the CE symbols to update the channel estimates and the noise variance estimates. The latency in the decode-encode chain can be reduced by removing RS decoder and RS encoder blocks. In this case, the Header Check Sequence (HCS) can be used to verify if the decoded header bits are correct. The estimates are updated only when the header bits are decoded without errors.

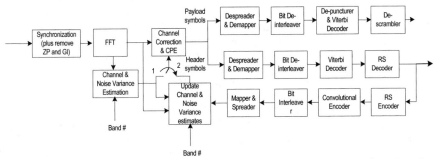

Figure 35.25: Block diagram of proposed TFI-OFDM receiver including the channel estimation, noise variance estimation and update channel estimates and noise variance estimates units.

In terms of complexity, the encoding chain in Figure 25.25 will not add any extra hardware cost since it can reuse existing hardware in the transmitter. Thus, compared to a standard receiver, the additional hardware cost is mainly due to two additional buffers. One buffer is to store $R_n(k)$ until $\hat{S}_n$ is available from the encoding chain. Another buffer is to hold payload symbols until the channel estimates and noise estimates are updated. The extra silicon area from these two buffers is marginal compared to overall baseband area: assuming that $R_n(k)$ use 8 bits in hardware, these two buffers will have a size of around 3Kbytes. In CMOS090 technology, the corresponding hardware area is around 0.2 mm^2.

We evaluated the performance improvement due to the estimate updates in the presence of pico-net interference through link-level simulations. The simulation parameters are same as the ones described in Section 35.6.8. Figure 35.26 shows the improvement in the performance of the receiver due to channel estimation update over the header symbols, in the presence of multi-path (CM1) and pico-net interference. Note that scaling by noise variance estimates is not applied in this particular case.

Figure 35.27 compares the performance of the 106.7 Mbps mode in the presence of multi-path (CM1) and pico-net interference with and without SNR scaling. The channel estimates are updated on the header symbols for both the cases. The legend NE5 represents the case when both the channel estimates and the noise variance estimates are updated over the header symbols. It can be observed that the in-band interference mitigation techniques together with the channel estimation and noise variance update scheme presented in this sub-section provides significant gains. In particular, we achieve a gain of about 4.0 dB at 1% PER in the case of one interferer and a gain of about 3.0 dB at 8% PER in the case of two interferers.

35.6.10 Multiple Antenna Receiver

Multiple antenna systems have found wide use in narrowband systems. Similar techniques can also be employed for UWB. Sources of gains with multiple-antenna systems come from three areas: diversity gain, fading gain and combining gain (noise reduction). While the difference in flat fading between two or more antennas is small for wideband systems compared to narrowband systems (see Figure 35.3), UWB can benefit from diversity and combining gains as much as narrow-band systems gain from these techniques. For example, a simple antenna selection per band provided only about 1.7 gain (fading gain) while antenna combining such as maximum ratio combing results in 5-6 dB gain for a 2 receive antenna system (3 dB from noise reduction gain and about 3 dB from diversity gain) [12]. Despite the need for multiple front ends, Maximal Ratio Combining (MRC) offers the greatest potential to improve the performance of UWB for current generation systems.

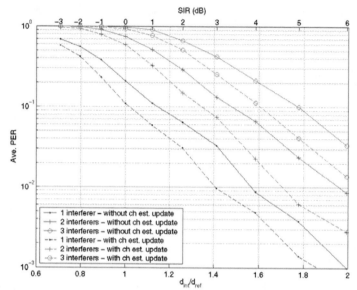

Figure 35.26: Average PER vs. d_{int}/d_{ref} (SIR) in the presence of pico-net interference with and without channel estimation update. SNR scaling is not used. Channel – CM1.

35.7 Overview of Next Generation High-speed UWB systems

The current Ecma specification provides a maximum PHY data rate of 480 Mbps. The effective payload from application point of view is somewhere about 60% of that due to MAC and PHY overheads, though some specific mode of PHY/MAC can provide a higher or lower efficiency. As evidenced by the past trends, UWB application requirements will most likely go beyond this maximum rate. Some vendors are already announcing proprietary modes with data rates up to 1 Gbps. Thus, next generation WiMedia systems (Gen II) will provide PHY data rates of 1-2 Gbps, with a possibility of supporting even higher data rates at lower ranges. There are four techniques that can be considered for Gen II, each having some pros and cons as described below.

35.7.1 Higher-order Constellation

The underlying constellation for current system is QPSK. Higher bit rates can be obtained by using higher constellation sizes such as 16-QAM. The maximum data rate can be increased to 960 Mbps using 16-QAM. However, the range that can be achieved by this mode is very low. Advanced receiver techniques such as MRC will have to be used to get better performance at these high data rates. Going beyond 1 Gbps will however be more difficult as this requires 64-QAM or even higher constellation sizes.

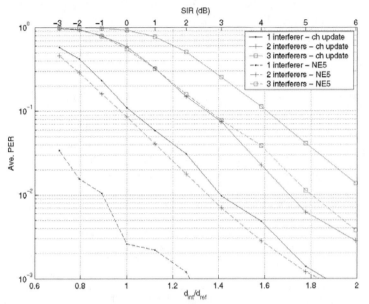

Figure 35.27: Average PER vs. d_{int}/d_{ref} (SIR) in the presence of piconet interference with and without SNR scaling; CM1 channel, $E_b/N_0 = 8.75$ dB, 106.67 Mbps mode. NE5 – With scaling using noise variance estimates.

35.7.2 MIMO

Spatial multiplexing in the form Multiple Input Multiple Output (MIMO) is another method that can provide robust performance at increased cost. Simulations have shown that 2-4m range can be achieved using MIMO QPSK (about 1 Gbps PHY rate). MIMO 16-QAM can provide about 2 Gbps data rates for shorter links.

35.7.3 Wider Bandwidth

Wider bandwidth in combination with MIMO and/or higher order modulation is another potential approach that is scalable for future systems. However, the main drawback of this option is that it requires high speed ADC/DAC and baseband processing modules. In addition, ensuring backwards compatibility becomes more challenging compared to using the current bandwidth.

35.7.4 Brute-force approach

Higher rates can be obtained by deploying several of the above and current WiMedia devices operating in different channels simultaneously. However, the increased cost and power may make this approach less attractive. In addition, a multi-channel MAC has to be employed to enable this feature.

35.8 Limitations on Capacity/Bit-rate and New Candidate Technologies

Due to the maximum power limitation imposed by regulatory bodies, UWB will find it difficult to push for higher bit rates at acceptable range with reasonable implementation complexity. Thus, for new emerging applications with much higher bit-rate in the order of few Gbps and beyond, alternative technologies need to be considered. This section briefly reviews a few candidate technologies.

35.8.1 60 GHz

60 GHz is an alternative band that has great potential for some applications. IEEE and Ecma have already formed task groups aiming to develop specifications for short-range high-speed wireless connectivity in this frequency band. The main drawback of 60 GHz is that at least 15 dBi antenna gain is required to make a link with acceptable range using current circuit technology that provides just about 10 dBm transmit power. Thus, some directionality will be needed. The following table presents a summary of the comparison of 60 GHz and UWB.

Table 35.4: Comparison of 60 GHz and UWB technologies.

	60 GHz	UWB
Power (Rate)	TX power limited by circuitry (13 dBm)	TX power limited by FCC (-2.7dBm @ 7.2 GHz BW)
	Large path loss (antenna gain is necessary)	Less path loss
	Antenna arrays give high gain and have small size.	
Multi-channel Support	High power in smaller BW allows easier channelization	Difficult to achieve multiple 3 Gbps channels
	Directional antennas allow many links in each channel	
Interference	60GHz band is not crowded	Many victims/interferers in 5 GHz band (e.g., 802.11a)
	Directional antennas reduce interference to/from others	

The 60 GHz band has wider bandwidth and relaxed transmit power limitations and therefore has the potential to meet the demand for high data rates, albeit, with some shortcomings such has the need for directionality. However, directionality does not necessarily mean that the user manually configures the antennae direction. Techniques such as sector switching and beam forming can be utilized to facilitate ease of communication with directionality. The rather small antenna sizes in higher frequencies also make these techniques very attractive.

35.8.2 Free-space Optics (FSO) and Terra-Hertz (THz) Band

Free-space optics (FSO) and Terra-Hertz (THz) band are attractive candidates due to the promise of abundant spectrum, relaxed regulatory constraints (at least for now) and less interference due to heavy directionality. One of the earliest standards of 802.11 is an infra red (IR) based physical layer for low data rates up to 2 Mbps. This and similar systems did not find popular commercial use due to the need for heavy directionality for acceptable performance. In order to address the need for heavy directionality, diffusing technologies and beam steering (e.g., servo for FSO) are being investigated. However, this will likely result in less mobility. While these approaches will enable some new applications down the road, they are not good candidates at the moment and are not matured enough for current and near-term applications. Nevertheless, further advances in implementation technologies can bring these within an affordable cost range and thus can become attractive in a decade or so.

35.8.3 Cognitive Radios

Though there is a perception of scarcity of spectrum, several measurements have indicated that most of the spectrum is not used at a given time and space. To meet the growing needs of commercial applications, the regulatory bodies are now relaxing the spectrum regulation, in which UWB is a good example. In addition they are also considering the use of intelligent wireless devices that allow a more efficient use of the spectrum. These devices are sometimes referred to as cognitive radios [13]. Cognitive radios sense the spectrum and transmit the appropriate waveform in a certain way so that the spectrum is utilized in a more effective manner. At the moment, an IEEE group, IEEE 802.22 is working on perhaps the first cognitive radio standard limited to vacant TV bands for regional area network applications (e.g., see [14]). The data rate supported is on the order of 20 Mbps. If the regulatory bodies relax the current regulations, then it is possible that cognitive radios can operate in much wider bandwidth. Thus, one could envision that future cognitive radios, if wide-bandwidth is available, can also be used for high-rate WPANs. However, several implementation challenges need to be addressed. For example, operating in the 300 MHz to 10 GHz bands poses a challenge in both antenna and RF circuits design to meet performance and commercialization (cost) requirements [15].

Acknowledgements

The authors would like to acknowledge Jun Yang, Alireza Seyedi, and Dong Wang of Philips Research North America for various stimulating discussions and ideas that directly and indirectly contributed to the contents of this chapter.

About the authors

Dagnachew Birru, PhD, (Dagnachew.Birru@philips.com) principal member research staff and project leader at Philips Research North America, has been with   Philips Research since 1992. Since 1998, he has been with the Wireless Communications and Networking Department of Philips Research Laboratories, Briarcliff Manor, NY. During this time, he worked on high-data rate WPAN (60GHz, UWB), cognitive radios, multi-standard channel decoder architectures and algorithms for digital TV reception. From 1992 to 1998, he was a Research Scientist at Philips Research Laboratories, Eindhoven, The Netherlands working on DSP core architectures and new sigma-delta modulation techniques and algorithms. He is the author and coauthor of several papers and patents. He has also contributed to technical groups such as ATSC (US digital TV), IEEE 802.15.3a, 802.22, WiMedia and Ecma. His current research interests include digital communications systems design for very high data-rate short-range wireless connectivity.

Vasanth R Gaddam (Vasanth.Gaddam@philips.com) received his B.E. in Electronics and Communications Engineering from Osmania University, Hyderabad, India in 1996 and M.S. in Electrical Engineering from Villanova University, Villanova, PA in 1998. Since 1998, he has been with the Wireless Communications and Networking department of Philips Research North America, Briarcliff Manor, NY, working on receiver enhancements and transmission standards development/improvements for terrestrial DTV, high data rate wireless LANs/PANs and WRANs. He has contributed to various technical groups involved in developing Physical layer (PHY) specifications such as ATSC, IEEE 802.15.3a and 802.22, and WiMedia. He has also co-authored several papers and patents. His research interests include error correction coding and digital signal processing for wireless communications systems.

35.9 References

[1] MBOA, "Multi-band OFDM Physical Layer Proposal", July 14, 2003.
http://www.ieee802.org/15/pub/2003/Jul03/03267r5P802-15_TG3a-Multi-band-OFDM-CFP-Presentation.ppt

[2] Standard ECMA 368, "High Rate Ultra Wideband PHY and MAC Standard, 1st Edition, Dec. 2005,
http://www.ecma-international.org/publications/files/ECMA-ST/ECMA-368.pdf

[3] J. del Prado Pavon, et. al., "The MBOA-WiMedia Specification for Ultra Wideband Distributed Networks", *IEEE Communications Magazine*, June 2006.

[4] J. Foerster, et. al. "High Rate WPAN Final High Rate WPAN", IEEE P802.15 Working Group for Wireless Personal Area Networks (WPANs), Feb. 2003.

[5] B. Muquet, Z. Wang, G. B. Giannakis, M. de Courville, P. Duhamel, "Cyclic Prefix or Zero-Padding for Multi-Carrier Transmissions?", *IEEE Transactions on Communications,* Vol. 50, No. 12, pp. 2136-2148, December 2002.

[6] IEEE 802.15 SG 3a channel modeling sub-committee, "Channel modeling sub-committee report final", Feb. 2003, IEEE P802.15-02/490r1-SG3a.

[7] A.A. Saleh and R. Valenzuela, "A statistical model for indoor multipath propagation", *IEEE Journal on Selected Areas in Comm.*, Vol. 5, Feb. 1987.

[8] A. Batra, et. al., "Design of Multiband OFDM System for Realistic Channel Environments", *IEEE Transactions on Microwave Theory and Techniques*, Vol. 52, No. 9, Sep. 2004.

[9] A. Ismail, A. Abidi, "A 3.1 to 8.2 GHz direct conversion receiver for MB-OFDM UWB communications", Solid-State Circuits Conference, 2005. Digest of Technical

Papers. ISSCC. 2005 IEEE International Publication, 6-10 Feb.2005, Vol. 1, pp. 208-593.

[10] B. Yang, K.B. Letaief, R.S. Cheng, Z. Cao, "Timing Recovery for OFDM Transmission", *IEEE J. Sel. Areas in Communications*, Vol. 18, No. 11, Nov. 2000, pp. 2278-2290.

[11] M. Ghosh and V. Gaddam, "Bluetooth interference cancellation for 802.11g WLAN receivers", *IEEE International Conference on Communications*, 2003.

[12] A. Seyedi, V. Gaddam, and D. Birru, "Performance of multi-band OFDM UWB system with multiple receive antennas", *IEEE Wireless Communications and Networking Conference*, Volume 2, 3-6 April 2006, pp. 792-797.

[13] J. Mitola, et. al., "Cognitive Radios: Making Software Radios more Personal," *IEEE Personal Communications*, Vol. 6, No. 4, August 1999.

[14] C. Cordeiro, K. Challapali, D. Birru, and S. Shankar, "IEEE 802.22: An Introduction to the First Wireless Standard based on Cognitive Radios", *J. of Communications*, pp. 38-47, Vol. 1, No. 1, April 2006.

[15] K. Challapali, D Birru, and S. Mangold, "Spectrum-Agile Radio for Broadband Applications", *EE Times*, Aug. 23, 2004.

## 36

Wireless Cities[a]

Foreword: A World Without Wires
Contributed by Steve Andrews, *BT Group's Chief of Mobility and Convergence*

Wireless working is mainstream. Whether on the go, at home at work or in the local coffee shop, it is increasingly possible to get the information we require, whenever we want it on the device we want it on. It is clear then that this is no longer a technology of the future. Wireless technology is radically changing the way local authorities, individuals and businesses work, talk and play. And it is doing so right now.

Simply put, there is a huge benefit for councils who understand the evolution to a wireless world, helping them to fulfil their vision of creating an e-enabled town or city. We have already seen it start happening in the U.S. and other parts of the world and the UK is now well underway. From Philadelphia to Westminster, municipal wireless networks are springing up in cities across the globe; ambitious but realistic projects which help council workers, communities and local businesses alike to mobilise and reap the benefits that wireless broadband networks can bring.

According to Gartner, Wi-Fi is accessible on 80% of professional PCs, as well as being in homes across the country, where people are increasingly getting comfortable with the wireless world. Wireless communications have taken a foothold throughout Europe and IDC research shows that two thirds of the European working population is equipped with mobile devices and predicts that during 2007 there will be 99.3 million mobile-enabled workers in Europe. Perhaps the most telling facts about this are that the world recently celebrated its 130,000th Wi-Fi hotspot which shows the reach and significance of wireless communications and that the Wi-Fi Alliance has already endorsed over 3,000 Wi-Fi products as fit-for-market. The wireless broadband world is here and now, but it is just the first step.

In Westminster, we have seen how Wi-Fi has improved staff productivity and more efficiently delivered services. In just one example, wireless CCTV cameras have led to a significant benefit to the general public by reducing crime.

In comparison to traditional fixed CCTV these new wireless cameras can be relocated to problem areas quickly and cheaply to allow wider and more flexible coverage. Plus the real-time footage can be viewed by the police through wireless handheld devices. This led Sir Simon Milton, Leader of the City Council, to say that making the city wireless was

---

[a] *British Telecom*

"one of the most exciting developments in Westminster's history" – a city, I'm sure you'll agree, not short of exciting developments in its past.

It is a similar story in Glasgow, in Milton Keynes, in Norfolk, in Sussex, in Cardiff, in Lewisham. Local councils are looking to create a wire-free working environment for citizens, business people and visitors, as well as using Wi-Fi to improve the flow of communication within the council itself.

Currently, BT has committed itself and significant investment to establishing Wireless Cities throughout the UK – where wireless technology creates wireless broadband coverage in the heart of a city. BT is working closely with each local authority to establish a network and applications to meet individual requirements. But this is only the start. Once the initial wireless cities have been rolled out others will follow and we are sure they will expand due to the huge benefits to the community.

This chapter sheds some light on the whys and wherefores of wireless cities. It demonstrates how Wi-Fi is making an impact right now for local authorities, businesses and citizens. It also sets out BT's vision – to keep people and communities better connected whether at home, in the workplace or out and about, with access to all their applications and information on their choice of device and utilising the best network available with a simple and seamless user experience. Great strides are being made in opening up access between local authorities and its citizens. Wi-Fi is at the heart of this enabling technology. This truly is an exciting time and one we are proud to be part of.

Unplugged and Recharged
Contributed by Councillor Paul Bettison, *Local Government Association e-Government Champion*

Local authorities across the UK are constantly looking to confirm their community's position at the forefront of the online digital revolution. As the percentage of workers on the road increases and wireless-enabled devices flood the market, councils need to recognise that a leading-edge technology environment is an expectation of big business and a prerequisite for attracting inward investment.

Simply put, councils want – and need – to emerge as leaders of a modern community on the move, fulfilling their vision of creating an e-enabled business and residential hub.

To answer this need, mobile wireless solutions are beginning to form a key layer in local authorities' business processes. They can help drive urban-generation through delivery of better and more cost-effective council services, as well as attracting businesses and visitors into city centres through the offer of ubiquitous broadband access.

From my point-of-view, the benefits of a wireless network in meeting the multiple agendas which local authorities must fulfil today are not difficult to envision. At least, they include creating an efficient workforce and attracting businesses to the local community and thereby providing the necessary environment for economic regeneration. At most they can improve access for a public who expect everything yesterday – be they business travellers, tourists or residents.

Specifically on the delivery of public services, councils can plan – and, with the likes of Westminster, Lewisham, Milton Keynes and Cardiff, are already seeing – both the cost savings and increased productivity on offer from a wireless-enabled workforce which has remote access to centrally-located business applications. From a harassed Licensing Officer desperately searching for an internet connection to an anxious boss needing urgent information from the Housing Officer out visiting local residents, wireless broadband technology can make it all much easier.

In more general terms, mobile working can help organisations of any size utilise their officers' time more effectively. Employees can stay in touch with the latest corporate information and communicate with the office cost-effectively, wherever they are at a Wi-Fi hotspot. Reducing employees' need to travel to and from the office means dead time is reduced; productivity, satisfaction and flexibility are increased and a better work-life balance and environmental improvements can be achieved. Not to mention the potential to release the value of real estate assets by setting up home practices. In addition, future residents will be able to use their Wi-Fi cameras, Wi-Fi phones, Wi-Fi enabled gaming consoles or Wi-Fi MP3 players wherever they are in a wireless community.

The historical problem is that, for many local authorities, it is too expensive to build their own wireless networks. However, working closely with established Wi-Fi providers such as BT, councils can look to the latest generation of high-speed broadband networks to provide the cohesiveness that a progressive city needs to move forward, making full functionality a reality.

Local authorities must keep pace with robust wireless technologies to optimise mobile working practices, while improving access for the public. Cost-effective public wireless broadband networks are essential to provide front-line workers with full access to back-office business applications, as well as empowering the general public. Suddenly the digital gap narrows and, in the meantime, the speed of Wi-Fi means increased speed of service delivery to its citizens, to whom councils are ultimately accountable.

36.1 Introduction

Just over four years ago, public wireless broadband (or Wi-Fi as it is more affectionately known) was a well-kept secret held by a few individuals in the know. Today, it is a completely different story.

The impact of Wi-Fi is now being felt across a multitude of areas including business, academia and home use but it is perhaps the scope of Wi-Fi for local authorities and its ability to connect entire communities where it becomes most exciting and where people can see a marked difference in the way they live their lives.

Wi-Fi can provide the network to glue together council services, making them more efficient and responsive and providing an additional platform with which to engage local residents. It can provide a more joined-up approach to delivering policy and ensure resources are targeted where they are most needed.

36.1.1 What is in it for local authorities?

With wireless technology becoming more popular and useable across an extensive audience, an obvious growth in communication opportunities emerges. Latest research from analyst firm, In-Stat, says there will be 248 deployments of municipal wireless networks worldwide by the end of 2006 and more than 1,500 by the end of 2010. Whilst it is true, and perhaps unsurprising, that most of this activity is taking place in the US, Britain is ramping up activity. In the US, hundreds of cities across the country are already offering or deploying wireless networks, including Boston, San Francisco and Chicago. Across the UK local authorities in Westminster, Cardiff, Norfolk, Milton Keynes and Sussex to name but a few are already making use of wireless technology to improve services and drive efficiency.

Whilst the next section of this chapter looks in more detail at the specific tangible benefits to local authorities, in general it is possible to place the benefits into three closely related categories – councils, residents and businesses.

36.2 The council worker and the resident
Contributed by *BT Local Government*

Local government nationwide is focused on providing the best amenities and delivering best value services to their citizens, whilst achieving Gershon's recommended efficiency savings. At the same time, councils face the challenge of finding ways of boosting their own staff efficiency and speed of response to situations, as well as decreasing crime, noise and pollution levels on the streets. By finding a solution to current communication problems and sourcing new opportunities to improve the citizen interface, councils can really position their city as a forward-thinker.

BT Wireless Cities can assist; it is a secure, reliable, fully managed wireless network service that has been designed to provide convenient, broadband speed access to back office systems, applications and information from virtually any corner of the community. In doing so, wireless broadband offers local authorities the opportunity to radically transform public service delivery in many ways. Wi-Fi can enable a mobile, multitasked local authority workforce to ensure real productivity gains, as well as reduce infrastructure costs and ongoing operational costs. If council workers are able to link on to the council's network from the street, it enables them to access information and resolve problems immediately, thus enabling local government activities to be more effective and less costly.

Benefits for the council worker

For local government, the benefits of Wi-Fi technology are significant across and within all departments. With a large proportion of the Council's employees working in the 'field' across the city or borough, there is much to be gained from a technology which can be accessed simply and securely whilst out of the office:

- Extension of in-office applications: out-of-office access to their usual IT network will result in a more productive, seamless workforce, as less time is lost in travelling back to the office to update systems;
- Broadening of data fields: field staff could access data from across Council departments, just as call centre staff do, which would enable the council worker to

deal with enquiries on a variety of subjects for the citizen they are sitting with, even if it is outside their immediate role of responsibility;

- Joined-up Government: other government agencies can work much more closely together, sharing knowledge and ensuring consistency;
- Street Event Management: leading to the creation of 'street rangers' with direct access to corporate applications, enabling resource mobilisation and real time event reporting;

Benefits for the residents

As well as the personal benefits which greater Wi-Fi access brings, citizens will benefit in many ways from a wireless city network.

For example:

- Community safety: wireless cameras are faster and easier to redeploy than conventional CCTV cameras, so new trouble spots or special events can be quickly addressed and repair or policing teams sent to the location immediately. Combined with noise monitors, evidence of criminal and anti-social behaviour can be enhanced;
- Parking and Traffic management: using the wireless CCTV cameras to deploy Parking Attendants to streets where significant congestion exists due to improper parking would clear trouble spots quickly. Additionally, parking meters using Wi-Fi are able to send data about their state which results in a faster response of repair. The more meters working, the less frustration and confusion for those wishing to park;
- Housing and social services: with appropriate access rights, individuals from different teams can fill in online forms for the rapid calculation of benefits, arranging a repairman to come to the home, make revisions to the care plan or changes to the home help diary and report any changes to the external environment on estates, all during one visit – making your home environment much more satisfying;
- Policing: Some city authorities have Street Guardians which provide additional Law Enforcement support. Having the ability to compile reports and access Town Hall databases creates a safer environment all round;
- E-tourism projects, where tourist location based content systems can be delivered to offer visitors real-time information during their stay in the City. With the high-speed core network acting as the enabler BT Wireless Cities can be progressively implemented to create a fully integrated mobile communications solution. We are already involved in such projects with Cardiff and Westminster City Councils and the London Borough of Lewisham.

Working alongside BT on a pilot scheme, Westminster City Council is set to transform council services ensuring residents and visitors can make the most of the city, whilst delivering significant cost savings. The pilot is also a key part of Westminster's One City vision, a five year programme to build strong communities, supported by excellent council services.

The installation of these wireless networks has been changing the way Westminster operates, from connecting street-based staff to their office systems and expanding the council's world-renowned CCTV system to run on wireless technology.

In particular, the pilot looked at how wireless CCTV cameras could help in the management of a 24-hour city. In comparison to traditional, fixed CCTV cameras, these new cameras can be relocated to problem areas quickly and cheaply to allow wider and more flexible coverage. Plus, the real-time footage can be viewed by police officers, Street Guardians and other council workers out on the street through their wireless handheld devices, so they can focus in on a crime scene and react immediately.

During the pilot, 38 cameras were installed across Westminster which resulted in over 58 incidents being captured over wireless CCTV. Indeed, the deterrent effect became clear almost immediately after installation with crime and anti-social behaviour on both pilot estates reduced and on one estate it was reduced to its lowest level since April 2004. Residents also reported feeling much safer walking through the area in the dark, knowing the CCTV cameras were helping to protect them.

Equally, the installation of the wireless networks enabled parking attendants being directed to problems on the street by trained operators viewing images from wireless CCTV cameras. As a result, staff were sent to 3556 parking contraventions, 211 obstructions, and moved on 1468 vehicles and the Public Perception surveys provided indirect evidence suggesting that 'smart deployment' had been effective, with residents and businesses perceiving more Parking Attendants on the streets when in fact there were less, and greater availability of parking spaces for loading and unloading.

To this end, the growth of Wi-Fi and the birth of Wireless Cities will ensure multi-layered benefits no matter what a resident's need is. The technology will mean businesses don't miss vital communications when staff are out of the office, encourage residents to feel safer on the streets and create cities which are key players in the 21st century global economy.

36.3 The business traveller and the mobile worker
Contributed by *Intel Corporation*

Research reveals that 88% of business travellers feel it is important for them to be instantly contactable during the day, with three quarters of them admitting that this need has increased over the past two years. As this 24/7 business trend expands, there's a clear need for technology to play a role in supporting the changing face of British business. Correspondingly, with a key task for local authorities being to help create an environment where business can flourish, it's clear that councils' can maximise their attractiveness to business by getting in tune with their wireless needs.

Figure 36.1: Wireless CCTV Deployment.

The research also highlights that, with UK workers wasting up to 689,000 hours a month waiting in airports alone, UK plc is losing out on more than £450 million a year thanks to this 'dead time'. Despite the fact that 53% of business travellers own Wi-Fi enabled laptops, only a quarter of them are currently using this 'downtime' to log onto their company networks whilst at the airport. Factor in other methods of travel be that train, bus or taxi and it's clear that this 'downtime' – and its associated cost – is likely to keep increasing unless business travellers review how they optimise their travelling time.

Most importantly for Wi-Fi providers and employers alike, the research also reveals that more than half (52%) of those questioned would like to be able to take advantage of this inevitable 'dead time' to log on to the internet to check their emails and download large documents. This shows a clear need and demand for out-of-the-office accessibility and wireless technology is a significant step on this path.

And the demand is not just coming from the 'jet-set'. Wi-Fi access is becoming ever more critical for a wide range of workers, for example:

- the travelling salesman who needs to update his sales figures for the weekly report;
- the home-based entrepreneur presenting some credentials to a potential buyer;
- the building site manager needing to download blueprints;
- the market researcher wanting to update their statistics on the corporate network;
- the recruitment consultant wanting to swot up on a company before meeting a candidate;
- the council worker wanting to report an urgent piece of information back to colleagues in the office;
- the management consultant out at a client site wanting to download a presentation;
- the journalist needing to file a story direct from a press conference;
- the medic needing to access and update patient information quickly and regardless of where their duties take them.

As these examples illustrate, we now live in a world where we are all under pressure to be more contactable. As such, whether we're the boss or the employee, the director or the PA, one of the chief drivers in achieving a change is the advancement, installation and uptake of technology which enables us to have the freedom and movement we need. As local authorities are tasked with providing local businesses with the climate they need to grow, develop and become successful, councils need to recognise the role which Wi-Fi can – and should – play in their infrastructure.

Figure 36.2: Mobile Worker Relaxing.

36.4 Wi-Fi changing the profile of gaming

Traditionally, fans of computer games have not had the greatest of reputations for sociability. The old cliché of the young male gamer spending hours in front of their games console communicating through monosyllabic grunts has all too frequently rang true – or at least that's what the stereotype would have you believe. All this is set to change, though, as gaming is brought out of the bedroom and into the outside world.

The emergence of Wi-Fi networks spanning the city and the growing popularity of handheld consoles such as Sony's Playstation Portable (PSP) and Nintendo's DS are part of the reason for this step change in the typical profile of the gamer and a transformation of the gaming environment.

Where wireless networks exist, people are venturing into town for the sole purpose of playing an online game. They can meet for a coffee, have a chat and then log on and play. Before the growing spread of wireless cities, gamers could only play online when they were in close proximity to a wireless access point or had a physical connection to the internet. Now gaming is more flexible.

Wireless cities are the perfect playgrounds for portable gaming. Such networks fit in perfectly with massively multiplayer online role playing games such as World of Warcraft which boasts a community of five million players worldwide. Such games have a strong social element with gamers able to talk to each other whilst playing. This interactivity has often been cited for the growing number of female gamers – who now account for around a third of all UK players.

So it is clear that gaming in the UK is undergoing a steady evolution. Wireless technology is helping to drive that change and, with it, consigning the traditional stereotypes of the gamer into the dustbin.

36.5 The entertainer

The growing popularity of Wi-Fi is not confined to just the business traveller or passing tourist. Wi-Fi providers are now realising that the future of wireless technology lies in making it less niche in the corporate mobile data market and more mainstream. All manner of wireless gadgets exist for our entertainment and ease of use in everyday life. The functionality of wireless today is beyond simply checking e-mail when on the move but extends in a much more creative and consumer-orientated way into our leisure time.

In particular, a new generation of people (children and adults alike) are discovering Wi-Fi enabled games consoles. Nintendo, for example, has taken the lead by creating the 'Nintendo Wi-Fi Connection' – a new era of gaming for fans worldwide.

The 'Nintendo Wi-Fi Connection' allows anyone with a Nintendo D.S. console to play against family and friends across the globe simply, safe and – best of all – for free. Gaming aficionados can visit any one of over 9,000 hotspots at locations across the UK and Ireland to play games. This means that a group of school children sitting in Westminster can play against a similar group of school children in Manchester, or a busy, working dad staying overnight in a London Hilton Hotel can use his Nintendo D.S. console to play against his son logging on at home.

The popularity of the 'Nintendo Wi-Fi Connection' is growing rapidly. In fact, during its first week of sales, 50% of those who purchased Mario Kart D.S. tried the online component. The Nintendo Wi-Fi Connection has now has seen in excess of two million unique players worldwide and more than 70 million individual game sessions in just over nine months of operation – a number expected to soar as more Wi-Fi enabled titles are released. All this is helping to change the solo element of gaming into a much more sociable activity. And it is not just the growing popularity of gaming which is driving entertainment without wires.

Today we can buy wireless digital cameras which can transfer images to other destinations without connection cables. There are even Wi-Fi wine glasses available which let people in long distance relationships feel more in touch with their other half. The idea is that when you and your partner raise the high-tech glasses they will glow warmly no matter how far apart you are so as to promote a shared, and hopefully romantic, drinking experience. A whole new chapter for wireless networks has been started. As Wi-Fi has become more ubiquitous in the business world it has spilt over into our leisure time and is playing a significant role in our entertainment. Wireless games, cameras and wine glass are only the beginning.

36.6 The Wi-Fi tourist – Giving your area a competitive edge

Expectations of UK tourism have fundamentally shifted in the last twenty years. Gone are the days of damp guest houses, bad customer service and inedible food. The need to compete with cheap flights to European destinations means that UK tourism is presented with a challenge to evolve and meet customer needs. Customer expectations around technology and tourism specifically have also changed and are where UK destinations can deliver an edge.

The majority of business travellers will grow used to a certain level of service during their frequent travelling which they will often want to see reflected when travelling for leisure. In short, business travel often sets the tone and expectation for leisure travel. In addition, technology is now firmly rooted in the household. According to government figures, an estimated 13.9 million households (57 per cent) in Great Britain could access the Internet from home between January and April 2006. In fact, it's likely that tourists visiting your area would have booked accommodation and travel online. The fact is that people are increasingly comfortable and reliant on technology during their day-to-day routine – something they want and need to see replicated wherever they are in the world.

So whether a seasoned business traveller or an American family on their first-ever trip to UK cities, the need to have a service that delivers regular access to information is still very much the same. And it's here where local authorities can give their area a competitive tourism boost through Wi-Fi.

People around the world want to be secure in the knowledge that they can write a quick email to their friends, book hotels at the last minute, work out local tourist hotspots they want to visit or send their relatives holiday photos to make them jealous.

This is where international roaming really fits in. As more people begin to appreciate the need for anytime, anywhere access, the demand for a wider range of Wi-Fi enabled locations increases correspondingly. To cater for this increase in tourist usage, Wi-Fi providers are joining forces with other providers offering similar, secure networks around the globe. For example, as a member of the Wireless Broadband Alliance – a collection of the leading telecommunications companies wanting to roam on each other's networks such as T-Mobile (USA and International), Portugal Telecom and Starhub (Singapore) – BT Openzone customers now have access to over 30,000 different hotspots across the globe.

On a more localised level, there is a clear offering available for tourists visiting a particular area – and it's not just the need to access their hotmail accounts to email their family back home. As one example, a Wi-Fi network allows visitors to use technology such as BT Softphone for cheap phone calls home, through Voice over the Internet (VoIP) technology. In addition, the installation of a wireless network can lead to clever and original e-tourism projects to position the city as forward-thinking, enhance the visitor experience and also benefit local businesses and services.

In this way, for other tourist industry stakeholders such as hotels, Wi-Fi can be an additional draw to attract visitors to them rather than competitors. Research has shown that 87% of frequent hotel users regard the provision of Wi-Fi as 'very important' during their stay and over half wouldn't stay in a hotel again if they were dissatisfied with the Wi-Fi service. It stands as tangible evidence that Wi-Fi is becoming so mainstream when an increasing number of tourists and businesses are putting Wi-Fi access on their checklist – along with more traditional needs such as convenient location, catering facilities and good transport links – when looking for a hotel or conference venue.

36.7 Wi-Fi art thou Romeo? – The Bard goes broadband

The historic birthplace of Shakespeare has recently embraced the latest Wi-Fi technology to help tourists find their way around the legendary landmarks. Thanks to the introduction of the Stratford Unplugged project, visitors to Stratford-upon-Avon can now hire a handheld organiser which will automatically point out the nearest literary hotspots as they wander round the town.

BT, Coventry University, Stratford University, Hewlett Packard and the Stratford Town Management Partnership joined forces to launch the Stratford Unplugged project and help raise awareness of the importance and real value of e-tourism. The online virtual tour guide provides tourist information and special offers from local businesses in Stratford-upon-Avon through a Personal Digital Assistant (PDA) for just £8 a day. Through the implementation of the UK's first Wi-Fi project for tourists, visitors will have the chance to have a new 21st century experience of the 16th century genius.

Working alongside several key businesses in the town, Stratford Unplugged will see a selection of hotspots installed across Stratford to give Wi-Fi coverage through BT Openzone across the centre. Visitors to the birthplace of England's most famous playwright can hire their PDA from the local Tourist Information Office or Thistle Hotel, giving them internet access throughout the day and an online guide to the heritage of Stratford, including ideas of which tourist hotspots to visit.

As part of the 12 month pilot, BT is rolling out around 20 Wi-Fi hotspots in a number of local businesses ranging from hotels, shops, restaurants, cafes and local tourist attractions, for example the Royal Shakespeare Theatre, Shakespeare's House (part of the Shakespeare Birthplace Trust) and the Brass-rubbing Centre. It's a win-win situation for businesses involved in this kind of e-tourism. The complete connectivity will enhance the tourist experience and provide an additional means of marketing for local businesses. Depending on where they are in town, tailored content is delivered at broadband speeds straight to the palm of the user's hand.

The benefits are clear. Whether visitors want to experience a taste of Shakespeare or a taste of Stratford hospitality, they can now find the right place to go and the best deals at the touch of a button.

But it's not just about e-tourism. In addition, local residents or business people with a wireless-enabled laptop or PDA can surf the internet or access their corporate networks within the Wi-Fi enabled areas covered by the project. Whether a business traveller needs to check out a website, respond to an email, download content-rich graphics or get at something saved on the corporate network, the Wi-Fi installations in Stratford can help them work any time, anywhere.

36.8 The application challenge
Contributed by *Nomad Wireless*

The past few years has seen a prolific rise in wireless technology. Hotspots in cafes and wireless home offices have demystified the technology and it has become increasingly common to see organisations seeking efficiencies through wireless technology.

The creation of outdoor wireless broadband networks within the UK is also becoming more prevalent. Such pervasive new wireless technologies can assist and empower local government in bridging the digital divide, enhancing safety and security, promoting local and economic development and transforming services to improve the quality of life for our citizens.

Figure 36.3: Wireless Access in Stratford Town.

However, a wireless network needs applications that can run over it to deliver services. This section provides an overview of the potential different types of applications which can be used with a wireless network.

36.8.1 Enterprise mobility for efficiency and effectiveness

A wide range of wireless applications are now possible. At the core are enterprise applications such as Voice over Internet Protocol, email, messaging, personal information management and internet access. Making such applications 'mobile' and delivering real-time information to the edge of the enterprise – to front line workers – makes individuals more efficient and effective. Decision making ability, enterprise communication and collaboration are improved.

The benefits of wireless mobility are even greater for the front line workers who spend most of their time outside the office. Transforming services through outdoor wireless networking for social care workers, environmental services officers, housing repair, building or planning inspectors is a clear target. These and other professionals spend many hours commuting to and from the office, and in transcribing handwritten notes to document their activities and update databases. They lose considerable time revisiting clients and issues because they lacked key information during on-site visits. These kinds of inefficiencies erode budgets and can take a heavy toll on service quality.

36.8.2 Creating an agile infrastructure

With the current focus on terrorism and crime, the public safety and security of residents is paramount for any local authority. Community safety teams responsible for the delivery of public safety need collaborative processes and underlying technologies to enable the sharing of information in real-time or near real-time. Hence, the rise of wireless CCTV in order to combat crime and reduce anti social behaviour by deploying mobile cameras rapidly in high crime areas is a paramount.

Wireless networks can also be used to support real-time monitoring of traffic. Advanced transportation information systems can also provide users with a single point of access to travel and information, help plan traffic control strategies and further develop intelligent parking applications.

36.8.3 Increasing the opportunities for all

Broadband and information communication technology is increasingly critical and pervasive in our working lives and our homes. The interaction between citizens and local government in the creation of better connected communities is vital. New access channels such as wireless information kiosks and Wi-Fi hotzones in libraries, community centres or extended schools create strategic areas within the community which allow residents to obtain information about local services. They allow them to use services such as paying bills online, providing direct interfaces to council services, reporting of street scene issues, access to local and international news, local tourist information, street maps and jobs.

In a digital world where the internet is extensively used within schools and society, it is imperative that local authorities do not disadvantage those who are unable to afford high speed internet services. Inclusion will only be fully achievable if local authorities overcome geography, culture and socio-economic obstacles so as to reach communities and individuals who may otherwise remain disenfranchised.

36.8.4 Sharing services

The efficiency drive in central departments and local services, aims to improve the performance of the public sector as a whole and provide better value for money. Whilst transformation of public services is about designing services around citizens' needs and choices, efficiency is about better use of resources. In many cases a major obstacle to this is the unavailability of the infrastructure required to enable access to resources (such as data centres) regardless of location. The potential development of one common infrastructure using wireless technologies to share services and the exchange data between multi-agency stakeholders is central to the success of partnership working.

It is clear that once the right infrastructure has been deployed, local authorities can deploy wireless applications with their individual business cases taking advantage of the enabling connectivity.

36.8.5 Examples of applications

- Mobile secure access to the Children's (Information Sharing) Index in compliance with Information Sharing Regulations 2007. Applications that will have access to the Index by schools' staff, childcare practitioners and the police.
- Mobile Benefits assessment (in co-operation with DWP). This is an example where joint working between central and local services may lead to both greater effectiveness

and efficiencies if supported by mobile technology. Benefits assessors and inspectors often work in areas of poor broadband coverage, and require secure access.

- Fraud prevention (mobile teams). Fraud prevention work is often dependent on the mobility and speed of access to the right information in the field.
- Business access to non domestic tax records, planning information, expert advice and accounts as well as other business support services would foster economic development in rural and sub-urban areas at present left out of the broadband revolution.
- Secure mobile access to shared infrastructure such as "Government Connect" resources and registration facilities would enhance the prospect for citizens and services take-up of these facilities, thus leading to potentially large cost savings.
- Mobile and secure access by 'Youth Offending Teams' to common data held across multiple agencies such as National Offender Management Service (NOMS), Youth Justice Board (YJB), Crown Prosecution Service (CPS), Courts and the Police.
- Support for schools in the delivery of wireless classroom access.

36.9 Frequently asked questions answered
Contributed by *Motorola*

36.9.1 Is Wi-Fi just for larger cities?

Wi-Fi is a sufficiently flexible technology that it can be provisioned by local authorities to flood city centres, business parks and small towns or villages with broadband access. However, to be cost effective, Wi-Fi in rural areas or remote regions is best implemented as part of a larger programme of broadband access provision.

36.9.2 When is the right time for me to roll out a wireless network?

Today, cities can quickly reap the benefits from a metro Wi-Fi network. Wi-Fi operates in unlicensed spectrum so there is no issue over license fees and there are millions of Wi-Fi enabled laptops and mobile devices ready for people to connect to services. Many local authorities have recognised that metro networks are critically important to present their region as a forward thinking, modern locality. On an organisational level, integrating Wi-Fi into IT systems enables local authorities to promote e-Government, operate more efficiently and improve public services.

Demand from the community is also strong. Consumers and businesses are increasingly au fait with the language and services of the digital economy and are increasingly using Wi-Fi hotspots to access those services. There is growing expectation that broadband access will be widely available whether they are connecting to services in their locality or travelling to cities for business or leisure. The availability of broadband is an important factor for many businesses assessing whether to open premises in a new area. In addition some innovative regions are using free or very low cost Wi-Fi to encourage companies to move to areas where urban renewal projects are underway.

Many local authorities are now investing in metro Wi-Fi technology because it provides a high quality yet affordable means to provide universal broadband access: it operates in the unlicensed spectrum, the network technology is cost effective to install and millions of laptops and mobile devices have been purchased that enable people to connect to services.

36.9.3 What's the difference between an 'owned' and 'operator' network?

An 'owned' network is commissioned by a local authority and once built the local authority oversees and operates the system. An operator network is constructed and run by a service provider such as BT for example.

Operator-run networks are become increasingly popular. There are several reasons for this. Key among these is that the operator has the depth of expertise and knowledge of the technology complemented by the economies of scale to quickly install systems and run them efficiently. Also, many local authorities agree revenue structures so that the operator funds the build and maintenance costs. The advantage of this approach is that the local community enjoys the advantage of broadband access without having to fund the deployment and maintenance of the network through its purse.

Many authorities are also electing to develop innovative partnership schemes where the network is run by a management company that's jointly owned by public and private bodies. Revenues are created through marketing access to business users and residents who may elect to buy premium services.

36.9.4 Paid-for versus free: what's the best solution?

The network profile will depend on exactly what each local authority wants to achieve. The objectives can be refined in conjunction with the technology supplier who will also advise on the funding options available and the optimum strategies for marketing the services. Many authorities expect to offer free services to their citizens and this requirement is often central to the tender.

36.9.5 What should local authorities advise citizens about wireless security?

Wi-Fi public access networks support the latest robust security protocols; so data sent and received by mobile devices is securely encrypted to prevent others from accessing it.

Due to widespread publicity about broadband security and virus attacks, complemented by many free security software offerings provided by Internet service providers (ISPs) such as BT, most consumers and virtually all businesses take steps to shield their computing assets. This said, local authorities may wish to remind citizens through regional publications and their web site of the need for online vigilance.

36.9.6 Are there any public health issues with wireless networks?

Local authorities are at liberty to reassure their citizens about the safety of Wi-Fi; the systems operate at a very low power of less than one watt per access point. Indeed, all Wi-

Fi access points installed in the UK will produce radio frequency fields well below the international guidelines, as verified by the Radio Communications Agency.

36.9.7 How can local authorities avoid restricting competition?

Local government can ensure there is a competitive open tender process for the franchise to operate a metro network within their locality. The tender for the network needs to be as open as possible, providing clear guidelines as to which submissions will be viewed the most favourably; judgement criteria may include for example, a commitment to free coverage in some areas, supplying service and support to the authority and ensuring that schools and universities have unrestricted access.

It's also possible to ensure, as part of the deal that they make wholesale data services available. This enables other communications companies to buy this space and market their own services, such as local ISPs. This facility encourages other providers to compete for business to drive more revenue from the infrastructure and lower the cost of premium service provision.

## 37

The Path to 4G and the Mobilization of the Internet

Philip Marshall[a]

With 3G having seen lack luster performance in mobilizing the Internet, the media and communications industries are migrating to more advanced technologies with the intention of improving the underlying performance and economics for delivering broadband services. Advanced technologies such as 3G-LTE, UMB, mobile WiMAX, are being touted as the migratory technologies towards 4G, with the support of technologies like 802.11n, and WiFi-mesh. As these technologies come to market it remains to be seen how well they perform relative to service demands. The open mobile Internet environment created with 4G will challenge the traditional service provider business models and dramatically lower the barriers for new entrants to penetrate the market with a variety of innovative solutions. It is competition derived from these innovative solutions that will drive the 4G market.

37.1 Introduction

As the media and communications industries grapple with 3G, and the mobilization of the Internet, 4G is emerging on the horizon with the promise of enabling the economic delivery of high bandwidth services to a plethora of cellphone, consumer electronic and computing devices. Traditionally 4G has been considered the realm of the mobile service providers, with the notion that they would migrate their legacy 3G networks to 4G once the technology was available and market demand evident. However advanced technologies like WiFi-mesh, 802.11n and WiMAX provide opportunities for fixed and broadband service providers and new entrants to offer portable and mobile Internet solutions that challenge the traditional position of mobile operators. The evolution towards 4G is illustrated in Figure 37.1, where the sweet spot for 4G is assumed to be an economic high bandwidth broadband solution that has ubiquitous connectivity for the purposes of mobility. Mobility is provided today with relatively low bandwidth services, and broadband services are economically delivered today in fixed and portable environments. Service providers with fixed solutions are implementing complementary WiFi, WiFi-mesh and WiMAX solutions to increase the portability of their broadband offerings. Mobile service providers are enhancing their 3G technology with solutions such as HSPA, CDMA2000 1xEV DO, and are looking towards technology standards such as 3G-LTE, UMB and mobile WiMAX.

---

[a] *Yankee Group*

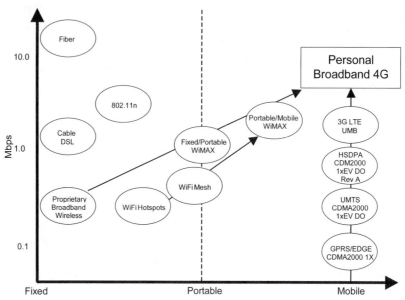

Figure 37.1: Technology migration towards the 4G vision comes from both broadband and mobile origins.

37.2 Mobilizing traditional and emerging media, communications and commerce business models

The mobile Internet has seen several false starts over the past decade, with early WAP solutions failing dismally and the "walled garden" approach to 3G proving untenable. The mobilization of the Internet with 4G technology aims to address these early market deficiencies by offering an open mobile Internet environment with broadband network capabilities. A variety of new and emerging Internet media and communications players are mobilizing their services and applications. Figure 37.2 illustrates these services and applications in terms of the following categories:

- **Carrier and communications centric:** Carrier centric includes the traditional facilities based service providers, and communications centric incorporates new entrant players offering VoIP services. It is anticipated that many of these players will gain sustainable intrinsic value from offering efficient media and service distribution capabilities through their networks.
- **Traditional and new media centric:** This includes companies offering traditional print, video and entertainment media, and new entrants offering digital media and user generated content.
- **Enterprise centric:** This includes the various players offering solutions to enterprises. The characteristics of these solutions vary between industry verticals, but ultimately drive the consumerization of enterprises, where employee and customer interactions are virtualized.

- **Commerce centric**: This category includes players like Amazon and eBay who offer a variety of eCommerce solutions. Key intrinsic value is derived from the scalable transaction capabilities that consumer centric players have, and their capabilities of supporting transactions spanning between the Internet and brick and mortar.
- **Community centric solutions:** These solutions are specifically designed to foster social networking. In the long term, social networks will act as change agents to traditional Internet and be applied to other business models, rather than creating a distinct category.
- **Search centric:** This incorporates players who provide Internet search and service discovery capabilities. Companies like Google and Yahoo! have capitalized on advertising supported Internet search capabilities. These capabilities will rapidly evolve to incorporate intelligent personalization and service discovery capabilities, and become a critical capability of the mobile Internet.
- **Connectivity centric:** This category includes a new breed of players that leverage existing residential WiFi access points, hotspots and possibly the resources of communication service providers and municipalities. Connectivity centric players are looking to build community networks that are capable of substituting traditional facilities based offerings.

Each of the solution categories shown in Figure 37.2 will capitalize on a mobile Internet by offering services with increased user centricity, and immediacy. As a consequence of the variety of existing and emerging business models, all players are vulnerable to disruption as the Internet is mobilized. For example, communication centric players like Skype threaten traditional carrier centric business models by offering low cost VoIP. At the same time the critical importance of interoperable communication capabilities in the other business model categories will see the commoditization of proprietary VoIP solutions, such as those offered by Skype.

37.3 Role of licensed and unlicensed technologies for the mobilization of the Internet

The mobilization of the Internet depends on adequate market competition to stimulate disruption of traditional business models. The various players illustrated in Figure 37.2 are poised to capitalize on mobile Internet, but are challenged by the significant control that mobile service providers have traditionally exercised over the market. This is illustrated fragmentation in mobile device operating system software and the control that service providers traditionally maintain over the delivery of Internet services to their devices. Figure 37.3 illustrates a Yankee Group 2006 device forecast, and shows that the lion's share of devices continue to be in the feature phone category, suggesting that they are customized to the specific service provider. The Yankee Group estimated that in 2006 there were nearly 4000 different proprietary mobile device profiles. This creates fragmentation in the device software and makes it challenging for players looking to achieve economies of scale with standardized mobile Internet solutions.

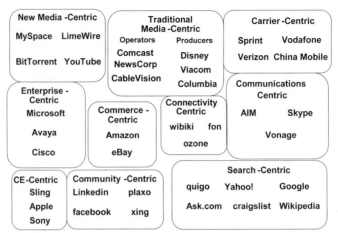

Figure 37.2: A collision of business models at the intersection of digital media, communications and the Internet.

With the mobile communications market approaching 3 billion subscribers, it represents an enormous market to attract players with disruptive business model innovations. For example, the aim for WiFi and WiMAX is to have standardized communications modules that can easily be integrated in a variety of cellular, consumer electronic and computing devices. If successful, this will dilute the control that mobile service providers traditionally have over the device.

The delivery value chain for mobile services is fragmented, largely as a consequence of the proprietary approach that traditional mobile service providers have used. This is illustrated by the large proportion of feature phones that are customized to specific service provider needs, (see Figure 37.3). It is also illustrated by the "walled garden" approach that service providers have traditionally adopted to manage and control service delivery. This fragmentation in service delivery has hindered the proliferation of the mobile Internet, and will slow the rate of migration to 4G. The lackluster performance of 3G coupled with diluted voice revenue opportunities is driving service providers towards strategies to create an open environment for service delivery. This is critical for the success of the mobile Internet.

Unlicensed technology like WiFi benefits from having less complicated value chain relative to technologies like 3G and 3G-LTE, which lowers the barriers for technology and market innovations. This is illustrated in Figure 37.4, where the positioning of licensed and unlicensed technologies are compared in the context a generic communications value chain components that include technology, regulatory, commercial and end user.

The regulatory environments for licensed and unlicensed technologies are significantly different. In the case of licensed mobile technologies, the regulatory regime is complex, and spectrum licenses are expensive and controlled by a small number of players. This contrasts unlicensed systems where the regulations do not control the user base. The regulations for unlicensed systems focus on parameters for the power spectral density envelope of transmitted radio signals to allow a large number of users to simultaneously use self installed systems. This effectively lowers the barriers for players to offer solutions in the unlicensed spectrum bands.

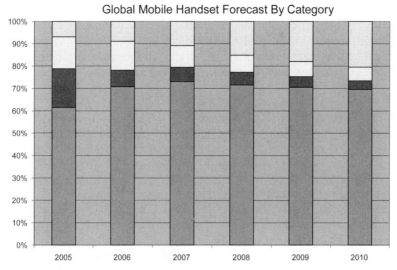

Global Mobile Handset Forecast By Category

■ Feature Phones ■ Voice Phones □ Browser Phones □ Advanced OS Phones

Voice Phone – Two way voice and SMS
Browser Phone – Adds Microbrowser to voice phone
Feature Phone - Adds features that further enhance wireless VASs (e.g., Java/color screen/MMS/camera/MP3 player).
Contains onboard storage of at least 200-300 KB. May contain dedicated applications processor. RTOS-based.
Advanced OS Phone - Adds advanced mobile OS (e.g., Palm/Symbian/Linux/Smartphone 2002), higher processing (at least
33 MHz) with dedicated applications processor and embedded storage of at least 8 MB.

Figure 37.3: The dominance of feature phones dramatically increases the fragmentation of software for delivering mobile Internet services.

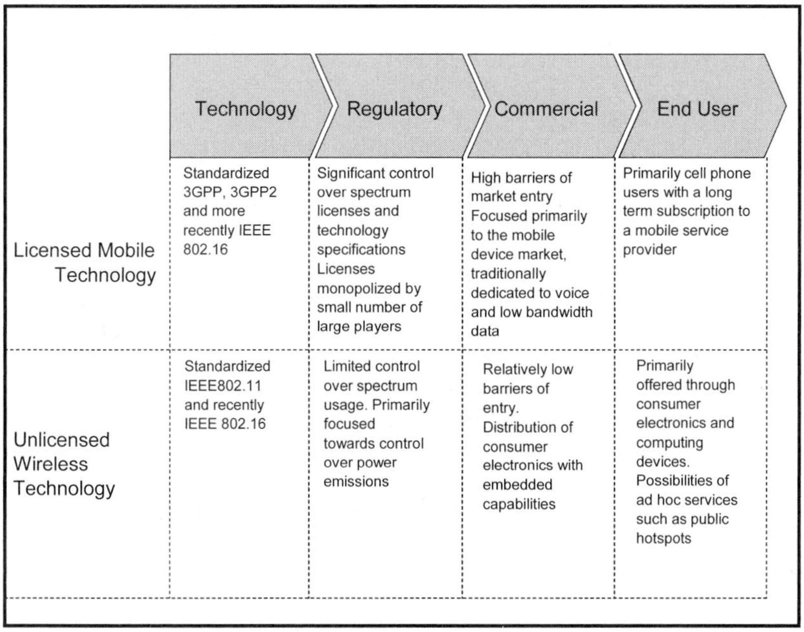

	Technology	Regulatory	Commercial	End User
Licensed Mobile Technology	Standardized 3GPP, 3GPP2 and more recently IEEE 802.16	Significant control over spectrum licenses and technology specifications Licenses monopolized by small number of large players	High barriers of market entry Focused primarily to the mobile device market, traditionally dedicated to voice and low bandwidth data	Primarily cell phone users with a long term subscription to a mobile service provider
Unlicensed Wireless Technology	Standardized IEEE802.11 and recently IEEE 802.16	Limited control over spectrum usage. Primarily focused towards control over power emissions	Relatively low barriers of entry. Distribution of consumer electronics with embedded capabilities	Primarily offered through consumer electronics and computing devices. Possibilities of ad hoc services such as public hotspots

Figure 37.4: A comparison of the licensed and unlicensed technology value chain.

Traditionally the commercial positioning of licensed mobile solutions like cellular and 3G has been vastly different to that of unlicensed WiFi systems. In the case of licensed mobile systems, service providers have capitalized on the high barriers to market entry and exercised a high degree of control over service distribution and the user experience. Licensed service providers come from a legacy of offering highly reliable voice and low bandwidth data services, and have developed their 3G service offerings, with limited success, using a similar approach.

In the case of technology domain shown in Figure 37.4, both licensed and unlicensed technologies are highly standardized and rely upon this standardization to drive market scale. However for unlicensed technologies, the low barriers to market entry generally necessitates that the rate of innovation is more rapid when compared with licensed technology. By way of comparison, the performance of unlicensed technologies double every two to three years, versus licensed technologies that see a doubling in performance every three to four years.

With unlicensed systems having lower barriers to market entry, residential WiFi offerings are essentially self installed consumer electronics offerings, and distributed through value added resellers and systems integrators in enterprise markets. This market environment has seen the rapid consolidation of the WiFi equipment vendors, rapid price erosion, and accelerated technology commoditization. Some players have capitalized on opportunities to offer WiFi into parallel markets, such as public hotspots. By capturing lock-in presence in key venues such as airports, hotels, casinos and coffee shops, these players have managed to monetize WiFi services.

37.3.1 WiFi-mesh and ad hoc networking

The notion of offering public WiFi services has been extended into outdoor installations, where players including local municipalities are looking to provide metro-zone and in many cases metro-wide wireless broadband services. Since WiFi uses low power transmitters in unlicensed radio spectrum, typical outdoor WiFi installations require between 40 and 100 access points per square mile, depending on coverage and interference conditions. With so many access points being deployed, a low cost technique is needed to provide connectivity to the Internet. Conventional techniques which have separate and distinct transmission links for each access point are prohibitively expensive.

A technique that is being widely adopted for outdoor WiFi solutions leverages WiFi-mesh technology. In a WiFi mesh architecture, each WiFi access point has the capabilities of supporting both access and transmission capabilities. The access capabilities provide connectivity to the end users, as is the case with conventional WiFi access points, and the transmission capabilities allow each access point to interconnect with other access points within close proximity. The result is that the access point network can provide a meshed transmission solution, which reduces the cost associated with implementing a large number of access points in metro-zone and metro-wide installments, see Figure 37.5. Currently mesh installations cost between $100 and $150 thousand US dollars per square mile to install, and therefore are well suited to high traffic areas. To economically deliver ubiquitous coverage, these systems require overlaid wireless technology such as 3G, LTE or WiMAX.

Current mesh network implementations are relatively static and tend to see the mesh circuits provisioned manually, or maintained with a high degree of control and manual intervention. While this approach enables WiFi installations to scale significantly relative to traditional installations, it typically results in the over provisioning of the network. Considerable performance improvements would be achieved if the mesh could adapt to changes in network usage patterns and interference conditions. Furthermore, the user devices themselves can potentially form part of the mesh network, particularly in areas where there is a low density of access points or a high system interference levels. Future architectures introduce ad hoc networking, where the mesh circuits use both access points and devices for transmission, and can adapt in an ad hoc fashion to the changes coverage, interference management and capacity requirements. Currently ad hoc networking solutions present a variety of technical challenges including the following:

- Identifying the need for access points versus areas where there is sufficient user device density for the ad hoc network to be self forming.
- Power management challenges and potential security challenges that emerge when user devices form part of the network
- Complex routing algorithm requirements, and
- MAC-Layer retooling.

As broadband wireless and mobile solutions become increasingly pervasive, service providers and technology vendors will continue to pursue a variety of alternative transmission solutions to economically deliver high bandwidth services. For example, future mobile base stations will support Ethernet based connectivity (as opposed to traditional transmission architectures based on incremental T1 or E1 connections) and flatter network architectures to allow for lower cost broadband connectivity. In addition, multi-service operators (MSO) are likely to integrate wireless capabilities in residential gateway and set-top box solutions, and integrate wireless access points to their cable plant using strand mounted access points. These strand mounted access points tap into the cable transmission networks using standard cable modem connectivity to offer low cost backhaul with modest capacity.

37.3.2 WiFi is an anchor point for Fixed Mobile Convergence

Fixed mobile converge is capturing market attention, with the promise of integrating wide area wireless mobile and local area WiFi networks for delivering VoIP and data services. These solutions are of particular interest to fixed service providers and cable companies who are eager to integrate mobile services in their bundled offerings.

The technical and commercial characteristics of FMC solutions depend on whether or not seamless mobility is required, and vary between enterprises, and consumer solutions offered by fixed and mobile service providers. For enterprises, converged solutions are normally aimed at augmenting their overall VoIP strategies. Mobile service providers are interested in converged solutions for reasons including improved network coverage, subscriber satisfaction, and the ability to offer advanced voice and other broadband services over WiFi. For many fixed and broadband service providers, mobile integration is key to their convergence strategies, with the WiFi being supported by the fixed infrastructure, and mobility enabled through wholesale agreements with mobile operators.

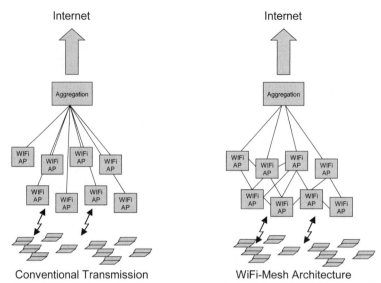

Figure 37.5: A comparison of conventional transmission with a WiFi-Mesh architecture.

For service providers offering integrated WLAN mobile solutions, there are a variety of integration points between the two environments. In principle, integration can occur at the customer facing, control and service centric, and network facing elements, illustrated in Figure 37.6.

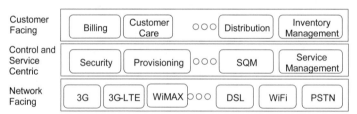

Figure 37.6: Integration points for converged WLAN/mobile service offerings.

The customer facing integration requirements depend largely on the positioning and associated synergies between existing and planned service offerings. For service providers to be effective in delivering converged 4G services, we believe that the customer facing solutions must be designed with the aim to integrate the plethora of Internet services and applications, which when mobilized, offers significant value to subscribers. The requires that standardized API interfaces be used to integrate with third party applications, and simplified business processes be established to enable the solutions to proliferate in the absence of complex implementation hurdles.

Service and control integration is necessary to ensure security and service continuity requirements are achieved. For example, a mobile operator with WiFi service might require authentication through the Home Location Register (HLR), or Home Subscriber Server (HSS) that it has in its mobile network, it may also require the capabilities for services, such as SMS to be delivered over both networks in a seamless manner.

If seamless mobility is required between the WLAN and mobile networks, network facing integration is necessary to facilitate inter-network handovers. While there are a variety of techniques used for seamless handovers including UMA and a variety of peer-to-peer softswitch techniques, integration for the purposes of seamless mobility in the 4G world will most likely be implemented with internet multimedia subsystem (IMS) infrastructure.

37.3.3 Finding an economic sweet spot for WiFi

By being an unlicensed low power wireless technology, WiFi initially found its sweet spot in extending broadband and Ethernet from a fixed into a local area wireless environment. By being unlicensed, it does not require the direct support of a licensed spectrum holder and can therefore be easily self installed by consumers and enterprises. The technology is highly standardized and has been promoted heavily by companies like Intel, with its Centrino campaign. Public WiFi has seen the expansion of WiFi into public areas and with the support of mesh type technologies allowed WiFi coverage to be extended to several square miles.

The economics for installations over large areas extending beyond several square miles is challenging. When ubiquitous coverage is required, WiFi solutions typically require between 40 and 100 access points per square mile at an average cost of between $100 and $150 thousand US dollars. This cost is manageable for WiFi solutions that capture sufficient user traffic over the area, but become prohibitively expensive as the coverage area extends beyond several square miles and in cases where full mobility solutions are required.

Licensed 3G, 3G-LTE and mobile WiMAX are capable of offering ubiquitous coverage within many square miles from a single cell site and therefore offer vastly superior economics in supporting broadband services over wide area environments. An analogy might be to consider the requirements for lighting a football stadium for night games. The equivalent of a macro-cell site would be the large elevated spot lights that are used today to cover the ground area. In the stadium itself it is impractical to use the same spotlights and a large number of low powered bulbs are used instead. These low power bulbs are equivalent to the WiFi access points in the analogy and are well suited to providing coverage in local areas. The combination of spotlights and low power bulbs ensures that the football stadium and grounds can be lit in an economically viable manner.

Economically efficient WiFi installations can either be targeted towards metro areas and indoor environments where there are high local traffic demands. In cases where users demand wide area service continuity, it is critical for WiFi to be integrated with a licensed broadband wireless technology like 3G, 3G-LTE and mobile WiMAX.

It is economically efficient to use WiFi in a similar manner, where it is well suited to support local areas that have high traffic requirements (a square mile or less), and integrate the WiFi service with an overlaid broadband wireless network such as 3G or WiMAX.

37.3.4 Municipal WiFi networks - more than a real estate play!

Municipalities worldwide have recognized the potential value of a WiFi-mesh technology in conjunction with the extensive rights-of-way (ROW), light pole and other structures that they control. They aim to offer a variety of services, including consumer access, public safety and commercial broadband over their networks. Municipalities have been partnering with WiFi-mesh vendors, systems integrators and Internet based service providers (like Google and Earthlink) to roll out their networks. While these networks do not scale to economically offer wide area broadband service, they are well suited to provide local and metro-area coverage. As wireless broadband networks become increasingly congested, the rights of way and structures owned by the municipalities will become increasingly valuable. Many of the sites enabled with WiFi will be integrated with macro-cellular sites using licensed technology like 3G or WiMAX.

While municipal networks aim to offer an alternative infrastructure for delivering broadband services, they are unlikely to gain massive scale without the direct involvement of telecom service providers. However they play an important role in the market in incubating new technologies like WiFi-mesh and strand mount solutions, and in grooming available structures and rights-of-way to support wireless broadband services operating in licensed radio spectrum.

37.4 Regulatory challenges for convergence and 4G

Traditional telecommunications regulations have been developed under the premise of distinct licensing regimes for fixed, mobile and broadband communications. Complex regimes have been established for specific aspects of network and service delivery, including radio spectrum ownership, interconnection tariffs, service rights and obligations, network roaming etc. The broadband mobilization of the Internet (4G), convergence of communications networks (fixed, mobile and broadband), and between media, the Internet and traditional telecommunications render traditional telecommunication regimes redundant, and call for new regulatory frameworks to anticipate the types of service offerings that are emerging in the marketplace.

Since there is a great deal of uncertainly surrounding the various services that are likely emerge with the convergence, it is challenging to establish regulatory regimes that enable fair and equitable use of scarce resources for the benefit the consumer. Furthermore, changes in regulatory regimes are hotly debated since traditional players have invested heavily in scarce resources such as radio spectrum, and in capital such communication network infrastructure. For example:

- Some regulators are proposing a network neutrality regime, which would essentially restrict network operators from prioritizing network traffic for commercial gain. Network neutrality is hotly debated, and opposed by network operators who fear that they will be disrupted by third party service providers, who would effectively have equivalent rights over the available networks.
- Radio spectrum allocation policies are also hotly debated, with incumbent service providers looking for regulators to place restrictions on the variety of services that can be offered in particular spectrum bands, while new entrants are seeking technologically neutral approaches.
- Perpetual battles rage between licensed spectrum holders and proponents of unlicensed bands, and both parties lobby for new spectrum allocations.

With significant uncertainties in the market, coupled with the complexities of introducing new policies, telecommunication regulation will continue to lag the market for the foreseeable future. This will create some uncertainty in the scalability and longevity of emerging disruptive business models, particularly those that are predicated on convergence between network technologies and traditional service categories.

37.5 The Emergence of Mobile WiMAX

Mobile WiMAX is promoted by telecom industry players, and equipment and silicon technology providers like Motorola, Nortel, Samsung and Intel as the next generation of wireless technology to come to market. It comes under a guise of a variety of names including WiMAX, IEEE 802.16e and WiBRO, (which is an early version of the mobile WiMAX technology).

As mobile WiMAX technology continues to be standardized, it is capturing significant market attention, particularly from those players who are interested in competing with 2.5G/3G technologies, or in offering residential and commercial broadband services in emerging markets with low tele-densities. In most markets, the success of WiMAX depends on its positioning with other technologies, most notably 2.5G/3G and its next generation migratory generics such as 3G, 3G-LTE and UMB. To capture initial success as a personal broadband solution, service providers must either position mobile WiMAX as a complement to 3G, or develop parallel markets that leverage on its anticipated positioning alongside WiFi in devices and other consumer electronic devices. In cases where service providers look for WiMAX to complement 3G, WiMAX devices will require 3G multi-modality and ultimately handover capabilities. In addition, WiMAX must interoperate with the 3G ecosystem, including IMS (Internet Multimedia Subsystem) and the proposed service delivery platform (SDP) framework, and where appropriate aim to leapfrog the legacy service delivery infrastructure, which 3G is evolving.

The long term success of WiMAX depends whether it is successful in creating a more definitive value proposition than merely being a 3G complement. It could provide a means for mitigating the traditional telecom technology lock-in controlled by a small number of industry players and enable non-traditional players such as Google and Newscorp to participate more actively in the mobile market. In addition, future positioning of WiMAX may see it being optimized for specific applications such as peer-to-group (P2G) and peer-to-peer (P2P) multimedia, mobile video, and hybrid P2P and application-to-peer (A2P) configurations.

Figure 37.6 illustrates the short-to-medium term positioning of mobile WiMAX relative to 2.5G/3G in enabling new market entrants and disruptive business models, improved network performance, and its technology supply chain, ecosystem and service continuity, and radio spectrum profiles.

Mobile WiMAX has significant potential to disrupt traditional telecom business models by enabling new market entrants not encumbered by traditional mass-market wireless mobile business models. Its business models are likely to be more closely aligned with WiFi than traditional 2.5G/3G mobile.

Network mobility and coverage for WiMAX will be inferior to that of 3G in the short to medium term since 3G is inherently backwardly compatible with 2.5G and therefore can benefit from the coverage offered by the legacy network. Compatibility and interoperability with 3G and possibly 2.5G is important for service continuity to be maintained for WiMAX in mobile operating environments.

Radio spectrum availability and consistency is a major challenge for WiMAX, with a variety of licensed frequency bands including 2.3, 2.5, 3.3 and 3.5GHz bands being touted. Although the bandwidth of the radio channels in these bands exceed that of 3G, in many markets the channels were originally allocated for fixed wireless operations and the notion of allowing mobile services is hotly debated. The current radio spectrum regime is a significant inhibitor to the market success of WiMAX. Service providers that have WiMAX spectrum are effectively the gatekeepers for WiMAX adoption in their respective markets, until new licenses are issued.

The market positioning of WiMAX goes beyond a technology discussion. Fixed and broadband service providers, and others which might include players like Google, see it as an opportunity to offer mobile solutions to stem churn, in the case of the fixed and broadband operators, or create new service distribution opportunities in the case of players like Google.

Vendors such as Intel are feverishly promoting WiMAX, and if successful will benefit from locking themselves in the personal broadband market. Adversaries, including vendors such as Qualcomm, and various mobile service providers are challenging the performance claims of WiMAX, and suggesting that the migration of 3G will be more fruitful. Mobile service providers that paid dearly for their 3G spectrum are challenging the suggestion of regulatory easement fro WiMAX to compete with 3G. Major equipment vendors such as Alcatel/Lucent, Ericsson, Huawei, and Nokia/Siemens are positioning themselves to support WiMAX, 3G and evolved 3G solutions such as LTE. Motorola and Nortel are betting heavily on WiMAX, largely because of their poor market positioning in 3G.

While WiMAX holds a great deal of promise as a next generation mobile technology, there are many hurdles that it must overcome before fulfilling this role. Furthermore, because the traditional telecom industry has a complex value chain with significant interdependencies, accelerated success of WiMAX will depend on it being focused on the aspects of the value chain that are most vulnerable to disruption. Presenting WiMAX solely as a superior technology is not enough. The industry must develop the entire technical and commercial ecosystem for WiMAX with the aim to leapfrog other incumbent telecom solutions.

37.5.1 Integration of WiMAX and WiFi

Companies like Intel are developing base band silicon products to integrate WiFi and WiMAX. They believe that at least 80 percent of gates used for WiFi baseband silicon can be reused for WiMAX, but more importantly that the existing base of WiFi usage and device availability significantly bolsters the supply chain for WiMAX device technology.

For WiMAX to be implemented in a WiFi enabled device, the base band technology can be shared, and separate radio frequency is used for the respective technologies, WiFi operates in the 2.4 and 5.8 GHz unlicensed band, and WiMAX being profiled to operate in a variety of bands including 2.3, 2.5, 3.5 licensed and 5.8 GHz unlicensed frequency bands. It is expected that other spectrum bands will be used for WiMAX in the future.

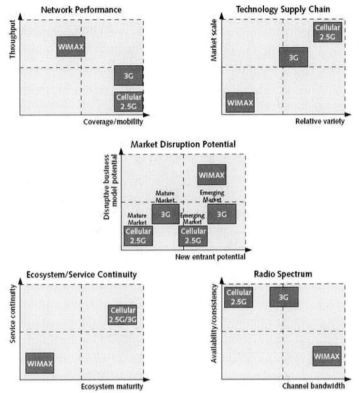

Figure 37.7: Short-to-Medium Term Positioning of WiMAX Relative to 2.5G/3G.

The integration of WiMAX with WiFi has the potential to disrupt traditional service provider business models. This is particularly the case when WiMAX is embedded in small form factor consumer electronic devices such as cameras and music and video devices. The devices can be distributed through independent channels, and service provides can offer a variety of subscriptions, including ad hoc approaches similar to that offered in the WiFi hotspot market. This is unlike the residential broadband and mobile subscription models currently offered by service providers today, where services is delivered through dedicated devices, and devices are often subsidized through fixed period subscriptions. As service providers develop new business models to capitalize in the integration of WiFi and WiMAX, they are faced with a variety of challenges. In particular:

- New service distribution approaches have the potential of cannibalizing existing subscription based business models;
- By relinquishing the control over device distribution, service providers face increased challenges in ensuring service quality is maintained, and;
- Independent device and service distribution models, coupled with the potential for disintermediation by Internet based service providers, disrupts traditional marketing, product positioning and brand strategies.

However, given that service providers are under tremendous pressure to change from their traditional subscription heavy business models, they will be ultimately driven towards open mobile Internet solutions enabled with technologies like mobile WiMAX.

37.6 Emerging Radio Technology Enhancements

With the performance of digital technologies being firmly rooted to Moore's Law, which implies a doubling of capabilities every 24 months, radio technologies tend to languish, with performance improvements tending to improve at a slower rate. There is a variety of radio technologies that offer the potential to enable considerable performance improvements include smart antenna technologies and cognitive radio architectures.

Currently broadband wireless systems are limited by their backhaul and transmission capabilities. Technologies such as WiFi-mesh and ad hoc networks, and other solutions that leverage DSL, cable strand-line and fiber plant are being applied to improve the economics for WiFi, WiMAX and 4G networks. As the economics for connectivity to base stations, access points and wirelessly enabled residential gateways are improved, smart antennas and cognitive radio solutions become more important.

37.6.1 Smart antennas

Traditional antenna technologies are essentially passive radio frequency arrays that are designed to transmit and receive signals within a predetermined coverage area. An omni-directional antenna provides equal gain through a 360 degree azimuth (horizontal) coverage area, and directional antennas localize the coverage area so that it is narrower than 360 degrees, as is illustrated in Figure 37.8. The advantage of omni-directional antennas is that they provide coverage over a 360 degree azimuth, the disadvantage is that propagating radio signals in all directions causes increased radio interference and therefore reduces the capacity of wireless systems. The traditional solution to this interference challenge is sectorization, which essentially involves using multiple directional antennas that each radiate over a narrower coverage radius. This approach requires dedicated equipment for each antenna and introduces some unnecessary interference as a consequence of the overlap between the radiation patterns of the directional antennas.

Smart antennas include active components that continuously estimate the location of the users and update the antenna patterns to efficiently provide directional coverage, while at the same time mitigating unnecessary interference. A smart antenna includes a system of antenna arrays and smart signal processing that can be used to identify the direction of arrival (DOA) of a radio signal from a user. Once the DOA of the signal is known, the digital signal processors calculate the necessary parameters for the pattern of the array to be adapted to suit the associated coverage requirements. The result is that the antenna adapts in real time to suit the necessary coverage requirements of the network, and in the process improves coverage, while at the same time dramatically reducing interference conditions.

Until recently smart antenna technologies have been prohibitively expensive to implement and challenged by the required processing requirements for implementation. Enhancements in digital signal processing (DSP) technologies that benefit from Moore's Law are enabling sophisticated smart antennas to be developed at a relatively low cost and contained in relatively small form factors. It is expected that these antennas will be developed for targeted commercial use over the next 24 to 36 months and will significant improve the performance of wireless networks such as WiFi in capacity limited environments. It will also improve coverage by creating higher antenna gain in the areas where active users are operating.

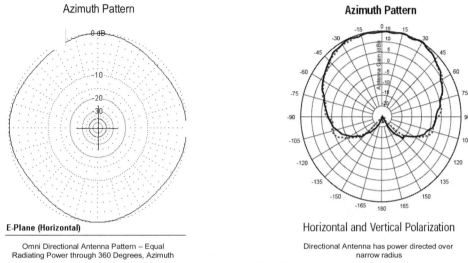

Azimuth Pattern

Azimuth Pattern

E-Plane (Horizontal)

Omni Directional Antenna Pattern – Equal
Radiating Power through 360 Degrees, Azimuth

Horizontal and Vertical Polarization

Directional Antenna has power directed over
narrow radius

Figure 37.8: Examples of omni and directional antenna patterns.

Figure 37.10: Hypothetical smart antenna pattern. The lobes in the antenna pattern point in the direction of specific users, or groups of users.

37.6.2 Cognitive Radio Technologies

While smart antenna technologies manage interference by controlling the direction that signals are radiated, cognitive radio systems manage the timing and operating frequency of radio transmissions. A cognitive radio continuously "listens" for unoccupied channels upon which it can transmit. These solutions are used extensively for military communications, particularly in hostile territories, but to date have not gained significant traction in unlicensed or commercial wireless systems.

Unlicensed and commercial wireless systems rely on structured and relatively static radio spectrum licensing. They use radio modulation techniques such as code division multiple access (CDMA) and orthogonal division multiple access (OFDMA) to coordinate radio resource allocations amongst users.

It is anticipated that cognitive radio technologies will take hold initially for systems that operate in unlicensed radio spectrum and then subsequently for systems. Unlicensed systems enable a heightened rate of innovation as a consequence of lower barriers to market entry, and an increasing need to better manage unlicensed spectrum allocations.

Cognitive radio and OFDM radio transmission is a powerful combination. The OFDM modulation scheme divides the radio spectrum into essentially sub-band tones with coordinated transmission timing. Cognitive radio technologies can be used to manage the transmitted power level of each of these tones depending on whether or not they are interfered. This is illustrated in Figure 37.11, where the time/frequency allocations shown on the left are adapted in the chart on the right in which interference is detected for certain tones. In this case the radio technology changes the tones used for user A and B so that the signal is not significantly impeded by the interference.

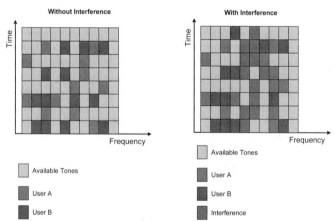

Figure 37.11: Simplified illustration of the powerful OFDM and intelligent radio combination.

An unlicensed system that has recently emerged is ultra-wide band (UWB), which essentially transmits its signal over extremely wide bandwidths at very low power levels using a combination of spread spectrum and OFDM modulation schemes. The bandwidth requirements for UWB means that they must occupy licensed spectrum bands, albeit at low power levels. The notion of allowing unlicensed systems to occupy licensed spectrum has created significant debate. These debates have tended to be associated with the ownership rights of the license holders, and the potential interference that UWB systems might cause to systems operating in licensed radio spectrum. To date UWB has been approved in several markets including the US and Japan, and is close to being approved in many Western European markets.

As UWB systems are deployed, they have the potential to create a precedent where unlicensed systems can coexist in licensed radio spectrum so long as it has a negligible impact on licensed system performance. Should cognitive radio solutions be approved and applied appropriately to UWB systems, there is the potential that the power limitations for UWB to be relaxed and the sub-channels that are used, be managed more carefully based on the occupancy of the licensed spectrum systems. There are a variety of technology vendors investigating these solutions with the hope that it will ultimately render the traditional static spectrum licensing regimes redundant and create new regimes that are not based on static licensing structures.

37.7 Conclusion

With a variety of vested interests at stake, the evolution towards 4G is complex, competitive and in many cases controversial. The underlying premise for 4G to mobilize the Internet disrupts traditional mobile communications. It creates opportunities for new players to mobilize a plethora of media and Internet services and applications, with relatively low barriers to market entry. This will significantly increase service deployment velocity and challenge traditional players in the market.

Technologies like WiFi and WiMAX offer an alternative to 3G and 3G-LTE for portable and ultimately mobile broadband services. By operating in unlicensed radio spectrum, technologies like WiFi (and WiMAX in the future) see a higher rate of innovation relative to traditional mobile technologies that are encumbered by spectrum licensing constraints. However current WiFi technologies lack adequate scale to support full mobile solutions as a consequence of unlicensed spectrum operations. WiMAX is currently being promoted as a licensed spectrum technology to address the scalability issues with WiFi, while at the same time proliferating a business model that is similar to that associated with public WiFi-type solutions.

Radio technologies have traditionally been the bottleneck for capacity on mobile networks. However the advancement in radio technology, combined with demand for low cost broadband services has created a bottleneck in transmission networks, which in many cases are based on legacy circuit switched architectures. Technologies like WiMAX and 3G-LTE address this issue with carrier grade Ethernet technologies. In the case of WiFi, meshing solutions are used to enable lower cost connectivity to the large number of low powered access points required for outdoor metro-wide operations.

In the future, commercial mesh transmission solutions will evolve to more flexible architectures (ad hoc network solutions) that enable the transmission paths to adapt to changes in operating conditions. Radio technologies such as OFDM and OFDMA will see significant performance improvements in both licensed and unlicensed operations, particularly when combined with cognitive radio and advanced antenna solutions. The precedence being set with OFDM based UWB operating in licensed radio spectrum could ultimately be a stimulus for other unlicensed systems (based on OFDM and cognitive radio) to operate and coexist in licensed spectrum bands.

## 38

All Internet is Local: Five Ways Public Ownership Solves the U.S. Broadband Problem

Becca Vargo Daggett[a]

The first wave of publicly owned information networks was in communities that already owned electric utilities. Today, cities of all kinds are being offered seemingly attractive deals from private companies that want to build new information networks. They would do well to also evaluate publicly owned alternatives. Public ownership means ownership by citizens, customers or the community. It provides communities with an ongoing voice in the design and operation of their information and communication infrastructure, and can ensure values that are not being enforced by federal regulators, including universal access and competition.

38.1 Introduction

Ten years after the 1996 Telecommunications Act, which was supposed to accelerate the introduction of high-speed communications systems, the U.S. has dropped from first to 15th in the world for the percentage of residents with high-speed Internet access.

Increasingly, local governments are stepping in where the private sector and federal government have failed. Hundreds of cities are currently debating strategies to develop citywide broadband networks. They share common goals – universal coverage, equitable access, increased competition, and more effective use of the new communications systems for municipal services, especially those related to public safety.

Their discussions often ignore or give short shrift to a crucial issue: who will own the information network?

Ownership matters. Public ownership of the physical infrastructure may be the only way to guarantee future competition. It is clearly the only way that communities can influence the design of their future information systems on an ongoing basis. And public ownership can allow a community to tap into the growing exchange of information to generate significant revenues, or cost savings, while enabling all households in the city to have affordable access.

As of mid-2006, more than 650 cities own telecommunications systems.[1] These range from downtown fiber optic networks that connect public buildings and major businesses, to

[a] *Institute for Local Self-Reliance*

citywide Wi-Fi networks that offer retail service to all residences and businesses. These publicly owned networks have proved remarkably successful in meeting the community's need for advanced services at fair prices.

This first wave of public ownership largely occurred in cities that already owned their electricity networks. That ownership was born a century ago out of public frustration at privately owned utilities' refusal to extend service beyond larger cities. Today, municipal electric utilities are expanding into broadband telecommunications, born of a similar frustration at telecommunications companies' slow response to the needs of small and rural communities.

More recently, communities without municipal electric utilities have begun exploring a governmental role in accelerating the deployment of high-speed information networks. These urban and suburban communities already have some level of high-speed Internet access through cable and telephone company networks. The incumbent suppliers vigorously oppose any municipal involvement, whether through public ownership or by facilitating a competitive network.

At the same time, companies that had been leasing space on incumbents' networks view municipal involvement as an opportunity to build their own networks, with public support. They offer cities what appear to be, at first impression, very attractive arrangements if the city grants them an exclusive contract.

Large cities – Philadelphia, San Francisco, Minneapolis, Boston, Houston, Seattle, and others – have become the front lines in the battle for affordable, high-speed information and communication networks.

So far, these larger cities have tended to choose privately owned, for-profit networks. They choose expedience over security. They choose the comfort of dependence rather than the risks and rewards of independence. They choose a small, guaranteed income via a franchise fee over the potentially large benefits, financial and otherwise, that stem from public ownership. We believe such a choice does a disservice to their households and businesses, as well as the local government itself.

38.2 What is Public Ownership?

Public ownership means ownership by citizens, customers, or the community. It comes in many different forms.

- *Municipal Networks* are owned by a local government entity. This may be the city itself, as in Saint Cloud, Florida, or a municipal utility, as in Moorhead, Minnesota.
- *Cooperative Networks* are customer-owned, as is the case with the Mountain Area Information Network in North Carolina.
- *Non-profit Networks* often are a partnership between a number of public and non-profit entities. OneCommunity (formerly OneCleveland), for example, is owned by a non-profit organization established through a partnership between a number of public and non-profit entities.

---

[1] Community Broadband Fact Sheet, American Public Power Association, July 2006.

- *Community Networks* consist of individual users owning the hardware and voluntarily participating in an ad-hoc network. Some are sponsored by non-profit organizations. Typically these networks offer free access. SoCalFreenet, NYCWireless, Seattle Wireless, and Ile Sans Fil in Montreal are all community networks.
- *Hybrid Networks.* Many networks are hybrids, building on the strengths of multiple partners. For example, REA-ALP Internet Services is a partnership between Runestone Electric Association, a rural electric cooperative, and Alexandria Light and Power, a municipal utility. The Urbana Project is a partnership between Champaign-Urbana Community Wireless Network[2] and the City of Urbana. Austin Wireless is a community wireless network, but operates some portions of its network in cooperation with the City of Austin.

38.3 An Astonishingly Brief History of Telecommunications Regulation

38.3.1 That Was Then

For the first century of telecommunications in the United States the public sector was deeply involved in the introduction and elaboration of both wired and wireless systems. Telephone networks were regulated monopolies. Companies received an exclusive franchise for a specific geographic location and a guaranteed profit, in return for which they had to provide universal coverage at affordable, fixed rates. Telephone and telegraph wired networks were declared common carriers, that is, open to all users on equal terms.

The wireless spectrum, used for radio and TV broadcasts, was regarded as a public asset and regulated by the federal government. Companies received licenses to use specific frequencies in defined areas based on a determination of "best public use." In return they had to abide by certain rules that protected the public interest, rules such as the "fairness doctrine" that required stations to allow access for opposing viewpoints. Broadcast licenses were limited in duration. Renewal depended on the licensees' living up to the rules, and their continuing demonstration that they served the "public interest, convenience and necessity." Congress also promoted competition by limiting the number of radio and TV stations a single company could own and the cross-ownership of newspapers and broadcasting stations in the same market.

When cable television was introduced as a way to deliver better reception and more than could be carried over the airwaves, companies received exclusive franchises to deliver non-broadcast television from local governments. In exchange they agreed to provide public benefits in the form of franchise fees and local programming.

The first computer networks emerged in the late 1960s, a result of federal research investments.[3] The Federal Communications Commission (FCC) issued its first regulation related to computer communications in 1971, when it ordered AT&T to allow competitors

---

[2] The Champaign-Urbana Community Wireless Network (CUWiN) also assists other community wireless networks around the world. For more information see "CUWiN: Wirelessing the Revolution with Open Source Mesh Wireless Technologies", *Government Technology Digital Communities*, November 1, 2006.
[3] The first data packet was sent in 1969 using the Department of Defense's ARPANET (Advanced Research Projects Agency Network).

to use the telephone network for data services without interference.[4] That order also prohibited AT&T itself from getting into the business, out of concern that the company would use its ownership of the network as an unfair advantage over competitors.[5] In 1980, the FCC allowed AT&T to begin offering data services, but still required the company to carry competitors' traffic on equal terms and without interference.[6]

These regulations facilitated a competitive and innovative market for services like voice mail, computer bulletin boards, and other "enhanced services."[7] Further public investment took computer networks to the next level. What we now know as the Internet began as a federal project, the National Science Foundation Network, in the mid-1980s.[8] The national backbone and regional networks it connected were developed with billions of state, university and federal dollars.[9]

38.3.2 This is Now

Over the last two decades, both the regulation and structure of wired and wireless telecommunications changed dramatically. The definition of public interest has been severely curtailed, as has the authority of local, state and federal governments to assert the public interest.

In 1984, Congress limited local authority to enforce cable franchise agreements. In 1987, the FCC eliminated the fairness doctrine. In 1994, the FCC began auctioning spectrum to the highest bidder, and the Internet backbone was turned over to private companies. In 1996, the Telecommunications Act lifted many of the restrictions on the concentration of media ownership, deregulated cable prices, and substantially deregulated the Baby Bells.

One of the few remaining areas of public involvement was the requirement that phone companies allow competing Internet service providers (ISPs) to connect to their customers via their networks. The 1996 Telecommunications Act contained this requirement. Its intent was to spur broadband, in this case DSL, infrastructure deployment. Cable companies were not covered by the same rules. They were not required to offer equitable access to their networks to ISPs.

As people began moving from dial-up to broadband, the different regulatory regimes for phone and cable companies became important. Local governments tried to rectify this inconsistent treatment of companies offering the same service – high-speed, always-on Internet access – by requiring their cable franchisees to become like the phone companies,

---

[4] Federal Communications Commission, *First Computer Inquiry*, Final Decision, 28 FCC 2d at ¶ 11.
[5] Robert Cannon, *The Legacy of the Federal Communication Commission's Computer Inquiries*, May 3, 2003. Harold Fedlman, *Debunking Some Telco Disinformation*, May 15, 2006.
[6] For a full timeline of the FCC's computer communications-related rulings and contemporary Internet developments, see Cybertelecom's *Computer Inquiries Timeline*.
[7] The FCC divided traffic over phone network into two categories. Basic telecommunications services were defined as transmission of an ordinary-language message from point-to-point, and were regulated as common carriers. Enhanced telecommunications services use common carrier transmission facilities, but use computer processing on the form or content of the transmitted information. For more definitions, see Cybertelecom's *Notes: Computer Inquiries*.
[8] Earlier networks facilitated the research that led to the development of NSFNET. But NSFNET, established in 1986, was the basis for the national Internet backbone we use today.
[9] Rijiv C. Shah and Jay P. Kesan, "The Privatization of the Internet's Backbone Network", Presented to the Association of Internet Researchers, Maastricht, Netherlands, October 16, 2002.

that is, common carriers and to allow other service providers to use their copper lines at a fair price.

Courts consistently struck down these local efforts, even when cities made open access a condition for renewing a cable franchise. The courts agreed with the FCC's position that Congress had preempted local authority on this issue. The U.S. Supreme Court upheld the FCC's position in its 2005 Brand X decision.[10]

Before the Supreme Court issued its decision, in 2003, the FCC ruled that telephone companies did not have to share the fiber optic portions of their networks. It was left to state governments to determine whether wholesale access rates for competing ISPs should be regulated. Most states chose not to regulate rates.[11] Almost immediately after the Brand X decision, the FCC extended its exemption from common carrier requirements to phone companies' data networks as well.

Today, neither cable nor phone companies are required to allow competing Internet service providers to use their networks (though some choose to do so).[12]

Meanwhile, technology is moving us into an era in which text, voice and video are carried over the same broadband networks. The FCC has used this as a further justification for deregulation, arguing that the existence of cable, phone, and satellite networks, and the emerging technologies of broadband transmission over power lines as well as the coming of terrestrial wireless, creates an adequate level of competition between network owners.

Such an argument is, at best premature. Approximately 98 percent of high-speed Internet connections come from cable or phone companies.[13] For most households, even in larger cities, the market is dominated by one cable company and one phone company. Many neighborhoods do not even have two choices, since not all areas of phone company networks are equipped to offer DSL. If they do offer DSL, it is at speeds of 1.5 Mbps or less, compared to 3 to 6 Mbps from cable, and with no capacity to support video.

Some ten percent of households cannot get high-speed Internet access from any provider at any price.

At the national level, the telecommunications industry is consolidating. Only slightly more competition exists in the telephone sector than in the days of Ma Bell. In 1984, AT&T was broken into eight regional "Baby Bells." Ensuing mergers and acquisitions have left us with just three: Verizon, AT&T,[14] and the much smaller Qwest. The two largest cellular phone companies, Verizon Wireless (majority owned by Verizon), and

---

[10] National Cable and Telecommunications Association v. Brand X Internet Services (04-277), 345 F.3d

[11] Federal Communications Commission, *FCC Adopts New Rules for Network Unbundling Obligations of Incumbent Local Phone Carriers*, February 20, 2003. For an overview of the federal court rulings leading to the FCC's decision, see *Written Statement of Michael K. Powell, Chairman, Federal Communications Commission*, before the Senate Committee on Commerce, Science and Transportation, January 14, 2003.

[12] "Specifically, the Commission determined that wireline broadband Internet access services are defined as information services functionally integrated with a telecommunications component. In the past, the Commission required facilities-based providers to offer that wireline broadband transmission component separately from their Internet service as a stand-alone service on a common-carrier basis, and thus classified that component as a telecommunications service. Today, the Commission eliminated this transmission component sharing requirement, created over the past three decades under very different technological and market conditions, finding it caused vendors to delay development and deployment of innovations to consumers." Federal Communications Commission, *FCC Eliminates Mandated Sharing Requirements on Incumbents' Wireline Broadband Internet Access Services*, August 5, 2005.

[13] S. Derek Turner, *Broadband Reality Check II*, Free Press, August 2006.

[14] SBC purchased AT&T and then took the older company's name, and is in the process of acquiring Bell South.

Cingular (soon to be wholly owned by AT&T) currently command more than half the market. Two cable companies, Comcast and Time Warner, control 47 percent of the cable television market.[15]

The lack of competition has slowed the expansion of the U.S. broadband market. We are 15th in the world in broadband penetration, according to the International Telecommunications Union, down from 4th in 2001. We perform even more poorly in the ITU's "digital opportunity" index, which considers price and capacity as well as other factors, coming in 21st after Estonia. Broadband subscribers in the U.S. pay twice as much as customers in Asia and Europe for one-twentieth the speed.

The Internet was invented in the U.S., but other countries are now taking the lead. For example, the private companies that own the Internet backbone in the U.S. have resisted upgrading to a new version of the Internet address system (IPv6) for nearly a decade. IPv6 greatly expands the pool of Internet addresses, allowing everything from cars to thermostats to have unique addresses, and allows for increased network security. In 2006, China converted to IPv6, and now the U.S. will have to follow its lead.[16]

FCC Commissioner Michael J. Copps recently wrote in the *Washington Post*: America's record in expanding broadband communication is so poor that it should be viewed as an outrage by every consumer and business person in the country. Too few of us have broadband connections, and those who do pay too much for service that is too slow. It's hurting our economy, and things are only going to get worse if we don't do something about it.

38.4 Why Public Ownership?

The stakes are high. Local governments are stepping in where state and federal policies of privatization and deregulation have failed. Despite a brief backlash against municipal broadband projects, it is increasingly accepted that cities have the authority to develop telecommunications plans. In elaborating such plans, they must take into account many factors, but the one that will have the greatest effect on competition, equity, and public benefits is the decision about who will own the network.

We propose five arguments for public ownership.

1. High-speed information and communication networks are essential public infrastructure.

Much of the infrastructure of the country – water, sewer, roads, airports, seaports – is publicly owned. Indeed, virtually all economists and economic development experts believe that public infrastructure is essential for improving productivity and maintaining competitiveness. Just as high quality road systems are needed to transport people and goods, high quality wired and wireless networks are needed to transport information. Both networks allow individuals and businesses in a community to connect to each other and the outside world.

---

[15] Free Press, *Who Owns the Media?*
[16] Robert X. Cringely, "The $200 Billion Lunch", *I, Cringely*, Public Broadcasting Service, November 2, 2006.

For over 100 years, cities have successfully built and managed public infrastructure like roads and water and sewer systems. Information networks are new kinds of infrastructure, but they are not outside the competencies of local government.

Public ownership of the physical network does not necessarily mean the city either manages the network or provides services. Benton Public Utility District in Washington State contracted for the construction of a fiber and wireless network, which it now manages as a wholesale only network. A half-dozen private companies sell retail services, including multiple ISPs and a home security company. UTOPIA (the Utah Telecommunications Open Infrastructure Agency) is financed and owned by a consortium of cities that contracted with a private company to build and manage the network, and has several providers of video, voice and Internet services, including AT&T.

Cities own roads, but they do not operate freight companies or deliver pizzas. Modern information infrastructure easily allows the transport layer (the road, or in this case the network hardware) to be separated from the service layer (the pizza delivery, or in this case Internet access or video services).[17]

A publicly owned network would not be a monopoly. Other networks would continue to exist. Indeed, as is explained in more detail below, the existence of publicly owned networks can raise the quality of services and the level of competition. As Franklin D. Roosevelt said, "the very fact that a community can, by vote of the electorate, create a yardstick of its own, will, in most cases, guarantee good service and low rates to its population. I might call the right of the people to own and operate their own utility something like this: a 'birch rod' in the cupboard to be taken out and used only when the 'child' gets beyond the point where a mere scolding does no good."[18]

2. Public ownership ensures competition.

---

[17] A more sophisticated comparison of the physical transportation network identifies another important factor. A network's design affects its function. Railroad networks, for example, can carry cars only whose wheels fit the tracks. A network using one width (gauge) for its tracks cannot carry cars from a network using a different gauge. Road networks, on the other hand, can carry two- or four- or 16-wheel vehicles of all different sizes and shapes.

Similarly, phone networks and computer networks are both telecommunications networks, but they function quite differently.

Traditional telephone networks are circuit-switched. They work by creating a temporary direct connection, a completed electronic circuit, between the users. Your voice is converted to electronic pulses that are sent through the connection. All the intelligence that allows services like call-waiting and caller ID are contained in the network itself, and can theoretically be used regardless of the type of phone plugged into the network.

Computer networks (high-speed data networks), on the other hand, are packet switched. They work by converting your voice (or email, or image) into digital format (1s and 0s), then breaking the digital communication into individual packets. Each packet is sent independently through the network. The packets are reassembled at the final destination. The intelligence is contained in devices that plug into the ends of the network.

A phone network is designed for phone service. A cable network is designed for video service. High-speed information networks, however, aren't designed for any particular service. Telephone calls, movies, photographs, and emails are all broken down into packets that look the same to the network. This flexibility means that modern high-speed networks are an unprecedented platform for innovation.

The nature of modern information networks is such that innovation takes place on the edge of the network. We don't think of municipal power networks as stifling innovation, because the innovation comes through the devices – computers, radios, security systems, etc. – plugged in at the edge of the network. The question is not what the network itself can do, but what users can do with the bandwidth supplied by the network.

[18] Works of Franklin D. Roosevelt, "The Portland Speech", A Campaign Address on Public Utilities and Development of Hydro-Electric Power, Portland, Oregon, September 21, 1932.

Tens of thousands of miles of fiber optic backbone cable have been laid by the private sector, but there is little incentive for the private sector to bring high-speed connections the "last mile" to homes and businesses (sometimes called the "first mile," to emphasize the fact that users are creators of content as well as consumers of content).

Owners of existing cable and phone networks have strong incentives to make use of their existing infrastructure for as long as possible. What's more, consolidation in the industry means that companies serving hundreds of markets make choices based on what is most advantageous for the corporation as a whole rather than any individual community.

Potential service providers seeking to compete with the incumbent cable and phone companies cannot use existing networks, or obtain access at rates that allow them to offer competitive services.[19] Thus, to reach customers, they must build their own network infrastructure. But here they face a significant barrier-to-entry. An overbuilder[20] faces the difficult challenge of having to simultaneously repay capital expenditures and compete for market share against incumbents that have already amortized their major capital investments.

A publicly owned, open access network could be open to all service providers on the same terms, thereby encouraging the entry of new service providers.[21] It would allow competing service providers to lease capacity on the network in order to sell services to customers. Customers could then choose Internet service providers according to the combination of price, speed and service that fits their needs. Competition would ensure fair rates, and if any service provider restricted what could be done with its connection, customers could choose a different service provider.

Cities establishing new, privately owned citywide networks can require the owner to allow fair access. But it is unclear whether these contractual obligations will be enforceable in the future. History indicates they may not. Cities negotiated cable franchise agreements that were later preempted by state and federal laws. Thus public ownership, which allows the public to establish the rules for using that infrastructure, may be the only way to ensure a network will provide open, nondiscriminatory access in the future.

3. Publicly owned networks can generate significant revenue.

Telecommunications networks are different from traditional public works like roads because they can be self-financing both in terms of initial construction costs and ongoing upgrades. Indeed, a growing body of data suggests an information network can be a very profitable investment, for the city and for its households and businesses.

Cities should welcome this prospect, given the strain on municipal budgets from increasing costs of public safety, health, welfare, and aging infrastructure.[22]

Saint Louis Park, Minnesota's publicly owned fiber network that connects public buildings has a five year payback period as a result of dramatically lowering operating costs below what it was paying for leased T-1 lines. In three years, the publicly owned

---

[19] The owner of the network charges independent service providers a higher access rate than they charge their in-house service provider.

[20] An entity that builds out a full, competing network.

[21] For more on open networks, see the International Network of e-Communities Declaration on Open Networks, signed October 2006.

[22] National League of Cities, *Advisory Council Trends and Changes Topics* and *About Cities: Trends in Our Nation's Cities*.

wireless network of Buffalo, Minnesota (population 10,000) generated over $150,000 in profits from a $750,000 investment.[23]

We offer the case of our hometown, Minneapolis, to demonstrate both the potential profitability of a publicly owned network and as a cautionary tale for cities tempted to use the public purse to allow private firms to capture those profits.

In April 2005, the City of Minneapolis issued a request for bids for a citywide wireless network. The City ruled out public ownership from the outset, insisting that given its weakened financial state, it could not afford the capital investment. In September 2006, the City announced the winning bidder, a small local company with gross sales of around $10 million in 2005.[24]

Since the company itself was far weaker than the City in terms of being able to finance a $10 million system, the City, under the terms of the 10-year franchise, agreed to purchase a minimum of $1.25 million in services each year, and likely much more. Part of this commitment, $2 million, will be prepaid before the network is launched.

The prepayment and the City's ongoing commitment to purchase services will enable the small, privately held company to finance the build-out. Indeed, the $2 million prepayment for services will cover about one-quarter of the cost of building the network.[25]

In other words, the City, which previously declared it lacked the financial wherewithal to finance the network, *is* financing the network. For the same amount of money the City could have owned the network, used subscriber revenues to pay operating expenses, and provided free services to itself.

As part of the agreement, five percent of net pretax revenue (that is, revenue after operating and debt expenses but before taxes) will go into a digital inclusion fund.[26] Minneapolis expects $4000 in the first year of operation and $1.7 million in the tenth year.

The company will receive 95 percent of the net income, over $32 million in pretax profits in the tenth year alone. Based on these numbers, the company would realize profits upwards of $130 million over ten years.

A city-owned, wholesale access only network would not have been as profitable as a private retail network. Our analysis of the City's numbers concludes that a publicly owned network might earn the City about $44 million over ten years, or roughly four times what it will receive from the contractual agreement (Figures 38.1 and 38.2).[27]

Some cities see public ownership not as an opportunity to make money, but as a way to strengthen the local economy. They view the city not narrowly as a municipal corporation trying to balance its internal books, but as a public corporation trying to maximize the total benefit to its community owners.

Saint Cloud, Florida, for example, chose to invest in citywide wireless in part to keep more money in their citizens' pockets. The $3 million capital expenditure is just 7 percent of the city's outstanding debt; its $300,000 annual operating costs represent just 1.5 percent

---

[23] Balhoff and Rowe, *Municipal Broadband: Digging Beneath the Surface*, September 2005. Blandin Foundation "Increasing Economic Vitality: A Community Guide to Broadband Development."

[24] Experian Business Reports, October 2, 2006.

[25] Bill Beck, Minneapolis Deputy Chief Information Officer, Minneapolis City Council Ways and Means Committee Meeting, September 1, 2006.

[26] The Digital Inclusion Fund will be managed by a private foundation. The fund will distribute money to organizations that promote technology literacy.

[27] *Is a Publicly Owned Network a Wise Public Investment?* ILSR, May 2006.

of the city's general fund expenditures in 2006.[28] The <u>free</u>, city supported service is saving the average household $450 per year – the amount they previously paid for broadband Internet access. That's more than the average household pays annually in local property taxes.[29]

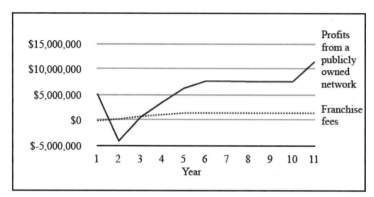

Figure 38.1: Community Benefits - Comparing Private and Public Networks.

	Years 1-5	Years 6-10	Year 11
Potential Net Revenue to a Privately Owned Retail Network	$55,110,000	$111,610,000	$30,950,000
Digital Inclusion Fund Share of Net Revenue to a Privately Owned Retail Network (5 percent of Net)	$2,755,500	$5,580,500	$1,547,500
Potential Profits from Publicly Owned, Wholesale Network	$12,175,300	$38,835,000	$14,641,600
Lost Public Revenue	$9,419,800	$33,254,500	$13,094,100
Estimates based on Wireless Minneapolis Business Case.			

Figure 38.2: Community Benefits - Comparing Private and Public Networks.

A publicly owned system can spur private competitors to lower their rates or improve their services, which will benefit all city households and businesses. The Clarksville, Tennessee Department of Electricity (CDE), for example, is asking local voters to approve a bond for $25 million to install a fiber to the home network. At launch, CDE's price will be lower than the current Internet provider, Charter. CDE fully expects that Charter will respond by lowering its rates, perhaps below that of the City's. And that's fine with the City. "That's not a bad thing," CDE General Manager Ken Spradlin says, "because not all our customers are going to choose to do business with us, but they are all our customers."[30]

---

[28] Saint Cloud, Florida, Annual Budget FY 2005/06.
[29] The average Saint Cloud household pays $300 annually in local property taxes. Jonathon Baltuch, MuniWireless Santa Clara, June 2006.
[30] "Charter, CDE go to battle over vote", *The Leaf Chronicle,* October 4, 2006.

4. Public ownership can ensure universal access

Society as a whole benefits when information and communication networks are accessible to everyone. More people on the network means more participants in online communities, and more customers for online products and services.

Publicly owned road, water and sewer, and sidewalk networks connect all households without discrimination. All have access to the same services, though they may purchase different amounts of these services based on household economics and need. A publicly owned telecommunications network similarly can choose to make a basic level of access available to everyone at a low cost, or offer free or subsidized access to some households.

Cities may be able to negotiate such requirements in initial contracts. But as pointed out above, federal and state intervention in cable franchises over the years demonstrates that local governments cannot count on retaining the authority to enforce these contracts.

Cities that choose private networks get one chance to set rules governing the network, in contract negotiations. After that, they rely on corporate good will.

Opponents of publicly owned information networks argue that the private sector is more responsive to customer demands than the public sector. Customers are not asking for high-capacity connections to the home right now, they argue, but once they do, the private sector will respond more quickly and efficiently than the public sector.

Yet the evidence indicates that the even the most aggressive telecommunications companies do not intend to serve everyone. Lower income areas, whether urban or rural, and sparsely populated areas, regardless of income are not attractive places for new investments.

The claims that access speeds have increased and prices have dropped are true only if phone and cable offerings are considered separately. Over the past five years, phone companies have not raised the average DSL speed of 1.5 Mbps available to most households. They have offered promotional prices, but often only to customers who buy home phone service, and they have created lower price tiers for 768 kbps or slower connections. Meanwhile, cable companies have kept prices the same, but increased advertised download speeds.

When these companies do invest, the incentive is to do so in ways that provide a quick return on investment. Consumer rights advocate Bruce Kushnick points out that over the last decade, states gave the Baby Bells tax breaks and deregulated some prices in exchange for their commitment to deploy high capacity, high speed fiber optic networks.[31] Instead of making investments in very high speed, very high capacity networks, however, the companies made lesser investments to add DSL to their existing copper networks. DSL offers much slower speeds, but it is almost immediately profitable. Yet even here, while the phone network is universal, DSL is not.

The single most reliable predictor of whether a household has a high-speed Internet connection is income.[32] Broadband data signals can travel only a limited distance over existing copper-based phone and cable networks, and companies are unwilling to invest in upgrades where average revenue per household is low. In rural areas, expensive satellite (upwards of $50 per month, plus hundreds of dollars for the dish[33]) is often the only

---

[31] Bruce Kushnick, *The $200 Billion Broadband Scandal*, TeleTruth, 2006.
[32] S. Derek Turner, *Broadband Reality Check II*, Free Press, August 2006.
[33] "With a Dish, Broadband Goes Rural," *New York Times*, November 14, 2006.

alternative to dial-up. In urban areas, a large percentage of households have access to cable modem service, typically at rates of more than $40 per month, but not DSL service, which provides slower speeds at a lower price.

Universal broadband access will be a long, long time coming from private companies, if it comes at all.

Consider the two highest profile projects currently underway. AT&T (formerly SBC) plans to run fiber to the streets of some 19 million homes in 13 states by 2009. The company will continue to use existing copper connections from the street to the home.[34] Verizon is spending $6 billion to run fiber directly to about 6 million homes by the end of 2006, and another 9 to 14 million homes by the end of 2010.

Combined, these deployments might reach about one-third of U.S. households in 2009, overwhelmingly located in communities of above-average income.[35]

Neither company plans to extend fiber to all their customers, ever, because "... there will be areas that are just not economic to offer fiber everywhere," says AT&T's Homezone managing director Ken Tysell.[36]

Phone companies are making this investment primarily to be able to offer video, a market dominated by cable television companies (less than one-third of households that subscribe to paid television do so through satellite rather than cable).[37] No cable companies have announced efforts either to connect fiber to subscribers' homes or to increase the capacity of their networks. Instead, they are packaging cable modem speeds that are slightly faster than current DSL offerings, along with video and voice over Internet protocol. A recent study from an industry supported research center indicates the capacity of these networks is strained as a result of these "triple play" packages.[38]

Most cities included full build-out and anti-redlining provisions in their cable franchise agreements. Cable must be available everywhere in the community. But phone companies are now lobbying at the state and federal level to create new franchising systems that would bypass local authorities and eliminate anti-redlining provisions.[39] Companies would be allowed to build out infrastructure only in the areas with the highest profit potential, that is, densely populated neighborhoods with higher incomes. Moreover, some of these proposals would allow cable companies to exit existing franchise agreements if a phone company began offering video services in any portion of the local market.

We are already seeing the results of the move to state-level franchising without build-out requirements.

---

[34]"AT&T is Calling to Ask About TV Service. Will Anyone Answer?" *New York Times*, July 2, 2006. About 1 million homes are in new developments and will get fiber to the premises.

[35] Broadband Everywhere, *A Picture is Worth a Thousand Words: How the Bell Business Model Leaves Much of America Behind*, April 4, 2006.

[36] *New York Times*, "AT&T is Calling to Ask About TV Service. Will Anyone Answer?" July 2, 2006.

[37] U.S. PIRG, *The Failure of Cable Deregulation: A Blueprint for Creating a Competitive, Pro-Consumer Cable Television Marketplace*, August 2003. Cable networks were essentially deregulated with the 1996 Telecommunications Act, which prohibited local franchising authorities in most places from regulating rates. Cable rates increased by over 50 percent from 1996 to 2003. While cable companies blamed increased programming costs for their rate hikes, 40 percent of the top cable channels are owned in whole or in part by cable companies or the conglomerates that own them.

[38] *Wall Street Journal*, "Cable Industry May Need to Spend Heavily on Broadband Upgrades," August 17, 2006.

[39] The Communications Opportunity Promotion and Enhancement (COPE) Act of 2006, which provides for national video franchising, passed the House in June 2006. The Senate version is being held in committee as of July 2006.

Communities with above average income have at least two competing providers of very high-speed networks – capable of providing video – while neighborhoods of lesser means are bypassed. Any infrastructure investments by the cable companies are in areas where they face competition from telephone companies. Lower-income and rural areas, many of which already have lesser networks, are ignored.

For example, of the Pennsylvania communities in which telephone companies have filed plans to upgrade their networks, 85 percent are above the state median income (Figure 3).[40] Meanwhile, Verizon is replacing its copper networks with fiber in certain Boston suburbs, but is reportedly trying to sell rather than upgrade its copper networks in Maine, New Hampshire and Vermont.[41] In New York, Syracuse is getting fiber to the home, but customers in Queens, New York City are being told there is no more capacity in their area. "You can't wire everything for unlimited capacity," said a Verizon spokesman. "It's more effective to engineer capacity to be a fixed percentage above the average use in a given day."[42]

	Communities Targeted for Fiber Deployment	Percent with Incomes Above State Median
Maryland	52	95
Massachusetts	39	97
New Hampshire	8	88
New Jersey	159	77
New York	97	96
Pennsylvania	145	88
Texas	41	90
Virginia	16	94
Source: Broadband Everywhere, A Picture is Worth a Thousand Words, April 2006.		

Figure 3: Existing and Planned Baby Bell Fiber Optic Deployments.

Private companies balance the price they charge against the number of households willing to subscribe at any given price. It makes no difference to the companies whether they generate $100,000 from 1000 people paying $100 per month, or 100,000 people paying $1.

---

[40] State Representative W. Curtis Thomas, "Redlining in the Digital World", *Philadelphia Daily News*, July 26, 2006.

[41] *Wall Street Journal,* May 10, 2006. *Boston Herald,* May 11, 2006.

[42] "Verizon says 'Come get DSL', then turns some away", *New York Post*, October 8, 2006.

5. Public ownership can ensure non-discriminatory networks.

Network neutrality is the term used to describe a network whose customers can use their broadband connections to access the content of their choosing, run the Internet applications of their choosing, and attach to their connection any devices of their choosing. This is possible because with Internet Protocol, bits are bits. Whatever you do with your internet connection – listen to radio programs, post your work on a web site, send pictures to family, or talk to friends in Canada – is broken down into little packets of data that move through the network in the same way.

With network neutrality, there can still be multiple tiers of service (i.e., $15 per month for 1 Mbps, $30 per month for 3 Mbps). What neutrality mean is that when customers pay for a connection with a certain level of service, they should be able to use that connection however they choose.

With the elimination of common carrier requirements and increased Internet traffic – both as a result of more people online and more bandwidth intensive applications like video – the debate over network neutrality has taken on new urgency.

Cable and phone companies have begun insisting they need to manage traffic in order to ensure "quality of service." A typically cited example is that X-rays shouldn't get tied up in network traffic created by someone downloading a movie.[43]

But reasonable traffic management can be incorporated into a network without changing the nature of the Internet. Just as emergency vehicles, like ambulances, can take priority on the roads, so emergency pieces of information, like X-rays, can be given priority over information highways.[44]

Private network owners argue that they need to charge differential rates in order to manage web traffic and provide "quality of service,"[45]. In reality they desire this ability to allow them to maximize their profits. Instead of offering faster or more affordable connections, they would charge you for what you do with your connection. For example, they can charge one rate to download video created by their own company, but a higher rate to download video from an independent filmmaker, and an even higher rate to post your own video for others to download. A digital book purchased from Amazon.com would download faster than the same book from your local bookstore or an independent author, just because the larger company can afford to pay for priority for its traffic.

---

[43] Not coincidentally, this also allows those who own the network to maximize revenue from their networks. Instead of offering faster or more affordable connections, they can charge you for what you do with your connection. For example, they can charge one rate to download video created by their own company, but a higher rate to download video from an independent filmmaker, and an even higher rate to post your own video for others to download. A digital book purchased from Amazon.com would download faster than the same book from your local bookstore or an independent author, just because the larger company can afford to pay for priority for its traffic.

Tom Rutledge, the Cablevision COO famously told a Wall Street conference in June about the contrast between his VoIP(Voice Over Internet Protocol) service with that of Vonage. "We actually prioritize the bits so that the voice product is a better product."

[44] Some have argued that allowing any prioritization of bits will lead to technologies that disguise low-priority bits as high-priority. See Sascha Meinrath and Viktor Pickard, "A New Net Neutrality."

[45] Quality of service is about guaranteeing a specific level of performance within a short time window. For most applications, quality of service doesn't matter. When browsing the web, a half-second delay is unlikely to cause you problems. With prerecorded audio or video, a ten second buffer ensures smooth viewing. The exceptions are voice over Internet Protocol, live video (e.g., teleconferencing), and online gaming. Edward W. Felten, "Nuts and Bolts of Network Neutrality", June 2006.

Those who own the network could make customer interaction with the Internet more like cable television. For example, AT&T's new service uses Internet technology, but won't allow users to browse just any content using the box on their televisions. According to the Wall Street Journal, "While the Homezone set top-box will be connected to the Internet, users won't be able to surf to any Web Site. They will only be able to download content from providers who have made deals with AT&T."

Publicly owned networks can ensure neutrality. Customers can be sure that any traffic management mechanisms are necessary and not simply to improve profitability. Communities can insist on neutrality from any service provider that uses the network, a form of local regulation they could not enforce if they were relying on privately owned networks. Or, if the market is large enough to support multiple service providers, they can leave neutrality to the market, knowing that unhappy customers can easily change service providers.

38.5 Broadband Access and Competition: Truth and Fiction

Do Americans have choices when it comes to broadband? Reports from the Federal Communications Commission would make you think so. A July 2006 Washington Post editorial cites statistics from the Commission. "More than 60 percent of Zip codes in the United States are served by four or more broadband providers that compete to give consumers what they want," they argue. Anyone who has tried to shop around for high-speed Internet access will find this assertion surprising.

In 1996, Congress required the FCC to report statistics on broadband penetration. Every other year, the FCC reported to Congress that the U.S. was making progress toward the goal, set forth in the 1996 Telecommunications Act, of making broadband available to all Americans. Yet each year, in international comparisons, the U.S. was falling further and further behind.

Finally, a frustrated and confused Congress asked the Government Accountability Office to evaluate the FCC's methodology. In May 2006, the GAO came to the same conclusion consumer advocates had reached years before. The FCC statistics are so flawed as to be useless in gauging broadband availability.[46]

The FCC, for example, does not distinguish between business and residential services. It counts a provider as offering service in the zip code even if it is offered only to businesses. If an ISP (Internet Service Provider) has a single business customer in a zip code, it is recorded as serving the entire community.

The FCC counts as competitive providers those ISPs who lease lines from the incumbent telephone company at retail rates. Given current federal rules, this is nonsense.

The FCC counts the ISPs who lease Qwest's facilities as competing service providers even though it is structurally impossible for them to compete on price. For example, an independent ISP in Minneapolis charges $20 per month for its services (i.e., email accounts, customer support), plus a "Qwest DSL monthly circuit rate" of $22 per month for a 1.5 Mbps/896 kbps connection. Qwest offers the same package for an introductory price of $32. Qwest imposes the same terms of service on all who use its lines, whether retail

---

[46] Government Accountability Office, "Broadband Deployment is Extensive throughout the United States, but It Is Difficult to Assess the Extent of Deployment Gaps in Rural America," GAO-06-426, May 2006.

customers or resellers. Thus an independent ISP cannot compete by, for example, by allowing customers to share its connection through a wireless router (something Qwest prohibits).

In removing common carrier requirements from phone and cable networks, the FCC argued that competition will come from so-called inter-modal, or network based, competition. They assume satellite, broadband over power lines (BPL), and terrestrial wireless will create more competition over time. But satellite represents just two percent of the broadband market, a figure that has changed little over time.[47] BPL is useful for power grid management and within building networking; there are only a handful of deployments providing Internet access to a few thousand homes in the U.S.[48] Both have limited potential to provide high-capacity connections.

Terrestrial wireless has emerged as the strongest competitor to wired networks. But whether it will be a competitor depends a lot on how it is deployed. WiMAX, which is being promoted as the future of wireless, relies on licensed spectrum that is auctioned to the highest bidder. Just a handful of companies, led by Sprint and Clearwire, hold rights to spectrum in the 2.3 GHz and 2.5 GHz bands (the two bands most likely to be used for WiMAX in the U.S.)

Increasingly, incumbent phone companies are also using unlicensed wireless. Local phone incumbent Embarq, a spin-off of Sprint, has a citywide Wi-Fi network in Henderson, Nevada. AT&T will soon be providing citywide Wi-Fi in Springfield, Illinois, where it is also the incumbent, on an exclusive franchise. Comcast, which controls about one-third of the U.S. cable market,[49] is an investor in Bel-Air Networks, which builds municipal wireless networks.

Unlicensed spectrum can be used by everyone. Thus, Wi-Fi is more open to competition than WiMAX. But the potential for interference between Wi-Fi networks, and other factors may give the first company into a community a de facto exclusive franchise.[50]

38.6 Evaluating "Public-Private Partnerships" and Other Private Business Models

The term "public-private partnership" is widely used to describe a bewildering variety of municipal broadband projects, projects as different as Philadelphia, where a private company will own and operate the network, and Saint Louis Park, where the city will own a fiber and wireless network and contract with a private company to manage and provide services over the wireless portion of the network.

It might be best simply to drop the term "public-private partnership" since it obscures more than it enlightens. What follows is an overview of business models in which the

---

[47] Satellite broadband is very, very expensive to provide. A satellite costs upwards of $250 million to launch. Unlike television, which can be beamed to an unlimited number of customers, Internet access requires a two-way connection and more satellite capacity must be added for additional customers. "With a Dish, Broadband Goes Rural," *New York Times*, November 14, 2006.

[48] BPL faces strong opposition from the amateur radio community because of the potential for interference.

[49] "Comcast, Cablevision target businesses for growth", *Reuters*, September 20, 2006.

[50] For a further discussion, see.

private sector owns the infrastructure, and an assessment of their risks and benefits to the public sector.

The Status Quo: The dominant business model for telecommunications networks in the United States is a network owned and operated by a private, for-profit company that is also the only or primary provider of monthly subscription services. This is true of your local phone and cable companies. They own the infrastructure, and you as a customer have no choice in who delivers the service. This is true even with long-distance service; if AT&T is your local telephone company, you cannot get Sprint long-distance without also paying AT&T for use of the line.

Cities have little regulatory authority over these networks. (As explained above, these networks are subject to few regulations at any level of government.) For example, they do not have the authority to require phone companies to expand their DSL coverage, nor can they include provisions related to equitable or affordable Internet access in their cable franchise agreements.

Franchise Model: A privately owned and operated, for-profit network that does not have the city as a major customer. The city grants the private company use of public assets for some period of time, and the company compensates the city for use of those assets.[51] Cities typically work with a company that applies for a franchise and do not issue a request for proposals (RFP), although some have done so as a way of soliciting competing offers.

One of the first wireless franchise agreements was in Anaheim, California. Earthlink will pay the city a fee for use of the public assets needed to support a Wi-Fi network. The city will not be an anchor tenant on Earthlink's network, because it is deploying a city-owned Wi-Fi system for municipal use. The franchise agreement does not include any requirements beyond the network providing a certain level of speed, coverage and reliability.

This model poses few risks, but also few benefits. It requires no public investment and little public involvement of any kind. The benefits are modest amounts of revenue from pole attachment fees, and the possibility of additional competition. The city has little influence over the network coverage quality of service, or the prices charged. Franchise models do nothing to overcome the digital divide between higher and lower income households.

Anchor Tenant Model: A privately owned network, with the city agreeing to become the anchor tenant by agreeing to buy a minimum annual level of services. The city grants the private company use of public assets (or assists in negotiating access from private entities[52]), and also agrees to be a major customer of the network (an anchor tenant). In exchange, the city is compensated for use of public assets. The agreement contains a public benefits section that may include a share of revenue or limited free access to the network.

---

[51] If the new network will provide television (i.e., if it is a fiber optic network rather than wireless), the company may be obligated to meet the same terms as the existing cable franchisee, providing public access channels and meeting build-out requirements.

[52] In some cities, investor-owned electric utilities own the light poles. Some wireless networks must also gain access to the roofs of buildings or other private assets.

One of the first anchor tenant models was in Minneapolis, as explained above. Under the terms of the contract, the City will pay the private owner of the network a minimum of $1.25 million annually for services over the 10-year life of the contract. The company will give five percent of net revenues to a digital inclusion fund managed by an outside foundation, and provide free access in selected parks and community technology centers.

The largest benefit of this model, in the eyes of many elected officials, is that the city does not have to finance construction of the network and assumes no responsibility for its ongoing operation. The city gains a new competing network to its incumbent phone and cable companies, and receives funds for public benefit projects.

This model, however, does have substantial risks. Since the city will rely on the network for its own internal communications and revenue for public projects, it cannot allow the network or the company that owns it to fail, even when its intervention contradicts the public interest. Consider the recent case involving Massport (Boston-Logan Airport). Massport entered into an agreement with a private company that would provide for-fee wireless Internet access throughout the airport and share a portion of its revenues with the airport. After the for-fee service was introduced, Massport tried to prevent airlines from offering their own free wireless Internet access in the airport. The conflict ended up at the FCC, which eventually ruled in favor of the airlines, on the grounds that landlords cannot prevent tenants from using legal technologies of their choosing.

Cities also face the possibility that state or federal legislation will preempt their authority to enforce these agreements at some future date, as has happened with cable franchise agreements.

38.7 The Dollars and Sense of Public Ownership

Every city that is seriously exploring a citywide broadband network should do a detailed economic and financial analysis. This will serve it well even if it should end up choosing a privately owned system because it will allow it to negotiate with the private company from an informed perspective.

The analyses can use different assumptions. Some of the issues involved are:

- Who will manage the network? This may be the entity that owns the network, or management may be contracted out.
- Will the network be for profit or not-for-profit?
- Will the owner of the network sell retail services only, wholesale access only, or a combination of the two?
- Will the city be a major customer?
- Will ongoing operations be supported by monthly subscriber fees, advertising revenue, sponsorships, municipal uses, or a combination of these?

A complete analysis requires that the city examine different ownership structures. A number of companies are offering to build networks at no up-front cost to the city. City officials should understand that although seemingly attractive for its convenience, such a model may not offer the city and its households and businesses the best long term benefits.

A financial analysis includes several key items:

Capital expenditures – Capital expenditures include wireless hardware and software, backhaul (the connection from wireless access points to the larger local network, which in turn connects to the global Internet network), network engineering and deployment. It also includes core network equipment (i.e., servers and routers). The city's existing assets – streetlights, electric poles, optical fiber connecting public buildings, etc. – can significantly affect the cost of a network.

Costs depend on the technology. Wi-Fi hot spots, like those found in cafes or homes, are inexpensive. Ongoing costs may be as much as ten times the capital investment, however, since each hot spot must be connected to a wired connection in the existing last-mile infrastructure.

More typical is the use of Wi-Fi mesh that reduces the number of wired connections in the network by allowing information to hop from one access point to another before reaching a wired connection. Wi-Fi mesh networks for municipal use only (public safety, meter reading, mobile municipal workforce) can be deployed for $100,000 or less per square mile. Residential service networks, typically designed to reach 90 to 95 percent of homes and businesses, can cost upwards of $200,000 per square mile.

Fiber to the home is the most expensive alternative, but it is also the longest-lived and the only "future proof" option. Estimates range from $600 to $3000 per home, depending on existing infrastructure and building density.

Operating expenditures – For municipal use only wireless networks, the rule of thumb is that operating expenditures are about 15 percent of capital expenditure annually. This includes 24-hour network operations, pole attachment fees and electricity, monthly equipment maintenance and software upgrades, and Internet bandwidth. For combination wireless networks, operating expenditures are about 30 percent of capital expenditure for a retail network, 15 to 20 percent for a wholesale network. The added costs include customer service, billing and marketing as appropriate for retail or wholesale customers.

For fiber to the home, annual operating costs will be around 5 percent of capital expenditure, though this may be slightly higher for smaller cities.
More detailed breakdowns vary by location. For example, average pole attachment fees are in the range of $36 annually in California, but $86 annually in Louisiana. Wi-Fi Access points with a single radio may draw $20 worth of power annually, while multi-radio deployments combined with high-powered wireless backhaul can draw five times more.

Wireless hardware maintenance will be in the range of 7 to 10 percent of equipment costs annually (though this may be higher for some backhaul components). Internet bandwidth consumption will depend on the number of subscribers and the average bandwidth use per subscriber, generally assumed to be 250 kbps to 500 kbps per user on average, and 1 Mbps per business on average.

Revenue – Monthly subscriptions are one of two major sources of revenue. Monthly rates depend on whether the network is wholesale only or retail. In a wholesale network, the city would be responsible for maintaining the network (or contracting for management) and relationships with companies that sell retail services. In a retail network, the city would be responsible for retail service and support, as well as all marketing and advertising. Gross wholesale revenue will typically be about one-quarter to one-third of gross retail revenue.

The wholesale rate that can sustain the network will depend not only capital expenditures and projected subscription rates, but also the division of responsibilities between the wholesaler and retailer(s).

Fiber to the premises can generate much higher revenues than wireless, because the networks can support television.

The other major revenue category is municipal use. Many cities currently budget for mobile computing, most often subscribing to cellular data services that are both slow (half the speed of a typical DSL or T-1 connection) and expensive ($60 per month). Within the city, the Wi-Fi network replaces these subscriptions, directly saving the city hundreds if not thousands of dollars each month. Other direct savings may come through replacing leased lines to public buildings with fiber or high-speed wireless connections that provide faster speeds at a lower price, or replacing local-use cellular phones with Wi-Fi phones. Cities that have invested in fiber connecting public buildings typically have a five to eight year payback relative to the expense of leased lines.

Advertising may be a source of revenue for wireless networks, but it would be unwise at this point for a municipality to count on that as anything other than an added benefit of perhaps one or two dollars per user, per month. To put it in context, that might be enough to cover the Internet bandwidth to support a free-to-the-user service, but little more.

The most challenging aspect of the evaluation will be to estimate second order effects. Some can be evaluated directly. For example, if the city has a choice between hiring a new building inspector or using wireless to improve the efficiency with the same number of inspectors, the salary of the inspector not hired can be credited as an avoided cost. But there is also a wide array of machine-to-machine communications (automated meter reading, wireless parking meters, traffic monitoring, etc.) that may improve provision of municipal services but do not directly reduce the city's expenditures.

If the city is planning to purchase these as communications services from a private network owner on a per unit basis, the value of the cost savings must be directly determined. Many of these are zero marginal cost applications, which is to say there is no additional cost beyond that of the hardware, that are essentially free to the city if it owns the network.

There are other, less tangible but very important benefits the city should take into account, including economic development, reducing the digital divide, and increasing municipal efficiency and service levels. The city should also take into account the citywide impact of reduced rates due to competition. Sometimes cities see this as a disadvantage. They worry that incumbents will reduce their rates below those of the city owned network. In no case of which we are aware, did this result in a city network's losing substantial amounts of money. Moreover, the city, by the nature of its mission and charter, should have a broader balance sheet. A drop in prices by incumbents by $10 per month translates into millions, perhaps tens of millions of dollars in collective savings to city households and businesses. That not only enriches their individual balance sheets, but the respending of a part of these savings will enrich municipal coffers as well.

38.7.1 Risk

Any financial analysis must analyze risk as well as return. There are two primary risks involved. One involves technology, the other subscription revenue.

Technology – The biggest decision cities must make is whether to deploy an inexpensive wireless network or invest in fiber to the premises. An all-wireless network has lower up front costs. The capital cost of a wireless network with fiber backhaul is as much as one-third higher, but leaves the city with a tangible asset with a lifespan of thirty years or more. A fiber to the premises network can cost ten to twenty times as much as wireless, but can carry all of a city's information and communication traffic for decades to come.

When it comes to the question of ownership, the most important part of the system for a city to own is the fiber infrastructure. However, many cities have chosen to own the Wi-Fi hardware because of its low investment and the fact that the investment can be paid off quickly. Standard depreciation for wireless components of a network is 5 years. On the other hand, it may be attractive for the City to contract with one or more private companies to install a wireless system and lease access to the City owned fiber network.

Households and businesses in cities that are touting low cost city-wide wireless are quickly learning there are often additional hardware costs. Although Wi-Fi is installed in most laptops, and Wi-Fi cards are widely available for desktop computers, many users will require additional equipment to connect to outdoor wireless from the interior of their homes or businesses. Often this has less to do with the strength of the signal from the wireless node than it does with the strength of the signal from the wireless connection in the user's computer. This is not a barrier to deploying a Wi-Fi network, but the cost of so-called customer premises equipment (currently around $100 but falling) and who will pay it must be factored into network planning.

A second decision is whether, if a city chooses wireless, it should commit to Wi-Fi with WiMAX on the horizon. The important difference between Wi-Fi and WiMAX is that the former uses unlicensed spectrum with power restrictions and smaller coverage areas, while the latter uses licensed spectrum that allows for higher power and therefore covers larger areas. Deploying WiMAX will not be an option for anyone, municipalities or otherwise, who does not hold licenses for spectrum in the bands that will most likely be used for WiMAX equipment in the U.S.

What is the risk of technological obsolescence? Fiber is, for all intents and purposes, a future-proof technology. The greatest expense is in installing the fiber. The electronic equipment used to "light-up" the fiber can be upgraded over time. Wi-Fi hardware is assumed to have a lifespan of five years, with software upgrades in the third year. Given that the useful life exceeds the payback period, and the investment itself is modest, the risk of obsolescence in Wi-Fi is minor.

A risk does arise if a city system depends on proprietary technologies. Wi-Fi is an open standard, meaning it is available free of charge to any one who uses a Wi-Fi device. Users interface with wireless mesh networks via Wi-Fi, but most hardware vendors rely on proprietary software for the network backhaul (the connection from the access point to the larger local network and the national and international Internet). There is no similar standard for mesh networking. Vendor bankruptcy, or even failure to invest in ongoing software development, could shorten the useful life of the wireless hardware.

Project pricing – Vendors in this field deem pricing to be proprietary information. They are unwilling to provide pricing information outside of a closed request for proposals process, and when they do so, are unwilling to itemize the bid components. This can make it very difficult for cities to estimate the actual costs for their specific circumstances. Under these circumstances, cities should insert contract provisions that shift the risk of cost overruns to the vendor.

Subscription rates – Subscription rates may not meet targets for any number of reasons, but increased competition is the most likely cause. Existing service providers may add services they previously did not offer or lower their prices in response to the new network. While this is problematic for private companies, it is no less a win for the policymakers that chose to build the new network. After all, regardless of whose customers they are, they are all constituents.

The city's options for dealing with this risk are substantially different between publicly owned and privately owned networks. If the city owns the network, the question is how much the city owned project can afford to lose while still generating a net benefit for the community.

The private sector benchmark is a return on investment of 30 percent or more within 5 years. Municipal projects also must recoup their original investment, but they have greater flexibility than privately owned networks both in the payback period, and in the willingness to accept indirect and community-wide benefits as part of the return.

For example, Saint Louis Park projects its network could lose between $240,000 and $1.4 million over five years if projected subscription rates are not obtained. On a per household basis, that is $2.31 to $13.48 per year. If competition drives prices down to $20 from the current $35 for DSL-equivalent service, and $30 from the current rates of $45 for cable equivalent service (both are rates that the new network will charge), households will save $180 per year. If the percentage of households with Internet access remains the same (an unlikely prospect, given the substantial reduction in price) the community as a whole will gain more than $12 million over five years, or more than eight times what it stands to lose in the worst case scenario.

38.7.2 A Note About Municipal "Failures"

City leaders considering municipal high-speed information networks may well come across reports, largely from anti-government think tanks, of municipal "failures." They should be cautious about taking these reports at face value. Here are some examples of such "failures."

Bristol Virginia Utilities (BVU): The Heartland Institute insists that BVU is a failure because "its operating budget is growing at an unexpected rate."
Facts: The Bristol City Council approved "OptiNet," a municipal fiber-to-the-home network operated by BVU, in 2001. In July 2003, OptiNet launched services after being delayed more than a year by legal challenges from the incumbent phone and cable companies. Throughout this time, it had to bear legal costs without revenue. Nevertheless, financial performance was 20 percent better than projected for the first, traditionally difficult start up years. Currently, the network is taking in more revenues than the sum of all its cash outlays, including debt service and interest.

Bristol is a town of 17,400 in Appalachia, a region hit hard by the decline of mining, farming and manufacturing. Median household income was $27,389 in 2000, one-third lower than the national median of $42,151. The City Council and BVU view OptiNet as an economic development tool. The network is credited with helping attract 700 new jobs in 2005. Cross Stone Products moved a 30-employee operation across the state line to take advantage of high-speed connection. Two technology companies, Northrup Grumman and CGI-AMS, are building data centers that will create 1500 high paying jobs.[53] The success of the network led neighboring Bristol, Tennessee to build its own fiber to the user network.

Cedar Falls Utility (CFU): The Heartland Institute cites CFU as a failure, principally for not generating enough free cash flow to finance its expansion.[54]
Facts: CFU started offering cable television in 1996 and high-speed Internet in 1997. Subscriber revenue has exceeded operating costs and debt service every year since 1997. No tax dollars have been used. Voters approved issuance of general obligation bonds to finance construction of the network, and CFU is on track to pay off all long-term debt by the end of 2011, five years ahead of schedule. In response to customer demand, additional bonds were issued to finance network expansion. Those bonds are also being repaid with subscription revenues.

CFU could be generating a significant profit, but being a public utility, it has opted to return most of its profits to its households in the form of lower rates. Cedar Falls residents pay $2 million less each year on cable and Internet access than the statewide average.[55] The community has also received services they would not otherwise have had. For example, CFU first offered high-speed Internet to businesses and residents in January 1997. The local cable company did not launch high-speed residential service until 2001, and high-speed business service in 2003.

DSL was not widely available in 2004. CFU built its fiber plant to the city's industrial park to attract businesses. The private cable company has yet to extend its infrastructure there. Since the fiber was installed, the industrial park has grown from 30 to 146 businesses, and employment from 1,400 to 4,300 people. And since 2005, CFU Internet subscribers also have access to a wireless network in the downtown area.

Muscatine (Iowa) Water and Power (MWP) – Heartland Institute views MWP as a failure because it has raised cable rates to cover its costs.
Facts: In 1996, after learning that incumbents TCI and U.S. West (now Qwest) had no plans to bring broadband into the community, Muscatine's business leaders recommended a municipal communications utility. Voters in Muscatine overwhelmingly approved the utility in 1997, with 94 percent voting in favor. Muscatine launched cable television service in March 1999, and Internet service later that year.

Despite predatory pricing and other anticompetitive behavior by the incumbent cable operator,1 Muscatine Power and Water successfully maintained its customer base by

---

[53] Paul Miller, "Bristol's broadband push," *Virginia Business Magazine*, November 2006.
[54] Ronald J. Rizzuto, *Iowa Municipal Communications Systems: The Financial Track Record*, Heartland Institute, September 10, 2005.
[55] CFU charges $34.50 per month for a 70-channel cable television package. It charges $40 per month for 3 Mbps Internet connections, compared to a state average of $50. All Internet subscribers also have access to the downtown Cedar Falls wireless network. It charges $69.50 per month for high-speed Internet and 70 channels of television, compared to a statewide average of $90. Businesses can get 10 Mbps connections for $80 to $150 per month, depending on the number of users.

providing higher quality services, including video on demand and wireless Internet access. In January 2003, the public utility bought the incumbent's (then Mediacom) assets in Muscatine and neighboring Fruitland.

The municipal utility has indeed raised its rates, but they remain below those of private providers elsewhere in the state.[56]

iProvo, Provo, Utah – The Reason Foundation calls this citywide fiber-to-the-premises network a failure because it has posted negative income in its first 18 months of operation. *Facts:* Fiber-to-the-premises requires a very large up front investment, and takes time to build, but the network will last for at least 20 years. It is normal to project losses for the first several years, during construction and while the customer base is built. This is equally true for the private sector. Verizon began offering its much publicized FiOS service in 2005, and expects to lose money on the investment until 2009.

Provo also had unexpected expenses. Like most networks built by public power utilities, the backbone of Provo's network was built to connect electrical substations, allowing for improved monitoring of the electrical grid. In 2001, the Utah legislature passed a bill making it possible for cities to build their own networks and sell wholesale access to private service providers. But the bill also imposed restrictions on the use of general or enterprise funds. Provo had used $2.3 million in power reserves to fund its network. The law was applied retroactively, and the city was given 10 months to repay the fund.

Finally, Provo made a single company, HomeNet, exclusive provider when the network was launched. After failing to meet subscription targets for a year, the company asked to be released from its contract and then filed for bankruptcy. In July 2005, the city added two new service providers and began meeting subscription targets, and is now on track to achieve its original goal of 10,000 subscribers in early-2007, and to begin breaking even sometime in 2008.

The network is not just for residential service. It also provides 100 Mbps connections to Provo's city buildings, fire stations, and schools, and improved reliability of the power grid. In the coming years, iProvo's fiber to the premises network can offer services the cable and phone networks are not capable of, such as distance learning courses with full-screen interactive video.

---

[56] City of Iowa City Cable TV Division, Frequently Asked Questions.

Epilogue

"Epilogue" – it sounds like the story is ending. But obviously the Wi-Fi story is continuing strong, evidenced by the contents of this book.

So let us consider this as not an "epilogue", but as just a brief pause to catch our breath. This book has covered so many of the topics that we know are important today. But based on our past experience, who really knows what future applications will be dreamed up? Who really knows which new technologies will prove to be important in the future evolution of Wi-Fi? It is very humbling to recall that back in the early and mid-1990s, when the IEEE 802.11 standards were originally being developed, the primary application on the minds of the key participants was not networking in the home, or wireless Internet access, or public hotspots, or voice over IP, or multimedia services, or city-wide wireless – but things like wireless bar code scanning and retail store inventory management. These "vertical" applications for Wi-Fi technology continue to be important today, but oh how far we have travelled.

So only an actual seer could predict the real future of Wi-Fi over the next 10 years. But one thing is clear: Wi-Fi will continue to play a role in our lives. Everything in technology has a finite lifespan – hardware products have a lifespan, software products have a lifespan – but the lifespan of a successful protocol, implemented in millions of devices worldwide, can be very, very long. Just consider TCP/IP, originally developed over 30 years ago, and we still use it every day each time we access the Internet. Wi-Fi has reached that level of universal, global presence that undoubtedly ensures it a long and healthy life.

The ability of Wi-Fi technology to expand in so many directions while maintaining backwards compatibility has been one key to its success – most obviously in the data rate leaping from 2 megabits to 11 to 54 to the hundreds of megabits possible with 802.11n – and the technology will certainly continue to evolve. This book has hopefully given you some insights into where we have been and where we may be headed. Of course, no one knows for sure – but it will certainly be a future with Wi-Fi, involving applications and technologies beyond anyone's dreams today.

Greg Ennis
Technical Director
Wi-Fi Alliance

Index